5th International Conference on Mechanical Engineering, Materials Science and Civil Engineering (ICMEMSCE 2017)

IOP Conference Series: Materials Science and Engineering
Volume 324

Kuala Lumpur, Malaysia
15 – 16 December 2017

ISBN: 978-1-5108-6110-7
ISSN: 1757-8981

Printed from e-media with permission by:

Curran Associates, Inc.
57 Morehouse Lane
Red Hook, NY 12571

Printed by Curran Associates, Inc. (2018)

For permission requests, please contact the Institute of Physics
at the address below.

Institute of Physics
Dirac House, Temple Back
Bristol BS1 6BE UK

Phone: 44 1 17 929 7481
Fax: 44 1 17 920 0979

techtracking@iop.org

Additional copies of this publication are available from:

Curran Associates, Inc.
57 Morehouse Lane
Red Hook, NY 12571 USA
Phone: 845-758-0400
Fax: 845-758-2633
Email: curran@proceedings.com
Web: www.proceedings.com

Table of contents

Volume 324

2017 the 5th International Conference on Mechanical Engineering, Materials Science and Civil Engineering

15–16 December 2017, Kuala Lumpur, Malaysia

Accepted papers received: 21 February 2018
Published online: 6 April 2018

Preface

Papers

IOP Conf. Series: Materials Science and Engineering **324** (2018) 011001 doi:10.1088/1757-899X/324/1/011001

Preface

Dear Distinguished Delegates and Guests,

The Program Committee warmly welcomes you to the 2017 5th International Conference on Mechanical Engineering, Materials Science and Civil Engineering (ICMEMSCE2017), held during December 15-16, 2017, in Kuala Lumpur, Malaysia.

The ICMEMSCE2017, which was held during 15-16 December 2017 in Kuala Lumpur, Malaysia, is the continuation of the activity of a group of research scientists and specialists who met in 2012 at the 1st ICMEMSCE2012 in Harbin, China. ICMEMSCE2017 is organized to gather members of our international community scientists so that researchers from around the world can present their leading-edge work, expanding our community's knowledge and insight into the significant challenges currently being addressed in mechanical engineering, materials science, civil engineering and other related topics. In keeping with the successful traditions established at previous conference series, all the submitted papers have been strictly peer-reviewed by the reviewers drawn from the program committee and external reviewers based on the subject matter of the papers, and only the work of the highest quality would be accepted. The book records the fully refereed papers presented at the conference, and will be very useful for those who are studying and working in the areas of mechanical engineering, materials science, civil engineering and other related topics worldwide.

The ICMEMSCE2017 required months of planning and hardworking, and each individual and institutional help has been very important for the success of this conference. The Program Committee of this conference is quite diverse and truly international, with committee members from USA, UK, Japan, France, China, South Korea, Australia, Poland, Turkey, Portugal, Hungary, Czech Republic, India, Iran, Brazil, Slovak Republic, Romania, Vietnam, Thailand, Malaysia and Sri Lanka. On behalf of the Program Committee, the Chair of ICMEMSCE2017 would like to acknowledge the valuable comments that we have received from the reviewers and thank to the authors for their contributions to the book of proceedings. We are grateful to all who have contributed to the success of ICMEMSCE2017, and do hope all the participants and other interested readers would benefit from the scientific and valuable proceedings.

On behalf of the program committee, I would like to especially thank all the editors from IOP Conference Series for their great support to ICMEMSCE2017. Without their excellent editorial work, ICMEMSCE2017 will not be published so timely and successfully.

Finally we wish all the delegates and attendees of ICMEMSCE2017 a unique, rewarding and enjoyable memory at ICMEMSCE2017 in Kuala Lumpur, Malaysia. We look forward to your participation in the 6th ICMEMSCE in 2018.

With our warmest regards,

Jingwei Zhao
Conference Chair of ICMEMSCE2017
University of Wollongong, Australia

IOP Conf. Series: Materials Science and Engineering **324** (2018) 011002 doi:10.1088/1757-899X/324/1/011002

ICMEMSCE2017 Committee

Conference Chair

Jingwei Zhao, University of Wollongong, Australia

Program Committee

Hyun-Do Yun, Chungnam National University, South Korea

Thanh-Lam Nguyen, Lac Hong University, Vietnam

Mou Chen, Nanjing University of Aeronautics and Astronautics, China

Keishi Matsuda, Matsuyama University, Japan

Jeremy (Zheng) Li, University of Bridgeport, USA

Ikuo Ihara, Nagaoka University of Technology, Japan

Norhaliza Abdul Wahab, Universiti Teknologi Malaysia, Malaysia

Abdul L Mohd Tobi, Universiti Tun Hussein Onn Malaysia, Malaysia

Aishah Najiah Dahnel, International Islamic University Malaysia, Malaysia

Ahmad Baharuddin Abdullah, Universiti Sains Malaysia, Malaysia

Toh Tien Choon, Universiti Tunku Abdul Rahman, Malaysia

Goh Kai Chen, Universiti Tun Hussein Onn Malaysia, Malaysia

Yusri Yusof, Universiti Tun Hussein Onn Malaysia, Malaysia

Mohammad Jawaid, Universiti Putra Malaysia, Malaysia

K.M. Muditha Dassanayake, University of Peradeniya, Sri Lanka

Michal Kelemen, Technical University of Kosice, Slovak Republic

Sangyoon Lee, Konkuk University, South Korea

IOP Conf. Series: Materials Science and Engineering **324** (2018) 011002 doi:10.1088/1757-899X/324/1/011002

IOP Conf. Series: Materials Science and Engineering **324** (2018) 011003 doi:10.1088/1757-899X/324/1/011003

Peer review statement

All papers published in this volume of *IOP Conference Series: Materials Science and Engineering* have been peer reviewed through processes administered by the proceedings Editors. Reviews were conducted by expert referees to the professional and scientific standards expected of a proceedings journal published by IOP Publishing.

Analysis of phase transformations in Inconel 738C alloy after regenerative heat treatment

N Kazantseva[1-2], D Davidov[1], N Vinogradova[1], I Ezhov[1] and N Stepanova[1]

1 Institute of Metal Physics, Russian Academy of Sciences, Ekaterinburg, Russia
2 Ural Federal University, Ekaterinburg, Russia
E-mail: kazantseva-11@mail.ru

Abstract. Study is based on the characterization of the chemical composition the phase transformations in Inconel 738C gas turbine blade after standard regenerative heat treatment. The microstructure and chemical composition were examined by scanning electron microscope and transmission electron microscope equipped with an energy dispersive X-ray spectrometer. It was found the degradation of microstructure of the blade feather. Redistribution of the chemical elements decreasing the corrosion resistance was observed inside the blade feather. The carbide transformation and sigma phase were found in the structure of the blade feather. It is found that the standard regenerative heat treatment of the IN738 operative gas turbine blade does not effect on carbides transformation, TCP σ-phase dissolution, and thus do not guarantee the full recovery of the IN738 gas turbine blade.

1. Introduction

Heat-resistant nickel alloys are composite structural materials. The complex chemical composition of these alloys is necessary for the formation of the structure, phase composition, and stability of the alloy under long-time exposure. Intermetallic γ'-phase based on Ni_3Al (γ'-phase, $L1_2$-type superstructure) and solid solution (γ-phase, FCC) are the main phases of the nickel superalloy. Cobalt, chromium, molybdenum, and tungsten are the strengthening elements of the nickel solid solution. In addition, chromium plays an active role in the protection of the alloys against oxidation. The volume fraction of the strengthening γ'-phase in the nickel super alloys depends on the aluminum content. Increase in the γ'-phase volume fraction at the operating temperature improves the long-term strength of the alloy [1].

In the carbon-containing nickel superalloy, MC-type carbides are formed – for example, NbC or TiC. During blade operation (with heating), the carbide reactions with the formation of the secondary $M_{23}C_6$-type carbides take place. Primary MC-type carbides and γ'-phase promote these reactions [2]. An important aspect of phase stability is the stability of the superalloy against the formation of excess intermetallic phases. The topologically close-packed phases (TCP-phases), such as the σ-phase, μ-phase or Laves phases are brittle and have unfavorable lamellar morphology [3].

The heat resistance of the nickel super alloy is determined by the complex interactions of several factors, such as:

- Growth dispersion of the crystalline structure and its perfection (small misorientations between the grain blocks);
- Volume fraction of the strengthening intermetallic γ'-phase, its morphology and dispersion distribution;
- Lattice mismatch between the γ-solid solution and intermetallic γ'-phase;

- Content of the intermetallic TCP-phases;
- Volume fraction and morphology of the carbide particles [4].

For heat resistance, the most important factor is the thermal stability of the strengthening γ′-phase. The diffusion processes take part in the phase formation, coagulation, and dissolution. In conditions of high stresses, effect of diffusion becomes especially important [5].

The Inconel 738C (IN738) super alloy, developed in 1968, is one of the important nickel based super alloys. This alloy shows the improved creep, hot corrosion, and oxidation resistance and is used in land-based gas turbines [6, 7]. The working temperature of the first stage INC738 gas turbine blade is about of 1100°C [8]. The exploitation of the gas turbine blade occurs under stressed thermo-mechanical conditions. Degradation process including of phase transformations occurs in structure of the blade under operation. Structure recovery under heat treatment is important process determining service lifetime of the gas turbine blade.

The main objective of this research is to study the degradation phenomena occurring in strengthened Inconel 738C alloy subjected to standard regenerative heat treatment.

2. Methodology

Standard regimes of the regenerative annealing were described in [3]. Service-exposed IN738 blade of the first stage of a gas turbine after a standard two steps regenerative annealing (1121 °C for 2 hr, air cool then 843 °C for 24 hr, air cool) was used for investigation. The annealing of the blade after standard exploitation time was done in accordance to the service plan. The study was carried out using an optical microscope Micromed MET and a scanning electron microscope JSM 6490 with the energy dispersive and wave micro-analyzer Oxford Inca. Small pieces were cut from the different parts of the blade for study.

3. Results and discussion

The X-ray diffractograms of the root and feather of the IN738 gas-turbine blade after regenerative annealing are presented in figure 1. In the sample cut from the blade root (Figure 1a), in addition to the γ-phase and γ′- phase lines, diffraction lines of MC and $M_{23}C_6$ carbides were found. The diffraction lines of the $M_{23}C_6$ carbide and the topologically close-packed (TCP) σ– phase were found in the sample cut from the blade feather (Figure 1b). This part of the blade usually operates at hot temperature and is more stressed.

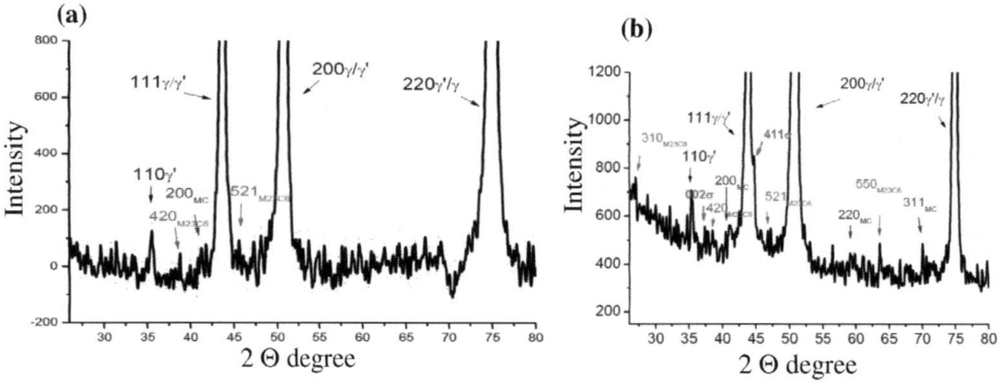

Figure 1. X-ray results of the IN738 blade: a- the root; b- the feather.

Structures of the root and feather of the blade are presented in figure 2. As can be seen from the figure 2a, carbides are observed inside the grains and on the grain boundaries of the sample cut from the blade root. Intergranular carbides have a well-defined globular shape of MC-type carbide. Carbides on the grain boundaries have a globular shape of MC-type carbide and an elongated shape which is typical for $M_{23}C_6$-type carbide. Figure 2b shows the results of optical microscopy of the sample cut from the blade feather. One can also see carbides inside the grains and on the grain boundaries. The precipitations of the irregular shape distributed along the grain boundaries and inside grains are associated with the σ-phase. The grain boundaries are completely covered with secondary $M_{23}C_6$-type carbides.

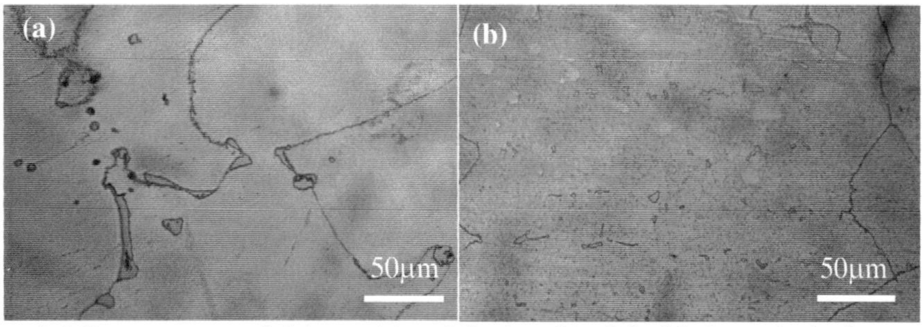

Figure 2. Microstructure of the root (a) and feather (b) of the blade, optical microscopy.

The SEM structures of the samples cut from the different parts of the studied blade are shown in figures 3-4.

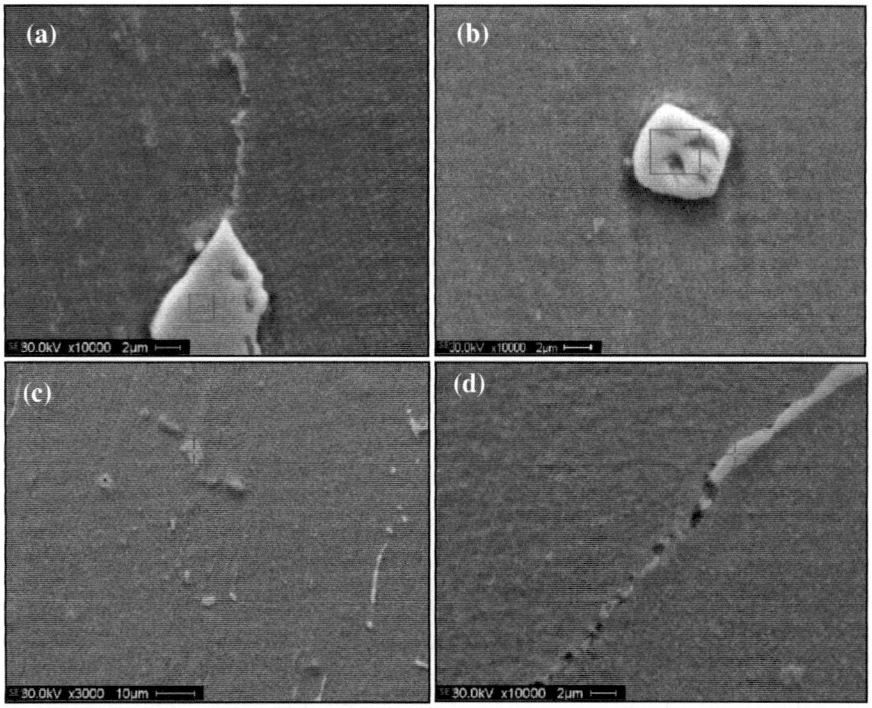

Figure 3. Microstructure of the $M_{23}C_6$ carbide (a) and MC carbide (b) in the blade root; the microstructure of the $M_{23}C_6$ carbide (c) and MC carbide (d) in the blade feather.

The EDS analysis supports these observations (Table 1-2). Sigma phase is enriched with nickel, tantalum, titanium, niobium, and molybdenum. MC-type carbide in the sample cut from the blade feather contains less tantalum content than that of the carbide MC-type in the blade root. EDS analysis of the primary MC carbide shows a high content of titanium, niobium, and tantalum; chromium is practical absent. Secondary carbide $M_{23}C_6$ has a high content of nickel, chromium, and cobalt. Grain boundary $M_{23}C_6$ carbides in the blade feather have approximately the same chemical composition as that in $M_{23}C_6$ carbides of the blade root.

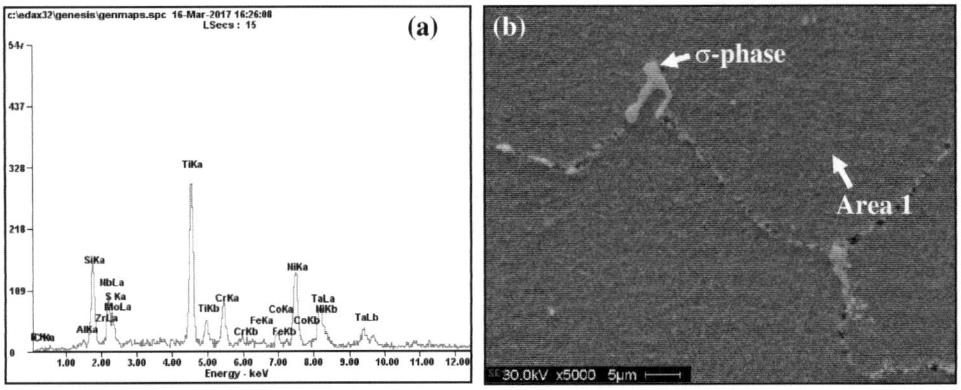

Figure 4. EDS spectrum (a) and the structure (b) of the σ-phase in the blade feather, SEM.

Table 1. Chemical composition of the phases in the blade root, wt. %.

Phases	Ni	Cr	Co	Mo	Ta	Al	Ti	C	Zr	Fe	Nb	W
Gamma prime	balance	13.66	7.42	1.67	4.93	2.49	3.67	2.02	0.89	0.47	1.70	2.06
MC carbide	balance	1.42	0.95	1.60	46.07	0.23	24.14	1.92	-	0.72	13.94	5.15
$M_{23}C_6$ carbide	balance	14.96	7.84	2.34	2.83	2.00	3.65	4.06	-	0.65	0.73	1.79

Table 2. Chemical composition of the phases in the blade feather, wt. %.

Phases	Ni	Cr	Co	Mo	Ta	Al	Ti	C	Zr	Fe	Nb
Gamma prime	balance	14.40	7.93	3.16	5.14	2.01	3.40	1.47	0.81	0.62	1.27
MC carbide	balance	2.32	0.92	14.79	21.06	0.52	22.59	1.53	2.69	-	19.07
$M_{23}C_6$ carbide	balance	13.62	7.72	2.97	4.32	2.27	4.36	1.36	0.82	0.22	2.50
Sigma phase	balance	4.14	1.50	13.66	18.91	0.66	19.81	1.14	2.76	-	16.02

It is known that $M_{23}C_6$ carbides in nickel super alloys are mainly deposited from the matrix during heat treatment and service (at 760–980 °C) [2]. $M_{23}C_6$ carbide may consist of chemical elements such as Cr, W and Mo. This type of carbide is formed due to super-saturation of carbon in the matrix and degeneration of the MC type carbide [2]. According to the literature data, hardening of the grain boundaries in heat-resistant nickel superalloy is achieved by MC type carbides based on Nb, Ti, W. To ensure a high heat resistance, carbides should have the globular shape, size of about 1 μm or less. MC carbides should be uniformly distributed along the grain boundaries without formation of a continuous grid. The probability of the formation of topologically close-packed (TCP) phases (σ, μ, Laves phases), as well as of M_6C or $M_{23}C_6$ carbides, leading to softening or embrittlement the alloy, should be minimized [3]. The type of carbide transformations and TCP phases depends on alloy doping and operating temperature. According to the literature, $M_{23}C_6$ carbides and σ-phase may form in the IN738 super alloy after the exploitation for a long time at high temperature [6, 9]. The appearance of σ-phase as well as $M_{23}C_6$ carbides lead to embrittlement the alloy. This process also points to the degradation of the alloy composition.

As can be seen from the Tables 1-2, the chemical composition of the gamma prime phase is different in the different parts of the blade. Comparison between Table 1 and 2 shows that the aluminum concentration of the gamma prime phase which is responsible for hardening in the blade feather is decreased in compare with that in blade root. Such decrease in aluminum content testifies the degradation process in the blade. The process of redistribution the chemical elements in the alloy may be explained by the Gorsky effect which was observed previously in the gas turbine blade after long service time in [5]. This effect deals with ascending diffusion under elastic-plastic conditions. Under stress, the atoms with small diameter move into the compressed regions and atoms with bigger diameters move into the stretched regions. In our case this process may occur because the different parts of the working gas turbine blade are under different stress-temperature conditions [10]. The upper edge of the blade feather is subjected to the greatest stresses and temperatures in comparison with that of the root part of the blade which is subjected the thermal load only. Service induced MC to $M_{23}C_6$ carbide transformation and formation of TCP σ-phase also indicate the redistribution of the chemical elements inside the material of the blade feather.

Conclusion

The following conclusions can be done from this study:

1. Phase transformations in the Inconel 738C alloy after standard regenerative heat treatment are studied. It is found the degradation of microstructure and chemical composition of the different parts of the IN738 gas turbine blade. It is suggested that the degradation process is caused by stress and high temperature.
2. Service induced MC to $M_{23}C_6$ carbide transformation and formation of TCP σ-phase indicate the redistribution of the chemical elements inside the material of the blade feather.
3. It is found that the standard regenerative heat treatment of the IN738 operative gas turbine blade does not effect on carbides transformation, TCP σ-phase dissolution, and thus do not guarantee the full recovery of the IN738 gas turbine blade.

Acknowledgments

The research was supported by the Russian Science Foundation, grant No 15-12-00001.

References

[1] Cahn RW, P A Siemers, J E Geiger and P Bardhan 1987 Acta Metallurgica 35 2737-2751.
[2] Kolagar A M, Tabrizi N, Cheraghzadeh M, Shahriari M S 2017 Case Studies in Engineering Failure Analysis 8 61–68.
[3] Chabina E V, Filonova E V, Lomberg B S, Bacragze M M 2012 Materials (russian) 6 1-11
[4] Crudden D J, Mottura A, Warnken N, Raeisinia B and Reed R C 2014 Acta Materialia 75 356 370.
[5] Kazantseva N V, Korolev A V, Davidov D I, Stepanova N N, Vinogradova N I, Rigmant M B 2013 Metallography 4 18-24.
[6] Jonšta P, Konečná K, Heide R, Gabčová M 2011 Z Metall-2011 5.18 -20 1-8.
[7] Zielinska M, Sieniawski J, Yavorska M, Motyka M 2011 Archives of metallurgy and materials 56(1) 193–197
[8] Rani S, Agrawal A K, Rastogi V 2017 Case Studies in Engineering Failure Analysis 8 1-10.
[9] Koul A K, Immarigeon J-P, Castillo R, Lowden P and Liburdi J 1988 Superalloys 755-764.
[10] Tapse S, Deshmukh A, Bhosale R, Gadhave S, Naidu M J 2017 International Conference on Ideas, Impact and Innovation in Mechanical Engineering (ICIIIME 2017) 5(6) 444-452.

Solar energy in buildings solved by building information modeling

B Chudikova[1] and M Faltejsek[2]

1, 2 VSB – Technical University of Ostrava, Faculty of Civil Engineering, Czech Republic

E-mail: blanka.chudikova.st@vsb.cz

Abstract. Building lead us to use renewable energy sources for all types of buildings. The use of solar energy is the alternatives that can be applied in a good ratio of space, price, and resultant benefits. Building Information Modelling is a modern and effective way of dealing with buildings with regard to all aspects of the life cycle. The basis is careful planning and simulation in the pre-investment phase, where it is possible to determine the effective result and influence the lifetime of the building and the cost of its operation. By simulating, analysing and insert a building model into its future environment where climate conditions and surrounding buildings play a role, it is possible to predict the usability of the solar energy and establish an ideal model. Solar systems also very affect the internal layout of buildings. Pre-investment phase analysis, with a view to future aspects, will ensure that the resulting building will be both low-energy and environmentally friendly.

1. Introduction

Solar thermal systems fall under the use of renewable energy sources, which are one of the much discussed topics, especially in connection with the implementation of the obligations laid down in Directive 2009/28 / EC. Above all, it is the fulfillment of the binding goal of producing 20% of energy consumption by renewable sources by 2020. [5].

Building Information Modelling (BIM) exponentially increases in connection with the global digitization of all branches, but above all with the emerging phenomenon Industry 4.0 respectively Construction 4.0,. This is a process through the entire building life cycle that connects individual building participants to working together on one information model. That is, creating the ideal design of a building, and then drawing on all the information to maximize the benefits. A comprehensive 3D model enriched with element-level information is an essential element of the BIM method. A lot of simulations and analysis using the building model, the surrounding area and the environment into which we place the building are also part of the process of ideal design by BIM. This can eliminate bad practices and predict problems that might arise during the lifetime of the construction. [1]

This paper deals specifically with the building of a nursery school on a real estate plot located in Ostrava, part of Mariánské Hory. The composition of the kindergarten is based on its shape and color. The building is designed in a compact rectangular shape because of the low number of thermal bonds and thus the lower energy demands of the building. [6] The roof is designed with a flat roof, which is a

IOP Conf. Series: Materials Science and Engineering **324** (2018) 012002 doi:10.1088/1757-899X/324/1/012002

suitable choice in terms of easier meeting the requirements of the thermal technology, as well as the location of the solar panels. Last but not least, the flat-roofed building appropriate better into the existing surrounding area and meets the potential conditions of the regulatory plan.

2. Description of the solar system

In solved building is designed active solar thermal system. The thermosolar system is characterized by the conversion of solar radiation into heat. The heat drain from collector dissipation is usually solved by a piping system in which is the heat transfer fluid circulates. The general requirements for these distributions make the system as short as possible and well insulated against heat losses. The water is heated in a well-insulated container and always indirectly, via the heat exchanger in the tank.

The connection of solar collectors to the distribution system in the building can be used for the preparation of hot water or for heating. The choice of the ideal type of solar collector also depends on the type of used connection. Solar thermal collectors are divided into flat, tubular and concentric. For civic amenities we can choose flat or tubular collectors.

The construction of flat solar collectors consists of a supporting frame, an insulation, a transparent layer, an absorber and a collector heat removal system. The flat collector achieves less efficiency and is more suitable for connection to the hot water system. The advantage of a flat collector is its affordable price.

The tube vacuum collector has a more complex design, consisting of a number of tubes. Larger tubes are then fitted with smaller tubes with heat transfer fluid. The space between the outer and inner tubes is filled with a vacuum. Vacuum collectors are especially desirable for the heating system as they are able to produce heat even in worse climatic conditions - in winter. Their disadvantage against flat collectors is their cost. [5]

3. Using BIM for designing solar panels

Building Information Modelling (BIM) should be a complex proposal of the building, including the technical equipment of the building, created using a 3D realistic model enriched with information. Solar systems have very specific parameters, and their proposal should be preceded by assessment of the building and knowledge of climatic conditions. [3] However, general proposals do not address the location of a building on a particular site in a building, and do not include different simulations of the future. If a site is surrounded by high-rise buildings or other elements, the design of the solar system may be affected by shadowing. Such a proposal is not effective and usually results in many additional or complete system design changes. In that case then costs and return on investment in solar panels increases and extend. High quality simulations of future phenomena and complex object status can prevent these problems.

3.1. Information in model

BIM is based on information. The data contained in the model reflects the future state of the building

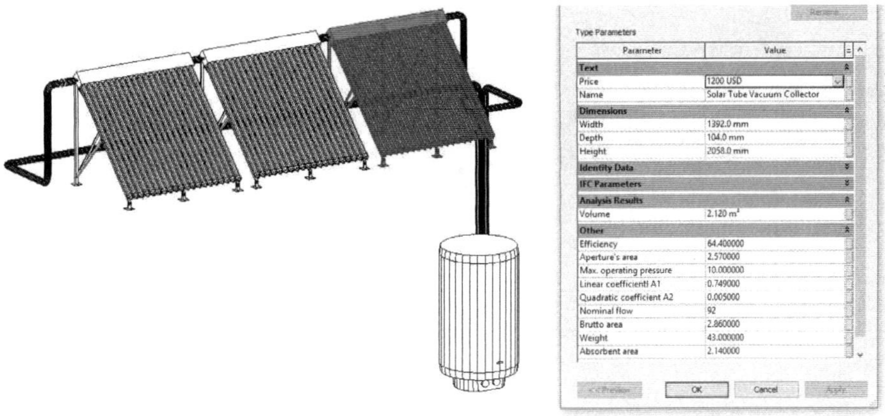

Figure 1 : Model full of information

and users who work with the model at various stages of the life cycle of the building and can use this information effectively. Comprehensive information and overview of the whole system can be generated from the model in a listing of elements including prices, materials and other specifications that we can define (Figure 1). Based on this information, we know the actual costs of acquiring the solar system, and if we link the BIM model to the CAFM system, we can use it, including enclosed documentation and information on revisions, inspections and possible repairs. The model carries information about the system's manufacturer, the warranty, or the life of the individual elements. [3]

3.2. Planning and using the model

Another benefit is the complex design of the actual state of the system. At the planning stage, before construction, we know a complete list of pipe fittings, lengths and other specifications for the proposed pipeline. The distribution, location or pipe fitting, for example, is designed with regard to the structural design of the building. By creating a complex building model, we know all the problem areas, the statically load-bearing structure or the management of other installations in the building, so we can modify the design into an ideal propose. [3]

We can use a real 3D model of construction to control collisions with other structures and installations in an object or for an ideal (efficient and economical) system design. The system also contains many valves and elements that must be installed on the pipeline. We immediately see potential design problems in the model and we can adjust, for example, the object or location of the system and individual fittings. In buildings with small technical rooms and without installation shafts, it is especially necessary for the design to be economical and appropriate located. These predictions of problems and their timely solution at the planning stage saves a lot of costs in the construction and operation of the building.

Possible changes to the building are very easy to put into the model and the documentation of the actual execution of the building is produced in parallel, which corresponds to the actual condition after completion of the construction. The model should serve throughout the lifetime of the building and, after the realization, be handed over to the operator of the building. [1]

4. Simulation of sunlight and influence of the surrounding area

In the classical case, the study of sunshine (shading) assesses in which daytime the apartment or outdoor space for recreation is sunny and whether in sufficient quantity. In our case, the sunlight study also allows us to assess the impact of neighboring buildings on the maximum possible sunlight of the solar collectors, thus determining their most suitable position on the roof of the building and thus the highest efficiency of the system. ArchiCAD software was used to study the sunlight, where the real situation was modeled and a simulation of the daylight flow of the object was created. From the shading of the surrounding buildings, the original design of the solar collectors was re-evaluated directly above the technical room to a more distant but less shaded site on the northwest side of the roof. [4]

Figure 2: Significant difference in shading depending on the location of the collectors on the roof

The picture shows complete shading of the solar collectors at 10:00 am, which greatly reduces the efficiency of the system and prolongs its return. Such a incorrect proposal, which could arise if the realistic model of a building built into the future building environment was not fully assessed, can lead to a non-economic and inefficient outcome. The added value of the information model of the buildings and the possibility to create real simulations of specific situations will in this case significantly affect the final design and balance.

4.1. Calculate the return on the system

A detailed calculation of the effect of the selected slope and azimuth on the return system for both types of solar collectors under consideration was performed. Thus, for a particular flat solar collector with real values of efficiency, losses and surfaces, as well as for a tubular vacuum collector. The calculation was made using the use of solar collectors for hot water using a simplified balance method. From the graphs that have been generated, it can be estimated that up to 14,000 CZK can be spared annually only with correct orientation and inclination of the collectors.

When buying a solar collector system with flat collectors, the payback time is 5 years under ideal conditions - ie slope 45° and azimuth 0° from the south. Under degraded conditions, the payback period will increase. In the most inappropriate orientation, returns up to 8.5 years. Vacuum collectors involve a larger investment and thus a longer payback period. Specifically 7.5 years with ideal orientation and in aggravated conditions up to 12 years. It is therefore appropriate to connect vacuum tube collectors to a heating system where the differences in payback time are not as huge against flat collectors.

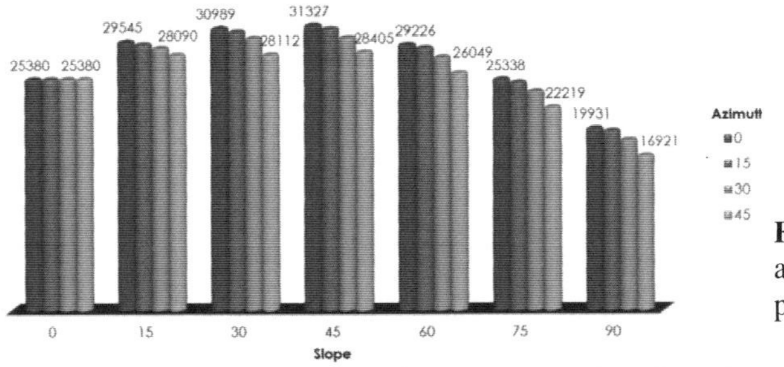

Figure 3: Influence of slope and azimuth on financial savings in CZK per year

Both payback time calculations, however, were based on the assumption that collectors receive the maximum possible supply of solar radiation. In the case of shading the surrounding buildings on average for 2 hours a day, the collector yield is decreasing. For a flat collector, the gain is reduced about 16% and the vacuum collector even has a loss of up to 23%.

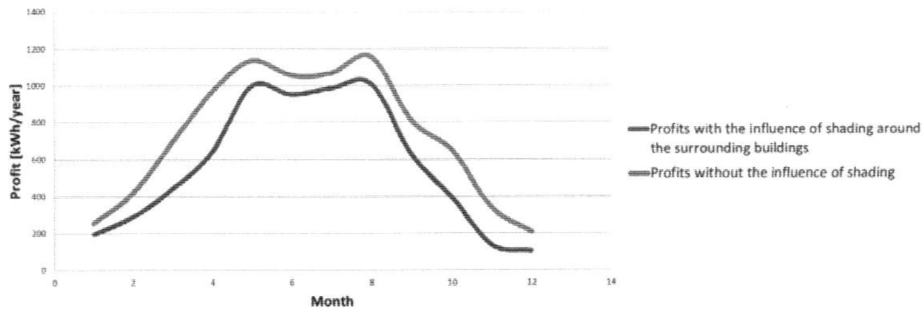

Figure 4: Influence of shading on profits from solar tube vacuum collector

5. Ideal solution

So how to proceed in designing an ideal solution? Firstly, it is important to choose a suitable location for solar collectors. Recognize whether we intend to use the solar system for hot water preparation or for heating. Based on this information, we select the appropriate slope of the collectors. The deviation from the South should be as small as possible, ideally to orient the collectors directly to the south. The calculations show that a 45 ° deviation from the South may result in up to 15% loss of profits. In our case we know that the solar system is connected to the hot water circuit and we choose the most profitable slope of the collectors 45 ° and the deviation from the south 0 °.[5]

The resulting type of collectors must be based on calculations and the current situation. From the calculated payback time, we show that the installation of flat collectors is better suited for this particular proposal building of civil equipments - kindergarten.

The non-last factor influencing the design of the system is the influence of the surrounding area on the shielding of the collectors. By using a 3D model inserted into a specific place intended for future building and creating a simulation of the sunshine, the original shorter design of the drain system was reject. By comparing the loss caused by shadowing and pipe losses in its longer lead, it has been found that a more efficient solution is to place the collectors in a more distant but sunny position over a longer period of time. The resulting pipeline losses, when properly insulated, do not have such a significant effect on the reduction of profits from the solar system, such as shading by buildings.
Based on the real model of the BIM model, we find more applications. Using ideal materials, manufacturers, or other specific parameters, and putting this information into the design phase at the project stage, the resulting construction costs can greatly influence. [2] The ideal proposal of solving the solar system itself is just one of the many benefits of the BIM method and can have a major impact on this particular example.

6. Conclusion

The use of solar radiation is one of the most widely used possibilities of using renewable resources, which are easily accessible to the public and relatively easy technical solutions. Their use and application can be found across all types of buildings, from family or apartment buildings to civic amenities. The design of such a system is practically dependent on the optimal placement solar collectors to the sun in the sky. The general rules for the slope of the panels, direct adjustment to the south and the requirement for the shortest piping system located best above the technical room, however, may not be the most effective solution in the final. Proper analysis of the situation and evaluation of various design alternatives in the planning phase allows us to choose the ideal and effective solution both in terms of technical and economic and cost-effective, with maximum system performance. Building Information Modelling (BIM) is a variant of how to create, evaluate and analysis different situations on a realistic model. It allows us to plan and model the actual future state of the object and installation with information about the dimensions, valuable and other specifications and to determine the ideal concept in the future location of the building, taking into account all factors, current and future. Last but not least, it eliminates errors and reduces costs throughout the life cycle of the construction.

7. References

[1] EASTMAN, Ch., TEICHOLZ, P., SACKS, R., LISTON, K. BIM Handbook, A Guide to Building Information Modeling for Owners, Managers, Designers, Engineers, and Contractors. New Jersey: John Wiley & Sons, Inc, 2008. ISBN: 978-0-470-18528-5.
[2] SMITH, Dana K. and Michael TARDIF. Building information modeling: a strategic implementation guide for architects, engineers, constructors, and real estate asset managers. Hoboken, NJ: Wiley, 2009, ISBN 978-0-470-25003-7.
[3] SANCHEZ A. X. and HAMPSON K. D. and VAUX S. Delivering Value with BIM, A whole-oflife approach. London a New York: Routledge, 2016. ISBN: 978-1-138-11899-7.
[4] WEI, X., BONENBERG, W., ZHOU, M., WANG, J., WANG, X. The case study of BIM in urban planning and design. *In: Advances in Intelligent Systems and Computing, vol. 600, pp 207-217, 2017. DOI: 10.1007/978-3-319-60450-3_20.*
[5] SIKULA, O., MERKA, V., HIRS. J., PLASEK, J. Maximization of solar gains of an air collector by simulations. *In: Applied Mechanics and Materials, vol. 284-287, pp 483-487, 2012. DOI: 10.4028/www.scientific.net/AMM.284-287.483.*
[6] LAUSOVA, L., SKOTNICOVA, I. Analysis of experimental measurements and numerical simulations of a heat field in the light weight building structure. *In: Advanced Materials Research, vol. 969, pp 33-38, 2014. DOI: 10.4028/www.scientific.net/AMR.969.33.*

Acknowledgement
The paper is supported from founds of the Conceptual development of science, research and innovation for 2017 allocated to the VŠB-TUO by the Ministry of Education, Youth and Sports of the Czech Republic.

Analysis of WBV on standing and seated passengers during off-peak operation in KL monorail

K Hasnan[1], Q Bakhsh[1], A Ahmed[1], D Ali[1], A R Jamali[1]
[1]Universiti Tun Hussein Onn Malaysia, 86400 Batu Pahat, Johor-Malaysia

Email: qadirquest@gmail.com

Abstract. In this study, the Whole-Body Vibration (WBV) was analyzed on the standing and seated passenger during off-peak operating hours when train was on the track. The experiments were conducted on two car train at one constant location (bogie-1, which is near to driver's cabin) during downward direction from KL sentral station towards Titiwangsa station. The aim of this study was to analyze that, in which posture of passenger's exposures the maximum level of WBV. Since, one passenger was performed the whole journey in standing posture whereas, the other passenger was in seated posture. The result obtained from experiments for the RMS accelerations (A_{rms}), maximum acceleration (A_{max}) and minimum acceleration (A_{min}) during the trip. As per standard ISO 2631-1, the daily vibration exposure (A_8), Vibration Dose value (VDV) and Crest Factor (CF) of this trip for both standing and sitting orientations were calculated. Results shows that the seated passenger was exposed to longer periods of continuous vibration as compared to the standing passenger. Whereas, the Vibration Dose value (VDV) value was greater than the action value as per ISO 2631-1 and within the limit values. The study concluded that whole body vibration transmitted towards both passengers either standing or seated during their journey. But in overall results comparison of both orientations, the seated passengers gained higher vibration than the standing one.

Keywords. Vibration, Monorail, Whole Body Vibration, Crest Factor, RMS acceleration

1. Introduction

In the field of railway engineering, the monorail is one of well famous transportation system in the field of railway, usually it is used in urban cities. The mono means one and rail means track or beam because it is a rail based transportation system, which is operated on a single rail or beam. In its design, the monorail considered as noiseless and more comfortable ride than the other trains or steel wheeled trains [1]. The monorails have no any road traffic interactions because usually they operated at elevated tracks or beams. Monorails are small or portable, it has the number of cars (2 to 8 cars) are coupled together in a series [2]. In monorail, each car contains two bogies at its front and rear axles. Each bogie equipped with the set of rubber wheels such as; load wheels used for propulsion, brake and guider wheels, which are used to hold or grip the train with track [3].

There are many factors that effects passenger ride comfort, in which the vibration is the fundamental or primary factor [4]. Many people are faced the whole-body vibration (WBV) in their occupational life, specially the drivers and passengers of various vehicles, such as trucks, cars, trains and buses [5]. In whole-body vibration the human body experienced with complex distribution of oscillatory waves and forces. It usually affects the human performance capability by discomfort or annoyance, influences a health and safety risk [6]. In vehicles like car, buses, trucks, trains and airplanes the persons sitting on seat either drivers or passengers are exposed to WBV and exposed to local acceleration at the hand, feet, back and head [7]. Usually, in vehicles they are drivers, passengers while riding on uneven surfaces and in case of machines they are operators. Also, the human body posture plays an important role in the magnitude of vibration transmission to their body [8]. The WBV usually cause health and safety risk such as; ride discomfort, badly disturb their performance, back pain, shoulder pain, nausea and other health conditions [9]. WBV also effects on many health risks to the human body such as; physiological system, psychomotor and psychological system [10]. Whenever, there is vibration transmission to whole body of human by means of the vibration source matting with the bottom back or feet of human is always considered as whole-body vibration [11].

In WBV, the humans are experienced with the six (6) type of different vibrational motions in three main orthogonal axes x, y, and z-axis, also the rotational motions around x, y, and z directions with the combination of individual motions [12]. Mainly in whole-body vibration exposure, the human body gained the vibration, which transmitted in their body from the seat or floor on that they are standing, or the contact between the machine and the operator from the driver seat [13]. When the drivers or passengers are seated on a fixed seat, the acceleration from the source then be will transmitted from the seat of that the body [14]. Among longitudinal and lateral vibrations, the vertical vibrations (in z-axis directions) are mainly affects the human body in case of WBV. These vertical vibrations are transmitted from the seat or floor to the buttocks and back of the persons along the vertebral axis through the seat pan and back [15]. The continues or long-term exposure to high amplitude WBV is strongly connected with the successive growth of lower back pain in human body [16].

In this study, the experiment was conducted to measure the whole-body vibration behavior for passengers in standing and sitting orientation during peak operation hours. The Human Vibration Meter (HVM-100) with accelerometer pad was used to record the vibration transmission from monorail to human body.

2. Design of experiment

In 2003, SCOMI launched their very first successful project so called KL Monorail with 2 car trains as the Malaysian first monorail system [17]. In its infrastructure and layout, 8.6 km long elevated track was built for KL monorail track with eleven (11) stations in total, which starts from Kuala Lumpur Sentral (KL Sentral) to Titiwangsa [18]. The experiment starts from KL sentral towards the last station Titiwangsa, this whole monorail line was divided into sections. The distance between the stations were named as section, therefore in eleven stations, total ten sections were designed as shown in Figure 1. As per KL monorail operating division, the peak operation hours are 6.30 AM to 9.00 AM and similarly in the evening 4.00 PM to 7.00 PM. Except these peak hours, the remaining all operating hours are Off-peak operation hours.

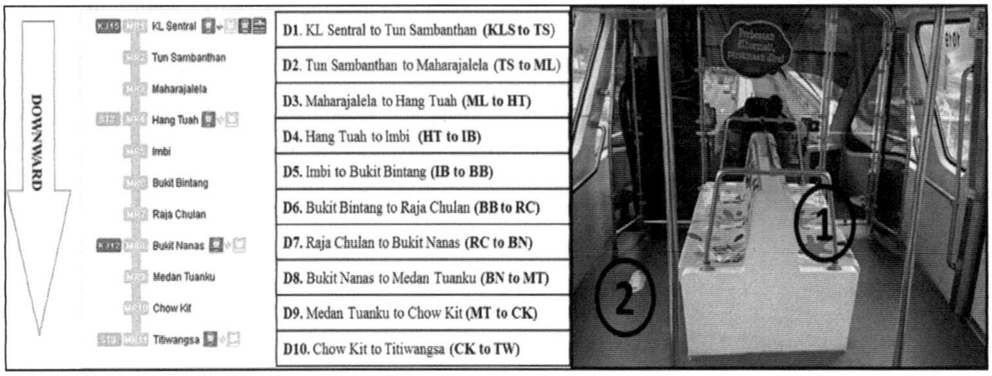

Figure 1. KL monorail stations, their subsections and location Bogie-1

2.1. Sitting on seat above bogie

In Figure 1, at position (1) the accelerometer pad was placed on the pan of seat and the passenger was seated on the accelerometer pad during the experiments.

2.2. *Standing above bogie*

In Figure 1, at position (2) the accelerometer pad was placed on the floor and the passenger was standing on it.

3. Results and Discussion

3.1. *RMS Acceleration (A_{rms})*

The Root Mean Square Acceleration (A_{rms}) was in peak value in standing position during MT to CK stations was obtained 0.389 m/s^2 in section D9, while during station RC and BN in the sitting position it was 0.418 m/s^2 in section D7. Moreover, the least Arms of this trip was generated 0.222 m/s^2 at section D6 in the standing position, whereas in sitting position it was 0.26 m/s^2 at section D6 from station BB to RC. The comparative results show that the seated passenger gained high vibration than the standing passenger as shown in Figure 2(a).

3.2. *Maximum Acceleration (Amax)*

The maximum acceleration (Amax) during standing position was generated 1.06 m/s^2 in section D9, while, in the sitting position it was 1.25 m/s^2 at section D10. Furthermore, during the standing position the lowest acceleration was generated 0.538 m/s^2 in section D6, since, in the seated passenger it was 0.598 m/s^2 in section D4. The comparison of both orientations is shown in Figure 2(b), in which it is clearly noticed that the seated passenger gained high vibration throughout the trip than the standing passenger except only in one section D1.

3.3. *Minimum Acceleration (Amin)*

The lowest minimum acceleration (Amin) 0.000772 m/s2 was transmitted to the standing passenger at section D6, while, 0.00042 m/s2 to seated passenger during section D6. Since, the highest Amin was developed during standing position 0.0355 m/s2 at section D7 and in seated passenger it was 0.0271 m/s2 at section D10. Figure 2(c) shows the comparison of both standing and sitting positions, in which the standing passenger gained highest vibration during eight sections (D1 to D8) than the seated passenger.

3.4. *Daily vibration exposure (A8)*

According to the equation; $\boldsymbol{A_8 = kA_w \left[\frac{T}{T_0}\right]^{\frac{1}{2}}}$ the daily vibration exposure 'A8' was calculated for all sections during standing and sitting position passengers, the comparative results are shown in Figure 2(d). The maximum A8 value in standing position was measured 0.0244 m/s2 in section D9 and in seated passenger the daily vibration exposure was 0.026 m/s2 in section D9.

3.5. *Vibration Dose Value (VDV)*

The VDV was calculated by the equation $\boldsymbol{VDV = \left[\int_0^T a^4(t)dt\right]^{\frac{1}{4}}}$ for the standing and sitting position. The maximum VDV in standing position was obtained 1.95 m/s1.75 in section D2, while in sitting orientation, it was 2.05 m/s1.75 in section D10. Similarly, the highest vibration is transmitted to the seated passenger than the standing one as shown in comparative graph in Figure 2(e).

3.6. *Crest Factor (CF)*

The CF for standing and seated passengers was calculated by equation $\boldsymbol{CF = \frac{a_{peak}}{a_{rms}}}$ the comparison graph is given in Figure 2(f). The maximum CF was obtained 5.718085 for standing passengers in section D8 and the for seated passengers the maximum CF value was 5.831325 during section D8. In both orientations the maximum crest factor value is less than 6, it means the ride was comfort as per standard.

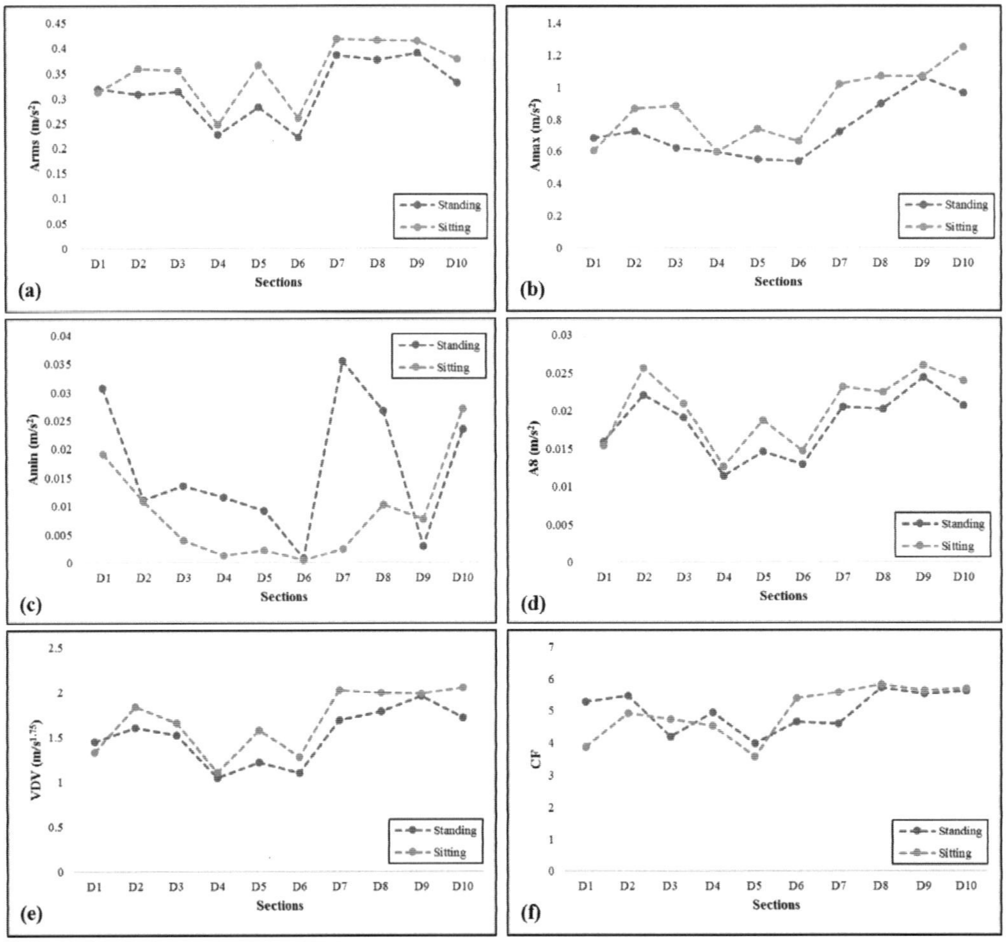

Figure 2. Results for Arms, Amax, Amin, A8, VDV and CF for Off-peak operation

4. Conclusion

The passengers who are seated on seat have experience of high vibration magnitudes because during the sitting condition the area in contact is high. While, in the case of the standing passengers the vibration transmission is less. Therefore, the results show that there is a difference in standing and sitting orientations. This is all because of the contact surface between the passengers and monorail body. Since, the overall daily vibration exposure of both orientations has difference but both values are under exposure limit value as per standard. In this study, the overall trip vibration exposure 'A8' was obtained during the sitting position 0.204 m/s^2 while, during the standing position it was 0.1819 m/s^2. As per the standard, VDV limit and action values should be 9.1 $m/s^{1.75}$ and 21.0 $m/s^{1.75}$, whereas in this study, the overall trip vibration dose value was 15.05 $m/s^{1.75}$ during standing position, and in sitting it was 16.81 $m/s^{1.75}$. Moreover, the crest factor value as per standard if exceeded from 6 that means the ride during that location or trip was not comfort. Therefore, it is concluded here in this study, that on same location (at bogie-1), same route (from KLS to TW station) and same operation hours (off-peak) the whole-body vibration magnitude that transferred to the passengers during sitting orientation is more than the standing orientation. Thus, it is concluded that the standing passengers has less exposure of WBV on their body as compared to the seated passengers.

IOP Conf. Series: Materials Science and Engineering **324** (2018) 012003 doi:10.1088/1757-899X/324/1/012003

5. Acknowledgement
The Authors would like to thank the Universiti Tun Hussein Onn Malaysia for financially supporting of this study.

6. References
[1] Kennedy R R 2010 Monorail Rapid Transit *Considering Monorail Rapid Transit for North American Cities.*

[2] Wang B 2003 Constructability Analysis of Monorail Project *Pre-Project Planning & Constructability Analysis.*

[3] Lee C H, Kim C W, Kawatani M, Nishimura N and Kamizono T 2005 Dynamic response analysis of monorail bridges under moving trains and riding comfort of trains *J. Engineering Structures*, ***27***(14), 1999–2013.

[4] Kim Y G, Choi S, Kim S W, Kim Y M and Park T W 2009 An Experimental Study on the Ride Comfort of the Korean High-Speed Train *J. Experimental Techniques*, ***33***(December), 30–37.

[5] Demić M, Lukić J and Milić Ž 2002 Some Aspects of the Investigation of Random Vibration Influence on Ride Comfort *J. Sound Vib;* **253**(1):109–28.

[6] ISO 1997 Mechanical vibration and shock-Evaluation of human exposure to whole-body vibration-Part 1 *International Standard Organization (ISO).*

[7] Griffin M J 1990 Handbook of Human Vibration *Elsevier Science Academic press UK.*

[8] Ismail A R, Nuawi M Z, How C W, Kamaruddin N F, Nor M J M and Makhtar N K 2010 Whole Body Vibration Exposure to Train Passenger *American Journal of Applied Scienes* **7**(3), 352-359.

[9] Mcphee B, Foster G and Long A 2001 A handbook on whole body vibration and exposure in mining *The Joint Coal Board Health & Safety Trust Sydney, Australia.*

[10] Sayed M E, Habashy S and Adawy M E 2012 Evaluation of Whole-Body-Vibration Exposure to Cairo Subway (Metro) Passengers *Global Advanced Research Journal of Engineering, Technology and Innovation*, ***1***(7), 168–178.

[11] Sylvester A A 2009 Evaluation operator whole body vibration and shock exposure in South African open cast mine *University of Pretoria. MS Thesis.*

[12] Rao S S 2010 Mechanical Vibrations *Pearson Education, (Fifth edition).*

[13] Farhana L 2008 The Effect of Whole Body Horizontal Vibration in Position Sense and Dynamic Stability of the Spine *University of Kansas. MS Thesis.*

[14] Falou W E, Duchêne ., Grabisch M, Hewson D, Langeron Y and Lino F 2003 Evaluation of driver discomfort during long-duration car driving *J. Applied ergonomics*, ***34***(3), 249-255.

[15] Cann A P, Salmoni A W and Eger T 2004 Predictors of whole-body vibration exposure by highway transport truck operators *J. Ergonomics*, **47**(13), 1432-1453.

[16] Limerick R B and Lynas D 2016 Long duration measurement of whole-body vibration exposures associated with surface coal mining equipment compared to previous short-duration measurements *J. of Occupational and Environmental Hygiene* **13**, 339-345.

[17] Scomi 2017 About Monorail: Scomi Rail *Retrieved March 26, 2017, from, http://www.scomirail.com.my/monorail.htm.*

[18] Prasarana 2014 Maintenance manual, Technical manual of bogie *Kuala Lumpur: Technology, Monorail Malaysia.*

IOP Conf. Series: Materials Science and Engineering **324** (2018) 012004 doi:10.1088/1757-899X/324/1/012004

Checkerboard deposition of lithium manganese oxide spinel ($LiMn_2O_4$) by RF magnetron sputtering on a stainless steel in all-solid-state thin film battery

T H Hsueh[1,2] , Y Q Yu[1], D J Jan[1], C H Su[1] and S M Chang[2]

1 Physics Division, Institute of Nuclear Energy Research (INER), Longtan District, Taoyuan City, Taiwan

2 Institute of Organic and Polymeric Materials, National Taipei University of Technology, Taipei, Taiwan

E-mail: tshsueh@iner.gov.tw

Abstract. All-solid-state thin film lithium batteries (TFLBs) are the most competitive low-power sources to be applied in various kinds of micro-electro-mechanical systems and have been draw a lot of attention in academic research. In this paper, the checkerboard deposition of all-solid-state TFLB was composed of thin film lithium metal anode, lithium phosphorus oxynitride (LiPON) solid electrolyte, and checkerboard deposition of lithium manganese oxide spinel ($LiMn_2O_4$) cathode. The LiPON and $LiMn_2O_4$ were deposited by a radio frequency magnetron sputtering system, and the lithium metal was deposited by a thermal evaporation coater. The electrochemical characterization of this lithium battery showed the first discharge capacity of 107.8 μAh and the capacity retention was achieved 95.5% after 150 charge-discharge cycles between 4.3V and 3V at a current density of 11 $\mu A/cm^2$ (0.5C). Obviously, the checkerboard of thin film increased the charge exchange rate; also this lithium battery exhibited high C-rate performance, with better capacity retention of 82% at 220 $\mu A/cm^2$ (10C).

1.Introduction

The thin film lithium battery (TFLB) is likely to be integrated with thin film solar cells and wireless signal network to be an energy harvesting module. Furthermore, the energy harvesting module is an inexhaustible power supply for internet of things (IoTs) applications and wearable devices [1].
According to related study for all-solid-state TFLBs that have significant advantages in high safety, long term storage and high cycle ability than that of the liquid electrolytes in batteries [2-6]. Among these solid electrolytes, lithium phosphorus oxynitride (LiPON) is the most popular and stable in all-solid-state TFLBs [7-9]. However, solid electrolyte has not been widely used in TFLB because of its ionic conductivity is too low to meet the required capacity at high power density. To overcome this drawback, some researchers focus on improving the ionic conductivity of the solid electrolytes, and succeeded in increasing the ionic conductivity, but the stability of these solid electrolytes is not good enough with the cathode or anode electrode [10-14]. Instead of above mentioned research, it was considered changing the morphology could create a larger area of solid electrolyte/solid active material contact, thus increasing the charge exchange rate. Here, we introduced a checkerboard deposition in all-solid-state lithium battery, which was made through radio-frequency magnetron sputtering of an amorphous $LiMn_2O_4$ thin film, and had changed its subsequent transformation to a crystalline phase by conventional thermal annealing. The checkerboard deposition of TFLB showed a discharge capacity of 107.8 μAh between 4.3V and 3V at a current density of 11 $\mu A/cm^2$ (0.5C).

Furthermore, the checkerboard deposition of thin film increased the charge exchange rate; also this lithium battery exhibited high C-rate performance, with better capacity retention of 76% at 220 μ A/cm^2 (10C).

2.Experimental details

All-solid-state TFLBs (on the basis of Li/LiPON/$LiMn_2O_4$) were fabricated by using a home-made radio frequency (RF) magnetron sputtering system to deposit thin-film $LiMn_2O_4$ electrodes from a 3 inches crystallized $LiMn_2O_4$ (99.9 wt.%) target. First, a flexible 100μm-thick of stainless steel was used as the substrate and current collector; the target-to-substrate distance was fixed at 70 mm. 1,000nm-thick of $LiMn_2O_4$ thin film was deposited by introducing an Ar gas into the sputtering chamber at a working pressure of 5×10^{-3} torr and a RF power of 91 W. The spinel-phase $LiMn_2O_4$ thin film was then subjected to conventional thermal annealing at 600 °C for 90min. Next, the solid-state LiPON electrolyte was deposited to a thickness of about 2,000nm by RF magnetron sputtering of a Li_3PO_4 target under nitrogen atmosphere. The ionic conductivity of LiPON was measured as ~1.5×10^{-6} S/cm. Finally, a Li metal film with the size of 22 mm × 22 mm was deposited on the LiPON electrolyte by a thermal evaporation coater, wherein a small evaporator (cylinder type, height 40 cm, and diameter 20 cm) was set up in an argon-filled glove box.

The film structure was characterized by field emission scanning electron microscope (FESEM, model Hitachi-S4800) (Japan) and X-ray Diffractometer (XRD, model Bruker-AXS D8 Advance) (USA); the cyclic voltammograms (CV) test was carried out between 3.6-4.3 V at a scan rate of 0.01mVs^{-1}, the frequency range of AC impedance was investigated from 1×10^{-3}Hz up to 2×10^5Hz and an AC voltage of 30mV was superimposed on a dc bias that was set at the 4.1V of each cell. The charge-discharge tests were conducted at a constant current of 55 μA between a voltage range of 3.0–4.3 V at 298K, while the high discharge current density tests were conducted at various discharge current between 5 and 200 μA on an electrochemical workstation (model VMP3) (USA). All measurements were conducted in an argon-filled glove box.

3.Results and discussion

Figure 1(a) and 1(b) show the top view SEM images of the layer and the checkerboard deposition of $LiMn_2O_4$ thin films, respectively. The layer deposition of $LiMn_2O_4$ thin film acting as the compared specimen is 22mm×22mm in area and the checkerboard deposition has a deposition of 1mm×1mm square with a distance of 0.1mm to each other under 22×22mm area. Both of the $LiMn_2O_4$ thin films are spinel structure without cracking or peeling after annealing, that indicates strong adhesion to the flexible stainless steel substrate achieved by RF magnetron sputtering. The ratio of active area between layer and checkerboard deposition of $LiMn_2O_4$ thin film is around 1.21:1.

IOP Conf. Series: Materials Science and Engineering **324** (2018) 012004 doi:10.1088/1757-899X/324/1/012004

Figure 1. SEM images of top view of (a) layer and (b) checkboard deposition of $LiMn_2O_4$ thin film.

Figure 2 shows the XRD patterns of $LiMn_2O_4$ thin films deposited on stainless steel substrates annealed at 650°C for 90min. As, all the XRD measurements were carried out with a scan rate of 1°/min in diffraction angle (2θ) range 10–70°. Both of the layer and checkboard films subsequently annealed at 650°C for 90min after deposition, the onset of crystallization was observed. These two samples exhibited similar intensity of characteristic peaks corresponding to (111), (311), (222) and (400) lines.

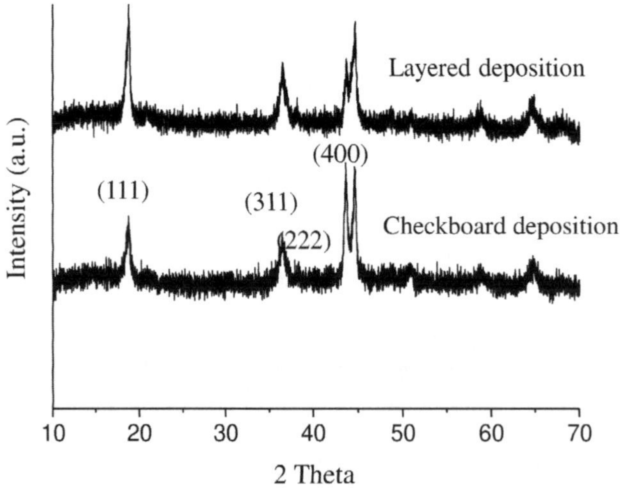

Figure 2. XRD patterns of layer and the checkerboard deposition of annealed $LiMn_2O_4$ thin films.

Cyclic voltammograms were investigated in this study. Figure 3 shows peak separation in the 4 V regions for the two-step de-/intercalation process of lithium ions into/out of the spinel-crystal lattice. The curves show two couples of redox peaks of the layer deposition of $LiMn_2O_4$ thin film were 3.97, 4.08 and 4.10, 4.18, respectively, and for the checkerboard deposition of $LiMn_2O_4$ thin film was 3.98, 4.04 and 4.11, 4.17, respectively. The potential differences of two pairs of redox peaks of the checkerboard deposition of $LiMn_2O_4$ thin film was smaller than the values of the layer deposition of $LiMn_2O_4$ thin films. This cyclic voltammograms indicate that the checkerboard deposition of $LiMn_2O_4$

thin films provide smaller resistance and clearer two couples of redox peaks than the layer deposition thin films.

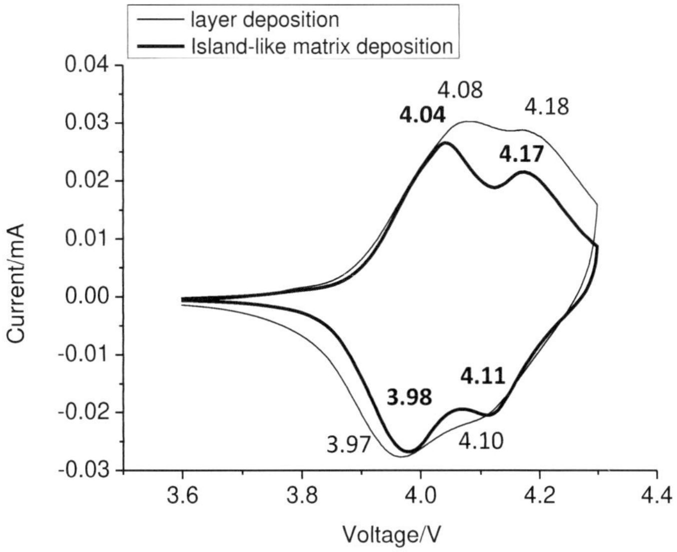

Figure 3. Cyclic voltammograms of the layer and the checkerboard deposition of $LiMn_2O_4$ thin films.

Figure 4 shows the Nyquist plots of the layer and the checkerboard deposition of $LiMn_2O_4$ thin films at room temperature. A single cell was tested at various frequencies ranging from 100 Hz to 100 MHz, with frequency decreased gradually; the Nyquist plots show two semicircles and a diagonal line. It is clear that the checkerboard deposition of $LiMn_2O_4$ electrode possesses low resistance of charge transfer (second semicircle), with resistance of 60 Ohm, but the value of the layer deposition $LiMn_2O_4$ electrode is up to 130 Ohm. The checkerboard $LiMn_2O_4$ electrode possesses smaller charge transfer resistance. This result is correspondent with the cyclic voltammograms mentioned above.

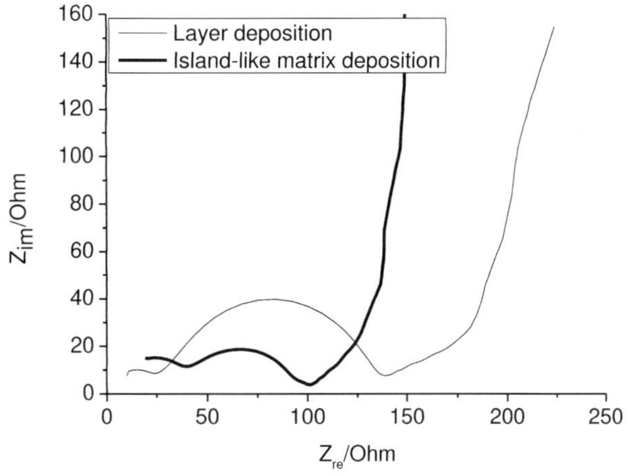

Figure 4. Nyquist plots of layer and checkerboard deposition of $LiMn_2O_4$ thin film.

Figure 5 displays the charge and discharge curves of the layer and the checkerboard deposition thin films. For the voltage region of 3.0–4.3 V, the characteristic charging and discharging plateaus of the checkerboard deposition of $LiMn_2O_4$ thin film slightly occur clearly than the layer deposition thin films. A discharge capacity of about 124μAh at a current density of 13 μA/cm2 (0.5C) was obtained for the layer deposition thin film and 108μAh at a current density of 11 μA/cm2 (0.5C) for the checkerboard deposition thin film. The ratio of discharge capacity between the layer and the checkerboard deposition of $LiMn_2O_4$ thin film is around 1.15:1. The data is very close to the ratio of active area between two type deposition thin films. Both of all-solid-state thin film batteries with the similar charge/discharge profile from the 1st to 3rd cycle between 4.3V and 3V means that the interfaces between LiPON/$LiMn_2O_4$ and Li/LiPON were stable.

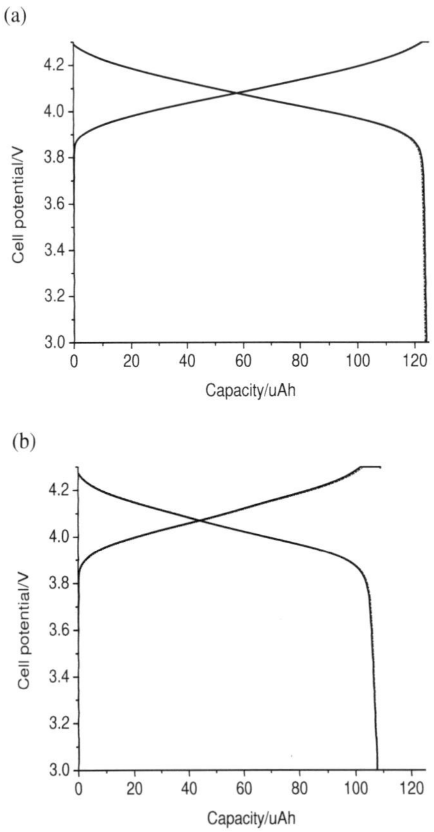

Figure 5. Charge and discharge curves of (a) layer deposition and (b) checkerboard deposition of $LiMn_2O_4$ thin film.

To confirm the all-solid-state of the layer and the checkerboard deposition of $LiMn_2O_4$ thin film lithium batteries were stable, the charge and discharge cycle test were investigated by an electrochemical workstation. Figure 6 shows the 1st discharge capacity of the checkerboard deposition of $LiMn_2O_4$ thin film lithium battery was 107.8 μAh between 4.3V and 3V at current density of 11 μA/cm2 (0.5C) and discharge capacity of about ~124 μAh (0.5C) was obtained for the first cycle of the layer deposition of $LiMn_2O_4$ thin film lithium battery. Within 3.0–4.3V charge/discharge cycle tests, the 150th discharge capacity was 102.9 and 117.2μAh, respectively. The results obtained after 150 cycles revealed that the checkerboard deposition of $LiMn_2O_4$ thin films retain over 95.5% of its initial discharge capacity and the layer deposition thin film battery was 94.5%. High capacity retention indicates a highly stable structure during long-term charging/discharging.

IOP Conf. Series: Materials Science and Engineering **324** (2018) 012004 doi:10.1088/1757-899X/324/1/012004

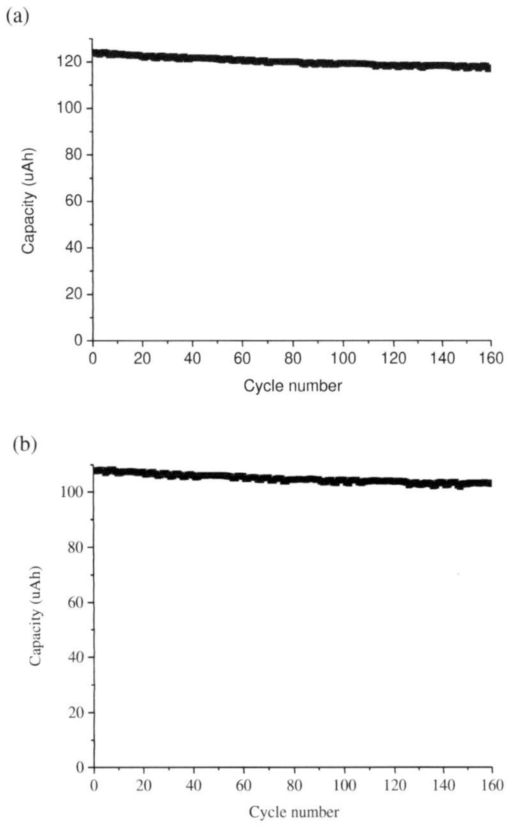

Figure 6. Capacity retention of Li/LiPON/$LiMn_2O_4$ cells cycled from 3.0 to 4.3 V.

Figure 7 shows the rate capability of the layer and the checkerboard $LiMn_2O_4$ thin films at room temperature. A single cell was discharged at various current densities ranging from 11 μA/cm^2 to 220 μA/cm^2, i.e., 0.5C to 10C for battery operation, with the gradual increase of discharge rate; the discharge capacity goes down by the restricted Li-ion diffusion. It is clear that the checkerboard $LiMn_2O_4$ electrode possesses excellent high current density capability, with capacity retention of 82% at 220 μA/cm^2 (10C) better than the capacity retention (70%) of the layer deposition thin film battery. By comparison with the layer sample, the checkerboard $LiMn_2O_4$ electrode possesses better rate capability because of the following factor - the larger specific surface area ensures a larger interfacial area for electrochemical reaction. The checkerboard $LiMn_2O_4$ electrode shows a discharge capacity of 84.2 μAh between 4.3V and 3V at a current density of 220 μA/cm2 (10C). The capacity retention is about 82.5%. However, the discharge capacity of the layer sample is only 75.9μAh at the same C-rate. The capacity retention is only 69%. Furthermore, the checkerboard sample almost recovers the capacity at a current density of 11 μA/cm^2 (0.5C) but the discharge capacity of the layer sample is slightly decreased when the current density goes down to 0.5C again.

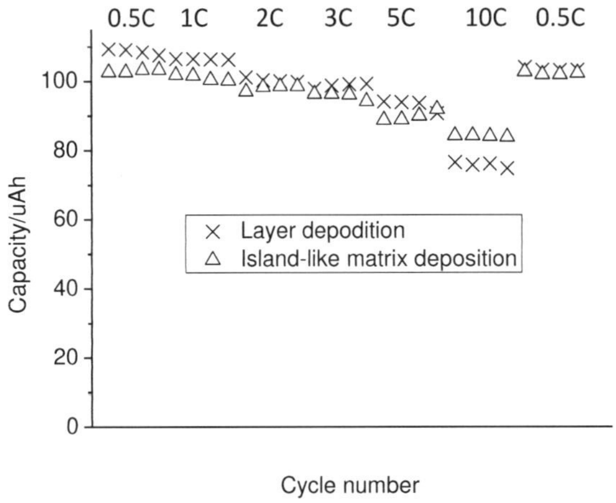

Figure 7. Rate capability of layered and checkerboard $LiMn_2O_4$ thin film.

From the results obtained above, the kinetics of Li^+ diffusion is normally the major reason for electrode polarization at a high discharge current; we believe that this checkerboard $LiMn_2O_4$ thin film electrode provides more lithium-transfer pathways as shown in figure 8. This can be attributed to the large surface area provided by the grains of the $LiMn_2O_4$ thin film, which allows a large number of Li^+ ions to be exchanged during conditions of high discharge current density. This large surface area also contributes to the reduction observed in the cathode polarization.

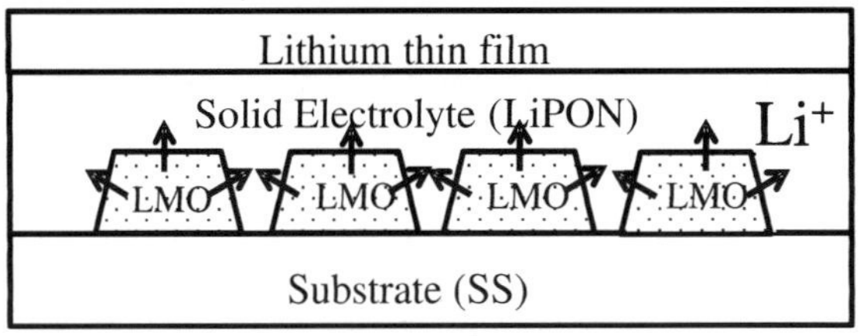

Figure 8. Scheme of the checkerboard deposition of $LiMn_2O_4$ thin films in all-solid-state lithium battery.

4.Conclusion

This study has demonstrated that checkerboard deposition of $LiMn_2O_4$ thin films in all-solid-state lithium-ion batteries can be successfully fabricated on a stainless steel substrate through the use of radio-frequency magnetron sputtering method and increase the area of solid electrolyte/solid active material contact, thus increasing the charge exchange rate. A single cell of such a battery demonstrated a discharge capacity of 107.8 μAh between 4.3V and 3V at a current density of 11 μA/cm2 (0.5C). It also proved to have a very stable cycling performance, retaining over 95 % of its initial capacity after 150 cycles. Similarly promising electrochemical properties were obtained at a high discharging current, with capacity retention of 82.5% at 220 μA/cm2 (10C). On the basis of these results, we believe that the checkerboard deposition of $LiMn_2O_4$ thin films in all-solid-state lithium batteries has great potential by some improvements for the future development of flexible lithium batteries capable of functioning equally well in high current density charge and discharge condition.

Acknowledgments
This work was supported by the Ministry of Science and Technology of Taiwan under Grant numbers MOST 106-3111-Y-042A-093 and MOST 106-3114-Y-042A-007.

References

1. Chamanian S, Ulusan H, Zorlu O, Baghaee S, Uysal-Biyikoglu E (2016) Wearable battery-less wireless sensor network with electromagnetic energy harvesting system. Sens Actuator A 249 : 77-84
2. Chiu KF, Chen CL (2011) Plasma assisted and manipulated deposition of thin film electrodes for micro batteries. Thin Solid Films 519 :4705–4708
3. Haruta M, Shiraki S, Ohsawa S, Suzuki T, Kumatani A, Takagi A, Shimizu R, Hitosugi T (2016) Preparation and in-situ characterization of well-defined solid electrolyte/electrode interfaces in thin-film lithium batteries. Solid State Ion 285: 118-121
4. Kato A, Hayashi A, Tatsumisago M (2016) Enhancing utilization of lithium metal electrodes in all-solid-state batteries by interface modification with gold thin films. J Power Sources 309 : 27-32
5. Lobe S, Dellen C, Finsterbusch M, Gehrke HG, Sebold D, Tsai CL, Uhlenbruck S, Guillon O (2016) Radio frequency magnetron sputtering of Li7La3Zr2O12 thin films for solid-state batteries. J Power Sources 307 : 684-689
6. Li D, Ma Z, Xu J, Li Y, Xie K (2014) High temperature property of all-solid-state thin film lithium battery using LiPON electrolyte. Mater Lett 134 : 237-239
7. Chiu KF, Chen CL, Chen BS, Leu HJ (2011) Modification of Electrolyte/Cathode Interfaces by Solid-State Electrolyte Thin Films. ECS Transactions 32: 67-75
8. Jouybari YH, Berkemeier F (2016) Enhancing Silicon Performance via LiPON Coating: A Prospective Anode for Lithium Ion Batteries. Electrochim Acta 217 : 171-180
9. Oukassi S, Giroud-Garampon C, Dubarry C, Ducros C (2016) All inorganicthin film electrochromic device using LiPON as the ionconductor. Sol Energy Mater Sol Cells 145 : 2-7
10. Kali R, Mukhopadhyay A (2014) Spark plasma sintered/synthesized dense and nanostructured materials for solid-state Li-ion batteries: Overview and perspective. J Power Sources 247:920–931
11. Takahashi K, Hattori K, Yamazaki T, Takada K, Matsuo M, Orimo S, Maekawa H, Takamura H (2013) All-solid-state lithium battery with LiBH4 solid electrolyte. J Power Sources 226:61–64
12. Yamamoto K, Iriyama Y, Asaka T, Hirayama T, Fujita H, Nonaka K, Miyahara K, Sugita Y, Ogumi Z (2012) Direct observation of lithium-ion movement around an in-situ-formed-negative-electrode/solid-state-electrolyte interface during initial charge–discharge reaction. Electrochem Commun 20:113–116
13. Richards WD, Miara LJ, Wang J, Kim JC, Ceder G (2016) Interface stability in solid-state batteries. Chem Mater 28:266–273
14. Yamada Y, Nozaki T, Kuriyama K, Kushida K (2013) Three hundred micron squared all-solid-state Li ion secondary battery fabricated by Si very large scale integration technology. J Alloy Compd 551:44–47

IOP Conf. Series: Materials Science and Engineering **324** (2018) 012005 doi:10.1088/1757-899X/324/1/012005

3D laser scanning in civil engineering - measurements of volume of earth masses

J A Pawłowicz[1] and E Szafranko[1] and J Harasymiuk[1]

[1] Institute of Building Engineering, Faculty of Geodesy, Geospatial and Civil Engineering, University of Warmia and Mazury in Olsztyn, Poland
E-mail: jopaw@uwm.edu.pl

Abstract. Considering the constant drive to improve and accelerate building processes as well as possible applications of the latest technological achievements in civil engineering practice, the author has proposed to use 3D laser scanning in the construction industry. For example, data achieved through a 3D laser scanning process will facilitate making inventories of parameters of buildings in a very short time, will enable one to check irregularly shaped masses of earth, heavy and practically impossible to calculate precisely using traditional techniques. The other part of the research, performed in the laboratory, consisted of measurements of a model mound of earth. All the measurements were made with a 3D SkanStation C10 laser scanner manufactured by Leica. The data were analyzed. The results suggest that there are great opportunities for using the laser scanning technology in civil engineering

1. Introduction

3D scanning is the state-of-the-art method based on laser technology, which creates endless opportunities for applications. Its main advantage is the rapid acquisition of huge amounts of data in a very short time, which makes it a much superior solution that any earlier measuring techniques. Terrestrial laser scanners easily find implementations in various branches of economy. The range of their capabilities means that they can be used in engineering geodesy, in specification of the parameters of buildings and building objects characterized by very complex shapes (e.g. historical buildings), in the research on deviations and deflections of building structures, but also in archeology or forensic science [1,2].

A scanner is an instrument which, once seated on a levelled and stable tripod, collects data while rotating around its own axis. Basically, it works by measuring the distance and angle between the device and the scanned object, using for this aim a laser beam which is reflected by an obstacle and returns to the instrument. The measured distance and angle allow us to visualize the scanned object by creating a point cloud with the set X, Y and Z coordinates, orientated in relation to the local set of coordinates of the laser. A cloud of points together with photographs used for reproduction of the texture of the building walls constitute a realistic visualization of the scanned building [3,4].

2. Measurements of volume of earth masses – laboratory assays

Construction works often inolve earthwork. The 3D laser scanning technology can also be successfully applied to control volumetric volume of earthen mounds and trenches. This part of the research was conducted in a laboratory. The objective was to determine what precison is attainable in measuring the volume of earthen material. The tests were carried out in the Laboratory for Testing Concrete, at the

UWM in Olsztyn. A portion of 0.005 m3 of medium grain aggregate was meausred and used to make a mound. The mound was then scanned at high resolution. The higher the resolution, the denser and more accurate the point cloud obtained.

The experiment was conducted in two stages. First, high resolution scanning was performed at three laser measuring stations. Next, during the second, control stage, the measurements were taken at four stations but with medium resolution. Once the observations were completed, the data were exported to a computer. They served as a basis for making a model of the mound in the form of a point cloud (Figure 1).

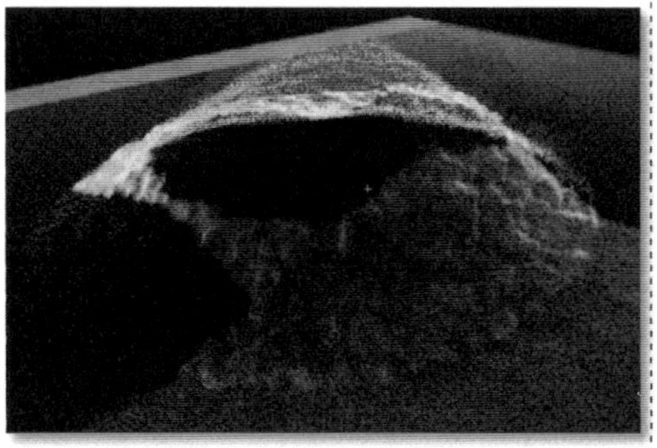

Figure 1. A model of a mound of aggregate in the form of a point cloud.

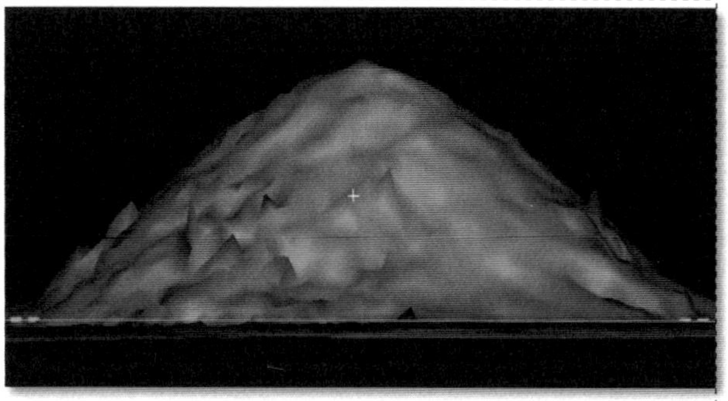

Figure 2. These two figures have been placed side-by-side to save space. Justify the caption.

The volumetric volume was measurable once the cloud was transformed into a 'mesh' model, but to achieve this a TIN (Triangulated Irregular Network) mesh had to be overlaid [5]. This is a mesh composed of triangles and each vertex of every triangle connects with points in a cloud, creating a surface. It is possible to modify the number of triangles in a model so as to reduce occlusions caused by matching noises appearing directly above the object to the Tin mesh (Figure 2). The default number of triangles is 105 of the basic model. Once the integration step is set, we can begin the volume of a mound. The result is given in m^3.

The volumetric volume was measured at three integration steps: 0.1m, 0.05m, 0.01m. The degree of integration attained through decreasing the number of triangles in the mesh was determined consequtively at 100%, 10%, 8%, 6%, 4%, 2%, 1%, 0.5%. The results are contained in the tables below (tab. 1 and tab. 2).

Table 1: Results from test 1.

Degree of diminished density of triangle mesh [%]	Model volume [m3]	Volume of the model	Difference in volume
Integration step – 0.10m			
100	0.005	0.0051668	-0.0001668
10		0.0050879	-0.0000879
8		0.0049888	0.0000112
6		0.0049749	0.0000251
4		0.0049803	0.0000197
2		0.0049701	0.0000299
1		0.0049674	0.0000326
0.5		0.0049576	0.0000424

	Integration step – 0.05m		
100	0.005	0.0050249	-0.0000249
10		0.0050509	-0.0000509
8		0.0050499	-0.0000499
6		0.005043	-0.0000430
4		0.0050411	-0.0000411
2		0.005021	-0.0000210
1		0.005018	-0.0000180
0.5		0.0050099	-0.0000099
	Integration step – 0.01m		
100	0.005	0.0050445	-0.0000445
10		0.004874	0.0001260
8		0.0049953	0.0000047
6		0.0050189	-0.0000189
4		0.0050182	-0.0000182
2		0.0050153	-0.0000153
1		0.0050087	-0.0000087
0.5		0.0050013	-0.0000013

Table 2: Results from test 2.

Degree of diminished density of triangle mesh [%]	Model volume [m3]	Volume of the model	Difference in capacity
	Integration step – 0.10m		
100	0.005	0.0050020	-0.0000020
10		0.0049558	0.0000442
8		0.0049517	0.0000483
6		0.0049456	0.0000544
4		0.0049327	0.0000673
2		0.004921	0.0000790
1		0.0049059	0.0000941
0.5		0.0049015	0.0000985
	Integration step – 0.05m		
100	0.005	0.0048004	0.0001996
10		0.0047998	0.0002002
8		0.0047986	0.0002014
6		0.0048001	0.0001999
4		0.0048011	0.0001989

2		0.0048201	0.0001799
1		0.0048047	0.0001953
0.5		0.0048021	0.0001979
Integration step – 0.01m			
100	0.005	0.0048644	0.0001356
10		0.0048443	0.0001557
8		0.0048447	0.0001553
6		0.0048459	0.0001541
4		0.0048446	0.0001554
2		0.004841	0.0001590
1		0.0048354	0.0001646
0.5		0.0048246	0.0001754

3. Summary and conclusions

Based on the results of tests, the smallest difference in the volume of the mound model was determined, relative to the actual volume of the aggregate, and the result was 0.0000013m3, which corresponds to a deviation in the range of 0.026% of the model value. The mean deviation of all the variants of possible configurations during the processing of data was 0.274% for test 1 and 2.740% for test 2. Obviously, the 3D laser scanning technology is suitable to measure volumes of earthen masses in civil engineering. Besides, a scan with a higher resolution of measurements yielded results more accurate by an order of values than observations with a laser device set at lower resolutions.

Despite the mentioned difficulties, it is doubtless that the 3D laser scanning technology proves useful in each of the analyzed situations. It is an exceptionally promising technology with endless applications in civil engineering, whose number and diversity are only limited by human imagination.

4. References

[1] Pawłowicz J. A., 2017 Importance of Laser Scanning Resolution in the Process of Recreating the Architectural Details of Historical Buildings, *WMCAUS, IOP Conference Series: Materials Science and Engineering,* Volume 245 (5)

[2] Rapp P., 2015 Methodology and examples of revalorization of wooden structures in historic buildings, *Wiadomosci Konserwatorskie – Journal of Heritage Conservation*, vol. 43, pp. 92-108,

[3] Kaspar M., Pospisil J., Stroner M., Kremen T. and Tejkal M., 2004 Laser Scanning in Civil Engineering and Land Surveying, *Vega s.r.o.*, Republika Czeska;

[4] Pawłowicz J. A., Bilko P., Sawczyński Sz. and Szafranko E., 2017 Diagnostic of Geometry Distortion of a Wooden Structure Based on a Point Cloud, *Geodetic Congress (BGC Geomatics)*, 2017 Baltic, IEEE, pp. 163-168

[5] Markley J. D., Stutzman J.R. and Harris E. N., 2008 Hybridization of Photogrammetry and Laser Scanning Technology for As-Built 3D CAD Models, *IEEE Aerospace Conference*, pp: 1 – 10,

Piezoelectric coefficients of the barium-strontium titanate film on a (111)-oriented substrate

V B Shirokov[1,2], R A Shakhovoy[3], A V Pan'kin[1,2] and V V Kalinchuk[1,2]

1 Southern Scientific Center of Russian Academy of Sciences, Rostov-on-Don, 344006, Russia
2 Southern Federal University, Rostov-on-Don, 344090, Russia
3 Sir Peter Mansfield Imaging Centre, School of Physics and Astronomy, University of Nottingham, Nottingham NG7 2RD, UK
E-mail: shirokov-vb@rambler.ru

Abstract. Within the framework of the phenomenological theory of phase transitions, we consider properties of the $Ba_{0.8}Sr_{0.2}TiO_3$ (BST80) film grown on a (111)-oriented cubic substrate. The film was revealed to exhibit three ferroelectric phases with different orientations of spontaneous polarization. It is shown that some piezoelectric moduli should demonstrate anomalous behavior near the phase boundaries.

Development of contemporary ferroelectric-based microelectronic devices [1,2,3] generally requires optimization of their operating parameters using numerical calculations. The cornerstone of the microelectronic components design is a selection of materials with required properties; therefore, modeling of electromechanical properties of ferroelectric thin films using non-linear thermodynamics [4] becomes a necessary stage of the material approbation. A special role in the thin-film technology is assigned to the substrate, whose properties and orientation significantly affects the features of the film.

Heteroepitaxial films are subjected to misfit strains from the substrate, which change the symmetry of the film. According to the Curie principle, the high-symmetry phase gets the symmetry group being the intersection (a general subgroup) of the symmetry groups of the bulk material and external action. Thus, under uniaxial pressure along the 4-fold cubic axis, the symmetry of the paraelectric cubic phase, $Pm3m(O_h^1)$, falls down to the $P4/mmm(D_{4h}^1)$ symmetry. Such situation occurs when the perovskite film is grown on the (001)-oriented cubic substrate. Types of ground states and misfit strain-temperature (u_m,T) phase diagrams of barium-strontium titanate (BST) thin films grown on the (001)-oriented cubic substrate are considered in ref. [5]. It is shown in other works [6,7] that material constants of analogous barium titanate (BT) film should exhibit anomalous behavior near the phase boundaries.

Under pressure along the cube spatial diagonal, the $Pm3m$ symmetry group falls down to the $R\bar{3}m(D_{3d}^5)$ group. Such symmetry lowering corresponds to the perovskite film grown on the (111)-oriented cubic substrate. Possible ground states and phase diagrams of the corresponding BST films are considered in ref. [8]. It is revealed that the $R\bar{3}m$ symmetry provides 13 possible low-symmetry phases [8]; however, only few of them can appear as a ground state. Fig. 1 demonstrates the misfit strain-concentration (u_m,x) phase diagram of the (111)-oriented BST film at $T=300$ K (x is the concentration of Ba in a solid solution). Along with the high-symmetry phase (*0*-phase), following phases occur:

- the *c*-phase of the C_{3v} symmetry and with the spontaneous polarization oriented perpendicular to the film surface ($p_x=0, p_y=0, p_z\neq 0$);

- the *a*-phase of the C_2 symmetry and with the polarization lying in the plane of the film ($p_x \neq 0, p_y = 0, p_z = 0$);
- the *r*-phase of the C_s symmetry with two non-zero components of the polarization ($p_x = 0, p_y \neq 0, p_z \neq 0$) oriented at an angle to the film surface.

We don't consider here the r_φ phase (left bottom corner of the diagram in Fig. 1) with the rotational order parameter.

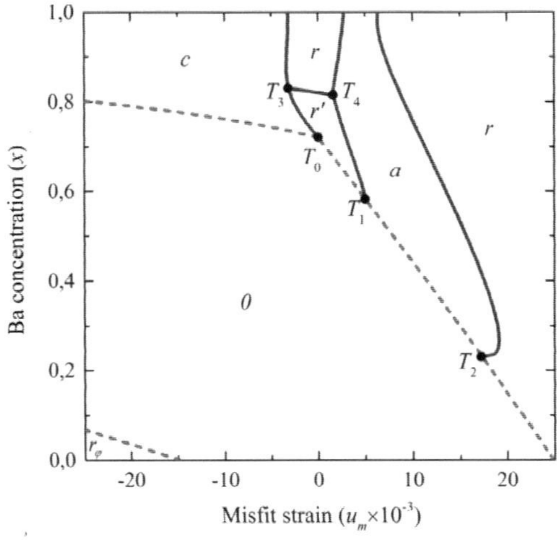

Fig. 1: (Colour on-line) Theoretical arrangement of phases occurring in the $Ba_xSr_{1-x}TiO_3$ film at $T = 300$ K on the misfit strain-concentration (u_m, x) plane. Dashed and solid lines correspond to the first- and second-order phase transition boundaries, respectively. Coordinates of points (u_m, x) are: $T_0(0, 0.72)$, $T_1(5.0\times10^{-3}, 0.58)$, $T_2(17.2\times10^{-3}, 0.23)$, $T_3(-3.2\times10^{-3}, 0.83)$, $T_4(1.6\times10^{-3}, 0.82)$.

We are interested here in the cross section of the diagram in Fig. 1 at $x = 0.8$, i.e. we focus below on the properties of the $Ba_{0.8}Sr_{0.2}TiO_3$ (BST80) film, where all three ferroelectric phases (*c*, *a*, and *r*) may appear. Coefficients of the thermodynamic potential for $x = 0.8$ may be calculated using an approach developed in ref. [5] or can be found in ref. [9]. Theoretical dependences of the polarization components on the misfit strain in different phases are shown in Fig. 2 by colored solid lines. The value of the tetragonality as a function of the misfit strain is shown by the dashed line.

One can see that at small strains the ground state of the film corresponds to the *r*-phase with the polarization vector ($0, p_y, p_z$). The in-plane component of the polarization tends to zero when increasing the clamping stress ($u_m < 0$), whereas the *z*-component increases towards the $r-c$ phase boundary ($u_m = -2.64\times10^{-3}$). In the *c*-phase ($u_m < -2.64\times10^{-3}$), the in-plane component disappears and p_z slowly decreases toward higher values of the clamping stress.

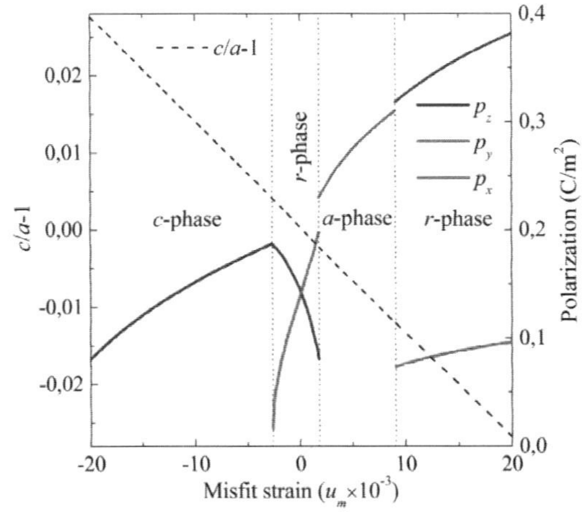

Fig. 2: (Colour on-line) Components of the polarization vector (solid lines) and the tetragonality (dashed line) of the $Ba_{0.8}Sr_{0.2}TiO_3$ film as a function of the misfit strain. Dotted vertical lines indicate phase boundaries.

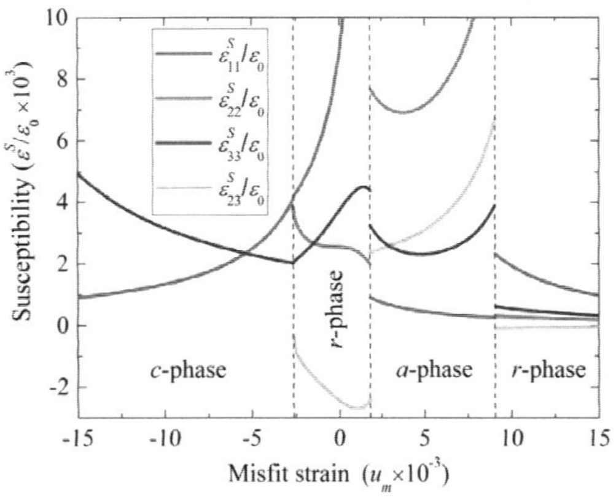

Fig. 3: (Colour on-line) Misfit strain dependences of the dielectric constant components in the (111)-oriented $Ba_{0.8}Sr_{0.2}TiO_3$ film.

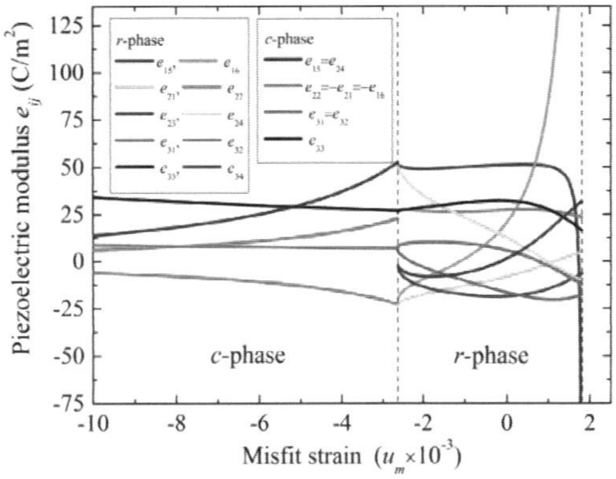

Fig. 4: (Colour on-line) Misfit strain dependences of piezoelectric moduli in the (111)-oriented $Ba_{0.8}Sr_{0.2}TiO_3$ film (*c*- and *r*-phases).

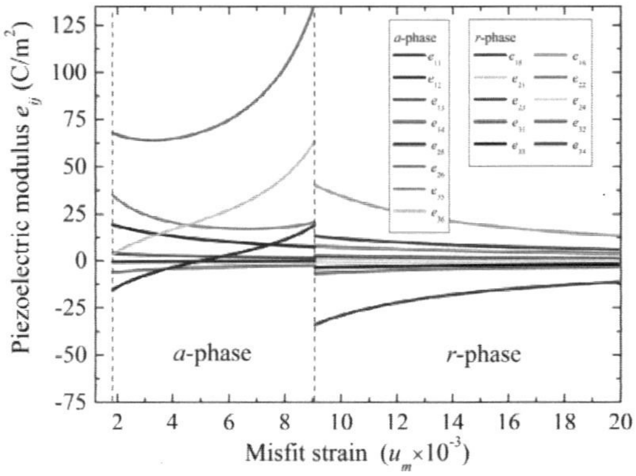

Fig. 5: (Colour on-line) Misfit strain dependences of piezoelectric moduli in the (111)-oriented $Ba_{0.8}Sr_{0.2}TiO_3$ film (*a*- and *r*-phases).

Under tensile stress, the *r*-phase is stable up to $u_m = 1.81\times10^{-3}$; however, growth of the strain causes an increase of the in-plane polarization, whereas p_z diminishes. At $1.81\times10^{-3} < u_m < 9.06\times10^{-3}$, both polarization components (p_y and p_z) disappear and the *a*-phase becomes stable with the polarization vector to be oriented along the *x*-axis. Increase of the tensile stress is accompanied by the growth of the in-plane polarization in the *a*-phase. Finally, at $u_m > 9.06\times10^{-3}$ the ground state again returns to the *r*-phase.

Theoretical misfit strain dependences of the dielectric constant components in the BST80 film are shown in Fig. 3. One can see that all components exhibit jumps on the first-order phase transition boundaries, although these jumps are almost negligible on the boundary between *c*- and *r*-phases. Note also that in the *c*-phase, increase of the ε_{33}^S component towards higher clamping stress is caused by approaching to the second-order phase transition boundary between the *c*-phase and the high-symmetry *0*-phase (see Fig. 1). Theoretical misfit strain dependences of piezoelectric moduli, e_{ij}, are shown in Figs. 4,5. One should note anomalous behavior of e_{15} and e_{16} moduli in the *r*-phase (Fig. 4), which exhibit abrupt change on the boundary with the *a*-phase. It is interesting that many of moduli change their sign within the *r*-phase. In the *a*-phase, three piezoelectric moduli, e_{25}, e_{26}, and e_{36}, exhibit anomalous behavior. One of them (e_{25}) changes its sign. In the *r*-phase occurring at large misfit strain ($u_m > 9.06\times10^{-3}$), all moduli change smoothly without any singularities.

Behavior of the piezoelectric moduli described above should be taken into account when developing microelectronic devices employing BST80 films on (111)-oriented cubic substrates. One may estimate the misfit strain and corresponding ground state of a real film from Fig. 2 using the tetragonality $c/a-1$ of the film, which can be determined from the X-ray data analysis.

This study was supported by the Russian Scientific Foundation (project no. 14-19-01676).

[1] Scott J F Science 2007 **315** 954
[2] Dawber M Rabe K M Scott J F Rev. Mod. Phys. 2005 **77** 1083
[3] Subramanyam G Cole M W Sun N X Kalkur T S Sbrockey N M Tompa G S Guo X. Chen C Alpay S P. Rossetti G A Dayal Jr K Chen L-Q and Schlom D G J. Appl. Phys. 2013 **114** 191301
[4] Khassaf H Khakpash N Sun F Sbrockey N M Tompa G S Kalkur T S Alpay S P Applied Physics Letters 2014 **104** 202902
[5] Shirokov V B Yuzyuk Yu I Dkhil B Lemanov V V Phys. Rev. B. 2009 **79** 144118
[6] Shirokov V Kalinchuk V Shakhovoy R Yuzyuk Y EPL 2014 **108** 47008
[7] Shirokov V B Kalinchuk V V Shakhovoy R A and Yuzyuk Yu I Phys. Solid State 2015 **57** 1535
[8] Shirokov V B Shakhovoy R A Razumnaya A G and Yuzyuk Yu I J. Appl. Phys. 2015 **118** 024101
[9] Shirokov V B Kalinchuk Yuzyuk Yu I Lemanov V V Phys. Solid State 2013 **55** 773

IOP Conf. Series: Materials Science and Engineering **324** (2018) 012007 doi:10.1088/1757-899X/324/1/012007

Thermal resistance approach to analyse temperature distribution in thick hollow FGM sphere: under dirichlet boundary condition

S M Shayak Ibna Faruqui[1], Al Arabi[2], M S Parvej[3] and Md. Atikur Rahman[4]

1, 3 & 4 Mechanical Engineering Department, Bangladesh University of Engineering and Technology, Dhaka, Bangladesh
2 Independent University, Bangladesh
E-mail: shayakbuet08@gmail.com

Abstract. A thick hollow FGM sphere with radially varying material property is considered under thermal loading. Thermal conductivity is assumed to vary radially following two different power functions separately and results for both have been analysed. Temperature distribution is obtained using Thermal Resistance Approach (TRA) where spatially varying thermal conductivity is assumed to be a discontinuous piece-wise function of radial distance rather than the continuous power functions. Whole body is considered to be constructed of layers having different material property but each layer is assumed to be homogeneous. Then temperature values are obtained considering each layer's thermal resistance and the total thermal resistance network. Results were compared with other established studies and results obtained by FEM.

1. Introduction

Functionally graded materials are a class of superior composite which are different from the traditional composite materials where variation of material properties along any direction is discontinuous but in FGMs this variation is continuous and smooth which allows researchers to tailor the gradient of material property variation with more flexibility. FGMs are used in various engineering fields because of their advantages over various metal alloys and composite materials to meet extreme material requirements. Many researchers have worked on FGMs since they were first introduced in Japan in 1984 [1]. Yahya et al [1] analysed the behaviour of thick FGM spheres under combined thermal and pressure loading. Jabbari et al [2] presented a general solution for the one dimensional steady state thermo-mechanical stresses in hollow thick porous FGM sphere. Jabbari et al [3] also developed an analytical method to obtain the solution for 2D transient thermal and mechanical stresses in a hollow FGM sphere and piezoelectric layers. S Faruqui et al [4] developed thermal resistance method for solving the temperature distribution problem in hollow FGM cylinders. S Karampour [6] presented solution for 2D thermal and mechanical stresses in poro-FGM spherical vessel. X L Peng and X F Li [7] analysed the thermo-elastic behaviour of cylindrical FGM vessels. Zimmerman R W and Lutz M P worked on thermal stresses and thermal expansion in uniformly heated FGM cylinders. In this paper temperature distribution in FGM sphere is analysed using thermal resistance approach. Results have been obtained for two different power functions of material gradient and also were compared with existing results in the literature. The advantage of this method over others is that results can be obtained for any type of material property variation where obtaining analytical solutions are much more complex due to nonlinearity.

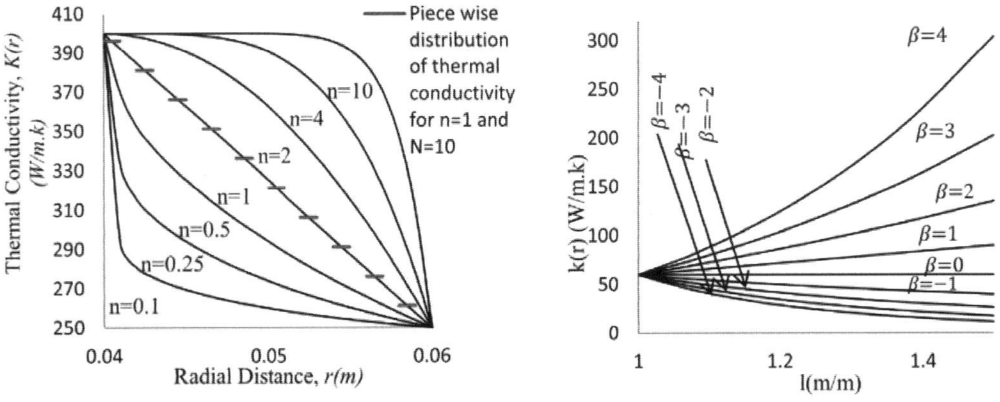

(a) (b)

Figure 1. (a)Radial distribution of $K_{fgm}(r)$ following power function equation (1) and an assumed piecewise distribution for Thermal Resistance Approach. (b) Radial distribution of $K_{fgm}(r)$ following equation (2).

2. Formulation of problem

A hollow spherical body is considered with inner radius r_i and outer radius r_o, thermal conductivity $K_{fgm}(r)$ is assumed to vary radially from k_i to $k_o$following any of the power functions of equation (1) and (2) where k_i and k_o are thermal conductivities of innermost and outermost surfaces respectively.

$$K_{fgm}(r) = k_i + (k_o - k_i)\left(\frac{r-r_i}{r_o-r_i}\right)^n \quad (1)$$

$$K_{fgm}(r) = k_c e^{\beta l} \quad (2)$$

In equation (1) "n" is the power law index and in equation (2) k_c is a material constant, "e" is the Euler constant, "β" is the inhomogeneity parameter and $l = \frac{r}{r_i}$.

2.1 Thermal problem and boundary condition

Taking spherical coordinate system into account and assuming the temperature of the spherical body varies only in radial direction and no internal heat generation the obtained governing equation is

$$\frac{d}{dr}\left(k_{fgm}(r) r^2 \frac{dt}{dr}\right) = 0 \quad (3)$$

Solutions were obtained under dirichlet boundary condition when temperature values at innermost and outermost surfaces of the body are known. These values were taken as $T(r_i) = T_i$ and $T(r_o) = T_o$ respectively.

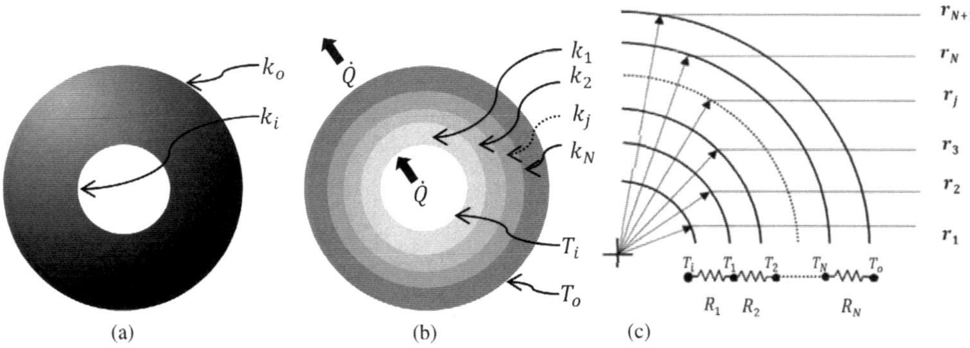

Figure 2. (a) Continuous distribution of k_{fgm}, (b) Layered distribution of k_{fgm} , (c) Thermal resistance network across the layers.

3. Proposed solution procedure (Thermal Resistance Approach)

For dirichlet boundary conditions a solution can be obtained by dividing the whole spherical body into layers as shown in Figure 2(b) and considering each layer a constituent element having homogenous material property throughout. Each element's thermal conductivity can be assigned to it from equation (1) or (2) by taking average of the values at innermost and outermost boundaries of that particular layer. The radial variation of thermal conductivity of the whole body is then assumed to be a piece-wise function of radial distance, r which is shown in Figure 1(a). If the spherical body is divided into 'N' number of layers or elements, each layer will have a thermal conductivity according to this piece-wise variation. After getting 'N' number of different thermal conductivities for 'N' different layers, thermal resistance can be computed for each of them. Thermal resistance R_j across the jth layer can be determined using the following equation

$$R_j = \frac{r_{j+1}-r_j}{4\pi r_{j+1} r_j K_j} \tag{4}$$

Where r_j and r_{j+1} are inner and outer radii of the jth layer respectively. The total thermal resistance across the inner and outer surface of the whole body can be calculated from equation (5) where R_1, R_2, R_3, etc. are the thermal resistances of different layers.

$$R_{total} = R_1 + R_2 + \cdots + R_j + \cdots + R_N \tag{5}$$

However, in this approach it is assumed that there is no heat loss or additional thermal contact resistance at the contact surface of two adjoin layers. Temperature values are obtained at each of the contact surfaces of constituent layers. Temperature at contact surface of the jth and $(j+1)^{th}$ element at $r = r_{j+1}$ is computed from (6). When $T(r_i) = T_i$ and $T(r_o) = T_o$,

$$T_{j+1} = \frac{(R_1+R_2+R_3+\cdots+R_{j-1}+R_j)T_o+(R_{j+1}+R_{j+2}+R_{j+3}+\cdots+R_{N-1}+R_N)T_i}{R_{total}} = \frac{\left(\sum_1^j R\right)T_o+\left(\sum_{j+1}^N R\right)T_i}{R_{total}} \tag{6}$$

3.1 Finite element analysis and validate FEM

A geometry specimen was modeled using commercial FEM solver ANSYS for a comparative study of the proposed method. Due to symmetry only one eight of a sphere was considered. The FEM model consisted of tota 1238 elements. Variation in material property was obtained by dividing the specimen into 5 homogeneous layers each layer having same but distinct material property. The meshing region as well as the obtained temperature distribution is shown in Figure 4(a).

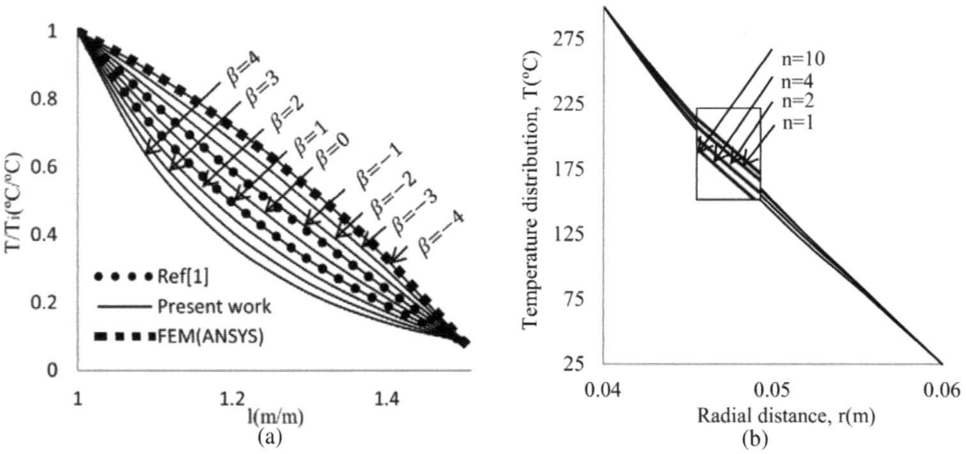

Figure 3. (a) Radial distribution of temperature for power function equation (2), (b) Radial distribution of temperature for power function equation (1).

4. Result, validation and discussion

Results were obtained considering radial distribution of thermal conductivity to follow equation (1) and (2) separately. Obtaining results for power function equation (2) the values of material and geometry parameters were taken as reference [1]. The material property was taken as $k_c = 60\frac{W}{m°C}$ and geometric properties, inner and outer radii of the sphere were taken as $r_i = 40mm$ and $r_o = 60mm$ respectively Boundary condition was $T(r_i) = 300°C$ & $T(r_o) = 25°C$. Results were obtained for inhomogeneity parameter $\beta = -4, -3, -2, -1, 0, 1, 2, 3$ *and* 4 as shown in figure 3(a). From the figure 3(a) it can also be observed that results obtained by thermal resistance approach are in well agreement with results obtained by Yahya et al [1] and FEM of this study using ANSYS solver. For power function equation (1) boundary condition and geometric parameters were kept same as previous and reference [1]. Material properties were taken as $k_i = 400 \frac{W}{m°C}$ and $k_o = 250 \frac{W}{m°C}$. Figure 1(a) shows the radial distribution of k_{fgm} for power law index $n = 0.1, 0.25, 0.5, 1, 2, 4$ *and* 10. Figure 3(b) shows obtained results for $n = 1, 2, 4, 10$ by thermal resistance approach. Table 1 contains a comparison among obtained results from thermal resistance approach and Yahya et al [1] for $\beta = 1$ & -1 and we can conclude that the results agree with each other. Figure 4(a) shows mesh and the radial temperature distribution obtained by FEM using ANSYS solver. Figure 4(b) is the radial distribution of thermal resistance for different values of β.

Table 1. Radial temperature distribution from this work and Yahya et al [1]

	$T/T(r_i)$					
	$\beta = 1$			$\beta = -1$		
r/r_i	This Work	Yahya[1] (Analytical)	Yahya[1] (FEM)	This Work	Yahya[1] (Analytical)	Yahya[1] (FEM)
1	1	1	1	1	1	1
1.0875	0.745182	0.74516	0.74574	0.810369	0.81036	0.81052
1.1875	0.520155	0.52008	0.52050	0.611509	0.61148	0.61162
1.2875	0.345443	0.34538	0.34568	0.428719	0.42870	0.42881
1.3875	0.207095	0.20707	0.20729	0.259595	0.25959	0.25968
1.5	0.083333	0.083333	0.08333	0.083333	0.083333	0.083333

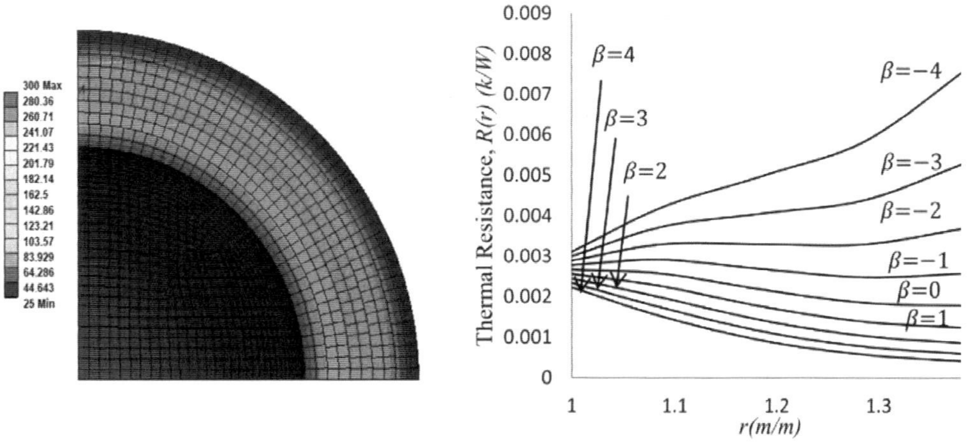

(a) (b)

Figure 4. (a) Temperature distribution obtained by FEM (ANSYS) for $\beta = -4$, (b) Radial distribution of thermal resistance for different

5. Conclusion

A solution to the temperature distribution problem of FGM hollow sphere is presented in this study for two different power law functions separately using thermal resistance approach. Solutions were obtained for different values of inhomogeneity parameter β for distribution of equation (2) and different values of power law indexes n for equation (1). Obtained results are also compared with some well-established data present in the literature and with results obtained from FEM using ANSYS solver. It is observed that results obtained in this study using thermal resistance approach comply very well with them.

6.Referance

[1] Yahya Bayat, M Ghannad and H Torabi 2011 Analytical and numerical analysis for the FGM thick sphere under combined pressure and temperature loading *Arch Appl Mech (2012) 82:229-242*

[2] M Jabbari, S Karampour and M R Eslami 2011 Radially symmetric steady state thermal and mechanical stresses of a poro FGM hollow sphere *international Scholarly Research Network ISRN Mechanical Engineering* Volume 2011, Article ID 305402, 7 pages

[3] M Jabbari, S M Mousavi and M A Kiani 2017 Functionally Graded hollow sphere with piezoelectric internal and external layers under asymmetric transient thermomechanical loads *Journal of Pressure Vessel Technology* Vol. 139 / 051207-1

[4] S Faruqui, A Arabi and M S Parvez 2017 Thermal Resistance Approach to analyze temperature distribution in hollow cylinders made of Functionally Graded Material (FGM): under dirichlet boundary condition *ICMIME-2017* RUET Rajshahi

[5] Yunus A. Cengel and Afshin J. Ghajar 2015 Heat and Mass Transfer Fundamentals and Applications (McGraw Hill) Chapter 3

[6] S. Karampour 2012 Dimensional two steady state thermal and mechanical stresses of a Poro-FGM spherical vessel *Procedia - Social and Behavioral Sciences* 46 (2012) 4880 – 4885

[7] X.L. Peng and X.F. Li 2010 Thermoelastic analysis of a cylindrical vessel of functionally graded materials *International Journal of Pressure Vessels and Piping*, 87 (2010) 203

[8] Zimmerman RW and Lutz MP 1999 Thermal stress and thermal expansion in a uniformly heated functionally graded cylinder *J. Therm. Stress* 22:177e88

IOP Conf. Series: Materials Science and Engineering **324** (2018) 012008 doi:10.1088/1757-899X/324/1/012008

Temperature Field Simulation of Powder Sintering Process with ANSYS

Hongxiu He, Jun Wang[a,*], Shuting Li, Zhilong Chen, Jinfeng Sun, Ying You

Center for Robotics Research, School of Mechanical Engineering,
Hubei University of Technology,Wuhan, Hubei,430068,China
[a]E-mail: junwang@mail.hbut.edu.cn

Abstract: Aiming at the "spheroidization phenomenon" in the laser sintering of metal powder and other quality problems of the forming parts due to the thermal effect, the finite element model of the three-dimensional transient metal powder was established by using the atomized iron powder as the research object. The simulation of the mobile heat source was realized by means of parametric design. The distribution of the temperature field during the sintering process under different laser power and different spot sizes was simulated by ANSYS software under the condition of fully considering the influence of heat conduction, thermal convection, thermal radiation and thermophysical parameters. The influence of these factors on the actual sintering process was also analyzed, which provides an effective way for forming quality control.

1.Introduction

As a modern manufacturing technology, selective laser sintering technology has its own unique advantages. In theory, it can make almost any shape of parts [1] . As one of the hot spots, metal laser selective sintering has been widely studied by scholars. The most common problems are the phenomenon of "spheroidization" and warpage. These are caused by the thermal coupling between the metal powder and the laser. Therefore, the study of the distribution of temperature field in the process of laser sintering and the relationship between forming characteristics and process parameters is the key to solving the problem[2].

2.Temperature field simulation

2.1 .Establish finite element model

The establishment of finite element model is divided into three steps, the size of the shape, the definition of material properties and the division of the grid. In order to meet the size of heat-affected zone during sintering, a rectangular block of size 100mm × 50mm × 6mm was established as the shape of the finite element model. Water atomized iron is regarded as a common engineering practice materials because of their excellent material properties in the water atomized iron powder. The grid is divided by considering the grid type, the grid size and the grid distribution. The solid70 hexahedron eight-node cell is used as meshing model[3], which can fill the heat loss due to the new added powder material[5]. The control grid size is 0.8mm, which subdivides the research object at the edge of the model.

2.2 .Determine the constraints

Constraints are the basis to determine the general solution or solution in the simulation analysis

process[4]. In the simulation of thermal analysis, there are a variety of heat transfer ways, such as radiation, conduction and convection, which are used to determine the following constraints.
The first one is the space temperature distribution in the material before simulation, which can be expressed by the equations (1) and (2) as ,

$$T(x,y,z,t)\big|_{t=0} = T_0 \tag{1}$$

$$T(x,y,z,t)\big|_{t=0} = \phi(x,y) \tag{2}$$

In the above two equaitons, the initial temperature of the object is the temperature distribution in the object, both of which indicate that the initial temperature distribution of the object is not uniform.
The second constraint is the temperature distribution of the space in contact with the experimental powder material, which can be expressed by equations (3) and (4).

$$\mathrm{T}\big|_{\Gamma} = \mathrm{T}_{\omega} \tag{3}$$

$$T\big|_{\Gamma} = f(x,y,t) \tag{4}$$

In Eqs. (3) and (4), T_{ω} is the initial boundary temperature of the space in contact with the powder material, and $f(x,y,t)$ is the expression of the temperature of the space boundary. Eq. (4) indicates that the temperature of the initial boundary of this space changes with time.
The third is the input and output of powder material boundary energy, which can be expressed by Eqs.(5) and (6).

$$-k\frac{\partial T}{\partial n}\bigg|_{\Gamma} = q \tag{5}$$

$$-k\frac{\partial T}{\partial n}\bigg|_{\Gamma} = g(x,y,t) \tag{6}$$

In Eqs.(5) and (6), q is the heat flux density value on the surface of the powder material, and $g(x,y,t)$ is the heat flux density function.
The fourth is the thermal convection between the powdered material and the molten liquid metal, which can be expressed by Eq.(7).

$$-k\frac{\partial T}{\partial n}\bigg|_{\Gamma} = a(T - T_f)\big|_{\Gamma} \tag{7}$$

In Eq.(7), a is the heat transfer coefficient between the powder material and the liquid metal and T_f is the liquid metal temperature.

2.3. Determine material parameters and process parameters

In the process of thermal analysis, it needs to define the material's specific heat capacity, thermal conductivity and other material parameters[6]. The final value of specific heat capacity and thermal conductivity are obtained from the experimental measurements as shown in Table 1.The process parameters are some reasonable experimental data to perform the experiment setup, as shown in Table 2.

Table 1 Thermal properties of the parameters

Temperature $/K$	273	573	673	773	873	1073	1173	1273	1373	1473	1573	1773
Coefficient of thermal conductivity $/W\cdot m^{-1}\cdot K^{-1}$	74.7	55.5	49.2	43	38.2	74.7	29.4	28.2	29.3	32.2	32.2	32.2
Specific Heat Capacity $/J\cdot kg^{-1}\cdot K^{-1}$	435.1	552.3	602.5	661.1	753.1	832.6	656.9	569	606.7	640.2	736.4	836.8

Table 2 Process parameters

Laser power P/W	Spot diameter D/mm	Scanning speed $V/mm\cdot s^{-1}$	Convection coefficient	Bed temperature $T/^{o}C$
215	0.4	50	10	20

2.4.Mobile Gaussian heat source simulation

During the process of laser sintering, the laser moves at a constant velocity. The energy is transmitted to the powder material in the form of heat flux during the movement. The laser's power density follows the Gaussian distribution, which can be expressed as follows,

$$q(x,y)=\frac{2AP}{\pi\omega^2}\exp(-2\frac{x^2+y^2}{\omega^2}) \tag{8}$$

In Eq.(9), $q(x,y)$ is the laser power density at point (x,y). A is the absorption of the laser energy by the powder material. P is the output power of the laser and ω is the laser power set in the experiment.

3.Results and analysis

3.1 Temperature field analysis under given conditions

After the process of model creation, meshing, material definition, unit type definition and definitive solution conditions, the temperature profile of the surface and side of the model at 0.44 s is obtained, as shown in Fig. 1 and Fig.2.

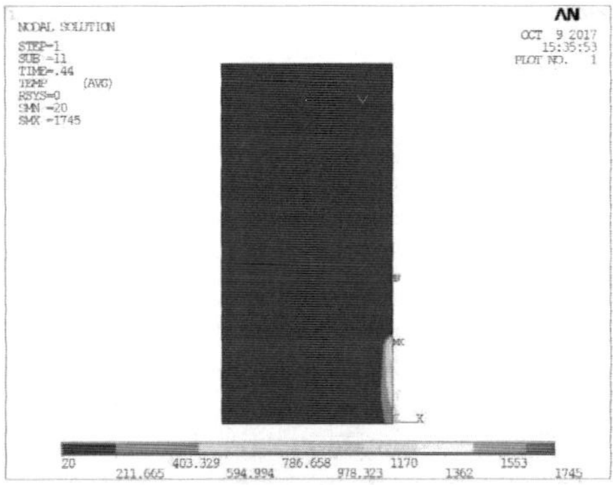

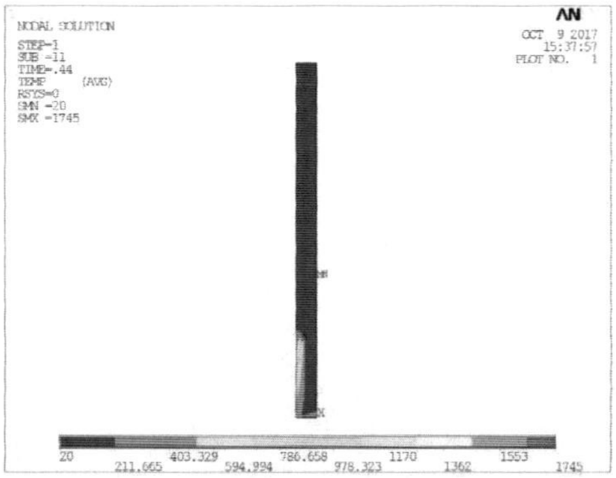

Figure 1 Temperature distribution in model surface **Figure 2** Temperature distribution in model side

From Fig. 1 and Fig. 2, the maximum temperature of the temperature field lags slightly behind the laser spot center because the laser contacts the water atomized iron powder at room temperature in the direction of the laser movement[7], which is different from the maximum temperature of the laser source at the laser spot center.

In order to find out the relationship between the quality of formed parts and the setting of laser parameters during the laser sintering process, the relationship between the quality of formed parts and the laser parameters during laser sintering process is discussed separately by changing the two intrinsic parameters of laser power and laser spot size.

3.2 Analysis of the temperature field under the change of laser power

The laser power is changed with other conditions remain unchanged. When the laser power is 100W and 400W, the temperature profile of the model at 0.44s is obtained, as shown in Figure 3 and Figure 4.

It can be seen from Figure 3 that when the laser power is 100W, the temperature at the center of the sintering spot is only 934.486°C due to insufficient energy compared with that shown in Figure 2 and Figure 4. In Figure 2 and Figure 4, the temperature does not reach the melting point of the powder material and the material can not be melted. Powder can not be sintered nor be shaped into parts.

As shown in Figure 4, when the laser power reaches 400W, the laser spot temperature is as high as 3293°C due to too much laser energy. On the one hand, the increase of laser power contributes to the increase of the temperature in the whole region, and the length of the linear region above the melting temperature of the iron increases[8]. On the other hand, the temperature of the spot center is much higher than the melting point of the iron material, the melting speed of the material is too fast and vaporization may even occur . At this moment, the melted material has too much fluidity, resulting in shear stress between the two layers of powder, which may cause warpage and deformation. It is not easily conducive to molding and may even have over-burning phenomenon. The adhesion between the layers may also affect the sintered surface of the uneven phenomenon.

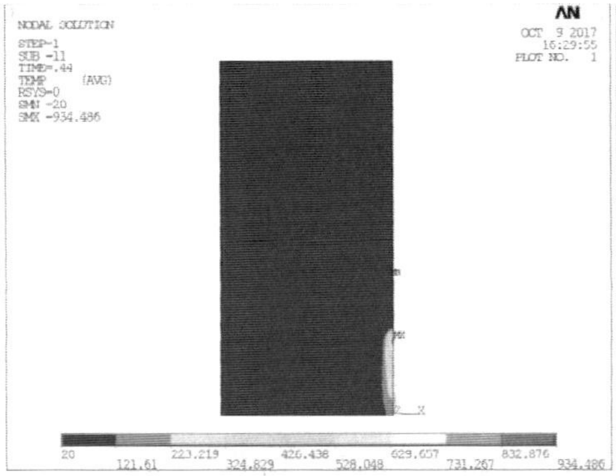

Figure 3 Temperature distribution at 100W

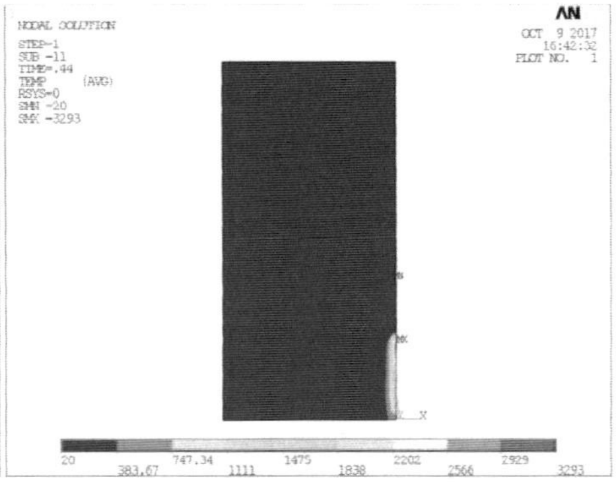

Figure 4 Temperature distribution at 400W

3.3 Analysis of the temperature field under the change of spot sizes

Under the given conditions, when the laser spot size is 0.2 mm and 0.8 mm, the temperature profile is obtained with only changing the laser spot sizes, as shown in Figure 5 and Figure 6.

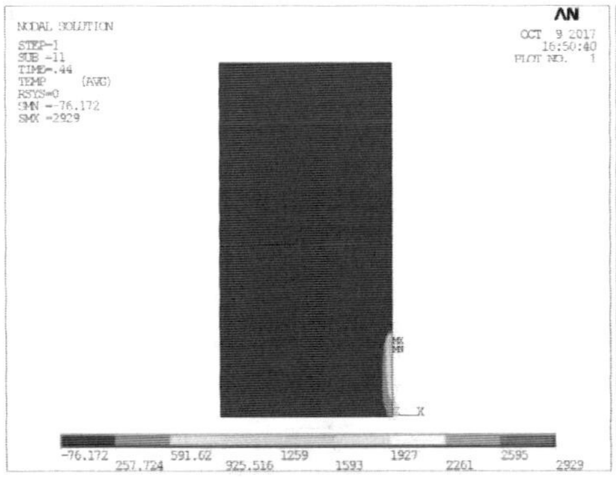

Figure 5 Temperature distribution at 0.2mm

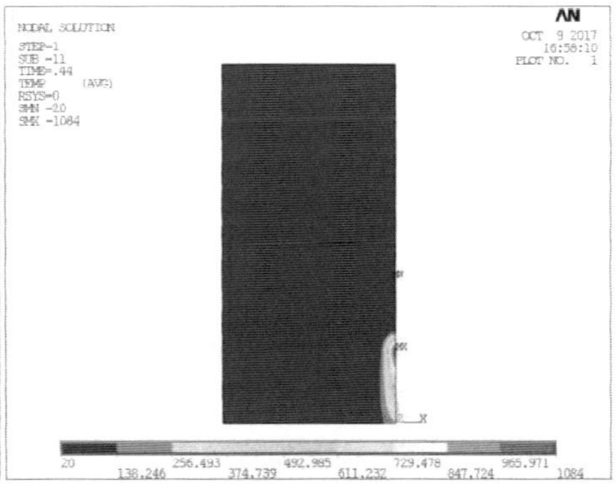

Figure 6 Temperature distribution at 0.8mm

As it can be seen from Figure 5, when the spot size is 0.2mm, the laser energy is densely distributed. Compared with Figure 1 and Figure 5, in the case of the same size of laser power and the same size of energy input, the center temperature is up to 2929°C and the metal powder can be melted to a liquid state faster. In addition, as the linear spot length is larger than the melting point of the iron material, the length of the linear region become larger than the melting point of the iron material which the spot size decreases. On the one hand, when the proportion of liquid phase increases, the liquid-phase sintering mechanism can conduct a more thorough, internal tissue adhesion and the quality of the molded parts become better[9]. On the other hand, the sintering center of the spot temperature is as high as 2929°C, which may cause the forming parts appearing as "ball phenomenon", "warp deformation" and other defects[10].

It can be seen from Figure 6 that when the spot size is 0.8mm, the spot center temperature decreases and the laser energy distribution is dispersed. Comparing Figure 1 and Figure 2, it can be seen that with the same size of laser power and same size of energy input, the temperature of the spot center is lower when the spot size is larger. The maximum temperature is 1084°C, which is lower than the melting point of iron material and the liquid phase sintering mechanism can not be carried out nor be shaped.

4.Conclusion

The three-dimensional transient temperature field of metal powder was numerically simulated. The water atomized iron powder was used as the research object. The finite element model was established and the Gaussian moving light source was loaded on the model surface. At the same time, the influence of convection heat transfer was considered. The simulated temperature field distribution of 0.44s was analyzed under different laser powers and different spot sizes. The results show that within a certain range, when the laser power increases, the spot size decreases and the sintering spot temperature increases, which can effectively improve the quality of the forming parts. When the laser power is too high, the spot size is too small and the workpiece appears "spheroidization", "warpage" and other defects, resulting in the quality of forming parts. This paper proposed some reasonable choices of laser power, spot size of the laser sintering process, which may benefit the research of laser sintering .

5. References

[1] Xu C T, Shen Y F, Gu D D and Yu C Y 2004 . *Cast,* 511-15
[2] Greco A and Maffezzoli A 2000 *J. Rapid Prototype* 2 24-40
[3] Guo H F, Zhou J Z and Hu Z R 2006 *Tool Technol* 40 13-17
[4] Dai D and Gu D 2016 *Int J Mach Tool & Manuf* 100 14-24
[5] Hu R X and Kang S T 2016 *Machinery Industry Press*
[6] Wang W F 2008 *North University of China.*
[7] Simchi and Pohl H r 2003 *Mat and Eng* A359119-28
[8] Gusarov A V, Laoui T and Froyen L，et al. 2003 *Int J Heat and Mass Transfer* 46 1103-09
[9] Ma C,Gu D,Dai D,et al.2015 *J Mat Research* 30 2816-28
[10] V. Morozov, V. G. Gusev, T. P. Dvoryaninova 2016 *Rus Eng Research*, 36 1004-07

Acknowledgments

The authors thank for the support of the Outstanding Youth Science and Technology Innovation Team Program of HuBei Provincial Department of Education Grant(T201505), the Key Project of Natural Science Foundation of Hubei Province Grant(2015CFA112) and the major project of technical innovation in Hubei Province Grant(2017AAA108).

IOP Conf. Series: Materials Science and Engineering **324** (2018) 012009 doi:10.1088/1757-899X/324/1/012009

An Optimized Trajectory Planning for Welding Robot

Zhilong Chen, Jun Wang*, Shuting Li, Jun Ren , Quan Wang, Qunchao Cheng, Wentao Li

Center for Robotics Research, School of Mechanical Engineering,
Hubei University of Technology,Wuhan, Hubei,430068,China
[a]E-mail: junwang@mail.hbut.edu.cn

Abstract: In order to improve the welding efficiency and quality, this paper studies the combined planning between welding parameters and space trajectory for welding robot and proposes a trajectory planning method with high real-time performance, strong controllability and small welding error. By adding the virtual joint at the end-effector, the appropriate virtual joint model is established and the welding process parameters are represented by the virtual joint variables. The trajectory planning is carried out in the robot joint space, which makes the control of the welding process parameters more intuitive and convenient. By using the virtual joint model combined with the B-spline curve affine invariant, the welding process parameters are indirectly controlled by controlling the motion curve of the real joint. To solve the optimal time solution as the goal, the welding process parameters and joint space trajectory joint planning are optimized.

1. Introduction

Since the 1960s, the manual welding has been replaced by welding robots which have a series of advantages such as flexible work, high production efficiency, and continuous work. So welding robots have been widely used in the mechanical processing, weapons industry, automobile manufacturing and other fields [1]. In the welding process, a reasonable space trajectory planning can improve the welding quality, welding efficiency and reduce welding costs, which is of great significance in the actual industrial production [2].

A variety of studies have been stated in the literature available on the trajectory planning of welding robots. K Abdel-Malek [3] presented an analytical formulation for flex barriers to motion inside the workspace of manipulator arms under the premise of accurate trajectory, which made the welding process successfully completed. TP Pachidis [4] used the visual sensor system for the overall path of the trajectory planning, the method is of high welding quality and low welding error; By identifying the groove characteristics, looking for welding groove and detecting the gap between the robot and the weld groove, Ke Zhang[5] adjusted the robot body and torch position, in which the method is of high precision and feasibility. Su Wang[6,7] provided the basis and algorithm support for the trajectory control of the sitting intersecting line weld by tracking the trajectory of the end point of the torch according to the Frente-Snow column vector, and developed the weighted fuzzy expert system to ride the riding Working angle and walking angle of welding torch for intersecting line welding robot Line planning, effectively completed the welding torch attitude control of the torch. Changliang Chen [8] developed the path control module to generate three B-spline curve welding paths, and introduced the point inversion module using particle swarm optimization to address the partition of path, which is required of the welding process and reduces the welding trajectory error.

Although a lot of research has been done on the trajectory planning of welding robots, few people has considered the interaction between spatial trajectories and other factors. Welding process parameters directly affect the welding quality, and welding process parameters such as welding work angle, welding walking angle are relevant to the welding torch position, so taking into account the welding process parameters and the spatial trajectory is the key to improve the quality of welding. In this paper, by constructing the virtual joints, the welding process parameters are represented by the joint variables of the virtual joints. By controlling the joints of the virtual joints, the welding process parameters are controlled to realize the real-time control of the trajectory during the welding process. The B-spline of three degree being adapted to trajectory interpolation guarantees the joint motion curve continuous and smooth, with its unique convex and affine deformation of indirect control of welding process parameters, which can meet the requirement of welding process.

2. Virtual joint establishment

Based on the principle of simplification, this paper first carries on the trajectory planning to the plane linear weld, and uses three virtual joints to realize the combined planning of the welding process parameter and the space trajectory, and draws lessons from other types of weld trajectory planning. The torch and the robot are rigidly connected, treat the torch as a connecting rod, and add three virtual joints at the end of it, as shown in figure 1.

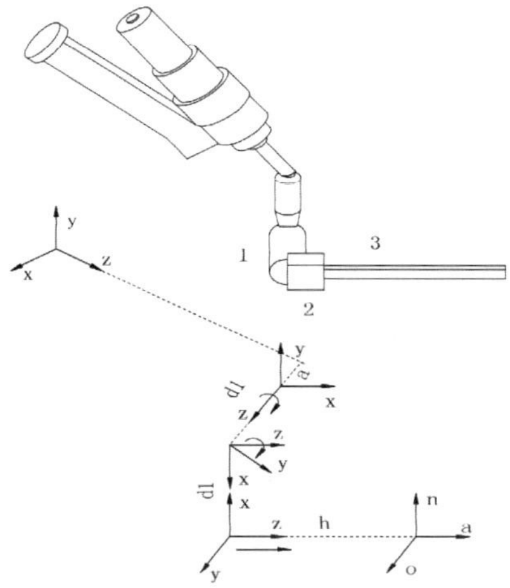

Figure 1. Virtual joint model and coordinate system of linear weld

In the figure 1, d1, d2 and h are the lengths of the links 1, 2 and 3, respectively, and a is the length of the connecting rod of the end-effector. The virtual joint 1 is a rotating joint located at the end of the torch with its axis perpendicular to the center of the torch. The virtual joint 2 is the rotation joint, at the other end of the connecting rod 1, the axial direction is perpendicular to the connecting rod 1; the virtual joint 3 is the moving joint, The direction of the axis coincides with the direction of the straight weld, perpendicular to the connecting rod 2.

Let θ_1, θ_2 and d be the joint variables of the virtual joints 1, 2 and 3, the end of the connecting rod 3 is fixed, d is the movement amount of the link 2 along the axis of the virtual joint 3, and the length of d1 and d2 is set to 0, d is the amount of movement of the torch in the axial direction of the virtual joint 3; the posture of the connecting rod 2 is kept constant due to the position of the end connecting rod 3 and the virtual joint 3 is the moving joint, θ_2 is the amount of rotation of the link 1 in the axial direction of the virtual joint 2, the length of the connecting rod 1 is 0, θ_2 is the rotation of the torch around the virtual joint 2 axis; Similarly, θ_1 is the amount of rotation of the torch in the axial direction of the virtual joint 1. So the torch walking angle, torch work angle can be represented by θ_1 and θ_2. For the plane straight seam welding torch walking speed, it can be written as:

$$v = d_s / t_s \tag{1}$$

Where v is the running speed of the torch, t_s is the unit time, and d_s is the joint variable in the unit time of the virtual joint 3, the running speed of the welding torch can be expressed as a function of d_s. It should be noted that θ_1 and θ_2 are the control of the welding process parameters are not the Euler angle in the posture of the torch. When the third virtual joint is the moving joint and the moving joint axis is parallel to an axis in the base system, θ_1 and θ_2 correspond to the Euler angle of the torch and equal. It can be seen that the establishment of the appropriate virtual joint model, by controlling the joint joints of virtual joints can be welded on the process parameters in real-time control, compared to the previous use of welding parameters to solve the gun position, and then use the gun attitude for trajectory planning, the method of using the virtual joint variable to control the welding process parameters is more intuitive, convenient.

3 Combined planning of process parameters and space trajectories for welding robots

3.1 Joint Kinematics Solution Based on Lie Group and Lie Algebra

Due to the particularity of virtual joint parameters, the solution of joint motion based on Lie group and Lie algebra will simplify the solving problem.

For n-type combined robots, the terminal link coordinate system, which is opposite to the base coordinate system, it can be expressed as[10]:

$$^{o}_{n}T = M_1 e^{S_1 x_1} M_2 e^{S_2 x_2} M_3 e^{S_3 x_3} \ldots M_n e^{S_n x_n} \tag{2}$$

$^{o}_{n}T$ is the arm transformation matrix and is a function of n joint variables.

For 9-type combined robots, it can be written as:

$$^{0}_{9}T = M_1 e^{S_1 x_1} M_2 e^{S_2 x_2} M_3 e^{S_3 x_3} \ldots M_6 e^{S_6 x_6}\, ^{6}_{7}T\, ^{7}_{8}T\, ^{8}_{9}T \tag{3}$$

As $^{0}_{9}T$ is known, $^{8}_{9}T$, $^{7}_{8}T$ and $^{6}_{7}T$ can be obtained by simple derivation, and the inverse kinematics problem of 9 joint robot can be transformed into the inverse kinematics problem of 6 joint robot. According to the method [9,10], the kinematic inverse of the 6-joint robot can be obtained, and the corresponding joint variables of six real joints can be obtained.

3.2 Joint trajectory planning method

After a series of key points of real joint, the trajectory planning is carried out in the joint space, and the selected key points are traversed by B-spline of three degree. B-spline of three degree are expressed as:

$$C(u) = \sum_{i=0}^{q} N_{i,p}(u) P_i, \ 0 \le u \le 1 \tag{4}$$

Where $\{P_i\}$ is the control point, $\{N_{i,p}(u)\}$ is B-spline of three degree basis function defined on the non-periodic node vector $U = \{0,0,0,0,u_4,u_5,u_6 \cdots u_{q+2},1,1,1,1\}$ .The normalization of t_i is done by the cumulative method, we can find the corresponding B-spline curve of the node vector, which can be written as:

$$\begin{cases} u_0 = u_1 = \ldots = u_3 = 0 \\ u_{q+3} = u_{q+4} = \ldots = u_{q+6} = 1 \\ u_i = u_{i-1} + |\Delta t_{i-4}| (\sum_{j=0}^{n-1} |\Delta t_i|)^{-1} \end{cases} \tag{5}$$

To solve the control vertices of the curve, we need to add four boundary conditions: the first two points of the curve coincide with the control vertex, the first two points of the speed of zero, which can be expressed as:

$$P_0 = d_0, P_n = d_n \tag{6}$$

$$P_0 = P_1, P_n = P_{n-1} \tag{7}$$

When the joint variable and the time series are known, the B-spline curve can be obtained [11,12]. If the joint variable sequence is known, the time node sequence is the variable to be determined, and in conjunction with equation (5), the B-spline curve can be written as a function of the variable u.

For the flat linear weld, when the weld direction along the axial direction of the base system, the gun walking angle θ_7 and the torch working angle θ_8 were the relative rotation of the torch around the other coordinate axes, due to the end of the connecting rod of the unchanged position, so we can control the posture of the torch by controlling the θ_7 and θ_8 .Since the virtual joint 9 is a moving joint, the welding robot always moves along the direction of the axis of the moving joint 9 in the direction of the straight line weld, and we can control speed by controlling the joint variable of the virtual joint 9 in the unit time. Therefore, the process parameters of the torch can be represented by the posture of the end of the torch for the linear weld. Assuming that $A = (\theta_1, \theta_2, \theta_3, \theta_4, \theta_5, \theta_6)$, $B = (\theta_7, \theta_8, \theta_9)$,when u increases from 0 to 1, the joints move from the starting point to the end position, and the 6 motion curve of the real joint is composed of a 6-dimensional B-spline curve, denoted *G (A, u)*, 3 virtual joint motion curve composed of three-dimensional curve, recorded as *H (B, u)*. The real joint variable to the virtual joint variable mapping can be obtained from equation (5), recorded as *f*. The *f* can be a continuous shot from the relationship between the joint variable and the position, which can be written as:

$$G(A,u) \xrightarrow{f} H(B,u) \tag{8}$$

Let P_i be the control vertices of H, P_i^1, P_i^2 be the control vertices of the first and second order curves of curve H, and use the convexity of the B-spline curve to control the trajectory constraints of the Cartesian space by controlling P_i^1 and P_i^2, which can be expressed as:

$$\begin{cases} \left|P_i^1\right| \le v_e \\ \left|P_i^2\right| \le a_e \end{cases} \tag{9}$$

Where v_e and a_e are the velocity and acceleration constraints of the joint space, respectively, P_i^1 and P_i^2 can be obtained from the De Boer-recursive formula, which can be written as:

$$C^{(c)}(u) = \sum_{j=i-5+c}^{i} P_j^{(c)} N_{j,k-c}(u) \quad c = 1,2,3 \tag{10}$$

$$P_j^{(c)} = (6-c)\frac{P_j^{l-1} - P_{j-1}^{l-1}}{u_{j+6-l} - u_j} \quad j = i-5+l,..i \quad l = 1,2,...c \tag{11}$$

It is also known that the relationship between the operating space velocity and the joint space velocity can be expressed by the Jacobi matrix, which can be expressed as:

$$\dot{B} = J(A)\dot{A} = J \circ A(u) \bullet \dot{A}(u) \tag{12}$$

When u increases from 0 to 1, each point of the curve $\dot{H}$ can be obtained by the corresponding transformation of the point of $\dot{G}$ through the linear transformation $J(A)$ .Since $J(A)$ is changing with u, the resulting curve is not a B-spline curve. However, the B-spline curve is the same as the control of the vertex shot transformation process by the affine invariance of the B-spline, which can be expressed as:

$$P_i^1 \xrightarrow{J \circ A(u)} P_i^{1'}, \quad i = 1,2 \cdots n \tag{13}$$

The motion curve *K* for controlling the vertices can be obtained by the following equation, which can be written as:

$$\dot{H} \xrightarrow{J} \dot{H}' \xrightarrow{P_i^1} K_i \tag{14}$$

IOP Conf. Series: Materials Science and Engineering **324** (2018) 012009 doi:10.1088/1757-899X/324/1/012009

we can see from the equation (14) that K is a B-spline curve, and the maximum value of K curve is smaller than $\max(P_i^{1'})$ by the convexity of B-spline curve. Obviously the maximum value of curve $\dot{H}$ is less than $\max(P_i^{1'})$, and the minimum value of the same curve $\dot{H}$ is greater than $\min(P_i^{1'})$.
On both sides of the time t on the guide, which can be written as:

$$\ddot{B} = J \circ A(u) \bullet \ddot{A}(u) + J \circ \dot{A}(u) \bullet \dot{A}(u) \tag{15}$$

As in equation (21), the curves of $\dot{A}(u)$ and $\ddot{A}(u)$ are B-spline curves, and are equally expressed as:

$$\max(\ddot{H}) \leq \max(P_i^{2'}) + \max(Q_i^{1'}) \tag{16}$$

$$\min(\ddot{H}) \leq \min(P_i^{2'}) + \min(Q_i^{1'}) \tag{17}$$

Where $P_i^{2'}$ is the new control vertex formed by the affine transformation $J \circ A(u)$ of the control vertex of $\ddot{H}$ and $Q_i^{1'}$ is the new control vertex formed by the affine transformation $J \circ A(u)$ of the control vertex of $\dot{H}$.
In order to solve the optimal time, joint planning of welding process parameters and space trajectories is carried out, and the optimal solution is solved by genetic algorithm. Because the problem relates to the constraints of the optimization, the general use of the penalty function to deal with, the cost function can be written as:

$$f(x) = \min \sum_{0}^{n-1} x_i + r_g \phi(x_i) \tag{18}$$

Where x_i is the time between two joint variables, r_g is the penalty factor, and $\phi(x_i)$ is the penalty function. Considering the requirements of the kinematic constraints of the joint space and the welding process, the constraints are expressed as:

$$\begin{aligned} s.t. \quad & Max(P_i^{1'}) \leq v_e^1,\ Min(P_i^{1'}) \geq -v_e^1 \\ & Max(P_i^{2'}) + Max(Q_i^{1'}) \leq a_e^1 \\ & Min(P_i^{2'}) + Min(Q_i^{1'}) \geq -a_e^1 \ \left|P_i^1\right| \leq v_e, \left|P_i^2\right| \leq a_e \end{aligned} \tag{19}$$

In the equation (19), v_e^1 and a_e^1 are the speed and acceleration constraints of the welding process parameters. $p_i^{1'}$, $Q_i^{1'}$ and $P_i^{2'}$ are a series of control vertices. The maximum and minimum values can be obtained by using the convexity of the B-spline curve. Limited by space, there is no detailed solution.

4 Simulation

Experiments were carried out using ABB robot studio simulation software for planar linear welds, the robot selected IRB1600 (capacity 8KG, reaching 1.45m) welding robot, and the torch was selected by Binzel air 22 type welding machine. Select a straight line on the workpiece as a welded straight seam to determine the end posture of the virtual link.

$$T_9^0 = \begin{bmatrix} 0 & -1 & 0 & 950 \\ -1 & 0 & 0 & 150 \\ 0 & 0 & -1 & 910 \\ 0 & 0 & 0 & 1 \end{bmatrix}$$

In the Cartesian space to select the key point (including the first two points a total of 6 points), the real joint space of the six key points can be solved by the above method, as shown in Table 1:

Table 1. Joint position sequence

point	joint 1	node 2	node 3	node 4	node 5	node 6
P0	-13.00	-4.59	-5.56	13.11	49.59	-16.00
P1	-4.40	-11.00	4.84	1.39	34.88	-12.00
P2	8.72	-14.00	14.16	-46.00	27.15	31.39
P3	20.39	-13.00	21.46	-88.00	36.42	69.23
P4	30.08	7.78	25.96	-108.00	54.99	86.62

P5	36.76	1.80	26.63	-116.00	74.83	94.30

Welding robot joint space kinematic constraints, welding process parameters as shown in Table 2 and Table 3:

Table 2. Joint space kinematic constraints

	joint1	joint2	joint3	joint4	joint5	joint6
speed v_e / (°/ s)	100	100	100	150	130	110
acceleration a_e/ (°/ s^2)	40	50	75	75	90	80

Table 3. Welding process requirements

	torch walking angle	torch working angle	torch walking speed
speed v_e^1/ (°/s)	5	5	10mm/s
acceleration a_e^1/ (°/s^2)	3	3	4mm/s^2

The principle of genetic algorithm is not described here, set the group M = 40, the termination algebra T = 300, the choice operator for the tournament method, the cyclic operator uses the cyclic crossing, PC = 0.8, the variation operator value is 0.02, the penalty factor is 50, using MATLAB software to write the program, find the optimal solution for [5.9375,6.5830,8.0174,7.0800,6.5586], the total time T = 34.1765s.

The joint point of the joint space and the corresponding time series are known, using MATLAB software on the key points of the B-spline interpolation, the joint motion curve, joint velocity curve, joint acceleration curve of six real joint are shown in figure 2~4:

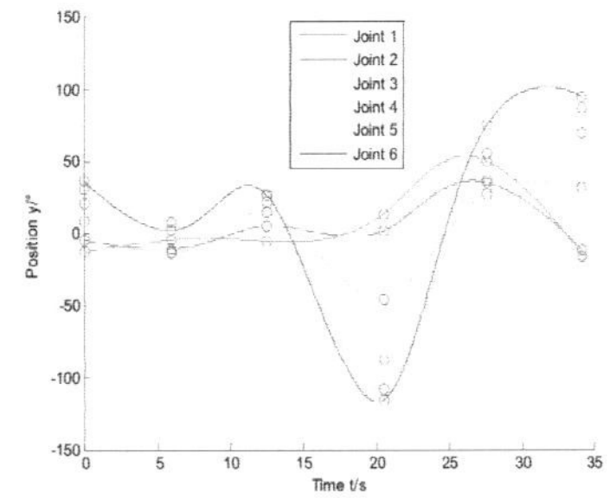

Figure 2. Joint movement curve

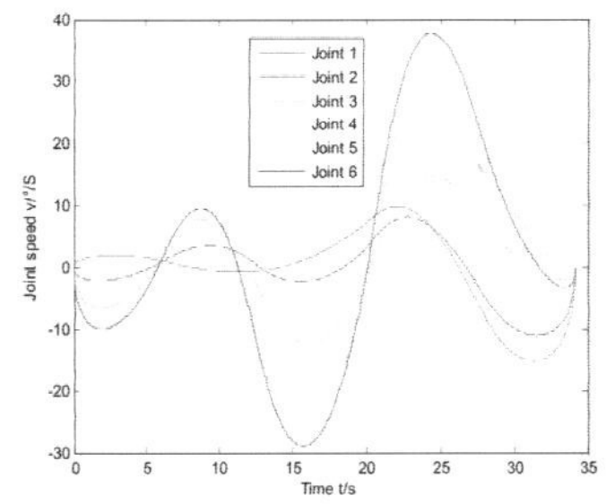

Figure 3. Joint speed curve

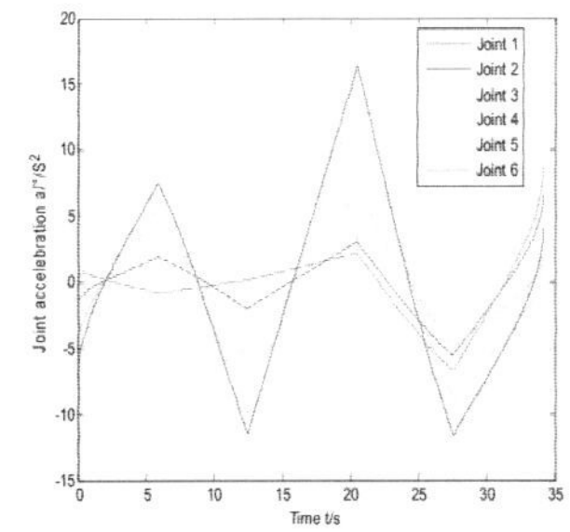

Figure 4. Joint acceleration curve

In figure 2, the blue circle on behalf of each joint must pass through the joint position sequence. The figure shows the joint motion curve through a given joint position sequence and the movement curve smooth and continuous; In figure 3, the speed curve of each joint smooth, the size of the fluctuation between (-30,40), much smaller than the set of the joint speed constraints, and the initial speed and stop speed is zero, to meet the set of initial conditions; In figure 4, the joint acceleration curve continuous, the value of the size Fluctuate between (-13, 16), less than the set joint acceleration constraint. For the welding process parameters of the test, the real joint motion curve can be programmed to write welding procedures, simulation experiments, figure 5 and figure 6 for the torch operation posture.

It can be seen from figure 5 and figure 6 that the working angle of the torch and the running angle of the torch change uniformly from 0 ° to 45 °, and there is no large mutation in the welding process. From the obtained time node, combined with the amount of change in position, a simple analysis shows that the results meet the requirements of the welding process parameters.

Figure 5. Variation of the working angle of the torch

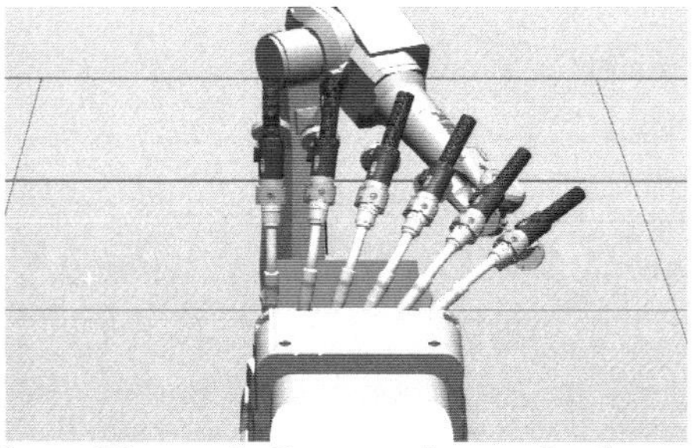

Figure 6. Gun running angle change map

5 Conclusion

By adding the virtual joint at the end-effector, the appropriate virtual joint model is established and the welding process parameters are represented by the virtual joint variables, so that the control of the welding process parameters is more intuitive. Reverse the key point of real joint movement curve by obtaining the key points of the real joint motion curve(the more complex the curve the more key points). Through the B-spline curve for trajectory interpolation, the welding process parameters are combined with the trajectory planning of the joint space, to meet the requirements of the welding process at the same time, so that the continuously joint speed and acceleration can reduce the mechanical vibration and joint wear.

6 References

[1] Zhu J M 2005 *manuf Eng* 12 42-7.
[2] Fu X L, He J P and Wang F X 2015 *Light Industry Machinery*, 33 110-13.
[3] Abdelmalek K and Yeh H J 2000 *Proce Institu of Mech Eng Part B J Eng Manu* 211 547-56.
[4] Pachidis T P and L ygouras J N 2007 *Journal of Intelligent & Robotic Systems* 48 307-31.
[5] Zhang K, Wu Y X, Jin X and Lv X Q 2005 *Chin J Mech Eng* 41 215-20.
[6] Wang S, Cai L L, Miao X G and Li X H 2010 *Weld Tech* 39 44-47.
[7] Wang S, Miao X G, Li X H 2010 *J. B Univ Aeronaut Astronaut* 36 771-75.
[8] Chen C, Hu S, He D, et al 2013 *Rob Comput-Integrat Manuf* 29 41-48.
[9] Lv S Z, Zhang D W and Liu H N 2010 *J. Mech Eng* 46 35-41.
[10] Park F C, Bobrow J E and Ploen S R 1995 *Intern J Rob Research,* 14(6):609-618.
[11] Zhou H M, Wand Y M, Liu Z G and Lu B H 2008 *J XI'AN JiaoTong Univ* 42 73-77.
[12] Zhu S Q, Liu S G, Wand X Y and Wand H F 2010 *J Mech Eng* 46 47-52.

Acknowledgments

The authors thank for the support of the Outstanding Youth Science and Technology Innovation Team Program of HuBei Provincial Department of Education Grant(T201505), the Key Project of Natural Science Foundation of Hubei Province Grant(2015CFA112) and the Major Project of Technical Innovation in Hubei Province Grant(2017AAA108).

IOP Conf. Series: Materials Science and Engineering **324** (2018) 012010 doi:10.1088/1757-899X/324/1/012010

Transient performance analysis of the master cylinder hydraulic system of a 6.3 MN fineblanking press

Guodong Yi and Jin Li

School of Mechanical Engineering, Zhejiang University, Hangzhou, 310027, China
E-mail: ygd@zju.edu.cn

Abstract. The master cylinder hydraulic system is the core component of the fineblanking press that seriously affects the machine performance. A key issue in the design of the master cylinder hydraulic system is dealing with the heavy shock loads in the fineblanking process. In this paper, an equivalent model of the master cylinder hydraulic system is established based on typical process parameters for practical fineblanking; then, the response characteristics of the master cylinder slider to the step changes in the load and control current are analyzed, and lastly, control strategies for the proportional valve are studied based on the impact of the control parameters on the kinetic stability of the slider. The results show that the kinetic stability of the slider is significantly affected by the step change of the control current, while it is slightly affected by the step change of the system load, which can be improved by adjusting the flow rate and opening time of the proportional valve.

1. Introduction

Fineblanking is a kind of rapid processing method for punching out high-precision and high-quality parts from sheet metal under a three-dimensional compressive stress state. Due to the high pressure and frequency of the fineblanking, the load on the master cylinder hydraulic system varies drastically and continuously during the process, which seriously affects the machining quality. Therefore, a key step in the fineblanking machine development process is establishing the simulation model for the master cylinder hydraulic system based on the fineblanking process and analyzing the dynamic performance of the system, especially the transient response under the step changes in the system load and control current, to lay the foundation for further design.

In hydraulic system modeling, some mathematical models are presented to study the dynamic characteristics of fluid in hydraulic pipes [1]-[3], and some simulation models based on commercial software are proposed for dynamic performance analysis of hydraulic systems based on different parameters [4],[5]. In hydraulic system analysis, the methods of finite element, finite volume, computational fluid dynamics, multi-body dynamics, and numerical computation are used to research hydraulic characteristics [6]-[11]. In hydraulic system control, electrical position feedback, built-in control rod, speed control, and real-time sensing and actuation are applied to the proportional valve, cylinder, and pump [12]-[15].

Due to the step change characteristics of the fineblanking load, it is necessary to study the transient performance of the master cylinder hydraulic system. On the basis of the abovementioned study, this study designed the typical process parameters of a B630T fineblanking press (rated load of 6.3 MN), established the equivalent models of the hydraulic components and master cylinder system, analyzed the response characteristics of the slider motion to the step change of the load and control signal, and studied the influence of the proportional valve parameters on the motion stability of the slider as well as the control strategy.

IOP Conf. Series: Materials Science and Engineering **324** (2018) 012010 doi:10.1088/1757-899X/324/1/012010

2. Equivalent model of the master cylinder hydraulic system

The master cylinder hydraulic system model is shown in Figure 1.

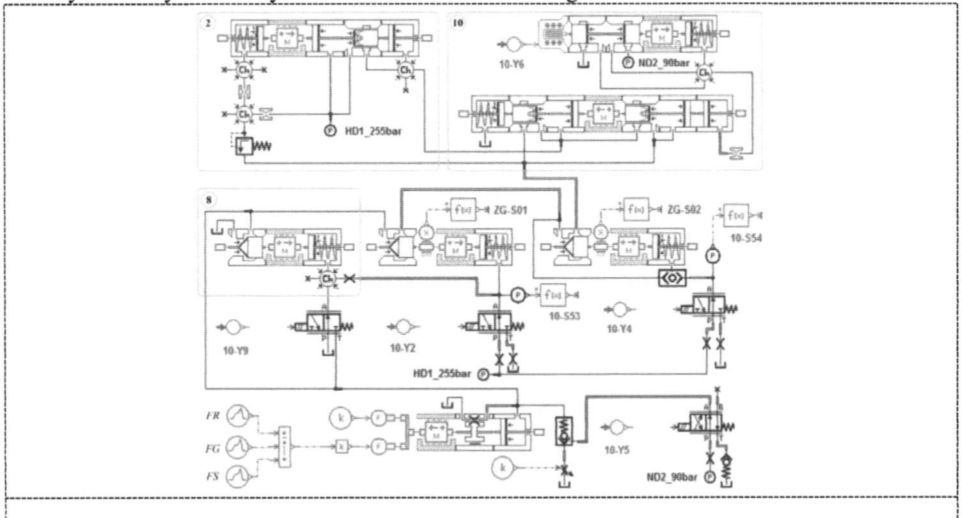

Figure 1. Master cylinder hydraulic system model.

3. Simulation and analysis of the transient performance of the master cylinder hydraulic system

3.1. Response of the master cylinder slider to the step change of the load

3.1.1. Simulation. At the end of the fineblanking process, the part is cut from the material, and the master cylinder load undergoes an approximate step change from large to small at this point. The simulation settings are set as follows: the load of the master cylinder is 3000 kN at the beginning of the simulation then jumps to 0 at 0.5 s; the control current of the proportional valve is constant; all solenoid valves are in the energized state. The simulation result is shown in Figure 2.

3.1.2. Analysis. In the simulation, when the load drops from 3000 N to 0 N in a very short time, the slider rushes forward 0.9 mm, and the proportional valve appears to be in the unstable state for 20 ms, with a rapid response in the inlet and outlet pressures. However, in practice, the load is unloaded step by step when the slider reaches the top dead center, and the solenoid valve is triggered at the same time to stop the slider quickly. Thus, the abovementioned simulation settings are extreme. Even so, the variation of the slider speed and acceleration are still controllable, and the master cylinder still runs stably under the step load using some optimized control methods.

3.2. Response of the master cylinder slider to the step change of the control signal

3.2.1. Simulation. At the beginning and the middle stage of the blanking, the slider speed is controlled by the proportional valve. At the end stage, the response of cartridge valve 5 is uncontrollable due to the damping effects of 10-Y2 and 10-Y4, so stopping the slider at top dead center mainly relies on the joint effect of proportional valves 10-Y5 and 10-Y9. Therefore, to obtain the system performance, it is necessary to analyze the response of the slider to the proportional valve control signal. The simulation settings are as follows: the master cylinder runs steadily under the load of 4000 kN; the control current of the proportional valve changes from 700 mA to 200 mA at 0.5 s; other solenoid valves are in the energized state. The simulation result is shown in Figure 3.

3.2.2. Analysis. When the control current jumps from 700 mA to 200 mA, it takes the slider 100 ms to reach the set position. However, because the blanking time is only 1.5 s, it is difficult to control the slider effectively using only the solenoid valve. A feasible solution to achieve the desired performance is collaborative control with other components.

IOP Conf. Series: Materials Science and Engineering **324** (2018) 012010 doi:10.1088/1757-899X/324/1/012010

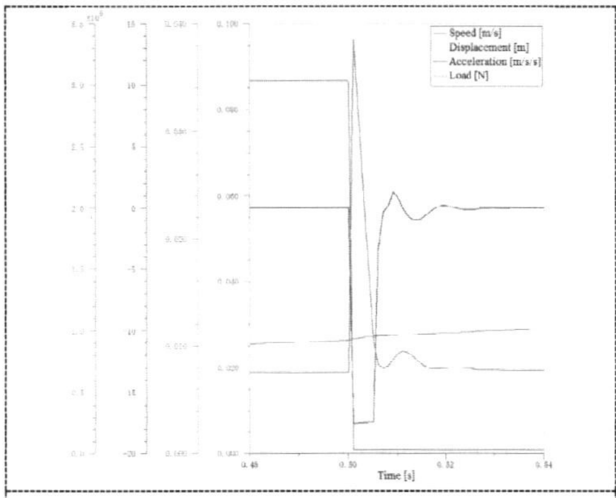

Figure 2. Response of the master cylinder slider to the step change of the load.

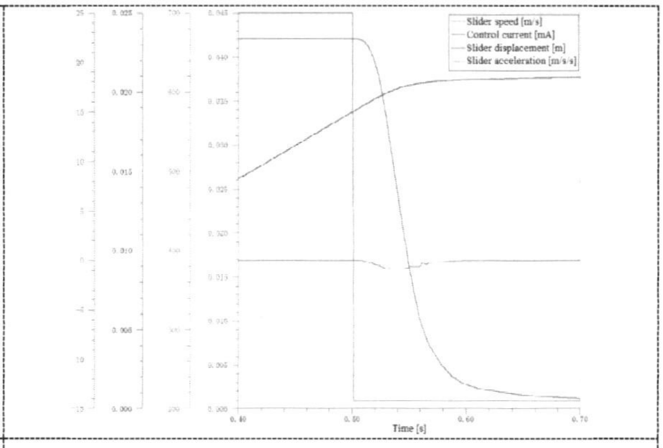

Figure 3. Response of the master cylinder slider to the step change of the control signal.

3.3. Control performance analysis of the pressure reducing cartridge valve

In the master cylinder hydraulic system, cartridge valve 3 and control cover 2 are used to keep the differential pressure between the inlet and outlet of the proportional valve within a certain range, which is adjusted by the direct relief valve on the cover. The stability of the slider speed is affected directly by the differential pressure between the inlet and outlet when the proportional valve control current is constant. Taking cartridge valve 3 and control cover 2 as the research objects, the factors influencing the pressure difference are analyzed. For ease of analysis, the proportional valve is assumed to be equivalent to the damping, and the load pressure is assumed to be equivalent to the pressure source.

If the orifice diameter is small, and it is easy to keep the differential pressure between the inlet and outlet stable, which means that the follow-up performance of the outlet pressure to the load pressure is better. The main influencing factor is the stroke of the cartridge valve spool. When the flow is small, the spool stroke is short, and when the flow is large, the spool stroke is long, which takes some time to complete.

The velocity of the valve spool can be adjusted by changing the size of the two orifices on the control cover, but this increases the risk of vibration. This variation increases the inlet pressure of the direct acting relief valve on the control cover, the differential pressure of the pressure reducing valve, and the flow of the load through the direct acting relief valve.

The opening speed of the proportional valve can be related to the change in the orifice diameter and the variations of the outlet pressure at three speeds. At speed 1, the orifice opens 1% in 10 ms and then to 100% in 50 ms; at speed 2, the orifice opens from 0 to 100% in 60 ms; at speed 3, the orifice opens with a step change.

3.4. Opening time control strategy for the proportional valve

At the moment the punch just touches the workpiece, the proportional valve of the master cylinder hydraulic system starts, reducing the impact load and optimizing the operating condition. In the transient analysis of the proportional valve opening, the master cylinder system is simplified as follows: the electric control pilot valve of the proportional valve is replaced by a pressure source; the step cylinder is replaced by a velocity source; and the times of turn-on and turn-off of the valves are ignored.

The master cylinder slider is driven by the step cylinder in the first 0.5 s and touches the work piece at 0.5 s, and then, the load rises from 0 to 1000 kN in 50 ms. The state of each valve is changed at 0.5 s, and the operating conditions of the components are shown in Figure 4. There are crests in the flow and velocity at position 1 because the proportional valve opening is constant at this moment, and the differential pressure between the two ports of the proportion valve is much larger than the design value due to the pressure valve lag, which results in the flow increasing. There are troughs in the pressure and velocity at position 2 because the slider has an initial speed at this moment and because the charging valve is closed, but the oil from the proportional valve has not arrived yet.

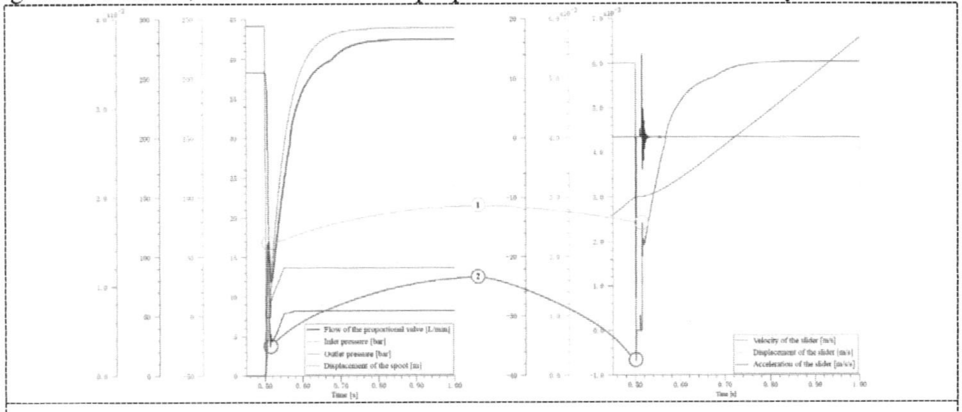

Figure 4. Operating conditions of the master cylinder and the proportional valve.

Based on to the flow crest at position 1 in Figure 4, the cartridge valve in the main oil circuit is opened at 0.49 s in advance, and the result is shown in Figure 5. The velocity and acceleration of the slider are significantly improved.

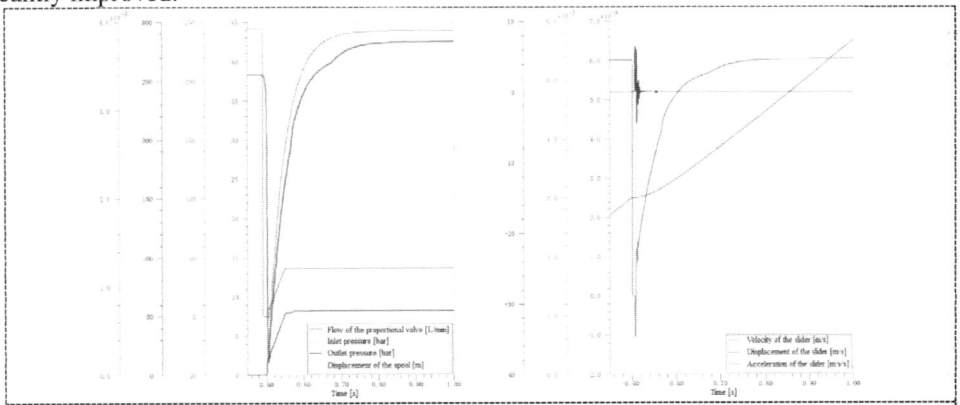

Figure 5. Operating conditions of the master cylinder and the proportional valve (the cartridge valve is opened in advance).

The negative velocity at position 2 in Figure 4 is due to the low instantaneous flow of the proportional valve. On the basis of the cartridge valve opening in advance, the proportional valve is opened 0.49 s in advance, and the result is shown in Figure 6. Although the negative speed is eliminated, the acceleration of the slider is increased.

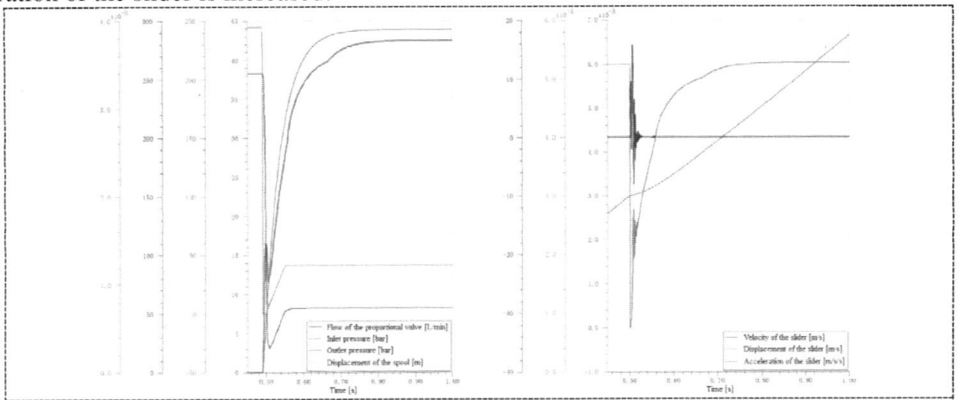

Figure 6. Operating conditions of the master cylinder and the proportional valve (the proportional valve is opened in advance).

In summary, the pressure variation in the cartridge valve is affected by changing the speed of flow and pressure. Therefore, stabilities of the slider velocity and displacement are difficult to take into account at the same time.

4. Conclusions

- The step change of the master cylinder hydraulic system load from large to small leads to a controllable variation in the slider velocity and acceleration, and some appropriate optimization controls can keep the master cylinder running stably.
- The stepped change of the control current of the master cylinder hydraulic system from large to small causes the slider to not be well-controlled when relying only on the solenoid valve, but the desired performance can be achieved by controlling other components.
- The stability of the slider velocity is directly affected by the differential pressure of the proportional valve, and a smaller orifice is conducive to maintaining a constant pressure differential, which causes better follow-up performance of the outlet pressure to the load pressure.
- The changing speeds of the flow and pressure affect follow-up performance of the pressure variations of the cartridge valve, which leads to difficulty taking the stabilities of velocity and displacement of the slider into account at the same time, but the velocity and acceleration of the slider are significantly improved by opening the cartridge valve in advance.

5. References

[1] Shahbodagh-Khan B, Khalili N and Alipour E 2015 *Comput. Geomech.* **69**, 411-423
[2] Liang J, Yuan X, Yuan Y, Chen Z and Li Y 2017 *Mech. Syst. Signal Process.* **85** 927-946
[3] Hu Q, Yang C, Zheng H and Chen M 2014 *Automation in Construction* **37** 211-216
[4] Andersson H, Nordin P, Borrvall T, Simonsson K, Hilding D, Schill M, Krus P and Leidermark D 2017 *Eng. Comput.* **33** 317-333
[5] Zhang M and Huang M 2011 *Proc. ICDMA* 465-468
[6] Ehlers W and Luo C 2017 *Comput. Methods Appl. Mech. Eng.* **315**, 348-368
[7] Zhu L, Peng S, Cheng X, Qi Y, Ge J, Yin C and Jen T 2016 *Soil Tillage Res.* **163** 168-175
[8] Xia Y, Zhu J, Wang K and Zhou Q 2016 *Int. J. Impact Eng.* **88** 139-152
[9] Wang C, Nilsson H, Yang J and Petit O 2017 *Comput. Phys. Commun.* **210** 1-9
[10] Igumnov L A and Metrikin V S 2015 *IFAC-PapersOnLine* **48** 640-644
[11] Yang Z, He Z, Li D, Cui X, Xue G and Zhao Z2015 *Sens. Actuators A Phys.* **236** 228-246
[12] Forental V I, Forental M V and Nazarov F M 2015 *Procedia Eng.* **129** 695-701
[13] Sun X, Qin B and Bo H 2017 *Ann. Nucl. Energy* **101** 151-158
[14] Gaspar J F, Kamarlouei M, Sinha A, Xu H, Calvário M, Faÿ F, Robles E and Soares C G 2016 *Renew. Energy* **97** 769-783
[15] Abu-Mahfouz A M, Hamam Y, Page P R, Djouani K and Kurien A 2016 *Procedia Eng.* **154** 99-106

Acknowledgments
The authors would like to thank the National Natural Science Foundation of China (Grant No. 51375438) for the supports given to this research.

IOP Conf. Series: Materials Science and Engineering **324** (2018) 012011 doi:10.1088/1757-899X/324/1/012011

Research on mining truck vibration control based on particle damping

SONG Liming[1],XIAO Wangqiang[2],Li Zeguang[1],GUO Haiquan[1] and YANG Zhe[1]
1 Inner Mongolia North Hauler Joint Stock CO.LTD, Baotou, China;
2 XIAMEN University, Xiamen, China
E-mail: slm340@163.com
Abstract: More and more attentions were got by people about the research on mining truck driving comfort. As the vibration transfer terminal, cab is one of the important part of mining truck vibration control. In this paper, based on particle damping technology and its application characteristics, through the discrete element modeling, DEM & FEM coupling simulation and analysis, lab test verification and actual test in the truck, particle damping technology was successfully used in driver's seat base of mining truck, cab vibration was reduced obviously, meanwhile applied research and method of particle damping technology in mining truck vibration control were provided.

1. Introduction

Mining truck is a kind of special transport vehicles, because of the product particularity and bad working condition, research on mining truck vibration control are still in a primary stage, which can only meet the minimum requirements of industry standards. With the development of mining truck technology and improvement of environment, health and safety awareness, more and more attentions were got about the research on mining truck driving comfort. Driver's seat is the mining truck vibration transfer terminal, it's also direct feeling source of driving comfort to driver. In this paper, particle damping technology was used in driver's seat base of mining truck, through discrete element modeling, DEM & FEM coupling analysis , lab test verification and actual test on the vehicle, cab vibration was reduced, meanwhile method of particle damping technology in mining truck vibration control were provided.

2. Theoretical basis

2.1. Particle damping technology

Particle damping technology is also called micro particle damping technology, which is a kind of vibration control technology in the 1990s. Filling a certain number of small particles in the closed space of vibrating body, Particle damping technology is a passive vibration control technology, which is based on the theory of energy dissipation mechanism. Damping effect is mainly caused by the friction and collision between particles filled in the cavity or between the particle and the damper wall, to realize mechanical structural vibration reduction by energy conversion.

2.2. Discrete element method

As a type of granular, particle is an another physical form relative to gas, solid and liquid. For this kind of discontinuity, motion equations can not be established similar to the solid' s or fluid' s, which movement is mainly controlled by the contact effect between particles. So there is a new numerical method, called discrete element method (DEM), which can solve the discontinuous medium problems. According to DEM, the interaction between particles is considered as a continuous dynamic process into balance. If the time step selected smaller enough, in every single time step, movement of a particle is only depends on itself and the particle direct contact with it. Contact constitutive relation and Newton's second law are applied in the DEM calculation, contact constitutive relation is used to get contact force and relative displacement between particles, Newton's second law is used to calculate the movement caused by contact and the mass.

2.3. Particles motion equations

Take, for instance, particle i and j contact, according to the Newton's second law, through the calculation of resultant force and torque of particle i mass center, the particles equations of motion in time t can be get as follows.

$$\begin{cases} \vec{\ddot{u}}_i^{(t)} m_i = \vec{F}_i^{(t)} = \sum_{j,j\neq i} \vec{F}_{ij}^{(t)} \\ \ddot{\theta}_i^{(t)} I_i = \vec{M}_i^{(t)} = \sum_{j,j\neq i} \vec{M}_{ij}^{(t)} \end{cases}$$

Where $\ddot{u}_i^{(t)}$, $\ddot{\theta}_i^{(t)}$ are the acceleration and the angular acceleration of particle i, m_i、I_i are the mass and the moment of particle i.

In calculation, the new location of particle can be obtained by central difference method.

$$\begin{cases} \vec{u}_i^{(t+\Delta t)} = \vec{u}_i^{(t)} + \vec{\dot{u}}_i^{(t+\Delta t/2)} \Delta t \\ \theta_i^{(t+\Delta t)} = \theta_i^{(t)} + \dot{\theta}_i^{(t+\Delta t/2)} \Delta t \end{cases}$$

Where Δt is time step.

3. Simulation and analysis of seat base

In general, finite element method (FEM) can be used in mechanical structure analysis and calculation, but for the issues related to particle movement, it is necessary to combine the DEM and FEM, that is DEM & FEM coupling analysis method. Particle movement characteristics were calculated in DEM, then the action of particles on damper wall in DEM can be passed into FEM model by coupling interface, to calculate influence of particles on structure. The DEM & FEM coupling analysis method process is shown in Fig. 3.1.

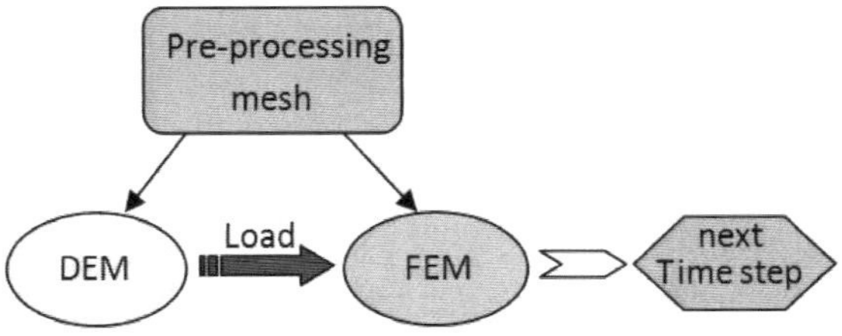

Fig. 3.1 DEM-FEM coupling analysis process

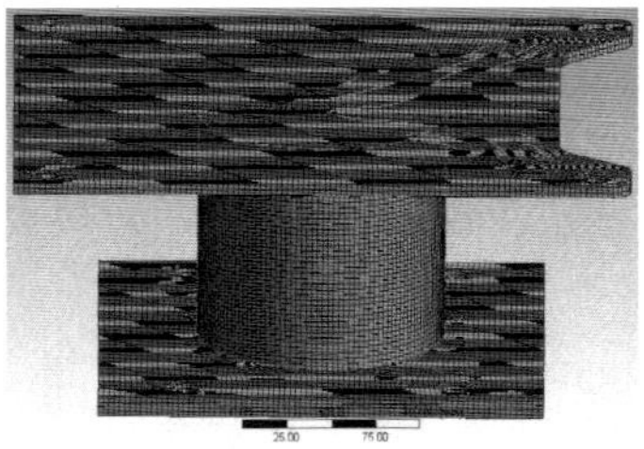

Fig. 3.2 FEM mesh model

3.1. Seat base modeling

According to the seat base structural characteristics, simplified and established the 3D model. Considering the element size and quality factors , meshed with hexahedral element. Fig. 3.2 shows the finite element mesh model.

3.2. DEM simulation and analysis

3.2.1. Time step

Time step is one of key problems of DEM, which should be determined firstly when calculating particle movement by DEM according to particle parameters, to track the collision and friction between particles. A small time step will increase the amount of calculation, also a big time step will affect the result accuracy. Usually Rayleigh rated time step was used, which formula shown as follow:

$$T_R = \pi R\left(\frac{\rho}{G}\right)^{\frac{1}{2}}(0.1631\nu + 0.8766)^{-1}$$

Where R, ρ, G, ν are particle radius, particle density, shear modulus and Poisson's ratio. In general, 20% ~ 40% of the Rayleigh rated time step was selected as a DEM time step.

3.2.2. Particle motion analysis

According to DEM simulation, analyzing particle motion at the beginning of the vibration of seat base. As shown in Fig. 3.3, at first, with the seat base vibration, due to the action of friction and force from inner wall of seat base, the upper particles began to vibration, and downward through bouncing back collide with other particles, rebound and slow down until collisions with the bottom of the base. During the process, the energy of the seat base is transferred to particles. The vibration energy is converted to heat by the inelastic collision and friction of particles.

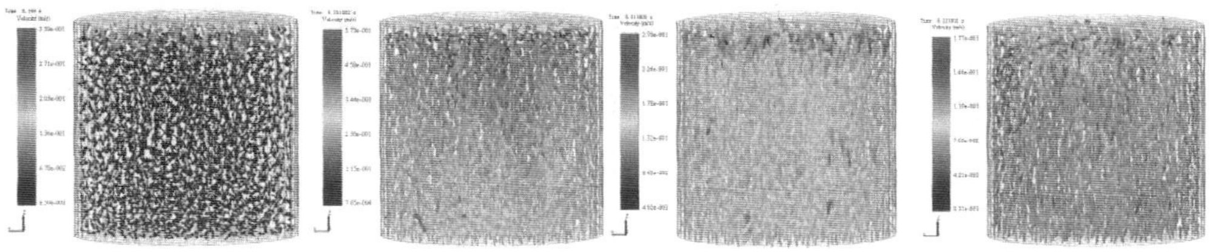

Fig. 3.3 Particle motion

As shown in Fig. 3.4, damping force of particles on the seat base, it can be seen that vibration start at 0.2s, the particles start from static to dynamic, damping force appears and is biggest. With base motion tending to steady, because of collision between particle and base wall, damping force decrease and display periodic changes.

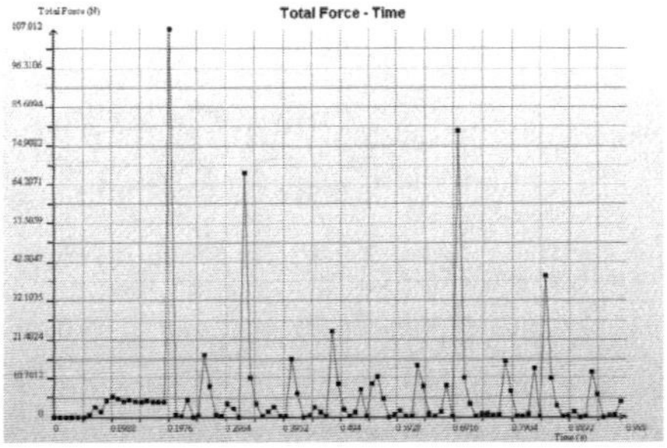

Fig. 3.4 Damping force of particles

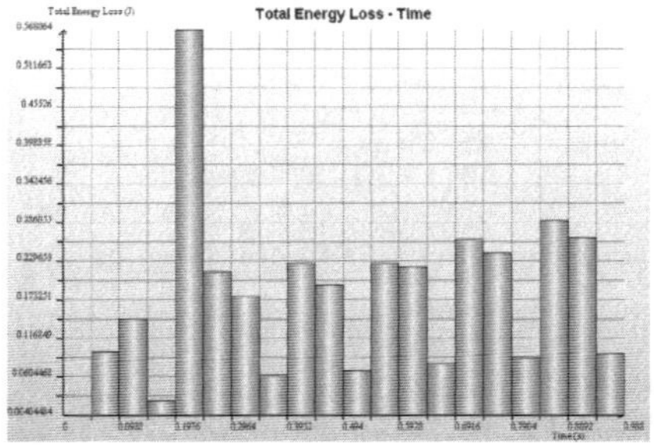

Fig. 3.5 Energy dissipation of particles

3.2.3. Energy dissipation

Particles vibration caused by Seat base, through collision and friction between particles or between the particle and the damper wall, system vibration is reduced. As shown in Fig. 3.5, same with trends of damping force, the largest energy dissipation appear in vibration beginning, then display periodic changes. Through the simulation at engine speed 750rpm and 2200rpm, energy dissipation comparative analysis is shown in Fig. 3.6, the higher engine speed, the more energy dissipation, the more obvious vibration reduction.

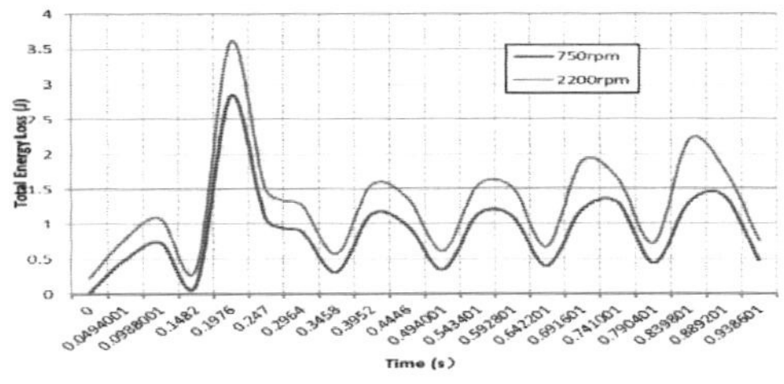

Fig. 3.6 Energy dissipation of particles

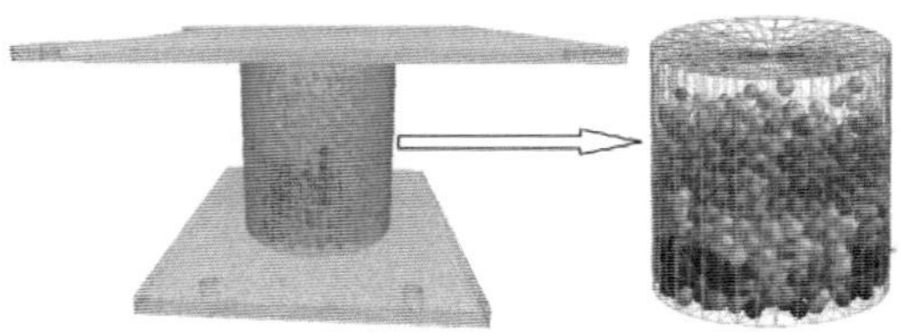

Fig. 3.7 Damped discrete element model

3.3. FEM-DEM coupling analysis

Analyze the influence of damping particles on the seat base vibration by FEM-DEM coupling analysis method, input the damping force of DEM calculation to the FEM model, build the damped discrete element model, which as shown in Fig. 3.7.

Through sine sweep excitation, get the seat base response curves with particles and without particles. As shown in Fig. 3.78 and Fig. 3.9, the peak of structural response reduce from 1.8744m/s^2 to 0.469m/s^2 at 750rpm, and the peak of response reduce from 2.53m/s^2 to 1.54m/s^2 at 2200rpm. The results indicate that vibration reduction is obvious with particles damping.

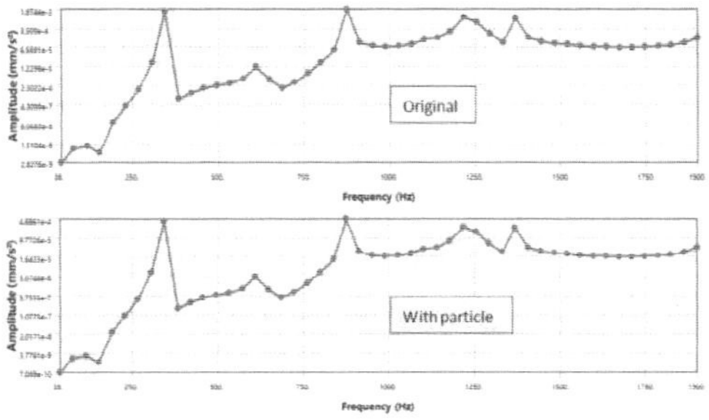

Fig. 3.8 Comparative response curve

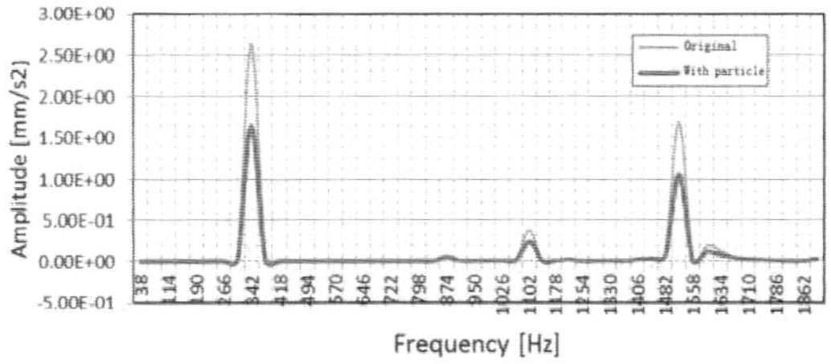

Fig. 3.9 Comparative response curve (2200rpm)

4. lab simulation and test

Fig. 4.1 Lab test

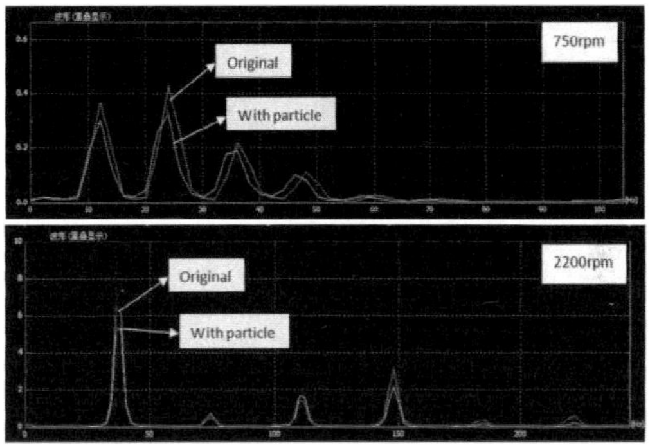

Fig. 4.2 Lab test results

As shown in Fig. 4.1, simulate seat base vibration at engine speed 750rpm and 2200rpm, obtain comparative test results with particles and without particles. From the results in Fig. 4.2 we can get conclusions: the vibration amplitude of seat base with particles reduced by nearly 30%, driving comfort is improved obviously; And, by comparing the vibration with particles at different engine speed, we can see that vibration reduction is more obvious at a higher speed, also the DEM simulation results are verified.

5. Actual test in the truck

Test in a truck with particles damper, select two points directly connected to seat base as the testing points, one is seat bottom (point 1) and another is cab floor (point 2), through vertical contrast test at different engine speed, analyze the effect of particles on vibration control.

As shown in Fig. 5.1 and Fig. 5.2 are the vertical vibration test results of point 1 at 750rpm and 2200rpm (where purple curve show results without particles, blue curve show results with particles, the same below.). Fig. 5.3 and Fig. 5.4 are the vertical vibration test results of point 2 at 750rpm and 2200rpm. According to the test results and data analysis, we can obtain: With the particle damping, the root mean square of acceleration decreased more than 30% at both point 1 and point 2, effect of cab vibration control is obvious; vibration reduction at 2200rpm are more better than 750rpm, verify the simulation results further.

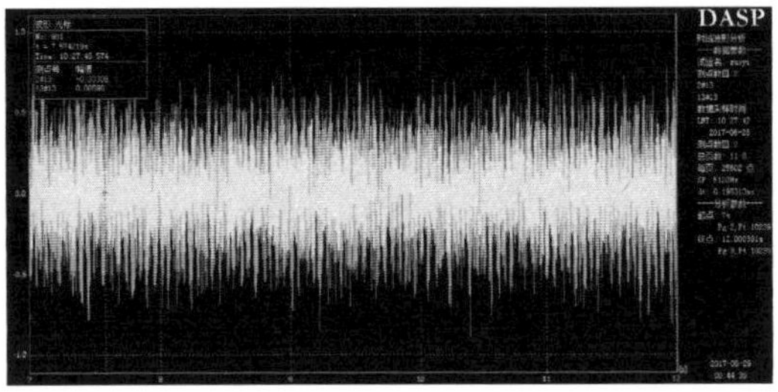

Fig. 5.1 Test results of point 1 at 750rpm

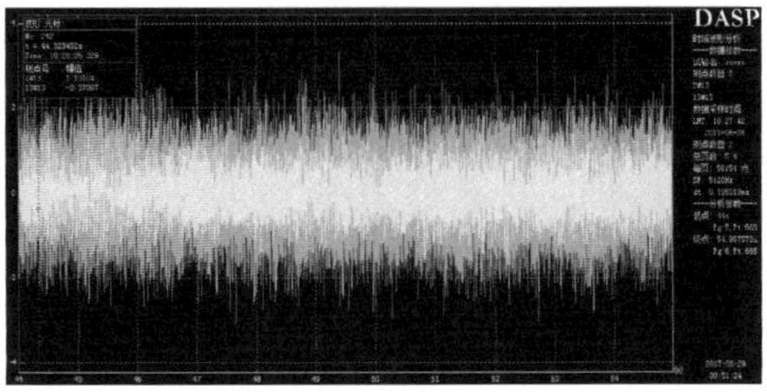

Fig. 5.2 Test results of point 1 at 2200rpm

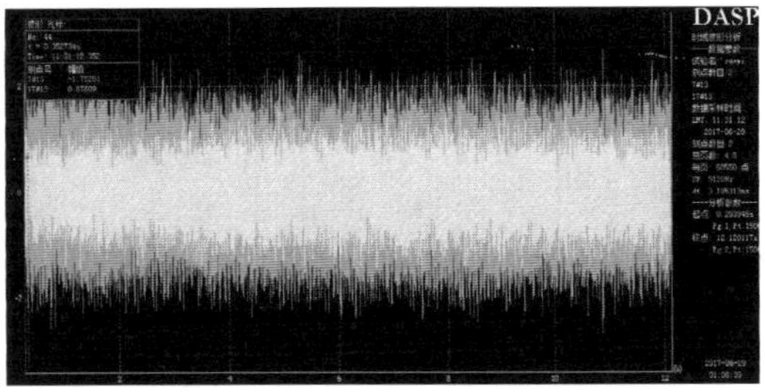

Fig. 5.3 Test results of point 2 at 750rpm

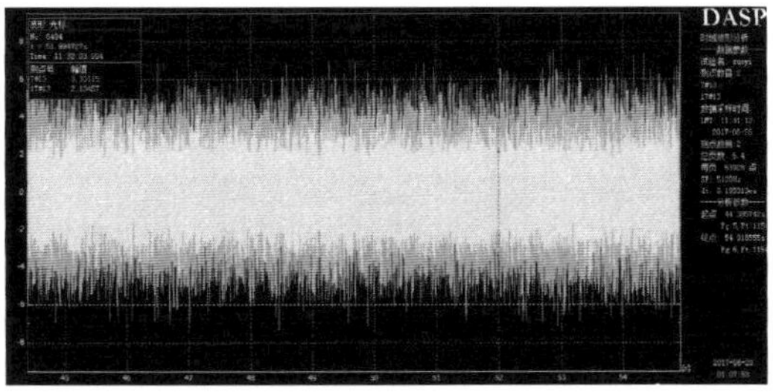

Fig. 5.4 Test results of point 2 at 2200rpm

6. Conclusion

With the development of mining truck technology, more and more attentions were got about the research on mining truck driving comfort. In this paper, based on particle damping technology and its application characteristics, through simulation and analysis , lab test verification and actual test in the truck, particle damping technology was successfully used in driver' s seat base of mining truck, cab vibration was reduced obviously, meanwhile applied research and method of particle damping technology in mining truck vibration control were provided.

References:

[1] HU Guoming. Discrete element method analysis and simulation of particles system[M]. Wuhan University of Technology Press, 2010.

[2] WANG Guoqiang, HAO Wanjun, WANG Jixin. The practice of discrete element method in EDEM[M]. Xi' an: Northwestern Polytechnical University Press, 2010.

[3] LU Zheng, Lv Xilin, YAN Weiming. A survey of particle damping technology[J]. Journal of vibration and shock, 2013, 32(7): 1-7.

[4] HU Li. Research on the Mechanics and Characteristics of Particle Damping [D]. Wuhan, Huazhong University of Science & Technology, 2008.

[5] KONG Xiangjun. Methods of Particle Damping and its Application in Mechanical Vibration Control [D]. Shenyang, Shenyang University of Technology, 2016.

[6] XIAO Wangqiang, HUANG Yuxiang, JIANG Hong, et al. Effect of powder material on vibration reduction of gear system in centrifugal field[J]. Powder Technology, 2016, 294: 146-158.

IOP Conf. Series: Materials Science and Engineering **324** (2018) 012012 doi:10.1088/1757-899X/324/1/012012

Alternative Constraint Handling Technique for Four-Bar Linkage Path Generation

S Sleesongsom[1] and S Bureerat[2]

[1]Department of Aeronautical Engineering and Commercial Pilot, International Academy of Aviation Industry, King Mongkut's Institute of Technology Ladkrabang, Bangkok, 10520, Thailand
[2]Sustainable and Infrastructure Research and Development Center, Department of Mechanical Engineering, Faculty of Engineering, KhonKaen University, KhonKaen City, 40002, Thailand
E-mail: suwins2000@yahoo.com

Abstract. This paper proposes an extension of a new concept for path generation from our previous work by adding a new constraint handling technique. The propose technique was initially designed for problems without prescribed timing by avoiding the timing constraint, while remain constraints are solving with a new constraint handling technique. The technique is one kind of penalty technique. The comparative study is optimisation of path generation problems are solved using self-adaptive population size teaching-learning based optimization (SAP-TLBO) and original TLBO. In this study, two traditional path generation test problem are used to test the proposed technique. The results show that the new technique can be applied with the path generation problem without prescribed timing and gives better results than the previous technique. Furthermore, SAP-TLBO outperforms the original one.

1. Introduction

In daily life and manufacturing, it is common to see many applications of four-bar linkage mechanisms. Steering wheels, oil wells, double wishbone suspensions, windshield wipers, automatic doors, and rock crushers are applications of such a mechanism. One task for designing a four-bar linkage is a path generation problem. This synthesis problem is aimed at finding significant link lengths of the mechanism to achieve a point on a coupler link moving along the desired path [1-16]. The problem is usually converted to be an optimization problem, which is posed to find the dimensions of a mechanism and some other parameters. The objective function of this problem is to minimize the errors between predefined target points and positions of one particular point on the connecting link, while the constraints are imposed for preventing an unusable mechanism solution. This design problem is the kinematic design of a mechanism involving position analysis. The path generation problem is of interest by many researchers due to the advantage of using such a mechanism such as work reported by Sleesongsom and Bureerat[16]. Almost all of the previous work tried to improve the design performance of the path synthesis problem in both the constraint handling [10-11, 16] and the performance of an optimizer [1-9, 16]. The path synthesis optimization problem normally composes of two types of constraints. The first constraint is the sequence of shortest link (crank) being able to rotate with a complete revolution in either direction, which agrees with the Grashof criterion. The second constraint is set to obtain unknown timings of the crank in a sequential order. From the

literature, it is found that a simple exterior penalty function technique is the most popular technique used to deal with constraints [1-7, 9-11], but it still lacks in efficiency as reported in [8, 11, 13, 16]. Very recent work in [16] proposed a new technique to neglect the second constraint from the optimization problem and a first constraint is still tackle with the traditional penalty technique, which is found to give the considerably better results than the traditional exterior penalty technique. The new technique can improve the design results but, in turn, increases computational time. The traditional path generation test problems used in this work are the problems without prescribed timing constraints. Researchers rarely studied and developed a new technique for constraint handling technique, therefore, the present work is conducted to show that a new technique proposed in [16] can be effectively used with a new constraint handling technique. The method is one kind of traditional penalty technique, which has been improved for optimal truss sizing [17]. It is an efficient technique for constraint handling. At the present study the efficiency of this technique for handling constraint of the path generation problem is needed to study. The optimizers used in this study are meta-heuristics (MHs) namely: Self-Adaptive Population Size Teaching-Learning Based Optimization (SAP-TLBO) [16], and original teaching - learning based optimization (TLBO) [18]. Performance testing of both optimizers for solving the mechanism synthesis is described with descriptive statistics, which focus on minimum (best), maximum (worse), mean value, and standard deviation of the obtained solutions.
The rest of this paper is organized as follows. Position analysis of a four-bar linkage, an optimization problem, and a constraint handling technique are briefly presented in Section 2. The numerical experiments are shown in Section 3 while the design results are given in Section 4. The conclusions of this study are drawn in Section 5.

2. Methodology

2.1. Position analysis

A model of a four-bar linkage, shown in Fig. 1, for position analysis is an important computation required for path synthesis. It is composed of four simple links, which are paired by 4 revolute joints and one link is set as a frame. It is a constrained chain mechanism fully operated by one actuator. The design variables of four-bar linkage lengths are r_1, r_2, r_3, and r_4 and other parameters, which shall give minimum error between the desire path (x_d, y_d) and actual trace points (x_P, y_P) on the coupler link. The coupler curve can be formed when a crank moves over a complete cycle. The coupler point coordinates in the global coordinate are shown in Fig.1 and can be expressed as

$$x_P = x_{O2} + r_2\cos(\theta_2 + \theta_1) + L_1\cos(\phi_0 + \theta_3 + \theta_1) \quad (1)$$

$$y_P = y_{O2} + r_2\sin(\theta_2 + \theta_1) + L_1\sin(\phi_0 + \theta_3 + \theta_1)$$

where x_{O2} and y_{O2} are the coordinates of the O_2 pin joint in the global coordinates.

ϕ_0 can be obtained by considering the couple link *BCP* using the cosine law, which is expressed as

$$\phi_0 = \cos^{-1}\left[\frac{L_1^2 + r_3^2 - L_2^2}{2L_1 r_3}\right] \quad (2)$$

The values of angles θ_3, θ_4, and γ for the known link lengths r_1, r_2, r_3, and r_4 at any crank angle (θ_2) can be obtained by common kinematic analysis given in [16].

2.2. Optimization problem and constraint handling

The path generation problem without prescribed timing is transformed to an optimization problem with an objective function being assigned to minimize the error between targeted and obtained coupler curves. The position error is the sum of squares of distances between each P_d and the corresponding P_o, where $\boldsymbol{P}_d$ is a set of predefined points, and $\boldsymbol{P}_a$ is a set of obtained points from an input design vector of linkage dimensions and other parameters. The members of a design solution **x** include r_1, r_2, r_3, r_4,

L_1, and L_2, the global position of O_2 (x_{O2}, y_{O2}), and the angle of frame 1 (θ_1). For the without prescribed timing problem, θ_2^i is undefined and included in the design variables. The optimization problem without prescribed timing is then written as

$$\min f(\mathbf{x}) = \sum_{i=1}^{N}\left[(x_{d,i} - x_{P,i})^2 + (y_{d,i} - y_{P,i})^2\right] \tag{3}$$

subject to

$$\theta_2^1 < \theta_2^2 < \ldots < \theta_2^i < \ldots < \theta_2^N \tag{4}$$

$$\min(r_1, r_2, r_3, r_4) = \text{crank } (r_2) \tag{5}$$

$$2\min(r_1, r_2, r_3, r_4) + 2\max(r_1, r_2, r_3, r_4) < (r_1 + r_2 + r_3 + r_4) \tag{6}$$

$$\mathbf{x}_l \leq \mathbf{x} \leq \mathbf{x}_u \tag{7}$$

where $\mathbf{x} = \{x_{O2}, y_{O2}, r_1, r_2, r_3, r_4, L_1, L_2, \theta_1, \theta_2^i\}^T$, N is the number of predefined points (x_d, y_d) on the prescribed or target curve, and $\mathbf{x}_l$ and $\mathbf{x}_u$ are lower and upper bounds of a design vector $\mathbf{x}$ respectively. The constraints (4) are set to obtain the values of θ_2^i being in ascending or descending order, which is considered if the problem is without prescribed timing. The constraints (5) and (6) correspond to the Grashof's criterion. Normally, an exterior penalty function is used to deal with constraints (5-7). It was however found to be inefficient for solving the above design problem [16], as a result, the unknowns θ_2^i are to be pre-determined and not treated as design variables for optimization. If the mechanism is a crank-rocker, which means constraint (5) and (6) are met. To tackle both constraints a new technique from [17] is used instead the traditional penalty constraint. Equation (5-6) is come from a common form as $g_i(\mathbf{x}) \leq g_{0,i}$. Then, a set of M input angle values (θ_2) equally spaced from 0 to 2π radian $T_2 = \{0, \ldots, \theta_2^i, \ldots, 2\pi\}$ is generated. With more number of T_2 and more efficient of the new constraint handling, better design results can be expected. Once the positions of point P corresponding to all values in T_2 are calculated, the objective function (3) is obtained as

$$f(\mathbf{x}) = \sum_{i=1}^{N} \min d_{ij}^2 \tag{8}$$

where $d_{ij}^2 = (x_{d,i} - x_{P,j})^2 + (y_{d,i} - y_{P,j})^2$ for $j = 1, \ldots, M$. The detail of this technique can see in [16].

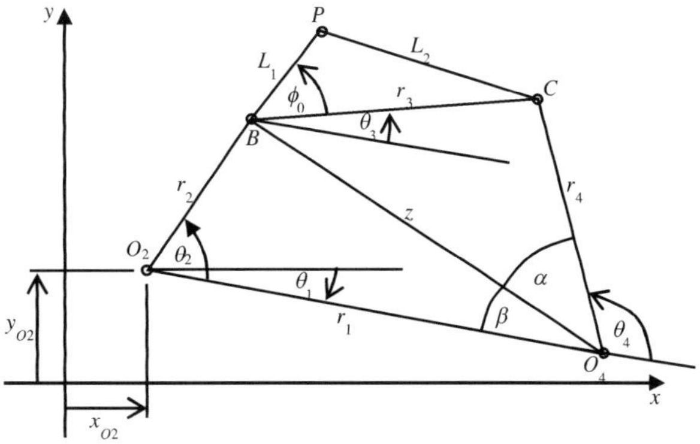

Figure 1. Four-bar linkage in the global coordinate system [16]

For detail of the constraint handling technique from [17] is showed here in brief. The constraint handling is similar technique with usual penalty technique, which is composed of two part is an objective function combine with constraint functions multiply with switching parameter. The design variable is invalided the constraint a large real switching parameter is added to modify the objective function value (ON state). Otherwise, it is an "OFF state" state, which means switching parameter is 0. With this technique the constraints can tackle, the solution is obtain is a feasible solution. If the switching parameter is too small or too large, it can effect to finding solution of an optimizer. The inefficient of the traditional technique is found by [8, 11, 13, 16]. Alternative penalty technique has been proposed by [18], it is found is an efficient technique. This idea is created the new equivalent objective function by using four constraint handling rules as follow.

1) Any feasible solution has a better equivalent objective function value than any infeasible one;
2) Between two feasible solutions, one with the better objective function value has a better equivalent objective function value;
3) Between two infeasible solutions, one with a similar constraint violation has better equivalent objective function value; and
4) As infeasible solution with small constraint violation ($c_i < c_O > 0$) is considered feasible in the optimisation process where c_O is a small positive constraint tolerance.

The fourth rule is using to exploring infeasible solution nearby constraint boundary if the feasible solution cannot find. In construction of a new objective unconstraint function from rules 1-3 can be defined as

$$f_p(\mathbf{x}, c_0) = \begin{cases} f' = \dfrac{f(\mathbf{x})}{f(\mathbf{x}) + k} & \text{if } \max|c_i(\mathbf{x})| \le c_0 \\ 1 + \max|c_i(\mathbf{x})| & \text{otherwise} \end{cases} \quad (9)$$

where k > 0 is a constant to be defined (for this paper k = 10), and c_0 is iteratively reducing to 0. Before using previous equation the constraint should rearrange firstly as follow.

$$c_i(\mathbf{x}) = g_i(\mathbf{x}) / g_{0,i} - 1 \le 0 \quad (10)$$

The equation (10) is in form parameterless.

3. Numerical Experiment

Two without prescribed timing optimisation problem of four-bar linkage path generation is used to study the performance of the proposed design technique. The first one is path generation with prescribed timing with straight line path in vertical direction, while the second is a problem with an elliptic path [6].

Case-1 : Path generation without prescribed timing

Design variables are

$\mathbf{x} = \{r_1, r_2, r_3, r_4, r_{px}, r_{py}, \theta_0, x_0, y_0, \theta_2^1, \theta_2^2, \theta_2^3, \theta_2^4, \theta_2^5, \theta_2^6\}^T$.

Target points are

r_d^i = {(20,20),(20,25),(20,30),(20,35),(20,40),(20,45)}

Limitation of design variables:

$5 \le r_1, r_2, r_3, r_4 \le 60$

$-60 \le L_1, L_2 \le 60$

$-60 \le x_O, y_O \le 60$

$0 \le \theta_0, x_O, y_O, \theta_2^1, \theta_2^2, \theta_2^3, \theta_2^4, \theta_2^5, \theta_2^6 \le 2\pi$.

Case-2 : Path generation without prescribed timing

Design variables are

$\mathbf{x} = \{r_1, r_2, r_3, r_4, L_1, L_2, \theta_0, x_O, y_O, \theta_2^1, \theta_2^2, \theta_2^3, \theta_2^4, \theta_2^5, \theta_2^6, \theta_2^7, \theta_2^8, \theta_2^9, \theta_2^{10}\}^T$.

Target points are:

r_d^i = {(20,10),(17.66,15.142),(11.736,17.878), (5,16.928), (0.60307,12.736), (0.60307,7.2638), (5,3.0718), (11.736, 2.1215),(17.66,4.8577), (20,10)}.

Limitation of design variables are:

$5 \leq r_1, r_2, r_3, r_4 \leq 80$
$-80 \leq L_1, L_2 \leq 80$
$-80 \leq x_0, y_0 \leq 80$
$0 \leq \theta_0, \theta_2^1, \theta_2^2, \theta_2^3, \theta_2^4, \theta_2^5, \theta_2^6, \theta_2^7, \theta_2^8, \theta_2^9, \theta_2^{10} \leq 2\pi$

In order to measure the performance of the new design technique and the SAP-TLBO and original TLBO are used to tackle the design problem. The parameter settings of those methods are detailed as follows:

1. Self-Adaptive Population Size Teaching-Learning Based Optimization (SAP-TLBO) [16]. IReset = 20, IRange = 5, δ = 1.
2. Teaching-learning-based optimization (TLBO) [18]: Parameter settings are not required.

The definitions of terms and variables can be found in their original sources. For solving four-bar linkage path synthesis, the population size n_P = 100 is used for all algorithms, while the maximum number of function evaluations which is a termination criterion is set to be 50,000. Each algorithm is run for 30 times for performance comparison of search consistency.

4. Design Results

The design results from the various optimizers with the new synthesis technique are shown in Table 1. The table shows the number of successful runs (a run which an optimiser can find a feasible solution), the mean objective function values from 30 optimization runs, the worse result (max), the best result (min), and the standard deviation (std). Fig. 2-3 show the best path traced by the coupler point and illustrates the kinematic diagram of the best linkage. In Case-1, there are 6 target points. It was found that SAP-TLBO gives the best result (error = 0.0000598) and the best mean objective value (error = 0.031747), which is better than TLBO. The result of Case-2 showed SAP-TLBO gives the better result than TLBO in both mean (error = 0.059718) and min (error = 0.002465). Furthermore, the best results are better than the best results by [16] (Case-1: error = 0.04760 (mean), 0.00088 (min) respectively, Case-2: error =1.644548 (mean), 0.027145 (best) respectively). The number of successful runs of all optimizers (present technique) is approximate equal to the previous work, but it higher than the traditional penalty technique [16]. The results concluded that SAP-TLBO with the new technique is better than those from using the traditional technique in both cases.

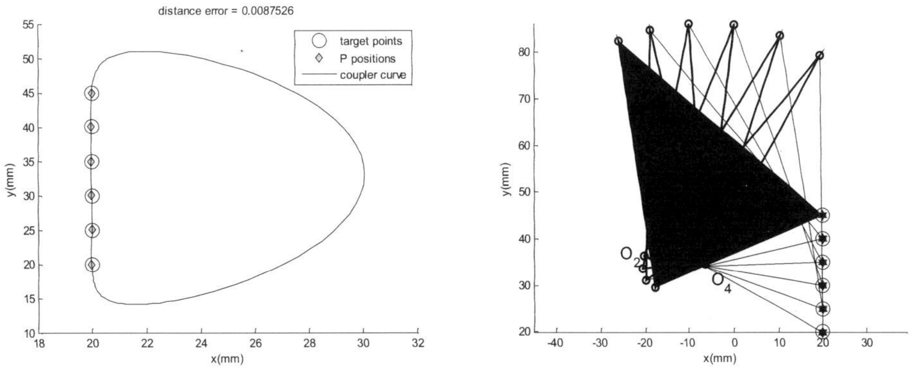

Figure 2. The best coupler curves and the best mechanism of Case-1.

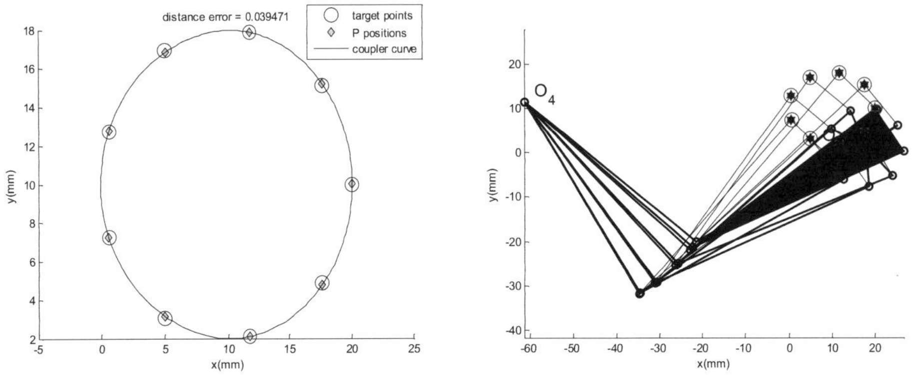

Figure 3. The best coupler curves and the best mechanism of Case-2.

Table 1. Comparative design results for Case-1 and Case-2 with a new technique.

Para-meters	Case-1		Case-2	
	TLBO	SAP-TLBO	TLBO	SAP-TLBO
r_1	12.3163	9.3329	80.0000	79.8358
r_2	5.0000	5.0546	8.6507	8.9499
r_3	37.6311	53.5740	66.3060	53.5336
r_4	32.7912	51.7039	47.0738	50.8890
L_1	-59.5259	-40.8228	14.9526	-11.9790
L_2	-52.4788	-59.3263	71.9939	52.6266
x_0	-34.5185	-15.7705	18.0731	17.8703
y_0	40.8644	34.1240	22.4629	1.1295
θ_0	0.0907	2.6586	94.9288	172.5849
mean	0.063749	**0.031747**	0.209332	**0.059718**
max	0.718959	0.206311	0.839434	0.352928
min	0.000433	**0.0000598**	0.003037	**0.002465**
std	0.160546	0.060032	0.302877	0.081523
Success.*	23	20	27	27

* Success. = no. of successful runs

5. Conclusions

This paper proposes a new technique to find the optimum parameters of a four-bar linkage for path synthesis. This technique is combing a technique to remove timing constraints from the optimization and a new penalty technique for handling Grashof's criterion. It is found that the new technique with SAP-TLBO is outperformed the original TLBO and the new penalty technique can upgrade the results from the previous study. However, it is recommended that the technique is an alternative technique for solving path generation problem without prescribed timing. For future work, a proper technique for handling Grashof's criterion should be investigated.

Acknowledgments
The authors are grateful for the financial support provided by King Mongkut's Institute of Technology Ladkrabang, the Thailand Research Fund (BRG5880014), and the Post-doctoral Program from Research Affairs, Graduate School, KhonKaen University (58225).

References
[1] Smaili A and Diab N 2007 *Mechanism and Machine Theory* **42**(1) 115–130
[2] Cabrera J A, Nadal F, Muñoz J P and Simon A 2007 *Mechanism and Machine Theory* **42**(7) 791-806
[3] Bulatović R R and Dordević S R 2009 *Mechanism and Machine Theory* **44**(1) 235–246
[4] Nariman-Zadeh N, Felezi M, Jamali A and Ganji M 2009 **44**(1) 180–191
[5] Acharyya S K and Mandal M 2009 *Mechanism and Machine Theory* **44**(9) 1784-1794
[6] Lin W Y 2010 *Mechanism and Machine Theory* **45**(8) 1096–1107
[7] Cabrera J A, Ortiz A, Nadal F and Castillo J J 2011 *Mechanism and Machine Theory* **46**(2) 127-141
[8] Peñuñuri F, Peón-Escalante R, Villanueva C and Pech-Oy D 2011 *Mechanism and Machine Theory* **46**(10) 1335–1349
[9] Khorshidi M, Soheilypour M, Peyro M, Atai A and ShariatPanahi M 2011*Mechanism and Machine Theory* **46**(10) 1453–1465
[10] Mateka S B and Gogate G R 2012 *Mechanism and Machine Theory* **52** 158–179
[11] Ebrahimi S and Payvandy P 2015 *Mechanism and Machine Theory* **85** 189–204
[12] Sleesongsom S and Bureerat S 2015 *Research and Applications in Mechanical Engineering***3**(2) 46-53
[13] Phukaokaew W, Sleesongsom S and Bureerat S 2016 *Proceeding of 7th TSME – ICOME conference*
[14] Heo J C and Yoon G H 2013 *Mechanism and Machine Theory* **64** 18-38
[15] Buskiewicz, J 2015 *Inverse Problem in Science and Engineering* **23** 818 -850
[16] Sleesongsom S and Bureerat S 2017 *Knowledge-Based Systems* **135** 180–191
[17] Bureerat S and Pholdee N 2016 Optimal truss sizing using an adaptive differential evolution algorithm *Journal of Computing in Civil Engineering* **30(2)** doi:10.1061/(ASCE)CP.1943-5487.0000487
[18] Rao R V, Savsani V J and Vakharia D P 2011 *Computer-Aided Design* **43**(3) 303-315

Output Error Analysis of Planar 2-DOF Five-bar Mechanism

Kejia Niu, Jun Wang[1,a], Kwun-Lon Ting[2,b], Fen Tao, Qunchao Cheng, Quan Wang, Kaiyang Zhang

[1]Center for Robotics Research, School of Mechanical Engineering,
Hubei University of Technology,Wuhan, Hubei,430068,China
[a]E-mail: junwang@mail.hbut.edu.cn

[2]Center for Manufacturing Research,Tennessee Technological University
Cookeville, Tennessee, 38501, USA
[b]E-mail: kting@tntech.edu

Abstract. Aiming at the mechanism error caused by clearance of planar 2-DOF Five-bar motion pair, the method of equivalent joint clearance of kinematic pair to virtual link is applied. The structural error model of revolute joint clearance is established based on the N-bar rotation laws and the concept of joint rotation space, The influence of the clearance of the moving pair is studied on the output error of the mechanis. and the calculation method and basis of the maximum error are given. The error rotation space of the mechanism under the influence of joint clearance is obtained. The results show that this method can accurately calculate the joint space error rotation space, which provides a new way to analyze the planar parallel mechanism error caused by joint space.

1. Introduction

For the planar 2-DOF Five bar parallel mechanism, its members can be regarded as rigid components, and the end errors of the output rod are derived from the rod length tolerance and joint clearance, while the joint clearance leads to the uncertainty of the output attitude of the end effector of the mechanism. However, the proper movement of the deputy clearance ensures the normal operation of institutions necessary conditions. The clearance is too small so that the mechanism may stuck and unable to run. And a large amount of space not only can not guarantee the accuracy of movement, but also causes the movement of side impact mechanism[1].Therefore, it is necessary to study the problem of kinematic pairs. Aiming at the precision analysis of planar parallel mechanism, many scholars have made a great deal of researches, Yu et al.[2] proposed a method of compensating clearance of the kinematic pair by optimizing the 3-RRR parallel mechanism. Zhang et al.[3] combined the principle of virtual work and optimization techniques and put forward to analyze the pose error of planar mechanism by numerical calculation method; Ting [4-5] studied the influence of the clearance error of the kinematic pair of the crank-slider mechanism and the four-bar mechanism with the end effector on the output uncertainty by establishing the kinematics model. Compared with the tedious and complicated numerical calculation and analysis method, it is simpler and More effective to establish the error model.

2. Model of joint clearance for planar 2-DOF Five-bar linkages

The planar 2-DOF Five-bar parallel mechanism includes two active input joints. The clearance error model is shown in Figure 1.One of the obvious characteristics of the rotating joint is that the diameter of the pin is different from the diameter of the hole in the rod. It is assumed that the pin and the hole are always in contact with each other in the research, There is no deformation or rupture between the pin and the hole in each joint, so the center distance between the pin and the hole is kept at a constant value. In the paper, the tiny distance caused by the rotating joint's clearance can be equivalent to a virtual link in the error model.The length of this virtual link is different from the radius between the hole in the rotating joint and the pin, which is also 1/2 of the distance of the rotating joint clearance. Each joint clearance adds a degree of freedom to the mechanism which results in an uncertainty of the mechanism's output orientation.

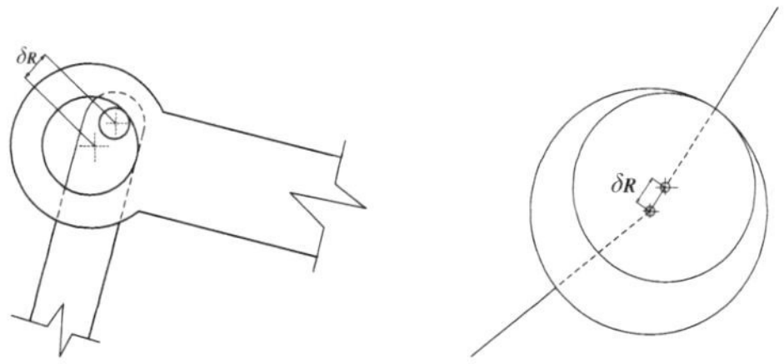

(a) (b)

Figure 1. Model of joint clearance for the revolute pair

3. N-bar rotatability laws and joint rotation space

a. N-bar rotatability laws

Based on the N-bar rotatability laws proposed by the Ting[6-7], Since a clearance link is essentially a short link ,a clearance link can have full revolution with respect to any link in the chain. The full rotatability of any clearance links is the key issue of the clearance link model.

Let lmax be the longest rod in the Five rods. Let lmin1 be the shortest rod. Let lmin2 be the second shortest rod, And the remaining two rod lengths are l_m and l_n respectively, and the following classification is made:

ClassI: $l_{\max}+l_{\min1}+l_{\min2} \le l_m+l_n$

ClassII: $l_{\max}+l_{\min1}+l_{\min2} \ge l_m+l_n$

ClassIII: $l_{\max}+l_{\min1}+l_{\min2} = l_m+l_n$

An N-bar linkages has full rotatability if and only if

1) The linkage belongs to classIchain, and

2) There is one and only one non-input joint between any pair of long links.

b. Joint rotation space

Joint rotation space is the input variable curve range. The movement of each joint is a single reciprocating cycle. For the whole mechanism, the motion of joint affects each other, and the motion state of the mechanism can be clearly expressed by the movement of all joints. A point in the rotation space of any joint of the N-bar linkage corresponds to a configuration of the mechanism.Therefore, the joint rotation space is a synthesis of the possible configurations of all mechanisms of an institution. The characteristics of the movements of a mechanism can be expressed in the form of a map of the characteristics of a mechanism so as to make the complex movement clearly and concisely presented.

c. Invariant link rotatability[8]

Invariant link rotatability suggests that in any planar single-loop chain formed by a set of links and revolve joints, the rotatability between any two links, whether they are adjoined or not, is independent of the order of connection. Because of the invatiant link rotatability, all clearance links can be grouped together. This reduces the complexity of the linkage.The independent rotation of each clearance link may be used to model the position uncertainty under worst as well as non-worst situations.

4. Influence of kinematic joint clearance on output error

The joint clearance is equivalent to a short rod, so that the Five-bar parallel mechanism is equivalent to a Ten-bar parallel mechanism, as shown in Figure 2.The number of degrees of freedom also increase from 2 to 7. However, with the design of the actual mechanism, the total length of all the equivalent clearance rod is much less than the sum of the rod length of the original ideal rod. Therefore, the equivalent parallel mechanism does not change the type of the mechanism.If both inputs are applied to the ideal connecting rod, the equivalent parallel mechanism has seven degrees of freedom, and the additional five degrees of freedom are due to the uncertainty of the joint space motion.It can be considered that because these five arbitrary degrees of freedom are applied to the five clearance links (short links), the parallel mechanism still has complete rotatability, which in turn leads to the existence of the entire mechanism error rotation space.

The ideal rod length of the planar Five-bar parallel mechanism is a, b, c, d, e, respectively, and the rod length of the equivalent clearance rod is δ,namely,

$$a+e<b+c+d \tag{1}$$

The equivalent joint clearance connecting rod,

$$a+(e+5\delta)<b+c+d \tag{2}$$

Therefore, according to the mechanism classification criterion mentioned above, the type of the planar parallel mechanism doesn't change after the equivalent clearance connecting rod is added, which ensures that the complete rotation between any original ideal connecting rod is stable and unchangeable.According to the Invariant link rotatability, given the input θ_2, θ_3, the lengths of l_1 and l_2 can be obtained from the input rod and the rod length of the frame,

$$l_1^2=a^2+e^2-2ae\cdot\cos\theta_2 \tag{3}$$

$$l_2^2=b^2+l_1^2-2bl_1\cdot\cos(\theta_3+\alpha) \tag{4}$$

The output range of the error between the two links of the mechanism can be obtained,

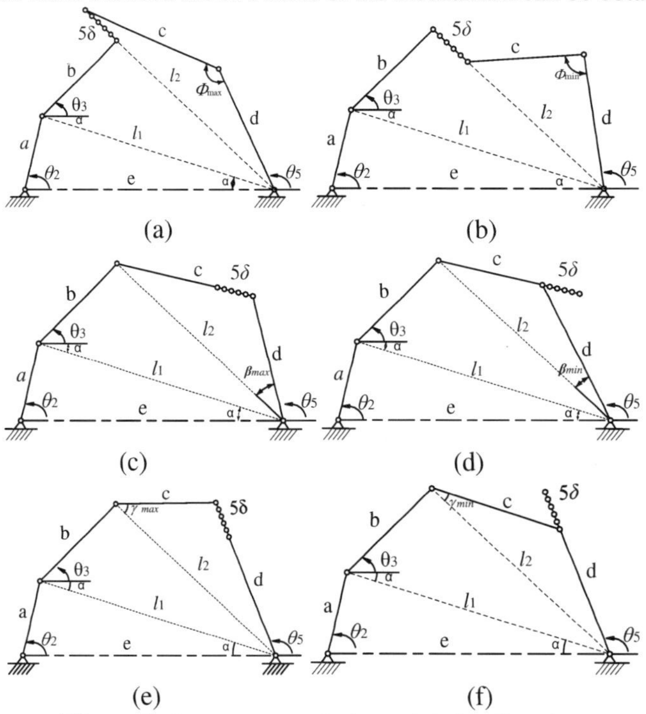

Figure 2. Output error of planar 2-DOF Five-bar

According to the joint rotatability ,The possible rotation range between links *c* and *d* is Φ_{max} and Φ_{min},

$$\phi_{\max,\min}=\arccos\left[\frac{c^2+d^2-(l_2\pm5\delta)^2}{2cd}\right] \tag{5}$$

According to the joint rotatability ,The possible rotation range between links *d* and l_2 is β_{max} and β_{min},

$$\beta_{\max,\min}=\arccos\left[\frac{d^2+l_2^2-(c\pm5\delta)^2}{2d\cdot l_2}\right] \tag{6}$$

According to the joint rotatability ,The possible rotation range between links c and l_2 is γ_{max} and γ_{min},

$$\gamma_{\max,\min} = \arccos\left[\frac{c^2 + l_2^2 - (d \pm 5\delta)^2}{2c \cdot l_2}\right] \tag{7}$$

Through the above calculation, according to any joint clearance on the direction of the output rod error is equivalent,Planar Five-bar parallel mechanism in the rotary joint space under the influence of the angular error between the rod is easy to get.

5. Solution of working space error caused by joint space

Based on the established error geometry model, according to the rotation mechanism of the link mechanism used in this paper, the mechanism shown in Figure 3.(a) can finally evolve into Figure 3.(b). Based on the given organization parameters, the error space of the equivalent planar parallel mechanism as shown in Figure4. can be precisely calculated. Table 1 shows the dimension parameters.

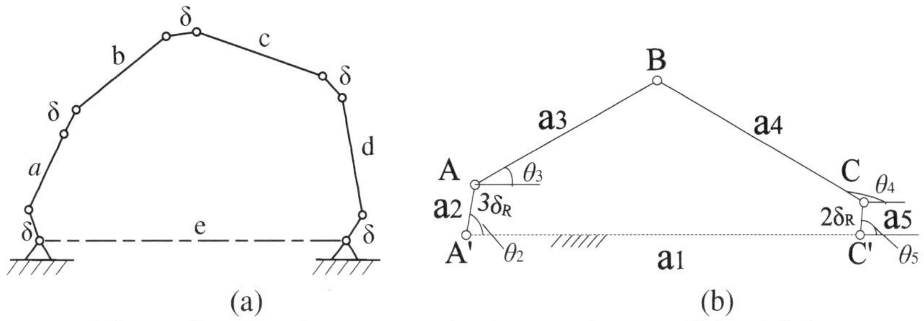

Figure 3. Joint clearance equivalent to Five parallel rod linkage

Table 1

δ_R(mm)	a1	a2	a3	a4	a5
0.76	304.80	2.29	152.40	274.32	1.52

The equivalent rod length of the rotating joint in the designed mechanism is estimated as δ_R = 0.76mm. Based on the theory of N-bar rotation mentioned above and the classification of Five-bar mechanism, as shown in Figure 3, the two shortest rods are all connected rods. The four rotating joints that are involved in the rotation are all rotating pairs. According to whether the input joint is applied, the following two situations exist:

a. The two input joints are not adjacent, as shown in Figure 4, $\theta_2$and θ_5 are input joints. When A'A and AB are collinear, the Five-bar linkages degenerate into a four-bar linkages. When A'A and AB coincidence, it also degenerates into a four-bar ring, the ring equation can be:

$$a_1 + a_5 \cdot e^{i\theta_5} + a_4 \cdot e^{i\theta_4} = (a_2 + a_3) \cdot e^{i\theta_2} \tag{8}$$

$$a_1 + a_5 \cdot e^{i\theta_5} + a_4 \cdot e^{i\theta_4} = |a_2 - a_3| \cdot e^{i\theta_2} \tag{9}$$

The above equation (8) can be obtained by Euler's formula:

$$\begin{cases} a_1 \sin 0 + a_5 \sin\theta_5 + a_4 \sin\theta_4 = (a_2 + a_3)\sin\theta_2 \\ a_1 \cos 0 + a_5 \cos\theta_5 + a_4 \cos\theta_4 = (a_2 + a_3)\cos\theta_2 \end{cases} \tag{10}$$

The above equation (9) can be obtained by Euler's formula:

$$\begin{cases} a_1 \sin 0 + a_5 \sin\theta_5 + a_4 \sin\theta_4 = |a_2 - a_3|\sin\theta_2 \\ a_1 \cos 0 + a_5 \cos\theta_5 + a_4 \cos\theta_4 = |a_2 - a_3|\cos\theta_2 \end{cases} \tag{11}$$

Eliminateθ_4, the above equation can be arranged as:

$$\begin{aligned} &[a_1^2 + a_2^2 + a_3^2 + 2a_2a_3 - a_2a_5\cos(\theta_2 - \theta_5) - 2a_3a_5\cos(\theta_2 - \theta_5) \\ &- 2a_1a_2\cos\theta_2 - 2a_1a_3\cos\theta_2 + 2a_1a_5\cos\theta_5 - a_4^2 + a_5^2] = 0 \end{aligned} \tag{12}$$

$$[a_1^2 - a_4^2 + a_5^2 + 2a_1a_5\cos\theta_5 + |a_2 - a_3|^2 - 2a_5 \cdot |a_2 - a_3| \cdot \cos(\theta_2 - \theta_5) - 2a_1 \cdot |a_2 - a_3| \cdot \cos\theta_2] = 0 \tag{13}$$

It can be seen from the above formula that the range of all input variables of the equivalent Five-bar mechanism A'ABCC' kinematic chain is determined by Eqs.(12)and(13). The range of all the input variables is called joint rotation space. The calculated error rotation space is shown in Figure 4(a).

b. When the two input joints are adjacent,the θ_4 and θ_5 are input joints. When A'A and AB are collinear, the Five-bar ring degenerates into a Four-bar ring. When A'A and AB coincide, it also degenerates into a Four-bar ring, the ring equation (8) (9) finishing as follows:

$$[a_1^2 - a_2^2 - 2a_2a_3 - a_3^2 + 2a_4a_5\cos(\theta_4 - \theta_5) + 2a_1a_5\cos\theta_5 + 2a_1a_4\cos\theta_4 + a_4^2 + a_5^2] = 0 \tag{14}$$

$$[a_1^2 - 2a_2^2 + 2a_2a_3 + a_5^2 + 2a_4a_5\cos(\theta_4 - \theta_5) + 2a_1a_5\cos\theta_5 + 2a_1a_4\cos\theta_4 - a_3^2 + a_4^2] = 0 \tag{15}$$

From equations (14) and (15), the error rotation space of two input joints can be calculated as shown in Figure 4(b).

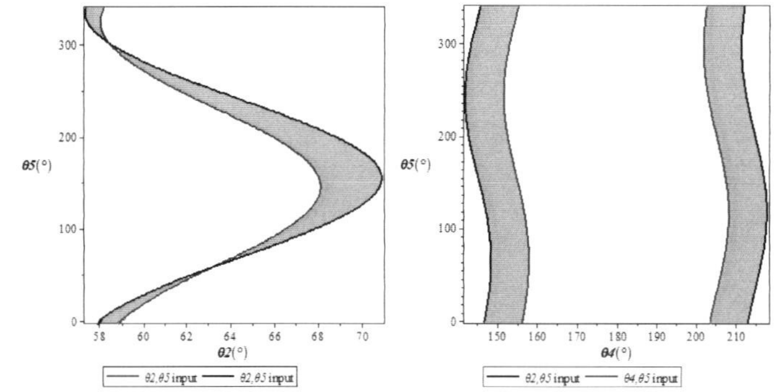

(a) (b)

Figure 4. Planar 2-DOF Five-bar error rotation space

6. Conclusion

In this paper, the output error of 2-DOF Five-bar planar parallel mechanism due to the uncertainty caused by the rotation of the joint was analyzed. The kinematic accessory joint space is equivalent to a virtual short rod. Establish 2-DOF Five-pole kinematic joint clearance error model. The influence of joint clearance on the output error of planar Five-bar mechanism is analyzed separately, and the calculation result of maximum error is obtained.An example of the length parameter of the mechanism is given. The accurate output rotation space due to the joint clearance is obtained. Due to the existence of joint space, the output error of planar parallel mechanism is inevitable. The method proposed in this paper selects reasonable joint clearance error value when designing planar parallel mechanism, and makes the error region within the controllable safety range has guiding significance.

7. Reference

[1] Shi Z C and Liu S H 1995 Mechanism Accuracy (Beijing:Higher Education Press) p 48-117.
[2] Yu Y Q and Tian H 2015 J. Optics Precis Eng. 23 1331-37
[3] Zhang Y, Huang X Z and Zhang X F,et al 2008 J. Northeast Univ (Nat Sci Ed) 29 1147-50.
[4] Ting K L, Hsu K L and Yu Z, et al. 2017 J. Mech & Mach Theory 111 66-75.
[5] Ting K L, Zhu J and Watkins D 2000 J. Mech & Mach Theory 35 391-401.
[6] Ting K L 1989 J.mech design ,vol.111 No.4 504-7.
[7] Ting K L and Liu Y W 1991 J.mech design 113(1):32-39.
[8] Shyu J.H.Ting K L 1994 J.mech design vol.116 No.1 343-34.
[9] Ting K L 2008 J.mech design 130 1404-04.

Acknowledgments

The authors thank for the support of the Outstanding Youth Science and Technology Innovation Team Program of HuBei Provincial Department of Education Grant(T201505), the Key Project of Natural Science Foundation of Hubei Province Grant(2015CFA112) and the Major Project of Technical Innovation in Hubei Province Grant(2017AAA108).

Combined effect of high curing temperature and crack width on chloride migration in reinforced concrete beams

L ELKEDROUCI[1], B DIAO[2], S PANG[3] and Y LI[4]

[1] School of Transportation Science and Engineering, Beihang University, Beijing 100083, China
[2] State Laboratory of Subtropical Building Science, South China University of Technology, Guangzhou 510640, China
E-mail : chinepekin0@gmail.com

Abstract. Deterioration of reinforced concrete structures is a serious concern in the construction engineering, largely due to chloride induced corrosion of reinforcement. Chloride penetration is markedly influenced by one or several major factors at the same time such as cuing in combination with different crack widths which have spectacular effect on reinforced concrete structures. This research presents the results of an experimental investigation involving reinforced concrete beams with three different crack widths ranging from 0 to 0.2mm, curing temperatures of 20°C or 40°C and water-to-cement of 0.5. Chloride content profiles were determined under non-steady state diffusion at 20°C. Based on the obtained results, higher chloride content was obtained under condition of high curing temperature in combination with large crack more than 0.1mm and there are no significant differences between narrow crack width (less than 0.1 mm) and beams without crack (0 mm).

1. Introduction

Reinforced concrete structures are wildly used nowadays in ocean environment construction where they usually provide satisfactory performance during their service life [1]. However, some structures suffer performance deterioration due to the chloride attack which may corrode reinforcement bars in RC structures and lead to cover cracking, spalling or in extreme cases collapse of the structure [2]. Transport of chloride ion through concrete is a complicated process [3] which can be impacted by several factors. Curing temperature to which reinforced concretes are subjected is one of the factors affecting concrete resistance [4]. Concrete resistance does not depend only on the material compositions or the water to cement ratio but also on the curing temperature which is required for cement hydration to attain the needed durability [5]. Higher curing temperature will accelerate hydration process of the cement paste [6], which results in high volume of pores creating an easy path for harmful substances such as chloride ion [7], and it has a substantial effect on the strength and durability of concrete [8]. High curing temperature plays an important role in identifying chloride ingress rate [9] and results in high chloride ion penetration [10]. High curing temperature in combination with other factors such as applied loads and concrete cover cracking can have enormous effects on chloride diffusion [11]. Cracks create preferential pathways for chloride ion penetration and other aggressive agents [12]. Besides, it reduces the strength of RC structure and induces loss of serviceability [13]. Increasing the crack width leads to increase chloride diffusion [14]. It is of great importance to study the effect of cracks on chloride diffusion through concrete [15].

The combined impacts of high curing temperature and different crack widths on chloride penetration of reinforced concrete beams were experimental investigated in this paper. Total 54 RC beams were initially cured under the temperatures of 20°C or 40°C, respectively. A non-steady state rapid chloride ion migration test was performed on total of 54 reinforced concrete beam with different crack widths under 20°C. Chloride contents were then tested and the comprehensive impacts of different crack widths and curing temperatures on chloride migration were discussed here.

2. Beam specimens and testing program

Total 54 reinforced concrete (RC) beam specimens were divided into two groups, one group of 27 RC beam specimens were cured under the temperature of 20°C and another group of 27 beam specimens were cured under the temperature of 40°C. The RC beam specimens were demoulded after cast 24 h and then cured under the temperature of 40°C or 20°C, respectively. At the age of 28 days, the RC beam specimens were loaded by three points bending to make different widths of flexural cracks. Then rapid chloride migration test was proceeded and the chloride content were tested for the RC beam specimens with different widths of cracks under high curing temperature or normal curing temperature. The details of the experiments were introduced in the following.

2.1. Beam specimens

Total 54 beam specimens were manufactured in prefabrication factory with ordinary Portland cement (42.5) and water-to-cement of 0.5. All beam specimens were designed with the same geometry sizes, same concrete materials and same steel bars. The length of beam specimens was 400mm, the width and the height of cross section was 100mm and 120mm, respectively. Two reinforcing steel bars of 340mm in length and 10mm in diameter were embedded at a cover depth of 30mm as illustrated in Figure 1.

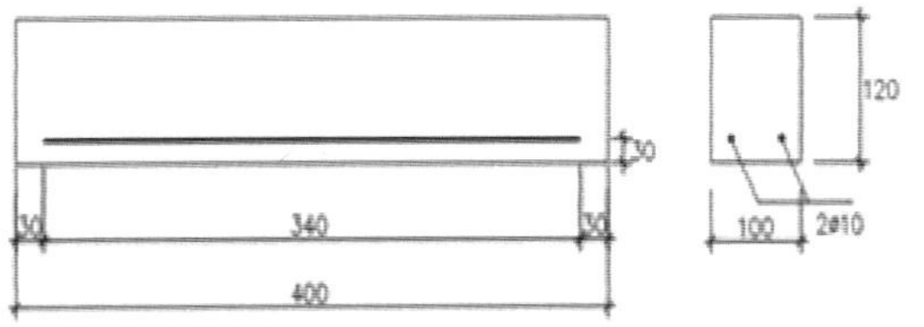

Figure 1. Geometry sizes and reinforcement bars of RC beam specimens.

2.2. Initial curing conditions

After demoulding, both groups of RC beam specimens were cured under two different curing temperatures of 40°C or 20°C, respectively, until the age of 28d. The group cured in high temperature chamber of 40°C was labelled HT for short. Another group cured in normal laboratory temperature (around 20°C) was labelled NT for short.

2.3. Crack width by three-points loading test

After initial curing of RC beam specimens finished, the three points loading were applied on the RC beam specimens at the age of 28d using a hydraulic universal testing machine. The schematic diagram of RC beam specimens was shown in Figure 2. The static loading of three points beading was applying to RC beam specimens until different widths of around bending cracks appeared, then unloaded. The maximum width of bending crack at tensile extreme fibres of RC beams are tested by crack width measuring and recording instrument, here the real flexural cracks were made by three points loading.

The RC beam specimens were classified into three types according to the testing results of the crack widths. First type of crack widths ranges from 0 to 0.1mm was named narrow cracks or named 'N' for short. Second type of crack widths range from 0.1 to 0.2mm was named wide cracks or named 'W' for short. Third type of cracks, as a reference, the beam specimen without crack was named '0'.

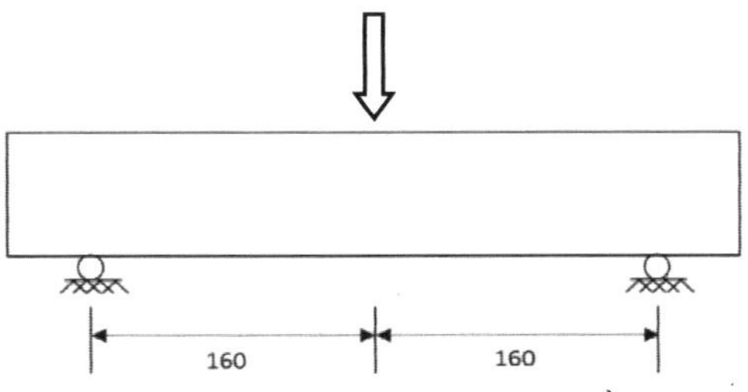

Figure 2. Schematic diagram of 3 points loading of beam specimen

2.4. Rapid Chloride Migration Test

The chloride penetration resistance of cracked RC beam specimens was tested by means of the non-steady-state rapid chloride migration (RCM) test, pushing chloride ion to penetrate through cracked concrete cover of RC beam specimens. The principal of RCM is similar to NTbuild492 [16], also called 'Nordtest' and was developed by several authors [17]. The schematic diagram of RCM test of RC beam developed here was shown in Figure 3, a gluing PVC pipe of 40mm high and 60mm diameter was put on tensile cracked or un-cracked surface and impenetrable all the other cracked area with epoxy resin to assure one way of penetration. To make sure the saturation of RC beam with water, it was immerged in pure water for 4 to 5 days except the covered PVC area. An electrical potential difference of 20 to 30 V was applied axially across the tensile cover depth, using steel tablet dipped in NACL solution as anode and steel bar as cathode which forced the chloride ion outside to migrate into concrete cover, as illustrated in Figure 3. The RCM test for each RC beam specimens lasted 24 hours.

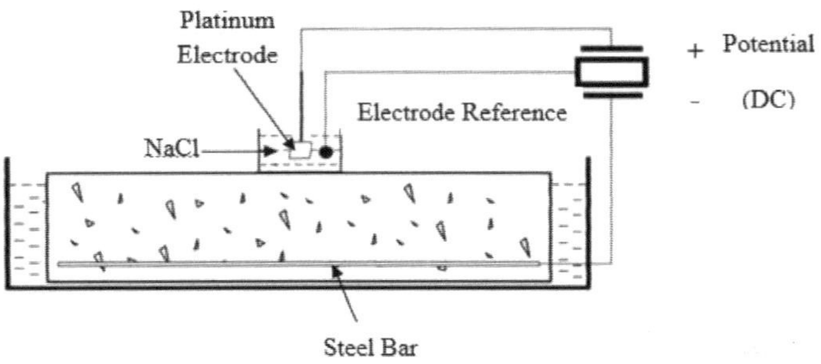

(a) (b)

Figure 3. Schematic diagram of RCM test of RC beam: (a) Photo of RCM test of RC beam, (b) Scheme of RCM test of RC beam.

2.5. Sampling and Chloride Testing

After rapid chloride migration test was performed, a cylinder sample with a depth of 30 mm was drilled at the cracked area which was covered by the PVC pipe in order to evaluate chloride penetration. Each drilled cylinder sample was then subdivided along the cover depth into 5 slices at a thickness of 6 mm. The sampling of concrete from RC beam was given in Figure 4. The slices were grounded into fine powdery particles to test free chloride content.

3. Results and discussion

3.1. Chloride content profiles

After rapid chloride migration test was performed, the results of total chloride content with three types of crack widths under high curing temperature or under normal curing temperature were shown in Figure 5a and 5b, respectively. Notice: Attention has to be drawn on chloride content of RC beam specimens

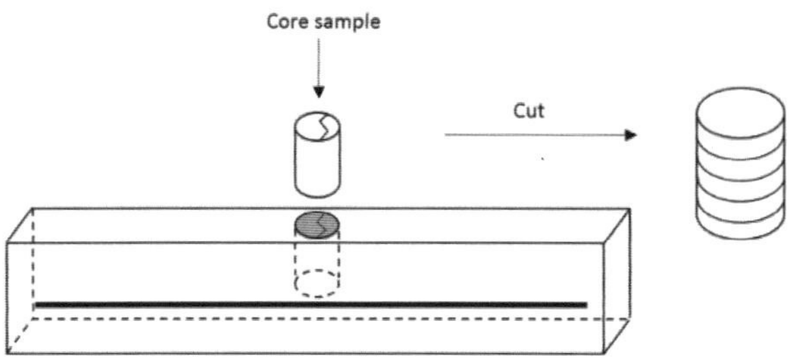

Figure 4. Schematic diagram core sample drilling and circle slices sub-cutting.

named '0' (without crack): beam specimens suffered a significant rate of chloride penetration that can be clarified by presence of micro-crack. As illustrated in Figure 5, the chloride content profiles show much higher values near the exposed surface and it increase significantly when increasing crack width in both considered curing temperatures. Both groups share the same variation trend of chloride content which is high at the beam exposure surface and steeply decreases reaching 10 mm of depth. Thereafter, chloride content decreasing rate deaccelerate for a depth range from 10 to 15 mm, then it becomes almost stable reaching the steel bar surface for all concrete beam specimens.

The effect of crack width at both curing temperatures groups was significant, where increasing crack width, imperatively, increase chloride content. High rate of chloride content was observed under condition of crack width large than 0.1mm, this indicate that crack width would be considered as one of the major factors affecting chloride incursion in concrete structures. For high curing temperature, chloride content distribution was higher than that of normal curing temperature. According to [9], curing at high temperature results in high volume of pores wider than 150 nm which leads to higher rate of chloride ion flow comparing to normal curing temperature; this point to that un increasing in chloride content due to an augmentation in curing temperature can be marked as one of major factors influencing chloride penetration in concrete. Chloride content is inversely proportional to depth, whenever depth increases, chloride content value decreases.

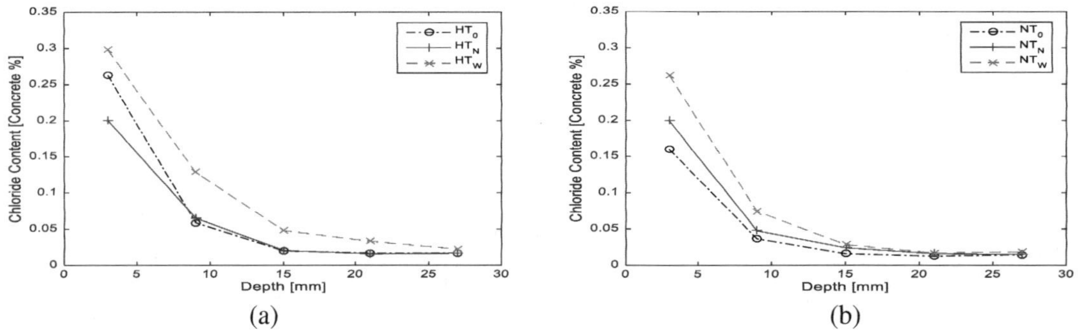

Figure 5. Chloride content along depth for two curing temperatures and different crack widths: (a): Chloride content for crack widths and HT, (b): Chloride content for crack widths and NT.

3.2. Comparison of Chloride Content Under Two Curing Temperatures

Figure 6 demonstrates a comparison of chloride concentration versus depth for similar crack widths in both considered curing temperatures. Usually, cracks form flow paths inciting higher concrete permeability and makes easier incursion of aggressive agent through concrete resulting in deterioration and loss of severability [18]. For the three considered crack widths, chloride content has the same tendency of distribution, increasing depth, chloride content value presents no substantial changes for both considered curing temperatures. Beam specimens with wider crack width 'W' display higher chloride concentration than that of 'N' and '0' specimens.

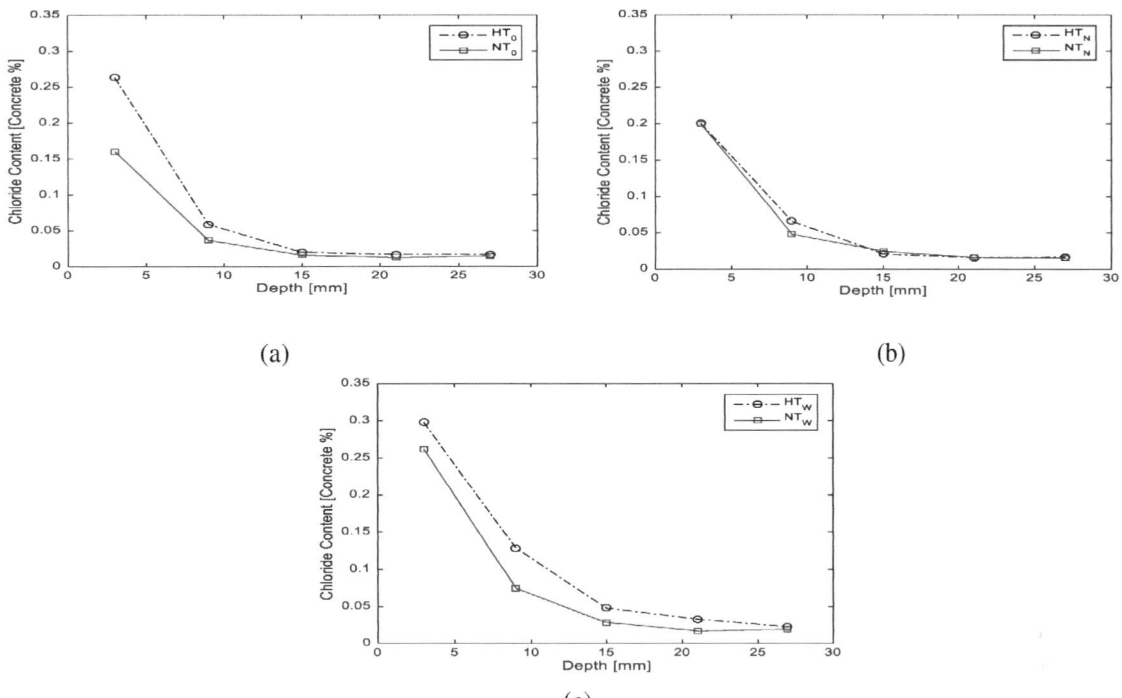

Figure 6. Chloride content under two curing temperatures and three crack widths: (a): no crack "0", (b): narrow crack "N", (c): wide crack "W".

3.3. Chloride Content at the Surface Steel Bar

The steel bar surface located at the maximum depth of concrete cover as shown in Figures (5, and 7) demonstrate the variation of chloride concentrations at the steel bar surface in both curing temperature groups for the three considered crack widths. The experimental data of Figure 7 were magnified 10 times in comparison with those of Figure 5. As it is illustrated in Figure 7, beam specimens with wide crack 'W' exhibit high chloride content values than crack width 'N' and '0', which leads to early corrosion of steel bar. The RC beam specimens named HT (curing in high temperature) display higher chloride content compared to those of NT (curing in normal temperature) RC beam specimens; this can be explained by the fact that curing in high curing temperature results in an acceleration in hydration

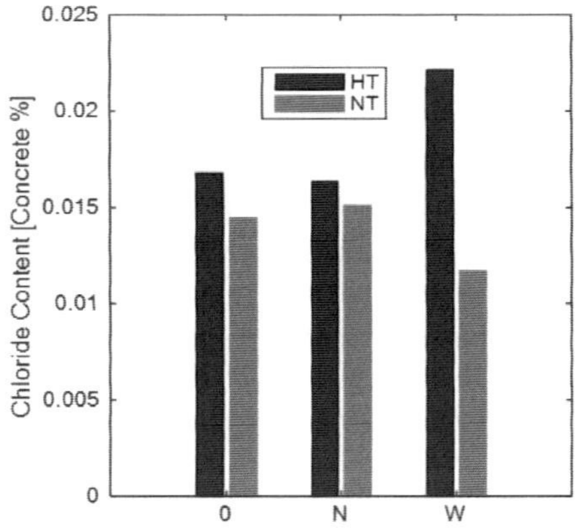

Figure 7. Chloride content with crack width for HT and NT State at the steel bar surface.

process of the cementitious material [19]. Also, it can be explained by the fact that raising the curing temperature leads to modify the porosity's characteristics by loss of free and absorbed water [20]. As it is shown in Figure 5, for the same curing temperature, wider crack results in high chloride content.

4. Conclusion

The ensuing conclusions are based on the test results stated in this paper:

Chloride concentration of reinforced concrete with different cracks is extremely dependent on the crack width. For both curing temperature, high values of chloride content were obtained in RC beams with crack width larger than 0.1 mm (wide crack) and much less in RC beams with crack width less than 0.1mm (narrow crack). This indicates that wide crack results in high chloride penetration rate.

Reinforced concrete beams under high curing temperature condition (40°C) displayed higher chloride content compared to normal curing temperature (20°C). This indicates that curing temperature have an obvious impact on chloride penetration.

For RC beam specimens with crack width 'N' (less than 0.1mm) and with crack width '0' (without crack), the chloride content profiles were almost equal for both curing temperatures. This shows that the effect of narrow crack width (less than 0.1mm) is nearly the same as without crack (0mm).

5. Reference

[1] Le T, Austin S, Lim S, Buswell R A, Law R, Gibb A G and Thorpe T 2012 Hardened properties of high-performance printing concrete. Cem. Con. Res. 42 pp 558-66
[2] Jin W, Yuan L, Wei Y S, Wang J and H L 2011 Durability Theory and Design Method of Concrete Structure under Chloride Environment
[3] Nilsson L O and Tang L 1996 Transport mechanisms in porous materials In the Modelling of Microstructure and Its Potential for Studying Transport Properties and Durability (Netherlands: Springer) pp 289-311
[4] Khatib J M and Mangat P S 2002 Influence of high-temperature and low-humidity curing on chloride penetration in blended cement concrete. Cem. Con. Res. 32 11 pp 1743-53
[5] Bonavetti V, Donza H, Rahhal V and Irassar E 2000 Influence of initial curing on the properties of concrete containing limestone blended cement. Cem. Con. Res. 30 5 pp 703-8
[6] Jaegermann C 1990 Effect of water-cement ratio and curing on chloride penetration into concrete exposed to mediterranean sea climate. Mat. J. 87 4 pp 333-9
[7] Goto S and Roy D M 1981 The effect of w/c ratio and curing temperature on the permeability of hardened cement paste. Cem. Con. Res. 11 4 pp 575-9
[8] Kjellsen K O, Detwiler R J and Gjørv O E 1990 Pore structure of plain cement pastes hydrated at different temperatures. Cem. Con. Res. 20 6 pp 927-933
[9] Detwiler R J, Fapohunda C A and Natale J 1994 Use of supplementary cementing materials to increase the resistance to chloride ion penetration of concretes cured at elevated temperatures. Mat. J. 91 1 pp 63-66
[10] François R and Arliguie G 1998 Influence of service cracking on reinforcement steel corrosion. J. Mat. Civ. E. 10 1 pp 14-20
[11] Otieno M B, Alexander M G and Beushausen H D 2010 Corrosion in cracked and uncracked concrete influence of crack width concrete quality and crack reopening. Mag. Con. Res. 62 6 pp 393-404
[12] Lu C, Jin W and Liu R 2011 Reinforcement corrosion-induced cover cracking and its time prediction for reinforced concrete structures. Corr. Sci. 53 4 pp 1337-47
[13] Pacheco J and Polder R 2012 Corrosion initiation and propagation in cracked concrete–a literature review. Adv. Mod. Con. Ser. L. (Netherlands: Springer) pp 85-93
[14] Wang H, Lu C, Jin W and Bai Y 2011 Effect of external loads on chloride transport in concrete. J. Mat. Civ. E. 23 7 pp 1043-49
[15] Build N T 1999 492 Nordtest method concrete mortar and cement based repair materials: Chloride migration coefficient from non-steady-state migration experiments (Finland: Espoo)
[16] Bogas J A and Gomes A 2015 Non-steady-state accelerated chloride penetration resistance of structural lightweight aggregate concrete. Cem. Con. Com. 60 pp 111-122

[17] Wang K, Jansen D C, Shah S P and Karr A F 1997 Permeability study of cracked concrete. Cem. Con. Res. 27 3 pp 381-393
[18] Kim J K, Han S H and Song Y C 2002 Effect of temperature and aging on the mechanical properties of concrete: Part I: experimental results. Cem. Con. Res. 32 7 pp 1087-94
[19] Care S 2008 Effect of temperature on porosity and on chloride diffusion in cement pastes. Con. Bui. Mat. 22 7 pp 1560-73

Acknowledgments
This work was financially supported by National Natural Science Foundation of China (51678021) and the Open Project of State Key Laboratory of Subtropical Building Science, South China University of Technology (2016ZA03).

IOP Conf. Series: Materials Science and Engineering **324** (2018) 012015 doi:10.1088/1757-899X/324/1/012015

Computational analysis for biodegradation of exogenously depolymerizable polymer

M Watanabe [1] and F Kawai [2]

1 Professor, Graduate School of Environmental and Life Science, Okayama University, Okayama, Japan
2 Specially-appointed Professor, Center for Fiber and Textile Science, Kyoto Institute of Technology, Kyoto, Japan
E-mail: watan-m@okayama-u.ac.jp

Abstract. This study shows that microbial growth and decay in a biodegradation process of exogenously depolymerizable polymer are controlled by consumption of monomer units. Experimental outcomes for residual polymer were incorporated in inverse analysis for a degradation rate. The Gauss-Newton method was applied to an inverse problem for two parameter values associated with the microbial population. A biodegradation process of polyethylene glycol was analyzed numerically, and numerical outcomes were obtained.

1. Introduction

A large amount of synthetic polymers have been produced since around med twentieth century. Those macromolecules would be nonexistent unless they were invented. However, some of those xenobiotic compounds are utilizable by microorganisms. Those synthetic materials have been accumulated, and they have become possible sources for carbon dioxide emission. So polymeric biodegradation should throughly be studied. There are two types of processes in polymeric biodegradation. The molecular reduction by terminal liberation of monomer units is the primary factor of a depolymerization process of exogenous type. Polyethylene (PE) and polyethylene glycol (PEG) are exogenously depolymerizable polymers. A mathematical model was formulated and numerical techniques were developed for PE biodegradation [1]. Those mathematical techniques were implemented in analyses of biodegradation processes of PEG [2]. A time-dependent degradation rate was considered in formulation of a PEG depolymerization process [3].

Unlike exogenously depolymerizable polymers, molecules of endogenously depolymerizable polymers break down randomly. Polyvinyl alcohol (PVA) and polylactic acid (PLA) are endogenously depolymerizable polymers. A mathematical model was proposed and numerical techniques were developed for an enzymatic degradation process of PVA [4]. Those techniques were implemented in a study of enzymatic hydrolysis of PLA, and outcomes for PVA and PLA were compared [5]. A time-dependent degradation rate was considered in formulatoin of an enzymatic hydrolysis of PLA [6]. A techniques developed for endogenously depolymerizable polymers were implemented in studies of exogenous type depolymerization processes of PEG [7] and PE [8]. Mathematical techniques for the PE biodegradation were implemented in study of PEG biodegradation processes [9]. A time-dependent degradation rate was considered in formulation of a PEG depolymerization process [10].

This study revisited a depolymerization process of PEG. In the following sections, mathematical model is described, numerical techniques used in an inverse analysis for a time factor of a degradation rate are illustrated, and numerical outcomes are introduced.

2. Mathematical formulation for exogenously depolymerizable polymers

Consider a microbial depolymerization process in a culture medium. Let M be the molecular weight, and let t be the time. Denote the weight distribution with respect to M at t by $w(t,M)$ [mg], and the residual of polymer consisting of molecules with molecular weight between A and B at time t by $v(t)$ [mg]. The residual $v(t)$ is expressed by the integral of $w(t,M)$ with respect to M,

$$v(t)=\int_A^B w(t,M)\,dM\,. \tag{1}$$

In particular, total weight $v(t)$ of the entire residual polymer at time t is expressed by

$$v(t)=\int_0^\infty w(t,M)\,dM\,. \tag{2}$$

Integral (1) is an appropriate approximation of integral (2) for suitable values of A and B.
Denote the population of viable microorganisms at time t by $\sigma(t)$. Equation

$$\frac{\partial w}{\partial t}=\sigma(t)\left[-\lambda(M)w+c(M)\int_M^\infty \lambda(K)d(K)w(t,K)\,dK\right]\quad\left(c(M)=Me^{\rho M},\ d(M)=\frac{e^{-\rho M}}{M\left(1-e^{-\rho M}\right)},\ \rho=\frac{\log 2}{L}\right) \tag{3}$$

was previously proposed [9 - 14]. Parameter L denotes the molecular weight of a monomer unit, *e.g.* PE: $L=28$ (CH_2CH_2), PEG: $L=44$ (CH_2CH_2O). The degradation rate per unit time, unit weight, and unit population was decomposed into a molecular factor and a time factor. Function $\lambda(M)$ denotes the molecular factor, whereas the microbial population $\sigma(t)$ is the time factor. Equations (2) lead to

$$v'(t)=\sigma(t)\int_0^M\left[-\lambda(M)w+c(M)\int_M^\infty \lambda(K)d(K)w(t,K)\,dK\right]dM\,.$$

In a previous study, equation

$$\frac{d\sigma}{dt}=k\left[1-h\frac{\sigma}{u(t)}\right], \tag{4}$$

was proposed for the microbial population, where $u(t)=-v'(t)$. In this study, equation (4) was replaced by

$$\frac{d\sigma}{dt}=ku(t)-h\sigma. \tag{5}$$

The term $ku(t)$ is the conversion of consumption of monomer units, and the term $-h\sigma$ is the reduction by viability loss. Suppose that $f_0(M)$ is an initial weight distribution, and that σ_0 is an initial population of viable microorganisms. system of equations (3), (5) and initial conditions $w(0,M)=f_0(M)$, $\sigma(0)=\sigma_0$ constitute an initial value problem.

3. Numerical results for inverse analysis of microbial population

The initial value problem associated with equations (3) and (5) is solvable for $w(t,M)$ and $\sigma(t)$ provided that the function $\lambda(M)$ is given and that values of parameters σ_0, h, , and are prescribed. To specify the function $\lambda(M)$ and values of parameters σ_0, h, , and , consider change of variables

$$\tau=\int_0^t \sigma(s)\,ds. \tag{6}$$

Denote $w(t,M)$, $\sigma(t)$, and $v(t)$ by $W(\tau,M)$, $S(\tau)$, and $V(\tau)$, respectively. Equations

$$\frac{\partial W}{\partial \tau} = -\lambda(M)W + c(M)\int_M^\infty \lambda(K)d(K)W(\tau,K)\,dK\,, \tag{7}$$

$$\frac{dS}{d\tau} = kU(\tau) - h, \qquad (U(\tau) = -V'(\tau)) \tag{8}$$

are derived from the equations (3) and (5), respectively. Note that the following equations hold.

$$V(\tau) = \int_0^\infty W(\tau,M)\,dM, \qquad V'(\tau) = \int_0^M \left[-\lambda(M)W + c(M)\int_M^\infty \lambda(K)d(K)W(\tau,K)\,dK \right] dM.$$

Suppose that $F_1(M)$ and $F_2(M)$ are weight distributions for $\tau = T_1$ and $\tau = T_2$ $(0 \le T_1 < T_2)$, respectively, equation (8) and conditions $W(T_1,M) = F_1(M)$, $W(T_2,M) = F_2(M)$ constitute an inverse problem for $\lambda(M)$ such that the solution of equation (7) satisfies those conditions.

Numerical techniques for the inverse problem were previously developed. Suppose that weight distributions $f_0(M)$, $f_1(M)$, ..., $f_m(M)$ for t_0, t_1, ..., t_m, respectively, are given, so that $w(t_0,M) = f_0(M)$, $w(t_1,M) = f_1(M)$, ..., $w(t_m,M) = f_m(M)$. A pair of weight distributions were assigned to the functions $F_1(M)$ and $F_2(M)$, and the inverse problem for $\lambda(M)$ was obtained numerically. Once $\lambda(M)$ was specified, equation (7) was solved for $W(\tau,M)$ with condition $W(0,M) = f_0(M)$. Once $W(\tau,M)$ was obtained, equations $V(\tau_i) = v(t_i)$ $(i = 0, 1, \ldots, m)$ were solved numerically, and values of τ_i $(i = 0, 1, \ldots, m)$ were obtained. Previous studies show that an exponential function $e^{-\mu\tau}$ approximates $V(\tau)$ with an appropriate value of μ. In this study, value $\mu = 0.1$ was set.

Suppose that $S(\tau,\sigma_0,k,h)$ is the solution of equation (8) with condition $S(0,\sigma_0,k,h) = \sigma_0$. Note that $S(\tau,\sigma_0,k,h) = \sigma_0 + k[v_0 - V(\tau)] - h\tau$, where $v_0 = V(0)$. The change of variables (6) leads to the equation $t = q(\tau,\sigma_0,k,h)$, where

$$q(\tau,\sigma_0,k,h) = \int_0^\tau \frac{dr}{S(r,\sigma_0,k,h)}. \tag{9}$$

Given m pairs of values of t and τ, (t_i,τ_i) $(i = 1, 2, \ldots, m)$, define functions $g_i(\sigma_0,k,h)$ $(i = 1, 2, \ldots, m)$ by $g_i(\sigma_0,k,h) = q(\tau_i,\sigma_0,k,h) - t_i$, and consider the system of equations for σ_0, k, and h,

$$g_i(\sigma_0,k,h) = 0 \quad (i = 1, 2, \ldots, m). \tag{10}$$

Values of residual PEG v_i at $t_i = i$ $(i = 0, 1, 2, \ldots, 6)$ (day), and v_7, and v_8 at 8 and 10 days, respectively, were reported [16]. In a previous study, some of those values were incorporated into analysis of an inverse problem derived from equation (4) [16]. Some of those values v_i $(i = 0, 1, 2, \ldots, 6)$ were also introduced into an inverse analysis in this study. Initial microbial population σ_0 was fixed ($\sigma_0 = 1.0$), and the Gauss-Newton method was applied to the least square problem for

$$\frac{1}{2}\sum_{i=1}^{5}\left[g_i(\sigma_0, k, h)\right]^2. \qquad (11)$$

With initial values $k = 0.1$ and $h = 2.0$, residuals $\left(\sqrt{(k_n - k_{n-1})^2 + (h_n - h_{n-1})^2}\right)$ between two successive approximations was reduced to a value that was less than 10^{-15} after fourteen iterations. Those final values are $k = 0.0493429362357295$ and $h = 1.01446006075672$. The graphs of the residual PEG $(t, v(t)) = (q(\tau, \sigma_0, k, h), V(\tau))$ and the microbial population $(t, \sigma(t)) = (q(\tau, \sigma_0, k, h), S(\tau, \sigma_0, k, h))$ were generated numerically for those values of h and k. Values of the microbial density [10^9/ml] (*S. terrae*) were obtained experimentally [14]. Those values and values of $\sigma(t) = S(\tau, \sigma_0, k, h)$ $(t = q(\tau, \sigma_0, k, h))$ were compared, and a linear conversion of $(t, \sigma(t)) = (q(\tau, \sigma_0, k, h), S(\tau, \sigma_0, k, h))$ according to a least square approximation is shown (Figure 2).

4. Conclusion

In previous studies [10 – 13], experimental outcomes before and after cultivation of microorganisms in culture media were incorporated into inverse analysis for $\lambda(M)$ and $\sigma(t)$. A recent study [16] demonstrated that an exogenous type microbial depolymerization process can be simulated with a set of residual polymer before after cultivation of microorganisms. The Gauss-Newton method [16] was applied to a nonlinear least square problem derived from the an equation for the microbial population. In this study, the Gauss-Newton method was applied to the square sum (11).

The symbiotic mixed culture E-1 consists of *S. terrae* and *Rhizobium* sp., and *S. terrae* was a primary PEG utilizing bacteria. The viable cell density of *S. terrae* was measured by colony counting [15]. Figure 3 (b) shows the transition of viable cell density of *S. terrae* and a conversion of the numerical result for the microbial population. Figure 2 (b) shows a notable disagreement between experimental results and numerical results for $3 \le t \le 6$.

Acknowledgement

The authors thank Dr. K. Yamamoto for his technical support. This work was supported by JSPS KAKENHI Grant Number 16K05276.

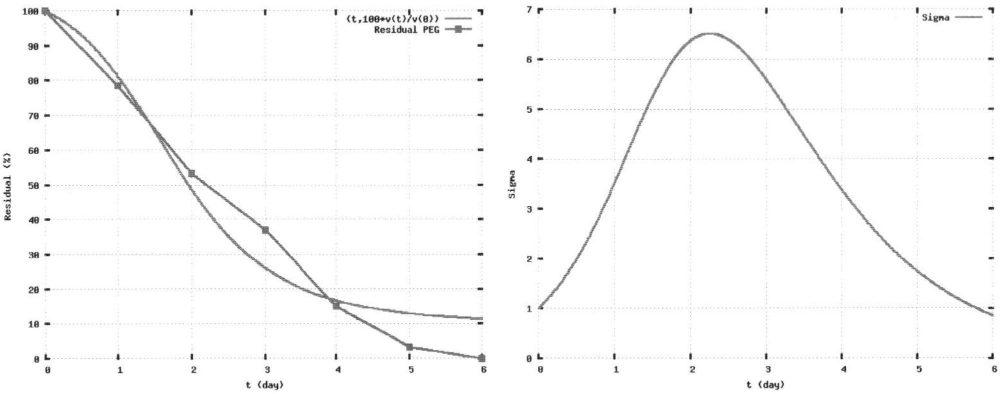

(a) (b)

Figure 1. (a) Residual PEG $(t, v(t)) = (q(\tau,\sigma_0,k,h), V(\tau))$. Experimental results [16] are also shown. (b) Microbial population $(t,\sigma(t)) = (q(\tau,\sigma_0,k,h), S(\tau,\sigma_0,k,h))$. Those curves were generated numerically for the values of σ_0, h, and $\sigma_0 = 1.0$ $k = 0.0493429362357295$, and $h = 1.01446006075672$.

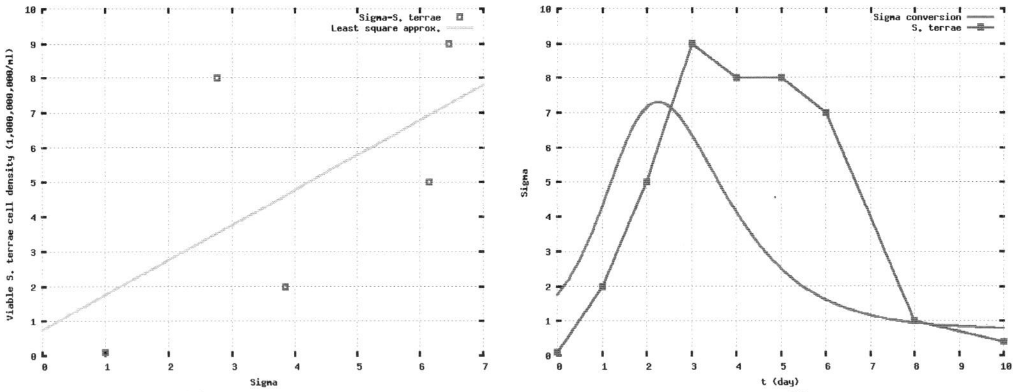

(a) (b)

Figure 2. (a) Values of the microbial density [10^9 /ml] of *S. terrae* and values of $\sigma(t) = S(\tau,\sigma_0,k,h)$ $(t = q(\tau,\sigma_0,k,h))$ (b) linear conversion of $(t,\sigma(t)) = (q(\tau,\sigma_0,k,h), S(\tau,\sigma_0,k,h))$ according to a least square approximation

References

[1] Masaji Watanabe, Fusako Kawai, Masaru Shibata, Shigeo Yokoyama, Yasuhiro Sudate, Shizue Hayashi, Analytical and computational techniques for exogenous depolymerization of xenobiotic polymers, Mathematical Biosciences 192 (2004) 19-37.

[2] M. Watanabe, F. Kawai, Numerical simulation of microbial depolymerization process of exogenous type, Proc. of 12th Computational Techniques and Applications Conference, CTAC-2004, Melbourne, Australia in September 2004, Editors: Rob May and A. J. Roberts, ANZIAM J. 46(E) pp.C1188--C1204, 2005. http://journal.austms.org.au/ojs/index.php/ANZIAMJ/article/view/1014

[3] M. Watanabe and F. Kawai, Effects of microbial population in degradation process of xenobiotic polymers, In P. Howlett, M. Nelson, and A. J. Roberts, editors, Proceedings of the 9th Biennial Engineering Mathematics and Applications Conference, EMAC-2009, volume 51 of ANZIAM J., pages C682--C696, September 2010.
http://journal.austms.org.au/ojs/index.php/ANZIAMJ/article/view/2433

[4] Masaji Watanabe, Fusako Kawai, Mathematical modelling and computational analysis of enzymatic degradation of xenobiotic polymers, Applied Mathematical Modelling 30 (2006) 1497-1514
[5] Masaji Watanabe, Fusako Kawai, Sadao Tsuboi, Shogo Nakatsu, Hitomi Ohara, Study on Enzymatic Hydrolysis of Polylactic Acid by Endogenous Depolymerization Model, Macromolecular Theory and Simulations 16 (2007) 619-626.
[6] M. Watanabe, F. Kawai, Modeling and analysis of biodegradation of xenobiotic polymers based on experimental results, Proceedings of the 8th Biennial Engineering Mathematics and Applications Conference, EMAC-2007, Editors: G. N. Mercer and A. J. Roberts, ANZIAM J. 49 (EMAC-2007) pp.C457--C474, March 2008.
http://journal.austms.org.au/ojs/index.php/ANZIAMJ/article/view/361
[7] M. Watanabe and F. Kawai, Study of biodegradation of xenobiotic polymers withh change of microbial population, In W. McLean and A. J. Roberts, editors, Proceedings of thc 15th Biennial Computational Techniques and Applications Conference, CTAC-2010, volume 52 of ANZIAM J., ages C410--C429, July 2011.
http://journal.austms.org.au/ojs/index.php/ANZIAMJ/article/view/3965
[8] Watanabe M, Kawai F (2012) Modeling Biodegradation of Polyethylene with Memoryless Behavior of Metabolic Consumption, J Bioremed Biodegrad 3: 146.
[9] M. Watanabe and F. Kawai, Study on microbial depolymerization processes of xenobiotic polymers with mathematical modelling and numerical simulation, In Mark Nelson, Mary Coupland, Harvinder Sidhu, Tara Hamilton, and A. J. Roberts, editors, Proceedings of the 10th Biennial Engineering Mathematics and Applications Conference, EMAC-2011, volume 53 of ANZIAM J., pages C203--C217, June 2012.
http://journal.austms.org.au/ojs/index.php/ANZIAMJ/article/view/5107
[10] Masaji Watanabe, Fusako Kawai, Simulation for microbial depolymerization processes of polyethylene glycol, Proceedings of the 2013 International Conference on Advances in Intelligent Systems in Bioinformatics, Editors: Dr. Ford Lumban Gaol, Dr. Benfino Soewito, Prof. Dr. Mohamed Bououdina, Prof. Dr. Mu-Song Chen, intel-13, Advances in Intelligent Systems Research, Atlantis Press, February 2014, 71 – 76. ISBN: 978-94-6252-000-4, ISSN: 1951-6851
http://www.atlantis-press.com/php/pub.php?publication=intel-13
[11] Masaji Watanabe, Fusako Kawai, Numerical Techniques for Inverse Problems from Modeling of Microbial Depolymerization Processes, International Journal of Applied Engineering Research ISSN 0973-4562 Volume 11, Number 8 (2016) pp 5461-5468.
http://www.ripublication.com/ijaer16/ijaerv11n8_18.pdf
[12] Masaji Watanabe, Fusako Kawai, Numerical Study of Microbial Depolymerization Process with the Newton-Raphson Method and the Newton's Method, International Journal of Engineering Research & Science (IJOER), Vol-2, Issue-2, February- 2016, Pages 119—130. ISSN: [2395-6992] http://ijoer.com/Paper-February-2016/IJOER-JAN-2016-55.pdf
[13] Masaji Watanabe, Fusako KawaI, Numerical Study of Inverse Problems in Modeling and Simulation of Microbial Depolymerization Processes,Volume 1, 2016, International Journal of Chemistry and Chemical Engineering Systems, 2016, 46-55. ISSN: 2367-9042
http://www.iaras.org/iaras/filedownloads/ijcces/2016/016-0006.pdf
[14] FUSAKO KAWAI, SHOGO ENOKIBARA, Symbiotic Degradation of Polyethylene Glycol (PEG) 20,000-Phthalate Polyester by Phthalate Ester- and PEG 20,000-Utilizing Bacteria, Journal of Fermentation and Bioengineering, Vol. 82, No. 6, 575-579, 1996.
[15] S. Gratton, A. S. Lawless, and N. K. Nichols, Approximate Gauss-Newton Method for Nonlinear Least Squares Problems, SIAM Journal of Optimization, 18(1), 2007, 106-132
[16] Masaji Watanabe and Fusako Kawai, Numerical Techniques for Analysis of Microbial Population with Residual Polymer in Microbial Depolymerization Process, 5th International Conference on Advances Engineering, Science, Technoogy and Sustainable Development (ESTSD-17) Sept. 28-30, 2017 Kuala Lumpur (Malaysia), ER0917409, 77-84.

IOP Conf. Series: Materials Science and Engineering **324** (2018) 012016 doi:10.1088/1757-899X/324/1/012016

Influence of the operating parameters of the needle-plate electrostatic precipitator on the size distribution of aerosol particles

P V Arsenov, A A Efimov, N V Protas, V V Ivanov

Department of Physical and Quantum Electronics, Moscow Institute of Physics and Technology, Dolgoprudny 141700, Russia

Abstract. The influence of the operating parameters (voltage and aerosol flow rate) of the needle-plate electrostatic precipitator (NP-ESP) on the size distribution of aerosol particles has been studied. The NP-ESP consists of a needle and a plate located in the plastic tube used as aerosol transport duct. Alumina (Al_2O_3) particles were synthesized by a spark discharge and used as a test aerosol with a size range from 25 to 500 nm. It was found that the average particle size decreases with increasing voltage and aerosol flow rate through the NP-ESP. It was also found that the average particle size can be reduced more than in 2 times in comparison with the initial size distribution at a voltage and aerosol flow rate through the NP-ESP are equal to 16 kV and 250 l/min, respectively.

1. Introduction

In recent years, there has been growing interest in the use of nanoparticles in science and technology [1]. Nanoparticles are widely used in electronics, optics, energy and medicine [2–4]. It is known that properties of nanoparticles mainly depend on their size [5]. Usually, many methods of producing particles, such as synthesis by spark discharge, does not significantly change the particle size during the synthesis process [6,7]. In this regard, there is a need to develop special equipment that allows to control the particle size after their production process. In this work, there has been developed the needle-plate electrostatic precipitator (NP-ESP) to separate particles by size. A distinctive feature of the developed equipment is a simple design, consisting of a needle and a plate in a dielectric tube. The influence of the parameters of the NP-ESP, namely, the voltage and the aerosol flow rete, on the particle size distribution is investigated.

2. Experimental

The scheme of the experiment is shown in figure 1a. Aerosol particles (Al_2O_3) were obtained using the multi-spark discharge generator (m-SDG) [8]. Then the aerosol particles were sent to the needle-plate electrostatic precipitator (NP-ESP). The particle size distribution was measured using an aerosol spectrometer (AS) at the outlet of the NP-ESP. Aerosol particles were obtained by the erosion of the electrodes made of Al in air at energy stored in the capacitor and the repetition rate of discharges is equal to 6 J and 1 Hz, respectively. A developed the NP-ESP consisted of a needle and a plate that were located in a dielectric tube with an internal diameter of 45 mm. A steel needle with a radius of curvature of about 40 μm was used as the corona electrode. The needle was located at a distance of 10 mm from the steel plate with a size of 100x30x2 mm. The distance between the needle and plate was chosen experimentally for achieving of the maximum current of a corona discharge. A corona discharge was created between the needle and the plate when a high voltage of 5 to 16 kV was applied to the needle. The source of high voltage was the source of VIDN-30 (up to 30 kV and 250 μA), and the corona discharge current was measured with an Agilent U1253B multimeter. The current-voltage characteristic of the NP-ESP is shown in figure 1b. Electrical breakdown of the gas gap length of 10 mm occurred at a voltage between the needle and the plate is more than 16 kV (figure 1b).

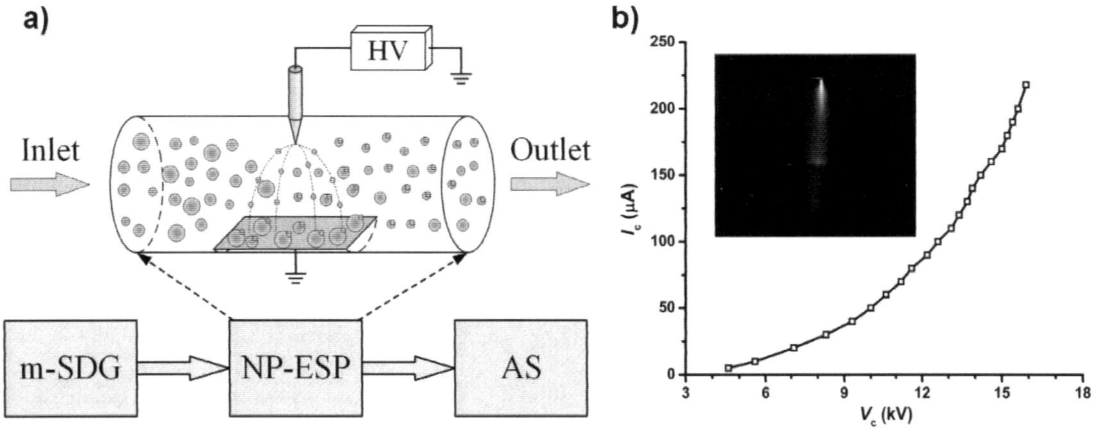

Figure 1 (a, b). The scheme of the experiment (m-SDG – multi-spark discharge generator, NP-ESP – needle-plate electrostatic precipitator and AS – aerosol spectrometer) (a). Voltage-current characteristics of the NP-ESP at needle-plate spacing is about 10 mm and negative corona polarity. The inset shows the appearance of the corona discharge (b).

Aerosol nanoparticles adsorbed charge due to collisions with ions or electrons. Then the particles were separated according to their size and charge in the field of corona discharge. The efficiency of separation E_s of aerosol nanoparticles in the NP-ESP was determined from equation (1):

$$E_s = (\frac{n_{in} - n_{out}}{n_{in}}) \cdot 100\% \quad (1)$$

where n_{in} and n_{out} – the particle number concentration measured when the NP-ESP was turned off and turned on, respectively.

Commercial aerosol spectrometer TSI SMPS 3936 was used to determine the size and concentration of aerosol particles. The particle size distribution was studied depending on the voltage between the needle and the plate V_c, and from the aerosol flow rate Q_a through the NP-ESP.

3. Results and discussions

Figure 2a shows the particle size distributions after passing the aerosol through the NP-ESP at a voltage between the needle and the plate V_c of 0, 8.5, 12.5 and 16.0 kV, respectively. The operating mode of the precipitator at V_c=0 kV means that the precipitator was turned off, and therefore the aerosol is the initial aerosol. Figure 2a shows that as V_c increases from 8.5 to 16.0 kV, the average particle size decreases from 117±8 nm to 87±6 nm, respectively. It should be noted that the initial aerosol at V_c=0 kV has an average particle size of 158±11 nm. Thus, it can be argued that a change in voltage between the needle and the plate V_c makes it possible to control the size of the particles separated in the NP-ESP. The result can be explained as follows, according to the theory of charging [9], it is known that large particles are charged more efficiently and gain a larger number of charges than small particles.

In this connection, it follows that larger particles in the electric field will have a higher electrical mobility (Z_p) in accordance with the equation (2):

$$Z_p = \frac{n_p e C_c}{3\pi\eta d_p}, \qquad (2)$$

where n_p – the charge per particle, e – the elementary unit of charge, C_c – Cunningham correction factor, η – the viscosity of the gas, d_p – the particle size.

As a result, in the electric field of the corona discharge, larger particles will precipitate more efficiently, while the smaller particles will leave the NP-ESP without deposition on the plate. This explanation is confirmed by the graph in figure 2b, which shows the dependence of the separation efficiency E_s on the particle size for different values of V_c. Figure 2b shows that large particles with a size of more than 200-300 nm are precipitate more effective, the separation efficiency of which exceeds 80%. However, the low efficiency of separation is observed for particles with sizes smaller than 200-300 nm because of their low charging efficiency. As a result, at the outlet from the NP-ESP an aerosol with a smaller average size is obtained than at the inlet to the NP-ESP.

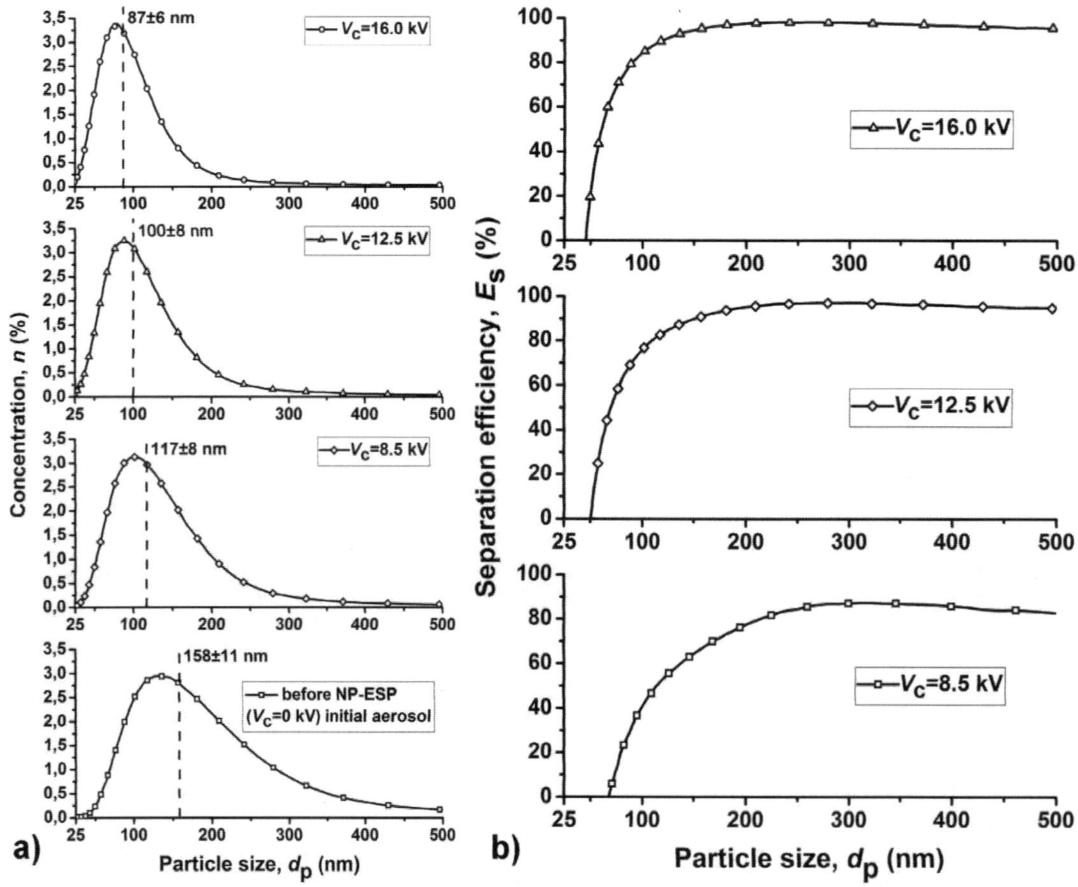

Figure 2 (a, b). The influence of the voltage between the needle and the plate V_c on the particle size distribution (a). The dependence of the separation efficiency E_s on the particle size for various voltages between needle and plate V_s (b). Aerosol flow rate Q_a is equal 33 l/min.

Figure 3 shows the particle size distribution depending on the aerosol flow rate Q_a through the NP-ESP. In figure 3, it is seen that when the aerosol flow rate Q_a are equal to 33 and 250 l/min, the average particle size in the initial aerosol are 156±11 nm and 95±7 nm, respectively. The decrease in the average size with the increase in aerosol flow rate is associated with a decrease in the particle concentration as a result of their dilution with a stream of pure gas.

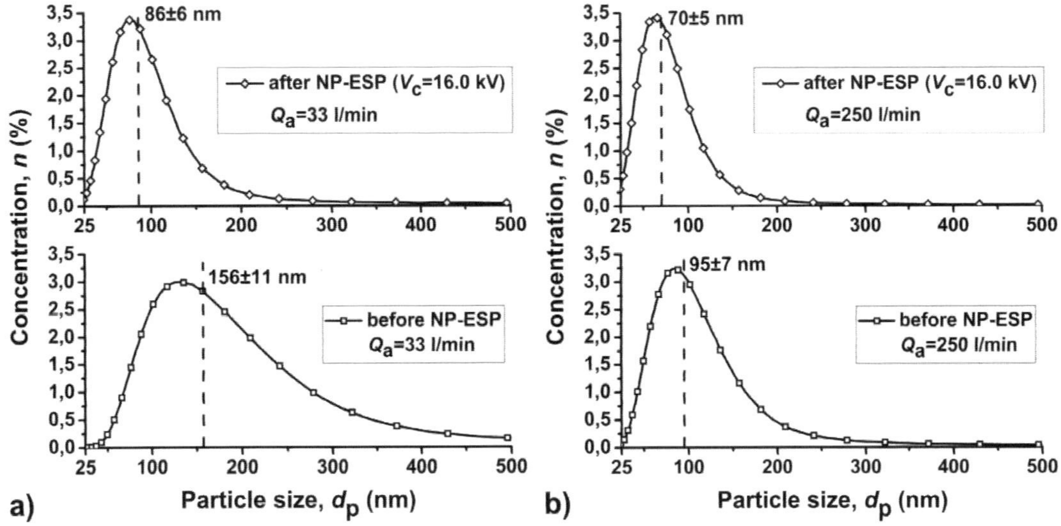

Figure 3 (a, b). The effect of aerosol flow rate Q_a through the NP-ESP on the particle size distribution, Q_a=33 l/min (a) and Q_a=250 l/min (b), respectively.

Figure 3 also shows that an additional reduction in the average particle size is achieved when the NP-ESP is turned on at V_s=16.0 kV. So, for example, with the aerosol flow rate Q_a of 33 and 250 l/min, the use of a precipitator at V_c=16.0 kV reduces the average particle size to 86±6 nm and to 70±5 nm, respectively. It can also be seen from the data in figure 3 that when the aerosol flow rate through the precipitator is low, Q_a=33 l/min, more efficient charge of particles is achieved, and as a result, the average particle size changes by a larger number than with the aerosol flow rate Q_a=250 l/min. Thus, the use of a precipitator can significantly change the average particle size at low aerosol flow rates. For example, it has been found that the average particle size can be reduced by more than 2 times (from 156±11 to 70±5 nm) compared to the initial size of the aerosol (figure 3a) at the V_c=16.0 kV and Q_a=250 l/min (figure 3b), respectively.

4. Conclusion

Based on the results of the experiments, it is established that the use of the developed NP-ESP makes it possible to reduce the average particles size of aerosol by more than 2 times in comparison with the initial aerosol. It is established that the NP-ESP performs the functions of an electrofilter-separator, precipitating large particles more than 200-300 nm from the aerosol stream. It is determined that the separation efficiency of the particles increases with increasing their size and increasing the voltage between the needle and the plate. The developed precipitator is a promising device for controlling the size of the synthesized particles, which is relevant for a number of applications in electronics, optics and medicine.

Acknowledgments

This work was supported by the Russian Science Foundation (project # 15-19-00190).

References

[1] Wang C, Friedlander S K and Mädler L 2005 Nanoparticle aerosol science and technology: an overview *China Particuology* **3** 243–54

[2] Kruis F E, Fissan H and Peled A 1998 Synthesis of nanoparticles in the gas phase for electronic, optical and magnetic applications—a review *J. Aerosol Sci.* **29** 511–35

[3] Raimondi F, Scherer G G, Kötz R and Wokaun A 2005 Nanoparticles in Energy Technology: Examples from Electrochemistry and Catalysis *Angew. Chem. Int. Ed.* **44** 2190–209
[4] Murthy S K 2007 Nanoparticles in modern medicine: State of the art and future challenges *Int. J. Nanomedicine* **2** 129–41
[5] Grassian V H 2008 When Size Really Matters: Size-Dependent Properties and Surface Chemistry of Metal and Metal Oxide Nanoparticles in Gas and Liquid Phase Environments *J. Phys. Chem. C* **112** 18303–13
[6] Efimov A, Sukharev V, Ivanov V and Lizunova A 2015 The influence of parameters of spark discharge generator on dimensional characteristics of synthesized TiO2 nanoparticles *Orient. J. Chem.* **31**
[7] Efimov A, Lizunova A, Sukharev V and Ivanov V 2016 Synthesis and Characterization of TiO2, Cu2O and Al2O3 Aerosol Nanoparticles Produced by the Multi-Spark Discharge Generator *Korean J. Mater. Res.* **26** 123–9
[8] Efimov A A, Ivanov V V, Bagazeev A V, Beketov I V, Volkov I A and Shcherbinin S V 2013 Generation of aerosol nanoparticles by the multi-spark discharge generator *Tech. Phys. Lett.* **39** 1053–6
[9] Hinds W C 1999 *Aerosol Technology: Properties, Behavior, and Measurement of Airborne Particles* (New York: Wiley-Interscience)

IOP Conf. Series: Materials Science and Engineering **324** (2018) 012017 doi:10.1088/1757-899X/324/1/012017

Unbonded prestressing tendons and their role in the construction of slender elements of buildings

M Mieszczak[1]

Ph D Student, Cracow University of Technology, Poland
E-mail address: malgorzatamieszczak16@gmail.com

Abstract. Steel unbonded tendons have been introduced in Europe for construction prestressing many years later than in the USA, Honkong, Singapore or Australia. In Poland, they appeared in the early 1990s. Despite their short application, in the last decade, the Institute of Materials and Building Constructions of the Cracow University of Technology in cooperation with the TCE Structural Design & Consulting company has developed and implemented several interesting and unique designs of building components, using the advantages of this type of prestressing. In the author's work, apart from the short description of these tendons, several selected (own and foreign) projects of unique character have been presented.

1. Introduction

Unbonded steel tendons were developed and introduced to prestressed structures in the 1960s, mainly in the USA. High labor costs and mass production of prestressing tendons by smelters, which are easy and fast the assembly has led to their widespread prevalence. The economic aspects have won with the basic drawback of unbounded tendons - lack of cement injection and therefore risk of losing the tendon if it is cut or if anchorage is damaged. In view of this threat, in Europe, this type of prestressing has encountered great difficulties in adaptation. In European countries, this prestressing found application many years later than in the USA, Asian countries or Australia. The scale of its use was also smaller. In Poland, such tendons were first used to reinforce the structure of a sewage tank in 1991. The first concrete slabs in buildings, prestressed this way, were created at the beginning of the present millennium.

2. Construction and characteristics of unbonded tendons

The unbounded tendon forms a 7ϕ5 mm steel strand placed in grease in a hard polyethylene tube (Figure 1). The tendons are delivered to the building site as prefabricated, in spools of a mass of about

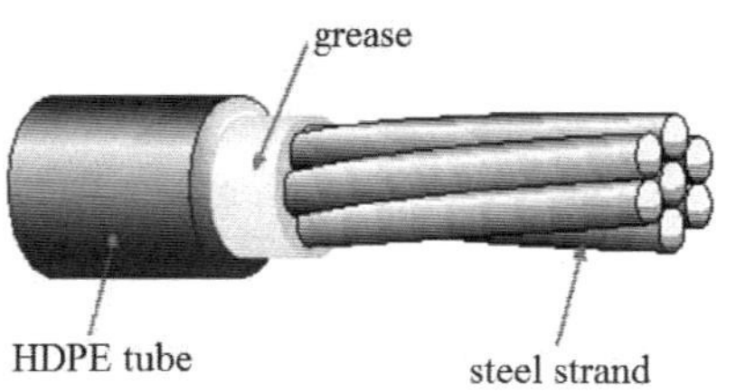

Figure 1. Unbonded tendon construction.

Figure 2. Unbonded tendon in the spool.

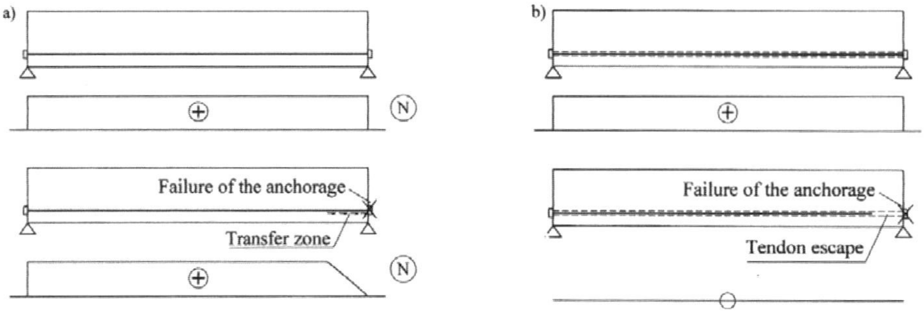

Figure 3. Bonded (a) and unbounded (b) tendon behaviour after anchorage failure.

2 tons (Figure 2). A low friction coefficient (<0.05), small diameter (about 20mm), small curvature radius (500-600mm), lightness and ease of assembly make them superior in particular types of construction, compared to the traditional systems of injected cables. Over the past decade, this new type of prestressing has led to the implementation of several interesting, unprecedented projects in Poland. In addition to several new applications, such as the elimination of thermal cracking in the walls of tanks fixed in the bottom slabs or post-tensioned slabs of vehicle scales, the tendons have been used in extremely thin floor slabs. At this point, it should be stressed that the threat connected with the possibility of damaging such a tendon in the element, e.g., due to its cut, drilling, damage of anchorage, etc. In contrast to a traditional tendon injected with cement paste, lack of bonding causes that the loss of force also includes the whole tendon, and not only the cross-section of the damage and its vicinity (Figures 3 and 4). Since the tendon has the ability to move freely in the tube, it is not possible to transfer force to concrete through adhesion and formation of a transmission zone. The ability to move the strand also causes that local cross-section effort does not induce local increase of stresses and strain in the prestressing steel. These increments cover the entire length of the tendon. Unbonded tendons are thus affected by larger elongations and the structures prestressed with their use are subject to larger deformations.

3. Post-tensioned slabs of large span

In recent years, several large-span slender post-tensioned slabs with the use of prestressing without bonding have been designed and erected in Poland. In the building of the Cultural and Artistic Center in Kozienice which was put into operation in 2015; due to the limited height of the building and the need to obtain as much space as possible, an opportunity of creating the thinnest construction panels possible was sought. Three post-tensioned floor slabs with the largest span and unprecedented slenderness have been designed (Figure 4 to 6). All three slabs were designed in post-tensioning technology, prestressed with unbounded tendons:

- slab Sl-1 +9.68m horizontally, with a span of 11.15m and thickness of 200mm located above the theater room. The ratio of span to depth is 55.8;
- slab Sl-2 +14.08m horizontally, with a span of 12.65m and thickness of 250mm, as flat roof over the theater room. The ratio of span to depth is 51.4;
- slab Sl-3 with projection dimensions of 17.65×19.6m and thickness of 350mm designed above a cinema hall. The ratio of span to depth is 50.4.

The largest slab Sl-3 over the cinema was the subject of computational analyses presented in the paper [1]. The constructional details of all three slabs, together with the results of measurements during the implementation, are given in the paper [2].

Here, reference should be made guidelines in literature regarding the maximum span and slenderness of such slabs. According to [3], the maximum span of one-way slabs, with a thickness of 250mm and a load over dead weight of 1.75 kN/m^2 (corresponding to flat roof Sl-2) should not exceed 9.4m for the simply supported slab and 10.8m for the inner span of the continuous slab (compare with the span of slab Sl-2). Meanwhile, for a thickness of 200mm and a load of over dead weight of 4.0kN/m^2

IOP Conf. Series: Materials Science and Engineering **324** (2018) 012017 doi:10.1088/1757-899X/324/1/012017

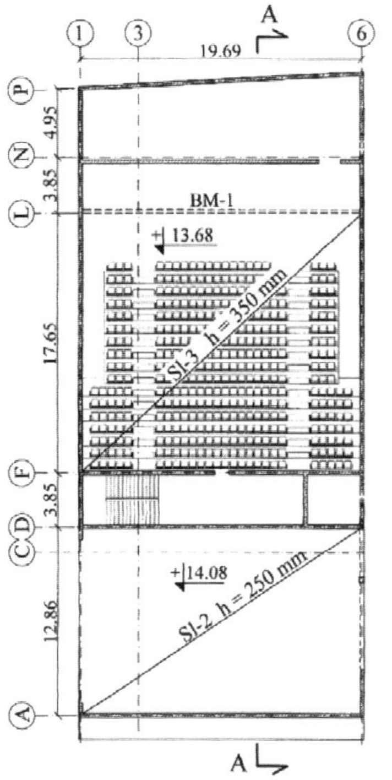

Figure 4. Plan of the building segment with post-tensioned slabs.

Figure 5. View of CKA in Kozienice..

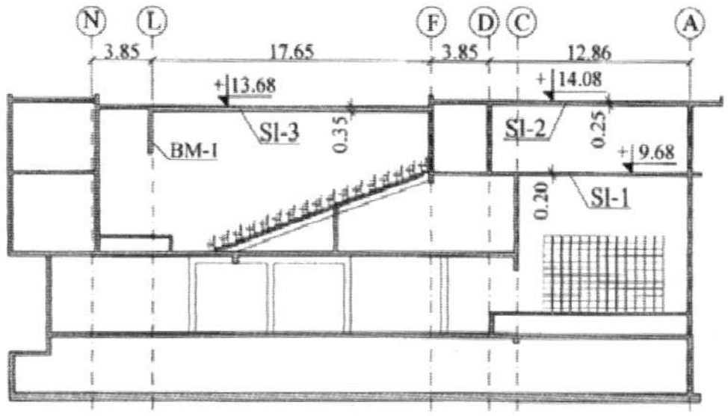

Figure 6. A-A cross section.

(corresponding to floor slab Sl-1), the recommended spans are 7.2m for the simply supported span and 8.0m for the inner span of the continuous slab (compare with span of slab Sl-1). In the case of two-way slab, with a load exceeding the dead weight of 1.75kN/m^2 and a thickness of 300mm, the largest suggested span for the simply supported slab is 13.2m. Meanwhile, slab Sl-3 over the cinema was 350mm thick but as large as 17.65m. The recommended maximum ratio of span to depth, with a minimum of two spans in each direction, is 42 for ceilings and 48 for flat roofs.

In their work [4], Khan and William have a different approach towards the maximum ratio of span to depth, depending on the construction of the ceiling and loads exceeding dead weight. For one-way solid slabs, the full ratio is 30÷45, ribbed slabs - 25÷35, flat solid slabs - 35÷45, waffle slabs - 20÷30. The upper value corresponds to a load exceeding dead weight - 2.5kN/m^2, and the lower value corresponds to a load of 15kN/m^2.

By comparing the geometrical parameters of the constructed slabs and the corresponding values recommended as the maximum, it is easy to see that the slabs are well above the recommended values. Despite this, they have shown surprisingly small deflections. For example, ceiling slab Sl-1 which, half a year after its execution, after all the components of the equipment on the ceiling, with fixed loads exceeding the dead weight of 3.9kN/m^2, was deflected by 11mm, i.e. 1/1014 of the span. Slab Sl-2, after 3 years of execution, is deflected by 24mm, i.e., 1/538 of the span, and the increase in deflection in time has significantly reduced.

Another post-tensioned slab with a large span constructed with the use of unbounded tendons is presented in [5]. A one-way simply supported slab over a concert hall, with a span of 21.3m and a thickness of 550mm, was designed. In order to reduce dead weight, lightweight internal relief inserts were used in the form of balls with a diameter of 315mm. This reduced the weight of the slab by 20%, reducing flexural stiffness by 10%, which resulted in a significant reduction in the expected deflections. Although the ratio of span to depth (38.7) is not impressive, the span for such a slab is a record value.

4. Church buildings

For the first time in Poland, prestressing with unbounded tendons has been used in two church projects. In the church of St. Peter and Paul in Bodzanow (near Wieliczka), a prestressed post-tensioned peripheral tie beam was used. The structure of the roof is built on a circular plan (Figure 7a) with a diameter of 25.5m. The wall-bearing construction is constituted by 12 columns evenly spaced on the circumference with a cross-section of 0.4×0.6m. The roof structure is constituted by 12 girders from glued timber with a cross-section of 300×700mm, pivotally attached at the peak and supported on reinforced concrete columns (Figure 7b). In order to avoid the use of additional steel ties running through the chords, or the construction of a massive reinforced concrete ring beam capable of

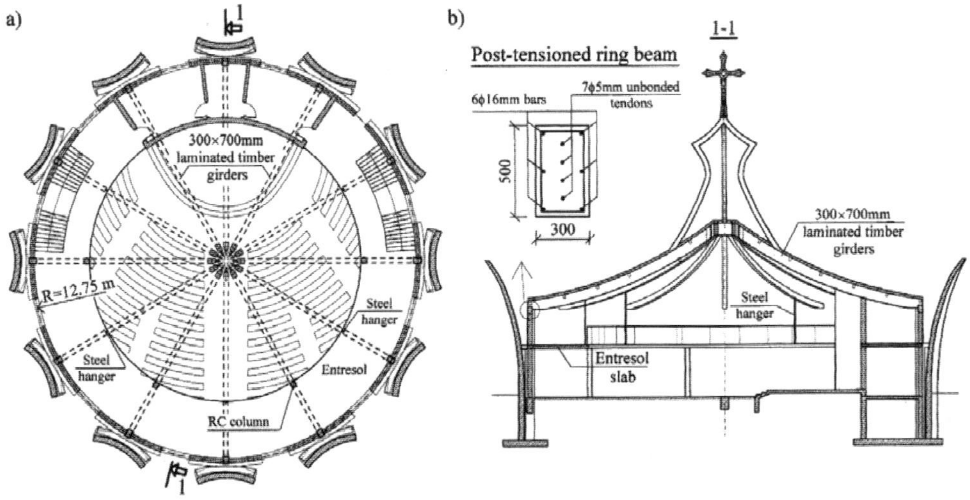

Figure 7. The church plan (a), vertical cross-section and post-tensioned ring beam (b).

transferring tensile forces from the roof with additional bending moments, a slender post-tensioned ring beam was constructed, with a cross-section of 300×500mm, hidden in the thickness of the wall. For the prestressing of the ring beam, 4 unbonded tendons were used, wrapping the entire circumference and anchored alternately on both sides of the circumference. The construction and technological details of the project are presented in the paper [6].

Another example of the use of unbonded tendons in churches is the church of St. Jack in Cracow, where, in the interior, a very thin, concrete structure of a mezzanine for the choir was used. A 3.0m wide and 250mm thick slab with a central span of 10.0m was prestressed along the ring with 10 unbonded tendons. This project is expected to be implemented and will be the subject of future publications.

5. Other, new use of unbonded tendons in a building

An interesting use of this type of prestressing in the construction of prefabricated Slimfloor slabs was proposed in [7]. Such ceilings are constructed of pre-tensioned prestressed concrete slabs supported on steel delta-shaped beams. The slabs are based on the bottom shelf of the beam, which is hidden in the ceiling. It is obvious that the spacing of the columns and the space between them are conditioned by the span of the heavily loaded and low beam. In the analyzed case [7], slabs with a span of 10.2m and a thickness of 320mm and an additional 60mm thickness of concrete overlay were used. Strength conditions, with the standard version of commonly used beams, allowed the use of a beam with a span of 7.2m. In the original version of the project, 2 unbonded tendons (Figure 8) with a parabolic route

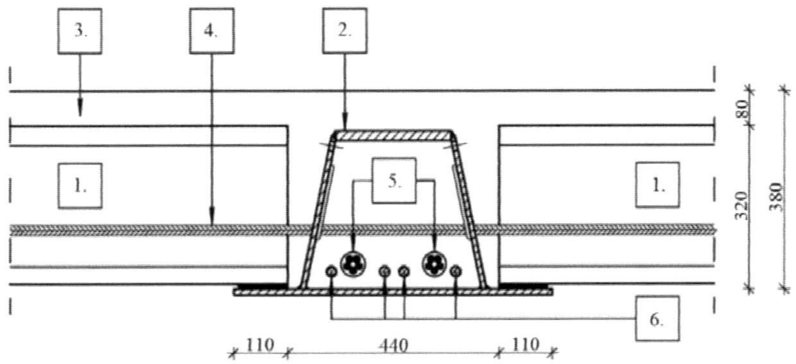

Figure 8. The church plan (a), vertical cross-section and post-tensioned ring beam (b) [7]. 1. Hollow core slab, 2. Delta-shaped steel profile, 3. Concrete slab with mesh reinfocement, 4. Transverese reinforcement, 5. Prestressing cables, 6. Fibre rebars.

were used inside the beam. This solution allowed to increase the span of the beam, and thus the span of the ceiling in the direction of the beam, from 7.2 to 9.0m.

6. Conclusions

The work presents some interesting examples of the use of unbonded tendons in post-tensioned elements in buildings in Poland. It should be emphasized that despite their relatively recent appearance in our country, they are currently used on a mass scale in the construction of buildings. The presented projects, however, are of innovative nature; they are a result of a scientific research and were created as an effect of the cooperation of an academic institution with a design company, under strict scientific supervision. Positive results obtained from the implementation are the basis for calibrating the long-standing calculation methods and contributing to the development of bold new projects.

7. References

[1] Mieszczak M 2013 Płaskie stropy sprężone cięgnami bez przyczepności (Flat concrete floors post-tensioned with 5 nbounded tendons) *3rd National Student Building Conference "Euroinżynier"* Cracow

[2] Szydłowski R, Mieszczak M 2015 Gdzie jest granica smukłości sprężonych płyt stropowych? O projekcie i badaniach stropów sprężonych w budynku Centrum Kulturalno-Artystycznego w Kozienicach (Where is the limit to span-depth ratio for prestressed slabs? About the design and research on prestresse slab in Art Gallery in Kozienice) in *Aktualne kierunki rozwoju teorii i praktyki konstrukcji sprężonych w Polsce* (*Current trends in the development of the theory and practice of prestressed structures in Poland*) Cracow University of Technology pp 67–78

[3] Fib-Bulletin No. 31 2005 *Post-tensioning in building* Lausanne p 109

[4] Khan S, Williams M 1994 *Post-tensioned Concrete Floors* Butterworth-Heinemann Bodmin-UK p 312

[5] Szydlowski R, Labuzek B 2017 Post-Tensioned Concrete Long-Span Slabs in Projects of Modern Building Construction *IOP Conf. Series: Materials Science and Engineering* 245 (2017) 022065

[6] Szydlowski R, Labuzek B, Turcz M 2017 Prestressed Ring Beam in the Church of St. Peter's and Paul's in Bodzanow, Design and Realization *IOP Conf. Series: Materials Science and Engineering* 245 (2017) 022066

[7] Derkowski W, Skalski P 2017 New Concept of Slimfloor with Prestressed Composite Beams *Procedia Engineering* 193 pp 176-183

IOP Conf. Series: Materials Science and Engineering **324** (2018) 012018 doi:10.1088/1757-899X/324/1/012018

Considerations on the use of elastic wheels to the urban transport vehicles

Ioan SEBESAN[1], **Sorin ARSENE**[2] and **Ion MANEA**[3]

[1, 2] Polytechnic University of Bucharest, Transport Faculty, Depart Rolling Stock Railway, Splaiul Independentei no. 313, Sector 6, Bucharest, Romania,
[3] SC Softronic SRL, R&D Department, Craiova, Calea Severinului No.40, Dolj, Romania
[1] ioan_sebesan@yahoo.com, [2] sorinarsene@gmail.com, [3] ion.manea@softronic.ro

Abstract. To minimize dynamic wheel-rail interaction efforts a condition is that the unassembled mass of the vehicle is as small as possible. The elastic wheel by its construction fulfills these conditions - she has interposed between the crown and the body of the wheel, the elastic rubber elements. In this way, it can be considered that the unsupported mass is represented only by the mass of the wheel crown. Additionally, this elasticity also has a reduction effect on rolling noise. This feature makes it suitable for use on urban transport vehicles.

1. Need and constructive type of the elastic wheels

The idea of breaking the metal chain that conveys shocks, vibrations and rolling noise from the wheel to the vehicle body by interposing an elastic element as close to the exciting source has been backed by the special benefits it offers.

Reduce the weight of the vehicle as much as possible which does not have an elastic suspension, make it as the perturbing influence of the irregularities of the path feel to feel only on these small masses and to be transmitted attenuated to the rest of the vehicle construction.

In fig. 1, is shown the wheel constructions that are more widely used to both in the light rolling stock (tram and metro), as well as the actual railroad. On these types of wheels, the rubber it as use are request at to compression and shear or only at compression.

The wheel of fig. 1a has two rubber rings with metal reinforcements, which are pressed to the mounting by bending the metal wings of the wheel center.

On the wheel of fig. 1b also uses two rubber rings, but without metal reinforcements.

The pre-compression of the rubber is achieved by tightening the screws, between two flanges of the rubber rings.

The particularity of this type of wheel is that it allows mounting / dismantling some to the transmission elements, such as the crown gear without moving the wheel hub.

For this purpose, the inner clamping flange is two halves.

The wheels of fig. 1c and 1d, patents of the German company Krupp-Bochum, have a very large spread and are called wheels with a rubber elastic ring.

In some constructive variants, the rubber ring is profiled in V, for greater safety towards transversal efforts. It is generally used for trams (fig. 1c).

The variant of fig. 1d has support collars on both the band and the center of the wheel which come in contact with larger axial deformations, limiting them to a maximum of 4 mm according to the traffic safety rules.

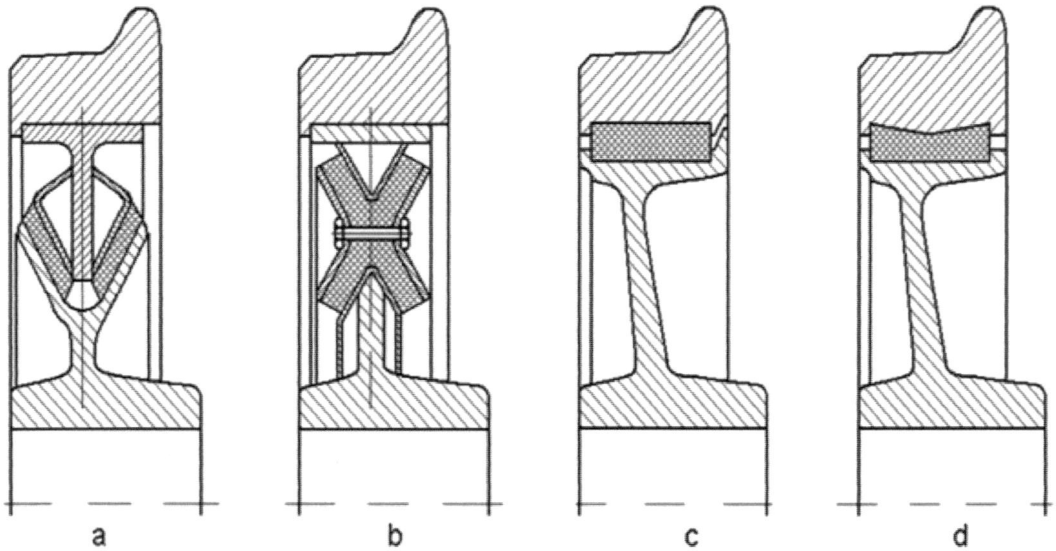

Figure 1. Elastic rolling wheels with the rubber requested mainly at compression.

In fig. 2 are some of the usual types of elastic wheels to which the rubber is only required for shear. It is the type of American conveyor wheel with very large distribution in North America and Canada where it is estimated that about 100,000 such wheels are in operation. And in Europe, companies that have taken US licenses use wheels of this type (Italy, Belgium, etc.).

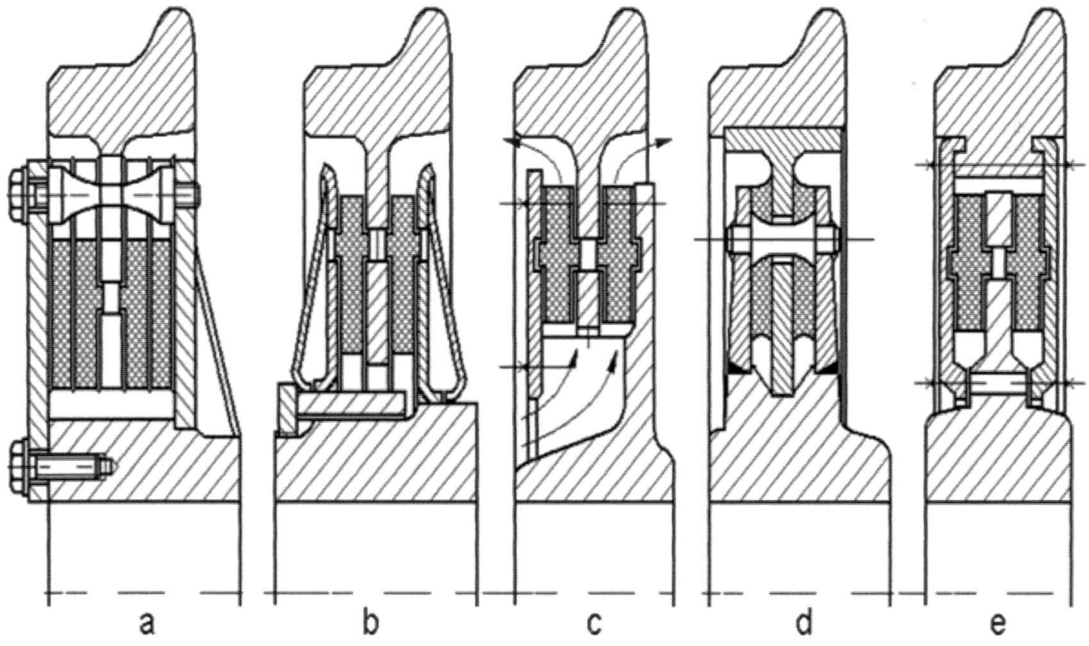

Figure 2. Elastic rolling wheels with the rubber requested mainly at to shear

In principle, the special band or wheel center has the shape of a disc that presses between two rubber discs. Pressing is done between metal flanges fixed to the wheel hub. In fig. 2a the bandage also acts as a center of the wheel and as a result, outside the section and the tread profile, it is executed as a disk in the central part. The piece is more expensive, but eliminates banding / deburring operations requiring special machines. If the thickness of the outer part in the form of bandage is higher, it is also possible to apply to these wheels after reaching the wear limit and machining as a wheel center as in fig. 2d.

Rubber discs can be made in one piece or in sectors, but in all cases they are provided with vulcanized metal fittings.

The tightening of the rubber disc is made under the press and the maintenance in this position is made with screws (fig. 2a) or with a central nut (fig. 2b) for wheels with a lower diameter and lower radial loads.

In general, these wheels, called wheels with two rubber discs, are removable for replacing the wheel-tire center or rubber discs.

There are also wheels of this type, such as a patent of German company Klöckner (fig. 2d).

On these wheels the rubber discs are vulcanized on the center of the wheel, and the press flanges are welded and hardened to the workpieces, also welded to the assembly.

It is worth noting the elastic wheel of fig. 2c, a patent of the Swedish SAB, in which the elastic ring consists of a number (generally 8 ... 10) of disk-shaped discs, the free space between these discs permitting air circulation which cools the rubber.

In fig. 2e shows the elastic wheel of Pirelli, to which the packing of the wheel is made with two bolted flanges, centered on the wheel band, which has a special machining. The radial and axial deformations of the wheel are limited between the clamping flanges and the wheel hub.

The centering of the component elements to all these types of wheels is made by the cylindrical projections of the metal fittings of the elastic members, protrusions which penetrate into the corresponding holes of the clamping flanges.

The centering is very good, the bandage or the center of the wheel-bandage, formerly turned, no longer requires, after the assembly of the wheel, than a small correction, which is generally done only by polishing the running surface.

Elastic wheels with two rubber discs generally have a lower stiffness in the radial direction (they are more elastic than the ones with a rubber ring described above but have the disadvantage of the tire-loaded tires) of a more labor-intensive technology maintenance.

For wheels with two rubber discs, according to SAB, the bearing capacity of the elastic wheels produced by this company (fig. 3) is presented.

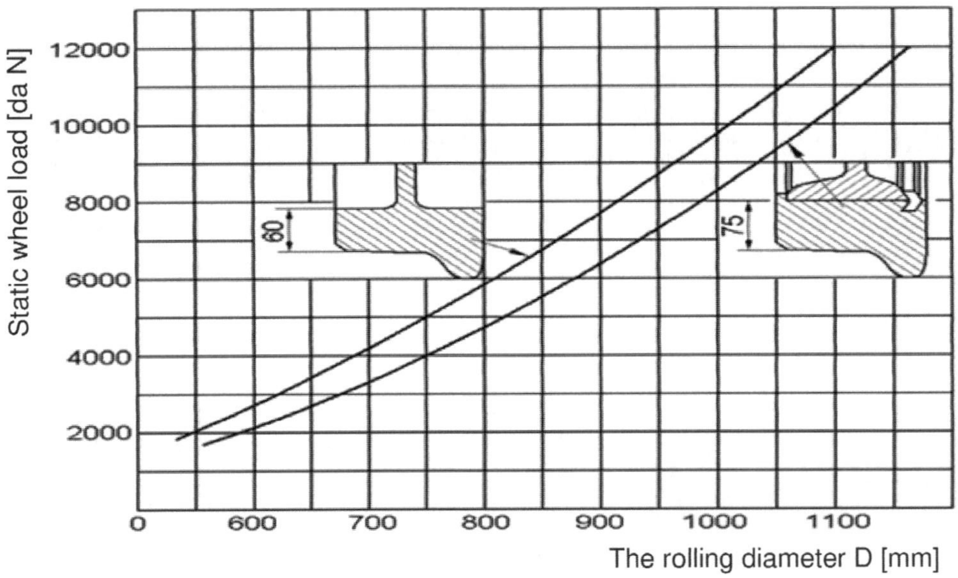

Figure 3. The bearing capacity of the elastic wheels produced by SAB

2. Rigidity of elastic wheels

The elastic wheels provide elasticity in three directions, namely: in the vertical direction (Z axis) in the transverse direction (Y-axis) and in the tangential direction, the acceleration or deceleration of the wheel rotation.

In fig. 4, the rigidity characteristics, as average values, for these three directions are indicated for the tires with a rubber ring (Bochum) and for the two wheels of rubber (Presidents Conference Committee).

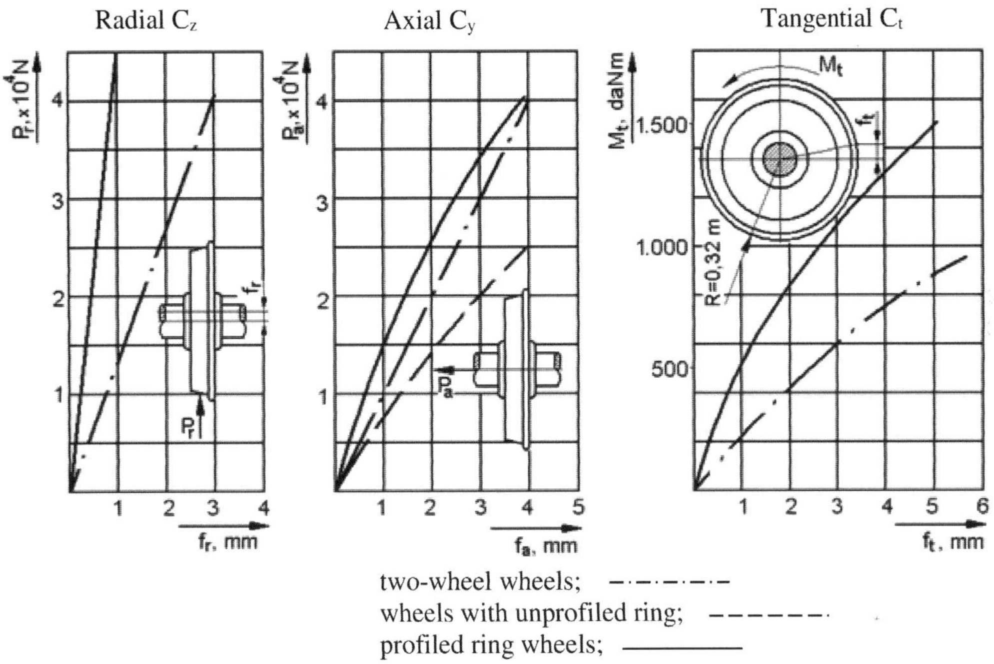

Figure 4. Rigidity of elastic rolling wheels (mean values)

Radically, wheels with two rubber discs are more elastic and, as such, better dampen shocks and vibrations taken from the track in that direction.

In the axial-transversal sense, the Bochum wheel may have two characteristics, as the rubber ring is profiled (see fig. 1c) or with a rectangular section (see fig. 1d). In the first case, the axial stiffness is higher than the wheels with two rubber discs.

If axial stresses (enrollment or curve guidance) are large and can cause axial deformations of the wheel (s) which endanger the safety of the movement or the correct entry on the trackside devices, the axial deformation is limited by the intervention of the shoulder and the center of the wheel fig. 1d). According to the German rules, the maximum admissible value for the axial deformation fa is 4 mm.

In the tangential sense, the wheels with a rubber ring are more rigid, as is the case with the other two directions.

The effect of tread rolling elasticity is manifested by the attenuation of shocks and vibrations transmitted to the suspended mass.

In fig. 5 indicate the vertical accelerations measured on the axle bearing housing in the case of standard (metallic) and elastic (Bochum) wheels on subway and railway vehicles. The higher the shock received from the tread of the wheel band, the higher the shock absorption factor transmitted to the center of the wheel and the wheel train, as well as the other vehicle organs (transmissions, motors ...).

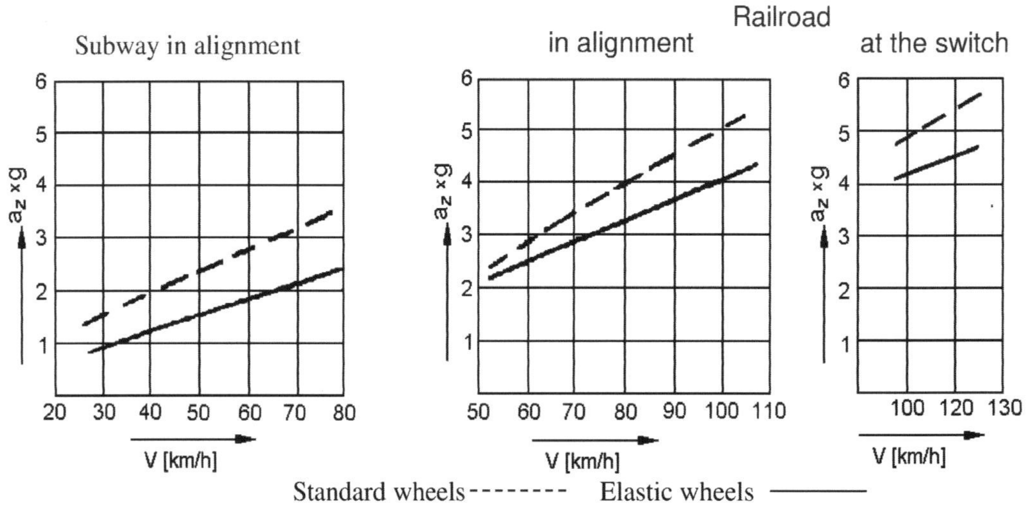

Figure 5. Elastic rolling wheel damping effect - Acceleration measured on the axle bearing (az) on elastic wheels and metal wheels

3. Conclusions

The quality of the running of the railway vehicles is dependent on both the elasticity and the resonant effect of the elastic wheels.
Related to these aspects, the paper presented the theoretical considerations and the results of experimental determinations carried out by the authors with elastic wheels used in subway vehicles.
For constructions used in light vehicles, the superiority of the elastic wheels results in the comparison of the entirely metal wheels. The high elasticity of these wheels also contributes to the reduction of the inertial forces with the rolling path.

Acknowledgments

This work was supported by a grant of the Romanian National Authority for Scientific Research and Innovation, CNCS–UEFISCDI, project number PN-III-P2-2.1-PTE-2016 -0008, within PNCDI III

4. References

[1] Sebeșan, I. Dinamica vehiculelor feroviare – Tratat, Ed. Matrix Rom, București,2011
[2] Gobel, E. Gremmifedern. Berlin, Springer Verlag, 1969
[3] Boutonnet,Y.C. La confort dans le transport ferroviare, Reven Generale des Chemins de Fer. Apr., 1981
[4] Anderson, C., Oscarsson, J. Dynamic train / Track interaction including state dependent track properties and flexible vehicle componentes, Vehicle System Dynamics 33,1999
[5] Ioan Sebesan, Marius Adrian Spiroiu, Sorin Arsene, Gabriel Popa, Gabriela Dinu Deformation Analysis of Clouth-type Springs of Railway Vehicles Suspension, Materiale Plastice, Volume 52, Issue 4, Pages 560-563, 2015

Optimization of the graph model of the water conduit network, based on the approach of search space reducing

Iakov S. Korovin[1] and Maxim G. Tkachenko[1]

1 Southern Federal University, Taganrog, Russia
E-mail: Korovin_yakov@mail.ru

Abstract. In this paper we present a heuristic approach, improving the efficiency of methods, used for creation of efficient architecture of water distribution networks. The essence of the approach is a procedure of search space reduction the by limiting the range of available pipe diameters that can be used for each edge of the network graph. In order to proceed the reduction, two opposite boundary scenarios for the distribution of flows are analysed, after which the resulting range is further narrowed by applying a flow rate limitation for each edge of the network. The first boundary scenario provides the most uniform distribution of the flow in the network, the opposite scenario created the net with the highest possible flow level. The parameters of both distributions are calculated by optimizing systems of quadratic functions in a confined space, which can be effectively performed with small time costs. This approach was used to modify the genetic algorithm (GA). The proposed GA provides a variable number of variants of each gene, according to the number of diameters in list, taking into account flow restrictions. The proposed approach was implemented to the evaluation of a well-known test network - the Hanoi water distribution network [1], the results of research were compared with a classical GA with an unlimited search space. On the test data, the proposed trip significantly reduced the search space and provided faster and more obvious convergence in comparison with the classical version of GA.

1. Introduction

Optimization of the structure of loop water distribution networks is a complex combinatorial problem, related to NP-hard (Non-defintistic Polynomial-time) class, since it is nonlinear, bounded, nonsmooth, nonconvex, and therefore multi-modal [2]. Although methods of mathematical programming such as linear and nonlinear ones are known to be applied to solve this problem [2-5], heuristic methods are currently relevant because of their ability to obtain quasi-optimal solutions for solving large-dimensional problems in conditions of limited working time. Genetic algorithms (GA) and a number of other algorithms are among the most common heuristic approaches used to design waterway networks. GA are widely used to create an effective net of conduits [2,3]. And also it is generally known that GA are based on the rules of evolution and natural selection.

Although heuristic approaches can cope with global optimization problems, they do not always provide finding the optimal solution. In addition to the apparently low possibility of task solved, another disadvantage is time, required for convergence of the method. Despite a significant number of works on optimization of convergence processes, heuristic methods are still relatively inefficient and take a lot of time when working with large graph models, including modeling the water distribution network. The low search efficiency is due apparently to the wide range of possible solutions that the algorithm should test. Since the search space is very large, general-purpose heuristic algorithms spend a significant amount of time evaluating unrealizable solutions. Consequently, the probability of finding the optimal solution decreases and the rate of convergence increases as the size of the search space increases. Thus, the actual task in this area is to develop methods for reducing the search space.

In this paper, we propose an approach to improve the efficiency of heuristic methods, used to optimize in the graph representation of the task of designing water distribution networks. The proposed approach is based on the restriction of the search space on the basis of the analysis of two opposite scenarios, the formation of a flow in the network with subsequent consideration of the constraints imposed on the flow velocity in the edges of the graph of the simulated network. The proposed approach is compared with the classical formulation of GA with an unlimited search space when analyzing data in the model of the Hanoi water distribution network (HWDN).

2. Formation of constraints in the graph model

In normalized graph model of water conduits (models without cycles), the flow distribution can be estimated by applying the equations of flow conservation in network nodes. From a practical point of view, the general procedure for this type of net is to determine speed limits of the flow. The limitations of the speed of the pipelines may vary depending on the material, diameter, and other considerations. Thus, sufficient flow rates can lead to increased pipe damage and significant energy losses. Low speeds, on the contrary, can lead to deposits and an increase in the total metal capacity and volume of the system. When speed limits are defined, the range of possible diameters that can be chosen for each edge of the graph is significantly reduced, which reduces the sophistication of net design. Nevertheless, in difference from normalized models, the flow distribution in loop networks is not known a priori and, as a consequence, this approach can not be applied.

The approach presented is built upon a reduction of the search space procedure by limiting the range of possible diameters that can be chosen for a specific edge of the net graph. The procedure consists of creating two opposite extreme flow propagation scenarios that satisfy the equations of nodal flow conservation and core requirements. The first scenario provides the most general distribution of the flow in the network that would satisfy the nodal requirements and flow restriction constraints. The result scheme will provide a network with high reliancy. The second (contrary) scenario is characterized by the highest accumulation of flow in several major pipes. This situation provides a thread distribution similar to that obtained for the spanning tree of the network graph.

The approach depicted in this paper, applied for calculating the distribution of extreme flows is the solution of the problem of optimization of quadratic functions (OQF) for each of them. The OQF problem is in general connected with the minimization or maximization of a quadratic function bounded by linear functions.

The objective function of the proposed OQP problem for the general distribution of the flow is to minimize the sum of the fluxes of squares of the fluxes in the edges of the network graph. These flow values must satisfy the law of flux conservation at network nodes. This set of constraints is linear. In general, the problem is presented by the equation below (1):

$$\sum_{i=1}^{n} Q_i^2 \rightarrow \min, \quad A \times Q = q, \quad Q \geq 0 \tag{1}$$

where: n is the number of edges in net graph, Q_i is the stream in the edge i, A is the matrix (m x n), and m is the number of vertices of the graph, and q is the constraint vector. The element of matrix A *[i, j]* is equal to 1 if the stream from vertex i is directed to vertex j, -1 if it is directed in the opposite direction, and 0 if vertex j is not connected with vertex i.

One of the drawbacks of this approach is that the direction of the flows must be determined in advance to perform the calculation. For complex networks, the number of possible combinations of the flow direction can be high, and the search for the correct version is a complex procedure. To delete this limit, it is proposed to double the number of edges of the graph by adding a dummy channel for each network line in such a way that for each network line two pipes with opposite flows are considered. Using this procedure, the number of edges in the graph increases to $2n$, and the matrix A will have dimension (m x $2n$). The solving of the OQF minimization task provides the correct flow directions and values that minimize the amount of network flows. This gives the distribution of the flow with the maximum variance (MD). The second scenario with the maximum accumulation of the flow, also called the maximum concentration scenario (MC), is obtained by maximizing the objective function and solving the equivalent problem of maximizing the OQP. The solution of these tasks defines 2 vectors, which limit the range of possible flows within each edge of the network graph. Given the restrictions on the flow rate, the pair of vectors that determine the range of possible diameters between the minimum (D_{min}, i) and the maximum (D_{max}, i) for each edge i can be calculated as follows:

$$D_{\min,i} = \sqrt{\frac{4 \times Min(Q_{MD,i}, Q_{MC,i})}{\pi \times U_M}} \quad (2)$$

$$D_{\max,i} = \sqrt{\frac{4 \times Max(Q_{MD,i}, Q_{MC,i})}{\pi \times U_M}} \quad (3)$$

3. Genetic algorithm with limitations

The approach proposed in this paper can be used in conjunction with various heuristic methods. In this article, the GA approach was chosen to test the effectiveness of the proposed methodology. GA is a simulation model of optimization, which is oriented to work with large complex systems. The paper uses GA, based on the classical model proposed in [2]. To implement the proposed approach, the model was changed and implemented in the form of software in the programming language C ++, the development environment of Qt Creator.

Algorithm [2] uses a scheme with integer coding. Each solution is encoded by a vector of n discrete variables (the pipe diameter sizes coincides a single edge of the net graph). Each chromosome is encoded by an integer value, that ranges from 1 to nd, i (the last diameter possible). This approach provides many advantages, since there are no restrictions on the number of possible pipe diameters that can coincide to a defined graph edge. In the classical formulation of GA, the number of possible diameters was the same for each communication line, and this value was equal to the total number of diameters in the pipe database. The same coding scheme is adopted in the proposed algorithm, although changes have been made allowing to vary the number of possible diameters for each edge of the graph. The method that we propose has an integer coding scheme and a variable number of possible options for each edge. The number of options depends on the number of possible diameters involved in the speed limits of each edge of the graph.

To test and compare the approach proposed with the classical GA, the initial population is formed randomly. This population evolves from generation to generation, conducting an iterative reproductive cycle. This cycle involves the use of three classical operators: selection, crossing-over and mutation. To use the selection operator, the suitability of each individual is estimated as the sum of the cost of the pipes making up the network, plus the penalty function used to account for the pressure deficit at the graph nodes, according to the formula

$$F(D) = \sum_{i=1}^{n} c_i \times L_i + p \sum_{j=1}^{N} (\max(hr_1 - h_1), 0) \quad (4)$$

where - c_i - the cost per meter of the pipe, depending on the diameter of D_i and the L_i length of the rib i, p - the coefficient of the penalty function, N - the number of nodes in the network, hr_j - the required pressure in node j, h_j - the operating pressure in node j, determined using the hydraulic model EPANET [6] .

The value of the coefficient of the penalty function can affect the accuracy of the solution, so it must be properly selected. To solve this problem, some researchers recommend various methods for processing constraints, such as the use of variable values or self-adaptive penalty functions [7]. However, in this paper, for simplicity and for comparing both approaches under the same conditions, the constant value of the penalty function coefficient is used. When simulating a penalty multiplier, a value was assigned to avoid decisions that violated pressure constraints at the nodes of the network. To calculate the effective values of nodal pressures for each individual, the EPANET hydraulic solver was used in the population, the source codes of which can be found in the public domain [7].

4. Modelling the operation of the algorithm

To simulate the GA and the proposed method, software was developed that implements these algorithms. As a result of the work, the application displays the parameters of the structure of the modeled waterway graph. For processing on the input of the application, both network information and a database of available pipe diameters are fed. Network information is imported from the EPANET input file. The database of pipes is stored in a text file of tabular format. The output module records the best solution found by the model (the resuting vector of pipe diameters and the cost of the net as a whole).
A block diagram of the software model is shown in Fig. 1.

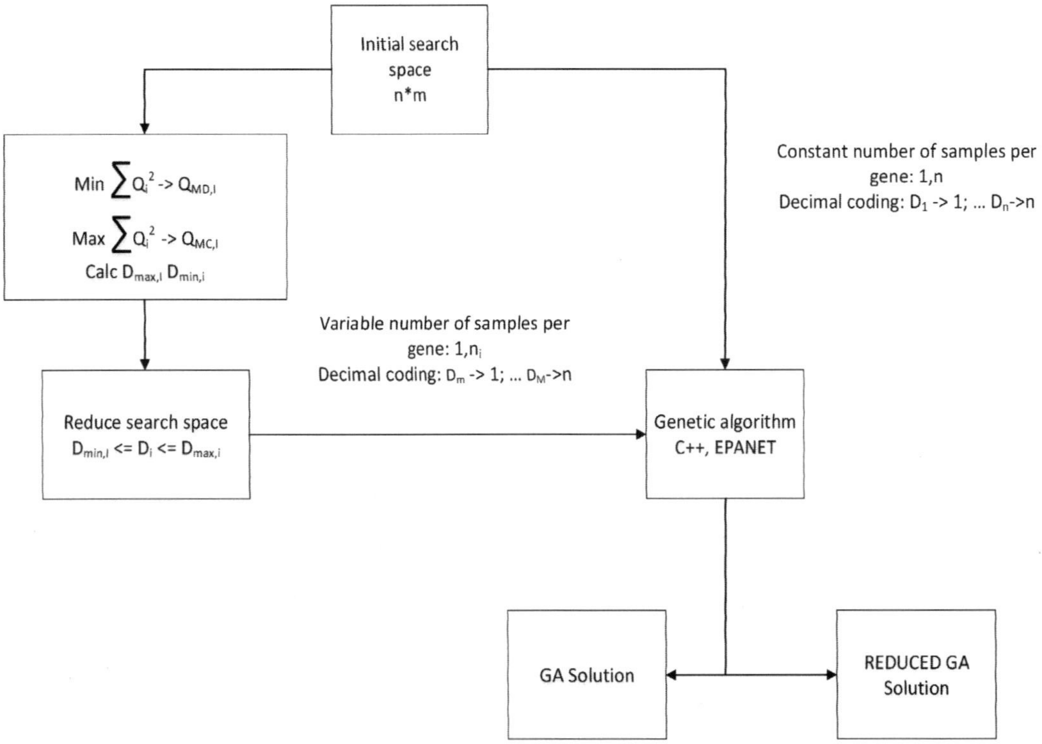

Figure 1. Block diagram of the software model.

The proposed approach was tested by a network test model to estimate the effectiveness of the classical GA approach with the limited GA model. For the test, the Hanoi waterway network was chosen, which was widely used in testing various algorithms for optimizing water conduit networks. The model can be considered as medium in size, with 32 nodes and 34 ribs and 3 loops. In the model, pumping units are not considered, since there is only one fixed source at a height of 100 m. The minimum pressure requirement for all nodes is fixed by a minimum height of 30 m. In this task, there is a set of 6 available diameters. The cost function is nonlinear. Pipe losses were calculated using the Hazen-Williams equation with a roughness factor C = 130. The values of the other parameters of the Hazen-Williams equation were established by default for the EPANET network analysis software [7].

To assess the effect of reducing the search space on the convergence rate and the accuracy of the solution, both algorithms were applied to solve the problem of designing a test network. For the subsequent comparison, in both algorithms the same input data and analysis parameters are selected. The size of the population was limited to 200 individuals. The number of generations was 300. The probability of crossing-over is Pcross = 0.85. The probability of mutation is Pmut = 0.08. The frequency of crossing the genes rcross = 0.3.
For simulation, the "leave-stable-delete-worst" reproduction plan is used, which leaves in the generation of the individual in the event that the value of their fitness function exceeds the level of the weakest suitable member of the population of ancestor. The least suitable member of the population is deleted and replaced by a descendant. The crossover operator implies that a pair of parent chromosomes exchange information to create a pair of chromosomes of the offspring that inherit their characteristics. Probability (the intersection of two chromosomes is determined by the input parameter Pcross.) In a homogeneous crossing-over, the chromosomes of the parents exchange their genetic information gene for a similar gene. The probability of gene exchange is determined by the frequency of crossing the genes (rcross) . To collect statistics, the application was run 20 times for both the proposed algorithm and for classical GA.

5. Simulation

The first stage in the calculation is determing the OQF, indicated in equation (1). The results of these calculations determine the range of flows for each channel of the network. Both flow rates represent the flow limits at each link in the network and, thus, reduce the search space. The results obtained are generated in Table 1. Table 1 contains the flow range, minimum and maximum diameters and the number of possible diameters compatible with the speed limits obtained as a result of calculating the OQF.

Table 1. Results of calculating OQF for the model.

Link	Q_{MD}	Q_{MC}	Dmin	Dmax (mm)
1	19,94	19,94	1016	1016
2	19,05	19,05	1016	1016
3	5326	6810	1016	1016
4	5196	6680	1016	1016
5	4471	5955	1016	1016
6	3466	4950	762	1016
7	2116	3600	609.6	1016
8	1566	3050	508	1016
9	1041	2525	406.4	1016
10	2000	2000	609.6	1016
11	1500	1500	508	1016
12	940	940	406.4	1016
13	1484	0	304.8	1016
14	2099	615	304.8	1016
15	2379	895	304.8	1016
16	2968	1205	508	1016
17	3833	2070	609.6	1016
18	5178	3415	762	1016
19	5238	3475	762	1016

20	7637	7915	1016	1016
21	1415	1415	508	1016
22	485	485	304.8	1016
23	4947	5225	1016	1016
24	2890	3065	609.6	1016
25	2070	2245	609.6	1016
26	992	1270	406.4	1016
27	92	370	304.8	762
28	278	0	304.8	762
29	1011	1115	406.4	1016
30	721	825	306.4	1016
31	361	465	304.8	1016
32	1	105	304.8	1016
33	104	0	304.8	1016
34	909	805	304.8	1016

To assess the efficiency of the proposed method with the classical GA algorithm, 20 iterations were performed for each method with the same input data and analysis process. The simulation performed gave the following results:

- GAL min cost 6 204 922;
- GA - max cost 6 573 272;
- GA - avr.cost 6,392,721;
- The proposed algorithm - min cost 6 122 136;
- The proposed algorithm - max cost 6 412 528;
- The proposed algorithm is avr.cost 6 241 235.

With the proposed methodology, an obvious reduction in search space was achieved. In the test network, there are six possible diameters in the database, and the number of edges is 34. The resulting number of alternative designs is 22.87 x 1026, while the search space in the restricted task is 4.35 x 1016. The speed limits also play an important role as Reduction of search space increases with decreasing speed range.

Regarding the rate of convergence, both algorithms converged quickly enough, although the proposed algorithm outperformed the classical GA (the algorithm converged on the 127th generation for the proposed algorithm and for the 312th in the case of GA). In the proposed algorithm, it was required to calculate slightly less than 25 thousand estimates, for the GA - more than 60 thousand. Both these quantities cover an insignificant part of the entire search space.

The proposed methods significantly reduced the search space and made it real to achieve much faster and more accurate convergence than the classical GA formulation. Both algorithms have found solutions close to the global optimum. Speaking on the test network, the best solution found by the proposed algorithm was 6 122 136, the average cost of the simulations was close to optimal. For GA, minimum solution found was 6,204,922, and also the average cost higher than the analogous parameters of the proposed algorithm. The solution found by the algorithm is comparable to the optimal solution found in the literature (6 056 000 [5]).

6. Conclusions

In this paper we presented a novel approach, based on limiting and reducing the overall search space in the design of water conduits using graph models. This approach minimizes search space by analysing two opposite scenarios for the architecture of extreme flows, and then applying restrictions on flow rates in the pipes. The proposed approach was used in conjunction with GA in order to improve its effectiveness.

IOP Conf. Series: Materials Science and Engineering **324** (2018) 012019 doi:10.1088/1757-899X/324/1/012019

Research evaluation process revealed that the proposed algorithm significantly reduced the search space and ensured faster and more accurate convergence than the classical GA formulation for the test net. It is forseen that for more complicated nets, the benefits, due to the new approach may be more significant. Extra vital advantage of the presented reduction is that it can be associated with other heuristic algorithms in the given subject area.

7. References

[1] *Hanoi water distribution network* 2012 Centre for Water Systems University of Exeter http://emps.exeter.ac.uk/engineering/research/cws/resources/benchmarks/layout/hanoi.php
[2] Reca J and Martmez J 2006 Genetic algorithms for the design of looped irrigation water distribution networks Water Resour. Res 42 W05416.
[3] Varma K and Narasimhan S 1997 *Optimal design of water distribution systems using NLP method* J Environ Eng 123 p 381-388.
[4] Lansey K E and Eusuff M M et al 2003 *Optimization of water distribution network design using the Shuffled Frog Leaping Algorithm* J. Water Resour. Plan. Manag 129 p 10-25.
[5] Montalvo I and Izquierdo J et al 2008 *Particle Swarm Optimization applied to the design of water supply systems* Comput. Math. Appl 56 p 769-776.
[6] EPANET 2017 *Software That Models the Hydraulic and Water Quality Behavior of Water Distribution Piping Systems* United States Environmental Protection Agency https://www.epa.gov/water-research/epanet
[7] Geem Z W and Kim J H et al 2001 *A new heuristic optimisation algorithm: Harmony search* Simulation 76 p 60-68.

Acknowledgments
This paper is published due the financial support of the Russian Science Foundation (RSF) via the grant No 15-19-00196.

IOP Conf. Series: Materials Science and Engineering **324** (2018) 012020 doi:10.1088/1757-899X/324/1/012020

The forming simulation of flexible glass with silt down draw method

Hou Yansheng[1,2,3], Cheng Jinshu[1,3], Kang Junfeng[4], Cui Jing[2]

1 State Key Laboratory of Silicate Materials for Architectures, Wuhan University of Technology, Wuhan, China;
2 Shanxi Industry Vocational College, Xianyang, China;
3 Hebei Shahe Glass Technology Research Institute, Shahe, China;
4 School of Materials Science and Engineering, University of Jinan, Jinan, China
E-mail: bshys@whut.edu.cn

Abstract. The slit down draw method is the main manufacturing process of flexible glass. In this study, Flow3DTM software was used to simulate the process of drawing and thinning glass slits during the slit down draw process. The influence of glass viscosity, initial plate thickness and initial plate speed on the glass spreading process was studied. The maximum pull-down force that the root can bear is linearly proportional to the viscosity, the initial thickness of 1.3837 power and the initial plate speed, respectively. The best way to improve the tensile strength of flexible glass is to increase the viscosity. Flexible glass was more easily to obtain with low viscosity, low thickness and low drawing speed.
Keywords: slit down draw method; flexible glass; drawing; simulation.

1. Introduction

Flexible glass is a new type of glass substrate material with a thickness of less than 0.1 mm. It has winding bending, high temperature, anti-aging, good light transmission and other characteristics. Flexible glass is a new type of thin film material for the next generation of flexible display [1-4], touch sensor [5], solar panel [6-9] and other electronic devices, and has attracted much attention from glass process researchers in recent years. Corning, Asahi Glass, Electric Glass and SCHOTT have flexible glass manufacturing technology. Recently, Cheng Jinshu's group at Wuhan University of Technology also succeeded in continuously drawing flexible glass with a thickness of 0.08mm through the slit down-pull method.

In the forming process of flexible glass, the slit down-draw method with simple equipment and thinner thickness of glass is the best flexible glass manufacturing process [10]. During the glass forming process, the glass liquid flows through the elongated slot made of platinum alloy and then made into the sheet by the nip roll [11,12]. In order to grasp the variation law of the sheet tensile, this paper adopts Flow3DTM software [13,14] to simulate the flexible glass slot down forming, and discusses the influence of initial sheet speed, initial viscosity and initial sheet thickness on the glass sheet spreading process.

2. Simulation method

2.1. Modeling

The geometry of the slit pull-down forming is shown in Figure 1. The width, length and depth are w_x, l and l_x, respectively. The molten glass liquid form a continuous glass plate into the drawing zone after flowing through the slit. In a section close to the slit, the glass sheet is stretched and thinned into a flexible glass by a downward tensile force. This stretched deformed area is called a "root" that is the simulation in this experiment on.

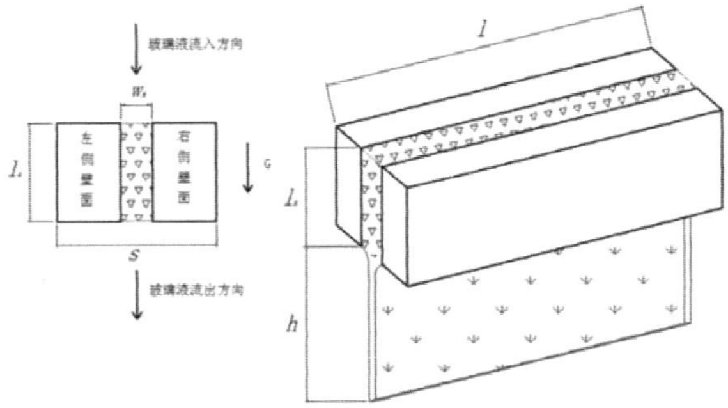

Figure 1. Slit down the tensile model.

2.2. Physical Models, Boundary Conditions and Initial Conditions

Parameters as shown in Table 1:

Table1. Modeling parameters of glass spread model

	Slit width w_0 (mm)	Glass viscosity η (Pa·s)	Entrance speed v_0 (mm/s)
Model I	1, 2, 3, 4	600	5.7
Model II	1	20, 50, 100, 200, 400, 600	5.7
Model III	1	600	2.85, 5.7, 11.4, 22.8

In order to facilitate the calculation of numerical simulation, the following assumptions and simplifications are made for the physical phenomena of the system:

1) Glass liquid is incompressible fluid, that is, constant density;
2) Other physical properties of glass (such as specific heat capacity, thermal conductivity) do not change with temperature, and defects such as bubbles, streaks, crystals and impurities are not considered;
3) The properties of platinum drain plate are constant;
4) Flow and heat transfer of glass liquid are all steady;
5) The effect of surface tension in the simulated area is not considered.

Physical Models: Gravity and Non-Inertial Reference Frame Models, Heat Transfer Models, Viscosity and Laminar Flow Models.

3. Results and Discussion

3.1. The maximum tensile force that the glass plate can withstand

According to the parameters of model I~III, three conditions of w_0 = 1～4 mm; η = 20～400 Pa·s; v_0 = 2.85~22.8 mm/s has been simulated in our present work. The simulation example shows in Figure 2, the glass plate formed immediately after glass emerges from the slit. The weight of the glass plate continues to increase, when the weakest parts of the root can't afford the glass plate gravity fracture simulation ended.

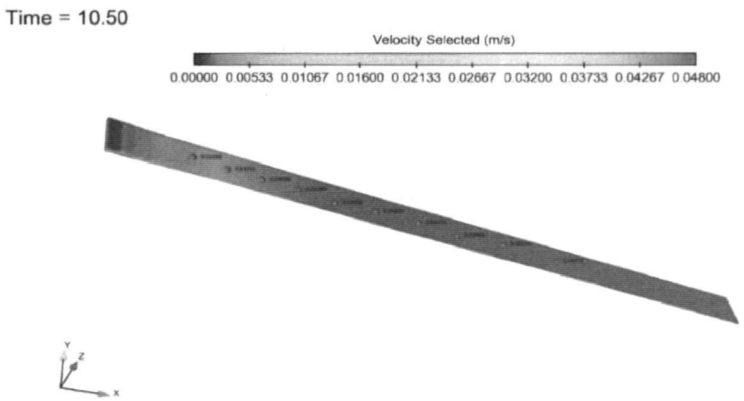

Figure 2. An example of simulation

Figure 3 shows the relationship between the maximum pull-down the glass plate fracture "root" can withstand and other factors. Figure 3(a) is the curve of the initial thickness and the maximum pull-down. The maximum pull-down force and 1.3837 times of the initial thickness have linear relationship: $f_{max} = 1.5018{w_0}^{1.3837} - 0.0672$; Figure 3(b) shows that the relationship between the viscosity of the glass and the maximum pull-down force. The viscosity η and the maximum pull-down force f_{max} approximate the linear relationship: $f_{max} = 0.021\eta + 0.1559$. Figure 3(c) shows the curve of the initial plate speed and the maximum pull-down. The maximum force f_{max} has the relationship of proportional with initial speed: $f_{max} = 0.317v_0 - 0.3191$

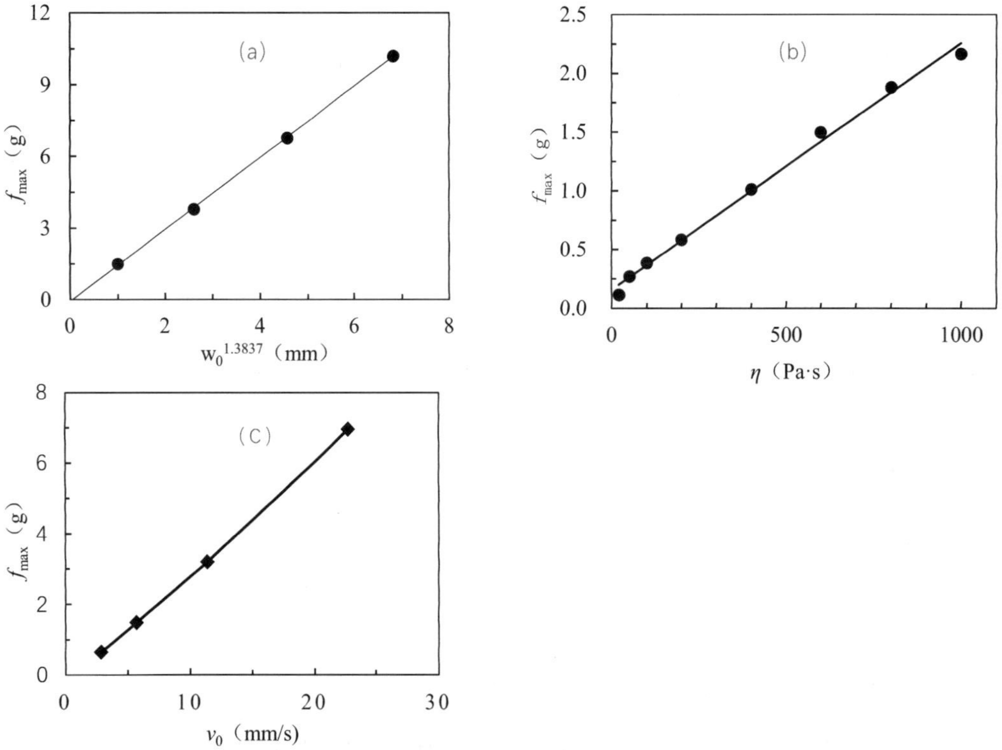

Figure 3. Relationship between f_{max} and w_0、f_{max} and η、f_{max} and v_0.

3.2. The combined effect of viscosity, initial plate thickness and initial plate speed on tensile stress

The maximum pull-down force that the root can bear is proportional to the initial plate thickness of 1.3837 power, the viscosity and the initial plate speed. Therefore, there are three ways to increase the tolerance of the pull force of glass plate: increasing initial thickness, initial viscosity and initial plate speed. The following discussion of the impact of changes of the three process parameters in the slit down-pull method.

The initial plate thickness of the glass is the same as the width of the slit. According to the previous study, the glass flow rate is proportional to the square of the slit width[15]:

$$Q = K(h_x {w_x}^2 / l_x \, \eta) \qquad (1\text{-}1)$$

Q is the glass flow, K is the coefficient, h_x is the glass level, w_x is the slit width, l_x is the slit depth, η is the glass viscosity.

The flow rate of the molten glass is also equal to the initial sheet speed v_0, initial thickness w_0 and width l of the product.

$$Q = v_0 \cdot w_0 \cdot l \qquad (1\text{-}2)$$

Obtain

$$v_0 = w_x K \cdot h_x / l_x \cdot \eta \cdot l \qquad (1\text{-}3)$$

In production, all of the coefficients K、h_x、l_x、η and l are constant, indicating that increasing the slit width causes a rapid increase in the initial plateau speed, as well as increasing the draw ratio of the glass plate, greatly increasing the difficulty of drawing and not suitable for flexibility glass drawing.

Increasing the viscosity of the glass leads to a decrease of the flow rate of the glass, so the viscosity has little effect on the pull-down force. Lower plate speeds also benefit from lower plate speeds. At the same time increase of the viscosity can reduce the glass temperature, and is good for the life of the drain plate. Therefore, it is advantageous to increase the viscosity of the flexible glass.

Initial speed is only proportional to the height of the glass surface, and no other coupling conditions that can be used as a separate process parameters.

3.3. Effect of initial state on plate surface velocity distribution during drawing

In addition to the pull-down force, the plate-speed distribution has a great impact on the flexible glass. Figure 4 is a glass plate velocity profile. The abscissa in the picture is the distance of the glass plate from the slit and the ordinate is the speed of the glass plate. All the curves have the same characteristics: the plate speed increases rapidly with increasing distance from the slit, but does not increase when the plate speed reaches a certain value. This is because when the glass sheet leaves the slit, the pull-down force is greater than the pull-out resistance in the reverse direction. The resultant drag force and pull-down force accelerate the glass sheet. However, as the distance increases, the cumulative drag resistance balances with the pull-down force plate speed is no longer increased.

Figure 4(a) shows the plate speed distribution curve with different initial plate thickness. The larger the initial thickness, the faster the glass sheet is stretched after stretching and the longer the tensile deformation area of the root.

Figure 4(b) is the viscosity of the board surface velocity distribution curve. The lower the viscosity, the faster the sheet speed of the glass sheet after stretching and the shorter the sheet root tensile deformation area.

Figure 4(c) is the plate speed distribution curve of different initial plate speed. The greater the initial board speed after stretch thinning the greater the glass plate speed, and rootstock tensile deformation zone longer.

Summary of the above characteristics: With the decreases of viscosity, increase of initial thickness and the initial plate speed, the glass plate after the balance rate also increased. The deformation distance decreases as the initial plate thickness, viscosity and initial plate speed. Therefore, flexible glass is suitable to be drawn at low viscosity, low thickness, low sheet speed process.

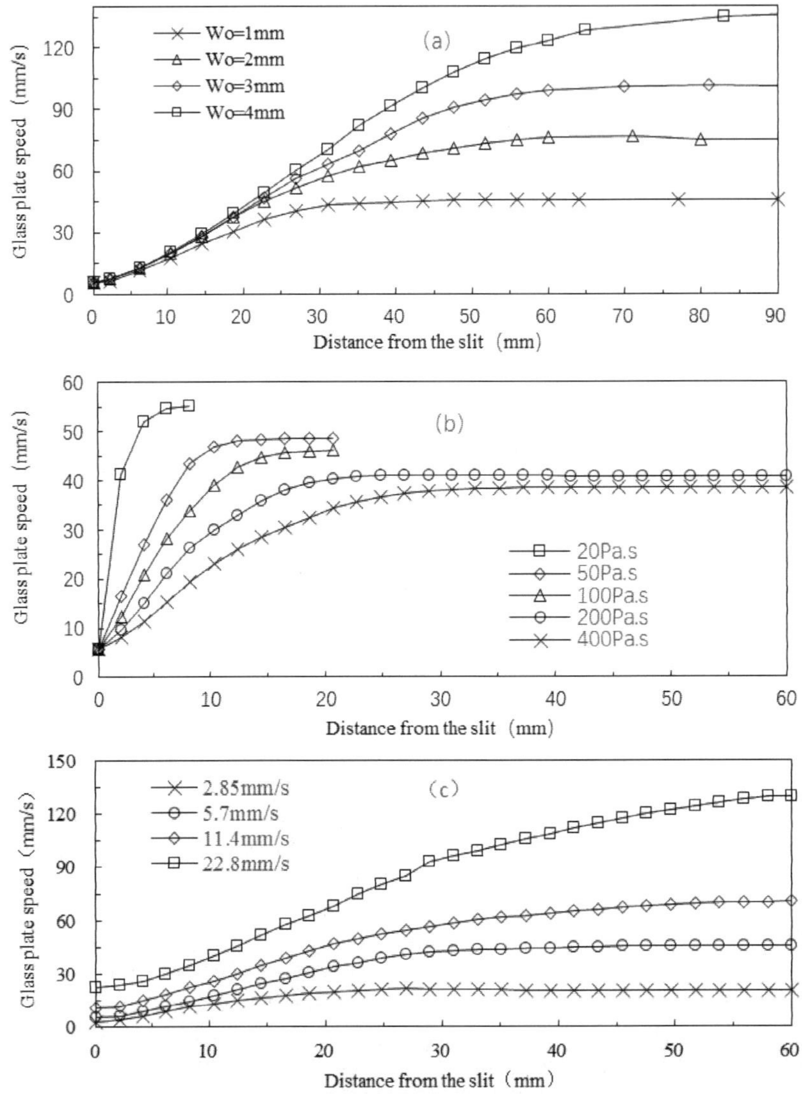

Figure 4. Velocity profile during glass deformation.

4. Conclusion

In this study, 3DTM software was used to simulate the influence of plate surface force and plate surface velocity during the pull-down of flexible glass slits. The study found:

(1)The maximum pull-down force that the plate root can bear is linearly proportional to the initial plate speed and the initial plate thickness of 1.3837, respectively.

(2)Among the three methods to increase the tensile strength of the glass plate, increasing the initial plate thickness of the glass plate is unfavorable for the drawn of glass; increasing the viscosity does not reduce the glass plate from being able to withstand the pull-down force and thinning the flexible glass favorable. The initial speed is not coupled with other conditions and can be used as a separate process parameter.

(3)As the initial thickness increases, the viscosity decreases, the initial plate speed increases, and the speed of the glass plate increases. As the initial plate thickness decreases, the viscosity decreases, the initial plate speed decreases and the deformation distance decreases.

IOP Conf. Series: Materials Science and Engineering **324** (2018) 012020 doi:10.1088/1757-899X/324/1/012020

References

[1] Hoehla S, Garner S, Hohmann M, et al. Active Matrix Color-LCD on 75 μm Thick Flexible Glass Substrates[J]. Journal of Display Technology, 2012, 8(6):309-316.
[2] Garner S M, Wu K W, Liao Y C, et al. Cholesteric Liquid Crystal Display With Flexible Glass Substrates[J]. Journal of Display Technology, 2013, 9(8):644-650.
[3] You H, Steckl A J. Lightweight electrowetting display on ultra-thin glass substrate[J]. Journal of the Society for Information Display, 2013, 21(5):192-197.
[4] Won S. Poly Si TFT on Microsheet[J]. Ecs Transactions, 2009, 25(3).
[5] Garner S M, He M, Lo P Y, et al. Electrophoretic Displays Fabricated on Ultra-Slim Flexible Glass Substrates[J]. Journal of Display Technology, 2012, 8(10):590-595.
[6] Liu C N. Development of flexible battery [J].Chinese Journal of Power Sources, 2014, 12(38):2215-2216.
[7] Sheehan S, Surolia P K, Byrne O, et al. Flexible glass substrate based dye sensitized solar cells[J]. Solar Energy Materials and Solar Cells, 2015, 132:237-244.
[8] Peng C Y, Dhakal T P, Garner S, et al. Fabrication of Cu_2ZnSnS_4, solar cell on a flexible glass substrate[J]. Thin Solid Films, 2014, 562(26):574-577.
[9] Fang Z Q, Zhu H L, Li Y Y, et al. Light management in flexible glass by wood cellulose coating[J]. Scientific Reports, 2014, 4:5842.
[10] Si M J, Guo W, Tian F, et al. The Current Research Status and Development Strategy on Flexible Glass[J]. Glass , 2016, 5:17-21.
[11] Xie J, Li X Q. Development and application of flexible thin glass[C].Bulletin the Chinese ceramic society. 2015.34:48-55.
[12] Wang C Y, Lu Q. Ultrathin Flexible Glass for Display [J]. Glass and Enamel, 2016, 44(3):43-46.
[13] Shao H, Mao Q, Pang S, et al. Numeric model of stretching process of float glass[J]. Journal of the Chinese Ceramic Society, 2004, 32(8):1025-1028.
[14] Pitbladdo R B. Overflow downdraw glass forming method and apparatus[J]. 2006.
[15] Hou Y S, Cheng J S , Kang J F, et al.Simulation of the glass melt flowing during the slit down draw process[J]. Materials Science and Engineering 274, 2017:012080

Acknowledgements

This work was funded by the National Key Research and Development Program of China. (No.2016YFB0303700)

Design of permanent magnet synchronous motor speed loop controller based on sliding mode control algorithm

Jiang Qiang[1] and Liao Meng-wei[2] and Luo Ming-jie[2]

1 Associate professor,Chongqing University of Technology,Chongqing,China
2 Master degree candidate,Chongqing University of Technology,Chongqing
E-mail:jq61@163. com

Abstract. The control performance of Permanent Magnet Synchronous Motor will be affected by the fluctuation or changes of mechanical parameters when PMSM is applied as driving motor in actual electric vehicle,and external disturbance would influence control robustness.To improve control dynamic quality and robustness of PMSM speed control system ,a new second order integral sliding mode control algorithm is introduced into PMSM vector control.The simulation results show that, compared with the traditional PID control,the modified control scheme optimized has better control precision and dynamic response ability and perform better with a stronger robustness facing external disturbance,it can effectively solve the traditional sliding mode variable structure control chattering problems as well.
Keywords:PMSM; SMC; FOC; second order integral

1. Introduction

Permanent magnet synchronous motor(PMSM) has the advantages of simple structure, small volume, high efficiency and reliability, but PMSM is a nonlinear object in nature, the performance of nonlinear control system will be influenced by the variation of motor parameters and external disturbance in the practical application.The sliding mode variable structure control strategy(SMC) can respond fast, be insensitive to parameter change, and has strong anti-interference ability. But in practical systems, chattering exists in SMC.

Employing A novel two order integral type sliding mode control method, based on Schiff and Jarno Lee $L=\frac{1}{2}s^2$, state error of control parameters in finite time convergence would be verified by the function.Results of a simulation based on MATLAB show the designed SMC can improve the

Project Supported by Scientific and Technological Research Program of Chongqing Municipal Education Commission（Grant No.KJ1400935） response time of the control system, relieve chattering problem easily which happened in the traditional sliding mode variable structure control system,.Meanwhile,the optimized control system has strong robustness and anti-interference ability.

2. Mathematical model of PMSM and vector control

Three phase permanent magnet synchronous motor (PMSM) is a multivariable, nonlinear and strongly coupled system. In order to facilitate the analysis easily, some assumptions about its conditions are given. The object applied in control algorithm proposed in this paper is the surface permanent magnet synchronous motor ,based on the vector control strategy $i_d = 0$ and the mathematical model of the motor can be simplified to the type below, the design of algorithm is based on it:

$$\begin{cases} u_d = -Lp_n\omega i_q \\ \dot{i}_q = -\dfrac{R_s}{L}i_q - \dfrac{p_n\psi_f}{L}\omega + \dfrac{u_q}{L} \\ \dot{\omega} = \dfrac{3p_n\psi_f}{2J} - \dfrac{B}{J}\omega - \dfrac{T_L}{J} \\ \dot{\theta} = \omega \end{cases} \tag{1}$$

3 .The design of control algorithm

The implementation process of sliding mode variable structure control of the sliding surface is to design the system according to the desired dynamic characteristics of system, making system state move from outside to the sliding surface of sliding surface, once the system reaches the sliding surface, the control effect will guarantee system along the sliding surface at the origin system.

Define the state error equation of the control system:

$$X = \omega^* - \omega_m \tag{2}$$

Where $\omega*$ is the reference speed set by the motor and the actual speed of the motor ω_m.

In this case $i_d = 0$ the formula (2) is derived and the formula (1) can be obtained together

$$\dot{X} = \omega^* - \frac{3p_n\psi_f}{2J}i_q + \frac{B}{J}\omega + \frac{T_L}{J} \tag{3}$$

In order to facilitate the design of control rates,Because some of the above formula can be set as a fixed value, let

$$\alpha = \frac{3p_n\psi_f}{2J} \qquad \beta = \frac{B}{J} \qquad \mu = \frac{T_L}{J} \tag{4}$$

So formula (3) can become:

$$\dot{X} = \omega^* - \alpha i_q + \beta\omega + \mu \tag{5}$$

According to the general form of integral sliding surface, the function of the sliding surface is defined as:

$$S = k_1 X + k_2 \int_0^t X dt \tag{6}$$

Among them, k_1 is the proportional gain of controller and k_2 is integral gain,.Moreover,both as design parameter must be greater than 0.

Derivation of formula (6) below:

$$\begin{aligned} \dot{S} &= k_1 \dot{X} + k_2 X \\ &= k_1(\dot{\omega}^* - \alpha i_q + \beta\omega + \mu) + k_2(\omega^* - \omega_m) \end{aligned} \tag{7}$$

According to the formula (7), in order to ensure the three-phase PMSM speed control system has better dynamic quality, because of $i_d = 0$,the control rate of the target object is designed as follows

$$\begin{aligned} &\text{Proportion al: } i_{q1} = \frac{1}{k_1\alpha}(k_2 X + k_1\beta\omega + k_1 \dot{\omega}) \\ &\text{Integral : } i_{q2} = \frac{1}{k_1\alpha}\int_0^t \left[x_1 \operatorname{sgn}(S) + x_2 \operatorname{sgn}(\dot{S}) + (r_1 S + r_2 \dot{S}) \right] dt \\ &i_q^* = i_{q1} + i_{q2} \end{aligned} \tag{8}$$

Where: x_1, x_2, r_1, r_2 the undetermined design parameters, and to ensure that the system converges to 0, x_1, x_2, r_1, $r_2 > 0$.

As can be seen from the formula (8), because the controller contains integral terms, on the one hand, it can weaken chattering phenomenon, on the other hand, it can also eliminate the steady-state error of the system and improve the quality of the control system.

According to the verification of the Lyapunov function, based on the control rate of speed error system design with sliding mode, according to the sliding mode reaching condition $S\dot{S} < 0$, easy to verify that the system can be converge gradually and tends to 0 in the designed controller.

4 .PMSM control system

In order to verify the validity of the sliding mode controller design, based on the $i_d = 0$ vector control strategy,build the overall simulation model of PMSM speed control system, which choose PI control for the current loop control and parameter, build the Simulink model.

According to the PMSM control system above, the control system mainly contains three parts, namely the speed loop, current loop and PWM algorithm, where speed loop has built-in designed two order integral sliding mode algorithm. Through giving specific speed value, the voltage value output is adjusted by the algorithm, so that the PMSM can realize the speed response more fast.

5 .Simulation and analysis

To verify the effect of designed control algorithm, make simulation over PMSM control system, motor parameters are set in the simulation as follows: polar number p_n =4, stator inductance $L_d = L_q = 8.5mH$, stator resistance $Rs = 2.875\Omega$, flux $\psi_f = 0.175Wb$, inertia $J = 0.003kg \cdot m^2$, damping coefficient $B = 0.008N \cdot m \cdot s$.

Settings of simulation condition: DC side voltage $U_{dc} = 311V$, PWM switching frequency $f_{PWM} = 10kHz$, cycle setting as $T_s = 10\mu s$.

Simulation process: set the reference speed $N^* = 1500 rad / \min$, that is, the dynamic response quality and speed when motor starts to achieve it . In the process of motor running, the initial load torque $T_L = 0N \cdot m$, in t=0.2 seconds increase load torque $T_L = 10N \cdot m$, which make changes in the external parameters perturbation effect on motor speed control system, to inspect anti interference ability of designed sliding mode controller. And the simulation experiments are carried out to see the dynamic characteristics of PMSM under PI control, traditional sliding mode control and two order sliding mode control respectively.
The simulation results are shown below

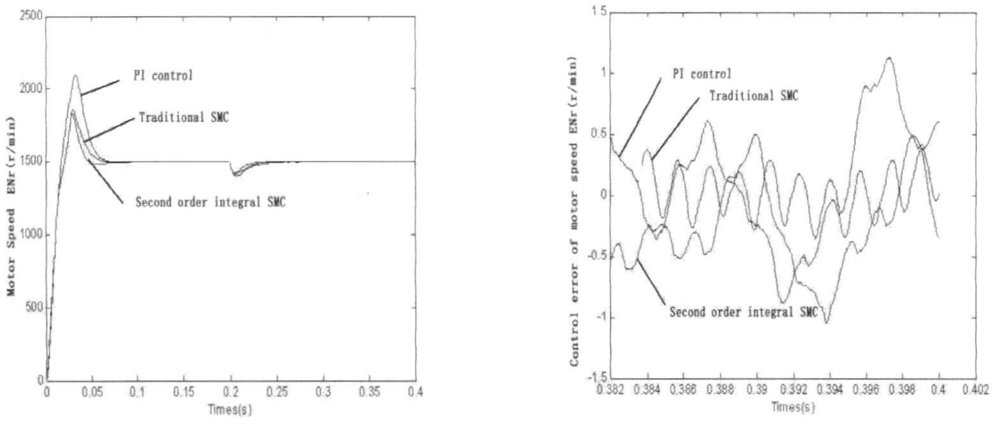

Figure1(a). Speed response curve of control system **Figure1(b).**Control system error curve

Figure1 shows the system characteristics under three control methods when the desired speed is 1500 . Among them, figure1(a) reflects the motor speed tracking characteristics; Figure1(b) reflects the control system error curve. The simulation results show that when the motor starts at the beginning of experiment, although motor speed appeared the overshoot, but still has faster dynamic response than PI control , after reaching the reference speed,system has better control precision. After 0.2 seconds increasing load torque suddenly, the motor can recover to the set speed value in a short time,which show that the designed speed loop sliding mode variable structure controller has good dynamic performance and anti-interference ability.

6. Conclusions

Aimed at PMSM speed loop control system, introducing the controller design method based on two order integral sliding mode control ,by tuning the proportional and integral parameters, effectively improves the dynamic response quality and anti-interference ability of PMSM speed control system . The simulation results show that the controller has obvious advantages over the PI control in speed tracking, and effectively suppresses the chattering phenomenon of the traditional sliding mode control, and enhances the robustness of the control system.

7.Reference

[1] Navaneethan.s.Jerome Jovitha 2015 *Speed control of PMSM using power reaching law based sliding mode controller* (WSEAS Transactions on systems and control vlo) p270-277

[2] Baik I C and Kim K H and Youn M J 2013 *Robust nonlinear speed control of motor using adaptive and sliding mode control techniques* ([J].IEEE Proceedings-Electric Power Application) p 145(4):369-376

[3] Andresscu G D and Pitic C I and Blaabjerg F 2008 *Combined flux observer with signal injection enhancement for wide speed range sensorless direct torque control of ipmsm drives* ([J].IEEE Transaction on Energy Conversion) p 23(2)： 393-402

[4] Ran L and Guangzhou Z 2009 *Position sensorless control for pmsm using sliding mode observer and phase-locked loop* ([C]//Preceedings of the IEEE 6th International Power Electronics and Motion Control Conference) p 1867-1870

[5] Chu J B and Hu Y W and Huang W X 2009 *An improved sliding mode observer for position sensorless vector control drive of pmsm* ([C]//Preceedings of the IEEE 6th International Power Electronics and Motion Control Conference) p 1898-1902

[6] Lu Tao and Yu Haisheng and Shan Bing and late Jieru 2015 *The maximum torque / adaptive sliding mode servo system of permanent magnet synchronous motor current control* ([J]. control theory and applications) p 02:251-255

[7] Jiang Hong and Han Junfeng 2014 *PMSM. Adaptive fuzzy sliding mode control of servo system* ([J]. micro motor) p 05:46-49

[8] Li Zheng and Hu Guangda and Cui Jiarui and Liu Guangyi 2014 *Integral sliding mode variable structure control of permanent magnet synchronous motor drive system* ([J]. proceedings of the Chinese society of electrical engineering) p 03:431-437

IOP Conf. Series: Materials Science and Engineering **324** (2018) 012022 doi:10.1088/1757-899X/324/1/012022

Research on Matching Method of Power Supply Parameters for Dual Energy Source Electric Vehicles

Q Jiang[1], M J Luo[2], S K Zhang[2]and M W Liao[2]

1 Associate Professor,Chongqing University of Technology, College of mechanical engineering, China.
2 Master,Chongqing University of Technology, College of mechanical engineering, china.
E-mail: jq61@163. com

Abstract. A new type of power source is proposed, which is based on the traffic signal matching method of the dual energy source power supply composed of the batteries and the supercapacitors. First ,analyzing the power characteristics is required to meet the excellent dynamic characteristics of EV, studying the energy characteristics is required to meet the mileage requirements and researching the physical boundary characteristics is required to meet the physical conditions of the power supply. Secondly, the parameter matching design with the highest energy efficiency is adopted to select the optimal parameter group with the method of matching deviation. Finally, the simulation analysis of the vehicle is carried out in MATLAB-Simulink, The mileage and energy efficiency of dual energy sources are analyzed in different parameter models, and the rationality of the matching method is verified.

1. Introduction

With the increase of on-board electrical appliances, pure electric vehicles are currently powered by batteries power supply. Not only does the batteries supply power to the motor, but it also needs to supply other high-power electrical equipment, which will lead to the batteries safety hazard or reduce the service life. At present, it is difficult to support the normal operation of electric vehicles on a single power supply, while The characteristics of super capacitor for long time and high power operation can meet the stringent requirements of electric vehicles on the power supply.

most researchers focus on the energy management systems of these vehicles, for example, *Soni*[1],*Rade*[2] et al studied the energy management strategy of dual energy source electric vehicles based on the determination rules, *Chen long*[3], *Kumar*[4] et al studied the energy management optimization strategy of dual energy source electric vehicles. However, few researchers have studied the parameter matching of dual energy sources. In this paper, the coupling of power battery and supercapacitor is used as the power supply. After analyzing the characteristics of dual power supply, the matching target is completed by the error matching method.

2. Analysis of power supply characteristics

Before matching the dual source parameters, a mature batteries pack of pure electric vehicle is used as the research model. This study takes the vehicle parameters of the model, the dynamic performance

Project Supported by Scientific and Technological Research Program of Chongqing Municipal Education Commission（Grant No.KJ1400935）

IOP Conf. Series: Materials Science and Engineering **324** (2018) 012022 doi:10.1088/1757-899X/324/1/012022

and mileage of the vehicle driving, and the physical conditions such as the size and quality of the on-board power supply are regard as the constraint conditions .

2.1. Power characteristic

Power characteristics of the power supply to meet the requirements of high performance dynamic characteristics of electric vehicles, determining the maximum power required for maximum speed, maximum climb and acceleration time of 0 ~ 100km/h .

$$P_{\max 1} = \frac{1}{3600\,\eta_e \eta_{PM} \eta_T}\left(mgfV_{\max} + \frac{C_a A}{21.15} V^3_{\max}\right) \tag{1}$$

$$P_{\max 2} = \frac{1}{3600\,\eta_e \eta_{PM} \eta_T}\left(mgfV \cos\theta_{\max} + mgV_i \sin\theta_{\max} \frac{C_a A V_i^3}{21.15}\right) \tag{2}$$

$$P_{\max 3} = \frac{1}{3600\,\eta_e \eta_{PM} \eta_T}\left(\frac{mgfV_t T}{1.5} ++ \frac{C_a A V_t^3}{21.15\times 2.5} + \sigma m \frac{V_t^3}{2\times T}\right) \tag{3}$$

η_e- Power discharge efficiency， **η_{PM}**- Permanent magnet synchronous motor efficiency， **η_T**- Drive efficiency of whole drive system， **m**- Fully loaded mass， **g**- Gravity coefficient， **f**-Coefficient of rolling resistance， **C_a** - Windward drag coefficient， **A**- Car windward area， **V_{max}**- Maximum speed,**V_i**- The speed of the climb， **θ_{max}**- Maximum gradeability, **V_t**- Speed after the acceleration,**T**-Acceleration time, σ - Correction coefficient of rotating mass;

The peak power of a dual energy source electric vehicle is powered by supercapacitors, the power of power supply for dual energy source electric vehicles is

$$N_{sc} P_{sc} \geq \max(P_{\max 1}, P_{\max 2}, P_{\max 3}) \tag{4}$$

N_{sc}- The number of supercapacitors， **P_{sc}**- The power of a single supercapacitors.

2.2. Energy characteristic

Endurance mileage is an important performance indicator for car manufacturers or users. In this case, the mileage is higher than 150km design of electric vehicles. Estimating the endurance of an electric vehicle by running the 100km power consumption of an electric vehicle at constant speed .

$$\frac{80\% \times \left(7.2\times 10^{-3} N_{sc} C_{sc} U_{sc}^{\ 2} + N_b C_b U_b\right)}{W} \geq \frac{150}{100} \tag{5}$$

C_{sc}- The capacity of a supercapacitor,**U_{sc}**- The voltage of a supercapacitor,**N_b**- Number of accumulators,**C_b**- A battery capacity,**U_b**-A batteriy voltage,**W**- The electric power consumed at a constant speed of 100 kilometres.

2.3. Physical characteristics

The quality and volume of the power supply affect the quality and volume of the vehicle, thus affecting the performance of the vehicle[5].

$$N_b m_b + N_{sc} m_{sc} \leq M_{\max} \tag{6}$$

$$N_b v_b + N_{sc} v_{sc} \leq V_{\max} \tag{7}$$

m_b- The quality of a single battery,**m_{sc}**- The quality of a single supercapacitor,**M_{max}**- Total power quality limits,**v_b**- The volume of a single battery, **v_{sc}**- The volume of a single supercapacitor,**V_{max}**-Total power supply limit.

2.4. Power parameter group

The limits of supercapacitors and accumulators in figure 1 are obtained through the calculation of (4) (5) (6) (7) under the condition of the power supply characteristics. There are 17,714 parameters in the black feasible region [Nb, Nsc]. How to determine the combination of supercapacitors and batteries is to optimize the selection of these parameters according to the driving conditions of electric vehicles andthe energy management strategy of the vehicle.

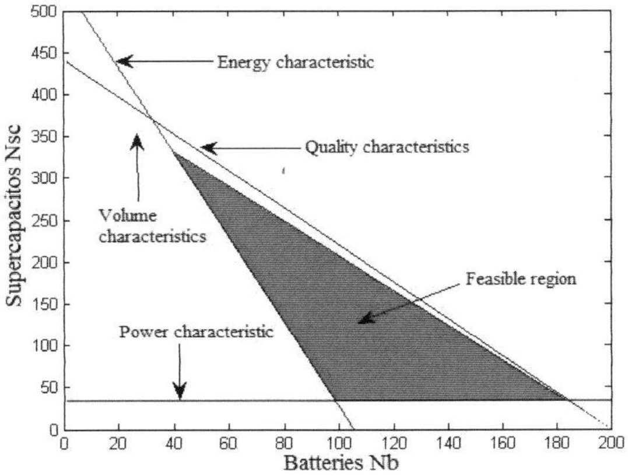

Figure1. Limit of supercapacitors and batteries.

3. Deviation matching method based on road condition

The matching parameters and energy management strategy of dual energy sources are strongly coupled[6]. The energy management strategy of a dual-energy electric vehicle can affect the distribution of energy in the power supply of supercapacitors and batteries. Because of the difference in operating conditions, too much energy is consumed in batteries or supercapacitors, and electric cars stop working, while batteries or supercapacitors save a lot of energy. This leads to the irrational use of power supplies This paper presents a method to obtain the optimal matching parameters of dual energy source power supply based on the driving condition in ensuring the power supply characteristics .The aim of this method is to make the supercapacitors and the accumulator at the same time reach the minimum energy value and improve the energy efficiency of the power supply.

3.1. Power conversion model

The UEDC cycle condition in figure 2 is regard as the actual driving condition. The powers of electric vehicles can be obtained according to the principle of the conservation of motor power.

$$P_p = \frac{u_a}{3600\eta_e \eta_{PM} \eta_T}\left(mgf + mgi + \frac{C_a A u_a^2}{21.25} + \sigma m \frac{du_a}{dt} \right)$$

P_p- Power required for power supply, **i**- Road grade, **u_a**- Running speed

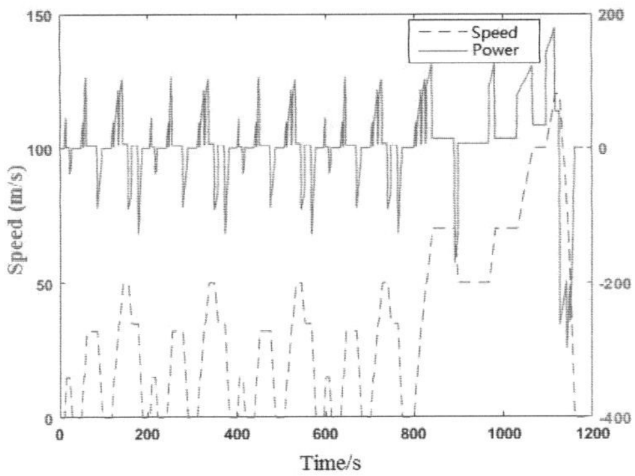

Figure2. Speed and power of NEDC.

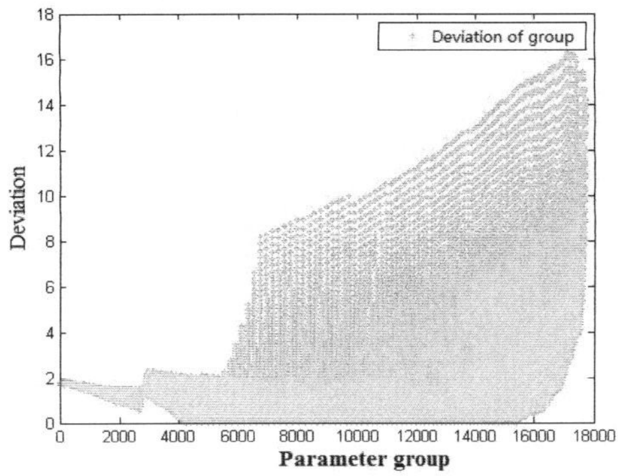

Figure 3. Matching deviation.

IOP Conf. Series: Materials Science and Engineering **324** (2018) 012022 doi:10.1088/1757-899X/324/1/012022

3.2. Power ratio of power supply under cyclic working condition

In this paper, the double power source electric vehicle energy management strategy is normal. High power charging and discharging is carried out by super capacitor, and the battery bears low power charge discharge, braking energy is recovery by supercapacitors. not only can it make the electric car with good dynamic performance, but also avoid the peak discharge of batteries. When the power of the power supply is less than the maximum power of the batteries, the batteries provides energy(Eb). When the power supply power is less than zero, the power of the supercapacitors is recovered(Ecov). When the power of the power supply is greater than the maximum power of the batteries, the supercapacitors provides energy(Esc).

$$P_{b\max} = C_b \bullet U_b \bullet k_{\max} \bullet N_b$$

P_{bmax}- The maximum discharge power of batteries,**K_{max}**- Maximum discharge ratio
Energy consumption is the product of power and time. The ratio of the energy of the batteries and the supercapacitors in the cycle condition is obtained.

$$K' = \frac{N_b C_b U_b}{\frac{1}{2} N_{sc} C_{sc} U_{sc} \times (3.6 \times 10^{-3})}$$

$$\delta = \left| K' - K \right|$$

The matching deviation can evaluate the operation of the parameter group under cyclic condition. The smaller the matching error is, the closer the dual energy source ratio in the matching parameter is to the dual energy source ratio required in the road condition.The matching deviation for the above mentioned 17714 sets is shown in figure 3.
The matching deviation of the 10313 parameter group [99,113] in the matching deviation is 2. 410-4, which is the smallest parameter group.It selects 99 batteries and 113 supercapacitors. Its main parameters are listed in table 1.

Table1. Mainparameters of power.

Unit of Parameter	Accumulator	Supercapacitors	Total
Quantity/(single)	99	113	212
Mass/(kg)	186. 12	96. 05	282. 17
Volume/($m^{2)}$	0. 085	0. 047	0. 132
Energy/(kwh)	25. 344	5. 706	31. 050
energy/(kwh/kg)	0. 136	5. 94×10^{-2}	0. 110

The method of matching deviation is to optimize the parameters of the driving condition and energy management strategy. First, the power of the power supply for a cycle condition is calculated. The power of the power supply is allocated by the energy management strategy to obtain the energy ratio of the supercapacitors and the batteries in the cycle condition. Finally, the matching deviation is calculated and the optimal parameter group is selected.

4. The simulation analysis

Building the vehicle model in MATLAB-Simulink, When the value of the SOC of the supercapacitors or the batteries is less than 20%, the electric vehicle power supply discharge ends in the NEDC cycle, and the simulation stops. A total of 99 batteries and 113 supercapacitors batteries were used. The input operation model is based on 1 cycle condition and 25 cycle conditions. The energy usage of the supercapacitors is shown in figure 4. The energy usage of the batteries is shown in figure 5.

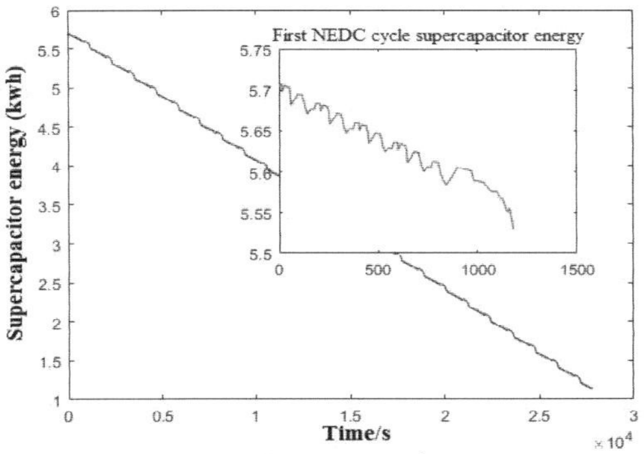

Figure4. Supercapacitorenergy.

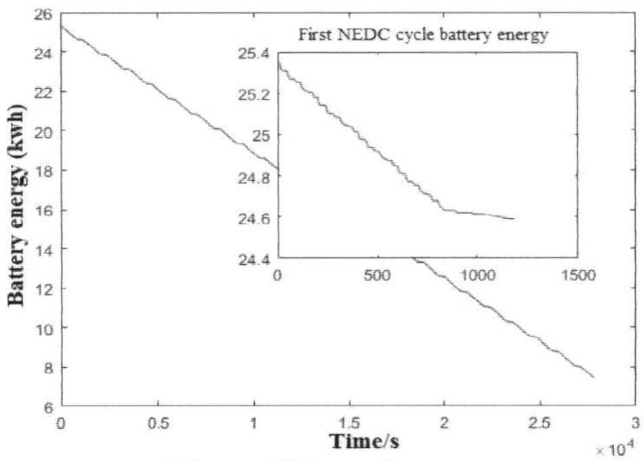

Figure5. Batteriesenergy.

The simulation stopped at 27784s. At this point, the supercapacitors is below the cutoff value of 1. 140 KWH, and the range is 256787m, which meets the design requirements. The batteries has 2. 37 KWH energy to use, and the power utilization efficiency is 92. 4%.

The NEDC was tested by taking the number 2131,10313, 15812. The test results are as table 2. The comparison test results of the three parameter groups are very straightforward to show that the smaller parameters of the match deviation are more efficient and having the longest range. The results show that the parameters matching method based on the road condition is reasonable and efficient.

Table 2.comparison of simulation result.

Parameter sequence number	Parameter set[N_b, N_{sc}]	Matching deviation	Simulation stop time/s	Range/ m	Energy efficiency/%
2131	[53,305]	1. 5	16859	155987	67. 1
10313	[99,113]	2.4×10^{-4}	27784	256787	92. 3
15812	[72,201]	0. 12	22157	191718	79. 8

5. Conclusion

This paper takes a mature electric car as the design prototype. A parameter matching method based on road condition is proposed to. The analysis results show that the parameter matching method is reasonable. Be up against a complex road in a region. after matching of the dual energy source ,And then the coupling mechanism of dual energy source is studied and the power control strategy is formulated. In view of the characteristics of multi terrain and multi road in our country, This method has certain reference value.

6. Reference

[1]SoniS and ChetanD2015 *International Conference on Energy Economics and Enviornment-1st IEEE Uttar Pradesh Section Conference(ICEEE)* p 1-7

[2]M R Rade2015*International Conference on Emerging Research in Erectronics Computer Science and Technology(ICERECT)* p 821-826

[3]Chen L,ZhouW J and Pan C F2016*Automotive Engineering* pp 422-428

[4]S R Kumar 2016A SRM Drive based Hybrid Electric Vehicle with Reduced Switch Converter *InternationalConference on Green Engineering and Technologies(ICGET)* p1-7

[5]Yu Y B,Wang Q N and Wang J X 2008*Journal of Jilin university (Engineering Edition)* pp 765-768

[6] Ebbseen S,Donitz C and Gzzellal 2012Particle swrm optimization for hybrid electric drive-train sizing*International Journal of Vehicle Design* p 181-192

Issues of Concrete in Silage Pits

S Rehacek[1], D Dobias[1], P Pokorny[1], D Citek[1] and J Kolisko[2]

[1] Research Engineers, Klokner Institute CTU in Prague, Prague, Czech Republic
[2] Head of department, Klokner Institute CTU in Prague, Prague, Czech Republic
E-mail: stanislav.rehacek@cvut.cz

Abstract. The article briefly describes issues of concrete in silage pits. It explains the basic principle of ensiling and the process of concrete degradation due to the action of acidic silage juices on the concrete surface. Typical signs of concrete degradation after using a silage pit for one year can be seen in the photographs.

1. Introduction

One of the methods of storing livestock feed is its preservation by so called ensiling. Ensiling is a fermentation process in which the feed (maximum dry matter content 45-50%) is stored in a silage pit, while being constantly compressed (e.g. by tamping down, rolling) and it is then airtightly sealed. This method of preservation, unlike hay drying, preserves the feed (silage) in a juicy state. The preservation is carried out by lactic acid fermentation of sugars contained in the silage feed without the access of air while retaining the nutrients and vitamins. In order to maintain the highest quality of silage, a sufficient amount of lactic acid, which is the basic preservative, must be produced as soon as possible. A pH of silage is generally lower than 4. The anaerobic environment in which no rotting processes occur is due to the acidic environment, the restriction of air access and the production of CO_2 produced by food respiration and microbial activity. Failure to comply with the technological discipline of silage preparation or the use of unsuitable fodder cause aerobic degradation - rot. This degradation destroys silage by biochemical processes with access of air. This process produces butyric, acetic or formic acids which make silage unusable (in addition, it contributes to the concrete degradation) [1, 2].

2. Methods of ensiling

Silage pits or silage troughs are most commonly used for silage in the Czech Republic, varying considerably from both the capacitive and the structural aspects. In terms of construction, pits can be divided into passable and impassable, aboveground, half-recessed and fully recessed, monolithic and prefabricated. The most common material for the construction of pit walls is concrete. Concrete or asphalt concrete are also used for the floor. After filling the trough with silage (maize, fodder crops, sunflower, etc.) and with admixture of additives (acids based additives and their salts, additives containing a source of sugar, probiotic additives etc.), the silage is properly compressed (using tractors, rollers) in order to force the air out. The silage surface is covered with a special tarp that is weighted down (usually with old tires). The cover is to prevent access of air and rainwater to silage. Other less usual methods of ensiling in the Czech Republic are: silage towers and silage put into bags or bales [1, 2].

Silage troughs in the Czech Republic are most often made out of monolithic or prefabricated reinforced concrete with a rectangular floor plan. The side length is 30-100 m, the side width is 15-30 m, and the wall height is 1.5-6 m. The wall thickness is usually 150 to 250 mm. The reinforced concrete floor has a thickness of 150-300 mm and is generally divided into units with a side length of 4-6 m, whereas the whole area is sloped in the direction of the entrance with a gradient of at least 1%, preferably 2%. The minimum compressive strength of the concrete class with respect to XC4 and XA3 exposure classes is C 30/37. Very often, the designer of the silage pits proposes concrete for XC4, XF4 or XA3 exposure classes. According to CSN EN 206 and CSN P 74 2304 the minimum requirements for C30/37 concrete of the above classes are the following [3 and 4]:

- maximum water coefficient w = 0.45
- minimum cement content = 360 kg/m^3
- minimum air content = 4%
- maximum pressure water seepage depth = 20 mm

3. Chemical action of silage juices on concrete

Silage juices are organic acids consisting mainly of lactic acid (butyric, acetic or formic acids are formed during inappropriate ensilage). The volume and acidity of juices fluctuates during fermentation and depends on the type of feed stored and its moisture content at the beginning of the ensilage process. The volume of silage juices further depends on the tightness of the plastic covering which prevents rainwater from flowing into silage. Rainwater leakage occurs especially along the walls. It is proven that the pH of silage juices ranges from 2.5-4. Due to the action of these very acidic juices, degradation of the surface layer of concrete occurs in a very short time (even after one year of use). The initial damage to the concrete is usually local, in places where acidic juice concentration usually occurs, usually into depths of 1-5 mm. Cement putty on the surface of concrete walls and the floor is broken down into non-cohesive grains and coarse aggregate grains are exposed.

The low resistance of concrete to the acidic environment of silage juices is due to its high alkalinity (the pH of young concrete is higher than 13). The reaction of acidic juices with cement hydration products causes type II corrosion. Type II corrosion takes place when solutions penetrating into the capillary pore structure of concrete react with the cement matrix which results in the formation of easily soluble compounds. The cement matrix is gradually degraded, losing its binding properties and ultimately reducing the mechanical properties of the concrete. In the initial phase, the portlandite $Ca(OH)_2$ is “attacked” first and subsequently the decomposition of the other components of the cement putty takes place, i.e. calcium hydrosilicates and hydroaluminates, while calcium salts are formed. The rate of concrete degradation depends on the solubility of these calcium salts and on the behaviour of the formed corrosion products layer. The more soluble the reaction products are and the faster they are washed out of the concrete, the faster the degradation process takes place. The speed of corrosion processes is dependent on whether the reaction products adhere to the surface of the concrete or are washed away. When adhering to the surface of the concrete, the reaction products help to slow down the access of acidic juices. During the silage removal, the mechanical action of the collection attachment of the loader rubs against the concrete wall and floor and thus accelerates the action of acids on the newly opened surface of cement putty [5, 6].

Effect of lactic acid on concrete can be described, for example, as follows:

$$Ca(OH)_2 + 2C_2H_4(OH)COOH + 3H_2O \rightarrow Ca[C_2H_4(OH)COO]_2.5H_2O \quad (1)$$

Effect of acetic acid on concrete can be described, for example, as follows:

$$Ca(OH)_2 + 2CH_3COOH \rightarrow Ca(CH3COO)_2.H_2O + H_2O \quad (2)$$

The usual method of protecting concrete against direct exposure to acidic environments is so-called secondary protection of concrete by coatings with organic substances (e.g. epoxy or polyurethane coatings). They guarantee long-term service life only in the case of a sufficient thickness (usually greater than 500 μm) and complex compactness. In the case of defects in the coating, the coating quickly becomes corroded and almost immediately delaminated. As a result, there is a significant reduction in the protection of the concrete structure. Defects in protective coatings on concrete components are frequent and are closely related to the quality of the technological application and implementation. This secondary protection also has very little resistance to additional mechanical damage.

4. Examples of concrete degradation of a silage pit

The following photos show the effects of acidic silage juices on concrete after only one year of use of two silage pits. The first silage pit 'figure 1' was made by a monolithic method from concrete C 30/37 XC4, XF4, XA3, S4 Dmax 22. The wall thickness is 200 mm and the height 5.5 m. The second silage pit was made from prefabricated panels 'figure 5'made of C 35/45 concrete XC4, XF4, XA3, S3 Dmax 22. The wall thickness is 150 mm and the height is 6 m. A coating was applied to the surface of the walls of both silage pits to provide secondary concrete protection. On the basis of a visual inspection of silage pits carried out after one year of use, it can be stated that the secondary protective coating failed in both structures and local concrete degradation due to acidic juices occurred in depth of about 5 mm 'figure 2 and 3'. In some places, coarse aggregate grains were exposed 'Figure 4 and 7'. Concrete degradation occurred especially in the area of wall footings and also in the wall area, especially up to a height of about 1.5 m 'figure 6'. An objectively effective method of protecting concrete structures against acids in full cross-section is to modify the used cement or concrete mix with suitable organomer components. Epoxy-based organomer components, copolymers based on acrylic acetate, or even specific acid-resistant polyethers can be used successfully. These groups differ significantly in binding properties with the C-S-H gel (if need be with the aggregate surface), the ability to fill in the pores, the effect they have on the mechanical properties of the concrete, their own resistance to acids (e.g. silage juice) and other aggressive substances such as petroleum products. However, not every such substance is suitable for application to a strongly alkaline environment of cement concrete.

Figure 1. View of partially removed silage and surface damage of a monolithic concrete wall.

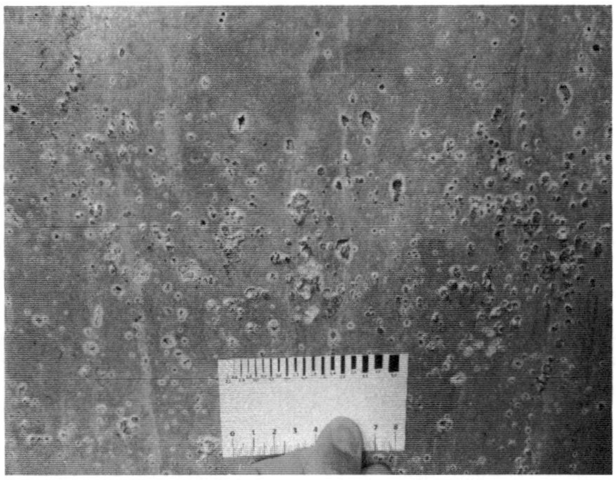

Figure 2. Small blisters in coating with degraded cement putty, depth max. 5 mm.

IOP Conf. Series: Materials Science and Engineering **324** (2018) 012023 doi:10.1088/1757-899X/324/1/012023

Figure 3. Small blisters and cracks in coating with degraded cement putty, depth max. 2 mm.

Figure 4. Exposed rough aggregate of the floor around the drainage channel.

Figure 5. Prefabricated panels with a non-resistant coating with a noticeable concrete degradation process along its height.

Figure 6. A degraded prefabricated panel footing to a depth of approx. 5 mm. There are obvious traces of the collection attachment of the loader on the panel - indicated by a red arrow.

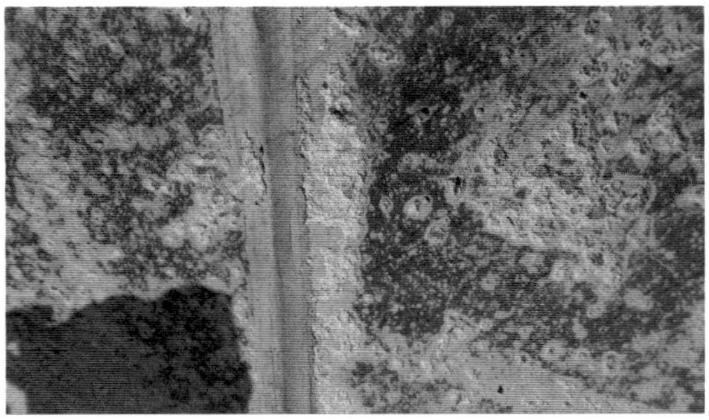

Figure 7. Detail of the degraded surface of the panel with exposed coarse aggregate grains.

5. Conclusion

From the above mentioned facts and illustrative examples of two silage pits, it is obvious that despite the type of concrete designed for the highest exposure class (XC4, XF4, XA3) the acidic silage juices cause relatively rapid degradation (decomposition of cement putty) of the concrete surface layer. In addition, in these cases an inappropriate coating was used that failed to provide secondary protection of the concrete. The highest degree of chemical attack of XA3 exposure class is intended for an environment with a pH $\geq$ 4, whereas the silage juice environment is more aggressive, the pH is usually 3-4. In addition to complying with the principles and requirements set out in CSN EN 206, secondary protection using organic acid-resistant coatings with a pH of 2.5-4 must be used to ensure the long-term service life of silage pits. Another option is the development of new concrete with exceptional resistance to aggressive substances. The aim of further research is to develop concrete with high tightness, compactness and resistance to aggressive substances throughout the cross-section of the structural element created by the application of polymer nanoparticle material directly to the concrete mix. The concrete thus created will not need to be provided with secondary protection and the concrete surface itself will be highly resistant to mechanical damage that the coatings used so far are poorly resistant to.

Acknowledgment

The paper was drawn up with the assistance of TACR (Technology Agency of the Czech Republic) grant TH02020690 Concretes with Exceptional Resistance to Aggressive Substances.

References

[1] www.agroweb.cz
[2] www.agroporadenstvi.cz
[3] ČSN EN 206 Concrete – Specification, performance, production and conformity
[4] ČSN P 74 0423 Concrete – Specification, performance, production and conformity – Additional information
[5] Al-Tamimi AK and Sonebi M 2003 Assessment of Self-Compacting Concrete Immersed in Acidic Solutions *Journal of Materials in Civil Engineering* **15** 354-357.
[6] Sernad A, Yazici H, Yigiter H and Baradan B 2007 Sulfuric Acid Resistance of High-volume Fly Ash Concrete *Building and Environment* **42** p 717-721.

IOP Conf. Series: Materials Science and Engineering **324** (2018) 012024 doi:10.1088/1757-899X/324/1/012024

Efforts of the occupant to change physical quality of residential unit through the change of building material at low cost flats in Jakarta

N Nurdiani[1]

1 Lecturer, Architecture Department, Faculty of Engineering, Bina Nusantara University, Jakarta, Indonesia 11480
E-mail: nnurdiani@binus.edu

Abstract. Low cost flats in Jakarta – Indonesia is provided by the government for low-income people in urban areas, in line with the program to redevelop or renew slum areas. Low cost flat is built with the minimum standard of building materials. The purpose of this study is to know efforts of the occupants to change of building materials at residential unit of low cost flats. The research was conducted by descriptive method at four of low cost housing in Jakarta: Rusuna Bendungan Hilir 1, Rusuna Tambora IIIA, Rusuna Bidara Cina, and Rusuna Sukapura. The results showed that physical changes which happened in low cost flats are aesthetic (residence paint color change), or improvement of physical quality of residential unit (change of building material), become dominant aspects done by residents in four rusuna.

1. Introduction

Jakarta as a government city, centre of economy, and services city in Indonesia, is still the main destination of migrants from various regions to seek a better life and improve their welfare. Immigrants who come and stay in Jakarta - Indonesia are not as found among workers in an industrial society. They are immigrants from the suburbs or the countryside who have limited skills and economic capacity, as well as low levels of education, so that they become low-income people living in the city with rural culture. They also tend to have lower levels of income and more work in the informal sector, so most of them lack the ability to obtain a decent house or dwelling. They live crowded into the village area around of city centre [1,2].

In order to improve the quality of neighborhoods and city kampong which crowded and slum, as well as to increase providing habitable housing for low income people in Jakarta, the Government of the Special Capital Region (Daerah Khusus Ibukota [DKI]) Jakarta provided urban renewal program, redevelopment program, or low cost housing program (Rumah Susun Sederhana/Rusuna/simple flats/Low cost flats).

The construction of low cost flats that has been started since 1983 is still regarded as a new form of modern living room by low-income people in Jakarta. Low cost flats which provided by the government or the private sector is a form of vertical housing which adopts modern housing for urban society in the developed countries. It massively built and designed with typical form and limited area in dwelling unit. The low cost flats are built with the minimum standard of building materials.

IOP Conf. Series: Materials Science and Engineering **324** (2018) 012024 doi:10.1088/1757-899X/324/1/012024

Currently the residents who stay in low cost flats have been done the change of building materials at their housing unit (see figure 1).

a b c

Figure 1. The change of building materials at housing unit of low cast flats.

The purpose of this study is to know efforts of the occupants to change building materials at residential unit of low cost flats.

Based on the results of previous studies, people who live in public housing (low cost flats) in Indonesia do require adaptation and adjustment process [2]. Adaptation process is actions to change of nonphysical factors of inhabitants (norms and behavior). Adjustment process is actions that did not only the changes of nonphysical factors of inhabitants, but also changes the physical factors of their residential unit, in order to realize the house which expected the residents. The physical changes are depending on the level of residents satisfaction on their residential unit [3,4,5].

2. Research methodology

The research was conducted by descriptive method through surveys and field observations at four of low cost flats in Jakarta. The population of this study is the earliest inhabitants who has owned dwelling unit, and stay in housing unit from the beginning until the time when the study was conducted.

The low cost flats (rusuna) which selected as the study cases are Rusuna Bendungan Hilir I in Central Jakarta, Rusuna Tambora IIIA in West Jakarta, Rusuna Bidara Cina in East Jakarta, and Rusuna Sukapura in North Jakarta.

The data were analyzed by descriptive approach to answer the research problem. Conclusion is prepared to answer questions about efforts of the occupants to change building materials at residential unit of low cost flats.

3. Results and discussion

The results of the analysis on the four rusuna owned by Rusuna Bendungan Hilir 1, Rusuna Tambora IIIA, Rusuna Bidara Cina and Rusuna Sukapura, show that there are similarities and differences in the types of physical changes made by residents in four rusuna case studies. The following is a descriptive description of the types of physical changes found in the rusuna belonging to the study case.

In Rusuna Bendungan Hilir 1, the pattern of physical changes of residential unit (adjustment) that occur and dominant done by residents, that is by changing the color of residential wall paint (done on 46% of units), or change building materials (26% of units). The next aspect that also happened in adjustment process related with the residential function, which is to improve the function of units in the rusuna, which is done by residents by (1) changing the function of the existing room (done on 14% of units) in the form of exchanging the function of the existing room or increase its functionality in the same area, and (2) expand the residential unit to a larger extent the size of the room, or add new function to this expansion area (12% of units), (3) renovation wall / door / window as it is damaged, or due to the expansion of residential (only 2% of units). This indicates that in Rusuna Bendungan Hilir 1 more adjustment to improve the physical condition of the dwelling than the effort to make improvements (performance) function of the dwelling or the room.

In Rusuna Tambora IIIA, the pattern of physical changes of residential unit (adjustment) that occur and dominant done by occupants are changes in the quality of units, that is by changing the building material (done on 40% of units) or changing the color of residential walls (30 % of units). The next aspect that happened in adjustment process related with the function of the dwelling, which is to improve the function of existing dwelling. The efforts made by the occupants to change the performance of the function of the dwelling is by (1) changing the function of the existing room (done on 26.7% of the dwelling) in the form of exchanging the existing room function or redefining its function, (2) can be expanded the size of the room, or add new function to this expansion area. The second way is only done on 3.3% of units. This means that on Rusuna Tambora IIIA, more adjustments to improve the physical condition of the dwelling than the effort to make improvements (performance) function of the dwelling or the room.

In Rusuna Bidara Cina, the pattern of physical changes of residential unit (adjustment) that occur and dominant done by residents, that is by changing the building material (done on 46% of units), or change the color of residential wall paint (27% of units). The next aspect that also happened in adjustment process related with the function of the dwelling, which is to improve the function of residential unit in the rusuna, which is done by residents through (1) changing the function of the existing room (done on 14% of units) in the form of exchanging the function of the existing room or increase its functionality in the same area, and (2) expand the residential unit to a larger extent the size of the room, or add new function to this expansion area (12% of units); (3) renovation the wall / door / window as it is damaged, or due to the expansion of residential (only 1% of units). This shows that in Rusuna Bidara Cina, more adjustments to improve the physical condition of the dwelling than the effort to make improvements on aspects of residential function or room.

In Rusuna Sukapura, the changing physical pattern of the residential unit which is done dominant by the occupants is the change of residential quality by changing the color of the residential wall paint (43% of units), or changing the building material (35% of units). The next aspect of the adjustment process is related with the aspect of the function of the dwelling, which is to improve the function of the existing residential in the rusuna, which residents do by changing the function of the existing room (done on 22% of units) in the form of exchanging the existing room function or add function in the same area. No occupancy expansion or demolition of walls / doors /windows by residents. This shows that in Rusuna Sukapura, more are doing adjustments to improve the physical condition of the dwelling than efforts to make improvements in aspects of residential function.

The results of the analysis of physical changes in the four rusuna properties illustrate that the dominant physical changes occur in four rusuna is aesthetic (residential paint color change), or improvement of physical quality of residential units (changes in building materials). Both types of change can be categorized to aesthetical and physical adjustment. While the other three types of physical changes (functional change, expansion, wall / door / window renovation) are physical changes related to the fulfillment of the residential function in the rusuna, so it can be categorized as functional adjustment.

4. Conclusion

From the comparison between the types of physical changes made by residents in four rusuna owned, namely at Rusuna Bendungan Hilir 1, Rusuna Tambora IIIA, and Rusuna Bidara Cina, and Rusuna Sukapura, it can be seen that physical changes are aesthetic (residence paint color change), or improvement of physical quality of residential unit (change of building material), become dominant aspect which is done by residents in four rusuna. Meanwhile, from the aspect of physical changes that are functional, almost all activities included in the physical changes related to the functional aspects (changes in the function of space, the expansion of occupancy, the renovation of walls/doors/windows) is done in three rusuna (Rusuna Bendungan Hilir 1, Rusuna Tambora IIIA, Rusuna Bidara China). While at Rusuna Sukapura, only the change of the function of the room is done by the residents, while the activities of expansion of residential or demolition of the wall/door /almost not done by residents in Rusuna Sukapura.

The above description shows that residents of Rusuna Sukapura are more able to control themselves against physical changes in their dwelling. This can be controlled because there is a manager role assisted by the agency to oversee the changes that may or may not be made by residents in the flats. So far only physical changes related to aesthetics and improvement of the quality of residential (change of paint color of residence and changes in building materials), or changes in space function (replacement of space functions) that many residents do at Rusuna Sukapura.

References

[1] Bahri S 2005 Rumah susun sebagai bentuk budaya bermukim masyarakat modern *Jurnal sistem teknik industri* **6** 97-102

[2] Nurdiani N 2012 Adjustment and self-help approach for improving housing unit quality in multi-storey housing *Applied mechanics and materials* **174-177** 3463-3466

[3] Mohit M A, Ibrahim M, and Rashid Y R 2010 Assessment of residential satisfaction in newly designed public low-cost housing in Kuala Lumpur Malaysia *Habitat International* **34** 18-27

[4] Awanyo L 2009 Meeting housing-space demand through in situ housing adjustments in the greater Accra metropolitan area Ghana *Environment and Planning C: Government and Policy* **27** 302-318

[5] Sinai I 2001Moving or improving: housing adjusment choice in Kumasi Ghana *Housing Studies* **16** 97-114

IOP Conf. Series: Materials Science and Engineering **324** (2018) 012025 doi:10.1088/1757-899X/324/1/012025

Application of Glass Fiber Waste Polypropylene Aggregate in Lightweight Concrete – thermal properties

D Citek[1], S Rehacek[1], Z Pavlik[2], J Kolisko[1], D Dobias[1] and M Pavlikova[2]

1 Klokner Institute - CTU in Prague, Prague, Czech Republic
2 CTU in Prague, Prague, Czech Republic
E-mail: david.citek@cvut.cz

Abstract. Actual paper focus on thermal properties of a sustainable lightweight concrete incorporating high volume of waste polypropylene aggregate as partial substitution of natural aggregate. In presented experiments a glass fiber reinforced polypropylene (GFPP) which is a by-product of PP tubes production, partially substituted fine natural silica aggregate in 10, 20, 30, 40 and 50 mass %. Results were compared with a reference concrete mix without plastic waste in order to quantify the effect of GFPP use on concrete properties. Main material physical parameters were studied (bulk density, matrix density without air content, and particle size distribution). Especially a thermal transport and storage properties of GFPP were examined in dependence on compaction time. For the developed lightweight concrete, thermal properties were accessed using transient impulse technique, where the measurement was done in dependence on moisture content (from the fully water saturated state to dry state). It was found that the tested lightweight concrete should be prospective construction material possessing improved thermal insulation function and the reuse of waste plastics in concrete composition was beneficial both from the environmental and financial point of view.

1. Introduction

The energy performance of a building is becoming increasingly important, because of environmental restrictions and rising costs of fuel end energy. Environmental aspects of constructions are more and more required by society nowadays. In order to design buildings with sufficient thermal attributes and optimum energy performance, specific attention must be paid not only to the design of building envelopes, but also to the thermal insulation of buildings subsoil [1]. To obtain buildings with high thermal resistance and sufficient thermal stability, the use of materials with suitable thermal insulation properties should be taken into account. The large growth of plastics consumption and consequently of plastic waste, requires new forms of recycling, avoiding landfill disposal [3]. In recent years in Europe, plastics recycling have increased to almost 30%. However landfilling and energy recovery still constitutes a large part of plastic waste disposal [4]. The construction industry, acknowledging the importance of environmental issues, has aim to find solutions that are able to combine economic aspects and growth and environmental conservation. Possible method should be the substitution of natural aggregates by recycled plastic aggregates in cementitious materials (mortars or concretes) [5]. From the literature, it appears, that incorporation of plastic waste as plastic aggregates in concrete leads into the significant improvement of thermal properties of developed composites [6, 7]. In that sense, this study is focused on the assessment of the lightweight concrete materials incorporating glass fiber reinforced polypropylene waste aggregates in terms of preferable thermal function and optimized energy performance in building construction.

2. Experimental research

Basic physical and thermal properties of GFPP aggregate were measured at first. Crushed waste glass fibre reinforced polypropylene (GFPP) used in this work was obtained from the PP tubes production Lightweight concrete was then prepared using silica sand, Portland cement CEM I 42.5 R, water and GFPP waste, which was used for the partial replacement of natural aggregate by 10, 20, 30, 40 and 50 mass %. The reference specimen without GFPP was prepared as well. The silica sand of fraction 0-2 mm was used. The water/cement ratio was 0.5 for all studied mixtures. The casted prismatic specimens (small beams) 40×40×160 mm and cubic specimens with side dimension of 70 mm were left for 1 day at laboratory conditions and then they were unmoulded and cured for 28 days in water until the tests. The mix proportions are summarized in Table 1.

Table 1: Composition of studied composites with GFPP waste

Mixture	Mass [g]				w/c
	Cement	Natural aggregate	Water	GFPP waste	
Ref	450	1 350	225	-	0.5
10 %	450	1 215	225	135	0.5
20 %	450	1 080	225	270	0.5
30 %	450	945	225	405	0.5
40 %	450	810	225	540	0.5
50 %	450	675	225	675	0.5

2.1. GFPP aggregate

In order to characterize GFPP, its powder (bulk) density, matrix density and grain-size distribution were determined. The average matrix density, measured using automatic helium pycnometer Pycnomatic ATC, was 940 kg/m^3. The grain-size distribution was analysed using standard sieve method with sieves of mesh dimensions 0.063- 63.0 mm. Obtained results showed that GFPP particles were smaller than 8 mm, what is beneficial for their incorporation into the composition of the lightweight concrete. The powder (bulk) density and thermal properties namely thermal conductivity λ (W/m·K), volumetric heat capacity Cv (J/m^3·K) and thermal diffusivity a (m^2/s) of GFPP aggregates were measured in dependence of compacting time, in order to get information on changes in thermal parameters based on aggregate processing. The compacting was done using a high-frequency vibrating table VSB-15. The highest compaction time corresponded to the steady-state specimen volume. To determine these thermal parameters, the commercially produced device ISOMET 2114 was used as a representative of transient impulse methods [8]. The measurement accuracy was 5% of reading + 0.001 W/m·K for the thermal conductivity in the range 0.015-0.70 W/m·K and 10% of reading for the thermal conductivity ranging from 0.70 W/m·K to 6.0 W/m·K. The accuracy of the measurement was 15% of reading + 1 × 10^3 J/m^3·K.

2.2. Thermal conductivity

In order to understand the influence of the plastic aggregate on improvement of thermal insulation properties of the developed lightweight composites, their thermal conductivity λ (W/m·K) was obtained using device ISOMET 2114. Measurement was performed in dependence on moisture content, from the fully water saturated state to dry state. Cubic lightweight concrete specimens were measured using surface probe (side with 70 mm).

2.3. Basic material properties

Not only basic material properties of the lightweight concrete specimens were examined. The bulk density, the matrix density and the total open porosity were determined. Composites were first dried in a vacuum drier at 60 °C. The matrix density was executed using automatic helium pycnometry as described above. The bulk density was given by gravimetric principle from measurement of specimen size and its dry mass according to EN 12390-7 [9]. On the basis of density measurement (bulk and matrix), the total open porosity value was calculated [10]. The relative expanded uncertainty of this applied measuring method was approx. 5 %.

3. Results and discussion

Generally, the plastic waste aggregate had significantly better thermal parameters than silica sand typically used in concrete production [11].Thermophysical properties of GFPP aggregate together with corresponding powder density are summarized in Table 2. With compaction of concrete mixture, due to the elimination of air gaps between the plastic particles, all measured thermal properties have increased. Table 3. shows basic material and physical properties of tested lightweight concretes.

Table 2: GFPP aggregates thermal properties

Compaction time [s]	Powder density [g/cm^3]	λ [W/m·K]	$C_v \times 10^6$ [J/m^{3}K]	$a \times 10^{-6}$ [m^2/s]
0	0.394	0.092	0.266	0.347
30	0.468	0.098	0.314	0.311
90	0.487	0.099	0.323	0.307
240	0.490	0.100	0.325	0.312

Table 3: Basic physical properties of tested composites

Material	Bulk density [kg/m^3]	Matrix density [kg/m^3]	Open porosity [%]
Ref	2 064	2 508	17.7
10 %	1 852	2 225	16.8
20 %	1 611	2 003	19.6
30 %	1 319	1 821	27.6
40 %	1 053	1 652	36.3
50 %	895	1 589	43.6

The bulk densities of the shown mixtures with incorporated GFPP aggregates reached values in the range 800-1900 kg/m^3. Therefore, they can be classified as lightweight concretes in classes LC 1.0 - LC 2.0 and LC 2.1 respectively, according to the standard EN 206-1.

Material porosity of specimens with 10% incorporated GFPP was lower compared to the reference specimens; for 20, 30, 40 and 50% replacement of silica aggregate the total open porosity increased with the rate of substituted plastic aggregate. Such behaviour can be explained taking into account the un-porous structure of this plastic waste, which decreased the total open porosity of cement composites when be used in a lower amount, whereas the workability and thus air imbibition in the mixes was not negatively affected. On the other hand, the higher amount of plastic waste aggregates led to the increase in porosity due to the bigger particles and to the general coarsening of composites inner structure [2]. Packing density of granular mix can also increase the open porosity. The moisture dependent thermal conductivity of lightweight concrete is introduced in Fig. 1.

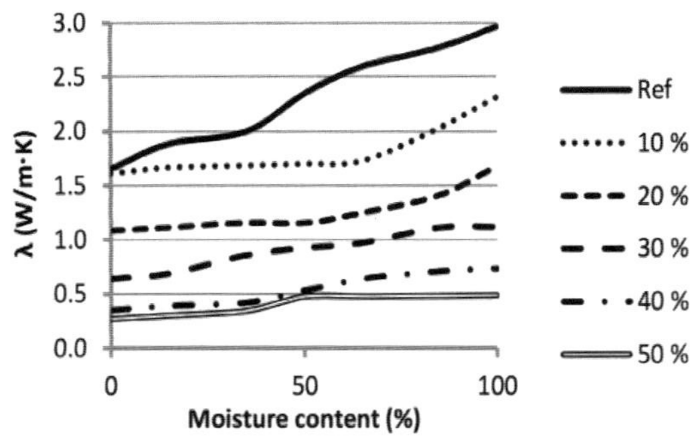

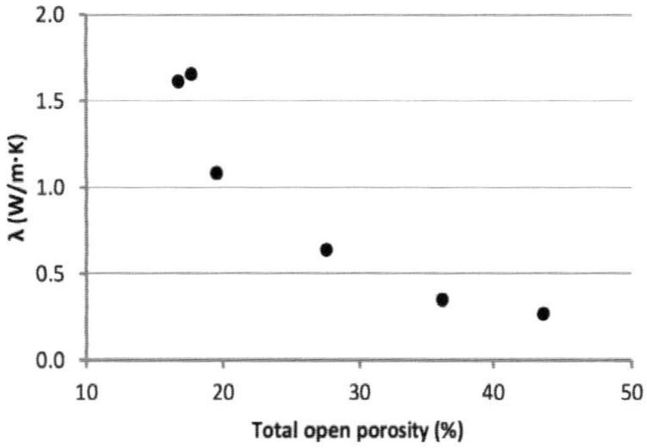

Figure 1. Thermal conductivity of lightweight concrete with GFPP in dependence on moisture content

Figure 2. Thermal conductivity of lightweight concrete with GFPP in dependence on the total open porosity (in the dry state)

The measured thermal conductivity has highly depended on moisture content. The effect of moisture content on the increase in thermal conductivity values was the highest for reference mix without GFPP use. In this case, the thermal conductivity has increased of about 74.4% for fully water saturated specimen compared to its dry value. It is quite logical, since GFPP granules were inert for water penetration and their inner encapsulated porosity contributed to the improvement of thermal conductivity without negative moisture effect. Regarding obtained data in Fig. 2, an exponential dependence of the thermal conductivity of researched composites in the dry state on their porosity should be see. The lowest contribution of moisture to the total thermal conductivity was observed for mix with 50 mass % of GFPP. Here, the difference between the dry and water saturated value of the thermal conductivity was about 38.1% only. The exception is the reference specimen, which at a higher porosity than composite with 10 % incorporated GFPP exhibited higher thermal conductivity. This is because the thermal conductivity in case of composites with plastic waste aggregates content does not affect only porosity, but also presence of plastic. Thermal conductivity is inversely related to the material weight.

4. Conclusions

Experimental research in the field of real application of the waste polypropylene in concrete was described in this paper. The assessment of thermal and basic material and physical properties of lightweight concrete with different amounts of glass fiber reinforced waste polypropylene was presented. The use of plastic particles (plastic aggregate) as sand substitution led to the increase in porosity and significant decrease in the bulk density. The results revealed the positive effect of GFPP aggregate incorporation on the moderation of heat transport even at presence of moisture. The use of plastic waste is an interesting way to extend its life in civil engineering applications providing lightweight concretes with enhanced thermal insulation properties. Because of environmental restrictions and rising costs of fuel end energy the energy performance of a building is becoming increasingly important nowadays. This is a way how to innovate view to design structures and safe environment.

Acknowledgments
The authors gratefully acknowledge the financial support received from Czech Scienty Society under project No. 17-04215S and project No. 17/162/OHK1/2T/31.

References

[1] Iucolano F., B. Liguori, B., Caputo, D., Colangelo, Cioffi, R., Recycled plastic aggregate in mortars composition: Effect on physical and mechanical properties. *Materials and Design*, 52, pp. 916-922, 2013.
[2] Z. Pavlík, M. Záleská, M. Pavlíková and R. Černý, "The thermal and mechanical performance of cement-based composites with enhanced thermal insulation properties", in Heat Transfer XIII: Simulation and Experiments in Heat and Mass Transfer, edited by B. Sundén and C.A. Brebia (WIT Press, Southampton, 2014), 251-260.
[3] B. Coppola, L. Courard, F. Michel and L. Incarnato, Composites Part B 99, 75-83 (2016).
[4] PlasticsEurope, Plastics – the Facts 2015, www.plasticseurope.org, consulted January 2017.
[5] A. M. da Silva, J. de Brito and R. Veiga, Constr. Build. Mater. 71, 226-236 (2014).
[6] J. L. Ruiz-Herrero, D. V. Nieto, A. Lopéz-Gil and M. Rodríguez-Pérez, Constr. Build. Mater. 104, 298-310 (2016).
[7] R. Sharma and P.P. Bansal, Use of different forms of waste plastic in concrete – a review, Journal of Cleane Production, vol. 112,2016, pp 473-482.
[8] Z. Pavlík, M. Záleská, M. Pavlíková and R. Černý, "The thermal and mechanical performance of cement-based composites with enhanced thermal insulation properties", in Heat Transfer XIII: Simulation and Experiments in Heat and Maşs Transfer, edited by B. Sundén and C.A. Brebia (WIT Press, Southampton, 2014), 251-260.
[10] M. Keppert, P. Reiterman, Z. Pavlík, M. Pavlíková, M. Jerman and R. Černý, Cement Wapno Beton 15, 187-193 (2010).
[11] H. Kiyohashi, S. Sasaki, H. Masuda, High Temp.-High Press. 35/36, 179-192 (2003/2004)

IOP Conf. Series: Materials Science and Engineering **324** (2018) 012026 doi:10.1088/1757-899X/324/1/012026

Experimental verification of concrete resistance against effect of low pH

D Dobias[1], S Rehacek[1], P Pokorny[1] D Citek[1] and J Kolisko[2]

[1] Research Engineers, Klokner Institute CTU in Prague, Prague, Czech Republic
[2] Head of department, Klokner Institute CTU in Prague, Prague, Czech Republic
E-mail: daniel.dobias@cvut.cz

Abstract. In the introductory part of this article, the principles of a concrete degradation by organic acids are mentioned, these acids occur, particularly in silage and haylage troughs, biogas stations, on concrete floors and grates in the vicinity of drinking basins with an addition of formic acid and also in fermenters and slurry reservoirs. In the experimental part, the first results of monitoring resistance of a concrete with a sealing admixture on the basis of styrene-acrylate against an effect of a low pH are presented. Additional accompanying tests are stated in the tested concretes.

1. Introduction

A method of preserving by so called ensiling is used for preservation of feed for farm animals in the Czech Republic, which is a fermentation process, in which the feed (maize, fodder plants, sunflowers, etc., a maximum dry weight is 45-50 %) is stored in silage pits under constant compaction (e.g. by packing down or rolling with tractors), in which the feed is closed hermetically. In this method of preservation, in contrast to making hay, is the feed (silage) stored in a sappy state. The most frequent material for the construction of the walls and the bottom of the silage pit is the concrete. In some cases, the asphalt concrete is used for making the bottom, which is more resistant than the concrete from the point of view of the resistance to the effect of acidic silage juices. Preservation of the feed takes place by the action of lactic fermentation of sugars contained in the ensiled feed without access of air, while the content of nutrients as well as vitamins is maintained. In order to maintain the highest quality of the silage, it is necessary to create a sufficient amount of lactic acid as soon as possible, which is a basic preservative agent. The pH of the silage is usually lower than 4. Due to the acidic environment, by limiting the access of air along with the production of CO_2, which is produced by food, respiration and microbial activity, an anaerobic environment is created, in which no putrid processes occur. If the technological discipline of the silage preparation is not observed or if the unsuitable fodder is used, there is an aerobic degradation - putrefaction. During this process, a butyric acid, acetic acid or formic acid comes into existence, which debases the silage (in addition it takes part in the degradation of the concrete) [1, 2].

With regard to the acidic environment, especially of the silage and haylage pits, biogas stations and concrete floors and grates in the vicinity of the drinking basins with water acidified with lactic acid, it is used the concrete in the degree of the environmental influence XC4, XF4, XA3. For these degrees of the environmental influence is determined the minimum strength class C 30/37. ČSN EN 206 and ČSN P 74 2304 determines for the concrete C 30/37 XC4, XF4, XA3 these minimum requirements [3 and 4]:

- Maximum water coefficient w = 0.45

- Minimum cement content = 360 kg/m^3
- Minimum air content = 4%
- Maximum depth of penetration with pressure water = 20 mm

In view of the fact that a larger amount of sulphates usually does not occur in the silage juices, it is not necessary to use sulphate-resisting cement according to ČSN EN 206 and ČSN P 74 2304.

2. Description of concrete degradation

The silage juices are organic acids constituted especially by lactic acid (in case of improper course of ensiling, the butyric acid, acetic acid or formic acid is formed). The volume and acidity of the juices fluctuate during fermentation and depends on the type of the stored feed and its moisture content at the start of ensiling. The volume of the silage juices further depends on the tightness of a cover plastic film, which should prevent from in-leak of rainwater into the silage. In-leak of rainwater occurs especially along the walls. It is substantiated, that the pH of the silage juices ranges from 2.5-4. Due to the influence of the effect of these very acidic juices, the degradation of the surface layer of the concrete occurs already in a very short time (and even after one year of use). The primary damage of the concrete is usually local, in areas, where there is the concentration of the acidic juices, usually to the depths of 1-5 mm. Cement lute on the surface of the concrete of the walls and the slab is disrupted to incohesive grains and there is the exposure of the coarse grains of the aggregate.

The low resistance of the concrete to the effect of the acidic environment of the silage juices is due to its high alkalinity (but the pH of the young concrete is higher than 13). The reaction of the acidic juices with cement hydratation products causes corrosion of the type II. The corrosion of the type II is characterized such that the solutions, which penetrate into capillary-pored structure of the concrete, react with a cement matrix and the result of these reactions is the formation of the easily soluble compounds. The cement matrix is decomposed gradually, loses its binding properties and as a result of this, the mechanical properties of the concrete are reduced. In the initial phase, the portlandite $Ca(OH)_2$ is "attacked" at first and subsequently the other components of the cement lute so called hydrosilicates and calcium hydroaluminates are decomposed to form calcium salts. The rate of the concrete depends on solubility of these calcium salts and on the behaviour of the layer of the corrosion product, which has arisen. The more are the reaction products soluble and the faster they are washed out of the concrete, the faster the degradation occurs. The rate of the corrosion processes is dependent on it, whether the reaction products cling to the surface of the concrete or whether they are washed away. In the case of clinging to the surface of the concrete, the reaction products help to slow down the access of the acidic juices. During withdrawal of silage, due to mechanical action of the withdrawal device of the loader, the abrasion against the concrete of the wall and the floor slab occurs and thus the acceleration of the acid's action on the newly opened surface of the cement lute [5 and 6].

It is evident from the literature [7, 8], that the development of the concrete admixtures warranting the increase in the resistance to the effect of acidic environment ($pH < 7$) is currently limited to modified polymer dispersions with low additions of substances with the ability to modify more often the physical properties of mixtures than the actual concrete after setting. Products, which are intended for application directly to the concrete, are very limited amount on the market (Czech Republic + EU). It can be said about all of these products that they are a certain form of analogy to the products intended for the surface protection of concrete products with the aim of creating a reliable and long-term effective insulation between the surface of the concrete and the acidic environment.

As the application admixtures warranting the increase in the concrete resistance to the acidic environment, these are always based on the polymer dispersion more often of copolymer frame. Primarily, the polymers and copolymers on an acrylate basis were tested (acrylates, styrene-acrylates, styrene-acrylate esters). Also the before available copolymers of butadiene-styrene were tested (currently elastomeric base), alternatively the copolymers of vinyl alcohol. At the present, in this case, mainly the copolymers of styrene-acrylate are used, since they primarily evince the highest resistance to the effect of the alkaline environment (problems of the strongly alkaline pH of the porous concrete solution) from all mentioned. Their use is now mainly as redispersible dispersion powders when specifying the recipe for the preparation of plasters for thermal insulation systems. The main task of dispersions is hardening of a conventional silicate binder (dispersion has binding properties) and the

IOP Conf. Series: Materials Science and Engineering **324** (2018) 012026 doi:10.1088/1757-899X/324/1/012026

reduction of an usual porosity of the concrete and thus decrease in an absorption capacity and permeability of the concrete for aqueous solutions. The organic connective dispersion base ensures better cohesiveness between the aggregate and the silicate binder during its sloughing due to the acidic solutions.

In our experimental works, we focused on the use of the sealing admixture on the basis of styrene-acrylate in various doses directly into the concrete mixture. The goal is to create the concrete with high tightness, density and resistance to the aggressive media over the whole cross section of the created constructional element.

3. Experimental part

In the experimental program, the test of the long-term monitoring of the effect of formic acid on five concretes with different composition was started, see table 1. The reason for using formic acid is frequent addition directly as the additive to silages and is also used to adjust the pH of water in the drinking basins. The pH of water in the drinking basins is then around 4. In order to achieve the highest density, the low water coefficient was used using the plasticizing admixture and the consistency of SI (subsidence 0 mm) was used intentionally. This consistency is used on some prefabricated components for the production of vibrocompressed goods intended for use in agriculture (troughs, grates, slabs, pillars, etc.). In some recipes, the sealing admixture was used with the goal to determine the influence of this sealing admixture on the resistance to the effect of the acidic environment. The dose of the sealing admixture was chosen with regard to the economic acceptability of the recipes. The parameters of both fresh concrete (temperature, volume weight, subsidence of cone), and the time course of the compressive strength (at the age of 1, 7, 28, 60 and 90 days) and the depth of penetration through pressure water (at the age of 28 days) were monitored (table 2). For each test term, three test specimens were made for each recipe.

Table 1. Composition of concretes.

Components	Recipe designation, doses in kg / m^3				
	I	**II**	**III**	**IV**	**V**
CEM I 42,5 R	390	200	200	200	390
Slag	0	190	190	190	0
Water	135	123	114	114	114
Sand 0/4	1110	1110	1110	1110	1110
Ballast 4/8	600	600	600	600	600
Plasticizer	2.7	2.7	2.7	2.7	2.7
Sealing admixture	0	0	2	7	7
Water coefficient, k-value for slag = 0.6	0.346	0.391	0.369	0.358	0.310
Water coefficient, k-value for slag = 1	0.346	0.315	0.297	0.310	0.310

The concrete compressive strength was determined at cubes with an edge of 150 mm according to the norm ČSN EN 12 390-3 [9, 10]. The depth of penetration through the pressure water was also performed on the cubes with the edge of 150 mm according to the norm ČSN EN 12 390-8 [11]. For the long-term monitoring of the concrete in the acidic environment, three cubes with the edge of 150 mm were made from each recipe. These cubes were placed in the 500 L tank with formic acid, whose pH is maintained at the value of 3 continuously (in practice, the pH ranges from 2.8-3.2), after 28 days of aging. The pH value is checked at weekly interval using a pH meter. The depth of degradation is observed on the cubes, the weight loss is monitored and the photo-documentation of the appearance change is performed.

Table 2. Properties of proposed concretes.

Properties	Recipe designation				
	I	**II**	**III**	**IV**	**V**
Subsidence of cone [mm]	0	0	0	0	0
Temperature of fresh concrete [°C]	22.5	21.2	21.0	21.3	23.0
Volume weight [kg/m^3]	2260	2250	2280	2260	2260
Depth of penetration [mm]	12	12	9	6	6
Compressive strength [MPa] at the age of:					
1 day	15.6	8.3	7.5	7.9	16.3
7 days	43.0	27.6	30.7	26.5	44.1
28 days	48.0	52.2	48.1	45.6	47.9
60 days	51.0	54.0	57.3	51.5	50.6
90 days	53.5	54.9	57.5	55.5	52.5

Note: For this type of the mixture is more suitable to use the VeBe test for the determination of the consistency of the fresh concrete.

A visual description of the concrete degradation of the test specimens (the cube with the edge of 150 mm) after 90 days of exposure:

- I and V – significant colour changes on the surface – "yellowing", surface degradation to the depth of about 1,5 mm ('figure 2')
- II to IV – small colour changes on the surface, surface degradation to the depth of about 0.5 mm ('figure 3')
- I x V – an influence of the sealing admixture was not observed
- II x III x IV – an influence of the sealing admixture was not observed

Figure 1. The surface of the reference specimen

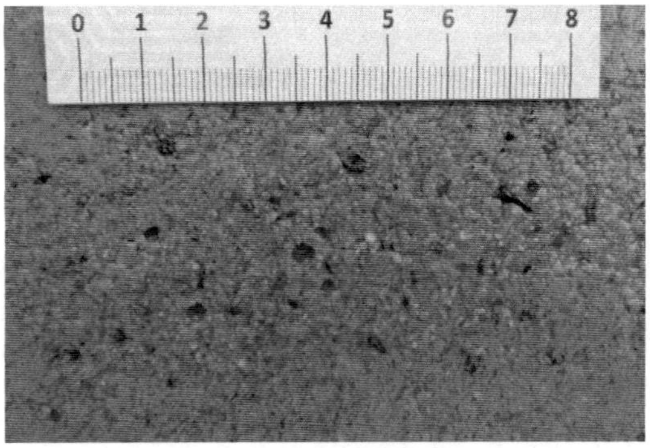

Figure 2. Surface of the test specimens I and V

Figure 3. Surface of the test specimens II, III, IV

4. Conclusion

Based on the results of the concrete compressive strength, it is possible to state that the concrete can be classified into the strength class C 30/37 and should satisfy for the classification into the degree of the environmental influence XA3. It should be noted that the results are, to a certain extent, influenced by the used dose of mixing water, which was processed so that the mixtures have the same workability. The slag has the positive influence on the reduction of mixing water dose, the influence of the sealing admixture on the decrease of mixing water is not significant (2-4 L/m^3). The rate of increase of compressive strength in the recipes with slag (II – IV) is slower (especially in the age of 1-7 days), and all recipes reach the similar strengths at the age of 28 days. Thanks to latent hydraulicity is the strength in the slag recipes at the age of 90 days higher approx. by 5 % in comparison with the recipes without slag. The influence of slag on the depth of penetration through the pressure water was not observed.

The sealing admixture has the influence in particular on reduction of the depth of penetration of pressure water up to 50 %. Already after two months, when the samples were exposed to the acidic environment (formic acid, pH = 3), the differences in the progress of degradation of the concrete surface are evident. In the concretes without slag, there was the significant colour change "yellowing" and washing out the concrete surface up to the depth of about 1 mm. The influence of the sealing admixture on reducing the degradation progress has not been demonstrated yet.

It can be stated that already after a relatively short exposure of the concretes in the simulated environment of the silage pits is evident that despite the design of the concrete in the highest degree of the environmental influence (XC4, XF4, XA3) there is the relatively rapid degradation (decomposition of cement lute) of the surface layer of the concrete due to the influence of the acidic silage juices. It is necessary to mention that the highest degree of the chemical influence of the environment XA3 is designated for the environment with the pH $\geq$ 4, where the environment of the silage juices is significantly more aggressive, the pH is usually 3-4. The concrete degradation test in the experimental tank will continue, with the goal to determine the rate and progress of degradation of the concrete and the comparison of the influence of the slag and the used sealing admixture and its dose.

5. References

[1] Gebrehanna M M, Gordon R J, Madani A, VanderZaag A C and Wood J D 2014 Silage Effluent Management: A Review *Journal of Environmental Management* **143** 113-122.
[2] Beddoe R E and Schmidt K 2009 Acid Attack on Concrete - Effect of Concrete Composition; Part 1 *Cement International* **7** 88–94.
[3] EN 206:2014 Concrete – Specification, performance, production and conformity
[4] ČSN P 74 0423 Concrete – Specification, performance, production and conformity – Additional information
[5] Al-Tamimi A K and Sonebi M 2003 Assessment of Self-Compacting Concrete Immersed in Acidic Solutions *Journal of Materials in Civil Engineering* **15** 354-357.
[6] Sernad A, Yazici H, Yigiter H and Baradan B 2007 Sulfuric Acid Resistance of High-volume Fly Ash Concrete *Building and Environment* **42** 717-721.
[7] Monteny J, De Belie N, and Taerwe L 2003 Resistance of Different Types of Concrete Mixtures to Sulfuric Acid *Materials and Structures* **36** 242-249.
[8] Sokolowska J J, Woyciechowski P and Adamczewski G 2013, Influence of Acidic Environments on Cement and Polymer-cement Concretes Degradation *Advanced Materials Research*, **687** 144-149.
[9] EN 12390-3:2009 Testing hardened concrete - Part 3: Compressive strength of test specimens.
[10] EN 12390-7:2009 Testing hardened concrete - Part 7: Density of hardened concrete.
[11] EN 12390-8:2009 Testing hardened concrete - Part 8: Depth of penetration of water under press.

Acknowledgment

The paper was drawn up with the assistance of TACR (Technology Agency of the Czech Republic) grant TH02020690 "The concrete with excellent resistance under aggressive media condition".

IOP Conf. Series: Materials Science and Engineering **324** (2018) 012027 doi:10.1088/1757-899X/324/1/012027

Influence of CuO nanoparticle on palm oil based alkyd resin preparation and its antimicrobial activity

Huei Ruey Ong[1]*, Md. Maksudur Rahman Khan[2], Ridzuan Ramli[3], Chi Shein Hong[2] and Rosli Mohd Yunus[2]

[1] Faculty of Engineering and Technology, DRB-HICOM University of Automotive Malaysia, 26607 Pekan, Pahang, Malaysia.
[2] Faculty of Chemical and Natural Resources Engineering, Universiti Malaysia Pahang, Lebuhraya Tun Razak, 26300 Gambang, Pahang, Malaysia.
[3] Malaysian Palm Oil Board (MPOB), No. 6, Persiaran Institusi, Bandar baru bangi, 43000 Kajang, Selangor, Malaysia.
*E-mail: hueiruey@dhu.edu.my; roi_rui86@hotmail.com

Abstract. An alkyd resin has been synthesized from palm oil that reacted with glycerol and phthalic anhydride by alcoholysis-polyesterification process and co-catalyzed by CuO nanoparticle. The CuO nanoparticle was pre-prepared in the glycerol via sol gel method, which creates a new reaction condition for resin preparation. The resins were characterized by fourier transform infrared spectroscopy (FTIR), where a new ester linkage bond (C-O-C) was noticed for resin sample. The antimicrobial activity and the curing behaviour of the resin were determined by Kirby-Bauer and differential scanning calorimeter technique. It was found that, the addition of CuO speeded up the reaction rate and played antimicrobial role. Moreover, it shortens the reaction time of alcoholysis and polyesterification process.

1. Introduction

In recent years, intensive research on the development of polymers from renewable resources has been triggered due to the environmental concerns [1]. Vegetable oils are one of the most favourable options to use for polymer development. This is due to its desirable properties such as cheap, ready availability and biodegradable [2, 3]. Alkyd resin, oil modified polyester which widely applied in paint, varnishes, lacquers, adhesive, composite making etc. [4-6] Alkyd is a mixture comprised of polyol, dicarboxylic acid or anhydride, and fatty acids. The common used polyol and diacid component in alkyd preparation was glycerol and phthalic anhydride, respectively [7]. Synthesis of alkyd involved a two-step process, Alcoholysis-polyesterification reaction, the first process which is known as alcoholysis referred as "monoglyceride process" [8]. This reaction consists of triglyceride (TG) and glycerol, which yield required amount of monoglyceride (MG) for later polyesterification process in alkyd synthesis. Simple Brönsted base catalyst commonly is used to conduct the reaction at 240-270 °C. Two kinds of catalyst system which practically used, homogeneous catalyst ($Ca(OH)_2$, KOH and NaOH) and heterogeneous catalyst (MgO, CaO, ZnO/alumina, calcium acetate and calcium carbonate). Homogeneous catalysed processes yield usually 40-60% of MG [9, 10] whereas the heterogeneous catalysed reaction yield 32-39% of MG [11]. Obtaining high yield of MG (at least > 40%) is important for the homogeneity of the reaction mixture and proceed to polyesterification [12]. Monoglycerides and diglycerides generated during alcoholysis take part in polymerization with diacid component. Furthermore, the reaction period required in acquiring maximum MG yield of heterogeneous catalysed is longer compared to homogeneous catalysed reaction. Lowering the alcoholysis period is crucial to yield a low cost process. Ferretti, et al. [9] used MgO heterogeneous catalyst in MG preparation and found that ~77% of MG was achieved after two hours of reaction. However, the utility of MgO in polyesterification has not been explored. It is well known that, nanosize particle with high surface area is a promising catalyst. Thus, the combination of nanosize catalyst in homogeneous catalysis system can induce a novel catalyst system for alkyd resin preparation. Apart from that, remaining of nano-catalyst in the matrix after polyesterification can act as a property enhancing agent. However, the catalyst act as an antimicrobial agent is not reported yet. In recent years, metal oxide nanoparticle such as zinc oxide (ZnO) [13], titanium dioxide (TiO_2) [14], copper oxide (CuO) [15] has raised the attention as antimicrobial agents as they exhibits good antimicrobial property with minimum concentration [16] and considered as toxic free material [17]. Among them, CuO attracts the most attention due to its catalytic reactivity and low cost [18, 19]. Usually, the inclusion of nanoparticle into polymer was through mechanical mixing technique [13, 15, 20]. Delgado, et al. [15] used copper (Cu) and CuO nanoparticles as antimicrobial agent in polypropylene. They found that, CuO possessed better antimicrobial property than Cu alone. Ebrahimiasl, et al. [13] also incorporated ZnO nanoparticle in polypyrrole/chitosan as antimicrobial agent where it effectively against all the studied bacteria. Our attempt is to study the combined system comprised of liquid bases and CuO nanoparticle in palm oil based alkyd preparation. Oil palm tree is one of the most planted agriculture crop in Malaysia and Malaysia is one of the top palm oil producers in the world [21, 22].

In the present study, a novel catalyst system comprising with NaOH and CuO nanoparticles has been studied for alkyd preparation. The CuO nanoparticle was synthesized in glycerol and the antimicrobial property of nanoparticle included resin has been inspected.

2. Experimental section

2.1. Materials

Copper (II) chloride salt ($CuCl_2 \bullet 2H_2O$), hydrazine monohydrate (64%), ethanol (99.9%), anhydrous methanol, sodium hydroxide, Methyl ethyl ketone peroxide (MEKP), cobalt octoate, glycerol, dimethyl sulfoxide (DMSO), copper (II) oxide (97%), copper (I) oxide (99.99%), *Escherichia coli* (ATCC 8739), *Pseudomonas aeruginosa* (ATCC 9027), *Staphylococcus aureus* (ATCC6538) and *Salmonella sp.* (ATCC 14028) were obtained from Sigma-Aldrich, Malaysia and used without further purification. Refined palm oil was provided by Malaysian Palm Oil Board (MPOB), Malaysia.

2.2. Preparation of CuO nanosol

The CuO nanosol was prepared by sol-gel method at ambient condition [18, 23-25]. Initially, the precursor was prepared by dissolving vital amount of $CuCl_2 \bullet 2H_2O$ in 44 mL of glycerol. The reducing agent solution at 7.060 mM was ready by mixing with ethanol. Then, ~6 mL of reducing agent was injected into the precursor solution and stired for 10 h. The concentration of copper precursor in the solution was maintained in the range of 100–300 mg/L and used for current study.

2.3. Preparation of alkyd resin

Alkyd was prepared by alcoholysis-polyesterification reaction involving palm oil, glycerol and phthalic anhydride. Initially, 100g of oil was heated to 240 °C and subsequently added 22.2 g of glycerol into the oil with constant stirring and inert condition. Afterward, 0.3 g of NaOH was introduced into the mixture. The end point of alcoholysis reaction was determined by methanol solubility test[26]. After the end point acquired, the reaction mixture was cool down to 140 °C. Then, fine phthalic anhydride (38.7 g) was supplied into the system to kick start the polyesterification reaction and the reaction was controlled under 240 °C and inert condition. During polyesterification reaction, acid value (AV) followed ASTM D 1639-90 was used to monitor the reaction progress [27]. The reaction mixture was quenched to room temperature when the AV reached ~ 10. For the preparation of alkyd resin catalysed by combined system, pure glycerol was substituted by the CuO nanosol in several concentrations, i.e. 0, 100, 200 and 300 mg/L and labelled as A0, A100, A200 and A300, respectively.

2.4. Calculation of extent of reaction

The acid value of in-process samples was withdrawn and determined by titrating method. The extent of the reaction, P with respect to the acid value was calculated using the following equation[28, 29]:

$$P = (C_0 - C_t) / C_0 \qquad (1)$$

where C_0 is the initial acid value and C_t is the acid value at time, t of the reaction.

2.5. Fourier transform infrared spectroscopy (FTIR)

The structural information of the sample was attained by FTIR Spectrometer, Imprestige–21 model, Schimadzu Co., Japan, equipped with an attenuated total reflectance (ATR) device in the wave number range 500–4500 cm^{-1} and resolution 4 cm^{-1}. The FTIR spectra were taken in a transmittance mode.

2.6. Antimicrobial test

The antimicrobial property of alkyd samples synthesized by A0 and A200 catalyst system was examined by Kirby-Bauer Method [30]. A number of microorganisms, such as *Escherichia coli* (ATCC 8739), *Pseudomonas aeruginosa* (ATCC 9027), *Staphylococcus aureus* (ATCC6538) and *Salmonella sp.* (ATCC 14028) was used. The cultures were maintained in a nutrient broth and subcultured at regular intervals. The samples were prepared (10 mg/mL) using 0.5% DMSO. The microbial cultures in nutrient agar plates were inoculated using spread plate technique and cork borer was used to make the hole on the agar. The samples (10 mg/mL) were loaded in the hole and incubated at 32.5 °C for 24 h. A blank sample was prepared separately as control. The diameter of the inhibition zone after 24 h incubation was measured.

2.7. Differential scanning calorimeter (DSC) measurement

The curing of the resin was monitored by DSC (model-TA-Q1000), under nitrogen purge gas. Prior to that, 1 g of resin was mixed with 0.04 g of MEKP and 0.02 g of cobalt octoate. The mixture was agitated manually and heated at 65 °C for 15 min. Thereafter, ~5 g of the mixture was put into a DSC sample pan and covered with an aluminum lid and closed tightly under pressure. The sample pan was placed in the DSC sample cell at ambient temperature, and an empty pan was used as reference cell. The DSC scan started from room temperature to 400 °C at 10 °C/min.

3. Results and Discussion

3.1. Preparation of alkyd resin

Methanol solubility method has been used to inspect the end point of alcoholysis reaction. Figure 1 exhibited the required time for end point of each catalyzed reaction. It can be seen that, combined catalyst system (A100, A200 and A300) significantly lowered the reaction period. Among them, A200 exhibited shortest reaction period. Beyond the 200 mg/L CuO limits, no significant performance was noticed. End point determination is important to control the alcoholysis reaction time, to avoid the intensification of polyglycerols formation [26].

Figure 2 exhibited the extent of reaction of different catalysed reactions. As seen, the extent of reaction increased as the reaction proceeded. It is noticed that, the early stage of reaction rate was faster than that of later stages. Comparable experiment curve were also reported elsewhere [27, 31-33]. These changes in extent of reaction during polyesterification is due to the different reactivity of primary and secondary –OH group in glycerol [34]. It is well known that, primary –OH react faster than that of secondary –OH. Thus, the fast reaction rate during early stage might be due to primary –OH, whilst secondary –OH at later stage [31]. In addition, 3D network in alkyd might be formed during later stages of the reaction. This might be due to alkyd chains cross-linking formation as the viscosity of the resin increased [35].

Apart from that, CuO addition is significantly affect the reaction during the early stage of the reaction. It can be seen that the initial reaction rates of combined system (A100, A200 and A300) are significantly higher compared to A0. The initial rates of A200 and A300 were almost alike. In addition, A200 possessed earliest equilibrium (60 min) compared with others (Figure 2). To this, A200 system has been designated as optimum.

3.2. IR analysis of alkyd resin

Figure 3 illustrated the FTIR spectra of oil and alkyd resin. It can be seen that, the oil spectrum was significantly different compared to alkyd. A wide band was observed at 3550-3200 cm^{-1} which assigned to O-H stretching in oil spectrum (Figure 3, i) [21, 36-40]. A characteristic absorption was detected at 3050-2800 cm^{-1} assigned to C-H stretching [41, 42]. The absorption bands at 1750-1720 cm^{-1} was endorsed to C=O. The C-O-C stretching in oil was noticed in the range of 1250-1050 cm^{-1} [43]. The C-H bending was observed at 1470-1440 cm^{-1}. Additionally, an characteristic peak at 722 cm^{-1} was noticed might be due to the existence of unsaturation in fatty acid [44].

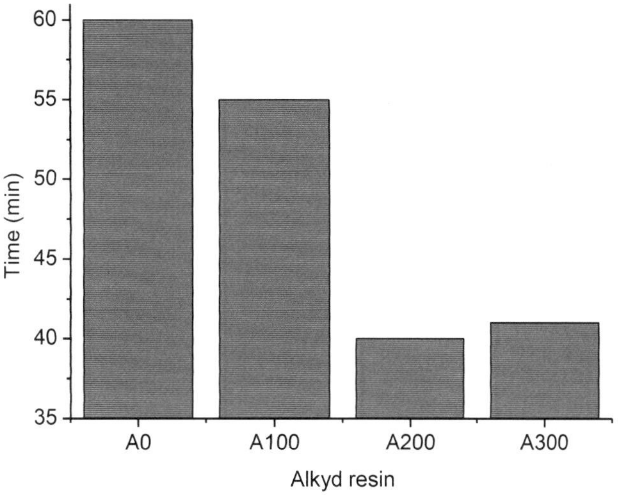

Figure 1. End point determination of alcoholysis for different catalyst system

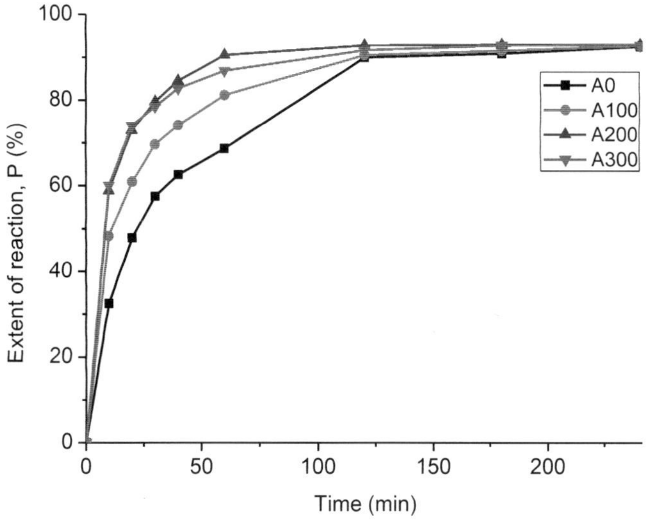

Figure 2. Progress of extent of reaction for different catalyst systems at polyesterification process (■A0, ●A100, ▲A200 and ▼A300).

The existence of ester linkages and other distinguish peaks in alkyd spectra was noticed (Figure 3, ii and iii). A new characteristic peak of C-O-C stretching at 1300-1250 cm^{-1} was appeared [33], which absent in oil spectrum. The C=O stretching of alkyd has shifted from 1748 cm^{-1} (oil's spectrum) to 1735 cm^{-1}. The C-H bending of aromatic ring of phthalate was visualized at 743 cm^{-1} [42]. It ensured the formation of alkyd resin [27, 41]. It is noticed that, the stretching frequencies of A0 (Figure 3, ii) and A200 (Figure 3, iii) samples were alike. Though, a slight shifts of C-O-C stretching (1270 cm^{-1} to 1266 cm^{-1}) and C-H bending of aromatic ring (743 cm^{-1} to 735 cm^{-1}) were detected. This phenomenon might be due to the interaction between polymer and nanoparticle [45].

3.3. Antimicrobial study

Figure 4 showed the antimicrobial property of A0 and A200 towards microorganism. The property of antimicrobial was determined based on the inhibition zone measurement. The inhibition zone of A200 sample were 23, 25, 15 and 40 for *Escherichia coli, Pseudomonas aeruginosa, Staphylococcus aureus* and *Salmonella sp.*, respectively, while A0 sample did not display any inhibition zones, indicating A200 have better antimicrobial property than A0. However, the bactericidal mechanism of nanoparticle toward microorganism is unknown. The disruption in the bacterial cell wall which resultant the cell death might be due to the binding between nanoparticle and the bacterial cell wall. Apart from that, the linkage can also form between nanoparticle with electron donor group like sulphur and nitrogen from bacterial molecule [46-48]. In addition, the bacterial cell metabolism and function like permeability and respiration might be disturbed by the copper-peptide linkage formed, which resultant the death of cell [48, 49]. In overall, the reduction rate of Gram negative bacteria is higher than Gram positive, due to their slimmer wall, which allow the easy absorption of the ions into the cell [50, 51].

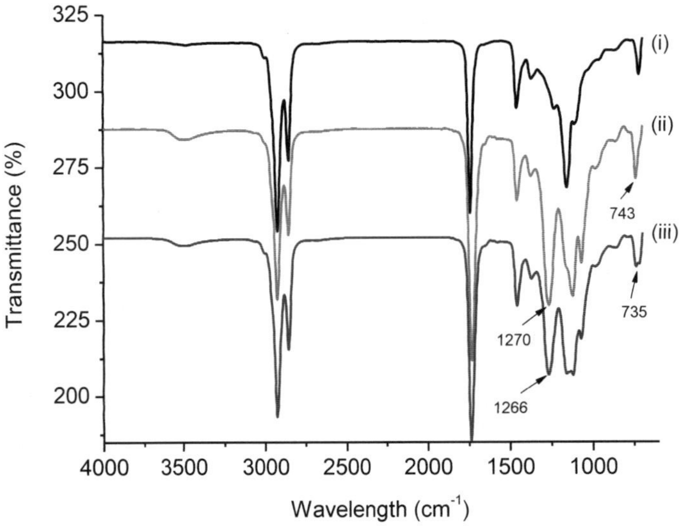

Figure 3. FTIR spectra of (i) oil, (ii) A0 and (iii) A200.

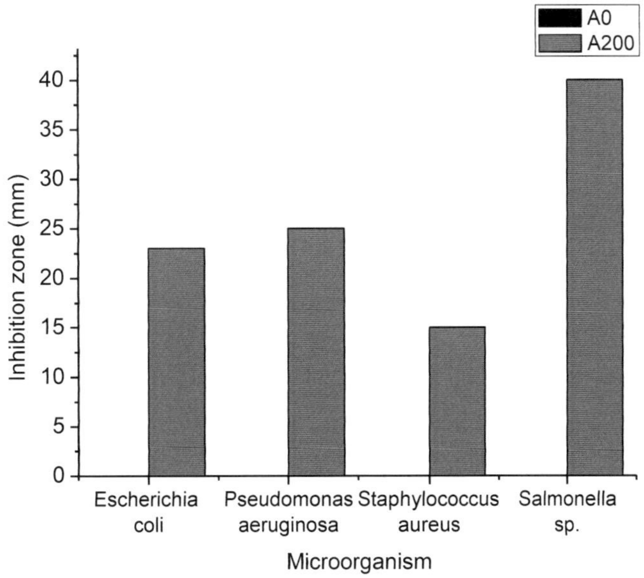

Figure 4. Antimicrobial activity of A0 and A200

3.4. Curing behavior

The DSC scans of the cure schedule for sample A0 and A200 were shown in Figure 5. A range of maximum cure temperature from 250 to 385 °C is depicted for the samples. The exothermic peak of sample A0 and A200 were 335 and 315 °C, respectively. It appears that CuO nanoparticles have an influence on the total heat released by the samples. Samples A200 have smaller exothermic heat compared to A0. The result obtained was in agreement with Ghaemy, et al. [52], Macan, et al. [53], which reported that nanoparticles able to lower the exothermic heat during the curing. This phenomenon occurred due to the absorption some of the heat by the nanoparticle [52]. Therefore, the presence of CuO nanoparticle in the alkyd resin can enhance the curing properties.

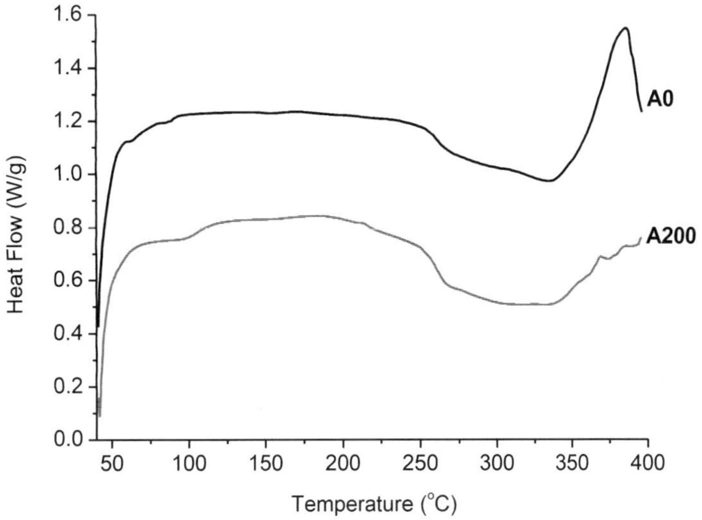

Figure 5. DSC curves for the curing of sample A0 and A200.

4. Conclusion

Palm oil based alkyd resin has been synthesized using the novel combined system. The addition of nanoparticle into the NaOH catalytic system improved the alcoholysis-polyesterification process performance. The antimicrobial activity of A200 was better than A0. The curing property of resin was improved by CuO nanoparticle via lowering the exothermic heat of the reaction. Therefore, it is evidenced that CuO nanoparticle can worked as co-catalyst for resin preparation simultaneously served as antimicrobial agent for alkyd.

Acknowledgements
The work was monetarily supported by DRB-HICOM University of Automotive Malaysia and Universiti Malaysia Pahang.

References

[1] Zhang C, Y Xia, R Chen, S Huh, P A Johnston, M R Kessler 2013 *Green Chem.* **15** 1477-1484
[2] Pathan S, S Ahmad 2013 *J. Mater. Chem. A* **1** 14227-14238
[3] Ong H R, R Ramli, M M R Khan, R M Yunus 2016 *Prog. Org. Coat.* **101** 245-252
[4] Meneghetti M R, S M P Meneghetti 2015 *Catal. Sci. Technol.* **5** 765-771
[5] Aulin C, G r Ström 2013 *Ind. Eng. Chem. Res.* **52** 2582-2589
[6] Dolui S K 2015 *ACS Sustainable Chem. Eng.* **3** 261-268
[7] Kalita H, S Selvakumar, A Jayasooriyamu, S Fernando, S Samanta, J Bahr, S Alam, M Sibi, J Vold, C Ulven 2014 *Green Chem.* **16** 1974-1986
[8] Ong H R, M M R Khan, R Ramli, R M Yunus, M W Rahman 2016 *New J. Chem.* **40** 8704-8709
[9] Ferretti C A, R N Olcese, C R Apesteguía, J I Di Cosimo 2009 *Ind. Eng. Chem. Res.* **48** 10387-10394
[10] Ferretti C A, A Soldano, C R Apesteguía, J I Di Cosimo 2010 *Chem. Eng. J.* **161** 346-354
[11] Hillion G, S Leporq, J-J Rouxel, R Stern, Reaction of vegetable oil with alcohol in presence of a catalyst, Institut Francais Du Petrole, 1999.
[12] Doulia D, S Rokotas, K Georgopoulou 2006 *Surf. Coat. Int. Pt. B-C* **89** 215-219
[13] Ebrahimiasl S, A Zakaria, A Kassim, S N Basri 2015 *Int. J. Nanomed.* **10** 217
[14] Kubacka A, M S Diez, D Rojo, R Bargiela, S Ciordia, I Zapico, J P Albar, C Barbas, V A M dos Santos, M Fernández-García 2014 *Sci. Report.* **4** 4134
[15] Delgado K, R Quijada, R Palma, H Palza 2011 *Lett. Appl. Microbiol.* **53** 50-54
[16] Brayner R, R Ferrari-Iliou, N Brivois, S Djediat, M F Benedetti, F Fiévet 2006 *Nano Lett.* **6** 866-870
[17] Gordon T, B Perlstein, O Houbara, I Felner, E Banin, S Margel 2011 *Colloids Surf., A* **374** 1-8
[18] Ong H R, M R Khan, R Ramli, R M Yunus 2014 *Appl. Mech. Mater.* **481** 21-26
[19] Chowdhury M N K, A F Ismail, M D H Beg, M R Khan, R J Gohari, M A R B Saidin, M R Islam, H R Ong 2016 *RSC Adv.* **6** 28937-28946
[20] Ozkan E, F T Ozkan, E Allan, I P Parkin 2015 *RSC Adv.* **5** 8806-8813
[21] Ong H R, R Prasad, M M R Khan, M N K Chowdhury 2012 *Appl. Mech. Mater.* **121-126** 493-498
[22] Ong H R, R P D.M., M M R Khan 2016 *Indian J. Chem. Technol.* **23** 39-46
[23] Ong H R, M M R Khan, R Ramli, Y Du, S Xi, R M Yunus 2015 *RSC Adv.* **5** 24544-24549
[24] Ong H R, M R Khan, M N K Chowdhury, A Yousuf, C K Cheng 2014 *Fuel* **120** 195-201
[25] Ong H R, M M R Khan, R Ramli, M W Rahman, R M Yunus 2015 *RSC Adv.* **5** 95894-95902
[26] Igwe I O, O Ogbobe 2000 *J. Appl. Polym. Sci.* **78** 1826-1832
[27] Ezeh I E, S A Umoren, E E Essien, A P Udoh 2012 *Ind. Crops Prod.* **36** 94-99
[28] Bobalek E G, M T Chiang 1964 *J. Appl. Polym. Sci.* **8** 1147-1168
[29] Bobalek E G, E R Moore, S S Levy, C C Lee 1964 *J. Appl. Polym. Sci.* **8** 625-657
[30] Ambika S, S Arunachalam, R Arun, K Premkumar 2013 *RSC Adv.* **3** 16456-16468
[31] Aigbodion A I, F E Okieimen 2001 *Ind. Crops Prod.* **13** 29-34
[32] Odetoye T E, D S Ogunniyi, G A Olatunji 2010 *Ind. Crops Prod.* **32** 225-230
[33] Odetoye T E, D S Ogunniyi, G A Olatunji 2012 *Prog. Org. Coat.* **73** 374-381
[34] Goldsmith H A 1948 *Ind. Eng. Chem.* **40** 1205-1211

[35] Aigbodion A I, F E Okieimen 1996 *Eur. Polym. J.* **32** 1105-1108
[36] Khan M M R, M W Rahman, H R Ong, A B Ismail, C K Cheng 2016 *Desalin. Water Treat.* **57** 14728-14738
[37] Ong H R, M R Khan, A Yousuf, N Jeyaratnam, D M Prasad 2015 *Pol. J. Chem. Technol.* **17** 41-47
[38] Ramli R, M M R Khan, R M Yunus, H R Ong, R M Halim, A A Aziz, Z Ibrahim, N H Zainal 2014 *Adv. Nanopart.* **3** 65
[39] Woon C W, M A Islam, B Ethiraj, H R Ong, C K Cheng, K F Chong, G Hedge, M Khan, M Rahman 2017 *ChemistrySelect* **2** 7637-7644
[40] Ong H R, G Hegde, V Chigrinov, M M R Khan 2016 *RSC Adv.* **6** 15094-15100
[41] Patel V C, J Varughese, P A Krishnamoorthy, R C Jain, A K Singh, M Ramamoorty 2008 *J. Appl. Polym. Sci.* **107** 1724-1729
[42] Cakić S M, I S Ristić, V M Jašo, R Ž Radičević, O Z Ilić, J K B Simendić 2012 *Prog. Org. Coat.* **73** 415-424
[43] Assanvo E F, P Gogoi, S K Dolui, S D Baruah 2015 *Ind. Crops Prod.* **65** 293-302
[44] Wu J, T Zhang, G Ma, P Li, L Ling, B Wang 2013 *J. Appl. Polym. Sci.* **130** 4201-4208
[45] Xu Y, D Chen, X Jiao, K Xue 2007 *J. Phys. Chem. C* **111** 16284-16289
[46] Morones J R, J L Elechiguerra, A Camacho, K Holt, J B Kouri, J T Ramírez, M J Yacaman 2005 *Nanotechnology* **16** 2346
[47] Applerot G, J Lellouche, A Lipovsky, Y Nitzan, R Lubart, A Gedanken, E Banin 2012 *Small* **8** 3326-3337
[48] Booshehri A Y, R Wang, R Xu 2015 *Chem. Eng. J.* **262** 999-1008
[49] Benov L T, I Fridovich 1994 *J. Biol. Chem.* **269** 25310-25314
[50] Rhim J-W, S-I Hong, H-M Park, P K Ng 2006 *J. Agric. Food Chem.* **54** 5814-5822
[51] Singh J, P K Dutta 2011 *Journal of Macromolecular Science, Part A* **48** 246-253
[52] Ghaemy M, M Bazzar, H Mighani 2011 *Chin. J. Polym. Sci.* **29** 141-148
[53] Macan J, H Ivanković, M Ivanković, H J Mencer 2004 *Thermochim. Acta* **414** 219-225

IOP Conf. Series: Materials Science and Engineering **324** (2018) 012028 doi:10.1088/1757-899X/324/1/012028

Adsorption of tobacco-specific nitrosamine 4-(methylnitrosamino) -1- (3-pyridyl)-1-butanone from aqueous solution with graphene aerogel

T G Xu*, R Shi, N Lu, J Zhang, R S Bai, Z D Yang and J Zhou

Beijing Third Class Tobacco Supervision Station, Beijing, 101121, P. R. China
Corresponding author, E-mail:xutongguang@tsinghua.org.cn

Abstract. The adsorption behavior of graphene aerogel in the 4-(Methylnitrosamino)-1-(3-pryidyl)-1-butanone (NNK) aqueous solution was studied. The adsorption kinetics fitted pseudo-second-order model with the rate constant (k_2) of 0.154 g/mg·h. The adsorption isotherm was investigated and fitted Langmuir and Freundlich models well, and the maximum adsorption capacity (q_m) was 59.66 mg/g estimated from Langmuir isotherm. Thermodynamic result indicated that the process of adsorption of NNK onto graphene aerogel was spontaneous and exothermic. Higher pH solution was favorable for NNK adsorption on graphene aerogel. The adsorption for NNK on graphene aerogel arose from the π-π interaction between them, and the high adsorption efficiency was resulted from the $-NO_2$ functional groups. The capability of graphene aerogel was maintained after repeated absorption-desorption cycles, which was benefit for convenient separating and recycling of graphene aerogel.

1. Introduction

4-(methylnitrosamino)-1-(3-pyridyl)-1-butanone (NNK), one carcinogenic tobacco alkaloid derivative, can cause tumors of the nasal mucosa, liver, and pancre to humans[1]. NNK is specifically formed in the tobacco and tobacco products.However, NNK was detected in source water and drinking water recently [2]. Thus, how to reduce the NNK in the environment water and wastewater is very important for the protection of the public health.Much efforts had been developed to reduce NNK in smoke and tobacco extracting solution by porous materials [3-5]. NNK could be adsorbed by zeolites by its N-N=O group inserted into the channels of adsorbent, and then catalytically degraded at high temperature by the fracture of N-NO bond. Moreover, incorporation of metal oxide in zeolites was able to further improve the adsorption and catalytic degradation of NNK [5].

In recent years, graphene, a novel 2D layered material, had received much scientific attentions with the great potential application in many fields. Among them, graphene showed excellent adsorption capacity in the removal of heavy metal ions from aqueous solutions, dye pollutants, fluoride ions, *etc.* [6-8].Comparing with other adsorbent, graphene exhibited the superiority of the selective adsorption ability to aromatic compound through strong π-π dispersion interaction. Consequently, graphene was expected to be a suitable adsorbent for the removal of pollutants with aromatic structures.

Aerogels was consisted of continuous nanopores forming open three dimensional networks, and in favour of providing effective mass transfer, which could further improve the adsorption behavior under suitable conditions [9,10]. Individual 2D graphene sheets were prone to stack in a parallel, resulting in the sacrifices of accessible surface area. However, 3D macroscopic assemblies could avoid the defect and exhibit better advantages, more paths with easy access and diffusion of ions and molecules could be provided in the application of catalysis, sensors, and so on. For this purpose, in this work, graphene aerogel was used as the adsorbent to remove NNK in aqueous solution. The adsorption capacity and behavior of graphene aerogels toward NNK, including adsorption kinetics, adsorption isotherms and thermodynamic parameters were systematically studied.

2. Experimental method

Graphite oxide(GO) was prepared by the modified Hummers' method from the natural graphite with concentrated H_2SO_4 and $KMnO_4$ as oxidant[11]. Typically, the graphene aerogel was prepared form GO suspensions with vitamin C as reducing agent. 100 mg of graphite oxide was added into 100 mL of distilled water to form homogeneous GO dispersion aqueous (2 mg/mL) under ultrasonic condition. Then, vitamin C was added to the dispersion with weight ratio of vitamin C to GO being 10:1. The dispersion was sealed in bottle and maintained at 95 °C for 2 h without stirring. The resulting sample was washed three times to remove impurities with water and ethanol and dried under vacuum condition.

The XRD patterns of the as-prepared sample were measured on a Bruker D8-advance powder diffractometer with Cu Kα radiation (λ = 1.5418 Å). Raman spectra were recorded with a HORIBA HR800 microscopic confocal spectrometer with a 514.5 nm argon laser. Field emission scanning electron micrographs of the as-prepared sample were observed on Hitachi SU-8010 field emission scanning electron microscope. Chemical compositions were analyzed with ULVAC-PHI, Quantera X-ray photoelectron spectroscopy (XPS). The charging effect was calibrated using the binding energy of C1s. Nitrogen adsorption/desorption was measured on a Micromeritics ASAP-2020 instrument at 77 K, the specific surface area of the samples was calculated by Brunauer-Emmett-Teller (BET) method.

Batch adsorption experiments were performed to evaluate the adsorption capacities of the graphene aerogel for NNK. The dependence of the initial solution concentration, solution pH, equilibration time and temperature dependent on NNK adsorption was investigated. 10 mg of graphene aerogel was added into 100 mL of NNK solution and shaken at room temperature with different NNK concentration (0.1-10 mg/L). The influence of pH in the range of 2-14 during the adsorption process was estimated with initial NNK concentration of 5 mg/L. The concentrations of NNK were determined on Spark Holland liquid chromatography coupled with TSQ Quantiva triple quadruple mass spectrometer.

3. Results and discussion

The XRD patterns of graphite and graphene aerogel was shown in Figure 1. As for the graphite, a sharp diffraction peak at 26.5° could be found which corresponded to the (d_{002}) interlayer spacing along the *c*-axis of 0.336 nm, disappeared in the grapheme aerogel. The relatively weak and broad diffraction peak at 25.1° was the typical diffraction peak of graphene, corresponding to the (002) interlayer spacing of 0.353 nm. It could be seen that the interlayer spacing of graphene aerogel was slightly higher than that of graphite sample, indicating the presence of residual oxygen-containing groups in graphene aerogel and the existence of π-π stacking between graphene sheets [12]. Apparently, the broad XRD peak of graphene aerogel reflected that the graphene sheets along their stacking direction was the poor ordering, and the framework of graphene aerogel was composed of few-layer stacked graphene sheets.

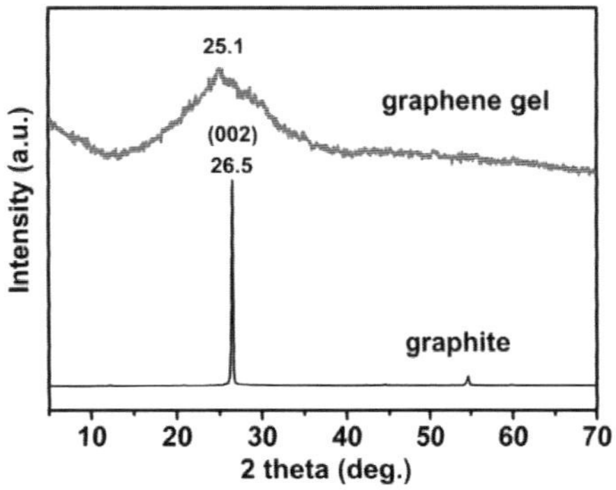

Figure 1. XRD patterns of graphene aerogel.

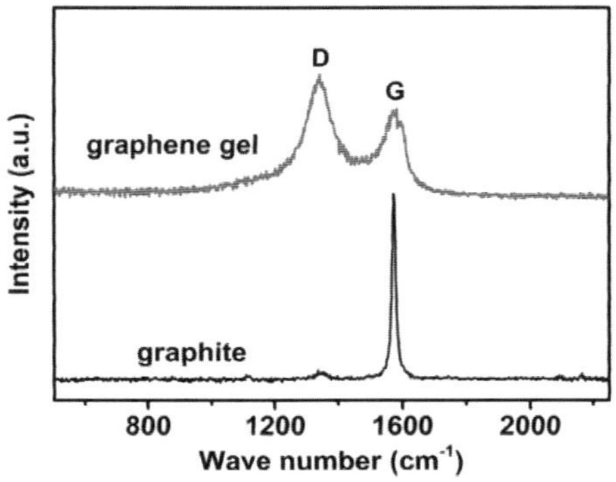

Figure 2. Raman spectra of graphene aerogel.

The Raman spectrum (514 nm excitation) of graphene aerogel sample was shown in Figure 2. As is shown, two Raman modes located at ca. 1350 and 1580 cm^{-1} in the spectrum of the graphite sample was found, showing the existence of D band and G band. It was believed that G band corresponded to the E_{2g} mode for sp^2 graphitic carbon, and D band was associated with amorphous carbon indicating the existence of structural defects. Therefore, the intensity ratio of I_D/I_G could reflect the disorder degree in crystal structure. The low I_D/I_G value (0.04) of pure graphite indicated the existence of intact graphite domains. For the graphene aerogel, the intensity ratio of I_D/I_G increased to 1.41 which suggested localized disorder defects formed within sp^2 carbon network [13]. Based on the XRD and Raman results, it was conceivable that both highly disordered and randomly arranged graphene sheets existed in graphene aerogel.

The surface functional groups of obtained graphene aerogel were identified by XPS to provide the direct proof for the reduction of GO. The survey spectrum showed that the graphene aerogel contained C, O elements with the binding energy of about 532.1 and 284.2 eV respectively (figure 3a). The deconvoluted XPS peaks of C 1s exhibited the presence of four different carbon groups: the non-oxygenated C (C–C), the hydroxyl groups (C–OH), the carbonyl groups (C=O), and carboxyl groups (O=C–O) [14], with the binding energies of 284.7 eV, 286.4 eV, 288.2 eV and 289.1 eV, respectively (figure 3b). The XPS spectra confirmed that there were still some oxygen functional groups on the surface of graphene aerogel, arising from the incomplete reduction reaction of GO dispersion aqueous under the present conditions.

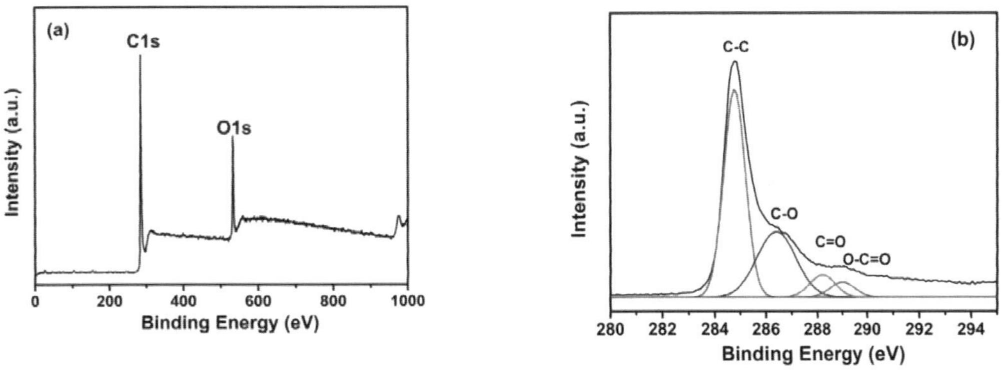

Figure 3. XPS spectra: (a) wide scan and (b) C1s spectra of graphene aerogel.

The morphological structure of graphene aerogel was characterized by SEM. The SEM images showed that graphene aerogel was composed of ultrathin grapheme nanosheets with a porous structure to form well-defined and interconnected structure (figure 4a). The partial overlaps and wrinkles of thin film were the evidence of graphene nanosheets. The sizes of the thin film ranged from several hundreds of nanometers to micrometers (figure 4b). In the 3D macroscopic graphene aerogel, flexible dispersed graphene nanosheets were in contact with each other to form the network structure, and then increasing the bonding interactions. As a result, self-assembly occurred to form 3D architechture and porous structure in graphene aerogel, which was driven by the increased hydrophobicity, π-π stacking interactions, and the cross-links between the remained oxygenated groups on the graphene sheets.

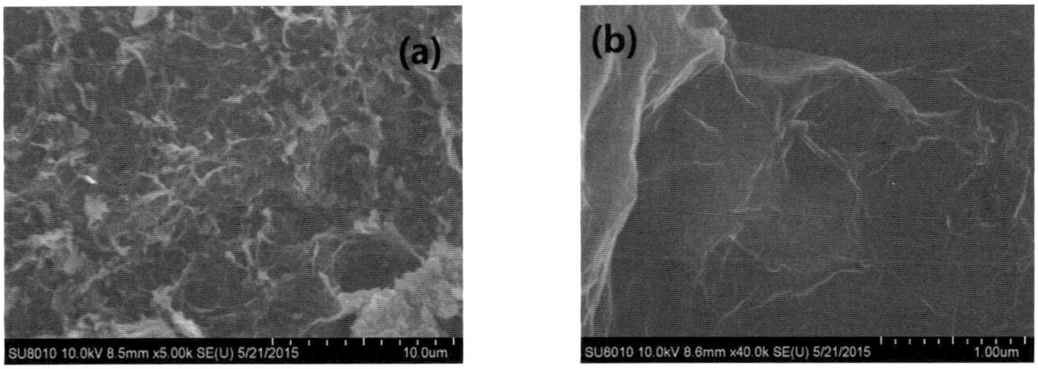

Figure 4. SEM images with different magnifications of graphene aerogel (a,b).

The specific surface area and porosity of the resulting graphene aerogel were revealed by nitrogen adsorption-desorption investigation (figure 5). The typical Type IV isotherm characteristic with adsorption hysteresis was observed in the adsorption-desorption isotherm of graphene aerogel, indicating the presence of mesoporosity. The pore size distribution curves suggested the presence of broad pores between 20 nm and 120 nm calculated from Barrett-Joyner-Halenda (BJH) model, which could be attributed to the interparticle pores from graphene sheets accumulation. It was worth mentioning that the specific surface area of the graphene aerogel was 133 m^2/g calculated from the Brunauer-Emmett-Teller (BET) method, which was lower than those of reported grapheme aerogrels. The relatively low specific surface area could be ascribed to the incomplete exfoliation and aggregation during the reduction reaction process.

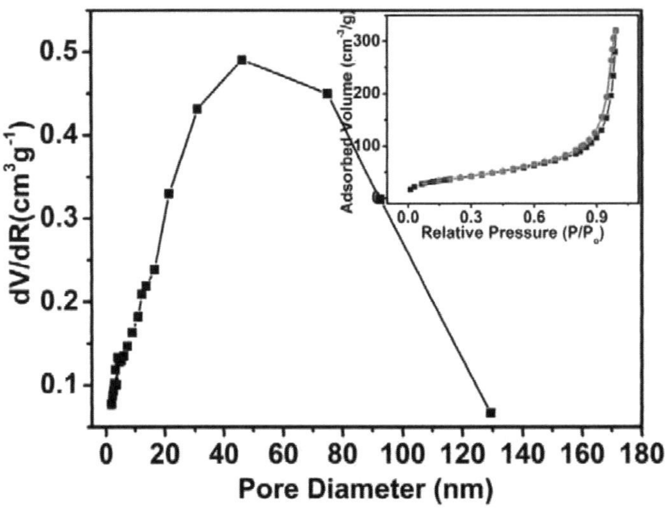

Figure 5. BJH pore size distribution curve and N_2 adsorption-desorption isotherm (inset) of graphene aerogel.

The macroporous structure with open pores and large pore volumes endow aerogels with good absorbing properties. The adsorption kinetics of graphene aerogel was investigated by detecting the decrease of NNK concentration. Figure 6 showed the effect of contact time of NNK adsorbed by graphene aerogel. As was shown in figure 6, the adsorption capacity increased quickly and then reached the adsorption equilibrium within 2 h, suggesting that the adsorption process of NNK on graphene aerogel was not a constant speed and might be divided into two stages: fast stage and slow stage. For the first 2 h, up to 90% of equilibrium absorption amounts of NNK in the solution were attained indicating most of the NNK molecule was absorbed to the surface of graphene aerogel at a fast ratebecause of the plenty of available adsorption sites on the graphene aerogel. There was no significant increase of adsorption capacity for graphene aerogel with the time prolonging to 4 h, which could be ascribed to the decrease of adsorption sites and the inhibition of absorbed NNK molecules with other molecules. The high adsorption rate at initial stages might be due to the existence of large number of adsorption sites, and then were saturated with the time prolonging. In addition, it was noticed that graphene aerogel floated on the surface of solution due to hydrophobic property during the absorption process. This property was important for the convenient separating and recycling of graphene aerogel.

The pseudo-second-order kinetics model was applied to analyze the kinetic adsorption of NNK on graphene aerogel. The pseudo-second-order model was described as [15]:

$$\frac{t}{q_t} = \frac{1}{k_2 q_e^2} + \frac{t}{q_e} \quad (1)$$

where q_e (mg/g) is the amount of NNK absorbed at equilibrium; q_t (mg/g) is the adsorption capacity at various time; k_2 is the pseudo-second-order rate constant (g/mg·h). Therefore, the values of q_e and k_2 were calculated from the slope and intercept of the linear plot of t/q_t. The calculated kinetic constant (k_2) and correlation coefficient (R^2) were given in Figure 6. It can be seen that the adsorption kinetic constant for NNK adsorption on graphene aerogel was 0.154 g/mg·h, and the pseudo-second order model was properly fitted with a high R^2 value of 0.989. Hence, the adsorption process of NNK on graphene aerogel could be described by the pseudo-second-order, implying the adsorption rate was depended on the concentration of adsorption sites of graphene aerogel. For the reason that external diffusion, adsorption, and internal particle diffusion were included in all the steps of adsorption for pseudo-second-order model, it could be concluded that the limiting step of adsorption rate was the chemisorption (i.e., valence forces) between graphene aerogel and NNK. The calculated adsorption amount ($q_{e,cal}$) of NNK on graphene aerogel obtained from the model was 11.53 mg/g, which was consistent with the experimental adsorption capacity ($q_{e,exp}$), implying that the adsorption for NNK was determined by graphene aerogel. Those results indicated that NNK uptake onto graphene aerogel was in accordance with the pseudo-second-order kinetic model.

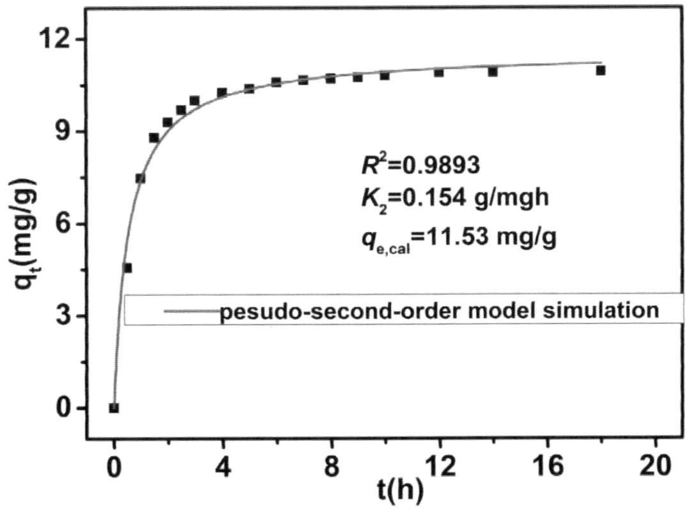

Figure 6. Effect of contact time on the adsorption of NNK by graphene aerogel (10 mg of graphene aerogel and 100 mL of 1 mg/L NNK at 298.15 K, pH = 7).

The adsorption isotherm was obtained with different initial concentrations of NNK (0.1 to 10 mg/L). In the present work, Langmuir and Freundlich model were used to describe the adsorption equilibrium. The expression for the Langmuir model was as follows:

$$\frac{c_e}{q_e} = \frac{1}{q_m} c_e + \frac{1}{q_m K_L} \qquad (2)$$

where c_e is the equilibrium concentration of NNK (mg/L); q_e is the adsorbed NNK amount on graphene aerogel (mg/g); q_m is Langmuir constant (maximum adsorption capacity, mg/g), it was assumed that monolayer coverage of the adsorption surface and no subsequent interaction among adsorbed substances in the Langmuir model, thus, q_m is corresponded to complete monolayer coverage; K_L is the Langmuir constant (L/mg), reflecting the affinity of the adsorbate with the adsorbent. A is obtained when q_m and K_L can be calculated from the straight line, which c_e/q_e was plotted against c_e.

Freundlich model, an empirical model allowing for multilayer adsorption, was described as:

$$\ln q_e = \frac{1}{n} \ln c_e + \ln K_F \qquad (3)$$

K_F is the Freundlich adsorption constant, an indicator of the adsorption capacity; 1/n is an empirical parameter of adsorption intensity. 0<1/n<1 suggested favorable adsorption and the adsorption capacity increased and new adsorption sites form [16].K_F and n can be calculated from the intercept and slope of the linear plot of q_e versus ln c_e.

The simulations of Langmuir and Freundlich isotherms at three different temperatures were shown in Figure 7. The adsorption capacity increased with the increasing equilibrium concentration of NNK and then reached equilibrium progressively, verifying the existence of surface diffusion and intra-particle diffusion in the adsorption process. Besides, the increase of NNK concentration could facilitate the diffusion of NNK molecules onto the graphene aerogel. The parameters of Langmuir and Freundlich adsorption isotherms of NNK on graphene aerogel at 298 K were given in table 1. It was noticed that the correlation coefficient R^2 values were fitted both Langmuir and Freundlich isotherms well (R^2>0.97). The Freundlich constant 1/n was between 0 and 1 for three temperatures, indicating that the adsorption was favorable under experimental conditions [16]. Moreover, the higher values of K_F at lower temperature indicated graphene aerogel possessed a higher adsorption capacity and affinity for NNK. the maximum adsorption capacity (q_m) at 298.15 K was determined to be 59.66 mg/g based on Langmuir model. It was approximately in a close agreement with the measured data. In addition, q_m calculated from Langmuir model decreased with the temperature increasing, indicating that lower temperature was in favour of NNK adsorption, which was consistent with the result of Freundlich model. Based on the results, it could be concluded that both Langmuir and Freundlich isotherms agreed well with the experimental data.

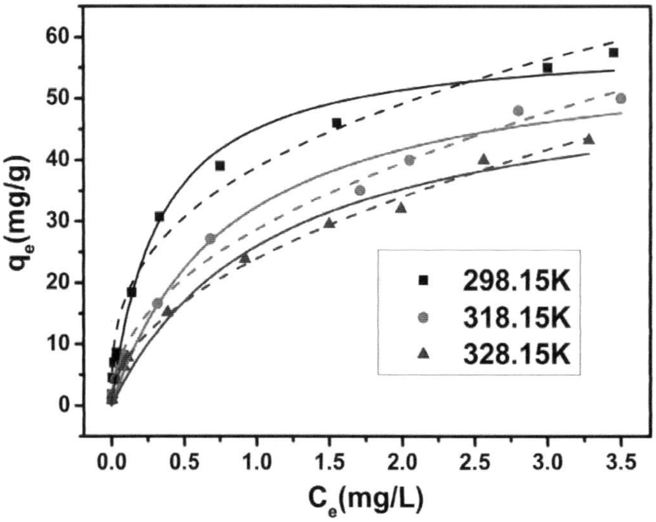

Figure 7. Adsorption isotherms of NNK by graphene aerogel at three different temperature (10 mg graphene aerogel, 100 mL NNK, contact time 8 h, pH = 7). Langmuir model simulation (solid lines); Freundlich model simulation (the dotted lines).

Table 1. Parameters of absorption isotherms of NNK on graphene aerogel.

	Langmuir			Freundlich		
T(K)	q_m(mg/g)	K_L(L/mg)	R^2	K_F	n	R^2
298.15	59.66	3.0921	0.9826	38.8053	2.9314	0.9858
318.15	58.76	1.2230	0.9800	28.6449	2.1477	0.9936
328.15	55.04	0.8913	0.9783	23.9339	1.9792	0.9965

To elucidate the NNK adsorption on graphene aerogel, the adsorption standard enthalpy change (ΔH°), standard entropy change (ΔS°) and free energy (ΔG°) were obtained from the adsorption of NNK by graphene aerogel at different temperatures. ΔH° and ΔS° reflected the adsorption process was endothermic or exothermic and spontaneous, The calculated value of ΔH° and ΔS° was -34.19 kJ/mol and -105.10 J/mol·K from the variation of the Langmuir constant (K_L), respectively (R^2 = 0.9921, as shown in table 2). According to the literature, the ΔH° value between 20.9 and 418.4 kJ/mol were commonly assumed to be associated with chemisorptions [17]. Hence, it was thought that NNK on graphene aerogel might proceed by chemisorption. The negative ΔH° value suggested the nature of adsorption was exothermic, supporting by the increase in the adsorption capacity of NNK with decline in temperature. Furthermore, the negative ΔS° indicated decreased randomness at the solid-liquid interface during the adsorption of NNK on graphene aerogel. The negative ΔG° (-2.86 kJ/mol at 298.15 K) suggested that the adsorption was a spontaneous process. In addition, the value of ΔG° became positive (0.29 kJ/mol) at 328.15 K, which indicated high temperature would prevent the adsorption of NNK on graphene aerogel.

Table 2. Thermodynamic parameters of NNK adsorption on graphene aerogel.

Thermodynamic constant	T(K)		
	298.15	318.15	328.15
ΔG° (kJ/mol)	-2.86	-0.76	0.29
ΔH° (kJ/mol)		-34.19	
ΔS° (J/mol·K)		-105.10	

The pH value was also one important factor to determine the adsorption behavior of an adsorbent, because it not only influenced the surface charge of the adsorbent, but also the degree of ionization of adsorbate[18].The results showed that the NNK molecule adsorption was strongly depended on the pH of solution, the adsorption capacity of graphene aerogel obviously increased in the pH range of 2.0 to 8.0. Further increasing pH from 8.0 to 14.0, the adsorption capacity of graphene aerogel slightly decreased and maintained stable for 40 mg/g at pH of 10.0. As reported, the reason for aromatic compounds adsorbed on graphene was originated from π-π dispersion interaction between the aromatic ring electrons of adsorbate and those on the graphene plane [19,20].Considering that pyridine ring in the molecular structure of NNK were presented, we suggested π-π interaction probably dominated the adsorption of NNK on graphene aerogel. Graphene aerogel was electrically neutral and presented a surface charge density close to zero over the whole pH value, which was proven by the zeta potential analysis. NNK was in molecular form in the basic pH range and started the protonation in around neutral range due to the pyridine ring in the molecular structure of NNK. Consequentially, NNK was mostly ionized to mono-cation after protonation in acidic and neutral pH range, which weakened the π-π interaction. Therefore, the reduction in adsorption capacity of graphene aerogel in acidic and neutral pH range was mainly owing to the weak π-π dispersion interaction between ionized NNK and graphene aerogel.

The recycling performance for the NNK sorption capacity of graphene aerogel was evaluated and the six cycle experiments were performed. Under the same experimental condition, there was no evident decreasing for the adsorption capacity of graphene aerogel. Even after 6 cycles using methanol aqueous solution as eluent, more than 90% of the NNK sorption capacity can be preserved, exhibiting a good reliability for environmental pollutants adsorption. The efficient adsorption performance indicated that graphene aerogel possessed great potential in environmental treatment.

4. Conclusion

In this work, the adsorption of graphene aerogel for removing NNK from aqueous solution was demonstrated. The maximum adsorption capacity of graphene aerogel for NNK was 59.66 mg/g estimated from Langmuir isotherm. The kinetics data could be well fitted with the pseudo-second-order model. The isotherms data were fitted with both Langmuir and Freundlich equation. The high pH and low temperature were favorable for the adsorption. The adsorption mechanism of NNK on graphene aerogel was deduced to be via π-π dispersion interaction. The pyridine ring in NNK molecular and hexagonal cells of the graphene aerogel facilitated π-π dispersion interaction between them.

5. References

[1] Hecht S S 1998 *Chem. Res. Toxicol.* **11** 559
[2] Wu M H, Qian Y C, Boyd J M, Leavey S, Hrudey S E, Krasner S W and Li X F 2014 *Environ. Sci. Technol.* **48** 1828
[3] Zhou Y, Lin W G, Wan M M, Yang J and Zhu J H 2012 *J. Mater. Chem.* **22** 23633
[4] Lin W G, Huang B C, Zhou B, Lv X L, Wang L J, Zheng S J, Wang W M, Tong Y G, Wang Y and Zhu J H 2014 *J. Porous. Mater.* **21** 311
[5] Xu Y, Liu H D, Zhu J H, Yun Z Y, Xu J H and Wei Y L 2004 *New Journal of Chemistry.* **28** 244
[6] Li Y, Zhang P, Du Q, Peng X, Liu T, Wang Z, Xia Y, Zhang W, Wang K, Zhu H and Wu, D 2011 *J. Colloid. Interface. Sci.* **363** 348
[7] Ramesha G K, Kumara A, V, Muralidhara H B and Sampath S 2011 *J. Collid. Interface. Sci.* **361** 270
[8] Deng X J, Lu L L, Li H W and Luo F 2010 *J. Hazard. Mater.* **183** 923
[9] Liang H W, Guan Q F, Chen L F, Zhu Z, Zhang W J and Yu S H 2012 *Angew. Chem. Int. Ed.* **51** 5101
[10] White R J, Budarin V, Luque R, Clark J H and Macquarrie D J 2009 *Chem. Soc. Rev.* **38** 3401
[11] Hummers W S and Offeman R E 1958 *J. Am. Chem. Soc.* **180** 1439
[12] Xu Y X, Sheng K X, Li C and Shi G Q 2010 *ACS. Nano.* **4** 4324

[13] Chandra V, Park J, Chun Y, Lee J W, Hwang I C and Kim K S 2010 *ACS. Nano.***4** 3979
[14] Wu X L, Wen T, Guo H L, Yang S B, Wang X K and Xu A W 2013 *ACS. Nano.***7** 3589
[15] Zhao G X, Li J X and Wang X K 2011 *Chem. Eng. J.***173** 185
[16] Hameed B H 2008 *J. Hazard. Mater.***154** 204
[17] Unlu N and Ersoz M. 2006 *J. Hazard. Mater.***136** 272
[18] Jimenez R M and Solache R M 2010 *J. Hazard. Mater.***180** 297
[19] Xu J, Wang L and Zhu Y F 2012 *Langmuir.***28** 8418
[20] Guo Y J, Deng L, Li J, Guo S J, Wang E and Dong S J 2011 *ACS. Nano.***5** 1282

IOP Conf. Series: Materials Science and Engineering **324** (2018) 012029 doi:10.1088/1757-899X/324/1/012029

The design and test of collecting device and film purge device joint work of residual film recovery machine based on Solidworks & Adams

Haozhou Zhai[1], Jianming Jian[1,2], Shulin Hou[1], Yunlong San[1], Wensong Guo[1] Yue Sun[1] and Mingqing Gao[1]

1 College of Engineering, China Agricultural University, Qinghuadonglu, Haidian Disrict, Beijing, China

2 Corresponding author. Email jamesjian@126.com

Abastract: The twine of residual film is an essential issue in the process of remnant residue recovery of the residual film recovery machine. It is difficult to clean up the residual film in the residual film recovery operation and to influence the subsequent film efficiency. Therefore, in response to this problem a composite tooth pocket residual film recovery device was designed. In this paper, the structure of the film recovery device design, theoretical analysis, simulation experiments, get the most appropriate film recovery device parameters. In addition, the residual film rate of the membrane is dramatically low, reaching about 1.3% only, and the operation of the whole machine is smoother, and the stability of the work is promoted. The operation of the film recovery device is very obvious. Lifting, in addition to the film rate has also been significantly improved to 93.88%

1 Introduction

With the promotion of mulching and planting technology, the mulching planting area in China is nearly 49.5 million acres, the usage amount of agricultural plastic film is about 2 million 600 thousand tons [1].The mulching technology of mulch has helped the growth of crop growthand has been widely applied to the crops of medium and late mature varieties, which can increase the yield [2-3]. As the cost of picking up the residual film is too high, the mainstream of the residual film picking way is by the machine now [4-6]. The membrane removal device is an important part of the residual film recycling machine. At work, a large number of residual film will be wrapped on the pick - up elastic tooth[7]. With the work going on, the residual film will gradually gather, and the ability to pick film will gradually decline[8]. The film removing device can be used to clean the residual film at the same time, which can improve the working capacity of the machine and ensure the stability of the residual film recycling machine. By using the method of combining theoretical analysis with experiments, the trajectory of pickup teeth as reducing residual film and elastic tooth winding for the purpose of optimal optimal target, developed in the optimal efficiency of membrane device set up in addition to the optimal efficiency of membrane device - scraping film removing device movement. Because the motion of the pick film missile has certain elasticity, the optimization design should be modified according to the actual effect, and further improve the performance.

IOP Conf. Series: Materials Science and Engineering **324** (2018) 012029 doi:10.1088/1757-899X/324/1/012029

2 The design of collecting device and the film purge device

2.1The working principle and the existing problems of collecting device

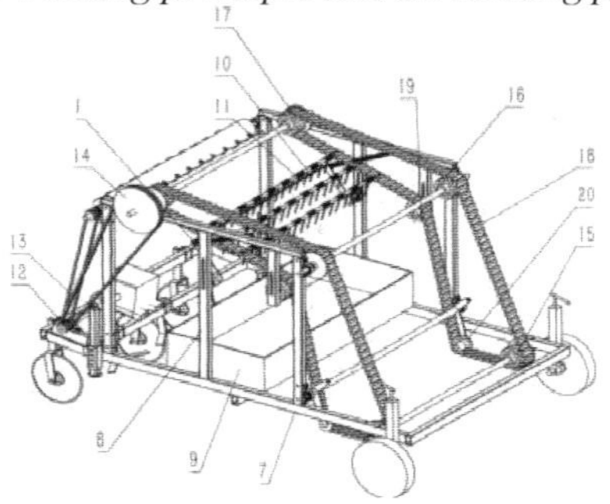

Figure 1. the structure of collecting device.
1. Film purge device 7. Chassis 8. Body frame clearner 9. Collector steel 10. Elastic teeth axle 11.Elastic teeth 12. 100mm belt pulley13. Conveyor belt 14. 280 mm belt pulley 15. Lower crawler wheel 16. upper crawler wheel 17. Back crawler wheel 18. Conveyor belt 19. lower crawler wheel 20. Tension pulley

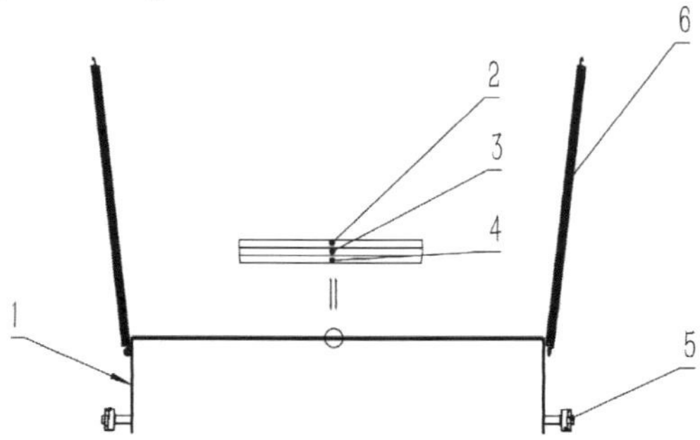

Figure 2. Structural diagram of purge device.
1.Alloy steel brace 2.The upper rubber 3.T9 high carbon plate 4.The lower rubber clearner 4.UCFL206 bearing 5.spring

In order to make the schematic diagram clearer, in figure 1 the elastic gear on the track is hidden, and only the gear shaft in the work is retained. The pick up mechanism of the machine consists of pick - up elastic teeth, conveyor belts, tension wheels and plum wheels. The pick up the work process: elastic teeth shaft connected by both ends and elastic teeth movement in the track, track and force conveying shaft under the gradually descending from the front running into soil to pick up the residual film in the soil, then reverse operation, gradually goes up when running, when running to the plastic film residue plastic film recycling machine in the middle, the most of the residual film drop into the film collector by gravity, the rest of film remaining on the teeth move to the next stage, causing some residual film drop to the soil, reduce the residual film recovery rate, increase the artificial picking up workload; another part of the residual film has been around in the residual membrane shells teeth, and it will gather more and more during work, and also causing teeth surface friction reduced, some residual film slipped from the teeth on the next collecting process.

2.2 The design of film purge device

The basic structure of the film purge device is determined according to the movement mode of the film spring of the residue film recovery machine, as shown in figure 2. The film removing plate made of two rubber cleaners and a high carbon steel plate compounded by bolts to form an integrated structure, the frame through the bearing is connected to the residual film recovery machine, film removing plate frame on both sides from the upper spring is connected to the rear of the frame residual film recovery machine. The spring device makes the film version in the elastic teeth scraping forward in the spring tension to follow teeth forward a long distance, increase the effective working distance of film removing plate. Due to the increase in the effective distance of film removing plate and scraping membrane on the two pieces of plate rubber blown film plate shape with the working process of the change in scratched the surface of the film version of the contact area and friction, greatly enhanced the ability of a film, the film will be relatively tight attachment can be successfully removed.

3 The assignments and force analysis of collecting device and the film purge device joint work

3.1 The assignments of collecting device and the film purge device joint work

The film purge device is installed at the front end of the pick-up spring teech of the residual film recycling machine, it aims to efficiently clean the residual film attached and wrapped on the teeth of the pick-up spring teeth. The upper end of a film purge device hanging by springs. The film removing plate moves in a circular motion under the thrust of the teeth, the forward and downward movement of residual film achieve the function of cleaning the residual film, in the movement to about 1/ 8 position clockwise circular trajectory, the plate run to bottom of elastic teeth, then the film purge device back to the initial position at this moment, reciprocated in addition to the membrane process.

After the cleaning process, the teeth with no residual film move to the below, take he residual film from soil again and it will drop to the film collector , by comparison with the teeth without cleaning, the residual film will be accumulated, excessive accumulation of residual film will affect the film picking the next tooth pick film work, even it's hard to pick the films. Therefore, film purge device ensures the stable and continuous operation of the residual film recovery machine, and greatly improves the recovery efficiency of the residual film.

3.2 Force analysis of joint work

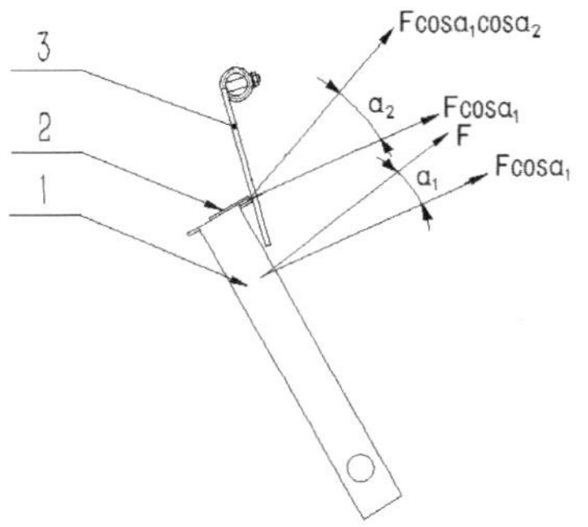

Figure 3. Force diagram of film purge device.
1.force direction of spring 2.removing plate 3.elastic tooth

In order to measure the force of purging residual film wrapping on the teeth, pulling down the film scraping rubber plate to complete the removal of residual film movement by using the electronic tension equipment, recording the force needed to achieve the desired analog measurement of scraping force. And the twenty groups of data measured by the electronic tension equipment will be summary and use the maximum value to represent the maximum tensile force data. On the other hand, the composite rubber and the elastic teeth on the film removing plate form greater friction in the process of movement, and the direction is opposite. The main condition that the residual film can be scraped is that the friction force is greater than the maximum tensile force.

$$f > F_{max} \qquad (3\text{-}1)$$

$$\begin{cases} f = \mu F\cos\alpha_1 \cos\alpha_2 \\ F = k(l_2 - l_1) \\ F_{max} = (m_1, m_2, m_3, ..., m_{18}, m_{19}, m_{20}) \end{cases} \qquad (3\text{-}2)$$

f —The frictional force between rubber and elastic teeth
μ —The coefficient of friction between rubber and elastic teeth
l_1—The initial length of spring
l_2 —The spring strength length
F —The spring tension
k —spring rate
F_{max} —Maximum pull of purging the film

m_1 to m_{20} —All the pulls of purging the film in the test.

α_1 —The angle between spring pull and film purging force directions

α_2 —The angle between elastic teeth and film purging force directions

Simultaneous equation（3-1）and（3-2）

$$k > \frac{F_{\max}}{\mu(l_2 - l_1)\cos\alpha_1\cos\alpha_2} \quad (3\text{-}3)$$

By type (3-3) can obtain the required minimum spring stiffness coefficient through access to relevant literature, and considering the influence of spring vibration during the movement of other factors will determine the scope of spring, so it is between 38mm~41mm, finally according to the actual production standard and the principle of economy is determined using 40mm tensile spring .

4.The working process simulation of film purge device

4.1 Solidworks *3 dimensional model design*

In this paper, ADAMS and Solidworks are used to complete the simulation test, and the simulation scheme is aimed at the simulation operation effect of the film purge device. The operation effect of the device is directly related to the effect of its pratical work. In this paper, Solidworks and Adams software are used to simulate the membrane removal device. Which draw each parts to a substantial proportion of 1:1 model by using Solidworks,then the CAD model into PARASOILD model and introduced into the dynamic simulation software ADAMS for dynamic simulation.

Figure 4. Solidworks 3D model

This experiment is aim at finding the most suitable plate height film elastic teeth to operate film removing, the simulation experiment will be working within the selected film removing plate distance of elastic teeth top from 2.0cm to 6.0cm height ,and they will be simulated in the simulation, the initial selection of the height is 2.0cm and 4.0cm and 6.0cm respectively in the preliminary test, the running speed of film board X, Y, Z sets are regarded as the standard, through the experimental results. The better performance range is from 2.0cm to 4.0cm, so the refinement test are operated in the second experiment, with 0.1cm as the interval, through simulation that the most appropriate height is 3.1cm.

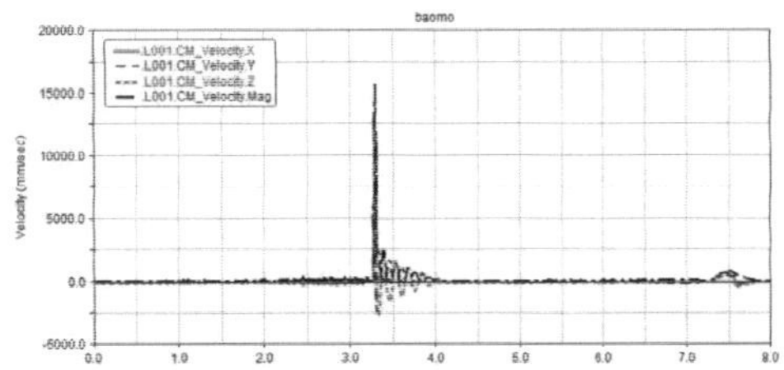

Figure 5. motion simulation of 6.0 cm

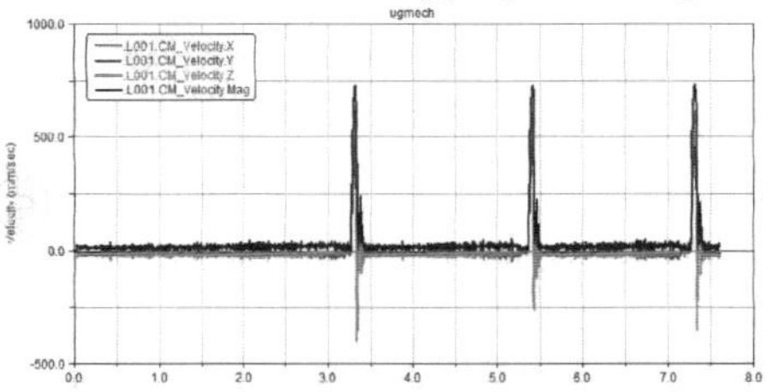

Figure 6. motion simulation of 3.1 cm

The figure 5 is representative of the height of 6.0 cm, due to the first film wiped action after springback wiped film face with elastic teeth cause rear operation cannot be carried out at a standstill, so the movement is stuck in the first session. When the height is adjusted to 3.1cm, the simulation results shown above, Velocity. X, Velocity. Y, Velocity. Z values have reached a relatively stable state at the same time period, Velocity. Mag values also showed a consistent, indicating that in this position, the membrane device has good stability, can work continuously in film motion to meet request. From the above analysis it can be concluded that when the elastic teeth top distance is in 3.1cm, the optimal motion effects of the velocity as shown in Figure 6, can smoothly complete the wiped film movement, and in second and third motion waveform showed a good stability, the results show that the film removing device installation location and elastic teeth movement has a high degree of fit.

5 Field test and analysis

5.1 Experimental environment

The field test of film purge device and collecting device was carried out in Liuquan potato film farmland Guan town Hebei Province, film thickness used for field is 0.008mm, flat terrain, and soil is in operation loam, soil hardness is 5.9kPa, the moisture content is 26.1%, the film had been broken, the experimental environment meet the most common cultivated land, so the working environment is generally representative. The power plant is supplied by Dongfanghong 40 tractor, with a speed of 3.0~4.0m/s and a working width of 1800mm.

Figure 7. Fields trial of residual film device

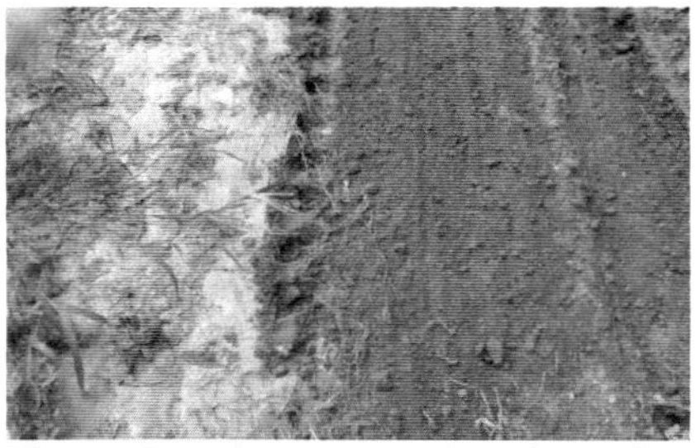

Figure 8. Enlargement of work results

5.2 Test scheme

In order to ensure the stability of the test and scientific comparation. The test will be divided into two schemes: install film purge device and uninstall film purge device . For each scheme is divided into 4 groups, in the experimental conditions permit in large extent to reduce the influence of random factors in the field, each test 2 times and take the average value, a single test with distance of 200m, taking into account the residual film recycling machine working efficiency and the applicable speed range of two factors, the operation speed is in the range of 2.0m/s ~ 3.5m/s test group A walking speed was 2.0m/s, test group B, walking speed is 2.5m/s, test group C, walking speed is 3.0m/s,test group D, walking speed is 3.5m/s.

The measurement method of residual film is: in each test residual plastic film left will be pick up by manual work and the film picking up by machine is also collected, then were washed and dried, weighed, and draw the final quality. The calculation formula of pickup rate is as follows:

$$S = \frac{m_u - m_c}{m_u} \times 100\%$$

S—Film cleaning rate

m_c —The quality of residual film after installing the film purging device

m_u —The quality of residual film before installing the film purging device

5.3 Test result

In the test, the film purge device is running closely to the design trajectory, the experimental results and simulation results are slightly different, but in the acceptable and reasonable range. When the velocity is between 2m/s ~ 3m/s, the membrane removal effect is good, and the recovery effect of residual film can be greatly improved. Among them, the running speed of 2.5m/s is the best, residual film recovery increased 6 percentages. At the same time, residual film recovery effect in speed 3.5m/s the residual film recovery appears to reduce the abnormally that residual film recovery machine was unstable in high speed operation, so it need to observed for further studies in the following work.

Table 1. The testing result of purge device.

Test serial number	Speed m/s	Twine film quality without device m_u / g	Twine film quality with device m_c / g	Rate of cleaning %	Film collecting rate with device %	Film collecting rate without device %
A	2.00	25.71	1.31	95.0	85.58	91.42
B	2.50	33.16	1.42	95.8	87.17	93.88
C	3.00	28.64	1.28	95.6	86.43	90.21
D	3.50	30.32	1.61	94.7	79.96	83.22

Table 2. Significant analysis of collecting experiment.

Difference source	SS	df	MS	F	P-value	F crit
Rows	116.81	1	116.82	11.20	0.0441	10.13
Error	31.28	3	10.43			
Total	148.09	4				

(Note: $p<0.01$ is extremely significant; $0.01<p<0.05$ is significant; $p>0.05$ is not significant)

Through the analysis data in the table 2, we get the experimental results F=11.20, P=0.0044<0.05, which shows that film purge device has remarkable effect on improving film collecting rate.

6. Conclusion

According to the circumstance that the residual film twines with the elastic teeth during the working process of film recovery machine, a following type film removing device is designed, its working principle is analyzed, and proved the feasibility for the device through the field test, suitable for elastic teeth type residual film recovery machine, the film recovery rate promotion effect is prominent, the quality improved obviously, so it is worth popularizing widely cultivated land in the application of residual film recovery machine.

Reference

1. China statistical bureau *China statistical yearbook national bureau of statstics of the people's republic of China Beijing* China Statistics Press 2015.
2. Hou S L Hu S Y Kong J M, Research status of residual film recycling machine in China *Transactions of the Chinese Society of Agricultural Engineering*, 2002, Vol. **2** pp 186-190.
3. Li M Y Ma S H The current research on the residual film recycling machine in China and suggestion Journal of Agricultural Mechanization Research, 2014,Vol. **36 (6)** pp 242-245,252.
4. Chen D W 2014 The residual film promotion of recycling machine *Agricultural Science&Technology and Equipment* Vol. **7 (7)** pp 34-35.
5. Qin W Popularization and suggestion of residual film recycling machine *Rural Science & Thchnology*, 2013 Vol.**7** pp 74-75.
6. Chang L R Pei Z L Cao L Review of mechanical technology for waste film recycling *Auhui Agricultural Science Bulletin*, 2013 Vol. **9** pp 153-155.
7. Mu D H Yao W Z Design and Experimental Research on elastic tooth pick up mechanism of residual film recycling machine *journal of agricultural mechanization research*, 2015 Vol.**11** pp 91-94.
8. Chen F Shi J X Zhao H J Optimum design of fixed cam residual film pickup mechanism *Transactions of the Chinese Society for Agricultural Machinery*, 2005 Vol. **3 (12)** pp 43-46.

IOP Conf. Series: Materials Science and Engineering **324** (2018) 012030 doi:10.1088/1757-899X/324/1/012030

Mechanical Design of Downhole Tractor Based on Two-Way Self-locking Mechanism

Delei Fang[1], Jianzhong Shang[1], Zirong Luo[1], Guoheng Wu[1] and Yiying Liu[1]

1National University of Defense Technology, College of Mechatronic Engineering and Automation, Changsha, 410073, China
E-mail: fangdelei365@126.com

Abstract. Based on the technology of horizontal well tractor, a kind of downhole tractor was developed which can realize Two-Way self-locking function. Aiming at the needs of horizontal well logging to realize the target of small size, high traction and high reliability, the tractor selects unique heart-shaped *CAM* as the locking mechanism. The motion principle of telescopic downhole tractor, the design of mechanical structure and locking principle of the locking mechanism are all analyzed. The mathematical expressions of traction are obtained by mechanical analysis of parallel support rod in the locking mechanism. The force analysis and contour design of the heart-shaped *CAM* are performed, which can lay the foundation for the development of tractor prototype.

1. Introduction

In recent years, the method of drilling horizontal wells and enhancing horizontal well length to increase oil and gas recovery factor has become an important way to increase production of oil and gas[1-2]. However, the development of large displacement horizontal well has raised higher and higher requirements for logging and other underground operations and traction technology of supporting instrument[3]. Well logging can't be placed into a horizontal well relying on gravity, and the logging instruments must overcome the devolution caused by the hole bending. For this kind of demand, the technology of horizontal well tractor has been studied extensively[4-5].

At present, there are various types of downhole traction robot. Based on the different driving modes, the tractors can be mostly divided into four types: the telescopic downhole tractor, the wheel type downhole tractor, high pressure jet downhole tractor and propeller propelling downhole tractor[6]. In the late 1990s, many companies in foreign countries developed horizontal well tractors capable of working independently in the mine[7]. In many products, the wheel type tractor developed by Statoil, Maritime Well Service (MWS) and Welltec 3 companies jointly, telescopic downhole tractor produced from SmarTract and the Sondex horizontal well tractor are most prominent[8].

This paper presents a kind of downhole tractor which can realize two-way self-locking function, based on the technology of horizontal well tractor and the locking principle in the retractable micro-pipe robot. The tractor selects unique heart-shaped *CAM* as the locking mechanism, aiming at the needs of horizontal well logging to realize the target of small size, high traction and high reliability. Firstly, the motion principle of telescopic downhole tractor is introduced in this paper. Then, the design of mechanical structure and locking principle of the locking mechanism are analyzed. The mathematical expressions of traction are obtained by the mechanical analysis of parallel support rod in the locking mechanism. Finally, the force analysis and contour design of the heart-shaped CAM are performed, which can provide the basis for the design and manufacture of the prototype of the tractor locking mechanism.

2. Motion Principle of Downhole Tractor

The basic motion unit of the tractor based on the two-way locking mechanism consists of two locking mechanisms and a telescopic actuator. The motion principle is shown in figure 1. Figure 1 (a) is the initial state, and the supporting arms of the two locking mechanisms are propped up and pressed against the wall. Then the telescopic mechamism expands, and the locking mechanism 1 locks and the front locking mechanism 2 unlocks, so the locking mechanism 2 is pushed forward, as shown in figure (b). Telescopic mechanism begins to bounce back after spreading to the back of range, the locking mechanism 2 in front locks. The locking mechanism 1 unlocks, and its body is drag to move forward, as shown in figure (c), When the above procedure is completed, the tractor returns to a state in figure 1 (a), and the system has been running forward a distance at this time. Repeating movement process of from figure (a) to figure (c), tractor will be able to move forward.

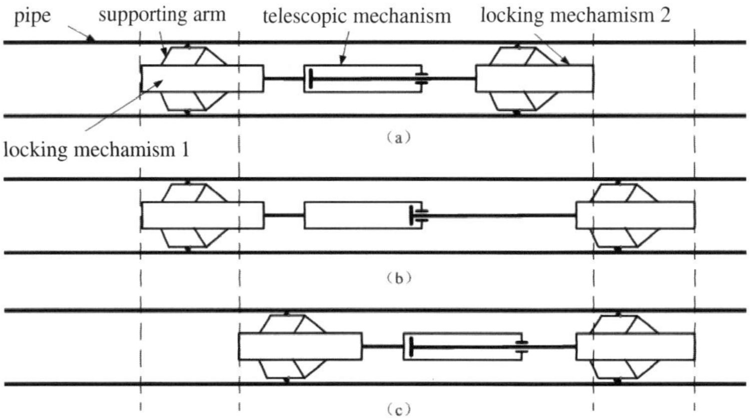

Figure 1 Motion principle of telescopic tractor

3. Design and Mechanical Analysis of Locking Mechanism

3.1 Design of the Two-way Locking Mechanism

The underground pipeline is filled with oil, sediment and so on. The realization of the telescopic tractor function has to set reasonable stress points, which is the use point of the supporting force of the tractor. In order to find the stress point in this complex environment, the two-way locking mechanism is designed. The locking mechanism is a key module that affects the ability of the tractor to function normally. Its function is to make the tractor only move in one direction, and the other direction is stuck. The schematic diagram of two-way locking mechanism is shown in figure 2.

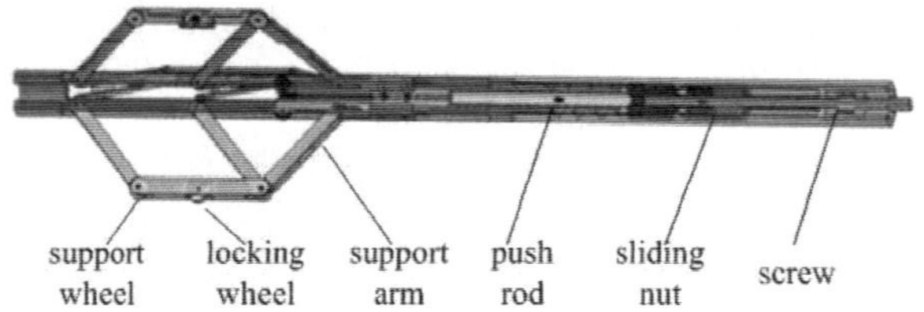

Figure 2 Schematic diagram of two-way locking mechanism

The heart-shaped locking wheel starts to work when the support wheel moves to the wall. The locking of different directions can be realized, if the heart-shaped locking wheel is in the different initial position. As shown in figure 3 (a), the initial position of the locking wheel is shown, When the tractor is moving to the left, the heart-shaped locking wheel turns clockwise, the contact force between the *CAM* and the wall is reduced, friction is reduced, and the tractor can move forward by overcoming the force of friction. When the tractor is moving to the right, the heart-shaped locking wheel turns counterclockwise, the contact force of the *CAM* and tube wall is increased, the contact force is increased, the mechanism locks. Similarly, as shown in figure 3 (b), the mechanism can achieve the right movement and the left locking movement.

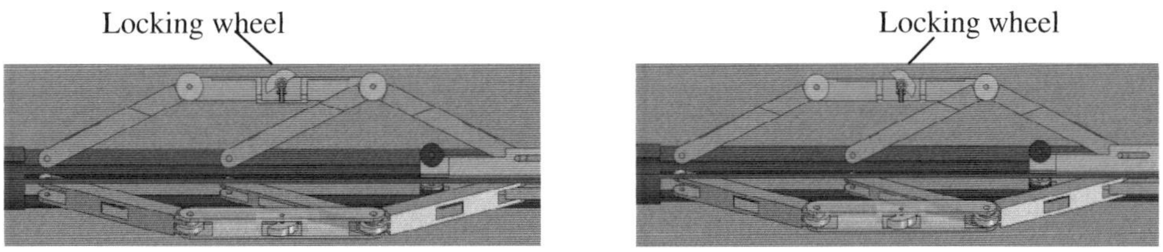

(a) left movement, right locking (b) right movement, left locking

Figure 3 Locking principle of two-way locking mechanism

3.2 Mechanical Analysis of Parallel Four-bar Mechanism

The forces of the parallel four-bar mechanism are shown in figure 4. The external forces of the mechanism are: N is the contact force of self-locking mechanism and pipe wall, fu is the friction force of self-locking mechanism and pipe wall, and F_s is the force of screw rod.

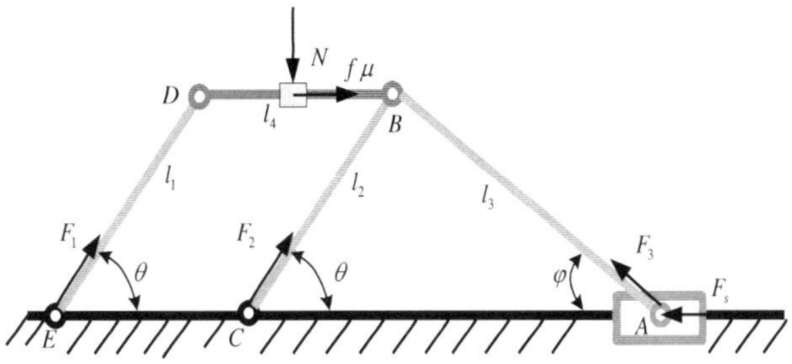

Figure 4. The force diagram of the four-bar mechanism

According to the tractor motion conditions, the friction force of the mechanism must be greater than the traction force,

$$fu \geq F_s \tag{1}$$

In the case of an expanding mechanism, the hypothetical equation (1) is satisfied, according to the force equilibrium condition,

$$\begin{cases} N = F_1 \cdot \sin\theta + F_2 \cdot \sin\theta + F_3 \cdot \sin\phi \\ fu + F_1 \cdot \cos\theta + F_2 \cdot \cos\theta = F_3 \cdot \cos\phi \\ F_1 \sin\theta = \lambda(F_2 . \sin\theta + F_3 . \sin\phi) \end{cases} \tag{2}$$

Applied the force balance condition of the ball screw,

$$F_S = F_3 \cdot \cos\phi \tag{3}$$

Combined (2) and (3), the forces of each support rod are

$$\begin{cases} F_1 = \dfrac{\lambda \cdot N}{(\lambda+1)\sin\theta} \\ F_2 = -\dfrac{N(-\sin\theta \cdot \cos\phi + \lambda \cdot \sin\phi \cdot \cos\theta + \sin\phi \cdot \sin\theta \cdot f\mu + \lambda \cdot \sin\phi \cdot \sin\theta \cdot f\mu)}{\sin\theta(\lambda+1)(\sin\theta\cos\phi+\cos\theta\sin\phi)} \\ F_3 = \dfrac{N(\sin\theta \cdot f\mu + \cos\theta)}{\sin\phi\cos\theta+\cos\phi\sin\theta} \end{cases} \tag{4}$$

Applied geometrical relation, and the rod thrust is

$$F_S \geq \frac{(fu(R-r) + \sqrt{l^2 - (R-r)^2})\sqrt{l_3^2 - (R-r)^2} F}{(R-r) fu(\sqrt{l_3^2 - (R-r)^2} + \sqrt{l^2 - (R-r)^2})} \tag{5}$$

It can be seen that the driving force is related to the pipe wall pressure, the length of the drive rod and the length of the support rod.

4. Profile Design of Heart-shaped *CAM*

When the wheel the trend of right movement is shown in figure 5 (a), the friction *fu* and the positive pressure *N* make the same direction torque on the center of rotation *B*, which can make the heart-shaped wheel run away from the wall.

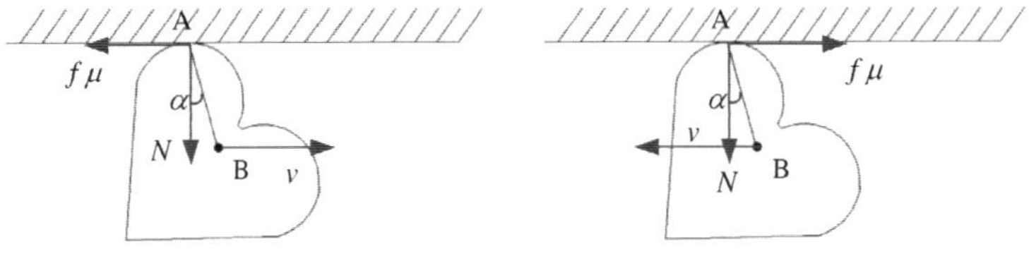

(a) the trend of right movement (b) the trend of left movement

Figure 5 Schematic diagram of the heart-shaped wheel

The heart-shaped wheel should be stuck, when it has the left movement trend as shown in figure 5 (b). Therefore, the direction is shown in Figure 5 (b).

As shown in figure 6, the distance between rotating center *B* and contact point *A* is *R*, after the time of Δt, the contact point is point *D*, the turning angle is $d\theta$. *AC* is the tangent direction of the point *A* and perpendicular to the *AB*. Because $d\theta$ is very small, it can be considered that *AC* is perpendicular to *BD*, BA is perpendicular to *AD*.

Hence the following equation is found in triangle *ACD*,

$$\angle CAD = \angle\alpha$$

$$\tan\alpha = \frac{DC}{AC} \tag{6}$$

Where, $DC = d\rho, AC = \rho \bullet d\theta$,

$$\tan\alpha = \frac{d\rho}{\rho \cdot d\theta} \leq \mu \tag{7}$$

The contour curve of heart-shaped wheel is

$$\theta = \frac{1}{\mu} \bullet \ln\frac{\rho}{\rho_0} \tag{8}$$

The maximum height of the mounting bracket is 19mm, and $\mu = 0.25$. The minimum distance of the center to the horizontal line is $1mm$. Solved equation,

$$\begin{cases} \rho \bullet \cos(\pi - \theta) = 6 \\ \theta = 4 \bullet \ln\dfrac{\rho}{15.53} \end{cases} \tag{9}$$

The optimal contour line of the heart-shaped locking wheel is shown in figure 7.

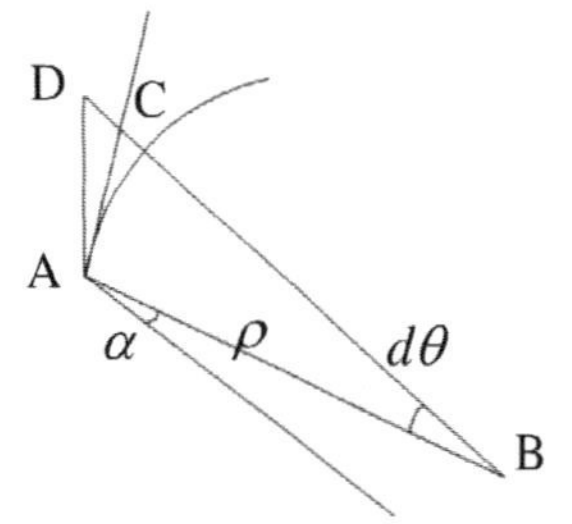

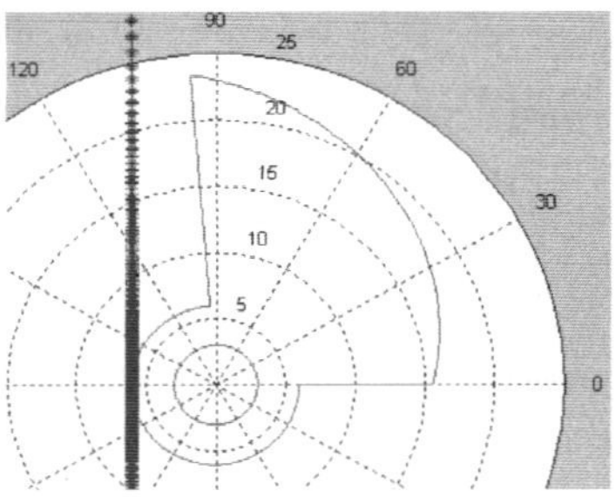

Figure 6 Diagram of heart-shaped shape **Figure7**. Contour of heart-shaped wheel

5. Conclusion

In this paper, a kind of downhole tractor was developed which can realize Two-Way self-locking function. The tractor selects unique heart-shaped *CAM* as the locking mechanism, which can meet the requirement of small size, high traction and high reliability. The motion principle of telescopic downhole tractor, the design of mechanical structure and locking principle of the locking mechanism were all analyzed. Finally the force analysis and contour design of the heart-shaped *CAM* are researched, which is helpful to develop the tractor prototype.

References

[1] A. Shukla, H. Karki, Application of Robotics in Onshore Oil and Gas industry, A review Part I, Robotics and Autonomous Systems, vol. 75, 2016, pp. 490–507.
[2] J. S. Cao, L. Lin, Y. Li, Research and Development Progress in Oil and Gas Pipeline Robotics, Oil & Gas Storage and Transportation, vol. 32, 2013, pp. 1-7.
[3] S. Roh, H. R. Choi. Differential-Drive In-pipe Robot for Moving Inside Urban Gas Pipelines [J]. IEEE Transactions On Robotics, 2005, 21(1): 1~17.
[4] Zheng T , Godage IS , Branson DT , Kang R , Guglielmino E , Medrano-Cerda GA . Octopus inspired walking robot: design, control and experimental validation. In: Robotics and automation (ICRA), 2013 IEEE international conference on. IEEE; 2013. pp. 816–21 .
[5] L. Q. Li, D. K. Zhou, G. Y. Zhang, Y. Q. Li, W. P. Song, Optimal design of horizontal well traction robot traction force regulating mechanism. Advanced Materials Research, vol. 1033, 2014, pp. 1350-1353.
[6] S. Gao, W. Sun, H. Ni, "Status and understandings of downhole tractor design" , Journal of Machine Design, vol. 31, 2014, pp. 1-8.
[7] Stolte C, Wu C, Carroll D, et al. Coiled tubing and well test experiences on sichuan horizontal shale gas wells[C]// International Petroleum Technology Conference, Beijing, China, March 26–28, 2013: 1–5.
[8] Billingham M, Etoukhy A M, Hashem M K, et al. Conveyance—down and out in the oil field[J]. Oilfield Review, 2011, 23(2): 18–31.

Acknowledgments

The work was supported by the National Natural Science Foundation of China (51675527).

IOP Conf. Series: Materials Science and Engineering **324** (2018) 012031 doi:10.1088/1757-899X/324/1/012031

Combined Experimental and Numerical Study on Fracture Behaviour of Low–Density Foamed Concrete

M Kozłowski[1] and M Kadela[2]

1 Silesian University of Technology, Faculty of Civil Engineering, Department of Structural Engineering, Akademicka 5, Gliwice, Poland
2 Building Research Institute (ITB), Filtrowa 1, Warsaw, Poland
E-mail: m.kadela@itb.pl

Abstract. The paper presents results of a combined experimental and numerical study on the fracture behaviour of low–density foamed concrete (< 1100 kg/m^3). A series of static tests was carried out on notched beams tested in three-point bending to determine the fracture properties of foamed concrete. Based on the load–displacement responses fracture energy and maximal tensile stress were evaluated. The paper also presents the results of numerical investigation of the fracture behaviour of notched beams using Extended Finite Element Method (XFEM). The numerical studies involve a simulation of fracture initiation and crack evolution in the notched beams subjected to three–point bending.

1. Introduction

Foamed concrete is a broadly recognised material in the building industry, mainly due to of its low self–weight and superb thermal and acoustic properties [1, 2]. It has been commonly used primarily as a thermal insulation of roofs and foundations, as backfill material for retaining walls and sound insulation of building partitions. In the last years, however, it has become a promising material for structural purposes [3]. This includes stabilization of poor soil conditions [4], sub–base layer in sandwich solutions for foundation slabs and industrial floors [5–9] and road–pavement systems [10].

Cracks of various shapes and forms may develop during construction phase and throughout the lifetime of concrete structures. Engineering practice shows that cracking of concrete is almost unavoidable, however, current technology can minimize the risk. Moreover, computation software used for simulations of fracture behaviour of concrete elements may also contribute to predict these risks. Numerical studies allows to reduce the costs of experiments and time to estimate the safety level of engineering structures and perform extensive simulations including various options to identify the optimal solution for a specific case.

Fracture behaviour of foamed concrete, especially of low–density (< 1800 kg/m^3), is still unknown phenomenon. Relatively limited number of studies report that the fracture energy of lightweight foamed concrete is below 25 N/m [10–16], which is only a fraction of what ordinary concrete shows [17].

2. Materials and Methods

An ordinary Portland cement CEM I 42.5R was used in current research. Water used was clean and drinkable. The procedure of preparation of foamed concrete samples and proportions of produced mixes are described in [18]. In total 20 specimens were fabricated with dimensions of 100×100×840 mm, five beams for the same mix resulting in varying density from approximately 500 to 1000 kg/m^3, see Table 1. A 3 mm wide notch was mechanically fabricated with a depth of 42 mm.

In this study three-point flexural tests were performed. The tests were performed on initially notched beams to obtain the fracture energy G_F and the maximal tensile stress at failure, according to [19, 20]. Further details regarding the test procedure were reported in previous studies [18].
Extended Finite Element Method (XFEM) is a development of the classical Finite Element Method, it allows for simulation of propagating cracks [21]. This approach does not require identification of preliminary cracks nor definition of crack paths. In addition, the crack domain does not require re-meshing during the crack propagation. A simple measure is used to detect the crack initiation. A crack starts forming when the maximal principal tensile stress exceeds the tensile strength of the material. The post–cracking behaviour (material softening) is based on the predefined value of fracture energy.
Three–point bending of foamed concrete beams with a notch were simulated using XFEM method [22]. Figure 1 shows a 2D model of the beams which represents the geometry and boundary conditions of the experiments. The loading was introduced in a displacement controlled manner through a steel roller on the upper surface of the beam at its mid-span. The loading allowed the crack to penetrate through the entire beam height. Contact properties between the steel roller and the beam were applied to represent the boundary conditions from experiments. Material parameters were set according to the experimental results, see Table 1. Rectangular mesh with the element size of 2×2 mm was applied in the model. The size of the elements in the vicinity of the notch were reduced to 1×1 mm.

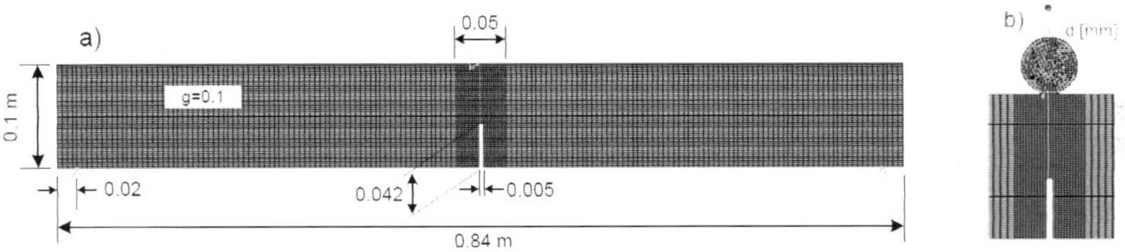

Figure 1. Geometry and boundary conditions of specimen (left), loading introduced as kinematic displacement of steel roller (right) [22].

3. Results and discussion

3.1. Experiments

The main objectives of the experiments were to investigate the influence of the density on the fracture energy and mechanical properties of foamed concrete. This includes Young's modulus and maximal tensile stress at initial cracking.

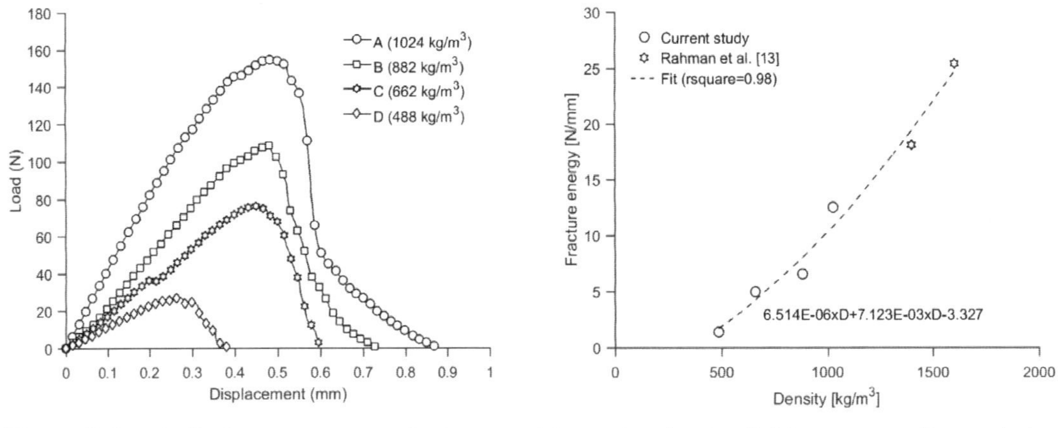

Figure 2. Load–displacement curves for beams with varying density (left), corresponding variation of fracture energy with density (right) [18].

Figure 2 (left) presents load–deflection curves for selected specimens of group A, B, C and D with different densities obtained from experimental study [18]. Structural responses, shown in Figure 2 (left), can be characterized by three particular phases which are common for all density groups. During the first phase the deflection increases linearly with the load, the notch is opening but no crack forms. A fracture process starts during the second phase where micro-cracks start forming and slow crack growth can be observed. During this phase a slight decrease of stiffness can be noticed. In the final phase (strain softening) rapid crack growth is noticed and corresponding vertical deflection increases. It can be noticed that specimens with higher density show faster increase of deformation during material softening phase thus present more brittle behaviour. It is clear from Figure 2 (left) that density affects stiffness, the value of maximal stress and ultimate displacement at failure. The specimens with higher density allow for greater displacement before failure.

Table 1. Test results from notched beams. The mean values of density, Young's modulus, tensile strength and fracture energy [18].

Material	Density (kg/m^3)	Young's modulus (GPa)	Poisson's ratio (-)	Maximal tensile stress (MPa)	Fracture energy (N/m)
A	1024	1.0	0.2	0.555	12.54
B	882	0.9		0.350	6.55
C	662	0.6		0.278	4.95
D	488	0.5		0.112	1.39

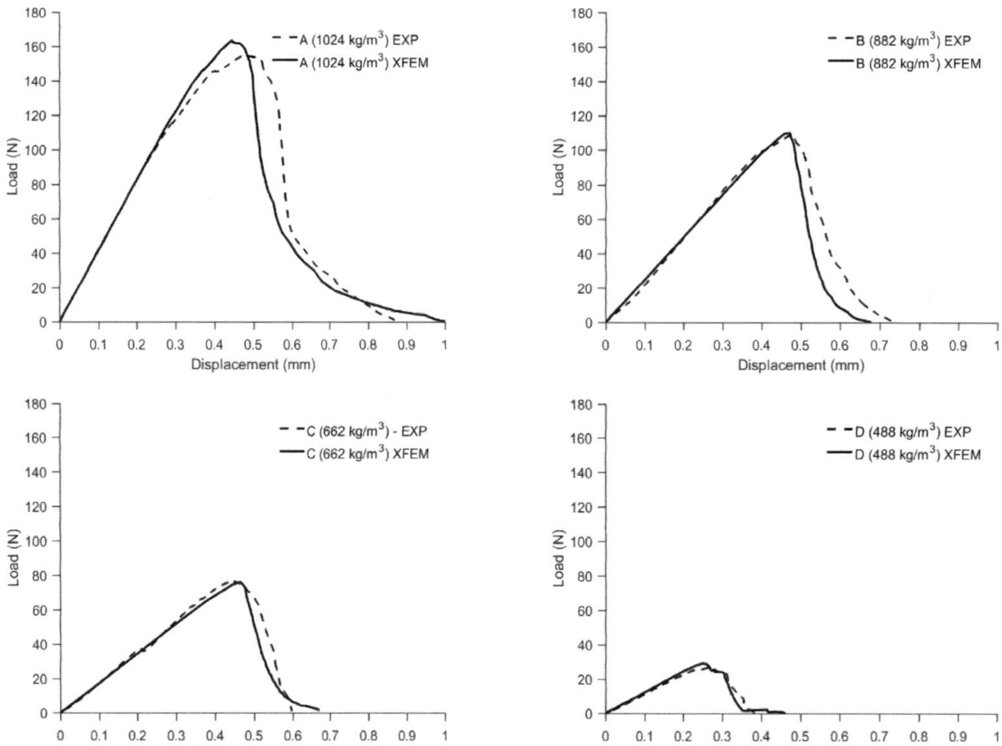

Figure 3. Comparison of experimental and numerical results for notched beams [22].

Figure 3 presents a comparison of load–deflection plots obtained from experiments and numerical simulations using XFEM method. The results present very good agreement regarding initial stiffness, maximal load and post–peak softening behaviour.

Based on the load–displacement behaviour, shown in Figure 2 (left), the value of fracture energy G_F was calculated for each specimen, according to [19, 20]. Figure 2 (right) shows a variation of the fracture energy with the density. It includes results of the current study and work by Rahman et. al [13]. The relation can be approximated with polynomial function shown in Figure 2 (right).

4. Conclusions and further study

Experiments and numerical study on notched foamed concrete beams in three-point bending were performed and simulated using the Extended Finite Element Method. The following conclusions have been drawn from the findings of the studies:

- an increase of the density of foamed concrete results in an increase of its stiffness, maximal tensile stress and fracture energy. For the notched beams with the density of 488-1024 kg/m^3 the mean values of fracture energy and maximal tensile stress obtained from experiments were 1.39-12.54 N/m and 0.112-0.555 MPa, respectively;
- XFEM method is suitable to represent fracture behaviour of notched foamed concrete beams tested in three–point bending. A comparison of the results from numerical models and experiments presents good correlation in terms of initial stiffness, peak load and post–cracking behaviour.

Based on the experimental studies and numerical simulations performed a number of topics is recommended to investigate in further studies:

- an investigation of fracture energy of foamed concrete with higher densities (>1000 kg/m^3);
- an experimental campaign involving the influence of addition of polypropylene fibres on the stiffness, strength and fracture characteristics of foamed concrete;
- numerical simulations of real application of foamed concrete as stabilization of poor soil conditions by an application of layer of foamed concrete as a subbase used in contact with subsoil.

References

[1] Van Deijk S 1999 *Foamed Concrete A Dutch View* 2-8

[2] Ramamurthy R, Kunhanandan N and Indu Siva Ranjani G 2009 A classification of studies on properties of foam concrete *Cem. and Concr. Compos.* **31** (6) 388-396

[3] Kearsley E P and Wainwright P J 2002 The effect of porosity on the strength of foamed concrete *Cem. and Concr. Res.* **32** (2) 233-239

[4] Drusa M, Fedorowicz L, Kadela M and Scherfel W 2011 Application of geotechnical models in the description of composite foamed concrete used in contact layer with the subsoil *Proc. 10th Inter. Slovak Geo. Conf. Geotechnical problems of engineering constructions (Bratislava)*

[5] Hulimka J, Krzywoń R and Knoppik-Wróbel A 2011 Use of foamed concrete in the structure of passive house foundation slab *Pro. 7th Int. Conf. on Ana. Mod. and New. Conc. In Concr. And Mas. Str. AMCM 2011*

[6] Hulimka J, Knoppik-Wróbel A, Krzywoń R and Rudišin R 2013 Possibilities of the structural use of foamed concrete on the example of slab foundation *Pro. of 9th Centr. Eur. Con. on Concr. Eng.* 67-74

[7] Hulimka J, Krzywoń R and Knoppik-Wróbel A 2017 Laboratory tests of foam concrete slabs reinforced with composite grid *Proc. of Int. Conf. on Ana. Mod. and New. Conc. In Concr. And Mas. Str. AMCM 2017 (Gliwice) Proc. Eng.* **193** 337-344

[8] Krzywoń R, Hulimka J and Jędrzejewska A Technical possibilities of foam concrete foundation slabs reinforcement *J. of Civ. Eng., Envir. And Arch.* **64** 341-350

[9] Kadela M, Kozłowski M 2016 Foamed Concrete Layer as Sub-structure of Industrial Concrete Floor *Proc. of World Multidisciplinary Civil Engineering-Architecture-Urban Planning Symposium, WMCAUS 2016 (Prague) Proc. Eng.* **161** 468-476

[10] Kadela M, Kozłowski M and Kukiełka A 2017 Application of Foamed Concrete in Road Pavement - Weak Soil System *Proc. of Int. Conf. on Ana. Mod. and New. Conc. In Concr. And Mas. Str. AMCM 2017 2017 (Gliwice) Proc. Eng.* **193** 439-446

[11] Mayer H K and van Mier J G M 2009 Fracture of Foamed Cementitious Materials: A Combined Experimental and Numerical Study *Proc. IUTAM Symp. on Mech. Prop. Of Cell. Mat. (Cachan)* 115-123
[12] Jaini Z M, Rum R H M and Boom K H 2017 Strength and fracture energy of foamed concrete incorporating rice husk ash and polypropylene mega-mesh 55 *Proc. Conf. Ser. Mat. Sci. and Eng.* **248** 012005
[13] Rahman N A, Jaini Z M and Zahir N N M 2015 Fracture energy of foamed concrete by means of the three-point bending tests on notched beam specimens *J. Earth of Sci.* **10** (15) 6562-6570
[14] Caplar R and Kulisic P 2015 An Experimental Study on the Fracture Energy of Foamed Concrete Using V-Notched Beams *Proc. Int. Civ. and Infrastruct. Eng. InCIEC 2014 (Singapore)* (Springer) 97-108
[15] Jaini Z M, Rahman N A, Rum R H M and Haurula M M 2017 Fracture Energy of Foamed Concrete: Numerical Modelling Using the Combined Finite-Discrete Element Method *MATEC Web Conf. Proc. Int. Symp. On Civ. and Environ. Eng. ISCEE 2016* **103** 02030
[16] Rahman N A, Jaini Z M, Rum R H M, Khairuddin S A A and Zamri M N 2017 Effect of Span-to-Depth Ratio on the Fracture Energy of Foam Concrete *Key Eng. Mat.* **730** 440-444
[17] Hengst R and Tressler R 1983 Fracture of foamed Portland cements *Cem. And Concr. Res.* **13** (1) 127-134
[18] Kozłowski M, Kadela M and Kukiełka A 2015 Fracture energy of foamed concrete based on three-point bending test on notched beams *Proc. of 7th Sci. Tech. Conf. Mat. Probl. In Civ. Eng. MATBUD 2015 (Cracow) Proc. Eng.* **108** 349-354
[19] RILEM FMC-50: Determination of the fracture energy of mortar and concrete by means of threepoint bend tests on notched beams.
[20] RILEM TC 187-SOC: Experimental Determination of the Stress-Crack Opening Curve for Concrete in Tension. Final report of RILEM
[21] T. Belytschko, T. Black, Elastic crack growth in finite elements with minimal remeshing. International Journal for Numerical Methods in Engineering 45 (1999) 601 – 620
[22] Kozłowski M, Kadela M and Gwóźdź-Lasoń M 2016 Numerical Fracture Analysis of Foamed Concrete Beam Using XFEM Method *Appl. Mech and Mat.* **837** 183-186

Acknowledgments
This work was supported by the on-going research project *Stabilization of weak soil by application of layer of foamed concrete used in contact with subsoil* (LIDER/022/537/L-4/NCBR/2013) financed by The National Centre for Research and Development (Poland) within the LIDER Programme.

IOP Conf. Series: Materials Science and Engineering **324** (2018) 012032 doi:10.1088/1757-899X/324/1/012032

The research of breaking rock with liquid-solid two-phase jet flow

X Z Cheng [1,2], F S Ren [1,3] and T C Fang [1]

1 School of Mechanical Science and Engineering, Northeast Petroleum University, Daqing Heilongjiang 163318, China
2 China National Petroleum Corporation, Beijing 100724, China
3 Corresponding author： renfushen@126.com

Abstracts. Particle impact drilling is an efficient way of breaking rock, which is mainly used in deep drilling and ultra-deep drilling. The differential equation was established based on the theory of Hertz and Newton's second law, through the analysis of particle impact rock, the depth of particles into the rock was obtained. The mathematical model was established based on the effect of water impact crack. The research results show when water jet speed is more than 40 m/s, rock stability coefficient is more than 1.0, the rock fracture appear. Through the experimental research of particle impact drilling facilities, analysis of cuttings and the crack size which was analyzed through Scanning electron microscope consistent with the theoretical calculation, the validity of the model was verified.

1. Introduction

Particle impact drilling is not only rapid penetration of high efficient rock, but also greatly reduced drilling cost. At present, the research of particle impact drilling is still in the experimental stage, its theoretical system needs further improvement. Scholars have conducted a series of research on particle impact drilling. Liu Jialiang et al[1] analyzed the damage of rock by numerical analysis of high pressure water jet impact rock. Through the destruction of high pressure water jet rock, Ni Hongjian et al[2] concluded that the failure form of the rock was tensile failure, and showed a step type. Li Gensheng et al[3] scanned the broken rock of high pressure water jet by scanning electron microscope, and observed the broken form of rock. Li Qinglu et al[4] He calculated the depth of the rock and the stress of the broken rock by studying the incident velocity of the particle. Kuang Chunyu[5], Ren Jianhua[6], Zhao Jian[7] et al established the particle impact rock model by ANSYS/LS-DYNA finite element modeling software, and analyzed the mechanism of particle impact breaking rock. Lu Xiaocong et al[8] used Hopkinson pressure bar equipment to impact sandstone. Li Liyun et al[9] obtained the change rule of the release strain energy and the dissipated energy through the analysis of the deformation energy of rock structure. Li Xibing et al[10] studied the stress distribution diagram

IOP Conf. Series: Materials Science and Engineering **324** (2018) 012032 doi:10.1088/1757-899X/324/1/012032

and the crack propagation law of rock under impact.

2. The effect of fluid-solid two-phase flow on rock

2.1. Mathematical modeling of single-particle impact rock

The motion equation and hertz equation are established for the impact of a single particle on the rock:

$$m_p z'' = -F'(t) \tag{1}$$

$$F(t) = (\frac{z}{k})^{3/2} \tag{2}$$

where $k = \left(\frac{9}{64} \frac{1}{R_a} \left(\frac{1-\mu_1}{E_1} + \frac{1-\mu_2}{E_2} \right)^2 \right)^{1/3}$; k is the material characteristic parameter; R_a is the radius of curvature; z is the center of gravity of the particle; $F(t)$ is the impact force of the particle on the rock; μ_1 is the poisson ratio of the particle; μ_2 is the poisson ratio of rock; E_1 is the elastic modulus of the particle; E_2 is the elastic modulus of rock.The derivative of equation (2) is substituted into (1), When the motion equation of the particle satisfies $z = 0$, $\frac{dz}{dt} = u_p$ So there is:

$$m_p u_p = -\frac{2}{5} k^{-3/2} z^{5/2} \tag{3}$$

where m_p is the mass of the particle, unit kg; u_p is the speed of the particle.

The maximum pressure depth $z_{\max}$ of the particle on the rock can be obtained:

$$z_{\max} = (\frac{5}{2} k^{3/2} m_p u_p)^{2/5} \tag{4}$$

The maximum impact of particles on the rock is:

$$F_{\max} = \frac{(\frac{5}{2} m_p u_p)^{3/2}}{k^{3/5}} \tag{5}$$

2.2 .The collision produces a crack

Using the formula of rock fracture mechanics, the crack length of rock is obtained:

$$K_1 = 1.12\sigma_y \sqrt{\pi a_e} \tag{6}$$

The equation (5) is substituted into the formula (6) to obtain the length of the crack a_e:

$$a_e = \frac{1}{\pi} \left(\frac{k^{3/5} K_1}{1.12(\frac{5}{2} m_p u_p)^{3/2}} \right)^2 \tag{7}$$

where K_1 is the first kind of strength factor; σ_y is stress; a_e is the length of the crack.

3. Mathematical modeling

The strength factors of the first and second types of rock can be obtained

$$K_1 = 5.51\bar{u}\sqrt{\pi a_0} + F(a)\sigma_{\max}\sqrt{\pi a_0} + 1.12\sigma\sqrt{\pi a_0} \tag{8}$$

$$K_{II} = 1.12\tau\sqrt{\pi a_0} \tag{9}$$

Where $\bar{u}$ is the average water pressure; K_1 is the first type of fracture strength factor; K_{II} is the second type of fracture strength factor; $a_0 = \frac{e}{\sin\beta}$; $\tau = F_1 \cos\alpha$; $F(a) = 1.122 - 1.4R_e + 7.33R_e^2 - 13.08R_e^3 + 14R_e^4$; $\sigma_{\max} = \frac{6M}{H^2}$; $\sigma = \frac{N\sin\beta}{H}$; $R_e = \frac{e}{H}$; $H = e + h \bullet \sin\theta_L$.

The crack initiation angle θ_L can be calculated as follow:

$$K_1 \sin\theta_L + K_{\text{II}}\left(3\cos\theta_L - 1\right) = 0 \quad (10)$$

The calculation method of crack fracture strength factor K_e can be expressed as follows:

$$K_e = \frac{1}{2}\cos\frac{\theta_L}{2}\left[K_1\left(1+\cos\theta_L\right) - 3K_{\text{II}}\sin\theta_L\right] \quad (11)$$

Therefore, based on the joint fracture strength factor and intact rock fracture toughness of the main control structure, the fracture mechanics expression of fracture rock stability coefficient is:

$$F_s = \frac{K_{1C}}{K_e} \quad (12)$$

where, the fracture toughness of the rock is K_{1C} .

In the process of operation, μ_L =0.5 and μ_V =0.8, fracture toughness of sandstone K_{1C} =9.8 $MPa \bullet M^{1/2}$, $\beta = 85^{\circ}$ and so on. Figure 7 is calculated for equation (12). According to Regulation 9.5.5 of the Code for Landslide Prevention and Control Engineering, the stable state of the rock containing cracks can be determined according to the stability factor of table 1.

Table 1. The stability coefficient of rock F_s .

Fs	Fs<1.0	1.0≤Fs<1.3	1.3≤Fs<1.5	Fs≥1.5
stable state	Unstable	Less stable	Basically stable	stable

According to the analysis of table 1, the stability of rock is unstable when the coefficient $F_S < 1.0$. The figure 1 shows that the stability of rock is unstable when the impact velocity of the particle is 40 m/s, the stability coefficient of the rock is 1. 0. When the particle velocity is greater than 40 m/s, the rock begins to fall off. When the velocity of the particle is less than 40 m/s, the stability coefficient of the rock increases exponentially, and the rock will not break and fall off.

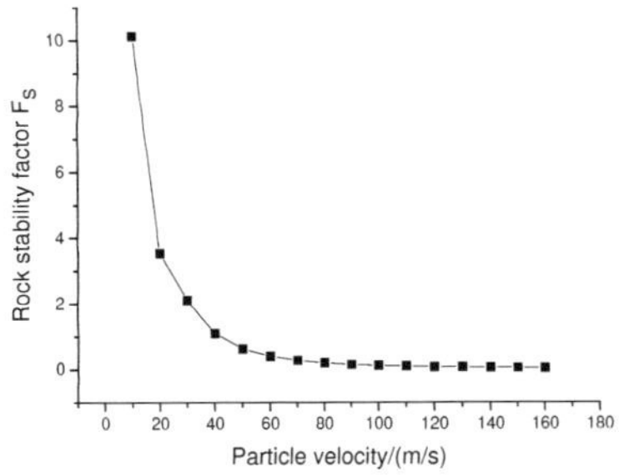

Figure 1. The velocity of particles and the stability coefficient of rock.

After each particle impact, the high pressure water acts on the rock, its crushing volume is:

IOP Conf. Series: Materials Science and Engineering **324** (2018) 012032 doi:10.1088/1757-899X/324/1/012032

$$V_{w}=h\cdot\left(\frac{H+e}{2}\right)\cdot\frac{\pi D}{n} \tag{13}$$

Since h is the depth at which particles invade the rock, substituting (4) into (13) yields:

$$V_{w}=\left(\frac{5}{2}k^{3/2}m_{p}u_{p}\right)^{2/5}\cdot\left(\frac{H+e}{2}\right)\cdot\frac{\pi D}{n} \tag{14}$$

where, D is the diameter of the particle.The volume relationship of rock with different particle velocities is shown in figure 2.

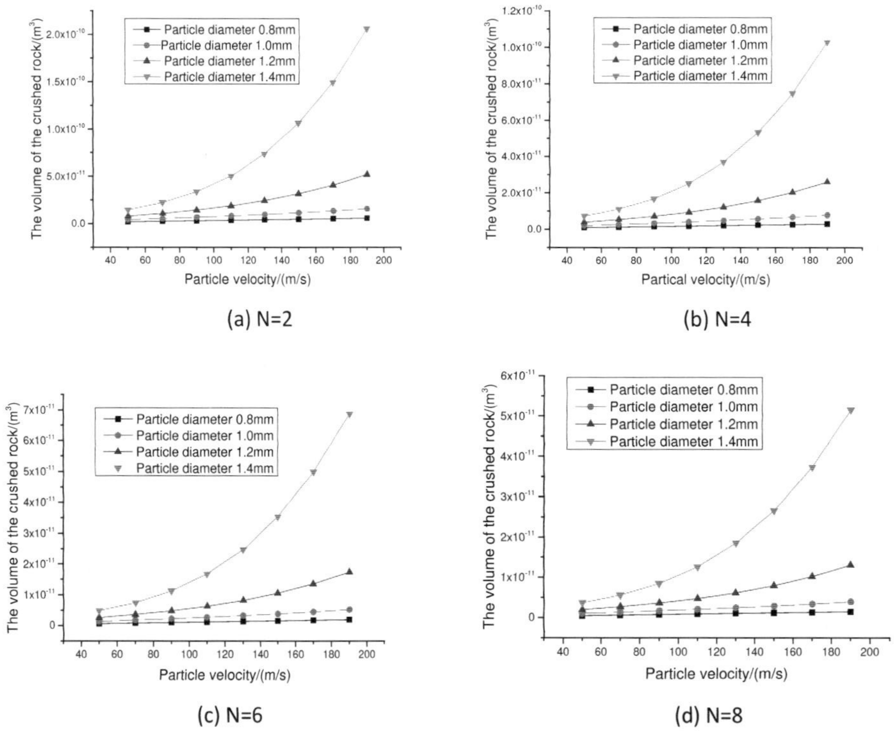

Figure 2. The volume of broken rock by water of different crack number.

It can be seen when the particle radius is more than 1 mm, the volume of water broken rock increases with the increase of particle velocity, and it increases linearly when the radius of the particle is less than 1 mm. The breaking volume of rock appears inflection point when the particle velocity is 150 m / s, the volume of broken rock is slow before the turning point, and the volume of broken rock increases rapidly after turning point.Through the above research, we can know that when the particle velocity is more than 150 m/s, the rock breaking effect reaches the best.

4. Experimental research

The diameter of the nozzle used is 4mm, the particle diameter is 0.8~1.4mm,the size of the granite is $250mm \times 250mm \times 100mm$ .In the experiment, the particles were added to the tank, and the volume of the particles in the high-pressure pipeline was controlled by adjusting the speed of the screw propeller. By adjusting the speed of high-pressure pump, control the speed of high pressure water in high-pressure pipeline, continuously for 30 seconds at different water speed, record the experimental data, observe the shape of rock debris, count the particle size distribution of the debris, and observe the rock fracture by scanning electron microscope. Table 2 is a set of parameters for the experiment.

Table 2. The distribution of the cuttings.

d<2mm	2mm<d<4mm	d>4mm
15%	60%	25%

During the experiment, the fragmentation of the rock is shown in figure 3, and the shape of the rock fracture pit is like an inverted funnel.

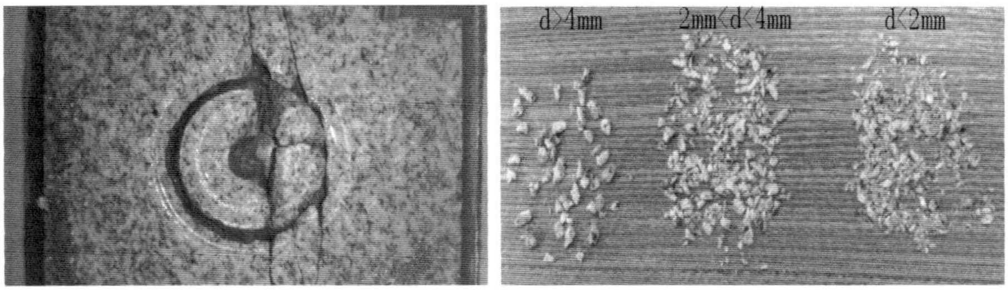

Figure 3. The picture of broken rock and cuttings.

5. Conclusions

By using the hertz theory to establish the mathematical model. When the impact velocity of water reaches 40 m / s, the stability coefficient is greater than 1.0, and the rock containing cracks begins to break.When the particle velocity reaches 150 m / s, the rock breaking effect is the best.Through the observation and study of rock debris in laboratory, the distribution proportion of cuttings diameter is obtained, and the volume is basically consistent with the theoretical analysis.

Acknowledgements

The authors would like to acknowledge financial support by the New Century Excellent Talent Training Fund of Heilongjiang Province in China, Project: 1254-NCET-005 and Postdoctoral Scientific Research Development Fund of Heilongjiang Province in China, Project: LBH-Q15018.

References

[1] Liu J L and Si H 2011 *China. J. Chongqing Univ.* **34** 40-6
[2] Ni H J, Wang R H and Zhang Y Q 2005 *China. Appl. Math. Mech.* **26** 102-5
[3] Li G S, Liao H L, Huang Z W and Shen Z H 2009 *China. J. Mech. Eng.* **45** 284-93
[4] LI Q L 2012 *Proc. 3th National Conf. on water dynamics and the 26th National Symp. on water dynamics-B Hydrodynamic basis* (China Shandong: Qingdao) pp171-76
[5] Kuang Y C, Zhu Z P, Jiang H J, Wu K S and Wang Q 2012 *China. Acta Petrol. Sin.* **33** 1059-63
[6] Ren J H, Xu Y J, Zhao J, Zhao X L and Zhang M 2012 *China. Chin. J. High. Pressure Phys.* **26** 89-94
[7] Zhao J, Wu X Z and Han L X 2013 *China. Drilling & Production Technology* **36** 1-6
[8] Lu X C, Xu J Y and Zhao D H 2011 *China. Eng. Mech.* **28** 138-44
[9] Li L Y, Xie H P and Ju Y 2011 *China. Eng. Mech.* **28** 35-40
[10] Jin J F, Li X B and Qiu C 2014 *China. Chin. J. Rock Mech. Eng.* **33** 1662-71

IOP Conf. Series: Materials Science and Engineering **324** (2018) 012033 doi:10.1088/1757-899X/324/1/012033

Analysis of titanium content in titanium tetrachloride solution

Xiaoguo Bi[1], Yingnan Dong[1*], Shanshan Li[1], Duojiao Guan[1], Jianyu Wang[1] and Meiling Tang[1]

1 Shenyang Institute of Engineering, Shenyang， P. R. China
E-mail: dyn4870@qq.com

Abstract. Strontium titanate, barium titan and lead titanate are new type of functional ceramic materials with good prospect, and titanium tetrachloride is a commonly in the production such products. Which excellent electrochemical performance of ferroelectric tempreature coefficient effect.In this article, three methods are used to calibrate the samples of titanium tetrachloride solution by back titration method, replacement titration method and gravimetric analysis method. The results show that the back titration method has many good points, for example, relatively simple operation, easy to judgment the titration end point, better accuracy and precision of analytical results, the relative standard deviation not less than 0.2%. So, it is the ideal of conventional analysis methods in the mass production.

1. Introduction

Strontium titanate ($SrTiO_3$) ultrafire power has already became a key material of electronic industry and ceramic industry .The performance of an application is surperior to the traditional materials,it in many areas (high voltage capacitor, grain boundary layer capacitor, pressure sensitive ceramics, temperature sensor, etc.). Strontium titanate powder is also a raw material of strontium titan ate monocrystalline. Strontium titanate single crystal is one of the perfect used widely high temperature superconducting single crystal substrate. It is well matched with the lattice of high temperature superconducting materials, such as Abacus, which has no twin structure and owns, good physical and mechanical properties. Using a variety of the membrane technology (such as magnetic sputtering, pulsed laser deposition, laser molecular beam epitaxy, etc.) producedly a variety of high temperature superconducting thin film (Y -, Bi -, La - system, etc.) own obtain some index date such as TCO acuity 90 ° K, JCO acuity 106 a/cm^2.

$SrTiO_3$ is also the preferred monocrystalline material for high-temperature superconducting junction technology (using twinning substrate or step substrate) and substrate step-cutting and heat treatment at a specific angle. Strontium titan ate and its derivative applications have a broad prospect, so the preparation of strontium titanate ultrafine powder is focused for a long time.

$SrTiO_3$, including of $BaTiO_3$ and $PbTiO_3$ and similar products, traditionally composed by using for solid phase .The cost of solid phase method is low , and the preparation process is simple and was big, but the particle size of powder and it was prepared by solid phase method, chemical homogeneity was poor, easy to cause pollution, and it was hard to get a single phase, which was far away from the demand of high purity materials. In recent years, liquid phase synthesis technology, including chemical co precipitation, sol gel is development rapidly, are generally four isopropyl titanium, titanium source four butyl titan ate and titanium tetrachloride, in which titanium tetrachloride is relatively cheap, is the industrial production for the most ideal choice. Therefore, the determination of

IOP Conf. Series: Materials Science and Engineering **324** (2018) 012033 doi:10.1088/1757-899X/324/1/012033

titanium tetrachloride water solution is one of the key technologies for $SrTiO_3$ or $BaTiO_3$, $PbTiO_3$ powder synthesis[1].

The determination of titanium tetrachloride ($TiCl_4$) water solution is rarely reported. The classic method of determination of titanium (HGB 3252-1960) is the oxidation reduction titration method of ammonium sulfate. Of this method measuring results are accurate and reliable, but the experiment has many steps, taking a long time, the oxidizing atmosphere requires. For this experiment it is difficult to control, and it requires special glass or instrument, it is hard to be used in the conventional test for the production process. Gravimetric analysis method[2] is also considered to be a more reliable method, but the gravimetric analysis method needs about three working days, which is either not suitable for mass production. EDTA-complexometric titration was reported by more researchers in the reference of titanium compound, titanium alloy and titanium content analysis. Although titanium and EDTA can form stable complexes, the reaction rate is slow. Therefore, the back titration method[3-5] or replacement titration method[6, 7] are usually adopted by researchers. The former uses lactic acid, malice acid, bitter almond acid, tart rate, citrate, phosphate, etc. As the release agent, which can quantitatively replace EDTA complexed with titanium. The latter is in the presence of hydrogen peroxide to makes the titanium and EDTA complex, titrate excess EDTA after a short time.

This article in view of the high purity titanium tetrachloride configuration solution method of calibration. According to the report in the reference the back titration method, the replacement titration method and the gravimetric analysis method three by the specific traget of this article in view for calibration samples of titanium tetrachloride solution. The results show that titanium tetrachloride solution method of calibration using the EDTA complexes with titanium ions, zinc standard solution titrate excess EDTA, has many good points, for example, relatively simple operation, easy to judgment the titration end point, better accuracy and precision of analytical results, the relative standard deviation (RSD) not less than 0.2%. So, it is the ideal of conventional analysis methods in the mass production.

2. Experiment

The chemical reagents used for the experiment were all analytically pure agents while ultrapure water was also used.

2.1 Back titration method

Add the 25.00 ml EDTA standard solution to the conical flask. Add 1ml hydrogen peroxide in the mixing condition, and quickly add 2.500 ml titanium tetrachloride solution and place 10 min. In stirring conditions, add 25 ml sodium acetate (2 mol/L) and a few drops of dimethylphenol orange solution, and use ammonia water (1 + 1) to the solution appearing red, and to pH=5.65 with hydrochloric acid (1 + 1). The concentration of $TiCl_4$ was calculated by using zinc standard solution to return the excess of EDTA and recording the volume of consumption.

2.2 Replacement titration method

(1) put 10.00 mL titanium standard solution into conical flask, (2) add 20.00 mL EDTA solution and 5 mL masking solution (sulphur succinct acid- acetyl acetone solution), (3) heat to boil for 2 min, (4) cooling to 60 ℃ ~ 70 ℃, (5) add 10 mL acetate buffer medium of pH 5.5, several drops PAN - Brij - 35 mixed indicator. (6) Titration with Zn standard solution from bright yellow to red (excluding reading). (7) Add 15 ml tartrate solution (release agent), (8) heat to boil 2 min, (9) cooling and 60 ℃ ~ 70 ℃. (10) Titration with Zn standard solution titration from bright yellow to purplish red, calculate the content of $TiCl_4$.

2.3 Gravimetric analysis method:

Excessive ammonia water was used to fully hydrolyze $TiCl_4$ completely into TiO_2. Ash free filter paper was used for filter washing. The constant heavy crucible was put into the maf furnace to reach constant weight, was calculated at last $TiCl_4$ content.

IOP Conf. Series: Materials Science and Engineering **324** (2018) 012033 doi:10.1088/1757-899X/324/1/012033

3. Results and discussion

The three batches of titanium tetrachloride aqueous solution were calibrated by back titration method, replacement titration method and gravimetric analysis method (parallel experiment 5 times). The experimental results were showed in table 1. From the results, the date obtained by the three methods are similar. It shows that the three methods all can be used to analyze the concentration of titanium tetrachloride solution. The back titration method and gravimetric analysis method have relatively high precision, RSD is no more than 0.2% and 0.3% respectively. The replacement titration method date is slightly higher, RSD is less than 1.0%. According to the reference and experimental results, the gravimetric analysis method is more reliable, but the process cycle is too long, which is recommended to be used as a comparison and reference analysis method in mass production. The RSD values of replacement titration method is higher than other two methods, So it is speculated that this is mainly due to the experiment process. For example, the replacement titration method operation is relatively complicated , and replacement which is not fully completed. Changing color is not sharp, is difficult to determine. By analysing the references, the main advantage of this method can undergo with the exist of a large number of interference ions titanium content analysis, this paper, analysis object is high purity titanium tetrachloride of aqueous solution. So the replacement titration method should not be the best choice.

Table 1. Analysis results of $TiCl_4$ concentration

$TiCl_4$ solution	Analysis method	Average value/ mol/L	Relative standard deviation
NO. 1	Back titration	2.186	0.11%
~2.2 mol/L	Replacement titration	2.184	0.82%
	Gravimetric analysis	2.188	0.24%
NO. 3	Back titration	2.025	0.18%
~2.0 mol/L	Replacement titration	2.030	0.95%
	Gravimetric analysis	2.028	0.20%
NO. 3	Back titration	1.174	0.16%
~1.8 mol/L	Replacement titration	1.171	0.78%
	Gravimetric analysis	1.175	0.26%

For the back titration method, complexation reaction of titanium and EDTA with in the presence of H_2O_2, are as follows: $TiO(H_2O_2)_2{}^{+} + Y4^{-} \rightleftharpoons TiOY_2 - (H_2O_2)$. The stable constant complex of the Ti-H_2O_2 -EDTA is 2.7×10^{20}. In this paper, the stability of the complex, H_2O_2 and excessive EDTA standard solution are added into aqueous solution, which can stop the hydrolysis of titanium, after a short time, zinc standard solution to titrate excessive EDTA standard solution for determining the content of titanium. The analysis method is relatively simple and easy to judge. The accuracy and precision of the analytical results are good, which is the ideal method for conventional analysis in mass production.

At the same time, the experimental data got by back titration method seems to have a very small systematic bias compared with the gravimetric analysis method in the experiment. This may be due to the fact that the hydrolysis of titanium tetrachloride which hydrolysis is quite complicated[8]. Although it is relatively stable under the high concentration conditions, very slow hydrolysis still exists. The hydrolysis product was negative in the analysis of the back titration method, which the hydrolysis was positive in the gravimetric analysis method. Therefore, regular back titration and gravimetric analysis method can be used to supervise the quality of titanium tetrachloride solution by comparing.

4. Conclusions

The back titration method, the replacement titration method and the gravimetric analysis method are used to calibrate the samples of titanium tetrachloride solution. The results show that the three methods can effectively analyze the concentration of $TiCl_4$ solution. Among the three methods, the back titration method has many good points, for example, relatively simple operation, easy to judgment the titration end point, better accuracy and precision of analytical results, the relative standard deviation not less than 0.2%. So, it is the ideal of conventional analysis methods in the mass production. The replacement titration method is more suitable for the analytical method in the complex composition sample and the gravimetric analysis can be used as the comparing experimental method.

Acknowledgments
Authors wishing to acknowledge assistance financial support from the fund of the Education Department of Liaoning Province (ZXLGL-1614).

References

[1] Xiaoguo Bi, Yingnan Dong and Yingjie Li 2017 Study on the Key Technology of High Purity Strontium Titanate Powder Synthesized from Oxalic Acid Co-sediment Precipitation *Mater. Sci. Eng.* **230** 012008
[2] Feng XiangQin, Tao XiangJun and Li PengFei 2015 Method for determining the Content of Barium Titanate *FUJIAN ANALYSIS & TESTING* **24** 34
[3] Zhang Lijun and Zhou Zhumin 1996 Determination of Titanium-to-Strontium Ratio in High Grade $SrTiO_3$ *Powder Bull. Chin. Ceramic. Soc.* **1** 54
[4] LU Yahong and Jiang Feng 2001 Rapid Determination of Lead and Titanium in High Purity Superfine Lead Titanate Powder *YUNNANMETALLURGY* **30(2)** 55
[5] YAN Qingzhi, YANG Feiyu and GE Changchun 2004 Study on the chemical analysis method of titanium and barium during the preparation of barium titanate powder *Metallurgical Analysis* **24** 47
[6] Wang Xianke, Li Yuping and Li Lifen 1997 ETERMINATION OF TITANIUM BY SELECTIVE CHELATOMETRIC TITRATION WITH TARTRATE AS RELEASING AGENT *HEBEI METALLURGY* **101** 50
[7] Tian Li, Tan Yongfang and Zhou Wude 2009 HE DETERMINATION OFTITANIA CONTENT IN THE CATALYST BY DISPLACEMENT TITRATION METHOD *QILU PETROCHEMICAL TECHNOLOGY* **37** 243
[8] Chen Ruicheng 1999 Study on the hydrolysis process of TiO_4 *Hydrometallurgy of China* **71** 1

IOP Conf. Series: Materials Science and Engineering **324** (2018) 012034 doi:10.1088/1757-899X/324/1/012034

Numerical analysis of strain localization for transversely isotropic model with non-coaxial flow rule

Ding Wei[1],Yu Cong-cong[1],Wu Chen-hui[1],Shu Zheng-yi[1]

1. Wuhan Second Ship Design and Research Institute ,Wuhan Hubei 430061
E-mail: dingwei1117@163.com

Abstract: To analyse the strain localization behavior of geomaterials, the forward Euler schemes and the tangent modulus matrix are formulated based on the transversely isotropic yield criterion with non-coaxial flow rule developed by Lade, the program code is implemented based on the user subroutine (UMAT) of ABAQUS. The influence of the material principal direction on the strain localization and the bearing capacity of the structure are investigated and analyzed. Numerical results show the validity and performance of the proposed model in simulating the strain localization behavior of geostructures.

1. Introduction

In the development of plastic mechanics, a basic assumption in classical plasticity mechanics is the law of orthogonal flow which means the direction of incremental plastic strain and the direction of stress are coaxial. With the further development of experimental studies, it has been found that the plastic strain increment deviates from the stress direction in the process of plastic deformation [1-2]. This phenomenon undermines the basic assumptions of many classical elastic-plastic constitutive models and makes many traditional elastic-plastic constitutive models unable to simulate the strain localization of soil bifurcation occurring in geomaterials. Therefore, many experts and scholars have carried out non-coaxial theory adding non-coaxial effects in the classical elastoplastic constitutive model, thus developing many non-coaxial models to study the anisotropic mechanical behavior of geomaterials. For example, Rudnicki and Rice [3] established a non-coaxial model to accurately predict the strain localization behavior of pressure-sensitive dilatant materials. Papamichos and Vardoulakis [4] successfully simulate the formation of sand shear bands in the compression process based on the elastoplastic non-coaxial model. Hashiguchi and Tsutsumi [5] used sub-loading surface model to simulate the bifurcation of sand and made an improvement for the bifurcation prediction. Huang et al [6] also conducted non-coaxial theory, and revised the bifurcation prediction based on non-coaxial model. Comparison of the predicted results and the experimental results showed a consistent outcome. The non-coaxial constitutive models formed by the combination of non-coaxial theory and classical constitutive model can well simulate the mechanical behavior of material strain localization. Therefore, a large number of non-coaxial elastic-plastic constitutive models were constructed by introducing classical elastoplastic constitutive models into the yield function or flow law to simulate the localized mechanical behavior of geomaterials [7-8]. Such models were more capable to predict the bifurcation of geomaterials and explain the anisotropy of geomaterials.

In order to further analyze the strain localization behavior of geomaterials and to predict the bifurcation behavior of materials more accurately. This paper simplified the transverse isotropic yield criterion of Lade [9-10], deduced the iterative scheme and the constitutive matrix of the corresponding forward Euler algorithm, and realized the corresponding program code through the UMAT interface of the finite element software ABAQUS. Numerical simulations were used to investigate the influence of

IOP Conf. Series: Materials Science and Engineering **324** (2018) 012034 doi:10.1088/1757-899X/324/1/012034

the main direction angle of the material on the ultimate bearing capacity and deformation localization of geotechnical structures and the effect of non-coaxial stress on the material.

2. Simplification of Lade Transversely Isotropic Model and Derivation of Finite Element Iteration

2.1. Transverse isotropic elastic constitutive descriptions

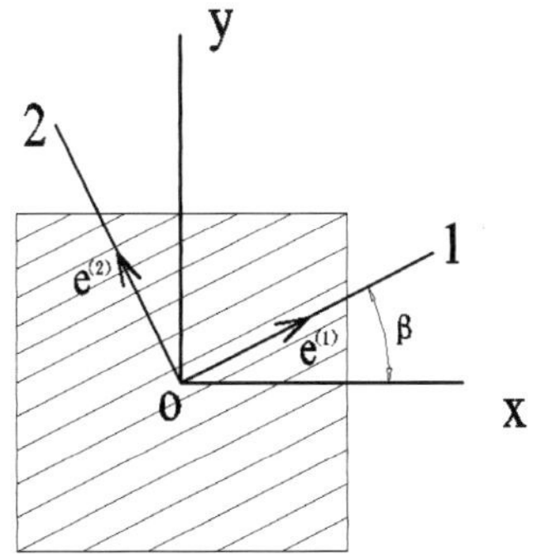

Fig 1. Material principal orientation

The main directions and axes of the transversely isotropic material are shown in the above diagram [11], β is the angle between the main direction of the material and the x-axis (hereinafter referred to as the main direction of the material). 1o2 indicates the local coordinate system; xoy indicates the overall coordinate system. 3-axis is perpendicular to 1o2 surface, 1o3 surface is isotropic, direction along 2-axis is the normal direction. $\boldsymbol{D}$ refers to the local coordinate system elastic constitutive matrix, then in the overall coordinate system the elastic constitutive matrix $\boldsymbol{D}_e$ is

$$\boldsymbol{D}_e = \boldsymbol{R}\boldsymbol{D}\boldsymbol{R}^{\mathrm{T}} \tag{1}$$

Where $\boldsymbol{R}$ is the transformation matrix:

$$\boldsymbol{R} = \begin{bmatrix} \cos^2\beta & \sin^2\beta & 0 & -2\sin\beta\cos\beta & 0 & 0 \\ \sin^2\beta & \cos^2\beta & 0 & 2\sin\beta\cos\beta & 0 & 0 \\ 0 & 0 & 1 & 0 & 0 & 0 \\ \sin\beta\cos\beta & -\sin\beta\cos\beta & 0 & \cos^2\beta-\sin^2\beta & 0 & 0 \\ 0 & 0 & 0 & 0 & \cos\beta & -\sin\beta \\ 0 & 0 & 0 & 0 & \sin\beta & \cos\beta \end{bmatrix} \tag{2}$$

$\boldsymbol{D}$ is showed in the following form:

$$\boldsymbol{D} = \begin{bmatrix} \frac{E_1(1-L\upsilon_2^2)}{(1+\upsilon_1)(1-\upsilon_1-2L\upsilon_2^2)} & \frac{E_1\upsilon_2}{1-\upsilon_1-2L\upsilon_2^2} & \frac{E_1(\upsilon_1+L\upsilon_2^2)}{(1+\upsilon_1)(1-\upsilon_1-2L\upsilon_2^2)} & 0 & 0 & 0 \\ \frac{E_1\upsilon_2}{1-\upsilon_1-2L\upsilon_2^2} & \frac{E_2(1-\upsilon_1)}{1-\upsilon_1-2L\upsilon_2^2} & \frac{E_1\upsilon_2}{1-\upsilon_1-2L\upsilon_2^2} & 0 & 0 & 0 \\ \frac{E_1(\upsilon_1+L\upsilon_2^2)}{(1+\upsilon_1)(1-\upsilon_1-2L\upsilon_2^2)} & \frac{E_1\upsilon_2}{1-\upsilon_1-2L\upsilon_2^2} & \frac{E_1(1-L\upsilon_2^2)}{(1+\upsilon_1)(1-\upsilon_1-2L\upsilon_2^2)} & 0 & 0 & 0 \\ 0 & 0 & 0 & G_2 & 0 & 0 \\ 0 & 0 & 0 & 0 & G_1 & 0 \\ 0 & 0 & 0 & 0 & 0 & G_2 \end{bmatrix} \tag{3}$$

The stress and strain satisfy the linear elasticity constitutive relation:

$$\varepsilon = \boldsymbol{D}_e \sigma \tag{4}$$

Where E_1, υ_1 and G_1 are elastic modulus, Poisson's ratio and shear modulus in the isotropic plane. E_2, υ_2 and G_2 are elastic modulus, Poisson's ratio and shear modulus perpendicular to the direction of the isotropic surface. Moreover $G_1 = E_1 / 2(1+\upsilon_1)$, $L = E_1 / E_2$, so there are only five independent material parameters, that is $E_1, \upsilon_1, E_2, \upsilon_2, G_2$.

2.2. Simplified Lade Transverse Isotropic Yield Criterion

Lade's method of anisotropy described by Pietruszczak and Morz in the form of tensor tensor [12] combined with the Lade-Kim isotropic yield criterion gives a suitable method for describing the extension of the transversely isotropic geomaterials Lade's criterion [9], the yield function is as follows:

$$f_p = f_p^{'}(\sigma) - f_p^{''}(W_p) \tag{5}$$

W_p indicates the plastic work, $f_p^{'}$ and $f_p^{''}$ represent stress-related terms and plastic work-related terms respectively in the yield function

$$f_p^{'}(\sigma) = \left(\frac{I_1^3}{I_3} - 27\right)\left(\frac{I_1}{p_a}\right)^m - \eta_0[(1+\Omega_1(1-3l_2^2)] \tag{6}$$

In equation (6), P_a is the atmospheric pressure, m and η_0 are both dimensionless constants, Ω_1 is the anisotropy intensity parameter, and l_2 is the loading direction. In order to improve the convergence of the Lade transverse isotropic model in the forward Euler algorithm and to reduce the computational parameters, the plastic work function is simplified, the parameters are integrated, only two dimensionless constants A and B are retained.

$$f_p^{''}(W_p) = A(\frac{W_p}{P_a})^B \tag{7}$$

Due to the experimental conditions, the law of associated flow is adopted, namely the plastic potential function is equal to the yield function. This paper simplifies the transverse isotropic model of Lade, deduces the elastic-plastic modulus tensor of the simplified finite element model, develops the constitutive integral algorithm, and realizes the corresponding program code based on UMAT interface of ABAQUS. The numerical simulations of plane strain biaxial compression tests are shown in the following section.

2.3. Elastic-plastic modulus tensor

Relationship between stress and strain increments:

$$d\sigma = \boldsymbol{D}_{ep} d\varepsilon = (\boldsymbol{D}_e - \boldsymbol{D}_p) d\varepsilon \tag{8}$$

$\boldsymbol{D}_{ep}$ refers to the plastic modulus tensor, expand as follows:

$$\boldsymbol{D}_{ep} = \boldsymbol{D}_e - \frac{\boldsymbol{D}_e(\frac{\partial g_p}{\partial \sigma})(\frac{\partial f_p}{\partial \sigma})^{\mathrm{T}} \boldsymbol{D}_e}{(\frac{\partial f_p}{\partial \sigma})^{\mathrm{T}} \boldsymbol{D}_e (\frac{\partial g_p}{\partial \sigma}) + H} \tag{9}$$

Where $\frac{\partial f_p}{\partial \sigma} = \frac{\partial f_p^{'}}{\partial \sigma}$, the specific expression in Eq. (9) is as follows:

$$\frac{\partial f_p^{'}}{\partial \sigma} = \frac{\partial f_p^{'}}{\partial I_1}\frac{\partial I_1}{\partial \sigma} + \frac{\partial f_p^{'}}{\partial I_3}\frac{\partial I_3}{\partial \sigma} + \frac{\partial f_p^{'}}{\partial l_2}\frac{\partial l_2}{\partial \sigma} \tag{10}$$

The flow direction consists of two parts. The first part $\frac{\partial f_p^{'}}{\partial l_2}\frac{\partial l_2}{\partial \sigma}$ contains the anisotropy tensor (which is not zero), so the direction of this part is different from that of stress increment, namely non-coaxial.

The other part $\frac{\partial f_p^{'}}{\partial I_1}\frac{\partial I_1}{\partial \boldsymbol{\sigma}}+\frac{\partial f_p^{'}}{\partial I_3}\frac{\partial I_3}{\partial \boldsymbol{\sigma}}$ is coaxial with the direction of stress increment, namely a co-axial term [13]. Shown as follows:

$$\frac{\partial f_p^{'}}{\partial I_1}=\frac{m}{I_1}f_p^{'}+\frac{3I_1^2}{I_3}\left(\frac{I_1}{p_a}\right)^m \tag{11}$$

$$\frac{\partial f_p^{'}}{\partial I_3}=-\frac{I_1^3}{I_3^2}\left(\frac{I_1}{p_a}\right)^m \tag{12}$$

$$\frac{\partial f_p^{'}}{\partial l_2}\frac{\partial l_2}{\partial \boldsymbol{\sigma}}=3\eta_0\Omega_1\frac{\partial l_2^2}{\partial \boldsymbol{\sigma}} \tag{13}$$

$$\frac{\partial l_2^2}{\partial \boldsymbol{\sigma}}=\frac{\frac{\partial[tr(\boldsymbol{M}\hat{\boldsymbol{\sigma}}^2)]}{\partial \boldsymbol{\sigma}}tr\hat{\boldsymbol{\sigma}}^2-\frac{\partial(tr\hat{\boldsymbol{\sigma}}^2)}{\partial \boldsymbol{\sigma}}tr(\boldsymbol{M}\hat{\boldsymbol{\sigma}}^2)}{(tr\hat{\boldsymbol{\sigma}}^2)^2} \tag{14}$$

$$\boldsymbol{M}=\begin{Bmatrix}-\sin\beta\\ \cos\beta\\ 0\end{Bmatrix}\begin{bmatrix}-\sin\beta & \cos\beta & 0\end{bmatrix}=\begin{bmatrix}\sin^2\beta & -\sin\beta\cos\beta & 0\\ -\sin\beta\cos\beta & \cos^2\beta & 0\\ 0 & 0 & 0\end{bmatrix} \tag{15}$$

Plastic power increment is $dW_p=d\lambda\frac{\partial g_p}{\partial \boldsymbol{\sigma}}$, $d\lambda$ refers to plastic multiplier, expand as follows:

$$d\lambda=\frac{\boldsymbol{D}_e(\frac{\partial f_p}{\partial \boldsymbol{\sigma}})^{\mathrm{T}}}{(\frac{\partial f_p}{\partial \boldsymbol{\sigma}})^{\mathrm{T}}\boldsymbol{D}_e(\frac{\partial g_p}{\partial \boldsymbol{\sigma}})+H}d\varepsilon \tag{16}$$

As a result of the associated flow law, in the above equation $\frac{\partial g_p}{\partial \boldsymbol{\sigma}}=\frac{\partial f_p}{\partial \boldsymbol{\sigma}}$, the hardening modulus H is obtained as follows:

$$H=-\frac{\partial f_p^{''}}{\partial W_p}\boldsymbol{\sigma}\frac{\partial g_p}{\partial \boldsymbol{\sigma}} \tag{17}$$

2.4. Constitutive integral algorithm

In this article, the stress and hardening parameters are updated using the forward Euler algorithm. In the forward Euler algorithm, the stress is replaced by an infinitely small elastoplastic stress-strain relationship by an incremental step. The initial value of the elastic-plastic constitutive matrix is determined from the initial stress state because the elastoplastic constitutive matrix is determined by the history of stress and strain, therefore this linear approximation can be accurate only at very small strain increments. So the strain is divided into smaller sub-strain increments:

$$\Delta\boldsymbol{\sigma}=\boldsymbol{C}^{ep}(\boldsymbol{\sigma}_0,\kappa_0)\Delta\varepsilon,\quad \delta\varepsilon=\Delta T\Delta\varepsilon=\frac{\Delta\varepsilon}{n} \tag{18}$$

ΔT is a fixed dimensionless time step, the finite increment of stress incremented is the sum of n incremental steps of the stress, $\delta\boldsymbol{\sigma}$ calculated as a forward Euler step:

$$\delta\boldsymbol{\sigma}_i=\boldsymbol{C}^{ep}(\boldsymbol{\sigma}_0+\Delta\boldsymbol{\sigma}_{i-1},\kappa_0+\Delta\kappa_{i-1})\delta\varepsilon \tag{19}$$

$$\delta\kappa_i=\delta\lambda_p(\boldsymbol{\sigma}_0+\Delta\boldsymbol{\sigma}_{i-1},\kappa_0+\Delta\kappa_{i-1},\delta\varepsilon)(\boldsymbol{\sigma}_0+\Delta\boldsymbol{\sigma}_{i-1})\frac{\partial g}{\partial(\boldsymbol{\sigma}_0+\Delta\boldsymbol{\sigma}_{i-1})} \tag{20}$$

Where:

$$\Delta\boldsymbol{\sigma}_{i-1}=\sum_{j=1}^{i-1}\delta\boldsymbol{\sigma}_j,\quad \Delta\kappa_{i-1}=\sum_{j=1}^{i-1}\delta\kappa_j \tag{21}$$

IOP Conf. Series: Materials Science and Engineering **324** (2018) 012034 doi:10.1088/1757-899X/324/1/012034

Forward Euclidean compared with other algorithms has the advantage of simple and direct and easy to implement, the process shown in Table 1:

Tab 1. Subincremental forward Euler scheme

Initial state $\boldsymbol{\sigma}_0$, κ_0 , $\Delta\varepsilon$, $\delta\varepsilon = \dfrac{\Delta\varepsilon}{m}$

Strain increment step $i=1,2,....,m$

$$\delta\boldsymbol{\sigma} = \boldsymbol{C}^{ep}(\boldsymbol{\sigma}_{i-1}, \kappa_{i-1})\delta\boldsymbol{\varepsilon}$$

$$\delta\kappa_i = \delta\lambda_p(\boldsymbol{\sigma}_{i-1}, \kappa_{i-1}, \delta\boldsymbol{\varepsilon})\boldsymbol{\sigma}_{i-1}\frac{\partial g}{\partial \boldsymbol{\sigma}_{i-1}}$$

$$\boldsymbol{\sigma}_i = \boldsymbol{\sigma}_{i-1} + \delta\boldsymbol{\sigma}_i$$

$$\kappa_i = \kappa_{i-1} + \delta\kappa_i$$

When $i=m$, stop calculating the sub-strain increment step

Ultimate state $\boldsymbol{\sigma}_i$, κ_i , $\boldsymbol{C}^{ep}(\boldsymbol{\sigma}_i, \kappa_i)$

3. **Numerical examples**

Based on ABAQUS we developed the corresponding user material subroutine (UMAT). The finite element model is a flat plate of 2m * 1m with a total of 20 * 10 units, using a planar strain 8-node 4 integral point reduction integral unit (CPE8R).

This paper numerically simulates the plane strain biaxial compression tests. The simulation process includes three loading steps:

1) The establishment of the initial isotropic stress field with the stress equals to 20kPa;
2) Isotropic consolidation, external load from 20kPa to 500kPa;
3) Maintain the confining pressure as 500kPa, impose the displacement load by moving the upper and lower boundary symmetrically, The first and second loading step model boundaries are shown in Figure 2a, and the third loading step is shown in Figure 2b.

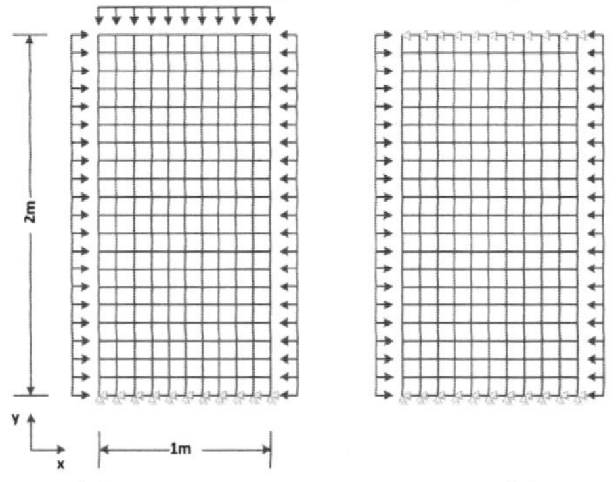

(a) (b)

Fig 2. Sketches of the model

The parameters [10] in the calculation are shown in table 2.

Table 2. Material parameters

E_1 /MPa	E_2 /MPa	υ_1	υ_2	G_2 /MPa	β	m	η_0	Ω_1	P_a /MPa	A	B
22.06	18.00	0.183	0.195	12.92	-	0.658	80.19	-0.10	0.101	68.597	0.343

Figure 3 shows the evolution of plastic work at $\beta = 30^\circ$ (unit: kPa), where u_y is the displacement applied to the upper boundary. It can be seen that during the formation of the shear band, a weaker shear band is formed in the direction of the weaker strength and then gradually developed into a stronger single shear band. At the same time another intersecting shear zone began to form. With the gradually development of the shear zone, eventually, a clear X-shaped shear band is formed as a whole. In this process, the plastic work in the direction of shear band is large. The anisotropy of strain localization is well reflected. The bifurcation of geomaterials is simulated and the material shear damage phenomenon is explained successfully.

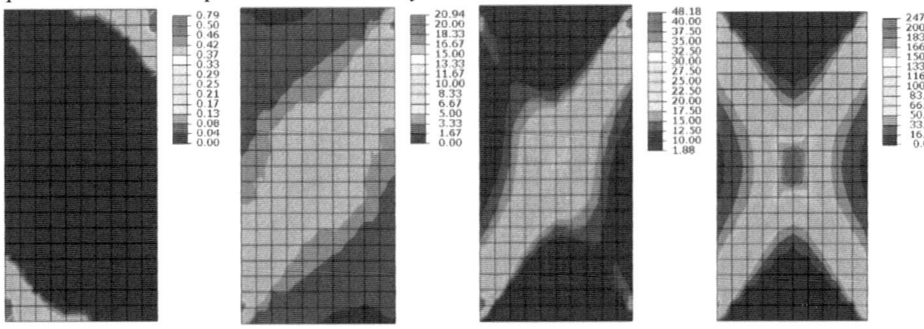

(a) $u_y = 0.08m$ (b) $u_y = 0.15m$ (c) $u_y = 0.20m$ (d) $u_y = 0.40m$

Fig 3. Variation of shear band with the increasing displacement

In order to investigate the influence of the main direction angle of the material on the shear band, that is, the influence of the angle between the bedding plane and the x-axis on the geotechnical material. Now, the changes of plastic work cloud when the material is compressed are numerically simulated with the angles of 0 °, 30 °, 60 ° and 90 °, respectively. Figure 4 shows the plastic work with the change of main direction when the upper boundary of the displacement applied to 0.4m. It can be seen that the two conjugate shear bands are basically symmetrical when $\beta = 0^\circ$ and $\beta = 90^\circ$, When $\beta = 30^\circ$ and $\beta = 60^\circ$, the two shear bands appear to be destructive along the weaker direction, and then become stronger in this direction. Combining the strain maps of different main directions it shows that the material failure modes are different when the main orientation angles are different, which shows the anisotropic characteristics of the material fracture modes. It can be seen from Fig. 4 that with the increase of β, the peak value of plastic work gradually decreases, and that the shear band width increases gradually as a whole. This is a better reflection of the evolution mode of material strain localization as β increases. It is a good predictor of the influence of different bedding angles on the failure mode of geomaterials, which can provide reference for explaining the phenomenon of shear-slip failure of geomaterials.

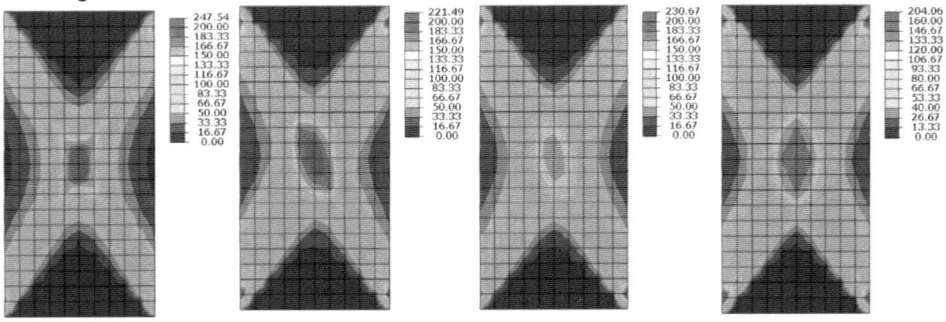

(a) $\beta = 0^\circ$ (b) $\beta = 30^\circ$ (c) $\beta = 60^\circ$ (d) $\beta = 90^\circ$

Fig 4. Distribution of plastic work with different material principal direction

Figure 5 shows the displacement curves of the bearing capacity under different main directions. It can be seen that the bearing capacity of $\beta = 30^\circ$ and $\beta = 60^\circ$ under the same displacement is larger at the initial loading. With the increase of loading displacement, the ultimate load-bearing relationship shows the highest bearing capacity of $\beta = 0^\circ$, the smallest bearing capacity of $\beta = 90^\circ$. Ultimately, the bearing capacity increases linearly with the increase of β.

IOP Conf. Series: Materials Science and Engineering **324** (2018) 012034 doi:10.1088/1757-899X/324/1/012034

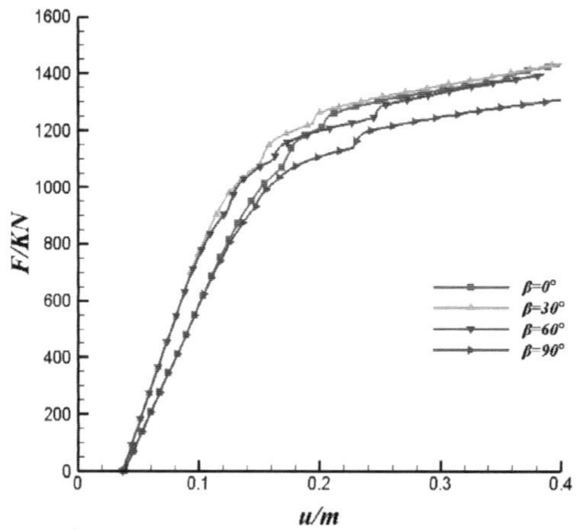

Fig 5. Variation of different material principal direction angle' bearing capacity with the displacement

From the plastic flow direction, we can see that $\frac{\partial f_p^{'}}{\partial I_2}\frac{\partial I_2}{\partial \sigma}$ is a non-coaxial term and $\frac{\partial f_p^{'}}{\partial I_1}\frac{\partial I_1}{\partial \sigma}+\frac{\partial f_p^{'}}{\partial I_3}\frac{\partial I_3}{\partial \sigma}$ is a coaxial term. To analyze the influence of non-coaxial part of the flow law during the loading process, take the middle node of the model for example. The curve of the ratio of the first invariant of the non-coaxial part to the coaxial part with node displacement of a node at $\beta = 30^{\circ}$ is shown in Fig. 6a. The ratio remains the same in elastic stage, combined with the corresponding time-step plastic power cloud chart (Figure 7), it can be seen that there is a faster scale-down process after the formation of the shear band. Fig. 6b shows more details of the process. Continue loading, the proportion gradually reduced, and tends to 0. The results show that after the plasticity is entered, the influence of non-coaxial stress on the material gradually decreases with loading, and finally its influence vanishes.

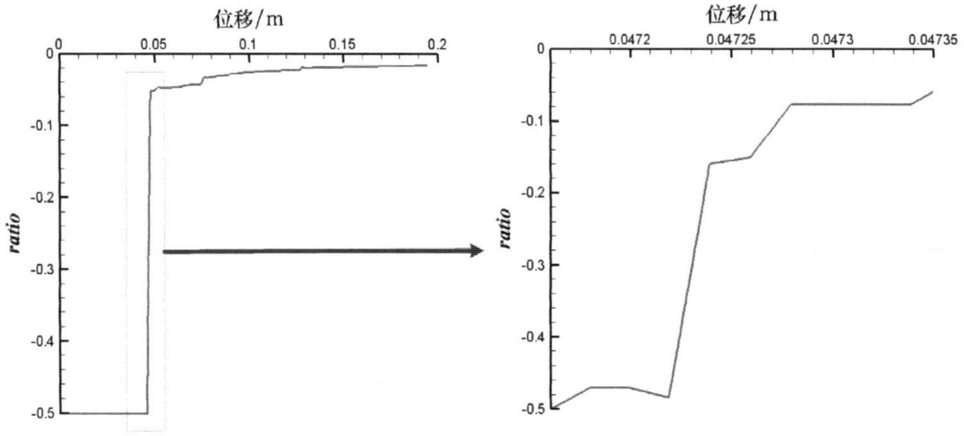

(a) (b)

Fig 6. Variation of non-coaxial percentage in the process of loading

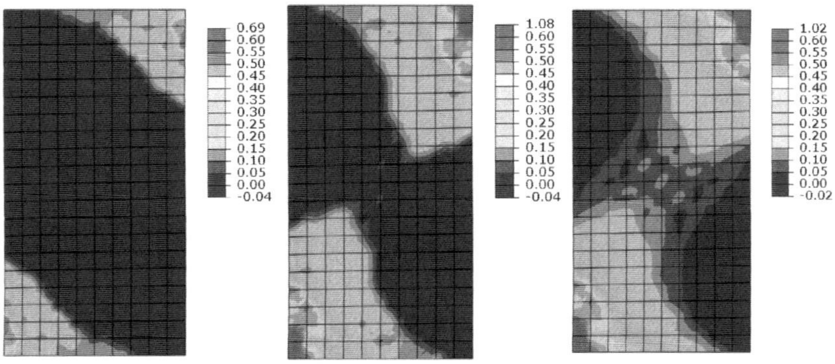

Fig 7. Variation of plastic work shear band in the beginning

4. Conclusions

In this paper, the anisotropic properties of the elastic and plastic phases of transversely isotropic geomaterials are considered. Based on the simplified Lade transversely isotropic elastoplastic model for transversely isotropic geomaterials derived the iterative scheme and the constitutive matrix of the corresponding forward Euler algorithm. Using UMAT interface based on ABAQUS numerically investigated the influence of the main direction angle of material on the shear band and the bearing capacity of the structure. The following conclusions are obtained:

1) The simplified Lade transversely isotropic elastic-plastic constitutive model can well simulate the strain localization of geomaterials. The main direction angle of the material directly affects the stiffness and ultimate bearing capacity of the geotechnical structure. The stiffness and bearing capacity of geotechnical structure can be fully and reasonably simulated with the evolution of bedding inclination, by reflecting the anisotropy of material strength.
2) The main direction angle of the material has a greater impact on the localized mode of strain. In the biaxial compression, for $\beta = 30^\circ$ and $\beta = 60^\circ$, the structure tends to preferentially form a significant shear zone in the weaker direction and destroy; for the case of $\beta = 0^\circ$ or $\beta = 90^\circ$, the structure breaks down by forming two conjugate X-shaped shear bands. With the increase of β, the peak value of plastic work gradually decreases. In general, the shear band width gradually increases, which better reflects the mode of material strain localization with β increasing.
3) After entering the plastic stage, with the loading, non-coaxial stress on the material gradually reduce the impact, and finally its influence vanishes.

References

[1] Arthur J R F, Dunstan T, Chua K S. Induced anisotropy in a sand[J]. Géotechnique, 1977, 27(27):13-30.
[2] Matsuoka H, Sakakibara K. A constitutive model for sands and clays evaluating principal stress rotation[J]. Soils & Foundations, 1987, 27(4):73-88.
[3] Rudnicki J W, Rice J R. Rice J.R.: Conditions for the localization of deformation in pressure-sensitive dilatant materials. J. Mech. Phys. Solids 23, 371-394[J]. Journal of the Mechanics & Physics of Solids, 1975, 23(6):371-394.
[4] Papamichos E, Vardoulakis I. Shear band formation in sand according to non-coaxial plasticity model[J]. Géotechnique, 1995, 45(4):649-661.
[5] Hashiguchi K, Tsutsumi S. Shear band formation analysis in soils by the subloading surface model with tangential stress rate effect[J]. International Journal of Plasticity, 2003, 19(10):1651-1677.
[6] Huang M, Lu X, Qian J. Non-coaxial elasto-plasticity model and bifurcation prediction of shear banding in sands[J]. International Journal for Numerical & Analytical Methods in Geomechanics, 2010, 34(9):906–919.
[7] Yao Y P, Kong Y X. Extended UH Model: Three-Dimensional Unified Hardening Model for Anisotropic Clays[J]. Journal of Engineering Mechanics, 2012, 138(7):853-866.
[8] Yeh G C K. Forced Vibrations of a Two-Degree-of-Freedom System with Combined Coulomb and Viscous Damping[J]. Journal of the Acoustical Society of America, 1966, 39(1):14-24.
[9] Lade, P. V., Failure criterion for cross-anisotropic soils[J]. Journal of geotechnical and geoenvironmental engineering, 2008, 134(1): 117-124.
[10] Lade, P. V., Jakobsen K P. Incrementalization of a single hardening constitutive model for frictional materials[J]. International Journal for Numerical & Analytical Methods in Geomechanics, 2002, 26(7):647-659.
[11] Chang J, Chu X, Xu Y. Finite-Element Analysis of Failure in Transversely Isotropic Geomaterials[J]. International Journal of Geomechanics, 2014: 04014096.
[12] Pietruszczak S, Mroz Z. Formulation of anisotropic failure criteria incorporating a microstructure tensor[J]. Computers and Geotechnics, 2000, 26(2): 105-112.
[13] Zhao J, Gao Z. Modelling non-coaxiality and strain localisation in sand: The role of fabric and its evolution[J]. 2015: ISBN 978-1-138-00148-0

IOP Conf. Series: Materials Science and Engineering **324** (2018) 012035 doi:10.1088/1757-899X/324/1/012035

Development of 300 mesh Soy Bean Crusher for Tofu Material Processing

E S Lee[1], P S Pratama[2], D Supeno[1], S W Jeong[1], J Y Byun[1], J H Woo[1], and C S Park[3] and W S Choi[1]

[1] Department of Bio-industrial Machinery Engineering, Pusan National University, 1268-50, Samnangjin-ro, Samnangjin-eup, Miryang, Rep. of Korea
[2] Life and Industry Convergence Research Institute, Pusan National University, 1268-50, Samnangjin-ro, Samnangjin-eup, Miryang, Rep. of Korea
[3] Keunwoo Tech Co. Ltd, 82-6, Geomdangongdan-ro, Buk-gu, Daegu, Rep. of Korea
E-mail: choi@pusan.ac.kr

Abstract. A machine such as bean crusher machine is subjected to different loads and vibration. Due to this vibration there will be certain deformations which affect the performance of the machine in adverse manner. This paper proposed a vibration analysis of bean crusher machine using ANSYS. The effect of vibration on the structure was studied in order to ensure the safety using finite element analysis. This research supports the machine designer to create a better product with lower cost and faster development time. To do this, firstly, using Inventor, a CAD model is prepared. Secondly, the analysis is to be carried out using ANSYS 15. The modal analysis and random vibration analysis of the structure was conducted. The analysis shows that the proposed design was successfully shows the minimum deformation when the vibration was applied in normal condition.

1. Introduction

Soybean is important for agriculture since it is one of the main food sources of nutrition for human. It is contains high quality protein, phytic acid, dietary minerals, and B vitamins. One of the most common food products made from soybean is tofu. The making process of tofu includes soaking of beans in the water, grinding the bean, filtering the grinded bean, boiling the bean extract, coagulating the extracted bean and finally moulding the tofu [1]. During tofu making process, inextricable dreg or soy pulp called okara is remained. In common, every 1 kilogram of dry beans makes into tofu produce about 1.1 kilogram of okara with around 80% moisture. It consists of high percentage of fiber and significant amounts of high quality oil and protein. However, until now okara is generally used as feed or fertilizer. Since the awareness of the dietary fiber importance in human health are growing, the utilization of whole beans as an alternative solution are increased [2]. Moreover, more than 90% of the water used during tofu making is wasted [3]. This waste water including coagulant recently has caused severe environmental issues. Some issues that caused by wastewater from tofu production are ecosystem contamination, certain organisms elimination, and decreased of water quality. Therefore wasted coagulant should be avoided. To produce whole beans tofu, the bean powder size should be less than 300 mesh. Conventional soybean pulverizer can only produce bean powder more than 100 mesh. Liquefied nitrogen is used to prevent the generation of oil in the pulverizer soybean during the pulverization. The surface hardening causes the wear of the pulverizer blade and the durability of the pulverizer.

In this research a bean crusher machine to produce the bean powder with size less than 300 mesh to produce high quality tofu. Furthermore, the temperature can be kept at 40 °C or less to prevent spoiling the beverage. A bean crusher machine is a mill used to grind hard soybean between two rotating abrasive surfaces. These surfaces are separated by some distance. When these two abrasive surfaces are far apart, the output material is bigger, and when the two abrasive surfaces are set closer together, the output material is smaller. According to the market demand, in this research a high speed bean crusher machine was developed. The structural analysis is one of important step in the development process such as in [4]. During operation, uneven load inside the crushing machine will lead the machine to the forced vibration. If the excitation vibration frequency and the resonance natural frequency of the frame structure are closed, the mechanical structure will produce local resonance and deformation [5]. Therefore, the simulation related to the vibration analysis is needed in development process.

2. System Design

The developed bean crusher machine is shown in Figure 1. The dimension of the machine was calculated optimally to maximize the capacity and the material strength. This machine powered by 220V 3.75KW induction motor. The nominal angular velocity of the crusher machine is 3000 rpm. Power transfer box transfer the energy generated by motor to the crush box through belt system. The belt conversion ratio is 1:2. The crush box is the most important part of bean crusher. It was made of steel. Three Nano inner crush tool inside the crush box rotate in different direction to crush the bean. The hopper was used as inlet for the crusher machine. Powder box was used to collect the bean powder processed by crush tool. The suction pipe then transports the bean powder to collector tank. Figure 2 shows the horizontal cut off schematics of the crusher machine. Since it consist of inner shaft and outer shaft, the Nano crush tool can rotate in two directions simultaneously. To reduce the noise occurred during operation, vibration proof rubber was attached to the base of each corner. To prevent degeneration of soybean grains with high oil content, a special cooling system was developed to maintain a low temperature of 40 degrees or less.

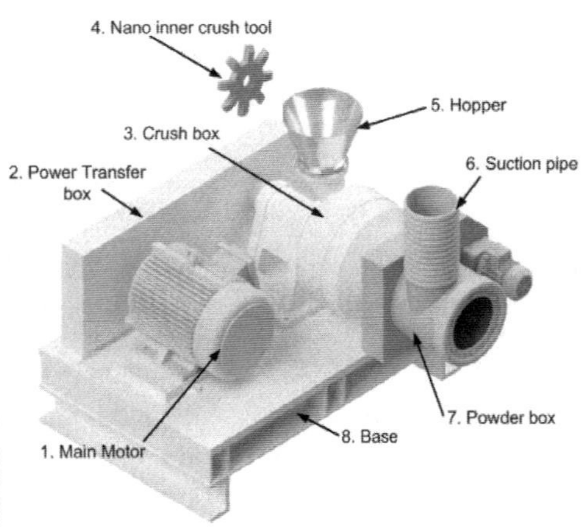

Figure 1. Developed system.

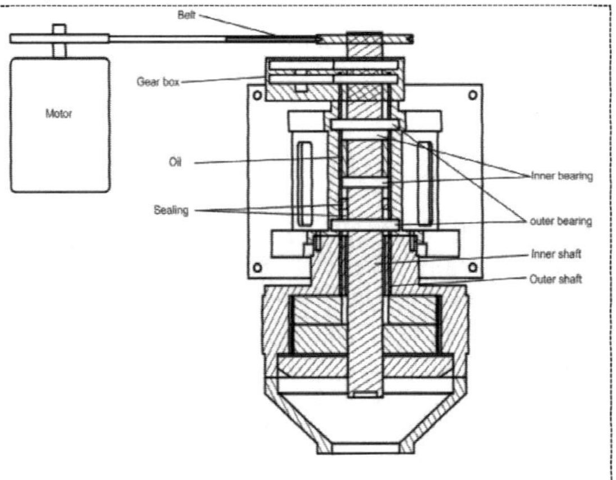

Figure 2. Schematics diagram of soy bean crusher machine.

IOP Conf. Series: Materials Science and Engineering **324** (2018) 012035 doi:10.1088/1757-899X/324/1/012035

3. Method

In this research the stress and modal analysis was done to evaluate the reduction gear performance. Stress analysis was done to clarify the design of reduction gear can withstand a specified load, using the specified amount of material or that satisfies some other optimality criterion. Modal analysis is a study of dynamics properties in frequency domain used to determine the natural frequency and vibration mode of a structure.

The static analysis and modal analysis process are consists of several steps. Firstly, the 3D model of important component was created. In this research focused on the crush box since this part affected by the operation condition. Secondly, the model was simplified to obtained geometry model. Thirdly, the mesh of the model was generated. Fourthly, the materials properties were defined. In this research the crush box is made of steel. Fifthly, the boundary condition was defined. Finally, solve the problem, visualized and read the results. In this research the 3D model was build using inventor software. The stress analysis was done using EDEM 2017 and modal analysis was done using ANSYS 15.

4. Result and Discussion

The calculation result of modal analysis is shown in figure 3. Modal analysis results in Table 1 shows the frames of 1 ~ 6 orders of natural frequency are between 338.55 ~ 1437.1 Hz range. Figure 3a~e shows the deformation effect on the frame when the natural frequency occurred. The scale is 1.

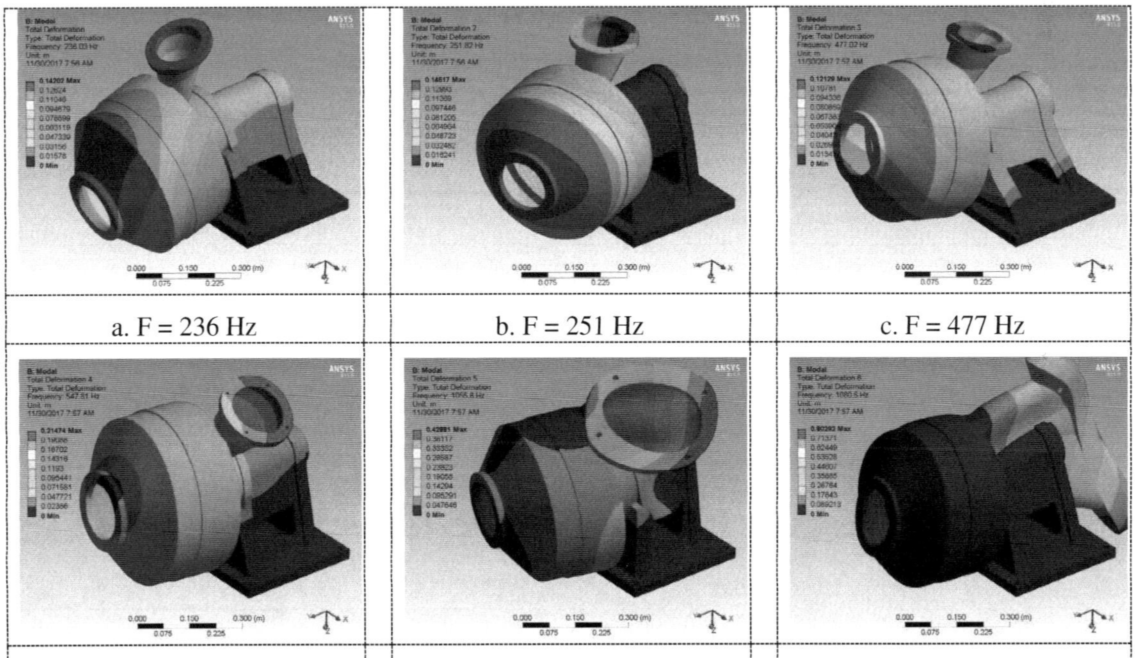

a. F = 236 Hz | b. F = 251 Hz | c. F = 477 Hz

d. F = 547 Hz | e. F = 1055 Hz | f. F = 1080 Hz

Figure 3. Modal analysis of bean crusher machine.

IOP Conf. Series: Materials Science and Engineering **324** (2018) 012035 doi:10.1088/1757-899X/324/1/012035

Table 1. Natural frequency and vibration mode.

No	Frequency	Characteristics of modal
1	236	Vertical across the Z axis at the front side
2	251	Horizontal across the X axis at the front side
3	477	Vertical across the Z axis at the front side
4	547	Horizontal across the X axis at the top side
5	1055	Horizontal across the X axis at the top side
6	1080	Vertical across the X axis at the top side

The bean crusher analysis based on discrete element method (DEM) using EDEM software is shown in figure 4. This figure shows the bean flow through the crusher machine at 1500 rpm motor speed. In this simulation the bean diameter size is 5 mm.

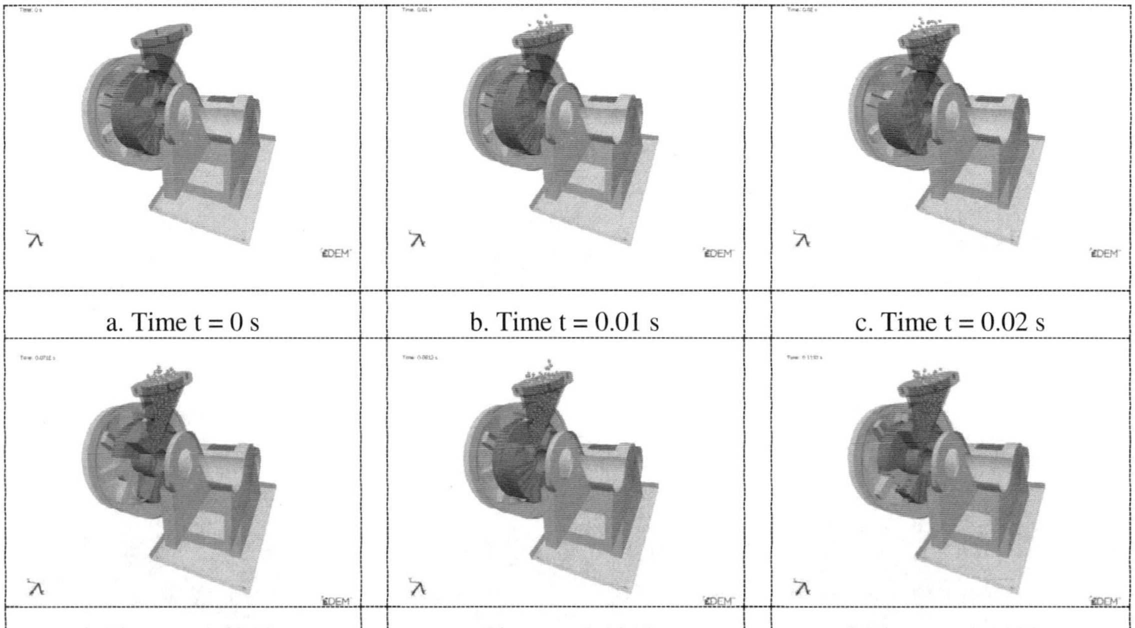

a. Time t = 0 s
b. Time t = 0.01 s
c. Time t = 0.02 s
d. Time t = 0.0712 s
e. Time t = 0.0812 s
f. Time t = 0.1132 s

Figure 4. Analysis of bean crusher machine based on discrete element method.

5. Conclusion

In this research, natural frequency analysis and random vibration analysis was done to the bean crusher machine. The simulation results show that the first order and second order natural frequency can be obtained from proposed modal analysis simulation. Modal analysis results show that the machine 1 ~ 6 orders modal natural frequencies range are 338.55 ~ 1437.1 Hz. This value should be considered during design process to avoid resonance condition. The behaviour of beans inside the bean crusher machine was investigated using discrete element method. The simulation shows that the bean flow the hopper inlet to the crush box.

Acknowledgments
This work was supported by Korea Institute of Planning and Evaluation for Technology in Food, Agriculture, Forestry and Fisheries (IPET).

References

[1] Rekha C R and Vijayalakshmi G 2013 Influence of processing parameters on the quality of soycurd (tofu) *J. Food Sci. Technol.* **50**(1) p 176.

[2] Kao F-J, Su N-W and Lee M-H. Development and Quality of Tofu Analogue Prepared from Whole Soybeans. Chemistry, Texture, and Flavor of Soy. ACS Symposium Series. 1059: American Chemical Society; 2010. p. 277.

[3] Faisal M, Gani A, Mulana F and Daimon H 2016 Effect of Organic Loading on Production of Methane Biogas From Tofu Wastewater Treated by Thermophilic Strirred Anaerob Reactor *Rasayan Journal Chemical* **9**(2) p 133.

[4] Choi W S, Pratama P S, Supeno D, Woo J H, Lee E S and Park C S 2017 Dynamic vibration characteristics of electric agricultural vehicle based on finite element method *Glob. J. Eng. Sci. Res.* **4** p 27.

[5] Wang W, Cheng S, Luo M and Zou J, editors. Experimental Modal Test for Structure Refinement of an Automotive Chassis. 2010 International Conference on Measuring Technology and Mechatronics Automation; 2010 13-14 March 2010.

IOP Conf. Series: Materials Science and Engineering **324** (2018) 012036 doi:10.1088/1757-899X/324/1/012036

Characteristics of Reduction Gear in Electric Agricultural Vehicle

W S Choi[1], P S Pratama[2], D Supeno[1], S W Jeong[1], J Y Byun[1], J H Woo[1], E S Lee[1] and C S Park[3]

[1] Department of Bio-industrial Machinery Engineering, Pusan National University, 1268-50, Samnangjin-ro, Samnangjin-eup, Miryang, Rep. of Korea
[2] Life and Industry Convergence Research Institute, Pusan National University, 1268-50, Samnangjin-ro, Samnangjin-eup, Miryang, Rep. of Korea
[3] Keunwoo Tech Co. Ltd, 82-6, Geomdangongdan-ro, Buk-gu, Daegu, Rep. of Korea
E-mail: choi@pusan.ac.kr

Abstract. In electric agricultural machine a reduction gear is needed to convert the high speed rotation motion generated by DC motor to lower speed rotation motion used by the vehicle. The reduction gear consists of several spur gears. Spur gears are the most easily visualized gears that transmit motion between two parallel shafts and easy to produce. The modelling and simulation of spur gears in DC motor reduction gear is important to predict the actual motion behaviour. A pair of spur gear tooth in action is generally subjected to two types of cyclic stress: contact stress and bending stress. The stress may not attain their maximum values at the same point of contact fatigue. These types of failure can be minimized by analysis of the problem during the design stage and creating proper tooth surface profile with proper manufacturing methods. To improve its life expectation in this study modal and stress analysis of reduction gear is simulated using ANSYS workbench based on finite element method (FEM). The modal analysis was done to understand reduction gear deformation behaviour when vibration occurs. FEM static stress analysis is also simulated on reduction gear to simulate the gear teeth bending stress and contact stress behaviour.

1. Introduction

Electric vehicles are an alternative solution to fuel engines for transportation vehicles due to advances in batteries, increasing oil prices, and the desire to reduce greenhouse gas emissions [1]. Many electric vehicles have been applied in industrial, logistics, and agricultural applications as material handling systems [2-5]. Electric vehicles are an important tool in agricultural modernization. With the rapid development of the economy, electric vehicles are widely used in agriculture. Electric agricultural vehicles are kind of mechanical and electrical integration of transportation vehicles. In this research, a small electric vehicle with lift and dump capabilities was developed. To achieve high quality and high productivity, a well-designed electric agricultural vehicle is needed.

One of important part to support an electric vehicle operation is a reduction gear. The reduction gear has been a research point of interest for many decades especially in agricultural vehicle. Gears can be defined as toothed member which is transmit power or motion between two shafts without any slip. Spur gears are the most recognized and common type of gears. Spur gears have their teeth parallel to the axis and are used exclusively to transmit rotary motion between two parallel shafts, while maintaining uniform speed and torque. They have high efficiency and excellent precision rating. The involute tooth form being the simplest to be manufactured permits high manufacturing tolerances to be attained.

There are several research related to the reduction gear. Wada [6] proposed method to investigate the failure condition on reduction gear operation. Badkas [7] proposed static and dynamic analysis of two spur gears. Jie [8] investigate the interfacial contact temperature based on FEM. Patil [9] studied about frictional tooth contact along line of a spur gear. Parra [10] investigates the gear condition based on the vibration occurs during operation. Li [11] and Zhao [12] proposed a fault diagnosis scheme for reduction geares using modified multi-scale symbolic dynamic entropy and mRMR feature selection. CAD object and computer software for simulation helps designer to analyzes, optimize design, reducing cost and time efficiency in physical prototype making [13].

To improve the strength of reduction gear, in this research some parameters of gear are modified. In this paper the static analysis was done to evaluate the stress and total deformation improvement. This analysis was done to prevent the gear failure caused by deformation during operation [14]. Furthermore, the modal analysis was performed to find out the natural frequency of the reduction gear.

2. System Design

The developed electric vehicle is shown in figure 1. This vehicle is basically a cart powered by an electric motor. The operator controls the vehicle velocity and direction from behind the vehicle through the control panel on the handle. The optimal dimensions of the vehicle were calculated to maximize the capacity and the material strength. The vehicle was designed with a total body weight of 330 kg. The frame is made of high-quality structural steel. The length of the vehicle is 1760 mm and the width is 830 mm. The cargo size is 1220 mm (L) x 800 mm (W) x 220 mm (H), and the total load capacity is 300 kg. With this amount of load space, bulky crops and vegetables can be loaded easily. Moreover, the cargo can be opened at the back, left, and right sides of the vehicle.

As shown in figure 2, this vehicle powered by two DA4A-29B4-ML57 24 V 400 W DC motors produced by Daehwa E/M Co. LTD from Korea. The rated angular velocity is 2900 RPM, the rated current is 23 A, the model size is 90 mm, and the shaft specification is powerful tooth cutting. To increase the torque and reduce the motor speed, a reduction gear was used. The D9K030US-S346 reduction gear was made by Daehwa E/M Co. LTD from Korea. The gear ratio is 30:1. The permissible overhang load is 55 kgf, and the permissible thrust load is 15 kgf. The maximum permissible torque is 300 kgf.cm. The output rotation direction is the same direction as the motor. Each motor was used to power the front wheels and rear wheels.

Figure 3 shows the failure condition of gear caused by contact stress and bending stress. The purpose of this paper is to improve the gear strength by modifying the gear parameter.

Figure 1. Developed electric vehicle.

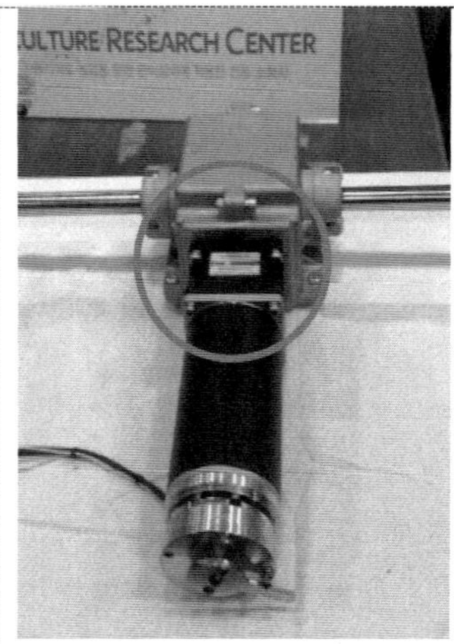

Figure 2. DC motor - reduction gear system.

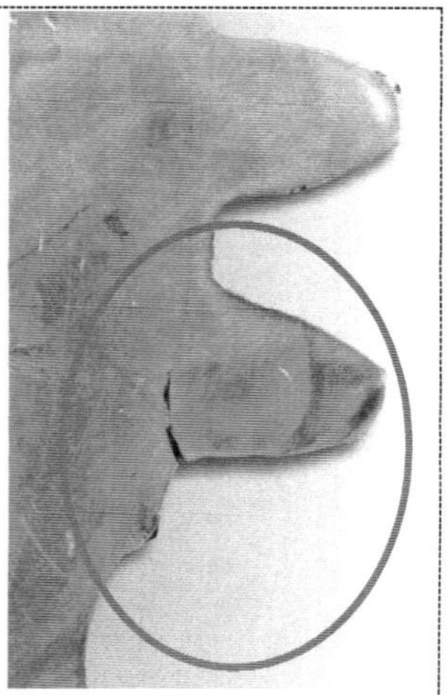

Figure 3. Failed gear.

IOP Conf. Series: Materials Science and Engineering **324** (2018) 012036 doi:10.1088/1757-899X/324/1/012036

3. Method

The structure of reduction gear used in this research as shown in figure 4 is consists of 5 pinions. To improve the strength of reduction gear, the pinion 3 and pinion 4 as shown in figure 5 are modified. The improvement are consists of gear shaft diameter increment and groove addition. The final form of reduction gear is shown in figure 6.

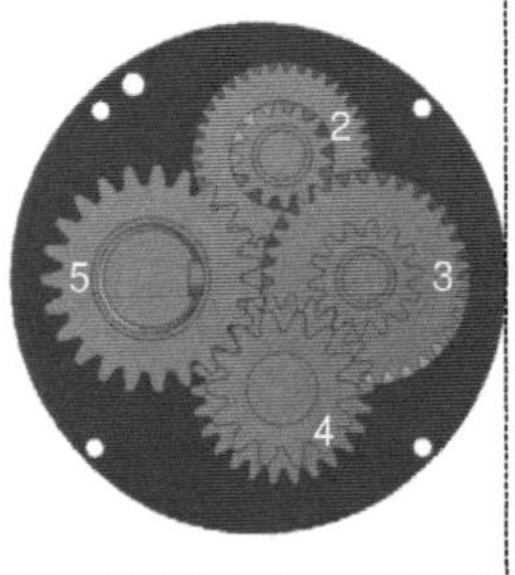

Figure 4. Front side view of reduction gear.

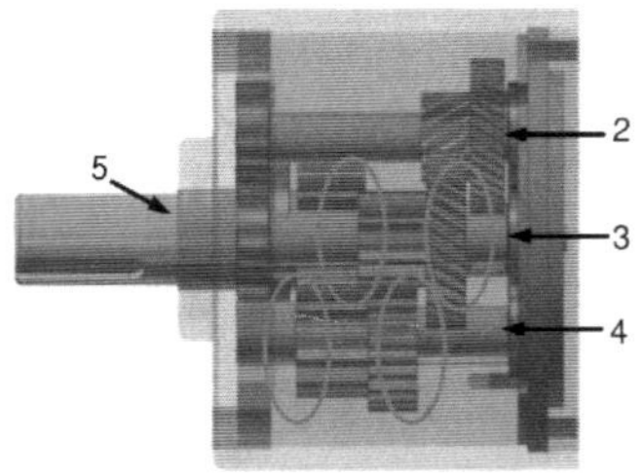

Figure 5. Right side view of reduction gear.

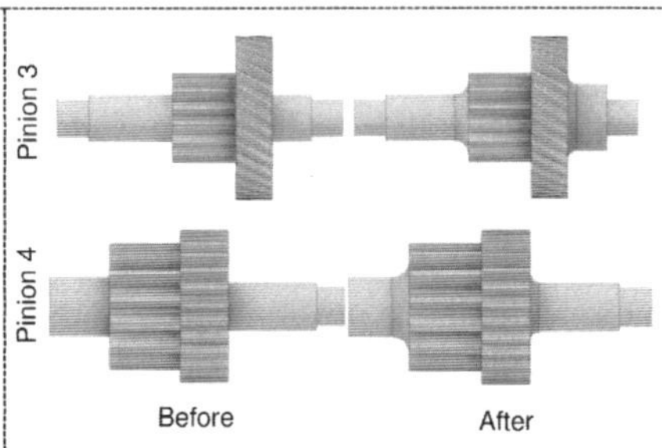

Figure 6. Modification of pinion 3 and pinion 4.

In this research the stress and modal analysis was done to evaluate the reduction gear performance. Stress analysis was done to clarify the design of reduction gear can withstand a specified load, using the specified amount of material or that satisfies some other optimality criterion. Modal analysis is a kind of linear analysis technology, used to determine the structure of the natural frequency and vibration mode. First natural frequency or resonance is one of the contributing factors for vibration and noise related problems that occur in structures and machineries.

The procedure of static analysis and modal analysis includes following steps as shown in figure 7. Firstly, the 3D model of important component of the reduction gear was created. In this research focused on the spur gears since this part affected by the loading condition. Secondly, the model was simplified to obtained geometry model. Thirdly, the mesh of the model was generated. Fourthly, the materials properties were defined. In this research the gear are made of AISI 4140. Fifthly, the boundary condition was defined. Finally, solve the problem, visualized and read the results. In this research the 3D model was build using inventor software and stress analysis and modal analysis was done using ANSYS 15.

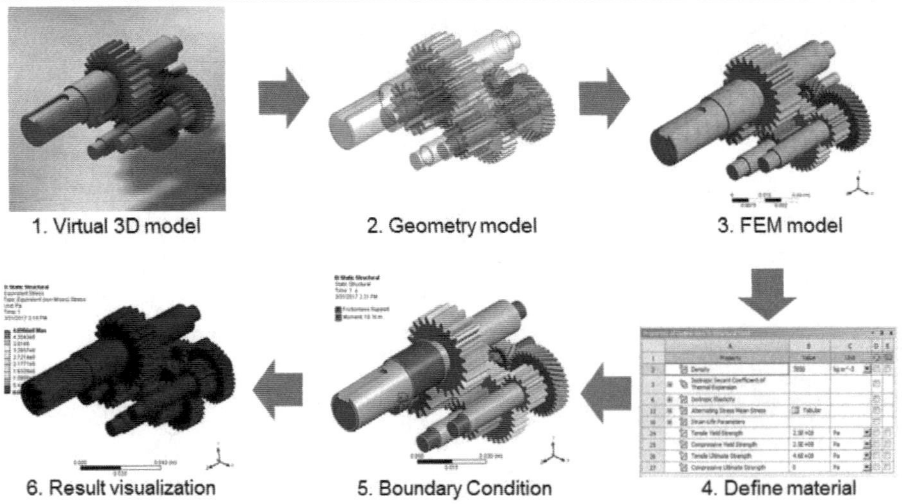

Figure 7. Simulation analysis procedure.

4. Result and Discussion

Figure 8 and figure 9 show the stress comparison before and after modification. The maximum stress occurred on the reduction gear before modification is 310.53MPa, and after is 288.42MPa. The safety factor of the gear can be improved from 1.33 to 1.43. The maximum deformation occurred on the reduction gear before modification is 30.36 μm, and after modification is 30.28 μm.

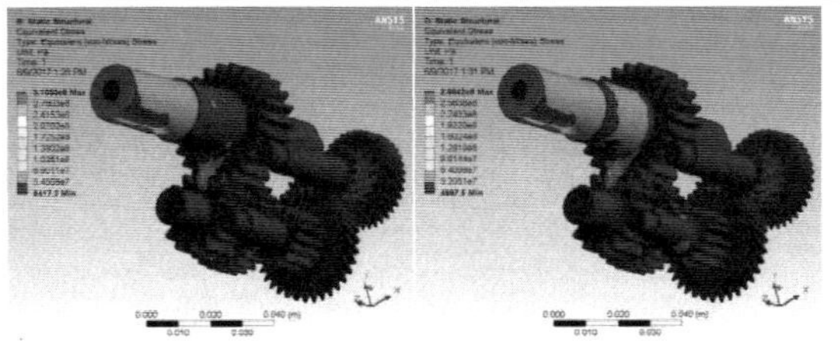

a. Before b. After

Figure 8. Von-Misses Stress of pinion 3.

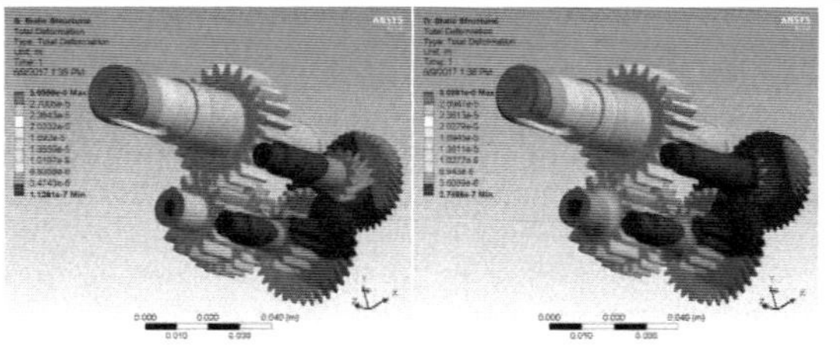

a. Before b. After

Figure 9. Total deformation of pinion 3.

Figure 10 shows the maximum stress occurred on the shaft 3 before modification is 158.47MPa, and after modification is 122.17MPa. The safety factor of shaft 3 can be improved from 2.618 to 3.396. Figure 11 shows the maximum stress occurred on the shaft 4 before modification is 310.53MPa and after modification is 288.42MPa. The safety factor of the gear can be improved from 1.33 to 1.43.

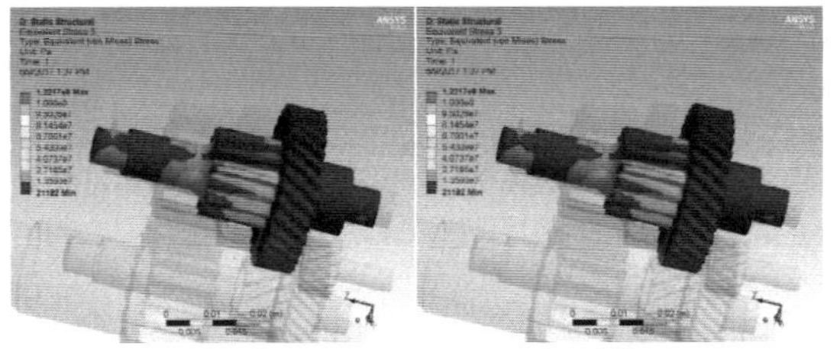

a. Before b. After

Figure 10. Von-Misses Stress of pinion 4.

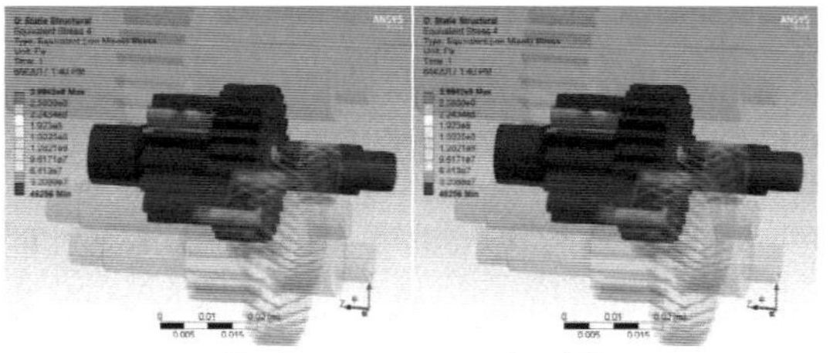

a. Before b. After

Figure 11. Total deformation of pinion 4.

Figure 12 shows the result of modal analysis of the gear. The simulation result shows that the natural frequency of the gear is between 4559.6 Hz and 7592.2 Hz. This is higher than the operational frequency of the gear which is 50Hz when the motor operate at 3000 rpm. This should be the key consideration to avoid produce resonance. Therefore, the operation of the gear only generated small deformation.

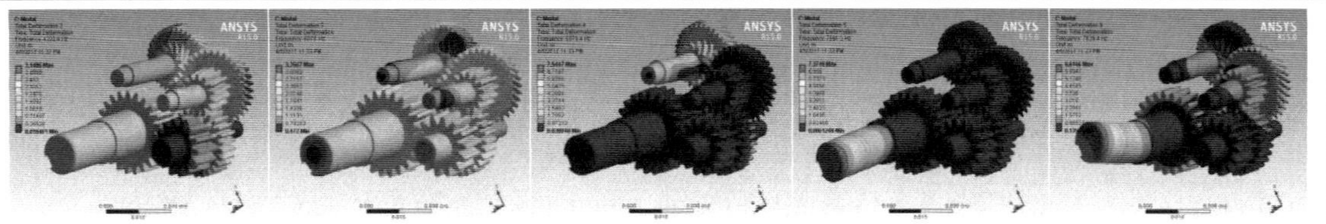

a. F=4559.6Hz b. F=6145.9Hz c. F=7063.8Hz d. F=7573.9Hz e. F=7592.2Hz

Figure 12. Modal analysis of reduction gear.

5. Conclusion

In this research the structure analysis was done to the DC motor reduction gear of electric agricultural vehicle. Based on this research following conclusion can be obtained:

1. By modifying shaft 3 and 4, the stress occurred in reduction gear can be reduced.
2. The safety factor of the gear can be improved from 1.33 to 1.43.
3. The natural frequency of reduction gear is between 4559.6~7592 Hz is higher than the operational frequency of the gear which is 50Hz.

Acknowledgments

This research was supported by the Ministry of Agriculture, Food and Rural Affairs Research Center of the Republic of Korea.

References

[1] Sperling D and Gordon D 2009 Two billion cars: driving toward sustainability *Disability & Society* p 22.

[2] Choi W S, Pratama P S, Supeno D, Woo J H, Lee E S and Park C S 2017 Dynamic vibration characteristics of electric agricultural vehicle based on finite element method *Glob. J. Eng. Sci. Res.* **4** p 27.

[3] Pratama P S, Gulakari A V, Setiawan Y D, Kim D H, Kim H K and Kim S B 2016 Trajectory tracking and fault detection algorithm for automatic guided vehicle based on multiple positioning modules *Int. J. Control. Autom. Syst.* **14** p 400.

[4] Pratama P S, Jeong S K, Park S S and Kim S B 2013 Moving Object Tracking and Avoidance Algorithm for Differential Driving AGV Based on Laser Measurement Technology *Int. J. Sci. Eng.* **4** p 78.

[5] Pratama P S, Luan B T, Tran T P, Kim H K and Kim S B. Trajectory Tracking Algorithm for Automatic Guided Vehicle Based on Adaptive Backstepping Control Method. In: Zelinka I, Duy V H, Cha J, editors. Lecture Notes in Electrical Engineering. Berlin Heidelberg: Springer-Verlag; 2013. p. 353.

[6] Wade D R and Larsen C G, editors. Measurement of Gearbox Surface Frequency Response Functions for HUMS Algorithm Improvement. American Helicopter Society 68th Annual Forum Proceedings; 2012 May; Forth Worth.

[7] Badkas S A and Ajmera N 2016 Static and dynamic analysis of spur gear *International Journal of Mechanical Engineering and Technology* **7** p 8.

[8] Jie L J L, Lei Z L Z and Qi Z Q Z 2009 Finite element modeling and analysis on interfacial contact temperature of gears based on ANSYS *2009 IEEE 10th International Conference on ComputerAided Industrial Design Conceptual Design* p 668.

[9] Patil S, Karuppanan S, Atanasovska I and Wahab A A 2014 Frictional Tooth Contact Analysis along Line of Action of a Spur Gear Using Finite Element Method *Procedia Materials Science* **5** p 1801.

[10] Parra J and Vicuna C M 2017 Two methods for modeling vibrations of planetary gearboxes including faults: Comparison and validation *Mechanical Systems and Signal Processing* **92** p 213.

[11] Li Y, Yang Y, Li G, Xu M and Huang W 2017 A fault diagnosis scheme for planetary gearboxes using modified multi-scale symbolic dynamic entropy and mRMR feature selection *Mechanical Systems and Signal Processing* **91** p 295.

[12] Zhao C and Feng Z 2017 Application of multi-domain sparse features for fault identification of planetary gearbox *Measurement* **104** p 169.

[13] Morozov A, Zou T, Saman M, Mousavi R, Setiawan Y D, Angeles J, et al. Design and Gear Shift Control of a Modular Swift-Shift Multi-Speed Transmission for EVs2015.

[14] Shipley E 1967 Gear failures *Machine Design* p.

IOP Conf. Series: Materials Science and Engineering **324** (2018) 012037 doi:10.1088/1757-899X/324/1/012037

Study on autogenous shrinkage characteristic and mechanism of ultra-high performance cementitous composite

Li Caixia[1], Sun Zhilin[2], Yang Liqi[3] and Zhang Guorong[4,5]

1 Zhangye municipal government and social capital cooperation management center office, Zhangye, China

2 Zhangye Traffic Construction Investment Co., Ltd., Zhangye, China

3 Chengde Road & Bridge Construction Corporation, Chengde, China

4 JSTI Group, Nanjing, China

5 State Key Laboratory of Safety and Health for In-service Long Span Bridges, Nanjing, China

Abstract. The self-designed vertical dial gauge holder was used in this paper. The influence factors of autogenous shrinkage of UHPCC, which include water-binder ratios, binder-aggregate ratios, sand ratios and fraction process of steel fiber, were mainly studied. The mechanism of autogenous shrinkage is explored by thermal analysis and pore structure analysis techniques. The results show that autogenous shrinkage of UHPCC increases with the decreasing of water-binder ratio. The autogenous shrinkage of 60d increases by 14.4% and 25.4%, when water-binder ratio decreases from 0.24 to 0.20, 0.16, respectively. Coarse aggregate has great positive effects on the autogenous shrinkage of UHPCC. Fine steel fibers have no obvious effects on the autogenous shrinkage of UHPCC. Non-evaporable water content decreases with the decreasing of water-binder ratio, which is positive effect on the autogenous shrinkage. However, the volume fraction of pores with the size range from 5 to 50 nm increases with the decreasing of water-binder ratio, which is negative effect on the autogenous shrinkage. Self-dessication effect is dominant factor in the autogenous shrinkage.

1. Introduction

Ultra-High Performance Cementitious Composites (UHPCC) is aimed at the problems of high cost, high energy consumption, complicated preparation process and difficult application in practical engineering for Reactive Powder Concrete (RPC) Using Portland cement and a variety of industrial waste as a matrix, with high strength coarse aggregate and short steel fibers as an enhancement to obtain a cement-based composite material with good flow properties, excellent mechanical properties and durability through the conventional preparation technology. Therefore, it has a wide range of application prospects in modern military protection projects as well as super high-rise, large span and some structural projects that have been subjected to dynamic loads for a long time.

Due to the low water-cement ratio of UHPCC, large amount of cementitious material, and the incorporation of superplasticizer, resulting in large autogenous shrinkage deformation, when the concrete is subject to internal or external constraints, it will produce tensile stress in concrete, it will emerge some cracks if the tensile stress beyond the tensile strength, the cracks will cause further significantly deteriorated influence to the mechanical properties and durability of cement-based materials .At present, the research on autogenous shrinkage of cement-based materials mainly focuses on high-strength concrete[1−4] and high-performance concrete[5−9], and few studies on UHPCC have been conducted. The research on the autogenous shrinkage that caused by water-binder ratio [10] , steel fiber content [11] has been done already. while fewer studies about sand rate, the ratio of binder-aggregate and other factors have been developed.Therefore, it is of great importance to study the self-shrinking properties of UHPCC and their related mechanisms systematically.

Vertical dial gauge holder method for the autogenous shrinkage (1d to 60d after molding) of ultra-high performance cementitious composites was used in this paper.The autogenous shrinkage properties of ultra-high performance cementitious composites were systematically discussed under the conditions of different water-cement ratios (0.16,0.20 and 0.24), different ratios of bindert-aggregate (1: 1, 1: 2 and 1: 3), different rates of sand (0.4, 0.5 and 0.6) different steel fiber volume contents (0%, 1%, 2% and 3%) .Besides, the mechanism of autogenous shrinkage was explored by comprehensive thermal analysis, Mercury method and Micro-analysis technology .

2. Experimental Study

2.1 Raw Material.

Cement (C): P · II 52.5 Portland cement produced in Konan Onoda, Nanjing, with a density of 3100kg /m^3; fly ash (FA): Class I fly ash produced by Jiangsu Nantong Power Plant, with a density of 2700kg / m^3; Silica fume (SF): Elkem Shanghai Company, with a density of 2100kg / m^3; Slag (SL): Nanjing Jiangnan Cement Co., Ltd. production S105 ground slag , with a density of 2870kg / m^3; fine aggregate: ordinary river sand, the apparent density is 2650kg / m^3; coarse aggregate: 5 ~ 10mm basalt crushed stone, the apparent density is 2950kg / m^3; steel fiber : Steel fiber: Copper alloy superfine-plated steel fiber produced by Ganzhou Grand Industrial Co., Ltd., with diameter of 0.17mm and length of 15mm; Water-reducing agent: polycarboxylate superplasticizer produced by Jiangsu Su Bot company, 50% solids, water reduction rate of 40% or more.

The chemical composition and physical properties of the cementitious materials are shown in table 1.

Table 1. The chemical composition and physical properties of the cementitious materials

Binder	Chemical composition /%							Specific surface area /(m^2/kg)
	SiO_2	Al_2O_3	Fe_2O_3	CaO	MgO	SO_3	IL	
C	16.52	8.33	3.37	69.52	0.85	1.53	1.59	310
FA	39.07	48.52	3.37	5.96	1.32	0.46	3.50	345
SF	95.01	0.82	1.86	0.35	1.24	0.32	1.58	20500
SL	32.07	14.68	0.97	35.81	9.30	2.51	0.58	480

2.2 Experiment Scheme.

A total of 16 UHPCCs were designed for autogenous shrinkage (from 1 day to 60 days after forming), of which W / B-0.16, W / B-0.20 and W / B-0.24 were used to study the effect of water- binder. B/A-1, B / A-2 and B / A-3 were used to study the effect of the ratio of binder- bone; SR-0.4, SR-0.5 and SR-0.6were used to study the effect of sand rate;FB-0,FB-1,FB -2 and FB-3 were used to study the influence of volume fraction of steel fiber. The amount of superplasticizer was 2% of the amount of cementitious material. The specific mixing proportion is shown in table 2.

Table 2. Mixing proportion of UHPCC

No.	Water-binder ratio	Proportion of cemen-titious material (%)				Binder-aggregate ratio	Sand ratio	Volume fraction of steel fiber (%)	90d compressive strength(MPa)
		C	FA	SF	SL				

W/B-0.16	0.16	60	10	10	20	1:2	0.5	2	150
W/B-0.20	0.20	60	10	10	20	1:2	0.5	2	130
W/B-0.24	0.24	60	10	10	20	1:2	0.5	2	118
B/A-1	0.16	60	10	10	20	1:1	0.5	0	88
B/A-2	0.16	60	10	10	20	1:2	0.5	0	121
B/A-3	0.16	60	10	10	20	1:3	0.5	0	112
SR-0.4	0.16	60	10	10	20	1:2	0.4	0	112
SR-0.5	0.16	60	10	10	20	1:2	0.5	0	119
SR-0.6	0.16	60	10	10	20	1:2	0.6	0	122
FB-0	0.16	60	10	10	20	1:1	1	0	103
FB-1	0.16	60	10	10	20	1:1	1	1	132
FB-2	0.16	60	10	10	20	1:1	1	2	151
FB-3	0.16	60	10	10	20	1:1	1	3	175

In addition, three groups of water-cement ratio (0.16,0.20 and 0.24), the hydration process and pore structure parameters of three mixed mineral admixtures (10% SF 、 10% FA and 20% SL instead of cement) at 7d, 14d age cementitious composites were studied by means of comprehensive thermal analysis (DSC-TG) and mercury intrusion porosimetry (MIP) and further explored the autogenous shrinkage mechanism of ultra-high performance cement-based composites. The mixing proportions of three group were shown in the table 3

Table 3. The experimental mixing proportions of DSC-TG、MIP

No.	water-binder ratio	Composition of cementitious materials (%)			
		C	SF	FA	SL
W/B-0.16	0.16	60	10	10	20
W/B-0.20	0.20	60	10	10	20
W/B-0.24	0.24	60	10	10	20

2.3 Test Methods.

2.3.1 Testing method of autogenous shrinking.

A self-designed vertical dial gauge holder device was used to test the autogenous shrinkage properties for UHPCC, shown in figure 1. One rack and chuck play a role in fixing the indicator and test pieces and ensuring full contact between them, the changes in the vertical length of cement-based composite materials were reflected through the changes in readings. For autogenous shrinkage test, to ensure the specimen and the surrounding environment has no moisture exchange is the most basic prerequisite, PVC pipe and plastic film was used to seal the test piece in this test.

autogenous shrinkage test specimens for ultra-high performance cement-based composites were prepared with a length of 400 mm and a diameter of 10 mm PVC pipe. Before the experiment, the organic sealant is used to bond the PVC pipe and the plywood. After the sealant is completely dried, the long-term autogenous shrinkage mold with the plywood as the bottom surface and the PVC pipe as the side surface is prepared. 1d after the specimen is molded, removing the bottom surface of the test pattern of the plywood, placing the specimen on the indicator frame, adjusting the height of the chuck, after the dial indicator is installed, starting the long-term autogenous shrinkage test, reading and recording corresponding age of the dial indicator value. Test environment temperature is 20 ± 2 °C, relative humidity is 60 ± 5%.

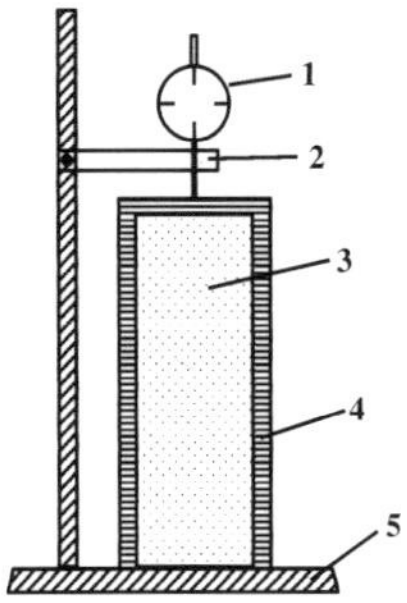

1- Dial gauge; 2-Collet; 3-UHPCC; 4-PVC pipe; 5-Hob

Figure 1. Testing apparatus of autogenous shrinkage of UHPCC

2.3.2 Comprehensive thermal analysis.

Netzsch STA449 F3 Synchrotron Thermal Analyzer (DSC / DTA-TG) of Germany resistant company was used in this test. In this experiment, the heating rate of synchronous thermal analyzer is 10 °C / min and heated to 1000 °C to get the DSC-TG curve. The process of UHPCC sample preparation: The fresh cement slurry was put into the plastic sealed bags, in order to eliminate the bleeding of the slurry and stratification, every 10min the plastic sealed bag was inverted several times, removing the sealed bag after the final seal.The sample is placed in 20 °C fog chamber and maintenanced immediately. By the age of 7d and 14d, the appropriate amount of hardened paste was crushed , immersed in anhydrous ethanol to stop the hydration, ground in an agate mortar to pass through a 200 mesh sieve to produce the desired sample.

2.3.3 Pore structure analysis.

The pore structure parameters of UHPCC were measured by mercury intrusion porosimetry (MIP). Autopore IV9510 automatic mercury porosimeter was used. The measured pore size ranged from 3.6nm to 400um. The maximum pressure was 60000 psia (414MPa). Sample Preparation Process: When the specimen is cured to 7 days and 14 days, the test sample is cut into small pieces of about 10mm × 10mm × 10mm with a small cutting machine, put it into anhydrous ethanol to stop hydration and soak for 24h .After that, it is put into a vacuum drying oven with the temperature of 60 ± 2 °C to dry the sample.

3. Results and Analysis

3.1 Autogenous Shrinkage Characteristic of UHPCC

3.1.1 Effect of water-binder ratio

Under the same conditions of other parameters (the content of basalt coarse aggregate and steel fiber volume is 2%), the autogenous shrinkage characteristics of UHPCC with different water-cement ratios of 0.16, 0.20 and 0.24 were studied. The results are shown in figure 2 . It can be found from the figure that the autogenous shrinkage characteristic shows a rapid growth in the early period and tend to be a steady growth in the late period. The three groups of 7d UHPCC autogenous shrinkage were 78.7% (W / B = 0.16), 59.7% (W / B = 0.20) and 60.2% (W / B = 0.24) , the main reason is that the internal self-dessication effect is determined by the hydration reaction rate of the cementitious material. The early hydration reaction rapidly causes the internal free water to be consumed rapidly and the self-dessication effect also develops rapidly. As the age increases, the speed of hydration reaction decreases and tends to stabilize, the self-dessication effect also decreases and tends to stabilize.

At the same time, it can also be seen that with the decrease of water-cement ratio, the autogenous shrinkage value increased obviously. The autogenous shrinkage values at 60d were 365 $\times 10^{-6}$(W / B = 0.16) and 333 × 10-6) and 291 $\times 10^{-6}$ (W / B = 0.24) respectively , and the autogenous shrinkage value of 60d increased by 14.4% and 25.4% respectively when the water-cement ratio decreased from 0.24 to 0.20 and 0.16 respectively. This is mainly affected by two factors: Firstly, the increase of water-cement ratio plays the role in thinning the pores, leading to an increase in capillary tension;

Secondly, with the water-cement ratio reduced, the internal free water decreases, with the hydration of cementitious material, material self-dessication occurs earlier in advance and with the water-cement ratio reduced the effect become more obvious.

3.1.2 Effect of binder-aggegate ratio

The the experimental of autogenous shrinkage characteristics were studied under the premise of keeping the water-cement ratio of 0.16 and the sand rate of 0.5 and the three groups of UHPCC are at different ratios of binder-aggregate (B / A-1: 1, B / A-1: 2 and B / A-1: 3) ,the test results were shown in figure 3.

It can be seen from the figure that UHPCC autogenous shrinkage also shows rapid growth in the early period and steady growth in the late period with the development of the age. At the same time, the volume fractions of the aggregates were 42.3%, 59.5% and 68.8% respectively when the ratios of binder-aggregate are 1:1,1:2 and 1:3 respectively. The autogenous shrinkage values at 60d were 556 $\times10^{-6}$ and 391 $\times10^{-6}$ and 344 $\times10^{-6}$, 60d autogenous shrinkage value decreases with the decrease of the ratio of binder-aggregate. This is because aggregates are a fixed-volume component of concrete that play a skeletal role in cement-based composites and have an important limiting effect on shrinkage and deformation. Decreasing the binder-aggregate ratio indicates that with the increase of the aggregate volume fractions in the cementitious composites, the corresponding limit effect is even more significant. In addition, as the volume fraction of aggregates increases, the amount of cementitious material per unit volume decreases and the self-dessication shrinkage resulting from hydration of the cementitious material decreases.

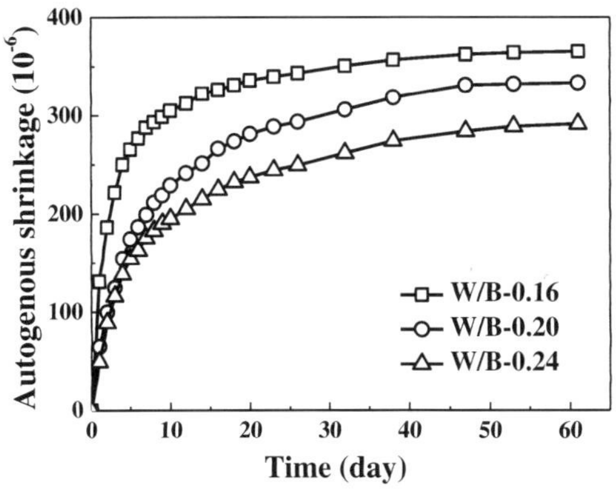

Figure 2. Effect of W/B

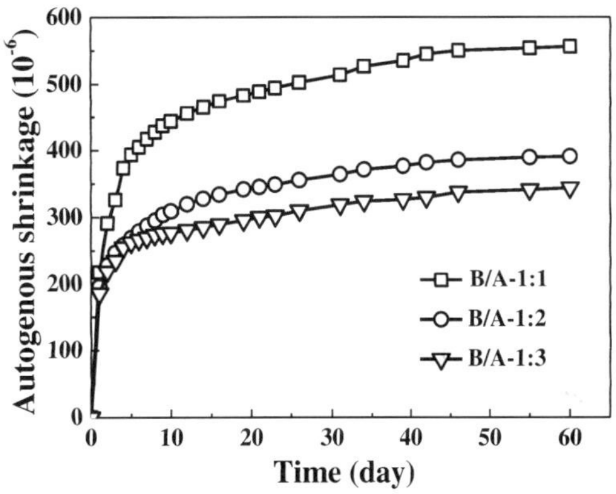

Figure 3. Effect of B/A

3.1.3 The impact of sand rate

Three different sets of sand rate (0.4, 0.5 and 0.6) UHPCC autogenous shrinkage characteristics were tested at binder-aggregate of 1:2. The results are shown in figure 4. The autogenous shrinkage increases with the increase of sand rate. The autogenous shrinkage values at 60d are 345 $\times10^{-6}$(SR = 0.4), 391 $\times10^{-6}$ (SR = 0.5) and 440 $\times10^{-6}$ (SR= 0.6), the autogenous shrinkage value of 60d increased by 13.3% and 15.9% respectively when the sand rate increased from 0.4 to 0.5 and 0.6. This is mainly due to the decrease of the overall elastic modulus of aggregates when the ratio of water-binder and the ratio of binder-aggregate are constant[12,13], thus weakening the skeleton effect of aggregates and the restraint effect of the shrinkage and deformation for cementitious composites is also weakened.

3.1.4 The influence of steel fiber volume

The autogenous shrinkage tests were carried out on UHPCC (non-aggregated volume) with water-cement ratio of 0.16 and steel fiber volume contents of 0%, 1%, 2% and 3% , the long-term autogenous shrinkage effect was also studied, the results were shown in figure 5. It can be seen from the figure that with the increase of steel fiber volume fraction, the UHPCC decreased slightly from long-term autogenous shrinkage. The autogenous shrinkage values at 60d were 670 $\times10^{-6}$ (FB-0%), The volume fraction of steel fiber increased from 0% to 1%, and the volume fraction of steel fiber increased from 0% to 1%, and the volume of steel fiber increased from 1% to 647 $\times10^{-6}$ FB-1%, 629 $\times$ 110^{-6} FB-2% and 606 $\times10^{-6}$ FB- % And 3% respectively, the autogenous shrinkage value at 60d decreased 3.4%, 6.1% and 9.6% respectively. The short-steel steel makes no difference to UHPCC after 1d molded for its long-term autogenous shrinkage inhibition effect.

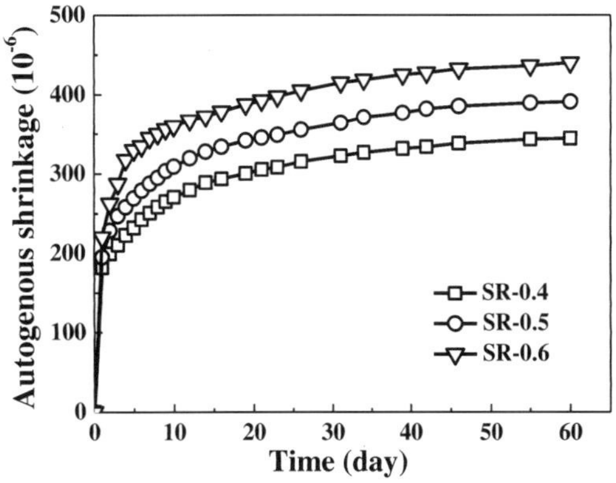

Figure 4. Effect of sand ratio

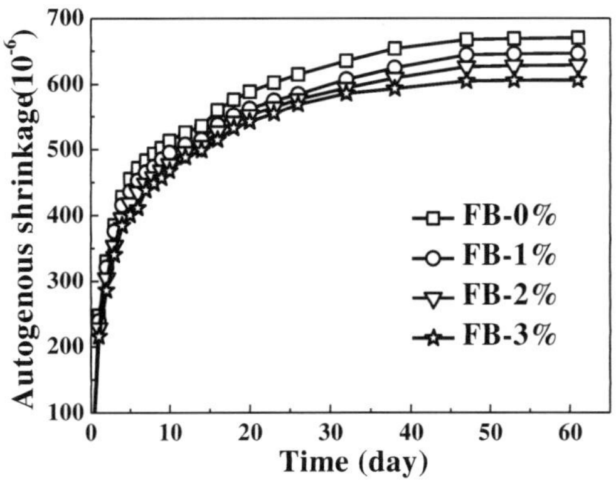

Figure 5. Effect of steel fiber volume

3.2 The mechanism research on UHPCC autogenous shrinkage

3.2.1 Comprehensive thermal analysis

Through a comprehensive thermal analysis test, the content of non-evaporated water was calculated to characterize the degree of cementitious material hydration. the content of non-evaporated water (W_n) can be calculated according to the following formula:

$$W_n = (m_1-m_2)/m_2\times(1-L)-L$$

Where m_1 is the mass of the sample corresponding to the TG curve at 105 ° C; m_2 is the mass of the sample corresponding to the TG curve at 1000 ° C; L is the loss on ignition (expressed in mass fraction) of the cementitious material at 105 ° C, it can also be converted according to the cement and mineral admixture mass fraction in cementitious material and the loss of each component. The calculated content of non-evaporated water of different water-cement ratio slurries is shown in table 4.

Table 4. Non-evaporated water content in the slurry of different water-cement ratio

No.	The content of non-evaporated water (%)	
	7d	14d
W/B-0.16	10.7	12.0
W/B-0.20	12.2	13.7
W/B-0.24	15.1	15.6

As can be seen from the table, under all ages, the content of non-evaporated water decreases with the decrease of the water-cement ratio, indicating that the smaller the water-cement ratio, the smaller the degree of hydration of the slurry. This is because the water-cement ratio is reduced, the amount of water that can be supplied internally to hydrate the cementitious materials is reduced. In addition, since the water-cement ratio is decreased, the capillary porosity of hydration products is low and the pore diameter of the pores is small, making it is difficult for the water that supplying to cement and hyrating the cementitious materials to move to reaction area, resulting in the reaction rate reduced, thereby reducing the degree of hydration of cementitious material[14].

3.2.2 Mercury Intrusion Method

The pore size distributions of three groups of cement-based composites with different ratios of water-binder and three mixed mineral admixtures are shown in figure 6 and figure 7, and the pore sizes are shown in figure 8. From figure 6 and figure 7, it can be seen that the most probable aperture of cement-based material decreases with the decrease of water-cement ratio. The most probable pore diameters are 0.24 are 12.25 nm and 17.11 nm and 21.10nm respectively when the water-cement ratio are 0.16, 0.20 and 0.24respectively , indicating that with the water-cement ratio decreases, the pore size is refined and the pore structure is improved. From Fig. 9, it is found that the porosity of cement-based composites decreases with the decrease of water-cement ratio. When the water-cement ratio decreases from 0.24 (18.13%) to 0.20 (13.35%) and 0.16 (10.08%) ,the 14d porosity decreased by 26.4% and 44.4% respectively. In addition, the porosity decreases with age, mainly due to the compact effect of the hydration product when it is hydrating.

During the autogenous shrinkage process, Self-dessication mainly occurs in the capillaries, and the internal liquid surface of the capillary water drops and the curvature of the meniscus becomes larger, resulting in a larger surface tension. The pore radius distribution of liquid surface, that is, the pore size distribution in the cement-based composite slurry, allows the autogenous shrinkage to be studied in conjunction with the pore size distribution of the material. Related literature[15] studies have shown that 5 ~ 50nm pores are the main area which self-contraction occurred. Therefore, the decrease of the water-cement ratio will refine the pore structure and increase the number of pores with 5 ~ 50 nm pore size, leading to the increase of capillary pressure under the self-dessication effect, thereby increasing the autogenous shrinkage of the cement-based composite.

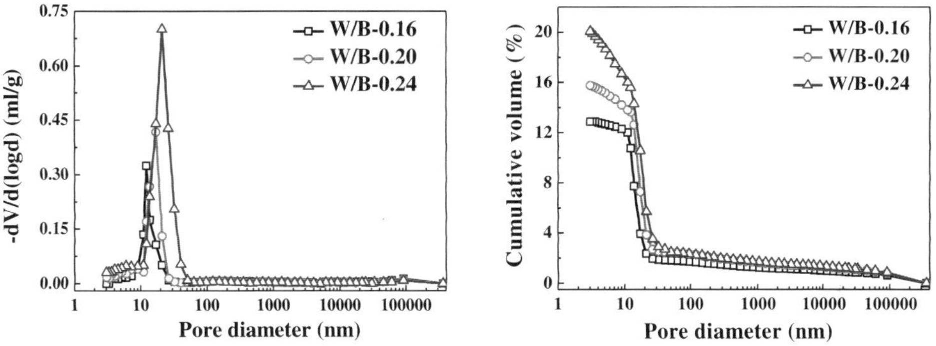

(a) Differential distribution curve (b) Cumulative distribution curve

Figure 6. Pore size distribution of paste with different W/B in 7d

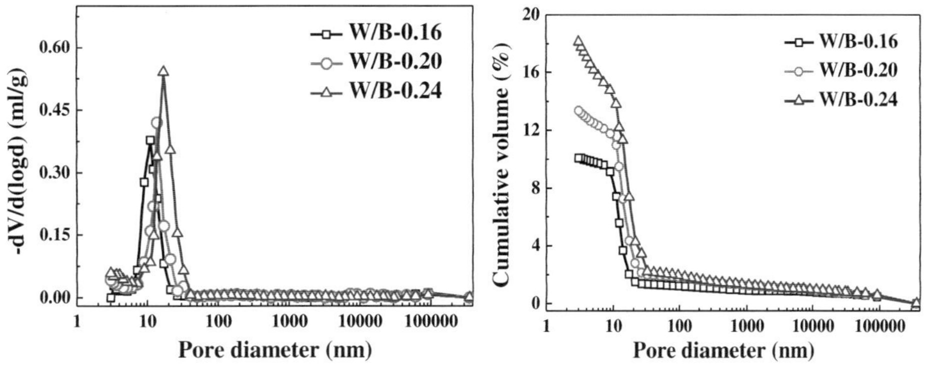

(a) Differential distribution curve (b) Cumulative distribution curve

Figure 7. Pore size distribution of paste with different W/B in 14d

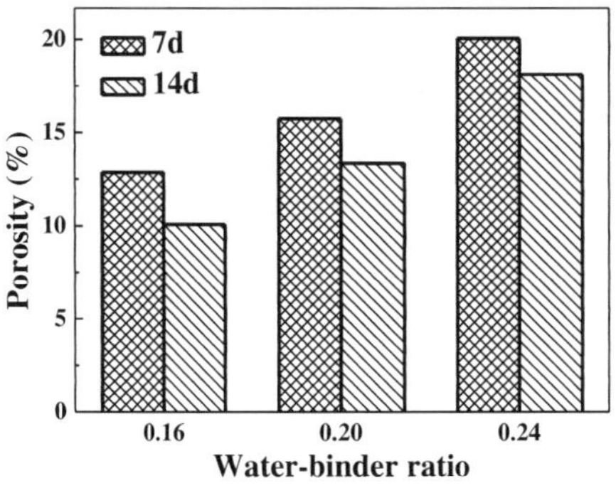

Figure 8. Effect of W/B on porosity of UHPCC

According to the test of non-evaporable water content and pore structure parameters, the water-binder ratio is decreased, the content of non-evaporated water is reduced, and the degree of hydration is reduced, which is a positive effect for autogenous shrinkage deformation. However, increasing the volume fraction of 5~50 nm pores increases the meniscus curvature, resulting in an increase in surface tension, which is manifested by an increase in the rate and extent of capillary negative pressure under seal conditions, an increase in self-dessication effect, which is a negative effect on autogenous shrinkage . The superposition of positive and negative effects shows that the autogenous shrinkage decreases with the increase of water-cement ratio, indicating that self-dessication effect plays a dominant role in the autogenous shrinkage factors.

4. Summary

(1) The self-designed vertical dial gauge holder measures 1d autogenous shrinkage value, with the water-cement ratio decreased, the autogenous shrinkage increased significantly, the water-cement ratio decreased from 0.24 to 0.20,0.16, 60d autogenous shrinkage value increased by 14.4% and 25.4% respectively.

(2) Aggregate addition significantly inhibited the autogenous shrinkage of UHPCC. 60d autogenous shrinkage decreased with the decrease of the binder-aggregate ratio. When the aggregate volume fraction increased from 42.3% to 59.5% and 68.8%, the autogenous shrinkage value decreased by 29.7% and 38.1% Volume stability has an extremely important role and significance.

(3) The smaller proportion of coarse aggregate in aggregates, the greater autogenous shrinkage value, the greater sand rate, the greater autogenous shrinkage. When the sand ratio increases from 0.4 to 0.5, 0.6, the autogenous shrinkage value Respectively, an increase of 13.3%, 15.9%. The inhibition effect on short fine steel fibers after 1d autogenous shrinkage is not obvious.

(4) Form DSC-TG and MIP microscopic analysis technology, decreasing oft water-binder ratio, the content of non-evaporating water decreased, the degree of hydration decreased, which is a positive effect for shrinkage deformation; however, 5 ~ 50nm capillary volume fraction increases, the surface tension increases, the capillary negative pressure increases , the speed and extent are increased, that is to say the self-dessication effect increases, which is negative for the autogenous shrinkage effect. The superposition of positive and negative effects shows that the autogenous shrinkage decreases with the increase of water-cement ratio, indicating that self-dessication effect plays a dominant role in the autogenous shrinkage factors.

IOP Conf. Series: Materials Science and Engineering **324** (2018) 012037 doi:10.1088/1757-899X/324/1/012037

References

[1] Igarashi S, Bentur A, Kovler K. Autogenous shrinkage and induced restraining stresses in high-strength concretes *J. Cement and Concrete Research*. 2000, 30(11):1701-07.

[2] X.M. Kang. Study on dry shrinkage and autogenous shrinkage performance of high strength concrete *J. Journal of Southwest University of Science and Technology.* 2012(1):37-39.

[3] L. Jin, Y. Yang. Experimental study on the autogenous shrinkage temperature dependence of high strength cement mortar *J. Journal of Lishui university*.2008(2):46-48.

[4] Min K, Jung H, Yang J, et al. Shrinkage characteristics of high-strength concrete for large underground space structures *J. Tunnelling and Underground Space Technology*.2010,25(2): 108-113.

[5] Z.H.He, C.X.Qian, G.F. Qian, et al. Influence of concrete components on internal relative humidity and autogenous shrinkage under equal strength *J. Function Material*.2011(2):222-225.

[6] Q.B. Yang. Study on autogenous shrinkage mechanism of high performance concrete *J. Journal of the Chinese Ceramic Society*. 2000(S1):72-75.

[7] H.S. Xu, Z.W. Jiang. Study on the change of relative humidity and autogenous shrinkage of high performance concrete *Journal of Chongqing Jianzhu University*.2004(2):121-125.

[8] Jiang Z, Sun Z, Wang P. Autogenous relative humidity change and autogenous shrinkage of high-performance cement pastes *J. Cement and Concrete Research*. 2005, 35(8): 1539-45.

[9] Yoo S W, Kwon S, Jung S H. Analysis technique for autogenous shrinkage in high performance concrete with mineral and chemical admixtures *J. Construction and Building Materials*. 2012, 34(0): 1-10.

[10] C. Wang. Y.W. Wang, X.C. Pu. Autogenous shrinkage characteristics and mechanism of ultra-low water cement concrete *J. Journal of Building Materials*.2010(1):75-79.

[11] H.B. Lin, C. Wang, Y.J. Zeng. Study on shrinkage deformation and cracking of steel fiber super high strength concrete *J. Concrete*.2011(5): 80-83.

[12] Y.L. Guo, Y. Li, D. Huo. Study on effect of aggregate and fiber amount on early autogenous shrinkage of concrete *J. China Concrete and Cement Products*. 2006(3): 50-52.

[13] H.J. Ba, W.M. Zhang. Effect of aggregate physical property on early autogenous shrinkage of commercial concrete *J. Ready Mixed Concrete*.2004(2): 10-14.

[14] C. Wang. Study on the preparation, structure and properties of super high performance concrete *D.Chongqing University*, 2005.

[15] Li Y, Bao J, Guo Y. The relationship between autogenous shrinkage and pore structure of cement paste with mineral admixtures *Construction and Building Materials*.2010, 24(10):1855-60.

Formability of Annealed Ni-Ti Shape Memory Alloy Sheet

K J Fann[1], J Y Su[1] and C H Chang[1]

[1] Department of Mechanical Engineering, National Chung Hsing University, Taichung 40227, Taiwan

kjfann@nchu.edu.tw

Abstract. Ni-Ti shape memory alloy has two specific properties, superelasiticity and shape memory effect, and thus is widely applied in diverse industries. To extend its application, this study attempts to investigate the strength and cold formability of its sheet blank, which is annealed at various temperatures, by hardness test and by Erichsen-like cupping test. As a result, the higher the annealing temperature, the lower the hardness, the lower the maximum punch load as the sheet blank fractured, and the lower the Erichsen-like index or the lower the formability. In general, the Ni-Ti sheet after annealing has an Erichsen-like index between 8 mm and 9 mm. This study has also confirmed via DSC that the Ni-Ti shape memory alloy possesses the austenitic phase and shows the superelasticity at room temperature.

1. Introduction

Recently Ni-Ti Shape memory alloy has been broadly used as biomedical material [1] not only due to its specific shape memory effect and superelasticity [2] but also because of good compatibility and well corrosion resistance in human bodies, such as coronary or vascular tents [3], bone clamps [4], dental braces [5], and so on for medical devices [6]. In the meantime, there are a variety of industrial applications implemented with Ni-Ti shape memory alloy [1] such as dampers, sensors, actuators, aligning pins, tube fittings and disc springs, and so on [7], by means of the particular properties of free recovery and constrained recovery based on shape memory effect.

The above mentioned shape memory effect is actually caused by phase transformation between austenite at high temperature and martensite at low temperature [8]. The shape formed in austenite at high temperature will be recovered, if the part is deformed in martensite at low temperature and the load is released, by raising the temperature, at which the austenite is stable. This shape recovery is called as free recovery. The superelasticity is in fact rubber-like behaviour, which is caused as well by phase transformation between austenite and martensite at high temperature where the austenite is table [9].

The parts made for the above mentioned applications are from Ni-Ti shape memory alloy tube or wire, mainly due to the material not having good machining properties, so that it must be formed into a parts. However, most of Ni-Ti shape memory alloy shows its superelasticity at room temperature and most of formed shape will spring back after forming. Therefore the forming process must be done in thermomechanical manner [10], in that Ni-Ti shape memory alloy part is formed in hot state, in which the ill-favored superelasticity will not present. On the other hand, such hot forming leads to limit the part shape only formed by tube or wire, not by sheet, which are usually formed at room temperature because of their geometrical characteristics to easily dissipate the heat. Furthermore, to secure or to memorize the shape of Ni-Ti shape memory alloy part, it needs further precipitation treatment at high temperature. Thus the tool sets used in the forming process are needed to clamp the parts to keep their shapes during the shape memory treatment at high temperature, which is a cost intensive process and needs lots of investment for the tool sets [11]. To overcome both above mentioned hindrances, Fann and Huang [12] therefore proposed a cold forming process, in that the shape memory alloys are first solid solutionized, then formed at room temperature, and thereafter aged for shape memory again at high temperature. Even this process would lose part of the shape formed at room temperature during the precipitation treatment, it is still worth to use it to form the shape memory alloy parts for saving the cost of the tool sets. However, Fann and Su [13] shows the solid solutionized Ni-Ti shape memory alloy sheet shows relatively low formability. Therefore this article is aimed to investigate the formability of Ni-Ti shape memory alloy sheets at room temperature, which are annealed at different temperatures [14].

2. Methods and Setup

2.1. Material

The material acquired for this study was a commercial Ni-rich (50.0-50.8at% Ni) Ni-Ti shape memory alloy sheet having a thickness of 0.9 mm. The chemical compositions of the alloy sheet was determined by a glow discharge spectrometer as shown in Table 1 and had 50.67at% Ni and 49.24at% Ti after converting weight percent (wt%) to atomic percent (at%). The content of Fe is as little as 0.09at% and can be regarded as a minimal contaminant during casting, so that the material can be considered as a pure binary Ni-Ti system.

Table 1. Chemical composition of Ni-Ti shape memory alloy sheet (wt%)

Ni	Ti	Fe
55.73	44.17	0.10

2.1.1. Determination of Transformation Temperatures. The characteristic temperatures of phase transformations occurring in the Ni-Ti shape memory alloy sheets were determined by a Differential Scanning Calorimetry (DSC) according to ASTM standard F2004 [15]. Samples were prepared by cutting them from the sheet blank, descaled with hydrofluoric-sulfuric acid and had a weight between 25 and 45 mg. By cooling the samples from 150°C to -150°C and heating from -150°C to 150°C with a rate of 10±0.5°C/min, the heat flow rate between the samples and a reference of empty aluminum pan was recorded. The characteristic transformation temperatures are then determined by drawing the tangents on the heating and cooling spikes from the DSC plot.

2.1.2. Annealing. The Ni-Ti shape memory alloy sheet was cut into a circular shape with a diameter of 36 mm by waterjet. Then the circular specimen was placed in a heated chamber at 300°C, 350°C, 400°C, 550°C, 600°C, and 650°C, respectively, for one hour and subsequently cooled in the still air at laboratory environment. Figure 1 shows the specimen having different colors in appearance after annealing.

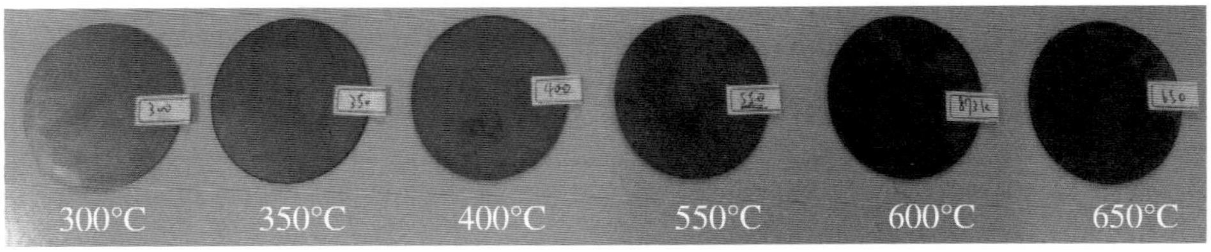

Figure 1. Ni-Ti shape memory alloy sheet blanks at different temperatures

2.1.3. Hardness Test. After annealing the mechanical property of the Ni-Ti shape memory alloy sheet blanks were determined by Rockwell hardness test according to ASTM standard D18 [16]. The average was taken from the Rockwell hardness measured at the mostly wide separated four points on each alloy sheet blank.

2.2. Erichsen-Like Cupping Test

Before executing any cold forming experiments of the Ni-Ti shape memory alloy sheets at room temperature, an Erichsen-like cupping test should be done to determine their process parameter. The Erichsen-like cupping test was actually served as a pre-test for preparing the tools of experiments by seeing how deep the Ni-Ti shape memory alloy sheet could be formed. Because the dimension for the cupping test deviated a little bit from the standard Erichsen cupping test [17], it was called in this study as Erichsen-like cupping test. Besides the punch diameter, the outer diameter of the blank was different to the standard as well. Figure 2 shows the schematic experiment arrangement used in the study, which was conducted on a computer universal testing machine. The sheet blank was clamped between the blankholder and the die by four M12 bolts with a fastening torque in 5 N·m, which induced a larger clamping force other than the statndard. Subsequently a hemispherical punch with a diameter of 30 mm descended with a traveling speed of 10 mm/min through the hole of the blankholder on the blank as well as into the cavity of the bottom die. Once the sheet blank could not endure the punch load and fractured, the punch stroke would be recorded as Erichsen-like index, which indicates the formability of the Ni-Ti shape memory alloy sheet.

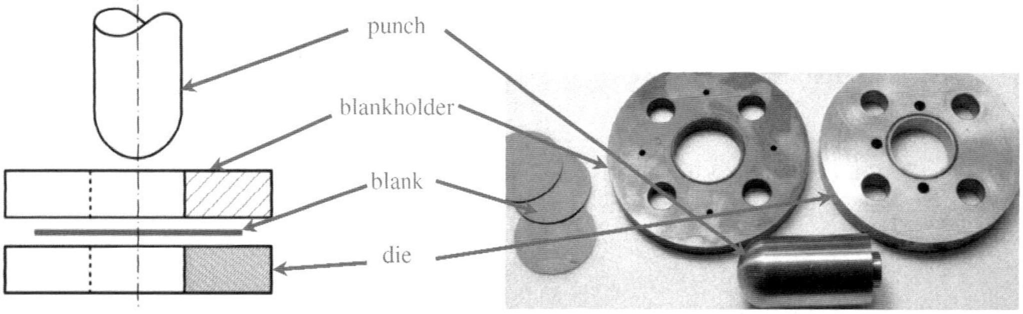

Figure 2. Schematic illustration of Erichsen-like cupping test

3. Results and Discussion

3.1. Transformation Temperatures

Figure 3 shows the DSC curve of the Ni-Ti shape memory alloy as received. On the upper part of the DSC curve, it can be seen that the martensitic start temperature (*Ms*) is about -53°C at the beginning of the spike, and martensitic finish temperature (*Mf*) should be lower than -150°C at the transition after the spike drops. On the bottom part of the DSC curve, the austenitic start temperature (*As*) is found about -47°C and the austenitic finish temperature (*Af*) is then about -19°C. This DSC curve shows that the Ni-Ti shape memory alloy formed at room temperature, which is higher than *Af*, is possessing the austenitic phase and showing the superelasticity.

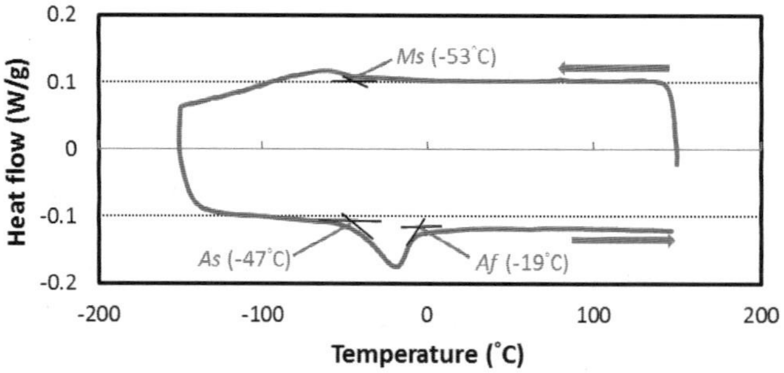

Figure 3. Transformation temperatures - martensitic start (*Ms*), austenitic start (*As*), and austenitic finish (*Af*) obtained for the Ni-Ti shape memory alloy as received.

3.2. Rockwell Hardness

Figure 4 shows the average HRC readings obtained from the Rockwell hardness tests on each Ni-Ti shape memory alloy sheet blanks annealed at different temperatures. It can be seen that the higher the annealing temperature, the lower the HRC value, even the hardness shown at 550°C departs from the trendline. With such high HRC values, the Ni-Ti shape memory alloy sheet shows very high strength and hard to deform.

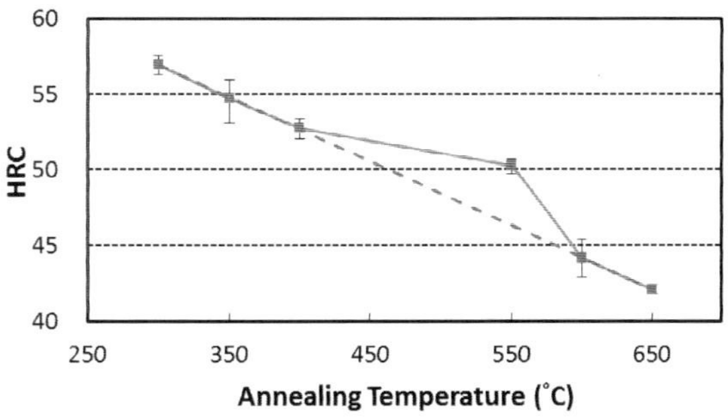

Figure 4. Rockwell hardness obtained on the Ni-Ti shape memory alloy annealed at different temperatures.

3.3. Erichsen-Like Cupping Test

Figure 5 shows the fractured specimen obtained from the Erichsen-like cupping test to each Ni-Ti shape memory alloy sheet blanks annealed at different temperatures. Most of them have the fractured location around the corner radius of the bottom die, while one of them has the fracture occurring at the final contact region of the punch to the sheet. The corner radius of the bottom die, which is not defined in the standard, is 1 mm in this study, which might cause the failure occurring there.

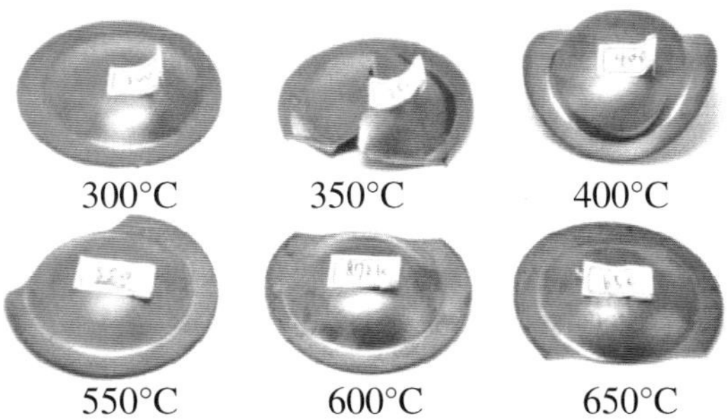

Figure 5. Fractured specimen after Erichsen-like cupping tests of the Ni-Ti shape memory alloy annealed at different temperatures.

Figure 6 shows the punch stroke-load curves obtained from the Erichsen-like cupping test to each Ni-Ti shape memory alloy sheet blanks annealed at different temperatures until the sheet blanks fractured. Even the HRC on the sheet annealed at 550°C higher than the trendline as shown in Figure 4, its maximum punch load still locates on the reasonable trail. As expected, the higher the annealing temperature, the lower the maximum punch load. However the Erichsen-like index shown in Figure 6, where the punch load droped, has different results, in which the higher the annealing temperature, the lower the index or the lower the formability, even the indices obtained from intermediate temperatures fluctuate slightly from the trendline. In general, the Erichsen-like index has between 8 mm and 9 mm.

4. Conclusions

This study has accomplished the hardness test and Erichsen-like cupping test at room temperature to investigate the formability of the Ni-rich Ni-Ti shape memory alloy sheet annealed at 300°C, 350°C, 400°C, 550°C, 600°C, and 650°C, respectively, for one hour and subsequently cooled in the still air.

As a result, the higher the annealing temperature, the lower the hardness and the lower the maximum punch load obtained from the Erichsen-like cupping test as the sheet blank fractured. However the Erichsen-like index shows that the higher the annealing temperature, the lower the index or the lower the formability. In general, the Ni-rich Ni-Ti shape memory alloy sheet after annealing has an Erichsen-like index between 8 mm and 9 mm.

This study has also confirmed by means of DSC that the Ni-Ti shape memory alloy acquired in this study possesses the austenitic phase and shows the superelasticity at room temperature, which causes significant spring back if it is formed cold.

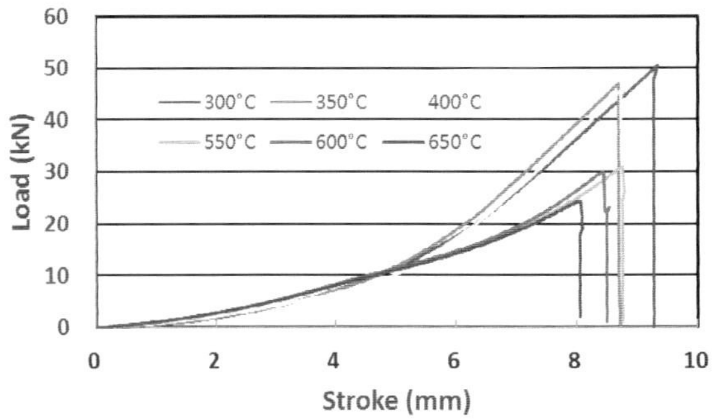

Figure 6. Punch stroke-load curves obtained from the Erichsen-like cupping tests to the Ni-Ti shape memory alloy annealed at different temperatures.

5. References

[1] Wayman CM 1980 *JOM* **32** 129
[2] Shabalovskaya SA 1995 *ICOMAT* 1199
[3] Sigwart U 1996 *Endoluminal Stenting* (London: W. B. Saunders)
[4] Kim DJ, Eun JP and Park JS 2012 *J. Korean Neurosurg. Soc.* **52** 21
[5] Duerig TW, Pelton AR and Stöckel D 1996 *Metall* **50** 569
[6] Duerig T, Pelton A and Stöckel D 1999 *Mater. Sci. Eng.* **A273–275** 149
[7] Hodgson DE, Wu MH and Biermann RJ 1990 *ASM Handbook, Volume 2: Properties and Selection: Nonferrous Alloys and Special-Purpose Materials* ASM Handbook Committee (Materials Park: ASM International) 897
[8] Chang LC and Read TA 1951 *T. Metall. Soc. AIME J. Metals* **189** 47
[9] Christian JW 1965 *The Theory of Transformations in Metals and Alloys* ed. Christian JW (Oxford: Elsevier) 1102
[10] Wayman CW and Duerig TW 1990 *Engineering Aspects of Shape Memory Alloys* ed. Duerig TW, Melton KN, Stöckel D and Wayman CM (London: Butterworth-Heinemann) 3
[11] Fann KJ and Hsu HC 2014 *Adv. Mater. Res.* **939** 430
[12] Fann KJ and Huang PM 2015 *Key Eng. Mater.* **661** 98
[13] Fann KJ and Su JY 2017 *Proc. Inter. Conf. Mater. Mech. Manuf. Eng. (ICMMME)*
[14] Huang X and Liu Y 2001 *Scripta Mater.* **45** 153
[15] American Society for Testing and Materials 2010 *Standard Test Method for Transformation Temperature of Nickel-Titanium Alloys by Thermal Analysis* (West Conshohocken: ASTM International) Designation: F2004-10.
[16] American Society for Testing and Materials 2017 *Standard Test Method for Rockwell Hardness of Metallic Materials* (West Conshohocken: ASTM International) Designation: E18-17.
[17] International Organization for Standardization 2013 *Metallic materials - Sheet and strip - Erichsen cupping test* ISO 20482:2013(E)

Acknowledgments

This research was financially supported by the Ministry of Science and Technology of the Republic of China with the grant coded MOST 105-2221-E-005-029.

IOP Conf. Series: Materials Science and Engineering **324** (2018) 012039 doi:10.1088/1757-899X/324/1/012039

Preparation of superhydrophobic copper surface by a novel silk-screen printing aided electrochemical machining method

X Y Yan[1], G X Chen[1] and J W Liu[2]

1 State Key Laboratory of Pulp and Paper Engineering, South China University of Technology, Guangzhou 510640, China
2 College of mechanical and electrical engineering, Guangdong University of Technology, Guangzhou 510090, China
E-mail: fejliu@scut.edu.cn

Abstract: A kind of superhydrophobic copper surface with micro-nanocomposite structure has been successfully fabricated by employing a silk-screen printing aided electrochemical machining method. At first silk-screen printing technology has been used to form a column point array mask, and then the microcolumn array would be fabricated by electrochemical machining (ECM) effect. In this study, the drop contact angles have been studied and scanning electron microscopy (SEM) has been used to study the surface characteristic of the workpiece. The experiment results show that the micro-nanocomposite structure with cylindrical array can be successfully fabricated on the metal surface. And the maximum contact angle is 151° when the fluoroalkylsilane ethanol solution was used to modify the machined surface in this study.

1. Introduction

There had been no reported about why lotus leaf and other plants and insects have hydrophobic surface before 1992. Afterwards, Barthlott et al. introduce the 'Lotus concept' in 1992, which is also called "lotus effect". Since then, Lotus leaves have widely attracted immense scientific attention due to their superior superhydrophobic properties. As for superhydrophobic surfaces, its water contact angle is higher than 150°, and roll angle is less than 10°. In fact, the double roughness structure is the origin of superhydrophobicity which is useful for floating ability on water surface. At the same time, epicuticular wax is also important, which is a typical low surface energy substance, so as to have superior superhydrophobic properties. Neinhuis et al. [1] demonstrated the super-water-repellent of the lotus leaf and attributed it to the double roughness surface structure with micro- and nanostructures, together with the hydrophobic properties of epicuticular wax.

Based on "lotus effect" theory, a series of fabrication methods (chemical substitution/deposition, electrochemical-etching, solution-immersion etc.) and hydrophobic agent (fluorooxysilane) have been reported for generating superhydrophobic surfaces. Cheng et al. [2] reported that the NiO/ZnO superhydrophobic surface on zinc was fabricated by chemical substitution/deposition and thermal annealing. Yang et al. [3] developed a simple and universal electrochemical-etching method to fabricate the superhydrophobic patterned surface on metal superhydrophobic substrates. Qu et al. [4] reported that superhydrophobic surfaces on engineering materials are prepared via a convenient solution-immersion method. And fluoroalkylsilane was used as a hydrophobic material to reduce the surface free energy. Up to now, more and more researchers have focused on their functions and applications in recent years, since there are many excellent properties besides superhydrophobicity. And it has been also considered for their practical applications in industry ultimately.

As we all know, metal is mostly corrosive, copper is no exception. To overcome this shortcoming, it is necessary to fabricate superhydrophobic copper surface with micro-nanocomposite structure to isolate with water, just because of the existence of water, causing copper corrosion. Zhang et al. [5] discovered that binary geometric structures at the micro- and nano-scale are fabricated on copper surfaces via simple sandblasting and surface oxidation process. He et al. [6] demonstrated that the micro/nano superhydrophobic surface on copper substrate has been successfully fabricated through a simple and effective electrodeposition method combined with annealing without low surface energy organics modification.

In this study, a silk screen printing technology was used to prepare the electrolytic mask on the copper surface, and a micro-nanocomposite structure of the copper surface can be obtained by electrochemical machining. The superhydrophobic copper surface was further modified by fluoroalkylsilane treatment. By analyzing the micro-nanocomposite structure topography, the mechanism of material erosion was studied during electrolytic machining. The formation conditions of superhydrophobic surface were analyzed by SEM and EDS. Meanwhile, the water contact angle of copper surface was measured by contact angle measurement.

2. Experimental and preparation

Copper substrates (99.9%) with a size of 4.0cm×4.0cm×0.02cm were obtained from Tianjin Kermel Chemical Reagent co., Ltd., China, which was selected as experimental material; screen printing plate; Heptadecafluorodecyl trimethoxy-Silane (98%, HTS) was purchased from En Fujia Technology Co., Ltd., Shanghai, China, which served as low surface modification materials. The experimental device is home-made electrolytic processing power with voltage of 0-60V and current of 0-100A, the electrolyte is the solution of sodium nitrate (5 wt %) and ferric chloride (1 wt %).

Figure 1 shows home-made electrolytic processing machine and machining process, before electrochemical machining, the copper surface was printed a mask layer, and then the non-printed areas were electrolytic corrosion during electrochemical machining process. Subsequently, the mask layer was rinsed. Finally, low surface energy was enhanced by fluoroalkylsilane ethanol solution immersion. Therefore, the superhydrophobic copper surface with micro-nanocomposite structure was obtained.

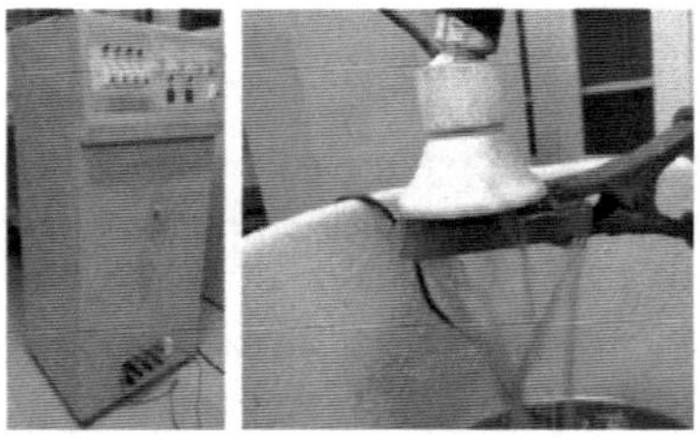

Figure 1. Home-made electrolytic processing machine and machining process.

3. Results and discussion

3.1 Effect of different dot area rates on the micro-nanocomposite structure

The screen printing plate was made by the number of screens with 100 lines / inch. And the screen printing dot area rates of 30%, 35%, 40%, 45% and 50% were obtained. The electrolytic mask was formed by printing on the copper surface shown in Figure 2.

Figure 2. Electrolytic mask printed out with different dot area rates.

The electrolysis voltage was 5V. The interval between the cathode and the anode was 3 mm. The copper with the mask was electrolyzed for 50s. And the obtained morphology is shown in Figure 3. By the way, the shape of Figure 3 (a-e) corresponds to the shape of Figure 2 (a-e). It can be seen from Figure 3 that the area that protected by the electrolytic mask was not electrolyzed, indicating that the printed electrolytic mask can play a protective role in the electrolytic process. The edge of the mask has a sharp shape, while most of the obtained upper surface of the microcolumn is close to a circle after the electrolysis, suggesting that the sharp shape is wiped out during the electrolysis. Meanwhile, as the shape of the tip is too small, it fails to be well protected copper. The height of the microcolumn was measured at a height of about 8 μm. Since the purpose of this paper is to prepare a micro-nanocomposite structure, the smaller the column is, the smaller the better. Comparing with Figure 3a and b, the shape of Figure 3a is not as round as that of Figure 3b, while the size of the microcolumn of Figure 3c-e is larger than that of Figure 3b. In conclusion, the screen printing plate with 35% dot area rate was selected as a mask version.

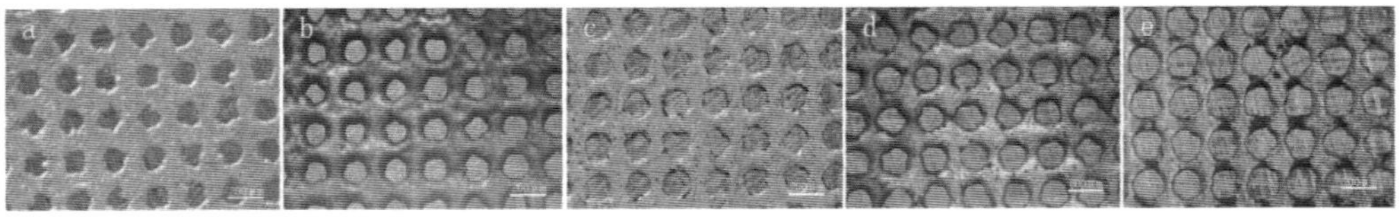

Figure 3. Different mask electrolysis obtained microcolumns.

3.2 Effect of different time on the micro-nanocomposite structure

In order to make the higher microcolumn, the electrolysis time was extended during the experiment. Figure4 shows that different shape was obtained by electrolysis for different times (30s, 50s, 60s, 70s, 90s) with corresponding height 5 μm, 8 μm, 9 μm, 10 μm, and 13 μm on the screen printing plate with 35% dot area rate. As can be seen from Figure 4, the shape of the microcolumn is good when the electrolysis is performed for 30s, and the size of the top and bottom of the microcolumn are similar, indicating that it is not obvious that stray corrosion during the electrolysis process, so the height of the microcolumn is only 5 μm. The height of the microcolumn reached to 9 μm, when the electrolysis was conducted for 60s, the side of the top and bottom of the microcolumn are not much difference. After the electrolysis for 70s, the height of the microcolumn reached to 10 μm, but the side of the top and bottom of the microcolumn are not the same, indicating stray corrosion is more obvious at this time. When the electrolysis carried out for 90s, although the height reached to 13 μm, the stray corrosion is too serious. At the same time, the top and bottom of the micro-column has much difference, and micro-column does not meet the requirements. Therefore, the choice of electrolysis 60s can get a higher degree of microcolumn.

IOP Conf. Series: Materials Science and Engineering **324** (2018) 012039 doi:10.1088/1757-899X/324/1/012039

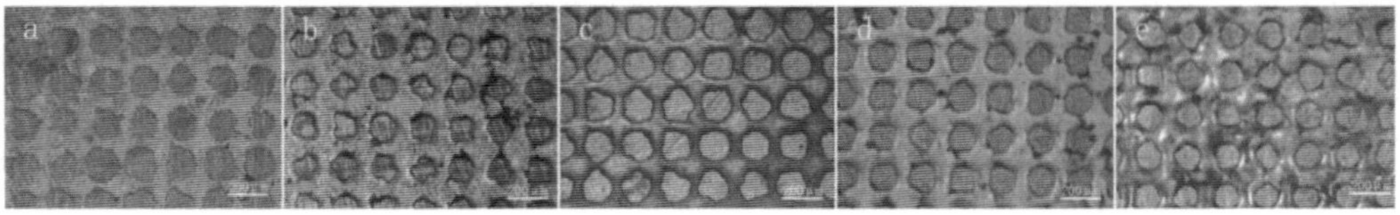

Figure 4. different electrolysis time microcolumn.

Based on above investigation, the copper surface with the micro-nano composite structure was obtained. Following, the copper surface was treated as low surface energy surface by fluoroalkylsilane ethanol solution (configurated by heptadecafluorodecyl trimethoxy-Silane and ethanol) immersion for 45 minutes. Then the surface was dried for 30 minutes. Finally, the surface of the superhydrophobic copper surface was obtained. The water contact angle of copper surface was measured by contact angle measurement. The results shown in Figure 5, the contact angle is 151 °, and rolling angle is 7 °, which was in line with the requirements of superhydrophobic.

Figure 5. The images of static contact angle on copper surface.

Figure 6 describes SEM magnified images of the copper surface. It can be clearly seen the edge of microcolumn, the side of microcolumn has a certain fine structure, because the electrolyte contained ferric chloride solution (1 wt%), and it was corroded the copper surface without mask protection during electrochemical machining process, resulting in the nano-scale structure was generated on the side of the microcolumn. And it is the condition to form a superhydrophobic surface. Therefore, superhydrophobicity can be achieved after fluorination.

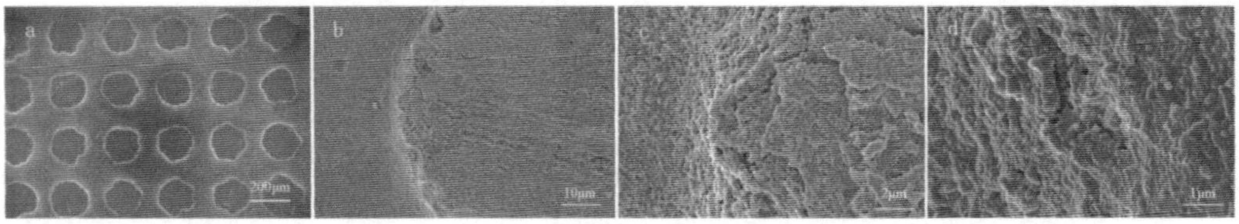

Figure 6. SEM image of copper surface with micro-nanocomposite structure.

The superhydrophobic copper surface element was further analyzed, and the results shown in Figure 7. It can be seen that the F and Si elements appeared on copper surface, indicating that the copper surface was adhered by fluoroalkylsilane successfully.

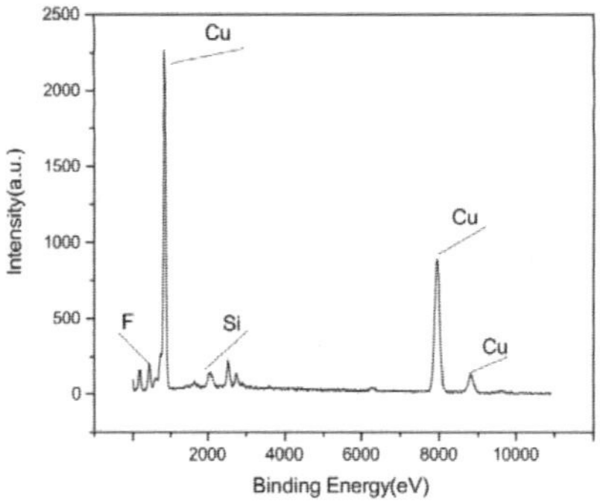

Figure 7. The EDS spectra of superhydrophobic copper surface.

4. Conclusion

In this study, the screen printing and electrolytic processing method was skillfully combined to successfully prepare a micro-nanocomposite structure on the copper surface, and the superhydrophobic effect was achieved, the experiment can be concluded as follows:

1. The screen printing method was used to quickly prepare the controllable electrolytic mask for the size and pattern. The microcolumn with the size of about 150μm and the height of about 9μm can be obtained by electrochemical machining.
2. The ferric chloride was added to the electrolyte. The electrolyzed side of microcolumns corroded by the ferric chloride, so as to the nano-structure formed.
3. The copper with a micro-nanocomposite structure has superior superhydrophobicity with the contact angle of 151°and rolling angle of 7° after fluoroalkylsilane treatment.

References

[1] Neinhuis C and Barthlott W. Characterization and distribution of water-repellent, self-cleaning plant surfaces [J]. *Ann. Bot. (London).* 1997, 79(6): 667-77.

[2] Cheng Y Y, Lu S X, Xu W G, Boukherroub B, Szunerits S and Liang W. Controlled fabrication of NiO/ZnO superhydrophobic surface on zinc substrate with corrosion and abrasion resistance [J]. *J. Alloys Compd.* 2017, 723: 225-36.

[3] Yang X L, Song J L, Liu J K, Liu X and Jin Z J. A twice electrochemical-etching method to fabricate superhydrophobic-superhydrophilic patterns for biomimetic fog harvest [J]. *Sci. Rep-UK.* 2017, 7(1): 1-12.

[4] Qu M, Zhang B, Song S, et al. Fabrication of superhydrophobic surfaces on engineering materials by a solution-immersion process [J]. *Adv. Funct. Mater.* 2007,17: 593-6.

[5] Zhang Y F, Yu X Q, Zhou Q H, Chen F and Li K G. Fabrication of superhydrophobic copper surface with ultra-low water roll angle [J]. *Appl. Surf. Sci.* 2010, 256: 1883-7.

[6] He G, Lu S X, Xu W G, Yu J Y, Wu B and Cui S. Fabrication of durable superhydrophobic electrodeposited tin surfaces with tremella-like structure on copper substrate [J]. *Surf. Coat. Tech.* 2017, 309: 590-9.

Acknowledgments

This research was supported by the National Natural Science Foundation of China (51675105 and U1134003), The Special Support Plan of the Guangdong province (2014TQ01X542) and the Fundamental Research Funds for the Central Universities (2015ZZ080).

Research on the tool holder mode in high speed machining

Zhao Zhenyu[1], Zhou Yongquan[1], Zhou Houming[2], Xu Xiaomei[1], Xiao Haibin[1]

[1]Shenzhen Institute of Information Technology, Shenzhen, China

[2]Xiangtan University, Xiangtan, China

E-mail:yxpzzy01@163.com

Abstract: High speed machining technology can improve the processing efficiency and precision, but also reduce the processing cost. Therefore, the technology is widely regarded in the industry. With the extensive application of high-speed machining technology, high-speed tool system has higher and higher requirements on the tool chuck. At present, in high speed precision machining, several new kinds of clip heads are as long as there are heat shrinkage tool-holder, high-precision spring chuck, hydraulic tool-holder, and the three-rib deformation chuck. Among them, the heat shrinkage tool-holder has the advantages of high precision, high clamping force, high bending rigidity and dynamic balance, etc., which are widely used. Therefore, it is of great significance to research the new requirements of the machining tool system. In order to adapt to the requirement of high speed machining precision machining technology, this paper expounds the common tool holder technology of high precision machining, and proposes how to select correctly tool clamping system in practice. The characteristics and existing problems are analyzed in the tool clamping system.
Keywords: tool holder, high speed machining, characteristic research;

1. Introduction

In recent years, the application of CNC machine tool has been popularized, and high speed machining technology is developing rapidly [1]. In order to achieve high speed, high efficiency, and high precision machining, all kinds of high speed machining centre spindle speed is not only as high as 10000-100000 r/min, and the spindle rotation accuracy is high. Its radial run-out is generally not more than 2 um. In addition to selecting high performance machining tools, a high performance tool clamping system is also selected.

In the application of high speed machining technology, it is often ignored how to select the tool holder method. This could mean a qualified parts and production of waste. The advantages and disadvantages of high speed machining depend greatly on the design of the tool clamping system. Therefore, it is very important to choose the tool clamping system which adapts to the high speed machining. The precision and reliability of the cutter holder should be highly regarded. The paper introduces the working principle and characteristics of heat shrinkage tool-holder with high speed machining, high precision spring chuck, hydraulic tool holder, static pressure expansion chuck and three-rib deformation chuck.

2. Technical analysis

2.1. Heat shrinkage tool-holder

The so-called hot tool holder is one of the methods used to realize the holder of milling cutter, drill bit, reamer and other kinds of tools. The clamping system is composed of the hilt monomer, and the center

of the hilt has a tool mounting hole [2]. Use the heating device to heat the tool installation part of the handle, and then insert the tool into it after the tool mounting hole expands. After cooling to normal temperature, the tool installation hole shrinks and holds the cutter, as shown in Figure 1.
Heat shrinkage tool-holder is a hot topic in recent years. The technology can provide the main reason is that hot charging handle is nearly perfect the handle. The handle of the radial run-out is small and large clamping force, and the use of this handle can keep the position of the machining tool precision, and can obtain high machining accuracy, surface roughness and long service life of machining tools.

Figure 1. Heat shrinkage tool-holder

The clamping principle of the tool system is to heat the shank of the tool by means of induction heating, as shown in Figure 2.When the temperature reaches 315℃~ 425℃, the minus tolerance shank diameter can fully expanded to tool petiole insertion. Insert the tool shank into the inner hole, then cool the handle and tighten the tool with the grip of the handle. The hot mounted handle has the following advantages.1) the shape of the head is slender.2) high pendulum precision.3) clamping force.4) can adapt to high speed rotation.5) facilitate access to workpieces.6) internal cooling can be adopted.

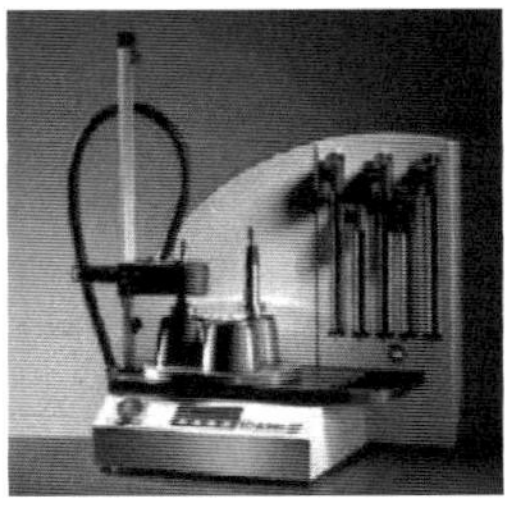

Figure 2. Induction heating device

2.2. *Spring chuck*

The spring chuck is a kind of cylindrical fixture used to hold a bit, or a milling cutter in a milling machine, as shown in Figure 3. It is a fixed device for fixing parts that need to be reworked.

Figure 3. Spring chuck

Before installing the spring chuck, the chuck must be loaded into the lock nut. The locking nut with the spring clip is gently screwed to the handle, the rod or the spindle. Be sure to clean the threaded part of the spring chuck, locking nut, the locating surface and the cone [3].

The working principle of the spring chuck is divided into four main parts, which are the spinning nut, pressing into the sleeve, reducing the inner diameter of the sleeve, and holding the cutter. Factors affecting the holder accuracy include the chuck body bore precision, in addition to accuracy of the sleeve, the sleeve outer cone accuracy, the accuracy of the holding hole and the thread accuracy. The nut and the sleeve contact surface accuracy and the sleeve are also important.

Figure 4. Working principle of spring chuck

The pressure of normal spring chuck is shown in Figure 4a. In the process of lock nut, the contact surface of the nut and sleeve will be relative. This not only causes the sleeve to be stressed, but also the contact surface to wear. It is difficult to obtain or maintain good holder accuracy. Japanese companies design and produce high-precision spring chuck, which is pressed into Figure 4b.The nut is divided into inner and outer parts, and the ball bearings are installed in the center. When the nut contacts the sleeve, the inner part of the nut stops rotating, and there is no relative motion between the sleeves. The locking force of the nut is completely transformed into the pressure of the sleeve, which allows the holder to get a larger clamping force and a higher clamping precision.

2.3. Hydraulic tool-holder

The hydraulic tool-holder can provide sufficient rigidity and dynamic balance, and can make the tool shank and the spindle of the holder head into a straight line, as shown in Figure 5. Hydraulic tool-holder can only hold round handle tool, and is not suitable for holding non-round handle cutter. If you use a hydraulic shank to hold a non-round shank tool, it will cause uneven deformation of the inner sleeve. If the use of time is too long, it is difficult to firmly clamp the tool [4].

Figure 5. Hydraulic tool-holder

Hydraulic tool-holder is a widely used tool holder. The kind of shank holder method is different from the traditional shank system, and is tighten only use a pressure bolt. When the bolt is tightened, it pushes the piston seal to generate a hydraulic oil pressure in the shank. The pressure is evenly transmitted from the circumferential direction to the steel expansion sleeve, which then clamps the tool, as shown in Figure 6. The tool clamping system can make the radial run-out error accuracy and repeat positioning accuracy controlled below 3μm. Due to the presence of high-pressure oil pressure in the shank, the built-in oil chamber structure and high pressure oil greatly increase the structural damping when the tool is clamped, which can effectively prevent the vibration of the tool and the spindle of the machine tool.

Hydraulic tool-holder is widely used in automobile manufacturing, electronics, mold, machinery and other departments, the current major equipment providers are Schunk, Sandvik, Kennametal and other internationally renowned brands.

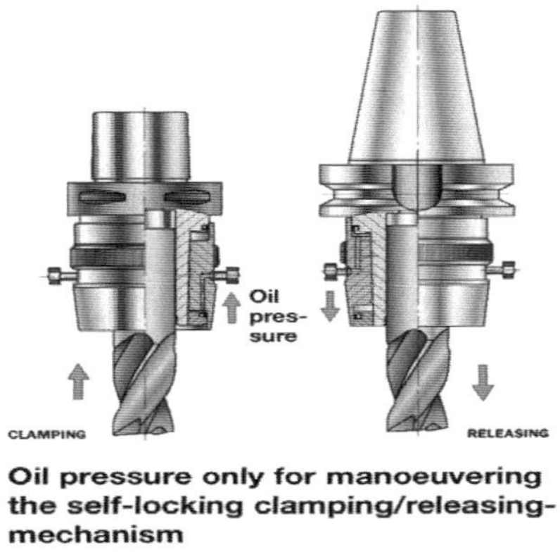

Figure 6. Hydraulic tool-holder working principle

2.4. The three-rib deformation chuck

The three-rib deformation chuck is used to clamp the cutter with its own deformation force [5-6]. The chuck is triangular in free state, as shown in Fig. 7a.The diameter of the inner circle of the triangular shape is less than the diameter of the shank of the holder, and a hydraulic pressure force device is used to apply external forces to the holder head, which can transform the clip head into a round hole, as shown in Fig. 7b.The diameter of the hole is slightly larger than the diameter of the shank, and the tool shank can be successfully inserted into the holder hole, as shown in Fig. 7c.Then remove the external force, the inner hole is redrawn into a triangular shape, the handle is tightened with a strong elastic resilience, as shown in Fig. 7d.

This kind of chuck has compact structure and good symmetry. Accuracy is high, holder rotation accuracy ≤ 3 μ m. Compared with the heat shrinkage tool-holder, tool handle is simple, and for different expansion coefficient of carbide handle and high-speed steel shank can be applied.

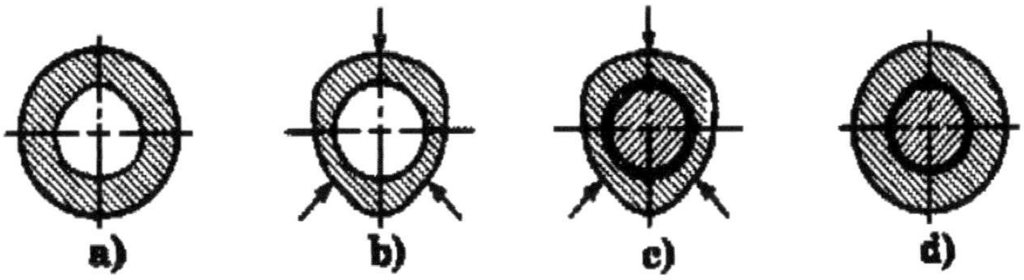

Figure 7. Working principle diagram of three rib deformation chuck

Through the previous analysis, the comparison of the properties of spring chuck, hydraulic holder, heat shrinkage tool-holder and three-rib deformation is shown in Table 1.

Table 1. The characteristics of various holder comparisons

Type	Milling Chuck	Spring chuck	Hydraulic tool-holder	Heat shrinkage tool-holder
Accuracy	0.01	0.003	0.0003	0.0005
Radial jump	0. 01mm~0. 02mm	0. 01mm~0. 02mm	0.002mm~0.005mm	0. 002mm~0. 005mm
Outer diameter	Very thick	Thinner	General	Very thin
Overhang length	Long	Long	Long	The longest
Torque	Good	Better	Best	Best

transmission				
Rigidity	Good	Better	good	Best
Blades material	High-speed steel, carbide	High-speed steel, carbide	High-speed steel, carbide	Carbide
Clamping force	The strongest	General	General	General
Dangerous	No	No	No	Yes
High-speed machining	Fit	Fit	Fit	Fit
Operability	Good	Better	Best	Very poor

3. Summary

At present, the new high-speed precision tool chuck is heat shrinkage tool-holder, high-precision spring chuck, hydraulic tool-holder and the three-rib deformation chuck in the most advanced high-tech CNC machine tool system. Although its share of value is small, it is an important tool system for connecting machine tool spindles and tools, which directly affects the quality of the finished product. Therefore, increasing the high-speed machining tool chuck theoretical research and manufacturing base for research is especially necessary.

4. References

[1] SCHALZ I H 2002 Development of high speed machining *Machinery Manufacturing and Automation* vol 2 pp 4-8
[2] CHEN S P and LI L 2008 Thermo Grip for Machining in high speed and exactitude *Manufacturing technology and machine tools* vol 2 pp 139-141
[3] WANG Z L 2013 New structure of collet chuck *Machine Building & Automation* vol 4 pp 82-83
[4] LIU Z H and ZHANG W M 2015 Comparative testing on milling for hydraulic and heat shrinkage tool-holders *Journal of China engineering machinery* vol 13 pp 457-460
[5] ZHOU Q Y and ZHOU Z X 2010 Research on tool chucks in high-performance NC machine tool *Machine tools and hydraulics* vol 38 pp 1-3
[6] SHEN J 2002 About a new precision tool clamping system *Manufacturing technology and machine tools* vol 5 pp 52-53

5. Acknowledgement

The authors would like to thank “Science plan project of Shenzhen” (JCYJ20130401095559824, GRCK20170424095629032, GJHZ20150316112419786, GRCK20160415111859786) and Guangdong province science and technology project (2015A050502006) for their financial support.

Durability and Strength of Sustainable Self-Consolidating Concrete Containing Fly Ash

O Mohamed[1] and W Al Hawat[2]

1 Professor of Structural Eng. and Materials, Abu Dhabi University, Abu Dhabi, UAE
E-mail: osama.mohamed@adu.ac.ae
2 Research Engineer, Abu Dhabi University, Abu Dhabi, UAE
E-mail: Waddah.alhawat@adu.ac.ae

Abstract. In this paper, the durability and strength of self-consolidating concrete (SCC) is assessed through development and testing of six binary mixes at fixed water-to-binder (w/b) ratio of 0.36. In each of the six SCC mixes, a different percentage of cement is replaced with fly ash. The development of compressive strength for each of the mixes is assessed by testing samples after 3, 7, and 28 days of curing. Durability of each of the six SCC mixes is assessed by measuring the charge passed in Rapid Chloride Permeability (RCP) test. Charge passed was measured in samples cured for 1, 3, 7, 14, 28, and 40 days of curing. All mixes out-performed the control mix in terms of resistance to chloride penetration. Binary mix in which 20% of cement is replaced with fly ash exhibited 28-day strength slightly surpassing the control mix.

1. Introduction

The use of self-consolidated concrete (SCC) in construction is gaining universal popularity due to its ability flow under its own weight and generally high quality of the finished product. The production of cement, the main ingredient in concrete, is accompanied by emission of significant amounts of CO_2 into the atmosphere. Therefore, decreasing the usage of cement through its partial replacement with more sustainable alternatives is a legitimate consideration. Fly ash, silica fume, and ground granulated blast furnace slag (GGBS), also known as supplementary cementitious materials (SCM), are industrial by-products that has been used extensively to improve fresh and hardened properties of concrete. Their reuse as partial cement replacements save precious landfill space, but may also enhance durability of concrete. Concrete structures that are in marine environments or in soils containing high levels of chloride ions are likely to get damaged when these chloride ions migrate through concrete voids and corrode reinforcing steel.

In this study, sustainable SCC mixes in which cement is partially replaced with fly ash were produced and tested to assess their compressive strength development and resistivity to chloride ion migration. Resistivity to chloride ions migration is assessed in this paper using the rapid chloride penetration (RCP) test in accordance with ASTM C1202 [1]. Concrete compressive strength was tested on standard 150 x 150 x 150 mm cubes after 3, 7, and 28 days of curing.

Mohamed [2] studied the durability and strength of SCC mixes containing various percentages of fly ash and chopped basalt fibres to assess compressive strength development and resistance to chloride penetration. Eight SCC mixes were developed in which the cement was partially replaced with 10%, 20%, 30% and 40% fly ash. In some mixes, 1% to 2% chopped basalt fibres were added. The compressive strength was determined after 3, 7 and 28 days of curing by testing standard 150x150x150 mm cubes in accordance with BS EN 12390-3:2009 [3]. Resistance to chloride penetration was determined using the RCP test conducted according to ASTM C1202 [1] on samples that have been cured for 7, 14, 28, 40 and 70 days. The highest 28-day compressive strength of 67.96 MPa was obtained on mixes in which 20% of the cement was replaced with fly ash (FA20). After each of the curing ages, the concrete resistance to chloride penetration improved as the FA dosage increased compared to the control mix.

Mohamed et al. [4] developed ANN models to predict the resistance of sustainable SCC mixes to chloride penetration. Model input parameters included water-to-binder (w/b) ratio, aggregate content, fly ash content, silica fume content, superplasticizer content, and number of curing days. The output data was the level of chloride penetration. The model was developed using one hidden layer with 7 neurons and ANN implemented the back-propagation method. A total number of 70 data sets was used to train the ANN model, and 20 data sets tested by the investigators were used to validate the accuracy of the ANN model. The ANN model with learning rate of 0.3, momentum of 0.5 and epochs of 5000 achieved an accuracy of 95%.

Ahari [5] investigated the compressive strength and chloride penetration resistance of 17 SCC binary, ternary and quaternary mixes containing various percentages of SCMs including GGBS and fly ash. The percentage of cement replacement with fly ash ranged from 18% to 36% while percentage of GGBS ranged from 8% to 18%. The compressive strength was tested after 7, 28 and 90 days of curing. Test results showed that replacing cement with 18% GGBS produced a binary mix with 28-day compressive strength of 54.6 MPa which exceeds the control mix. The highest compressive strength of 56.6 MPa for binary mixes was obtained when 36% of the cement was replaced with fly ash. The chloride penetration resistance was tested after 28 days and after 90 days of curing. The chloride penetration resistance of the 36% fly ash mix was better than the control mix, after 28 days of curing.

Khodair [6] studied the effect of high volume of fly ash and slag as cement replacement of the properties of SCC. 20 SCC mixes were developed with different ranges of fly and slag at a constant w/b ratio of 0.38. Nine samples were tested for compressive strength was tested at 3-, 14-, and days using a 100 X 200 mm cylinder according to ASTM C39 [7]. The permeability of concrete was tested at 90 days using the RCP test that is based on ASTM C1202 standards [1]. The results indicated that using SCMs as a replacement for cement lowered the 28-day compressive strength compared to the control mix. The maximum compressive was obtained by replacing the cement with both 25% fly ash and 25% slag. All mixes containing SCMs had low permeability according to ASTM C1202 [1].

Singh [8] investigated the durability of self-consolidating concrete containing iron slag (IS) as replacement of fine aggregates. The percentage of IS that replaced the fine aggregate ranged from 0% to 40%. The compressive strength was tested after 28, 91 and 365 days of curing. The permeability of SCC mixes was measured using RCP test after 7, 28, 91, and 365 days of curing. The results showed that the use of IS increased the compressive strength of SCC compared to the control mix. The highest 28-day compressive strength was 40 MPa, achieved by replacing fine aggregates with 40% of IS. Increasing the percentage of IS decreased the permeability of concrete and improved resistance to chloride penetration compared to the control mix as indicated by the very low passing charge.

Gesoglu [9] studied the fresh and hardened properties of SCC containing fly ash combined with marble powder and limestone filler. The SCC mixes were produced as binary and ternary mixes at a constant w/b ratio of 0.35. RCP tests were conducted on six 100 x 200 mm standard cylindrical specimens according to AASHTO T277 [10]. The compression test was conducted on standard 150 x 150 x 150 mm cubes according to ASTM C39 [8]. The compressive strength was measured after 28 days and after 90 days of curing, while the chloride permeability test was conducted on samples that has been cured for 90 days. The results showed that high replacement percentage of cement by fillers affected the fresh properties negatively, while replacement of cement with fly ash enhanced concrete fresh properties. Fly ash mixes produced lower 28 days compressive strength compared to the control mix due to the slow hydration process.

Beglarigale and Yazici [11] studied the effect of GGBS and steel microfibers on reducing the deleterious effects of alkali-silica reaction (ASR) on concrete. The investigators noted that adding GGBS and steel fibers to tested samples showed a reduction on expansion associated with ASR compared to control samples. Other studies on 50-year old concrete panels have shown that expansion due to ASR may result in significant damage and loss of compressive strength in the long term [12].

2. Experimental Program

The purpose of this study is to investigate the resistance to chloride penetration of SCC mixes in which cement is partially replaced with various percentages of fly ash. A total of six SCC mixes were produced with various percentages of fly ash at a constant w/b ratio of 0.36. While the fixed w/b is in a way limitation of this study, w/b = 0.36 is a typical average ratio for SCC nowadays in many parts of the world. Total binder content is 480 kg/m^3 and the total natural course gravel content is 800 kg/m^3. Fine aggregates content is a combination of 582.4 kg/m^{3}black sand and 313.6 kg/m^3 dune sand. The superplasticiser dosage was maintained at 7.2 kg/m^3 in each of the six trial mixes. The six binary SCC mixes were created by partial replacement of cement with 10%, 15%, 20%, and 25%, 30%, or 40% fly ash.

The compressive strength was determined by testing standard 15x 150x150 mm cubes in accordance the BS EN 12390-3:2009 [9]. Samples for compression test were cured for 3, 7 and 28 days. The resistance of the SCC mixes to chloride penetration was assessed using the RCP test conducted on cores extracted from standard 150 x 150 x 150 mm cylinders. The RCP test was conducted in accordance with ASTM C1202 [1] on samples cured for 1, 3, 7, 14, 28 and 40 days of curing. The chloride penetration level is determined by the currents the current (I0, I30, etc.) and total passing charge, Q, measured in coulombs using Equation 1.

$$Q = 900\,(\,I_0 + 2I_{30} + 2I_{60} + \cdots + 2I_{300} + 2I_{330} + 2I_{360}) \qquad (1)$$

Where I_0, I_{30}, I_{60}, ... are the Current (Ampere) immediately after the voltage is applied, at 30 minutes and at 60 minutes. The passing charge computed from Equation 1 is converted to concrete penetrability rating according to Table 1.

Table 1. Chloride Ion Penetrability Based on Charge Passed

Charge Passed (Coulombs)	Chloride Ion Penetrability
> 4000	High
2000 – 4000	Moderate
1000 – 2000	Low
100 – 1000	Very Low
< 100	Negligible

3. Results and Discussion

The fly ash mixes were tested for the compressive strength after 3, 7, and 28 days of curing. Table 2 shows the compressive strength of fly ash mixes at different curing ages. The mixture name FA10 means a binary mix in which 10% of the cement in the control mix was replacing with fly ash, and so on. Table 2 and Figure 1 show that FA20, the binary mix in which 20% of the cement was replaced with fly ash is the optimum mix in terms of compressive strength as it marginally surpassed the 28-day compressive strength of the control mix. This is remarkable noting the notoriously slow reactivity of fly ash. Table 2 also shows that fly ash replacement below the optimum ratio of 20% (10% and 15%) or above the optimum ratio (30% and 40%) produce lower 28-day compressive strength compared to the control mix.

Figure 1 shows that the compressive strength of binary fly ash mixes coverage to the strength of control mix over time, except for FA20 which slightly exceeded the control mix after 28-days of curing.

Table 2. Compressive Strength of Binary Fly Ash Mixes

Mixture Type	Compressive Strength (MPa)		
	3d	7d	28d
Control Mix	51	61.48	66.08
FA 10	46.3	54.3	61.325
FA 15	42	50.3	62.5
FA 20	46.225	50.025	67.96
FA 25	38.6	43	61.7
FA 30	38.6	46.4	56.5
FA 40	31.43	36.7	55.75

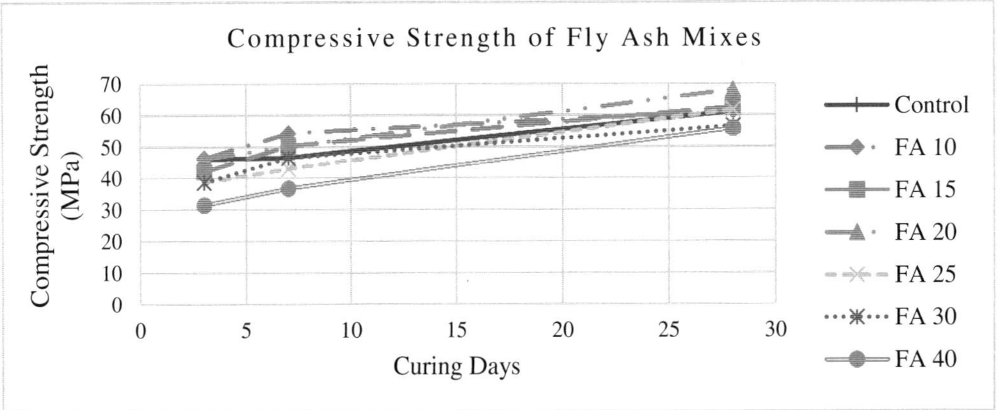

Figure 1. Compressive Strength of Fly Ash Mixes

Figure 2 shows the passing charge versus curing days for binary fly ash mixes. The passing charge was lower than the control mix for all replacement ratios and decreases with increase in cement replacement ratio. The higher the replacement ratio, the lower the passing charge.

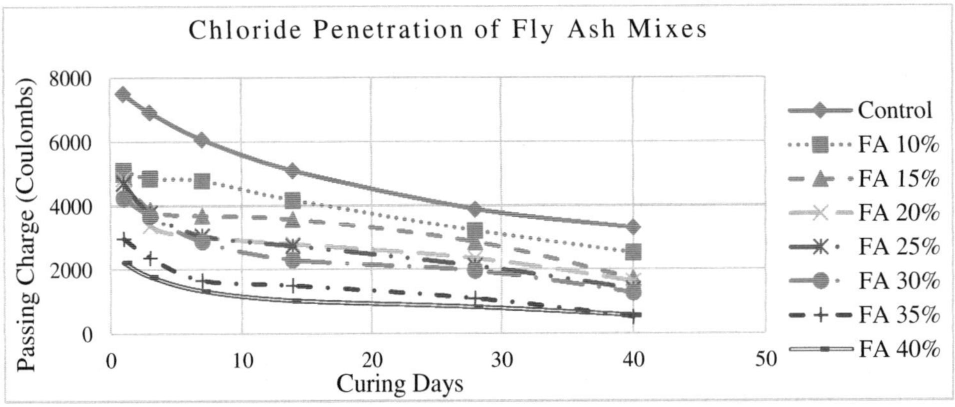

Figure 2. Chloride Penetration of Fly Ash Mixes

The best resistance to chloride penetration in fly ash mixes was exhibited by FA40 after 14 days of curing where the passing charge classifies as very low. FA20, the mix that produced the highest compressive strength of fly ash mixes, exhibited low chloride ion permeability after 40 days of curing.

4. Conclusions

In this paper, compressive strength development and chloride penetration resistance of six binary self-consolidating concrete mixes were assessed through an experimental program. Binary mixes contain cement that is partially replaced with various percentages of fly ash. The compressive strength was tested after 3, 7, and 28 days of curing. The chloride permeability of concrete was tested after 1, 3, 7, 14, 28, and 40 days of curing. The following observations were made:

- Partial replacement of cement with fly ash improves resistance to chloride penetration resistance compared to a control mix. This applied to cement replacement ratios examined in this study and after each of the curing days examined in this study as well.
- Binary mix FA20 in which 20% of the cement was replaced with fly ash produced the maximum 28-day compressive strength, slightly surpassing the control mix and higher than all other fly ash mixes. The chloride ions permeability for this mix was rated as *moderate* after 3 days of curing and rated as *low* after 40 days of curing.

Acknowledgement
The authors gratefully acknowledge the financial support of Abu Dhabi Department of Education and Knowledge (ADEK) through the ADEK Award for Research Excellence (AARE) 2017.

5. References

[1] ASTM C1202-17, (2017) Standard Test Method for Electrical Indication of Concrete & apos;s Ability to Resist Chloride Ion Penetration, *ASTM International*
[2] Mohamed O, Al-Hawat W, 2016 Influence of fly ash and basalt fibres on strength and chloride penetration resistance of self-consolidating concrete *Materials Science Forum* **866** 3-8
[3] BS EN 12390-3:2009, Testing hardened concrete Part 3. Compressive Strength of Test Specimens. *British Standards Institution*
[4] Mohamed O, Ati M, Al Hawat W, 2015 Using neural networks to predict chloride penetration of sustainable self-consolidating concrete, *Civil-Comp Press.*
[5] Ahari R, Erdem T, Ramyar K, 2015 Permeability properties of self-consolidating concrete containing various supplementary cementitious materials *Const. and Build. Materials* **79** 326-336
[6] Khodair Y, Bommareddy B, 2017 Self-consolidating concrete using recycled concrete aggregate and high volume of fly ash, and slag *Construction and Building Materials* **153** 307-316
[7] ASTM C 39/ C39M-10, 2010 Standard Test Method for Compressive Strength of Cylindrical Concrete Specimens, *American Society for Testing and Materials*, Pennsylvania, PA
[8] Singh G, Siddique R, 2016 Effect of iron slag as partial replacement of fine aggregates on the durability characteristics of self-compacting concrete *Const. and Building Materials* **128** 88-95
[9] Gesoglu M, Güneyisi E. Kocabag M, Bayram V, Mermerdas K, 2012 Fresh and hardened characteristics of self-compacting concretes made with combined use of marble powder, limestone filler, and fly ash *Construction and Building Materials* **37** 160-170
[10] ASSHTO T277-89, 1990 Rapid determination of the chloride permeability of concrete. *American association of states highway and transportation officials*, Standard Specifications-Part II, Washington, DC.
[11] Beglarigale A Halit Yazici H, 2014 Mitigation of detrimental effect of alkali-silica reaction in cement-based composites by combination of steel microfibers and ground granulated blast-furnace slag *J. of Mat. in Civil Engineering* **16**(12) doi.10.1061/(ASCE)MT.1943-5533.0001005
[12] Mohamed O A, Rens K L, and Stalnaker J J, 2001 Time effect of Alkali-aggregate reaction on performance of concrete, *Journal of Materials in Civil Engineering* **13**(2) 143-151

The Comprehensive Evaluation Method of Supervision Risk in Electricity Transaction Based on Unascertained Rational Number

Wang Haining[1,2], Wang Lei[1,2], Zhang Qian[1,2], Zheng Zongqiang[1,3], Zhou Hongyu[1,2] and Gao Chuncheng[1,2]

[1] Nari Group Corporation/State Grid Electric Power Research Institute Nanjing China

[2] Beijing Kedong Electric Power Control System Co., Ltd. Beijing China

[3] Nari Technology Co., Ltd. Nanjing China

E-mail:hbkjdx2009@126.com

Abstract. For the uncertain problems in the comprehensive evaluation of supervision risk in electricity transaction, this paper uses the unidentified rational numbers to evaluation the supervision risk, to obtain the possible result and corresponding credibility of evaluation and realize the quantification of risk indexes. The model can draw the risk degree of various indexes, which makes it easier for the electricity transaction supervisors to identify the transaction risk and determine the risk level, assisting the decision-making and realizing the effective supervision of the risk. The results of the case analysis verify the effectiveness of the model.

1. Introduction

With the establishment of the united and mutual power market[1-2], the objects and methods of electricity transaction are more complicated, and there is greater uncertainty in market behaviors, resulting in increasing supervision risk in electricity transaction. In order to avoid the deteriorating supervision risk in electricity transaction, affect the market order and bring more risk to the electricity transaction, it is urgent to research the supervision risk in electricity transaction.

At present, fuzzy comprehensive evaluation method [3-4]is widely used to evaluate the supervision risk in electricity transaction[5-6]. However, the evaluation objects of the method are most the power market operation efficiency, the power market operation rules and so on which are highly certainty. Relative to the supervision risk in electricity transaction with greater uncertainty, the fuzzy comprehensive evaluation method does not exactly reflect the uncertainty in the evaluation which is embodied in two aspects. Firstly, uncertainties in market behaviors lead to greater uncertainty in the comprehensive evaluation of supervision[7] risk in electricity transaction. Secondly, when determining the true state of things and the number relation, policy makers are with uncertainty in purely subjective understanding due to insufficient information.

In fact, the information which is provided by any system has both state factors and behavioral factors is mainly all essentially unidentified information .For the information, we must consider its uncertainty and can not simplify it to the determining information to deal with it. Therefore, aiming at the uncertainty of supervision risk in electricity transaction, this paper constructs the comprehensive evaluation method of supervision risk in electricity transaction based on unascertained rational number[8].

2. The comprehensive evaluation model of supervision risk in electricity transaction based on unascertained rational number

This paper builds the warning index system for transaction supervision risk that is applicable to the united and mutual power market. It contains 4 first class indexes namely market synergies, trading plans, contracts and settlements and energy efficiency including 10 second indexes which are shown in table 1.

Table 1. The warning index system for transaction supervision risk that is applicable to the united and mutual power market

the first class index	**the second class index**
market coordination	provincial trading volume
	the ratio of transaction volume in the total electricity
	clean energy efficiency
trading plan	the completion rate of base electricity plan
	the implement balanced rate of base electricity plan
contract and settlement	the rate of contract record
	the rate of settlement completion
	the rate of settlement prompt
energy efficiency	the rate of clean energy consumption
	the year-on-year growth rate of coal saving calculation

(1) the overall thinking

1)Conduct a comprehensive credibility model for power market experts, drawing the comprehensive credibility of the experts involved in the evaluation;

2)The second class index are marked by each expert;

3)According to the credibility of experts and the grade, calculate the scores of the two class index and the score is belong to unascertained rational number;

4)Finally, combining the weight of second class indexes and the scores, calculate the scores of first class indexes. By that analogy, calculate the score of the general goal.

(2) Conducting the comprehensive credibility model for power market experts

According to the expert title, academic qualifications and working years, determine the credibility of the experts which is shown in table 2.

Table 2. Expert Credibility Assessment Standard Classification

Item	Job title			Education				Working-age (year)		
Category	Senior Title	Intermediate	Junior title	Master's degree	Undergraduate	Specialist	High school	⩾ 20	20~ 10	<10
Confidence interval	[8,10]	[4,7]	[1,3]	[8,10]	[7,10]	[3,6]	[1,3]	[8,10]	[4,7]	[1,3]

$\varepsilon_i (i=1,2,3)$ is the job title, education, working-age respectively, and then the credibility of experts is β_i that can be expressed by the formula (1).

$$\beta_i = \frac{\sum_{i=1}^{3} \varepsilon_i}{30} \tag{1}$$

The closer the value of β_i is to 1, the more trustworthy this expert is and the more accurate the judgment is. Conversely, the smaller β_i is, the less likely it is to believe this expert.

The credibility of experts $B_1, B_2, \dots B_n$ involved in comprehensive evaluation are $\bar{\alpha}_1, \bar{\alpha}_2, \wedge, \bar{\alpha}_n$ 。

The comprehensive credibility of expert B_i for expert group $B_1, B_2, \dots B_n$ is α_i.

$$\alpha_i = \frac{\bar{\alpha}_i}{\sum_{k=1}^{n} \bar{\alpha}_k} \quad (i = 1, 2, \wedge, n) \tag{2}$$

The comprehensive credibility of expert group $B_1, B_2, \dots B_n$ is α.

$$\alpha = \frac{1}{n}(\bar{\alpha}_1 + \bar{\alpha}_2 + \wedge + \bar{\alpha}_n) \tag{3}$$

(3) The uncertainty quantization of expertise

This paper uses the unascertained rational number to quantify the evaluation information of experts. $A_1, A_2, \dots A_m$ are the assessment of A, and the experts $B_1, B_2, \dots B_n$ use the percentile system to rate the indicator factors, drawing the scoring table.

The scores of factor A are $C_{i1}, C_{i2}, \dots, C_{in}$, respectively, and the comprehensive credibility of experts are $\alpha_1, \alpha_2, \dots, \alpha_n$,.Now, arrange the $C_{i1}, C_{i2}, \dots, C_{in}$ into $C_{ij1}, C_{ij2}, \dots, C_{ijn}$, then the paper draws the unascertained rational number of the factor A.

$$f_i(x) = \begin{cases} \alpha_{ij1} & x=C_{ij1} \\ \alpha_{ij2} & x=C_{ij2} \\ \dots & \dots \\ \alpha_{ijk_i} & x=C_{ijk_i} \\ 0 & x \notin \{C_{ij1}, C_{ij2}, \dots, C_{ijk}\} 且 x \in G \end{cases} \tag{4}$$

In the formula,

$f_i(x)$ is the unascertained rational number of the factor A.

$i=1,2,..,m$;

IOP Conf. Series: Materials Science and Engineering **324** (2018) 012042 doi:10.1088/1757-899X/324/1/012042

A_{ij1}, A_{ij2},…, A_{ijk} are the sum of experts' comprehensive credibility which the score are C_{ij1}, C_{ij2},…, C_{ijk}.

The weight of factor A_i is w_i, so the uncertain quantizer of object A is C.

$$C = \sum_{i=1}^{m} w_i f_i(x) \qquad (5)$$

3. The analysis of case

The paper supposes that there are three power market experts involved in the evaluation. Through the model of experts' comprehensive reliability, the comprehensive reliability of experts are 0.5,0.3,0.2 respectively. The scores of second class index in market coordination index are shown in table 3.

Table 3. The scores of second class index in market coordination index

	Expert 1	Expert 2	Expert 3
provincial trading volume	75	80	78
the ratio of transaction volume in the total electricity	65	60	66
clean energy efficiency	40	50	35

The scores of market coordination index are shown in table 4.

Table 4. The scores of market coordination index

score	40.43	52.33	61.63	75.10
credibility	0.105	0.270	0.365	0.260

The higher the score, the greater the risk. Table 4 shows that the probability of 60 or more is 62.5%, indicating that in the electricity market transaction at this stage, the market coordination that the first class index has certain risks. So the relevant person should strengthen the supervision of market coordination, make early warning response to the indicator and put forward countermeasures to avoid greater risks.

4. Conclusion

The comprehensive evaluation method of supervision risk in electricity transaction based on unascertained rational number in this paper solves the problem of uncertainty in the current supervision risk evaluation in power transaction. At the same time, the model facilitates the risk supervisor to identify and judge the risk level, avoiding the ever-worsening of trading risk, and instruct power supervisors to take appropriate measures to effectively monitor the risks.

5. References

[1] Wen Xu, Yan Wei, Wang Junmei, et al. Regional Energy-Saving Electricity Market and Stochastic Programming Power Purchasing Model Considering Assessment on Energy Consumption in Trans-Provincial Transaction [J]. *Power System Technology*,2013,**2**:500-506.

[2] Zeng Bo, Li Yingzi, Zhang Jianhua, et al. Comprehensive Evaluation Model and Method for Smart Distribution Network Planning Under New Electricity Market Layout[J]. *Power System Technology*,2016,**40(11)**:3309-3315.

[3] Zhang Xinjie, Ge Shaoyun, Liu Hong, et al. Comprehensive assessment system and method of smart distribution gird[J]. *Power System Technology*, 2014, **38(1)**:3642-3649.

[4] Yu Hongyan, Ji Shenjia. Estimation and Comparison on prediction accuracy of weight array model based on FAHP[J]. *Statistics and Decision*,2017,**23**:74-78.

[5] Li Hongxia, Ren Jiahe. Study on the effectiveness of emergency signs based on ordering relation and fuzzy comprehensive evaluation[J]. *Journal of Xi'an University of Science and Technology*,2017,**5**:706-711.
[6] Wang Fei, Du Lijuan. Low-carbon Economy Comprehensive Evaluation Based on Entropy Weight and Grey Correlation and TOPSIS [J]. *Science and Technology Management Research*,2013,**7**: 48-51.
[7] Li Cunbin, Yu Xinyi, Wang Zhijie. Risk Assessment on Photovoltaic Power Generation Project by Grey Correlation Analysis and TOPSIS Method[J]. *Power System Technology*, 2013,**6**:1514-1519.
[8] Zhao Zitong, Shen Junhui, Xu Peng, et al. The application of unascertained mathematics method to the extraction of integrated deformation modulus value in intensive altered rock zone[J]. *Journal of Chengdu University of Technology(Science & Technology Edition)*,2016,**3**:119-124.

6. Acknowledgments

SGEPRI-KJB-KJ(2017)2941,Technical project of State Grid Corporation of China "Research on Key Technologies of Mobile Application and Big Data Analysis for Power Trading Platform"

IOP Conf. Series: Materials Science and Engineering **324** (2018) 012043 doi:10.1088/1757-899X/324/1/012043

Study on the key role of hierarchical microstructure for strength and plasticity in a lath martensitic steel

Ming Yang [1, 2]**, Shao-lei Long**[2]**, Yi-long Liang**[2]

1Faculty of Material Science and Engineering, Kunming University of science and Technology, Kunming 650093, PR China
2College of Materials and Metallurgy, Guizhou University, Guiyang 550025, PR China
E-mail: 429428817@qq.com

Abstract. In this paper, the effect of substructure of lath martensite on the mechanical properties was discussed in detail. Results indicated that prior austenite grain, packet and block increase with the increasing of quenching temperature. A good linear relationship exists between the packet, block and prior austenite, which reveal that the size of packet, block depends on prior austenite grain. However, lath is increased with not determined by prior austenite grain. Based on the EBSD analysis, the large ratio of the low angle orientation boundaries determines the better plasticity is obtained in coarse grain. Therefore, the refining of martensite lath or the increase of the low angle orientation plays an important role on improving the plasticity in lath martensite steel.

1. Introduction

Low-carbon martensite steel belongs to an important family of advanced high strength or ultra-high strength steels due to an excellent combination of strength, plasticity and toughness. Recently, extensive researches have attempted to build the relationship between the multi-level microstructures and the mechanical property. These results indicated that microstructure is the effective control unit for strength and toughness of martensite, such as prior austenite size (dr) [1-3], packet size (dp) [4-7] and block size (db) [8-10]. However, with the plasticity many investigations attended to plastic deformation processing and mechanisms, few reports concerned with the effective control unit of macroscopic plasticity in low-carbon martensite steels. The work aims at revealing the relationship between macroscopic plasticity and micro-structure and providing a new way in the design of new material with the high strength and high plasticity based on 20CrNi2Mo steel.

2. Experimental procedure

2.1. Experimental material

The low-carbon 20CrNi2Mo steel was utilized in the present study. The chemical composition of the tested steel in weight percent is: C 0.20, Si 0.25, Mn 0.66, P 0.01, Cr 0.64, Cu 0.02, Ni 1.69, and rest Fe. The tested steel was rolled into Φ 80mm round bars, and forged into Φ17mm round bars. Finally, three set of samples were heated up 1000°C, 1100°C and 1200°C for full austenitization and quenched into 5wt% iced brine, then tempering at 250°C for 2 h, respectively. Therefore, the martensite multi-level microstructures are determined via the different quenching temperature.

2.2. Microstructure and mechanical property

The martensite multi-level microstructures, consisting of prior austenite grain, packet, block and lath, were observed by means of OM, FESEM, EBSD and TEM.The specimens were polished and etched in a super-saturation picric acid for optical microscopy (Leica, DMI5000M), to observe the prior austenite grain. And the martensite packets of tested steel were measured by scanning electron microscopy (ZEISS, SUPRA40) after etched from 4% Nital solution. In order to analysis martensite blocks, EBSD images with an orientation imaging microscopy system were used to present it. The samples for EBSD were made in the way of mechanical and vibrating polishing for 10 h, from the VibroMetTM2 vibratory polisher. The lath width was statistically measured by a high-resolution transmission electron microscopy (Tecnai G2 F20 S-TWIN, HR-TEM).The uniaxial tension test of the tested steel was executed on a MTS system.The dimension of tensile samples was prepared from the round bars with the size of Φ7×115.

3. Results and discussions

3.1. Quantitative microstructure analysis

A quantitative microstructure analysis of the various quenching temperature is listed in Table 1 from about averaging 400 grains, 300 packets, 300 blocks and 150 laths. Result shows that the different size of prior austenite grain is obtained based on the different quenching temperature from 15.3μm to 90.3μm. Meanwhile, according to the result from Table 1 and Fig.1, the martensite packets and blocks are increased from 9.8μm to 31.6μm and from1.4μm to 2.78μm, respectively, with the increase of quenching temperature in consistent with prior austenite grain. But the lath width is decreased from 0.342μm to 0.201μm.

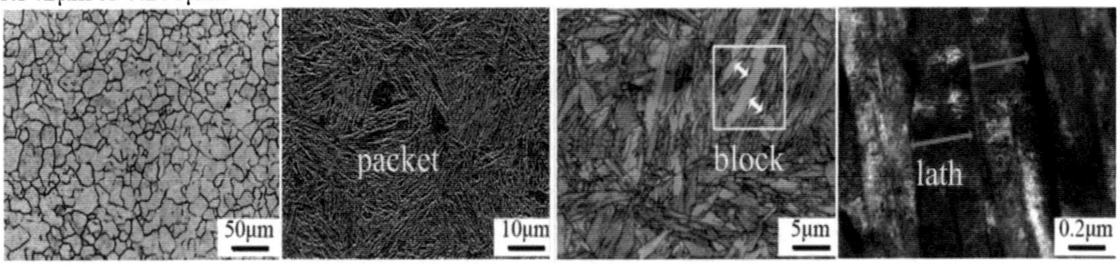

Figure 1. The observation of the martensite microstructure for B1 sample

Table 1. The result of the martensite microstructure for tested steel

No.	Grain dr/μm	Packet dp/μm	Block db/μm	Lath dl/μm
B1(1000℃)	15.3	9.8	1.4	0.342
B2(1100℃)	19.9	11.7	1.64	0.275
B3(1200℃)	90.3	31.6	2.78	0.21

3.2. Effect of martensite microstructures on the yield strength

The martensite packet is a lath groups with the same habit plain in a prior austenite grain, while the block represents a lath groups with the same orientation in a packet. It suggests that the martensite packet and block may depend on the prior austenite grain. The result in Table 1 reveals that the packet and block are increased by 3.22 times and 1.99 times with the coarsening of grains. Furthermore, the following linear relationships in Fig.2 (a) between microstructural parameters can be obtained for the steel, which demonstrate that the packet and block depend on the prior austenite grain, the equation (1) and (2) can describe well the relationship.

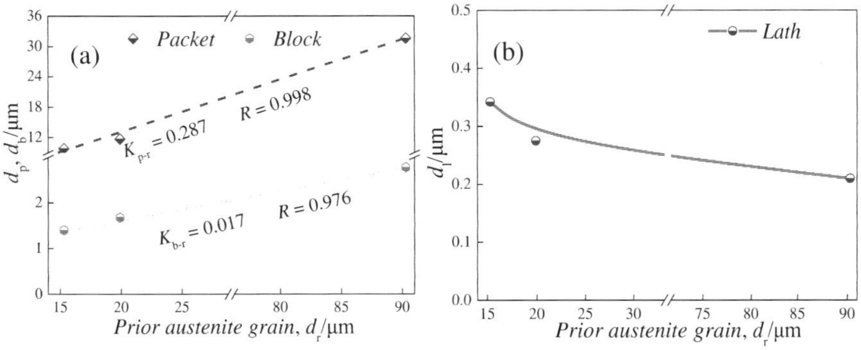

Figure 2. Relationship between (a) martensite packet, block, (b) lath and prior austenite grain

$$d_p = 0.287 d_r \tag{1}$$

$$d_b = 0.017 d_r \tag{2}$$

However, Fig.2 (b) shows that lath is not determined by the prior austenite grain. The previous works had demonstrated that dl depends on the nucleation rate and martensite start temperature (Ms) [8], with independence on the prior austenite grain. As well, the higher nucleation rate is determined.

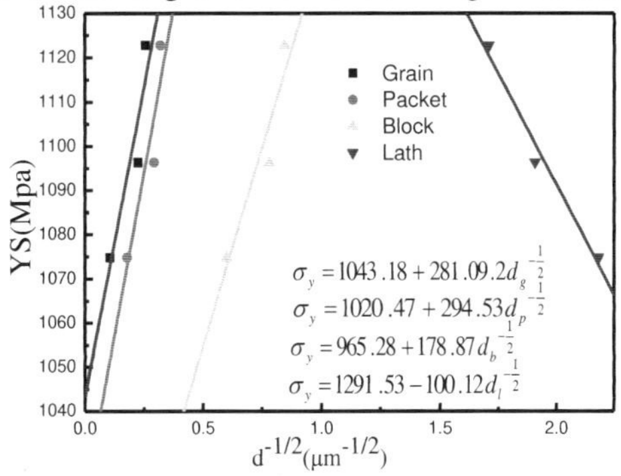

Figure 3. Plots of the yield strength vs. reciprocal square root of substructure

In addition, as shown in Fig.3, Hall–Petch relations are presented for the various substructures. Consequently, the block should be the effective grain size due to the far lower size than the packet size and the prior austenite grain size

3.3. Effect of prior austenite grain on the crystal orientation

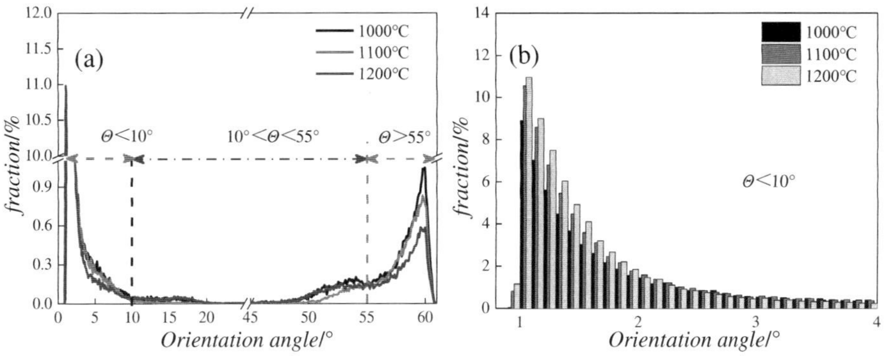

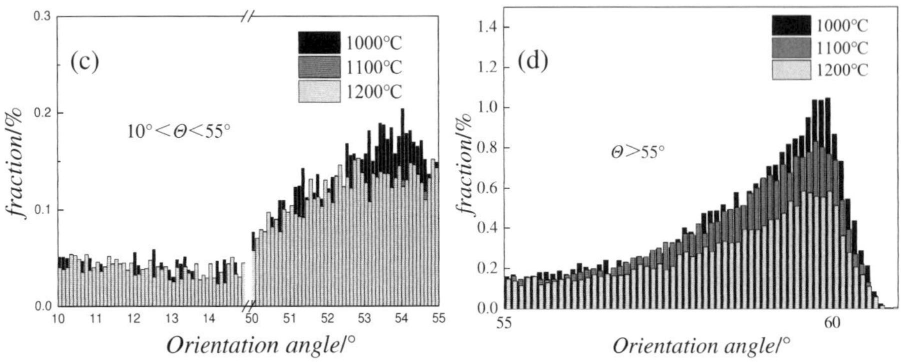

Figure 4. The distribution of orientation from different processes for tested steel, (a) The whole, (b) Θ<10°, (c) 10°<Θ<55° and (c) Θ>55°

In this paper, the martensite orientation is proposed by Tango software, as shown in Fig.4. It is found that the orientation distributions for tested steel are same with the range of 0.8°-62.5°. And the orientation distributions are divided into 3 sections, such as Θ<10°, 10°<Θ<55° and Θ>55°, which are used to represent the boundaries of lath, grain, packet and block [8]. Further observation, the orientations of Θ<10° and Θ>55° in fine grain are different from coarse grain: the number of Θ<10° in coarse grain is higher than fine grain, and the number of Θ>55° is lower than fine grain. It is because the size and number of the multi-level microstructure for different heats are different.

3.4. Effect of martensite microstructures on the plasticity

The mechanical properties for tested steel was obtained by the uniaxial tension test and showed in Fig. 5. And the higher strength is obtained in fine grain because of the more high-angle orientation. In addition, in low-carbon lath martensite steel, the packet and block play more important role in strength.

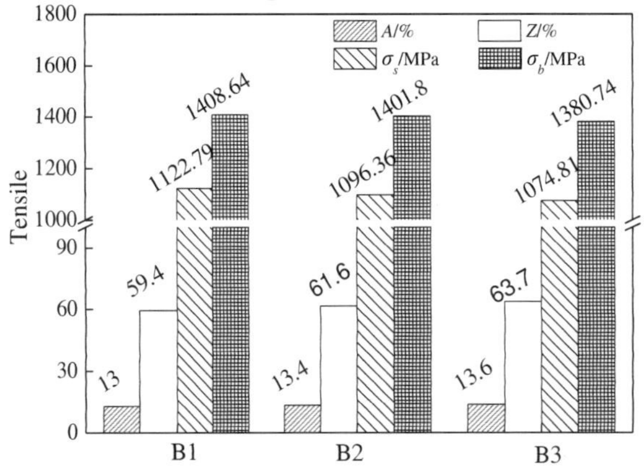

Figure 5. The mean value of the mechanical properties for tested steel

However, the plastic parameters of A% and Z% increase with the coarsening of prior austenite grain from Fig.5, this is contrary to the conventional understanding The increase amplitude of A% and Z% is obtained with 4.6%, and 7.2%. It can be illustrated from two points [12]: (1) In finer grain, the micro-crack is easy to form and propagate along the grain boundaries for the more high-angle boundaries. On the contrary, the more low-angle boundaries decrease the crack initiation, and it is difficult for the crack propagation along the grain boundaries. Therefore, the crack will encounter more laths to improve the plasticity; (2) In lath martensite steel, the fracture is controlled by the strain or the crack propagation. Thus, the prior austenite grain is not effective grain of plasticity.

4. Conclusions

In the present work, the effect of substructure of lath martensite on the mechanical properties was discussed in detail. The main conclusions can be summarized in the following:
(1) Prior austenite grain size, packet and block width increased as the quenching temperature increased, a pretty linear relationship meet between packet width, block width and prior austenite grain size, respectively. The results indicated that the width of block and packet in lath martensite are dependence on the prior austenite grain size.
(2)Yield strength with prior austenite grain size, packet size and block width obeyed Hall-Petch relationship regardless of the variation of quenching temperature. The block should be the effective grain size due to the far lower size than the packet size and the prior austenite grain size
(3) Based on the EBSD analysis, the orientation distribution of tested steels corresponding to different microstructures showed that more high angle boundaries exists in fine microstructure, while more low angle boundaries exists in coarsen microstructure.
(4)The low angle boundary can increase the compatible deformation capability of multi level microstructures, improving the plasticity of lath martensite.

Reference

[1] Grange.R. A. Strengthening steel by austenite grain refinement [J]. Trans. ASM, 1966, 59: 26-48.
[2] Krauss G. Martensite in steel: strength and structure. [J]. Mater. Sci. Eng, A. 1999, 273: 40-57.
[3] Antti J. Kaijalainen, Pasi P. Suikkanen, Teijo J. Limnell, et al. Effect of austenite grain structure on the strength and toughness of direct-quenched martensite. [J]. Alloys Compd, 2012.
[4] Roberts M J. Effect of transformation substructure on the strength and toughness of Fe− Mn alloys [J]. Metall. Trans. A, 1970, 1(12): 3287-3294.
[5] Swarr T, Krauss G. The effect of structure on the deformation of as-quenched and tempered martensite in a Fe-0.2 pct C alloy [J]. Metall. Trans. A , 1976, 7(1): 41-48
[6] Tomita Y, Okabayashi K. Effect of microstructure on strength and toughness of heat-treated low alloy structural steels [J]. Metall. Trans. A, 1986, 17(7): 1203-1209.
[7] Inoue T, Matsuda S, Okamura Y,AokiK.The fracture of a low carbon tempered martensite. Transactions of JIM, 1970, 11(1):36-43.
[8] Morito S, Tanaka H, KonishiR,et al. The morphology and crystallography of lathmartensite in Fe-C alloys [J]. Acta Mater, 2003, 51(6): 1789-1799.
[9] Zengmin Shi, Kai Liu, Maoqiu Wang .Effect of tensile deformation of austenite on the morphology and strength of lath martensite. [J]. Met Mater Ini, 2012, Vol.18 (2), pp.317-320.
[10] Luo Z J, Shen J C, Hang S U, et al. Effect of Substructure on Toughness of Lath Martensite/Bainite Mixed Structure in Low-Carbon Steels [J]. J Iron Steel Res Ini, 2010, 17(11):40-48.
[11] Maresca F, Kouznetsova V G, Geers M G D. On the role of interlath retained austenite in the deformation of lath martensite[J]. Modelling Simul.Mater.Sci.Eng, 2014, 22(4): 045011.
[12] Liang Y.L, Long S.L, Xu P, et al. The important role of martensite laths to fracture toughness for the ductile fracture controlled by the strain in EA4T axle steel [J]. Mater. Sci. Eng, A.2017, 695:154-164.

Acknowledgments

The research documented in this work was financially supported by the Joint Foundation ofGuizhou province, China (Grant No. [2017] 7244), the National Natural Science Foundationof China (Grant No. 51461006), and the Natural Science Foundation of Guizhou province,China (Grant No. [2014] 2003)

Design and Checking Analysis of Injection Mold for a Plastic Cup

Xuebing Li

Guidaojiaotong Polytechnic Institute, Shenyang 110023,China
251336431@qq.com

Abstract. A special injection mold was designed for the structural characteristics of a plastic cup part. The mold was simulated by Moldflow software and verified by calculating the stripping force, the pulling force and the clamping force of the mold so that to determine the appropriate injection parameters. It has been proved that the injection mold is effective and practical in the actual producing and can meet the quality requirements during the course of using it, which solved some problems for injection molding of this kind of parts and can provide some reference for the production of other products in the same industry.

1.Introduction

With the continuous development of various engineering plastics, injection molding is used more extensively in the manufacturing field to make various products for meeting requirements. In order to produce high quality injection products economically, we must consider the problem about injection mold comprehensively. The quality of injection mold directly affects the efficiency, quality and cost of molded products. It plays a very important role in the forming of injection products. Not only the surface quality, precision of plastic products, are completely determined by the mold, but also the internal quality and efficiency of forming products are also affected by the mold. So how to fast and standardize designing injection mold has become the primary problem to develop the technology.

Here the plastic cup in this article was made of polypropylene including antioxidants and UV Absorbents. Its shape and structure are shown in Figure 1.

2.Designing injection mold

The drawing of injection mold to this cup is shown in Figure 2. . According to the actual situation of the plastic part, the parting surface is set as the location shown in Figure 3. . In order to avoid the occurrence of sharp corners between plate and the slide block, the side should be rounding at the bottom of the parting surface. Due to repeated contact and collision between mainstream , high temperature plastic melt and nozzle of injection machine, the main channel is often designed as a detachable gate sleeve, which is made of steel T10A and hardened to 50 ~ 55HRC. Steel T10A is suitable for manufacturing various tools which has a high wear resistance , a certain toughness and a sharp edge cutting under poor conditions.

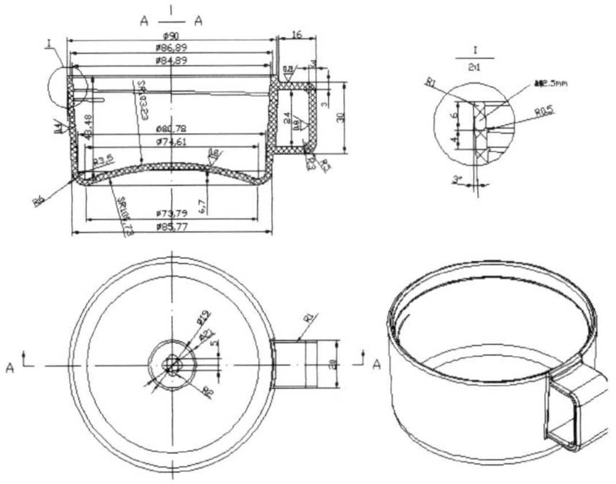

Figure 1. part drawing of plastic cup

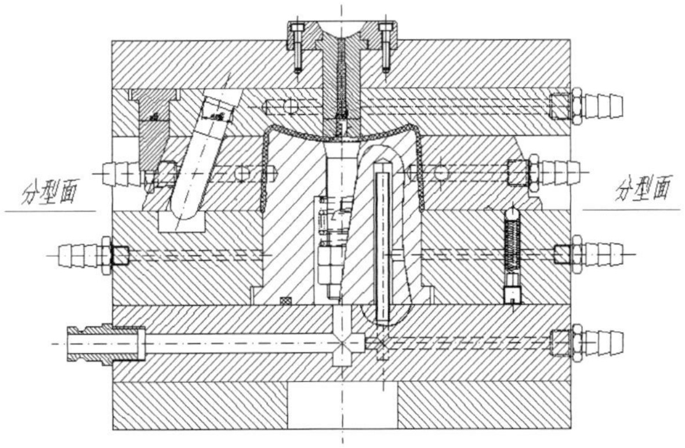

Figure 3. Mold parting

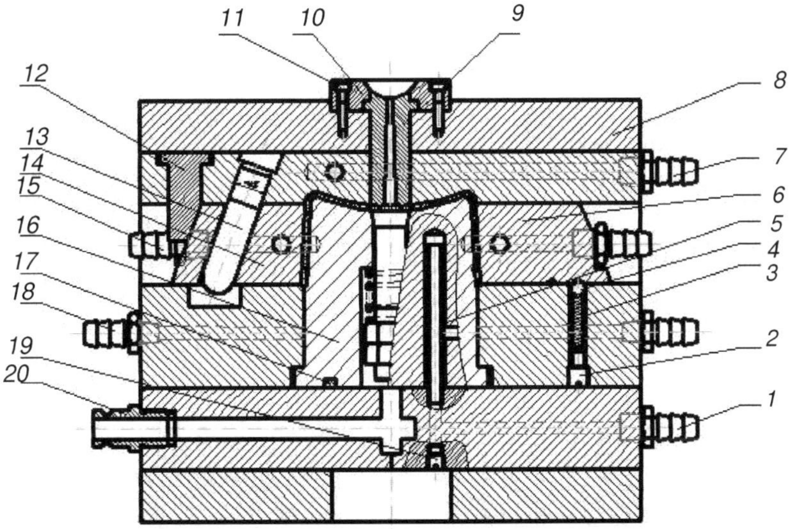

Figure 2. Mold structure drawing

1. Water injection well choke 2. Stopper 3.Spring 4. Steel ball 5. Thin wall pipe 6. Right slider 7. Water injection well choke 8. Base plate of fixed mold 9. Spruebush 10.Locating ring 11. Hexagon socket head cap screws 12. Wedge block 13. Inclined guide column 14.Left slider 15. Water injection well choke 16.Core 17. Sealing gasket 18. Water injection well choke 19. Plug screw 20. Interface to pressure pump

3.Calculation and analysis

(1)Calculating demoulding force

According to $\delta\ d^{-1}=0.0278<0.05$, Here $\delta=2.5$mm, $d=90$mm; Therefore, the plastic part is a thin-walled plastic part and because the section of the part is circular ring ,so

$$F = 2\pi\delta_1 ESl\cos\varphi(f - \tan\varphi)(1-\mu)^{-1}k_2^{-1} + 0.1A \qquad (1)$$

where S is average shrinkage rate of plastic (%), E is modulus of elasticity of plastic (MPa), l is the tolerance length of the plastic to the core (mm), ϕ is demoulding slope of mold core (°), μ is Poisson's ratio of plastic, δ_1 is the average wall thickness of a plastic part (mm), f is the static friction between the plastic and the core, f =0.1~0.2, A is the projection area of the mould core in the blind hole in the demoulding direction and k_2=1+fsinϕ cosϕ ≈ 1.

The material of this plastic part is PP and the relevant parameters are S=1.75%, E=950MPa, l=40mm, μ=0.392, A=276.46mm2, ϕ=3 degree. It can be calculated from (1) that the demoulding force is approximately 81.65KN.

(2) calculation of core pulling force

The core pulling force can be obtained in Mould design and manufacture; that is,

$$Q = lhp_2(f_2\cos\theta - \sin\theta) \qquad (2)$$

where l is the circumference of the section shape of a movable core packaged with a plastic part. h is the depth of the cavity part. p_2 is extrusion pressure per unit area that plastic parts to core (MPa), generally 8~12MPa. f_2 is the friction coefficient between plastic and steel generally 0.1~0.2. θ is the draft angle of a side hole or side concave(°). It can be calculated from (2) that the core pulling force is approximately 125189.1N.

(3)Checking of the clamping force for the mold

When the melt high pressure plastic fills the cavity, it will generate a great thrust, which should be less than the nominal clamping force of the injection machine.Otherwise, it will generate overflow phenomenon. That is,

$$F \geq pA \qquad (3)$$

where F is the nominal clamping force of the injection machine(N). P is the injected pressure into the cavity during injection. Here P=25MPa. A is the sum of the vertical projection area both the plastic and the pouring system on the parting surface (mm^2),Here $A = nA_1 + A_2, n = 1$. A_1 is Projection area of a single plastic part on a parting surface (mm^2). A_2 is the projection area of a runner (including a gate) on a parting surface. It can be calculated from (3) that the thrust force in the mold is approximately1506.9KN. So the clamping force of the injection molding machine can meet the requirements.

(4) Analysis of plastic parts by Moldflow

IOP Conf. Series: Materials Science and Engineering **324** (2018) 012044 doi:10.1088/1757-899X/324/1/012044

1) preprocessing

(1) importing the model, analyzing the type, and select the material

Import the cup model, choose the mesh type as the surface model, double-click the fill bar on the left side of the task bar, select the analysis type as "gate location", then select "TaiwanPP" as the material and 6331 as brand, take the"immediate analysis". As shown in Figure 4., the blue display is the best gate location from the results of the analysis.

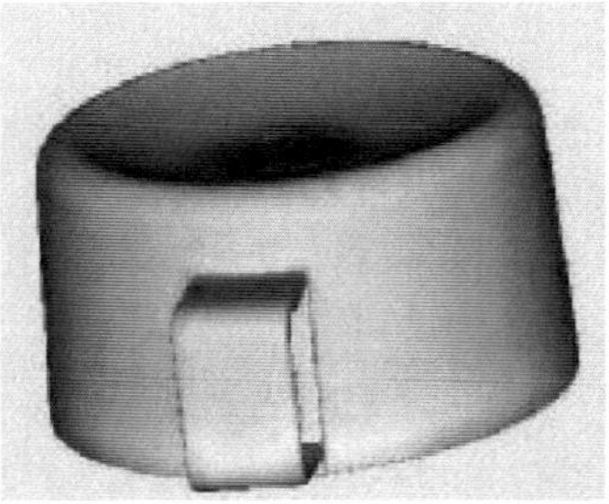

Figure 4. optimal gate position

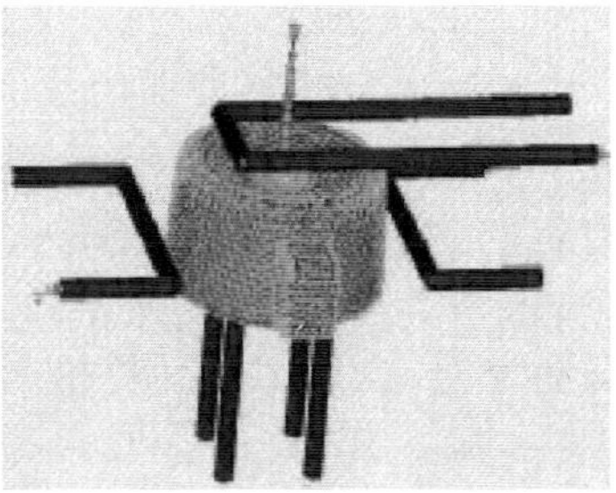

Figure 5. cooling loop

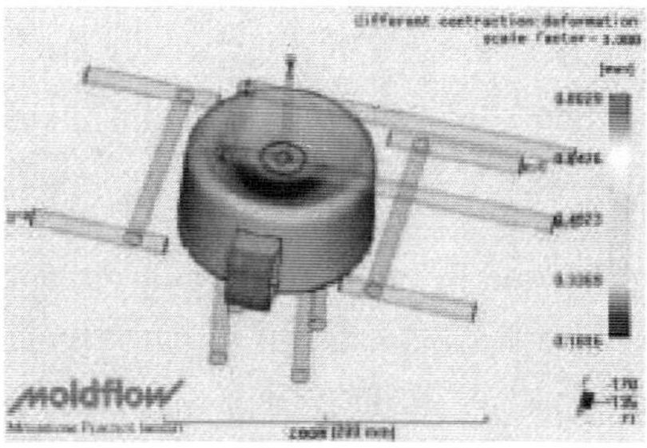

Figure 6. Deformable and contraction

(2) creating the gating system and cooling system

Place injection position at the center of the cup's bottom. Choose gate location center as the mainstream position. Set the entrance diameter as 3mm, taper angle as 3 degrees, the length as 52mm. Create a cooling system by hand, place the inlet and outlet of the water and the inlet of the coolant, so as to finish creating the gating system and cooling system. As shown in Figure 5..

2)Output analysis Results

Carry out analyzing the cup by "cooling + flowing + warping" in Moldflow. Found from the analysis, the largest clamping force of injection molding machine is 7 tons, the maximum injection pressure is 1.8MPa,the filling time is 1.5s. When the filling stage is 1.48s and the flowing rate is 33.46 cubic centimeters per second, switch speed and pressure. Holding pressure begins from 1.5s , at 11.49s pressure releases and at 28.14s pressure ends. The surface temperature of mold is set as 60 degrees Celsius. The ejection condition is set as 100 degrees Celsius and injection pressure is set as 120MPa.

It can be seen from the report: 1) The deformation caused by different cooling is different from each other, and lengthening the cooling water channel at the slide block, as shown in Figure 6., will make the deformation shrink. 2) When the filling time is 1.5s, rotate the plastic part and choose different points on parts. As shown in Figure 7., it's ok for the changed time between the filling time is not more than 0.2S. 3) The position which generate the gas holes is on the parting surface or in the gap between the left and right slider, or the edge of the cup .Where it is easy to exhaust, so there is no problem. 4) Although it is inevitable to get rid of the weld marks. There are still some weld marks in the position of the handle .We can lessen the weld mark to make it unobvious by increasing the speed, temperature and pressure. 5) The desire to the clamping force is much less than the maximum clamping force that the injection molding machine owns, so it is feasible. 6) finding the temperature deviation on the wall of loop tubes is very low shows that the cooling effect is very good.

4. conclusions

The results show that the various calculations of theories and advantages in mature CAD and CAE software, taking the mold design, product design and experimental verification before manufacturing into a comprehensive consideration, can reduce the cost on material and time caused by

IOP Conf. Series: Materials Science and Engineering **324** (2018) 012044 doi:10.1088/1757-899X/324/1/012044

some unreasonable design, can increase the reliability of the product design. It can help avoid the defects on designing, shorten the cycle for producing.

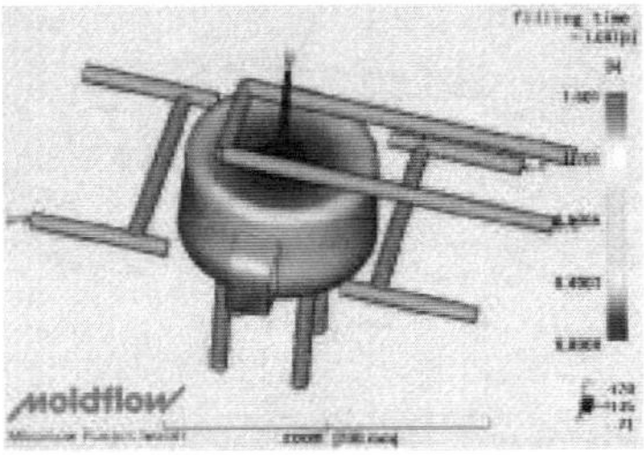

Figure7. Filling time

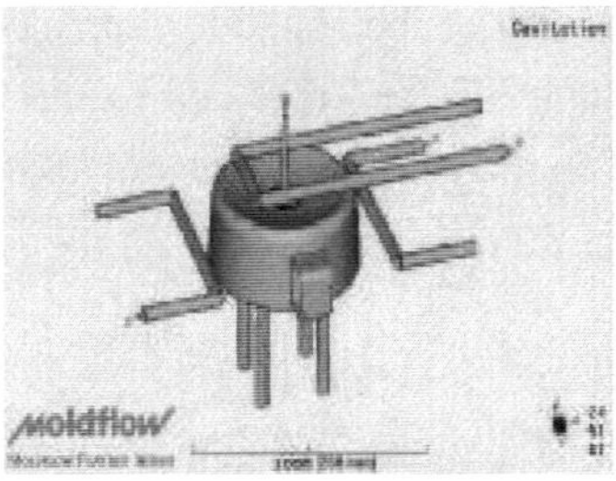

Figure 8. Cavitation

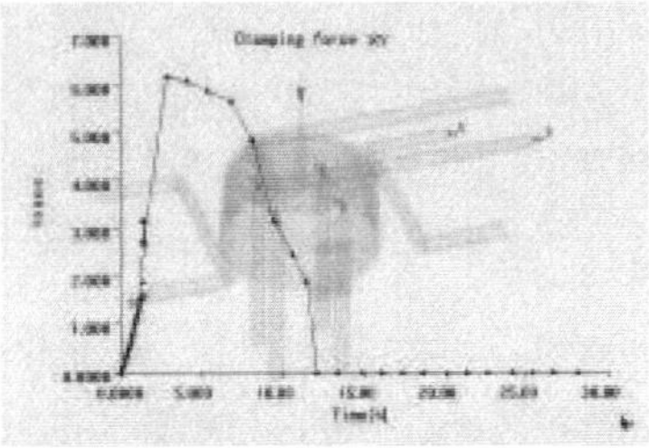

Figure9. Clamping force

The method here for design and checking analysis to this injection mold can provide some reference during the injection processing. But there are still some deficiencies in the study of some parameters for this mold. So further research must be needed in some related aspects.

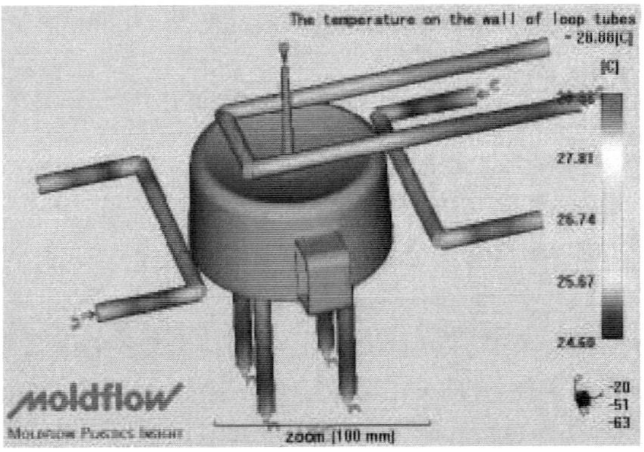

Figure 10. The temperature on the wall of loop tubes

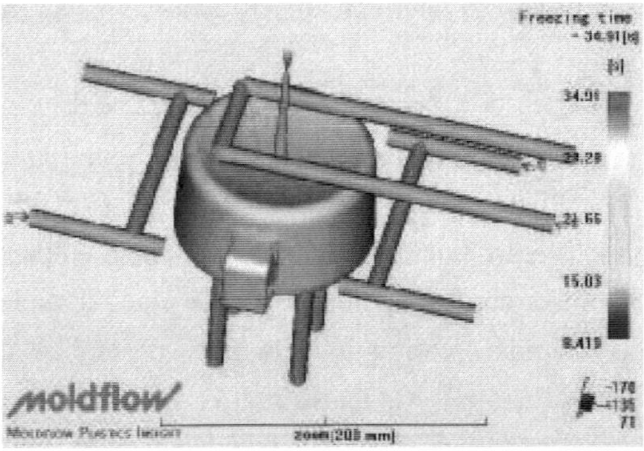

Figure 11. Freezing time

References

[1] CAO Jiang-dong, LU Li-xin. Optimum Design of Injection Mould of Hairdryer Shell Based on Moldflow , J. PLASTICS， 2010, 39(1)： 122-125.

[2] Liu Bin, Wu Xi. Analysis and Design Method of Conformal Cooling Channel of Injection Mould, J. Engineering Plastics Application， 2017, 45(8)： 123-127.

[3] SONG Ke, HU Qing-chun, JIANG Xiao-ping. Optimization Design of Injection Mold Cooling Channels Arrangement, J. China Plastics Industry， 2014, 42(12)： 45-48.

[4] ZHAO Zhenfeng ,SHI Xianzhang ,WANG Jing,SHEN Changyu,LIU Chuntai. Experimental analysis of pipe network for injection mold , J. CIESC JOURNAL， 2009, 60(12)： 3148-3151.

[5] LI Xi-ping, ZHAO Guo-qun, GUAN Yan-jin, LI Hui-ping. Analysis and optimal design of heating and cooling process for rapid heating cycle injection mold, J. JOURNAL OF PLASTICITY ENGINEERING， 2009, 16(1)： 196-200.

[6] Jeppson R w． Analysis of Flow in Pipe Networks． Boston：Butterworth publishers， 1976： 1-86

IOP Conf. Series: Materials Science and Engineering **324** (2018) 012045 doi:10.1088/1757-899X/324/1/012045

Analysis of the effectiveness of various cross-sections in large-span post-tensioned ceilings

K Bednarz[1]

Ph D Student, Cracow University of Technology, Poland
E-mail address: katarzyna.maria.bednarz@gmail.com

Abstract. The correct construction of large span, slim post-tensioned concrete slabs is conditioned by an appropriate cross-section selection. It is generally accepted that the thinnest slab can be constructed using the full cross-section as the largest compression stress storage. However, completely different cross-sections may help to overcome large spans. The paper presents the results of the computational analysis of several types of cross-sections (full, with internal relieving inserts and ribbed) in the application to a post-tensioned slab with a span of 15.0m. Based on the results presented, appropriate conclusions were drawn.

1. Introduction

Post-tensioned concrete slabs prestressed with unbounded tendons have been used in the world for several decades. They were initially applied in the USA, Hongkong, Singapore, Australia, and later also in European countries. The architectural trend that has been going on for many years is aimed at creating large, support-free spaces in buildings. On the other hand, obtaining the highest possible height of rooms makes it necessary to use the thinnest floors possible. All of this forces designers to thorough analysis and refinement of constructional solutions that have been used for years and to look for new varieties.

Improvements in the efficiency (increase in span) of ceilings are also sought in ceilings made from pre-tensioned hollow core slabs. A larger span can be obtained using an additional concrete topping [1], or using additional unbounded tendons inside the steel beams supporting the pre-tensioned concrete hollow core slabs [2]. An interesting solution of a semi-prefabricated ceiling on prestressed concrete boards, increasing the spread of traditional reinforced concrete floors at the same thickness, was published in paper [3].

Undoubtedly, much better results in the construction of large span ceilings can be obtained in post-tensioned concrete technology, using unbounded post-tensioning. These tendons with small diameters, a low friction coefficient and small curvature radiuses allow to shape long, high-curved routes. In addition to applications in ceilings, their advantages allow shaping slender rod structures in buildings, e.g. [4]. In Poland, in recent years, several very thin prestressed ceiling slabs with unbounded tendons have been designed and successfully constructed [5].

There are many types of cross-sections as well as design variations of floor slabs. Each of them has its advantages and disadvantages. It is difficult to say unequivocally that one is better, and another is worse. Its usefulness and effectiveness will depend on the conditions of use. In this paper, the results of the analysis of three types of cross-sections will be presented: full, with internal relieving inserts, and ribbed, in application to a one-way simply supported slab with a span of 15.0m. Post-tensioned one-way concrete slabs are more often used in buildings due to their simpler execution in relation to two-way slabs.

IOP Conf. Series: Materials Science and Engineering **324** (2018) 012045 doi:10.1088/1757-899X/324/1/012045

2. Design assumptions

Computational analysis has been applied to the one-way flat roof slab shown in Figure 1. The slab is freely supported on four walls with a thickness of 250mm. The axial spacing of the walls in the shorter direction is 15.0m while in the longer 50.0m. By omitting the end zones, the slab therefore works unidirectionally. The slab thickness was originally set to 400mm (1/37.5 of the span).

The following slab loads were assumed: slab self-weight - 25kN/m^3, layers on the roof – 1.5kN/m^2, load of devices on the roof and under the slab – 2.5kN/m^2 (live load), snow – 1.0 kN/m^2 (live load).

Figure 2 shows the prestressing profiles adopted from the principle of using the maximum tendon sag. For this purpose, a minimal structural distance of the tendons from the lower edge in the middle of the span (50mm) and prestressing in the middle of the cross-section's center of gravity over the supports have been assumed. For cross-sections with two axes of symmetry (table 1 - cross-sections 1, 2 and 3) a 150mm sag was obtained, for ribbed cross-sections (with a raised center of gravity) 199mm were obtained for cross-sections no. 4 and 5 and 229mm for cross-section no. 6.

It was assumed that the slab will be made of concrete class C30/37 according to [6]. For prestressing, commonly used for unbounded tendons 0.6" strands with a cross-sectional area of 150mm^2 and steel strength of 1860MPa were assumed. The value of initial force in tendon is 220kN.

Two design situations were considered:

- the initial design situation (just after prestressing) - the dead load of the slab was taken into account and the prestressing was reduced by immediate losses of 10% and the increasing factor equal to 1.05 [6],
- final design situation - all operating loads, prestressing after time-dependent losses assumed as 20% of the initial force and a decreasing factor equal to 0.95 were taken into account.

As a criterion for correctness of the cross-section's work, apart from limiting deflection, boundary stresses in concrete were assumed. Maximum stresses were limited to:

- for compression in the initial design situation - $0.6f_{ck}(t_0)$, where $f_{ck}(t_0)$ means the characteristic compressive strength of concrete during prestressing. $f_{ck}(t_0)$ was assumed as $0.8f_{ck}$. Ultimately, the limit compressive stresses amount to $\sigma_{cc,lim} = 0.6 \times 0.8 \times 30 = 14.4$MPa.
- for tension in final design situation $-f_{ctm}$ that is the average concrete tensile strength of 2.9MPa. In the initial situation, this value was reduced by 20%. This limitation results from the necessity of ensuring lack of cracking. The cross-sections cracked at the bottom, resulting in a significant reduction in stiffness, cause a radical increase in deflections. However, cracking the top in the initial stage is not dangerous for the work of the structure, but it requires the use of an appropriate upper ordinary reinforcement in the slab.

The total long-time deflection of the ceiling was limited to 1/300 of the span, i.e., to 50mm. The long-time deflection was determined in accordance with [7], using the increasing factor to the deflection

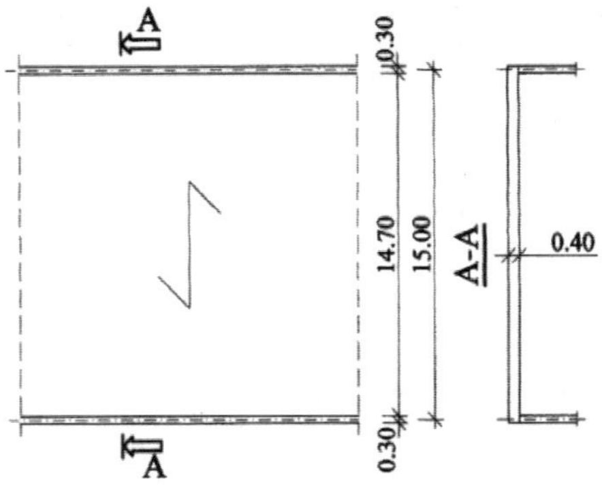

Figure 1. Slab geometry.

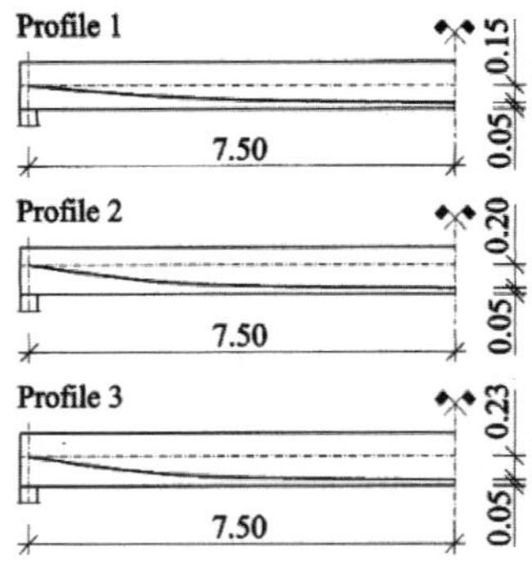

Figure 2. Tendons profiles.

IOP Conf. Series: Materials Science and Engineering **324** (2018) 012045 doi:10.1088/1757-899X/324/1/012045

calculated from the elastic analysis. The value of this coefficient is 3.0 for dead loads and prestressing and 1.5 for live loads.

Four cross-sections were considered in the analysis:

- solid cross-section with a height of 400mm (no. 1 in table 1),
- cross-section with a height of 400 mm with relieving inserts *Slim Line* of the *Cobiax* system (no. 2), with a height of 260mm and a diameter of 315mm. The spacing of the inserts was assumed to be 500mm in the transverse direction and 350mm in the load-bearing direction. The reduction of the slab's weight due to the use of inserts amounted to 226kg/m^2, i.e., 22.6% while the reduction of the moments of inertia 11.4%,
- ribbed cross-section with a height of 400mm (no. 4 and 5) with a 150mm-thick slab and ribs with a width of 185mm in a 500mm spacing,
- ribbed cross-section with a height of 450mm (no. 6) with a 150mm-thick slab and ribs with a width of 185mm in a 500mm spacing.

Table 1 Analyzed cross-sections as well as their properties.

Cross-section		Tendon profile (see fig.2)	Cross -section area	Moment of inertia	Slab weight reduction		Moment of inertia reduction
No.	Shape		m^2	m^4	kg/m^2	%	%
1	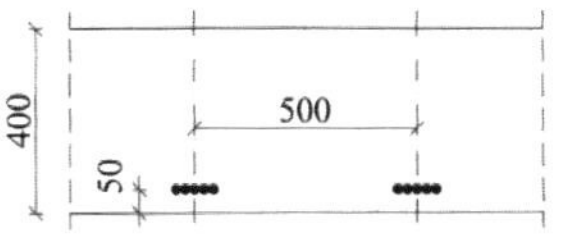	1	0.400	0.00533	-	-	-
2	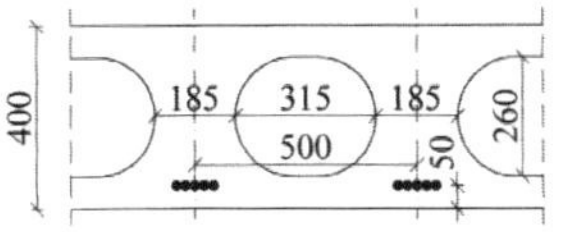	1	0.2652	0.00472	226	22.6	11.4
3	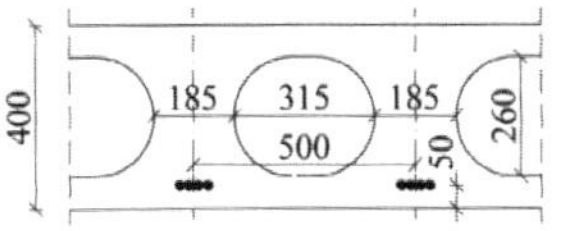	1	0.2652	0.00472	226	22.6	11.4
4	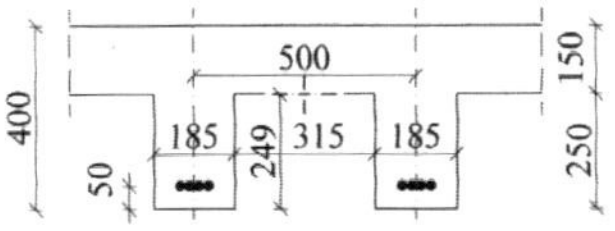	2	0.242	0.00305	394	39.4	42.8
5	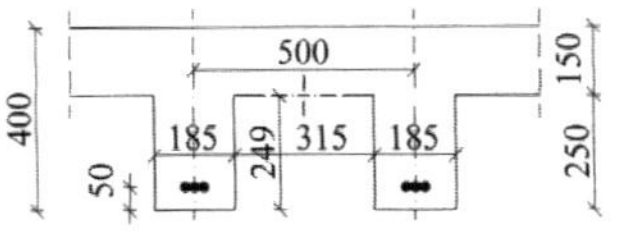	2	0.242	0.00305	394	39.4	42.8
6	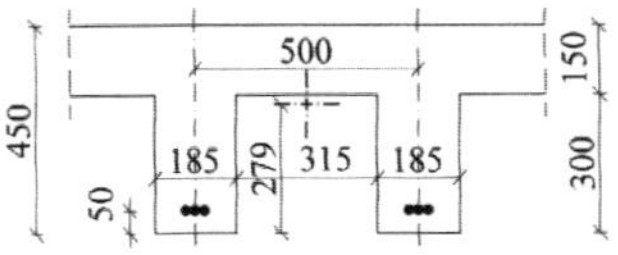	3	0.261	0.00434	473	42.0	42.8

3. Calculation

In order to perform the calculations, the slab calculation model was performed in the Finite Element Method (Figure 3a). Linear hinged support was assumed on the perimeter, with only one longest edge deprived of the possibility to move horizontally. Surface 4-node, square finite elements with 500mm

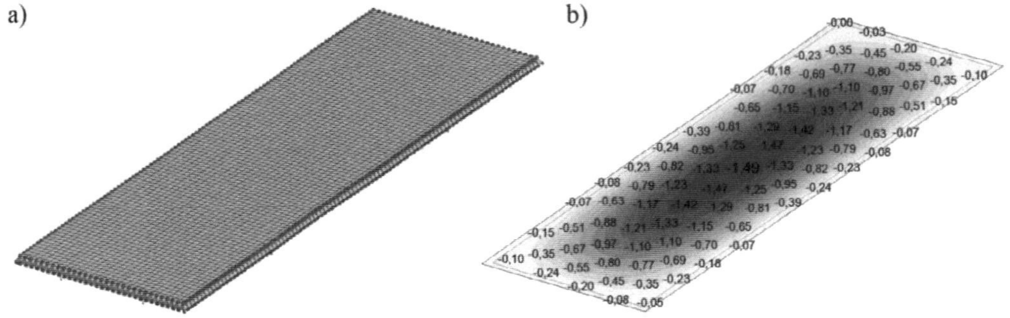

Figure 3. FEM model of analyzed slab (a), elastic slab deflection for cross-section no. 1 (b).

sides were used. The prestressing load was applied as a substitute horizontal load on the edges and a vertical surface load determined according to the formula:

$$w = \frac{8P \cdot \delta}{L^2} \tag{1}$$

where P is the force in the tendon, δ is the tendon sag and L is the tendon length. Computational analysis was carried out in the range of linear-elastic concrete work.

4. Calculation results

The results of the computational analysis carried out for all 6 considered cross-sections are summarized in table 2. In section no. 1 (solid slab) correct results were obtained by using 5 prestressing strands every 500mm. The use of relieving inserts in cross-section no. 2, with identical prestressing, allowed to reduce the deflection (in relation to the solid cross-section no. 1) from 44.7 to 23.4 mm. However, the correct deflection value (44.7mm) can be obtained by reducing the number of tendons in the bundle from 5 to 4 (cross-section no. 3). The positive effect of internal inserts relieving both the deflection of the ceiling and the amount of prestressing. This is due to a much smaller reduction in cross-section stiffness (11.4%) in relation to the slab's weight reduction (22.6%).
In the case of a ribbed cross-section (no. 4) with a similar height (400mm) and with 4 tendons in the rib (as in the case of a properly working cross-section with relieving inserts), a considerable exceeding of the compressive stresses was obtained in the initial design situation (19.9 in relation to 14.4MPa). Due to the large drop in the moment of inertia (42.8%) and surface area (39.4%), this cross-section proved to be too weak to transfer such a prestressing. Subsequently, the number of strands in this cross-section was reduced in the rib to 3 (section 5), obtaining the appropriate stresses in the initial design situation. Such prestressing, however, turned out to be too small in a situation of use where tensile stresses of 4.25 MPa were obtained. This means cracking of the cross-section. In the case of a cracked cross-section, one should expect long-time deflections significantly greater than 3 times the

Table 2 Values of concrete stresses and slab deflection.

Cross-section No.	Initial design situation		Final design situation		Elastic deflection	Long-time deflection
	Upper stress	Bottom stress	Upper stress	Bottom stress		
	MPa		MPa		mm	mm
1	3.83	6.39	10.2	-1.93	14.9	44.7
2	3.68	11.8	10.4	1.99	7.8	23.4
3	4.69	7.67	11.2	-1.42	14.9	44.7
4	-1.22	19.9	-	-	-	-
5	1.06	11.7	9.14	-4.25	1.26	63.0
6	0.85	10.5	7.13	-1.72	6.1	18.3

elastic deflection. Assuming an analogy to cracked reinforced concrete cross-sections, this deflection can be up to 5 times greater than elastic deflection, i.e., it can be 63mm. This is already a value much greater than the 50mm limit. This cross-section proved to be too weak to the assumed geometrical conditions and ceiling loads. Its height has been increased to 450mm (cross-section 6). For such a cross-section, correct values were obtained with 3 tendons in the rib. The use of ribbed cross-section and increasing its height from 400 to 450mm allowed to reduce the number of strands in the bundle with 5 for solid cross-section and 4 for cross-section with internal relief inserts up to 3.

5. Conclusions

Based on the performed computational analysis, it was found that:

- Although the solid cross-section constitutes the largest compressive stress storage (important in case of prestressed structures), the use of internal relieving inserts with the same prestressing allows to reduce the slab's deflection. The same effect (as to deflections) can be achieved by reducing the amount of prestressing (in the analyzed case by 20%). Although it is commonly recognized that the full cross-section, due to its highest compressive stress storage capacity, allows the smallest thickness of the slab, the internal relieving inserts seem to be a good solution when overcoming a large span with slender slabs. Their advantage lies in a much smaller reduction in the stiffness of the cross-section in relation to the slab's weight reduction.
- Ribbed cross-sections, due to a significant reduction in stiffness and surface area, are much weaker compared to solid ones and those with internal relief inserts. Although they allow a significant reduction in the amount of prestressing (in the analyzed case - by 20 and 40% respectively), this occurs at the expense of increasing the height.

6. References

[1] Derkowski W, Surma M 2014 Influence of concrete topping on the work of prestressed hollow core slabs on flexible supports *Proceedings of 4th International fib Congress 2014: Improving Performance of Concrete Structures* (Mumbai, India, 10-14 February 2014) pp 339–341

[2] Derkowski W, Skalski P 2017 New Concept of Slimfloor with Prestressed Composite Beams *Procedia Engineering* 193 pp 176-183

[3] Szydlowski R, Szreniawa M 2017 New concept of semi-precast concrete slab on pre-tensioned concrete boards *IOP Conf. Series: Materials Science and Engineering* 245 (2017) 022090

[4] Szydlowski R, Labuzek B, Turcz M 2017 Prestressed Ring Beam in the Church of St. Peter's and Paul's in Bodzanow, Design and Realization *IOP Conf. Series: Materials Science and Engineering* 245 (2017) 022066

[5] Szydlowski R, Labuzek B 2017 Post-Tensioned Concrete Long-Span Slabs in Projects of Modern Building Construction *IOP Conf. Series: Materials Science and Engineering* 245 (2017) 022065

[6] EN 1992-1-1: Eurocode 2: *Design of concrete structures – Part 1-1: General rules and rules for building* (European Committee of Standardization, December 2004)

[7] Concrete Society 2005 Post-tensioned concrete floors *Design Handbook, Technical Report No. 43* (Cromwell Press Wiltshire UK) p. 110

IOP Conf. Series: Materials Science and Engineering 324 (2018) 012046 doi:10.1088/1757-899X/324/1/012046

The Speech multi features fusion perceptual hash algorithm based on tensor decomposition

Y B Huang[1] ,M H Fan[1] and Q Y Zhang[2]

1 College of Physics and Electronic Engineering, Northwest Normal University, Lanzhou, Gansu, China
2 School of Computer and Communication, Lanzhou University of Technology, Lanzhou, Gansu, China
E-mail:huang_yibo@foxmail.com

Abstract. With constant progress in modern speech communication technologies, the speech data is prone to be attacked by the noise or maliciously tampered. In order to make the speech perception hash algorithm has strong robustness and high efficiency, this paper put forward a speech perception hash algorithm based on the tensor decomposition and multi features is proposed. This algorithm analyses the speech perception feature acquires each speech component wavelet packet decomposition. LPCC, LSP and ISP feature of each speech component are extracted to constitute the speech feature tensor. Speech authentication is done by generating the hash values through feature matrix quantification which use mid-value. Experimental results showing that the proposed algorithm is robust for content to maintain operations compared with similar algorithms. It is able to resist the attack of the common background noise. Also, the algorithm is highly efficiency in terms of arithmetic, and is able to meet the real-time requirements of speech communication and complete the speech authentication quickly.

1. Introduction

Speech messages as human communication means is the most natural, effective and convenient. Hash function is based on the perception of human psychology multimedia information processing theory, multimedia data is unidirectional mapped to a data set of multimedia perception summary sets. Only the perceptions of simple and fast hash function to meet multimedia massive data analysis applications. Therefore, hash function is very suitable for application in the field of information security[1].

At present, the main methods of speech perception hash feature extraction and processing are as follows: Mel frequency cepstrum coefficient(MFCC), Hilbert transformation, linear prediction coefficient, the buck with energy[2], the energy ratio of wavelet features[3] and spectrum energy feature[4] can also be used as the characteristics of the speech perception hash values. In literature [5], a perceptual hashing algorithm based on Mel cepstral coefficient combined with linear prediction cepstral coefficient is proposed, which has good robustness and tampering localization, but it is not effective in differentiating and content preserving operations.The literature [6] proposes a speech-aware hashing authentication algorithm based on the MFCC correlation coefficient, which considers security, it has good robustness and discrimination, but is not efficient. The literature [7] Hilbert transform spectrum estimation method is used to implement robust speech feature extraction, construct hash function perception. In literature [8], a perceptual hash algorithm based on RT and DCT is proposed, which has good robustness and high computational efficiency, but is poor in distinction. The literature [9] proposes a multi-format audio perceptual hashing algorithm based on dual-tree complex wavelet transform, which has higher efficiency and achieves audio authentication in five different audio formats, including original domain and compressed domain, and achieves small-scale Tamper detection and positioning, but without considering the security issues, weak robustness.

The perceptual hashing must satisfy discrimination and robustness. Discrimination is an indicator which should reflect the data content in a unique way. The speech have different content should yield different hash sequences. Robustness is used to evaluate the identity of the speech after the content maintain operation, such as low-pass filtering, volume improvement, volume reduction, resample. Robustness is used to evaluate whether the speech after the operation is the same as the original speech.

2. Methodology

2.1. Tensor decomposition

Tensor in the field of information processing has been widely used. Tensor can be considered as a product of vector space, and it is a higher order generalization of vector and matrix. The order of the tensor can be expressed as $X \in R^{N_1 \times N_2 \times \cdots \times N_M}$. The tensor decomposition includes two kinds of decomposition methods: CANDECOMP/PARAFAC and Tucker[10]. CANDECOMP/PARAFAC decomposition and Tucker decomposition are the high order generalization of the singular value decomposition in the matrix decomposition[11].

2.2. Wavelet packet transform

Wavelet Packet Transform (WPT) as the further expansion of wavelet analysis theory. Wavelet packet decomposition can reflect the feature and nature of the signal, very suitable for the analysis and processing of speech signal types of non-stationary signal [12].

2.3. LPCC Feature Coefficient Extraction

The recurrence relation between LPC coefficient and LPCC coefficient is below:

$$\begin{cases} c_0 = a_1 \\ c_n = a_n + \sum_{k=1}^{n-1} \frac{k}{n} c_k a_{n-k}, & 1 \le n \le N \\ c_n = \sum_{k=1}^{n-1} \frac{k}{n} c_k a_{n-k}, & n > N \end{cases} \tag{1}$$

c_0 is the DC component, a_n is the LPC coefficient, c_n is the LPCC coefficient.

2.4. LSP feature parameter extraction

Linear spectral pair (LSP) is a mostly used as the feature vector. LPS can be obtained by solving the conjugate complex root of P + 1 order symmetric and antisymmetric polynomials.

$$P(z) = A(z) + z^{-(p+1)} A(z^{-1}) \tag{2}$$

$$Q(z) = A(z) - z^{-(p+1)} A(z^{-1}) \tag{3}$$

The roots of the above two polynomials are on the unit circle and appear alternately. Where $P(z)$ has a root of $z = -1$, $Q(z)$ has a root of $z = 1$. $P(z)$ and $Q(z)$ have $M/2$ conjugate complex roots located on the unit circle. So get a new polynomial.

$$P(z) = (1 + z^{-1}) \prod_{i=1}^{p/2} (1 - 2\cos\omega_i z^{-1} + z^{-2}) \tag{4}$$

$$Q(z) = (1 - z^{-1}) \prod_{i=1}^{p/2} (1 - 2\cos\omega_i z^{-1} + z^{-2}) \tag{5}$$

Where $\cos\omega_i, i = 1, 2, \cdots, p/2$ is the representation of the LSP feature in the cosine domain.

2.5. ISP feature parameter extraction

Immittance Spectral Pairs (ISP) is an important equivalent representation of linear prediction feature, which can effectively reduce the computational complexity and ensure the calculation precision at the same time. According to the transfer function of linear prediction filter $A(z)=\sum_{i=0}^{p} a_i z^{-i}$, $a_0=1$.

$$F(z)=A_p(z)+A_p^{\#}(z)=A_p(z)+z^{-p}A_p(z^{-1}) \tag{6}$$

$$G(z)=A_p(z)-A_p^{\#}(z)=A_p(z)-z^{-p}A_p(z^{-1}) \tag{7}$$

3. Speech perception hash authentication scheme

The flow chart of speech perception Hash authentication is shown in figure 1.

Step 1: preprocessing, framing and windowing: conduct pre-emphasis on the speech in the speech library to be tested, enhance the useful frequency spectrum of high frequency, reduce the edge effect and eliminate noise.in order to eliminate the inter frame loss during framing, conduct framing and add the Hamming window for speech $x(t)$; during framing, the frame length is L; when the frame moves at $L/2$, $s(n)$ can be obtained; later, add Hamming window for $s(n)$ to obtain $s_w(n)$. n is the frame number.

Step 2: construction of speech feature tensor and decomposition of speech tensor: carry out wavelet package decomposition for the speech frame, calculate the MFCC and ΔMFCC features of each frame, conduct the tensor construction of feature feature to obtain the speech feature tensor χ. Carry out Tucker decomposition for the feature tensor χ to obtain the low-dimensional core tensor G;

Step 3: quantization: Reconstruct the core tensor G to form the two-dimensional feature matrix, calculate the sum of each column of matrixes.

$$R_h(j)=\sum_{i=1}^{r} H^{(n)}{}_{ij} \quad 1\le j\le k \tag{8}$$

In the formula above, $H^{(n)}{}_{ij}$ signifies the feature feature in row j and line i; k is the number of rows of feature matrix. Quantize the feature formed and row matrix to form the Hash value $h(j)$ of speech segment, $\hat{R}_h$ is the mid-value.

$$h(j)=\begin{cases}1 & R_h(j)>\hat{R}_h \quad 1\le j\le k\\ 0 & other\end{cases} \tag{9}$$

Step 4: calculation and matching of perception Hash distance: suppose that there are two speech segments α and β; define the Hash mathematic distance is $D_h(:,:)$, which is shown below:

$$D_h\left(H_\alpha,H_\beta\right)=\sum_m\left|h_\alpha(j)-h_\beta(j)\right|, j=1,2,\cdots,n; \tag{10}$$

if the perception contents of the two speech segments α and β are the same.

$$D_h(H_\alpha,H_\beta)\le\tau \tag{11}$$

if the perception contents of the two speech segments α and β are different.

$$D_h(H_\alpha,H_\beta)>\tau \tag{12}$$

In the formulas above, τ is the matching threshold.

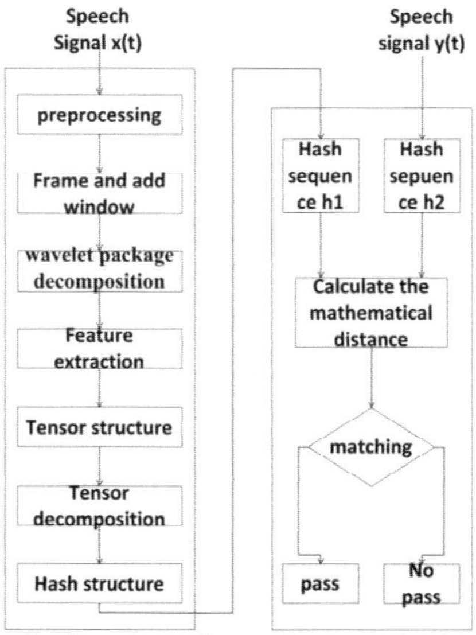

Figure 1. The flowchart of speech perception hash authentication based on tensor decomposition

4. Experimental analysis

4.1. The experimental environment

As for the experiment in the paper, the MATLAB 2010b simulation realization is adopted. The speech of the length 4 seconds 1189 segments, in which different content of speech contains Chinese and English and the same content of different people read speech. Speech parameters used for the sampling rate 16000Hz, bit rate is 256kbps, the number of channels is mono, sampling precision is 16bit, and format is wav. Frame length is 20ms, frame shift is 10ms.

4.2. Discrimination

The perception hash gain 816003 bit error rate data .The obtained bit error rate of the normal distribution as shown in figure 2.

The bit error rate of perception hash value of different speech content mainly obey normal distribution, where probability distribution parameter is μ (μ=0.4971),and Standard deviation σ=0.0274.

As can be seen from table 1, the algorithm deterrent rate in this paper has no more obvious improvement than the deterrent rate based on LPC speech perception hash algorithm.

4.3. Robustness

The average bit error rate can be obtained due to the attack to speech library.

as shown in figure 2, FRR and FAR curve has obtain BER, extracted from the content of the same speech perception of the hash value and BER are below the threshold value of 0.35, the experimental results show that this algorithm has higher robustness. And FRR - FAR curve in the figure is not crossed, this algorithm also has good distinguishability and robustness. It can accurately identify their content and content of malicious operation. According to table 2 shows that when the threshold τ= 0.35, FAR=3.8185e-008.

As can be seen from the figure 3, discriminant threshold of algorithm in this paper has arrange between 0.33 and 0.41.As can be seen from table 2, In addition to 20dB Gaussian white noise , the average bit error rate of the above several content keeping operation are below decision threshold 0.35, increasing and decrease the volume does not change the channel model. Therefore, the coefficients greater change will not happen, so adjust the volume does not produce the error rate. Due to the algorithm using parameters normalization which can improve the resistance to Gaussian white noise, the algorithm robustness in the paper against noise is great.

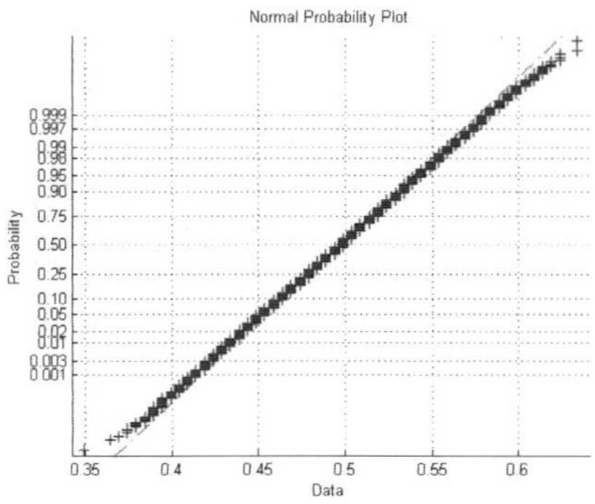

Figure 2. The algorithm of normal distribution graphics

Table.1 The threshold false accepted rate of the algorithm

	proposed algorithm
τ=0.25	8.5838e-020
τ=0.30	2.9505e-013
τ=0.35	3.8185e-008

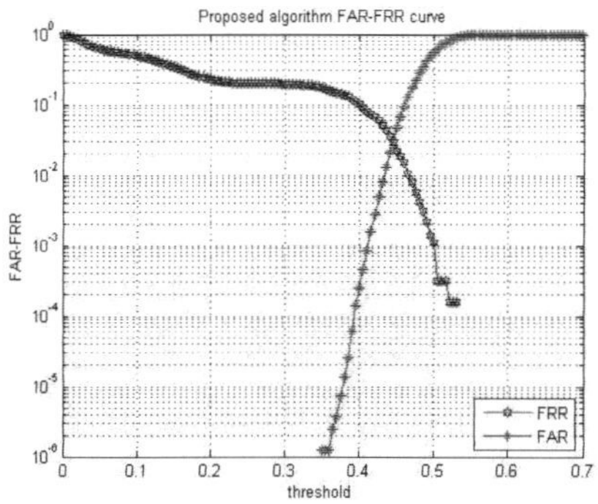

Figure 3. The algorithm FRR - FAR curve

Table.2 The matching rate of speech authentication after the being kept operating content %

	The proposed algorithm	Average bit rate
-50% volume	100	0.0497
+50% volume	100	0.1211
Resampling 8kHZ-16kHZ	100	0.1295
Resampling 32kHZ-16kHZ	100	0.0518
20dB Gaussian white noise	72.3	0.3215
30dB Gaussian white noise	87.1	0.2942
4kHz low-pass filtering	99.7	0.1978

5. Conclusion

In order to solve the balance of authentication efficiency, the content remain operational robustness and the discriminative.an efficient and robust speech authentication algorithm based on perceptive hash of tensor decomposition model is proposed. Through theoretical analysis and experimental analysis, the proposed algorithm has good anti-collision performance and robustness which can effectively resist the content preservation attacks and common background noise. The algorithm can authenticate speech information and speaker information more accurately.

We only consider the matching of speech in the speech library when the algorithm is built. Personal speech habits are different, and feature extraction is complex. Post-research needs to consider the extraction the unique characteristics of personal speech, the speech authentication which is not related to speaking time, the establishment of hash value database and so on. Therefore, the speech perception hash algorithm, which is unrelated to the speaker is the direction to be studied.

REFERENCES

[1] Z Q Liu, Q Li, J R Liu and X M Niu, 2012, ICDIP.Proc. of spie, A Novel Evaluation Method For Perceptual Hash Algorithm In Authentication Mode,Vol.8334 83444-1.

[2] M Ramona and G Peeters, 2011, Acoustics, Speech and Signal Processing (ICASSP), 2011 IEEE International Conference on, Audio identification based on spectral modeling of bark-bands energy and synchronization through onset detection, 477-480.

[3] J Z Wang, T L Xu, K G Xuan, 2011, Computer Applications and Software, Audio Digital Watermarking Algorithm Based on Energy Ratio of Wavelet Coefficient, 28(5): 288-291.

[4] J J Deng, W G Wan, S Ram, X Q Yu and X Q Pan, 2011, Smart and Sustainable City (ICSSC 2011), IET International Conference on, An audio fingerprinting system based on spectral energy structure, 6-8 July:1-4.

[5]Y B Huang, Q Y Zhang, Z T Yuan, Z P Yang, 2015, Huazhong University of Science and Technology (Natural Science Edition),The Robust Hash Algorithm of Speech Perception Based on the Integration of Adaptive MFCC and LPCC, 43(2), 124-128.

[6] J F Li, W Tao, H X Wang, 2015, Journal of Beijing University of Posts and Telecommunications, Perceptual Hashing Based on Correlation Coefficient of MFCC for Speech Authentication, (2) 89-93.

[7] H Zhao, H Liu, K Zhao and Y Yang, 2011, International Journal of Digital Content Technology and its Applications, Robust Speech Feature Extraction Using the Hilbert Transform Spectrum Estimation Method, 5(12):85-95.

[8]J F Li, T Wu, 2015, Proc of 2015 11th international Conference on computational intelligence and security. Shenzhen; IEEE, Perceptual audio hashing using RT and DCT in wavelet domain, 363-366.

[9]Q Y Zhang, P F Xing, Y B Huang, R H Dong, Z P Yang. 2016, Journal of Beijing University of Posts and Telecommunications, Perceptual Hashing Algorithm for Multi-Format Audio, 39 (4) 77-82.

[10] B W Bader and Kolda T G, 2006, Scient. Comput., Efficient MATLAB computations with sparse and factored tensors. SIAM J. 30: 205-231.

[11]J Li, L X Jin and G N Li, 2013. J. Electron. Inform. Technol., Hyper-spectral remote sensing image compression based on nonnegative tensor factorizations in discrete wavelet domain, 35: 489-493.

[12]Y Yang and S Nagarajaiah, 2014, Signal Process., Blind identification of damage in time-varying systems using independent component analysis with wavelet transform, 47: 3-20.

Acknowledgment

The authors would like to thank for the support by the National Nature Science Foundation of China (No. 61363078), the Natural Science Foundation of Gansu Province of China (No. 1606RJYA274).

Load-bearing Characters Analysis of Large Diameter Rock-Socketed Filling Piles Based on Self-Balanced Method

Wu tongqing[1,2], Li liang[3*],Liu xinjian[4],Xu nianchun[1,2],Mao Tian[3]

1 School of Civil Engineering and Architecture, Chongqing University of Science & Technology, Chongqing, 401331,China
2 Chongqing Key Laboratory of Energy Engineering Mechanics & disaster prevention and mitigation,Chongqing,401331,China
3 National Engineering Research Center for Inland Waterway Regulation, Chongqing Jiaotong University, Chongqing ,400074, China
4 Chongqing Construction Science Research Institute, Chongqing, 400020, China
E-mail: 634986248@qq.com

Abstract. Self-balanced method is carried out on the large diameter rock-socketed filling piles of high-pile wharf at Inland River, to explore the distribution laws of load-displacement curve, pile internal force, pile tip friction resistance and pile side friction resistance under load force. The results showed that: the tip resistance of S1 and S2 test piles accounted for 53.4% and 53.6% of the pile bearing capacity, respectively, while the total side friction resistance accounted for 46.6% and 46.4% of the pile bearing capacity, respectively; both the pile tip friction resistance and pile side friction resistance can be fully played, and reach to the design requirements. The reasonability of large diameter rock-socketed filling design is verified through test analysis, which can provide basis for the optimization of high-pile wharf structural type, thus reducing the wharf project cost, and also providing reference for the similar large diameter rock-socketed filling piles of high-pile wharf at Inland River.

1. Introduction

In recent years, along with the large-scale construction of river ports, large diameter rock-socketed filling piles with higher carrying capacity has been generally used as the pile foundation of high-pile wharf, so how to accurately determine its carrying capacity is particularly important. Self - balanced method can accurately test the bearing capacity value of the foundation pile, which was first proposed by Japanese scholars Nakayama and Fujiseki [1]. In the 1980s, J.Osterberg [2] applied it successfully in the bridge steel piles for the first time, and formed a new type of pile foundation test technique, so it is also known as J.Osterberg test pile method, and then widely used and promoted in Europe, Singapore and other places; Li Guangxin et al [3] introduced this method into China, and carried out a large number of theoretical and model experimental researches, but no actual project site test was conducted; Followed by this, Shi Peidong et al [4] introduced the promotion and application conditions of this method in foreign countries, and discussed existing problems in domestic applications; Gong Weiming et al [5] committed to the theoretical research and engineering application of self-balancing method; the industry standard was developed after years of engineering application [5], in addition, this test technology was applied in Hong Kong-Zhuhai-Macao Bridge [6-7] and other major projects. This method has conducted a lot of researches in regards of the basic

IOP Conf. Series: Materials Science and Engineering **324** (2018) 012047 doi:10.1088/1757-899X/324/1/012047

theoretical research and practical engineering application of rock-socketed filling piles foundation, Gao Ping [8] and Zhang Mingli et al [9] studied test pile self-balanced test under the condition of soft rock; Xu Xibin et al [10] conducted comparison analysis between the actual pile test and indoor model tests of rock-socketed filling piles.

In this paper, it aims to analyze the load - displacement, tip resistance, side friction and ultimate load value based on the measured data obtained from the self-balanced method of high-pile wharf rock-socketed filling piles foundation, which can verify the reasonability of pile foundation design, and provide basis for the optimization of high-pile wharf structural type, as well as provide reference for the similar large diameter rock-socketed filling piles of high-pile wharf at Inland River.

2. Self - Balanced Method Test on Large Diameter of Rock-socketed Filling Piles

2.1. Overview of test pile

The high pile wharf piles foundation at a certain Inland River adopts reinforced concrete large diameter rock-socketed filling piles with the diameter of 2.0m or 2.3m. Two piles with the pile foundation diameter of 2cm shall be selected to have pile test, and the parameters are shown in Table 1, the engineering geological conditions are shown in Figure 1. The strength of pile concrete is 35MPa after 28 days curing, the elastic modulus is 3.5×10^4MPa.

Table 1. Test pile parameters

No	Diameter/mm	Length/m	Top elevation/m	Bottom elevation /m	Load box bottom elevation/ m
S1	2000	25.7	179.7	154.0	155.0
S2	2000	19.7	179.7	160.0	160.5

Through the laboratory test on the main rock-soil layer sample of S1 and S2 test pile geological drilling and sampling, the various rock layers parameters at pile test position are shown in Table 2. Combined with the practical engineering and test analysis on covering soil, the ultimate friction resistance of pile side takes 21.45kPa.

Table 2. various rock layers parameters

Lithology	Medium-weathered mudstone	Medium-weathered sandstone
Compressive strength of natural uniaxial /KPa	8.03×10^3	25.67×10^3
Compressive strength of saturated uniaxial /KPa	4.62×10^3	19.75×10^3
Natural gravity /kN.m^{-3}	25.3	24.5
Softening coefficient	0.58	0.8

2.2. Test pile load test system diagram

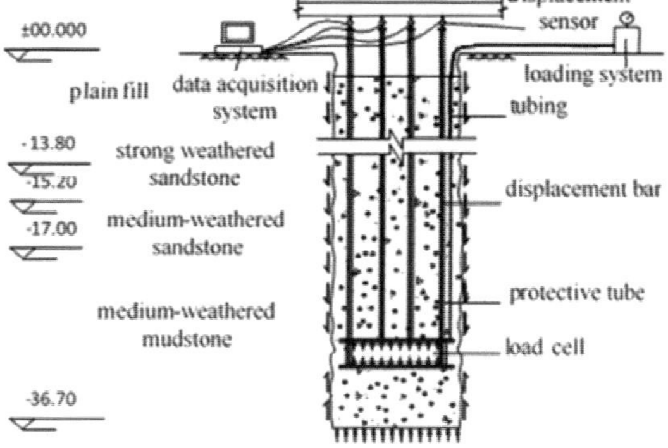

Figure 1. The test system of self-balanced method

The test system of self-balanced method is shown in Figure 1. It sets an inbuilt load box at the self-balancing point of pile, and then apply the load to load box through hydraulic device, the displacement sensor on displacement rod that lead to the ground is used to detect the deformation, the strain meter is arranged at the boundary of rock-soil layer to detect the pile strain, and the quantity of strain meter should be increased or reduced according to different rock-soil layer thickness. Thus, load - displacement (up, down, tip) curve and the stress-strain curves of strain meter under different load values can also be drawn the according to readings, so as to analyze the pile foundation bearing capacity, pile body force and pile side friction resistance distribution law along with changes of depth.

2.3. Test pile load

In accordance with the requirements of test procedures [5], slow maintain load method is adopted to conduct gradual loading, 1/15 of estimated limit value is regarded as each level load value, and the initial estimated load value is 30000 kN. For the test pile S1, when the load is loaded to 32000kN, the upper pile will be lifted, test pile is destroyed, and the load is terminated, so the load value for the upper and the lower parts of S1 test pile shall be 15000 kN; for test pile S2, when the load is loaded to 24000 kN, the upper pile will be lifted, test pile is destroyed, and the load is terminated, so the load value for the upper and the lower parts of S1 test pile shall be 11000kN. Test pile estimated load rating please see Table 3.

Table 3. Estimated gradation loading of test pile

Loading gradation	S1 Corresponding loading/MN	S2 Corresponding loading/MN	Loading gradation	S1 Corresponding loading/MN	S2 Corresponding loading/MN
1	2×2	2×2	9	2×10	2×10
2	2×3	2×3	10	2×11	2×11
3	2×4	2×4	11	2×12	2×12
4	2×5	2×5	12	2×13	2×13
5	2×6	2×6	13	2×14	2×0
6	2×7	2×7	14	2×15	
7	2×8	2×8	15	2×16	
8	2×9	2×9	16	2×0	

Note: In the table, "0" in S1 and S2 corresponding loading indicates unloading.

3. Results of test pile

3.1. The curve results of test pile Q-S

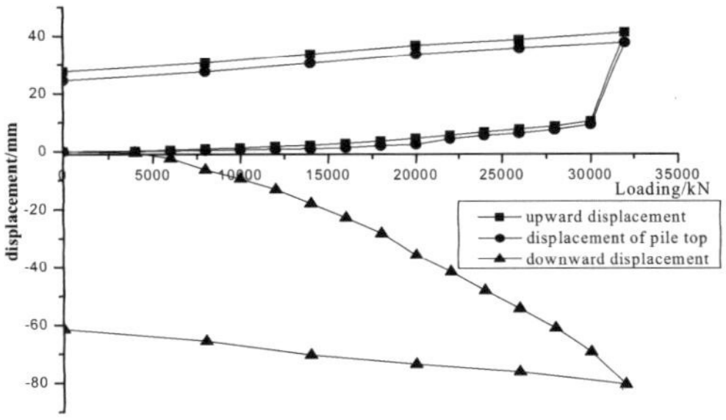

Figure 2. Q-S curve of S1 test pile

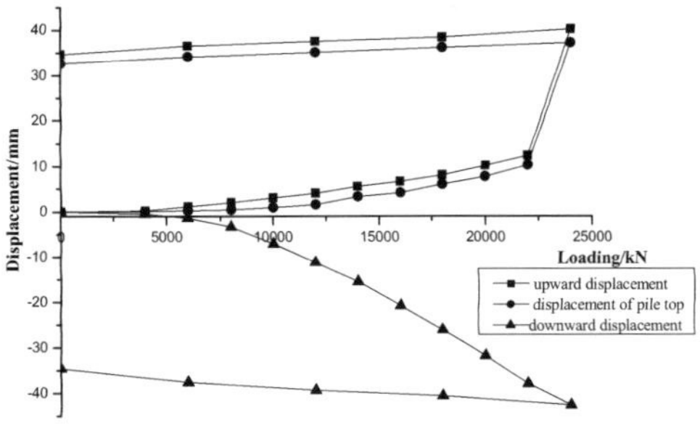

Figure 3. Q-S curve of S2 test pile

The Q-S curve result of test pile S1 and S2 is shown in Figure 2 and Figure 3. It is known from the figures that: the Q-S curve of both the two test piles had a slowly varying type, of which, the maximum displacement of the upper and lower section of the pile is 41.15mm and 80.58mm respectively for the maximum loading value of S1 test pile; while the maximum displacement of the upper and lower section of the pile is 40.21mm and 42.60mm respectively for the maximum loading value of S2 test pile. Due to the different geological conditions, pile length and the influence of balance point selection of both S1 and S2 test piles, the upper and lower section of the QS curve of test pile is not completely symmetrical; this phenomenon is quite common in practical engineering testing.

3.2. Test pile axial force and pile side friction test results

The axial force can be calculated through elastic modulus of steel and concrete, as well as their cross-sectional areas by a strain gauge, and the changes curve of axial force of each test pile under different load values along with the depth is shown in Fig 4. The figure shows that the axial force of pile at the same cross-section increases gradually along with the increase of load value; under the condition of different levels of preloading, the axial force distribution law is similar, and the axial force amplitude is close; but under the condition of same loading, axial force transmitted and decreased from the bottom to top along with the loading box position; in addition, the axial force at pile tip approaches to zero.

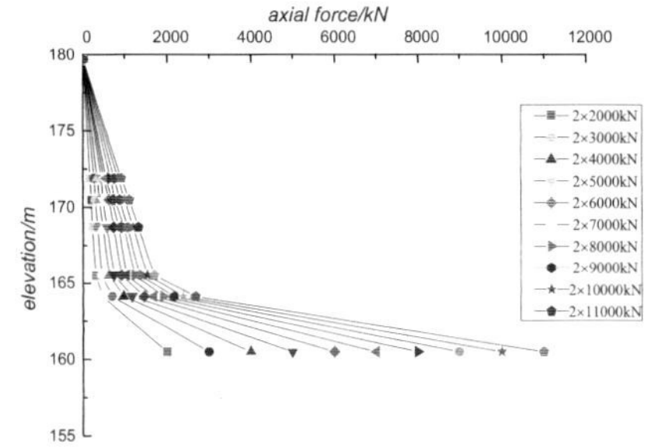

Figure 4. Distribution curve of S2 test pile axial force

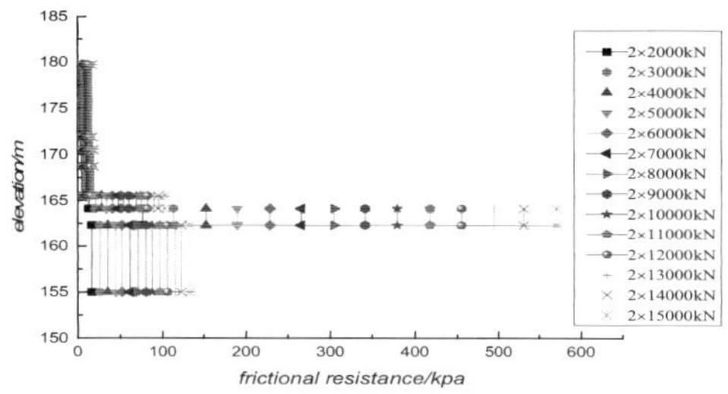

Figure 5. Distribution curve of S1 test piles side friction resistance

Pile side friction resistance can be obtained with the ratio between the axial force difference of two adjacent steel strain gauges and the pile side area, under different levels of loading, the distribution curve of average pile side friction resistance of different sections is shown in Figure 5. The results showed that based on the differences in the nature of each rock-soil layer at a test pile, the pile side friction gradually increases with the increase of load value, and the increase amplitude is similar; however, the friction resistance of S1 test piles suddenly increased at the elevation of 164.12~162.32m (medium-weathered sandstone), the maximum friction resistance reaches to 570KPa, indicating that the pile side friction resistance is greatly influenced by the nature of rock-soil layer, the higher strength the pile geotechnical layer, the larger the role of pile side friction resistance; The rock-soil layer of pile side friction resistance at loading box plays the role first, and then the degree of role decreases from near to far, the minimum value is at the pile tip. While designing the rock-socketed pile, the upper part of side friction resistance value can be reduced appropriately.

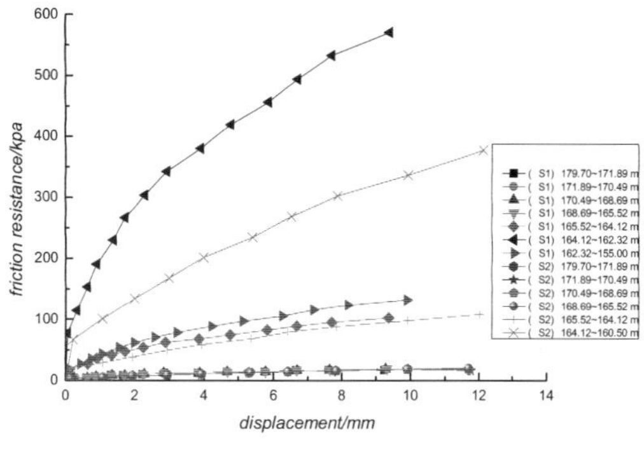

Figure 6. Each rock-soil layer friction resistance of S1, S2 test pile

From Figure 6 we know that, the amount of pile compression and displacement increases along with the progressively increase of load, so does the side friction resistance of each rock layer, and the maximum side friction resistance is near the loading box; soil layer of friction resistance at the distance pile showed linear homogeneous changes; influenced by the axial force distribution and Poisson ratio effect, friction resistance of side rock layer increases along with the load, but not in linear changes. When the soil displacement of S1 test pile reaches to about 8mm, pile side friction resistance is stabilized. The side total resistance of S1 and S2 piles is 13099kN and 9522kN, respectively.

3.3. Pile tip resistance of test pile

As known from Figure 7 that the pile tip resistance increases along with the increase of load, so does the increase amplitude. The resistance close to zero when the load at pile tip is relatively small; tip resistance of embedded rock piles S1 and S2 test pile accounted for 53.4% and 53.6% of the pile bearing capacity, while the total side resistance accounted for 46.6% and 46.4% of the pile bearing capacity. Each test pile can give full play to the tip resistance and side friction resistance, and meet the design requirements.

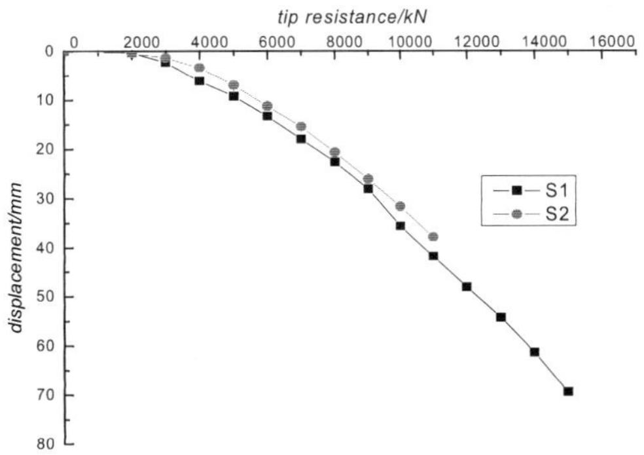

Figure 7. Tip resistance distribution curve of S1, S2 test pile

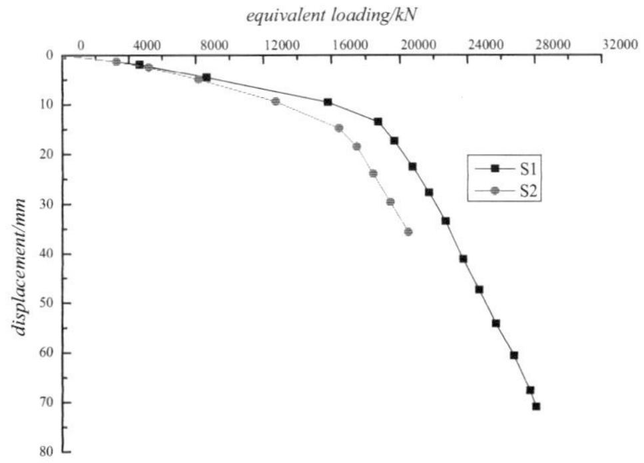

Figure 8. Equivalent conversion curve of S1 and S2 piles

3.4. Determination of the ultimate bearing capacity

Through equivalent conversion calculation, and based on the conversion between each rock-soil layer friction resistance and displacement measured value, we can obtain each test pile's loading equivalent conversion curve, as shown in Figure 8. The figure shows: The load - displacement curve changes of S1, S2 test piles are more gentle, no significant turning point; so the ultimate load value of the last level is taken. In addition, the limit bearing capacity of S1, S2 piles is 28099kN and 20522kN, respectively, and the corresponding displacement is 70.96mm and 35.74mm, respectively.

4. Conclusion

The application and research of self-balanced method at the rock-socketed filling piles foundations of high-pile wharf at Inland River can be summarized in the following points:

Pile bearing capacity of rock-socketed filling piles in this project has met the design requirements. The vertical bearing properties of foundation piles indicate that, the pile tip resistance of rock-socketed pile S1 and S2 test piles accounted for 53.4% and 53.6% of the pile bearing capacity, while the total pile side resistance accounted for 46.6% and 46.4% of the pile bearing capacity, each test pile can give full play to their tip resistance and side friction resistance.

The soil-level friction resistance of pile side showed a linear and uniform distribution, influenced by the axial force distribution and Poisson's ratio effect, the friction resistance of tip rock-level friction continued to increase along with the load, but not in linear changes. When the soil displacement of S1 test pile reached to about 8mm, pile side friction resistance is stabilized. During the rock-socketed filling pile design, it is recommend that the value of upper side friction resistance value can be appropriately reduced.

Through the self-balanced method that conducted on the bearing property of rock-socketed filling piles of high-pile wharf at Inland River, it has verified the design reasonability and provided basis for the optimization of high-pile wharf type, which can not only reduce the construction cost of pier, but also provide reference for the similar large diameter rock-socketed filling piles of high-pile wharf at Inland River.

Acknowledgments
This project is supported by Fundamental and Advanced Research Projects of Chongqing (cstc2015jcyjA30002,cstc2015jcyjA30003), Research Foundation of Chongqing University of Science & Technology (CK2015Z29),Chongqing Municipal Commission of Urban-rural Development (2015-1-39),Scientific and Technological Research Program of Chongqing Municipal Education Commission (KJ1601336).

References
[1] Nakayama.J, Fujiseki.Y. A pile load testing method. Japanese Patent No.1973～27007.(in Japanese)
[2] Osterberg J. New device for load testing driven piles and drilled shaft separates friction and end bearing[J]. *Piling and deep foundations*,1989,1 (989): 421-427.
[3] LIU Chaogang, FANG Lei HUANG Feng,et al. Study on the Osterberg pile loading test by using similitude method [J]. *Uranium Geology*,1996, 12(6): 369-374.
[4] SHI Peidong, YU Xingfu, CAO Jianmin, et al. Application of Osterberg load test method in China and abroad and relevant problems[J]. *Building Construction,* 2007, 29(8): 573–576.
[5] Transport Industry Standard in the People's Republic of China. JT/T 738-2009,Static loading test of foundation pile-self-balanced method [S]. *China Communications Press* ,2009.
[6] Wang Zhongwen,Liu Zhifeng,Luo Yongchuan. Self- balanced method for large-diameter bored piles of Hong Kong-Zhuhai-Macao Bridge [J]. *Chinese Journal of Geotechnical Engineering*, 2013,35(S2): 1216-1219
[7] Hu Xiaobo,Liu Renyang,Xia Mingliang,et al. Studies on a self-balance test method for GMS Mekong Bridge pile foundations[J]. *China Civil Engineering Journal*,2015, 48(s1):171-180.
[8] Gao Ping,Yang Jianmin,Yang Xu. The design of the pile foundation in soft rock strata verification by self-balanced static load test[J].*Highway Engineering*,2012,37(3):234- 238.
[9] ZHANG Mingli,SUN Xuexian,GUO Nan,et al. O-cell experimental study on bearing capacity of rock-socketed piles in weathered soft rock region [J]. *Railway standard design*, 2014,58(5):39-44.
[10] XU Xibin，ZHOU Liang，WU Tongqing,et al. Test research on large-diameter rock-socketed cast-in-place piles at river port [J]. *Port & Waterway Engineering*, 2015,499(1):175-180.

IOP Conf. Series: Materials Science and Engineering **324** (2018) 012048 doi:10.1088/1757-899X/324/1/012048

Kinematics analysis of a 3-DOF hydraulic driven parallel mechanism

Bing Zhang [1,*], **Jiandong Li** [1], **Ziliang Jiang** [1] ,**Cuntang Wang** [1]and **Zhentao Zhang** [2]

[1]School of mechanical Engineering, Jiangsu University, No.301 Xuefu Road, Jingkou District, Zhenjiang, Jiangsu, China, 212013
[2]Zhuhai RULEBIT intelligent Robot Technology Company, No.9 Pinggong Road, Nanping Science and Technology Park, Xiangzhou District, Zhuhai, China, 519000
**Corresponding author: zhangbing@ujs.edu.cn*

***Abstract*:** In order to provide the basis about the selection of parts for the mechanical design of a 3-DOF parallel mechanism, such as hydraulic cylinders, valves and so on. Based on coordinate transformation and vector guidance method, the kinematics of the hydraulic drive 3-DOF parallel mechanism is analyzed at first. Secondly, deducing the displacement of each hydraulic cylinder according to the position of the upper platform on parallel mechanism. Then, The ADAMS is used to build virtual prototype of 3-DOF parallel mechanism, the validity about the kinematics of the parallel mechanism is verified by the co-simulation of MATLAB / SIMULINK and ADAMS, which provide the theoretical support for dynamics analysis.

1 Introduction

Parallel mechanism is widely used in testing equipment, machine tools, sensors and entertainment equipment because of its large stiffness, fast response, high precision and high load carrying capacity[1-2]. In view of the drive system, it can be divided into hydraulic drive, pneumatic drive, motor drive and so on[3]. Hydraulic drive is suitable for the testing equipment of heavy equipment because of its high power density, large output characteristics, such as vehicles, tanks and other heavy engineering machineries[4-5]. Kinematics analysis is the basis to analyze the dynamics, singularity and workspace of parallel mechanisms, mainly including the kinematics forward and inverse solution, speed and acceleration analysis and so on[6-7]. Kinematic inverse solution gives the pose of platform to solve the position parameter of driving system, kinematic forward solution is the opposite. Velocity and acceleration analysis refers to the relationship between the velocity and acceleration of the platform with the velocity and acceleration of driving mechanism, and they can be linked by the Jacobi matrix[8].

Recently, few degree of freedom parallel mechanism has become research hotspot, which has broad prospects in the field of industrial applications because of its low cost. At present, 3-DOF parallel mechanisms mostly evolve from Delta mechanism, while 3-RPS spatial parallel mechanism is proposed by foreign scholars. Generally speaking, 3-DOF parallel mechanisms are fewer models and inadequate research depth. The schematic diagram of mechanism about the novel 3-DOF parallel mechanism researched in this paper is shown in Fig. 1.

As can be seen from figure 1, the compositions of 3-DOF parallel mechanism researched in this paper are upper platform, lower platform, three sets of drive hydraulic cylinders ($1^{\#}$、 $2^{\#}$ and $3^{\#}$) and three sets of folding combinations (upper folding are a1d1, a2d2 and a3d3; the corresponding lower folding are d1b4, d2b5, d3b6).

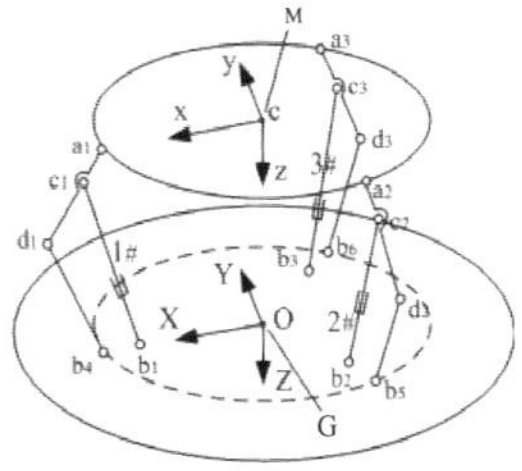

Figure 1. Schematic diagram of three degree of freedom parallel mechanism

Three degrees of freedom movements that the 3-DOF can achieve are: when the hydraulic cylinder $1^{\#}$ stays in the middle position, the cylinder $2^{\#}$ and $3^{\#}$ do differential motion, achieving the roll motion. When the three hydraulic cylinders run synchronously in middle position, the platform achieves the heave movement. When the hydraulic cylinder $1^{\#}$ does reciprocate motion in the middle position, and the other two hydraulic cylinders run synchronously with differential motion of cylinder $1^{\#}$ to achieve pitch movement at the same time.

2 Kinematic analysis of parallel mechanism

In order to express spatial motion of parallel mechanism clearly, two coordinates systems are needed. They are body coordinate system{M} and static coordinate system{G} that respectively shown in Fig. 1. Spatial motion pose parameter of parallel mechanism upper platform is defined as $q=[q_1\ q_2\ q_3\ q_4\ q_5\ q_6]^T$; Let $q_r=[q_1\ q_2\ q_3]^T$ be the angle of body coordinate system relative to static coordinate system; $c=[q_4\ q_5\ q_6]^T$ is the position of upper platform center in static coordinate system. Meanwhile, let $q_0=[q_1\ q_2\ q_6]^T$ be the generalized coordinate of 3-DOF parallel mechanism that is researched, respectively represent roll, pitch and heave motion of parallel mechanism.

The system rotation transformation matrix can be obtained after three axes rotation transformation as follows

$$\boldsymbol{R}=\begin{bmatrix} cq_3cq_2 & cq_3sq_2sq_1 - sq_3cq_1 & sq_3sq_1+cq_3sq_2sq_1 \\ sq_3cq_2 & cq_3cq_1 - sq_3sq_2sq_1 & sq_3sq_2cq_1 - cq_3sq_1 \\ -sq_2 & cq_2sq_1 & cq_2cq_1 \end{bmatrix} \tag{1}$$

where, $cq_1=\cos q_1$, $sq_1=\sin q_1$, and the others are same defined. Therefore, the transformation matrix from body coordinate system to static coordinate system can be given as

$$\boldsymbol{T}=\begin{bmatrix} \boldsymbol{R} & \boldsymbol{c} \\ \boldsymbol{0} & 1 \end{bmatrix} \tag{2}$$

According to the triangle cosine theorem, the length b1c1 and total length of hydraulic cylinder $1^{\#}$ can be derived, and the hydraulic cylinders $2^{\#}$ and $3^{\#}$ are calculated Similarly. The geometric relationship about $1^{\#}$ compounds of driving chain of 3-DOF parallel mechanism is shown in figure 2.

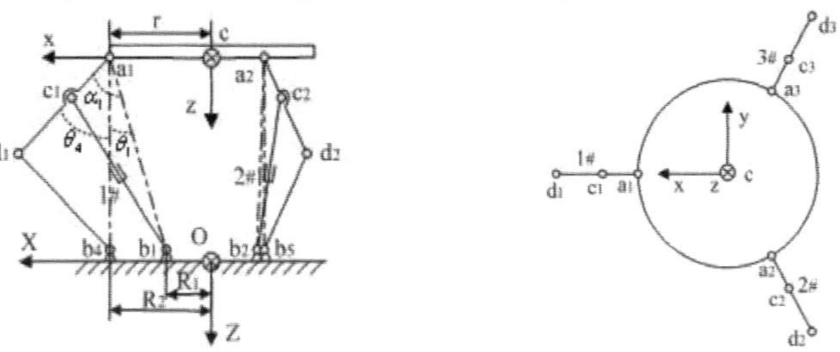

(a) Side view of 3-DOF parallel mechanism (b) Top view of 3-DOF parallel mechanism

Figure 2. Schematic of kinematic inverse solution of parallel mechanism

If homogeneous coordinate matrix $\boldsymbol{A}$ in body coordinate system of upper platform hinge point $a_i(i=1,2,3)$, homogeneous coordinate matrix $\boldsymbol{B_1}$ and $\boldsymbol{B_2}$ in static coordinate system of lower platform hinge point $b_i (i=1,2,3)$, $b_{i+3} (i=1,2,3)$ are

$$A=\begin{bmatrix} a_1 & a_2 & a_3 \\ 1 & 1 & 1 \end{bmatrix} \quad B_1=\begin{bmatrix} b_1 & b_2 & b_3 \\ 1 & 1 & 1 \end{bmatrix} \quad B_2=\begin{bmatrix} b_4 & b_5 & b_6 \\ 1 & 1 & 1 \end{bmatrix} \tag{3}$$

Through coordinate transformation, the coordinate matrix G in the static coordinate system of upper platform hinge points can be derived as

$$G=TA=\begin{bmatrix} a_1^g & a_2^g & a_3^g \\ 1 & 1 & 1 \end{bmatrix}=\begin{bmatrix} R & c \\ 0 & 1 \end{bmatrix}\begin{bmatrix} a_1 & a_2 & a \\ 1 & 1 & 1 \end{bmatrix} \tag{4}$$

where, a_i^g is coordinate matrix of upper hinge point a_i in static system, $a_i^g=\begin{bmatrix} x_i^g & y_i^g & z_i^g \end{bmatrix}^T$ $(i=1,2,3)$; a_i is coordinate matrix of upper hinge point a_i in body system, $a_i=\begin{bmatrix} x_i & y_i & z_i \end{bmatrix}^T$ $(i=1,2,3)$; R is coordinate rotation matrix; c is coordinate origin translation matrix of body coordinate system.

The upper hinge can only move in one plane due to the limitation of the three folds, for example, point a_1 can only move in the plane determined by point b_1, b_4 and d_1. Thus three constraint equations can be described as

$$y_1^g=0, y_2^g=\sqrt{3}x_2^g, y_3^g=-\sqrt{3}x_3^g \tag{5}$$

Combining formula (3) and (4), q_3, q_4, q_5 can be derived as

$$q_3=\arctan\frac{sq_1sq_2}{cq_1+cq_2}, q_4=\frac{-r}{2}(cq_1cq_3+sq_1sq_2sq_3-cq_2cq_3), q_5=-cq_2sq_3\cdot r \tag{6}$$

At this time, the six parameters q_1, q_2, q_3, q_4, q_5, q_6 of motion platform pose have already been solved, so transform matrix T is known. The coordinates of the upper hinge point a_i $(i=1,2,3)$ in the static coordinate system can be derived by formula (4). The expression of the position vector between the upper hinge point a_i and the lower hinge point b_i in the static coordinate system can be given by

$$L_{a_ib_i}=a_i^g-b_i=Ra_i+c-b_i \ (i=1,2,3) \ , \ L_{a_ib_{i+3}}=a_i^g-b_{i+3}=Ra_i+c-b_{i+3} \ (i=1,2,3) \tag{7}$$

Then the distance between the upper and lower hinge points can be expressed as

$$l_{a_ib_i}=\left\|L_{a_ib_i}\right\|, l_{a_ib_{i+3}}=\left\|L_{a_ib_{i+3}}\right\| \ (i=1,2,3) \tag{8}$$

So the angle θ_i $(i=1,2,3,4,5,6)$ in Fig. 2(a) can be derived as

$$\theta_i=arc\cos\left(\frac{l_{a_ib_i}^2+l_{a_ib_{i+3}}^2-l_{b_ib_{i+3}}^2}{2l_{a_ib_i}l_{a_ib_{i+3}}}\right) (i=1,2,3), \theta_{i+3}=arc\cos\left(\frac{l_{a_id_i}^2+l_{a_ib_{i+3}}^2-l_{d_ib_{i+3}}^2}{2l_{a_id_i}l_{a_ib_{i+3}}}\right) (i=1,2,3) \tag{9}$$

And $\alpha_i (i=1,2,3)$ can be defined as

$$\alpha_i=\theta_i+\theta_{i+3}(i=1,2,3) \tag{10}$$

According to the cosine theorem, the length of hydraulic cylinder in $\Delta a_ib_ic_i (i=1,2,3)$ can be derived as

$$l_i=l_{b_ic_i}=\sqrt{l_{a_ib_i}^2+l_{a_ic_i}^2-2l_{a_ib_i}l_{a_ic_i}\cos\alpha_i} \ (i=1,2,3) \tag{11}$$

If the initial length of hydraulic cylinder is $l_{i0} (i=1,2,3)$, then the piston rod movement of hydraulic cylinder can be described by

$$\Delta l_i=l_i-l_{i0} (i=1,2,3) \tag{12}$$

3 Co-simulation verification of parallel mechanism system

It is a low cost, high precision and high efficiency method to verify the mathematical model of system by using virtual prototype instead of physical prototype. In this paper, the kinematics model, dynamic model and control strategy of 3-DOF parallel mechanism will be established in MATLAB/SIMULINK; The virtual prototype of parallel mechanism is established in ADAMS software. To verify the feasibility and effectiveness of the kinematics model, dynamic model and control algorithm of 3-DOF parallel mechanism built above by co-simulation.

The parallel mechanism is in initial state while three cylinders are in middle position, and three groups of driving and supporting structures are evenly distribute, the structural parameters are shown in table 1 when the parallel mechanism is in middle position.

Table 1 Table of the 3-DOF parallel robot's structural parameters (m)

Radius of upper hinge circle r	Radius of lower hinge circle R_1	Radius of lower hinge circle R_2	Fold vertical distance between upper and lower point	Length of upper fold $L_{a_i d_i}$	Length of lower fold $L_{d_i b_{i+3}}$	Distance between fold and hinge point of cylinder $L_{a_i c_i}$
2.000	1.600	2.000	2.000	1.154	1.154	0.450

The process of kinematic model verification is shown as: the displacement of hydraulic cylinder is obtained by using inverse kinematic with input pose of parallel mechanism, and the output pose can be obtained by using motion virtual prototype of displacement driven platform. The validity of kinematic model is evaluated by the error between input and output pose, its principle is shown in figure 3.

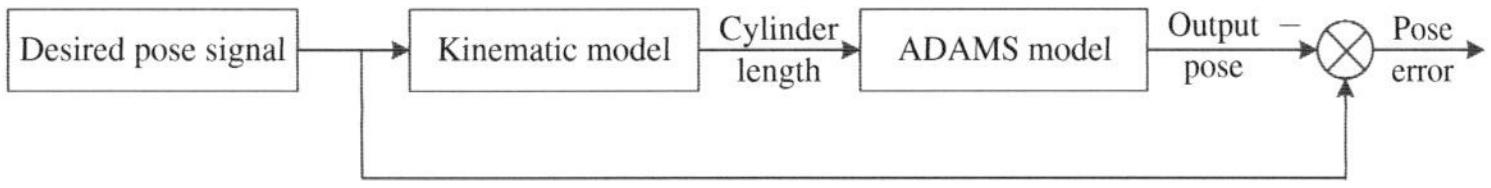

Figure 3. The verification principle of kinematics

The desired input pose signals are $q_1 = 8^{\circ}$, $q_2 = 8^{\circ}$, $q_6 = 0.28\text{m}$. So the results of system simulation are shown in figure 4.

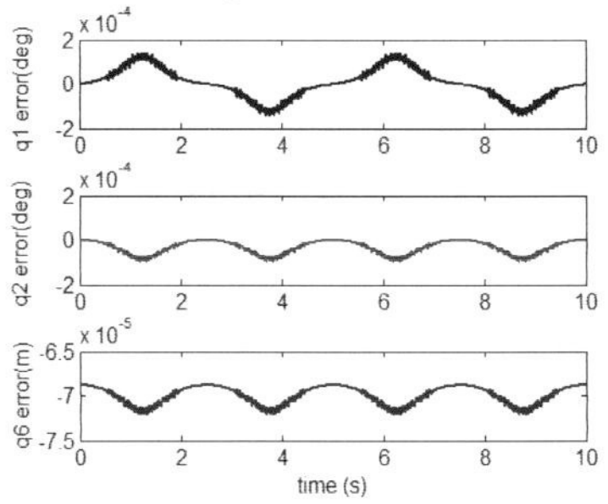

(a) Pose error of roll motion

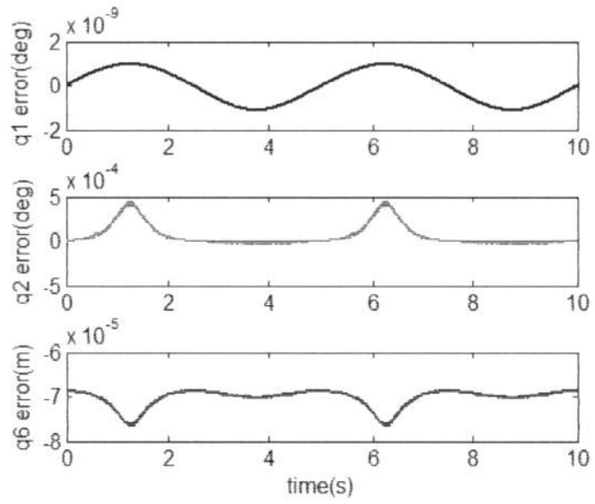

(b) Pose error of pitch motion

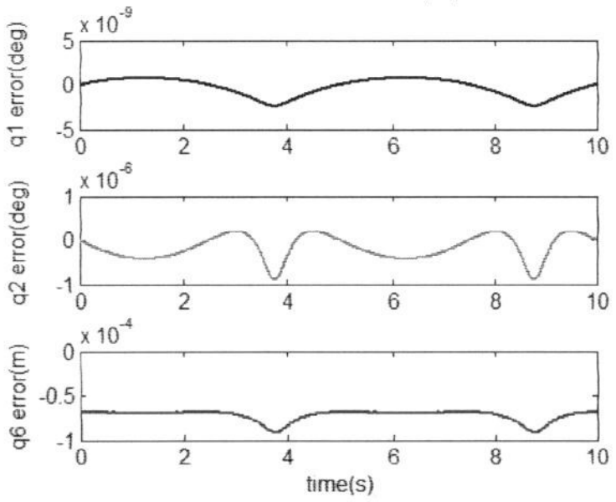

(c) Pose error of heave motion

Figure 4. Validation error of kinematic model

According to the simulation results, the roll motion error is $1.48^{\circ} \times 10^{-4}$, that is 0.00185%; The input of pitch motion and heave motion is zero, and its errors are $9.85^{\circ} \times 10^{-5}$ and $7.22 \times 10^{-5}\text{m}$ respectively; The pitch motion error is 4.42×10^{-4}, that is 0.0055%; The errors of roll motion and the heave motion at zero input are respectively $9.82^{\circ} \times 10^{-9}$ and $7.65^{\circ} \times 10^{-5}$; The heave motion error is $9.21 \times 10^{-5}\text{m}$, that is 0.0033%; The errors of roll motion and the pitch motion are respectively 7.65×10^{-10} and $8.97^{\circ} \times 10^{-7}$. From the simulation results, it can be concluded that the kinematic model established to analyze the parallel mechanism is accurate.

Table 2 Cylinder length range during the three degrees of freedom motion

	roll	pitch	heavy
Cylinder 1#	-4.32×10^{-5}m~1.78×10^{-3}m	–0.180m~0.162m	–0.175m~0.168m
Cylinder 2#	–0.155m~0.144m	–0.088m~0.087m	–0.175m~0.168m
Cylinder 3#	–0.155m~0.144m	–0.088m~0.087m	–0.175m~0.168m

According to table 2, the final motion range of hydraulic cylinder can be obtained as follows: cylinder $1^{\#}$: $-180\text{mm}\sim168\text{mm}$; Cylinder $2^{\#}$ and $3^{\#}$: $-175\text{mm}\sim168\text{mm}$, which can be ready for the selection of hydraulic power mechanism.

4 Conclusion

The kinematics of a novel 3-DOF parallel mechanism was established using the derivative methods, and the high precision of model was verified by co-simulation in using MATLAB/SIMULINK and ADAMS software. The length of each hydraulic cylinder was gained by calculating the kinematics model under the given work condition, which guides the choice of power mechanism. And the inverse kinematic model supplies the fundament for the joint control of parallel mechanism. There is a slight translational movement along the axis x in roll and pitch movements because of the existence of implicated movement.

5 Acknowledgement

This work was supported by the Starting Foundation of Jiangsu University Advanced Talent (grant number 14JDG048) and the Postdoctoral Research Funding Plan of Jiangsu Province (grant number 1402093C).

References

[1] Chablat D and Wenger P. Architecture optimization of a 3-DOF translational parallel mechanism for machining applications, the orthoglide[J]. Robotics & Automation IEEE Transactions, 2003, 19 (3): 403-10.
[2] Xu Qingsong. New Flexure Parallel-Kinematic Micropositioning System With Large Workspace[J]. IEEE Transactions on Robotics, 2012, 28 (2): 478-91.
[3] Stump E and Kumar V. Workspaces of Cable-Actuated Parallel Manipulators[J]. Journal of Mechanical Design, 2006 128 (1): 159-167.
[4] Yang Chifu, He Jingfeng, Han Junwei and Liu Xiaochu. Real-time state estimation for spatial six-degree-of-freedom linearly actuated parallel robots[J]. Mechatronics, 2009 19 (6): 1026-33.
[5] Yang Chifu, Huang Qitao and Han Junwei. Computed force and velocity control for spatial multi-DOF electro-hydraulic parallel manipulator[J]. Mechatronics, 2012 22 (6): 715-22.
[6] Yang Xiaolong, Wu Hongtao, Li Yao and Chen Bai. A dual quaternion solution to the forward kinematics of a class of six-DOF parallel robots with full or reductant actuation[J]. Mechanism & Machine Theory, 2017 107: 27-36.
[7] Cai Kunhai, Tian Yuanfeng, Wang Fujun, Zhang Dai and Liu Xianping, B Shirinzadeh. Design and control of a 6-degree-of-freedom precision positioning system[J]. Robotics and Computer-Integrated Manufacturing, 2017 44: 77-96.
[8] Wang Liping, Xu Huayang, Guan Liwen and Zhi Yu. A novel 3-PUU parallel mechanism and its kinematic issues[J]. Robotics and Computer-Integrated Manufacturing, 2016 42: 86-102.

Analysis of the Mean Absolute Error (MAE) and the Root Mean Square Error (RMSE) in Assessing Rounding Model

Weijie Wang [1] and Yanmin Lu [1]

1 CFETS Information Technology (Shanghai) Co., Ltd. China Foreign Exchange Trade System, Shanghai, China
E-mail: wangweijie@chinamoney.com.cn

Abstract. Most existing Collaborative Filtering (CF) algorithms predict a rating as the preference of an active user toward a given item, which is always a decimal fraction. Meanwhile, the actual ratings in most data sets are integers. In this paper, we discuss and demonstrate why rounding can bring different influences to these two metrics; prove that rounding is necessary in post-processing of the predicted ratings, eliminate of model prediction bias, improving the accuracy of the prediction. In addition, we also propose two new rounding approaches based on the predicted rating probability distribution, which can be used to round the predicted rating to an optimal integer rating, and get better prediction accuracy compared to the Basic Rounding approach. Extensive experiments on different data sets validate the correctness of our analysis and the effectiveness of our proposed rounding approaches.

1. Introduction

Collaborative Filtering (CF) systems have attracted lots of research attentions to overcome the information overload problem. Many recent algorithms, such as neighborhood methods, matrix factorization, various probabilistic models, clustering techniques, graph-based approaches and machine learning techniques, based on, e.g., association rule mining [4], have been proposed which aimed to improve the accuracy of prediction rating. For evaluation, Mean Absolute Error (MAE) and Root Mean Square Error (RMSE) are widely adopted in many recommendation systems to measure the difference between the predicted scores and users' actual ratings, such as Slope One [8], Recommending Based on Rating Frequencies [5] and SVD model [9].

In general, CF algorithms output a decimal number as the predicted preference of an active user toward a given item. However, the actual ratings in most systems are integers. According to our experiment, it is necessary to round the decimal predicted scores to the proper integers, as this not only reduces the MAE values, but also improves the precision and quality of recommendation. Few preliminary works [3] have observed this fact. However, this issue has not been seriously studied. In fact, many recent popular CF algorithms, such as Recommending Based on Rating Frequencies [5], get very inspiring accuracy attributed to the fact that the original predicted ratings are integers.

In this paper, we present a comprehensive study on analyzing and demonstrating how and why rounding influences MAE and RMSE values. We show that the nonuniform distributions of ratings in training sets bias the predicted scores. Based on our empirical findings, we further propose two approaches to round the predicted rating to the most likely integer rating by considering the predicted ratings' probability distribution of each rating level. Extensive experiments show that our proposed approaches achieve inspiring performance in MAE metric.

The rest of this paper is organized as follows. Section 2 briefly reviews some related works. Section 3 analyzes the influence of rounding on MAE and RMSE, and proves rounding is necessary in postprocessing of the predicted ratings using classical matrix factorization model. The empirical process of proof and our proposed rounding approaches are presented in Section 4. Section 5 presents the extensive experimental results of our proposed approaches compared to the baseline rounding approach. Section 6 concludes this paper and discusses our potential future works.

2. Related Works

In most CF algorithms, the rounding of predicted scores has seldom been seriously studied. Thomas Hofmann [6] and Rong Jin [7] take the user's rating trend into account and translate the ratings into normalized intervals by employing Gaussian and Decoupling normalization. Both of them avoid the issue of rounding the predicted rating to the integer rating.

A preliminary work [3] has observed that rounding can effectively improve the prediction accuracy and reduce the value of MAE. In [3], a rounding algorithm is proposed by considering the users' trend degree, departure degree and judgment degree. However, the computational complexity of this algorithm is very high when the rating matrix is very sparse (which usually is the case in most recommendation systems) and the number of the prediction rating is very huge. Most importantly, it is not illustrated why rounding can improve the precision of predicted ratings and reduce the MAE values.

The basic rounding approach which rounds the predicted rating to the nearest integer is the simplest solution for rounding the predicted rating to integer; however, it lacks effective rationale and is indemonstrable. In our work, we take it as the baseline rounding approach to validate the effectiveness of our proposed approaches.

In response, in this paper, we analyze and demonstrate how and why rounding predicted ratings to integer can influence the performance of the CF algorithms in terms of MAE and RMSE metrics, and prove that rounding is necessary in post-processing of the predicted ratings and improving the performance and quality of the recommendation. Based on our findings, we further propose two approaches to round the predicted rating to the most likely integer rating. Our experiments show that the proposed approaches get inspiring performance in MAE metric.

The three notable contributions of this paper are:

- First, we analyze and demonstrate why the rounding will bring different influences for MAE value and RMSE value, indeed reduce the MAE value, increase the RMSE values in some specific context.
- Second, we prove that rounding is necessary in post-processing of the predicted ratings and improving the performance and quality of the recommendation.
- Third, we propose two rounding approaches based on the predicted ratings' probability distribution of each rating level, which can be used to round the predicted rating to the most likely integer rating and offer better performance both in MAE and RMSE metrics.

3. The Influence of Rounding Prediction

This section, we demonstrate why and how rounding can affect the performance of recommendation systems in terms of MAE and RMSE metrics. A rating denotes the actual rating given by a user to an item . To distinguish predicted ratings from actual ones, we use the notation to denote the predicted score of user to item . The actual ratings are integers lying in .

For evaluation and analysis, we use 100k-MovieLens rating database, detailed description of the dataset can be seen in Section 5. As CF algorithm, here we employ Matrix Factorization (MF) model to analyze the probability distribution of predicted ratings. MF technique has become a popular choice for implementing CF and a survey can be found [9].

3.1. Evaluation Metrics

Mean Absolute Error (MAE) and RMSE (Root Mean Square Error) have widely been used in evaluating the accuracy of a recommender system, given by:

$$MAE = \frac{\sum_{n=1}^{N} |\hat{r}_n - r_n|}{N} \qquad RMSE = \sqrt{\frac{\sum_{n=1}^{N} (\hat{r}_n - r_n)^2}{N}}$$

where $\hat{r}_n$ means the prediction rating; r_n means the true rating in testing data set; N is the number of rating prediction pairs between the testing data and prediction result.

3.2. Performance of Rounding and Analysis

The Basic Rounding operator is defined as follows:

$$BasicRound(x) = \begin{cases} 1 & if\ x < 1 \\ \lfloor x \rfloor & if\ x - \lfloor x \rfloor < 0.5\ \&\ x \in [1, R] \\ \lceil x \rceil & if\ \lceil x \rceil - x \geq 0.5\ \&\ x \in [1, R] \\ R & if\ x > R \end{cases}$$

where $\lfloor x \rfloor$ is the floor integer of x (the largest integer less than or equal to x) and $\lceil x \rceil$ is the ceiling integer of x (the least integer greater than or equal to x).

As depicted in Fig. 1, the MAE values decrease when we round the predicted rating to the nearest neighbor integer with the basic rounding operator. While the RMSE values increase after rounding, especially when the training set ratio is larger than 0.5.

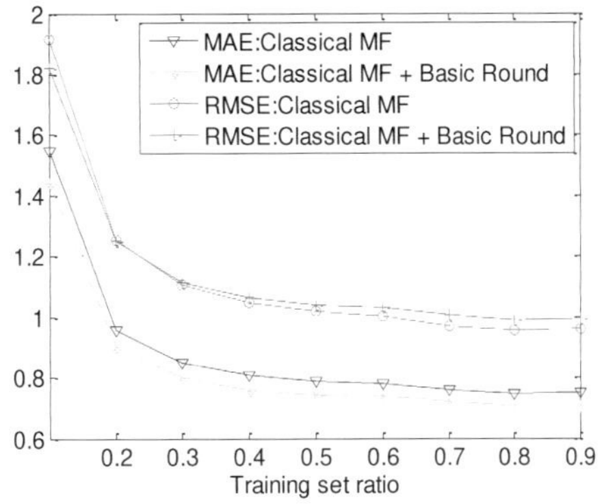

Fig. 1. Rounding the predicted ratings on MovieLens

- Reasons Analysis

In order to demonstrate why MAE values will decrease when predicted ratings, which predicted by classical Matrix Factorization model, rounded to integers, we analyzed the data set and all predicted ratings.

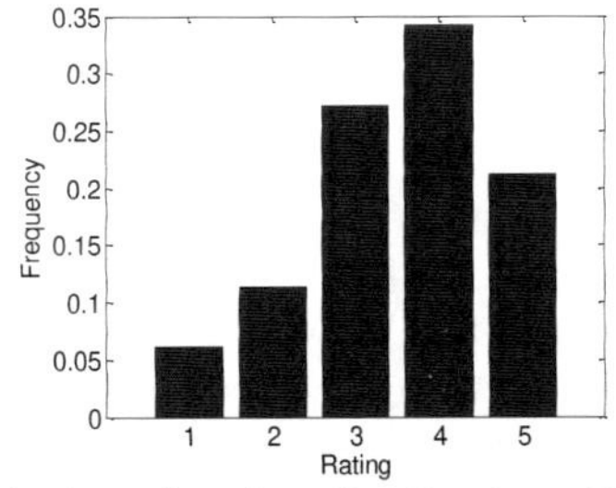

Fig. 2. Actual rating distribution of MovieLens

Fig. 2 illustrates the distribution of ratings in the training set for 5 different ratings (1~5). As can be seen in Fig. 3, the vast majority of the ratings are 3 and 4. In other words, users tend to rate 3 and 4 more frequently. As the result, the ratings of 3 and 4 have more influence on the final prediction model which is learned by all ratings in the training set.

We conduct experiments on MovieLens with the training set ratio of 0.8 and compute the difference between predicted ratings and actual ratings for all training samples. Fig. 3(a) illustrates the probability distribution of the prediction errors. Obviously the distribution follows a unimodal distribution, with a mode at -0.2. This means that the overall predicted ratings are lower than actual ratings by about 0.2.

In addition, we satiated the distributions of the prediction errors for 5 different rating levels (1~5) in the training set individually. Interestingly, as shown in Fig. 3(b-f), for different rating levels, the distributions of the prediction errors are quite different from each other. Indeed, although each distribution follows a unimodal distribution, their modes are materially different which range from xx to xx. It reflects that most of the modes of these distributions curve are not at the right value (0). One of the explanations for this effect is that among the training rating during the MF model learning, the majority ratings that used to train the model are rating 3 and rating 4. That cause the predicted ratings, which actual ratings is higher than 4, are lower than its actual value. Similarly, the predicted ratings, which actual ratings is lower than 3, are higher than its actual value.

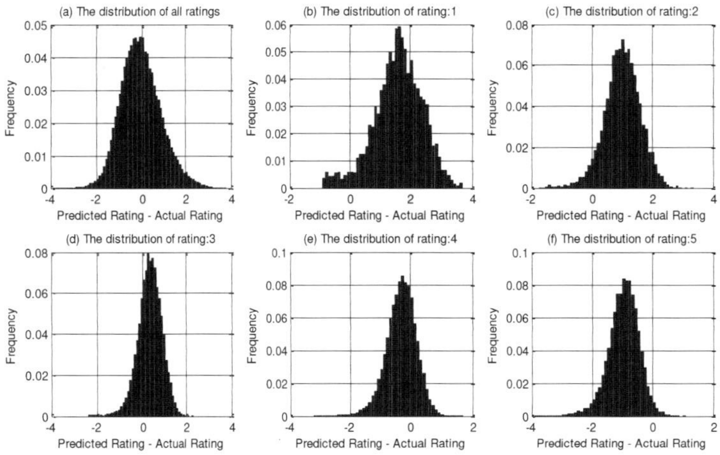

Fig. 3. Distribution of prediction errors in training set on MovieLens

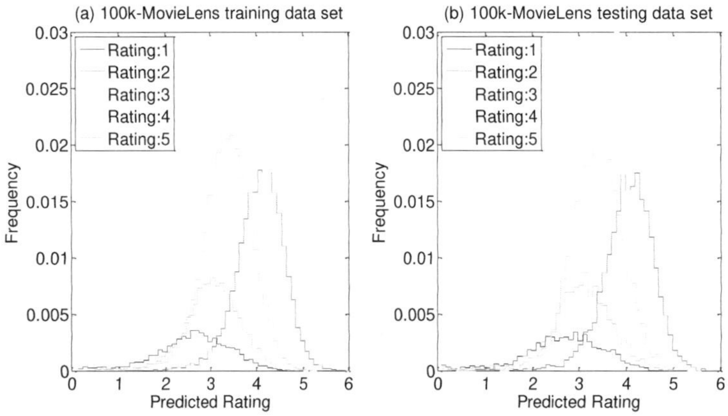

Fig. 4. All predicted ratings' probability distribution

We combine these distributions curve to one Figure as shown in Fig. 4(a). The x-axis represents the predicted scores, and the yaxis represents the percentage of the predicted ratings in training set. Fig. 4(a) shows each distribution of the each rating levels' predicted ratings in training set. For the actual rating of 1, the mode of its distribution is 2.2, which means most of the predicted rating are predicted to be around 2.2. This actually results in quite large prediction errors. Similarly, for the actual ratings of 2, 3, 4, and 5, the modes of their distributions are 2.9, 3.2, 3.8, and 4.3 respectively.

For the probability distribution function, we denote by f_r the probability function of the actual ratings of predicted ratings belonging to integer r. Naturally, $f_r(x)$ is the probability of the actual rating of the predicted rating x belonging to integer r.

According to Fig. 4(a), for a predicted rating x, supposing $\lfloor x \rfloor$ is the nearest integer of x, we have $x \in [\lfloor x \rfloor - 0.5, \lceil x \rceil + 0.5]$, and $f_{\lfloor x \rfloor}(x) = \max_{r \in \{1, L, R\}} \{f_r(x)\}$.

In summary, from the above analysis, we can see rounding can definitely decrease the value of MAE. It should be noted that we assume the distribution of predicted ratings in test set keeps identical with the training set. Fig. 4(b) shows the distribution of predicted ratings on test set. Compared with Fig. 4(a), we can see the distributions of predicted ratings on test set and training are almost the same.

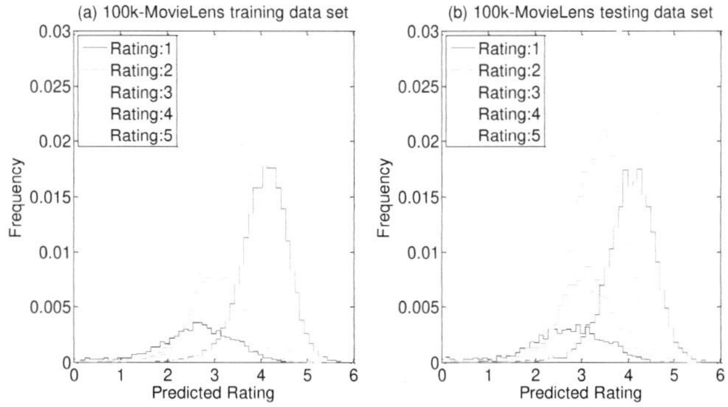

Figure 5. All predicted ratings' probability distribution

4. Proving and Our Approaches

In this section, we present the detailed analysis and discussion on the performance of MAE and RMSE in the rounding classical MF model. First, we theoretically prove why rounding bring different influences for MAE and RMSE. Second, we further propose two new rounding approaches to improve the performance of recommendation systems in term of MAE metric.

4.1. MAE Performance in Rounding Model

There exist two ways to round the predicted ratings to proper integers with the purpose of decreasing the MAE. One way is to round a rating to its neighbor, i.e. rounding a predicted rating x to the floor integer $\lfloor x \rfloor$ or the ceiling integer $\lceil x \rceil$. The other way is rounding a rating to the optimal integer. This allows a rating to be rounded to any integer, as long as this integer could decrease MAE the most. In the following, we discuss these two possible rounding approaches respectively.

- **Rounding to the Neighbor Integers**

Here, we first consider rounding the predicted ratings to their neighbor integers. It should be noted that we discuss not a signal predicted rating, but the all predicted rating which has same value.

Theorem 1: Given a predicted rating x, rounding x to at most one of its two neighbor integers ($\lfloor x \rfloor$ or $\lceil x \rceil$) can reduce the value of MAE after rounding.

Proof: For any predicted rating x, we have $x \in [\lfloor x \rfloor, \lceil x \rceil]$. If we round all predicted ratings with the value of x to $\lfloor x \rfloor$, the decrement of MAE is calculated as follows:

$$Dec_{MAE}(x, \lfloor x \rfloor) = \frac{\sum_{r=1}^{r=R} f_r(x) * |x-r| - \sum_{r=1}^{r=R} f_r(x) * |\lfloor x \rfloor - r|}{\sum_{r=1}^{r=R} f_r(x)} = \frac{\left[\sum_{r=1}^{r=\lfloor x \rfloor} f_r(x) - \sum_{r=\lceil x \rceil}^{r=R} f_r(x)\right] * (x - \lfloor x \rfloor)}{\sum_{r=1}^{r=R} f_r(x)} \quad (1)$$

Similarly, if we round these predicted ratings to $\lceil x \rceil$, the decrement of MAE will be:

$$Dec_{MAE}(x, \lceil x \rceil) = \frac{\left[\sum_{r=\lceil x \rceil}^{r=R} f_r(x) - \sum_{r=1}^{r=\lfloor x \rfloor} f_r(x)\right] * (\lceil x \rceil - x)}{\sum_{r=1}^{r=R} f_r(x)} \quad (2)$$

It is important to note that the sign of Equation (1) (2) depends on the sign of $\sum_{r=1}^{r=\lfloor x \rfloor} f_r(x) - \sum_{r=\lceil x \rceil}^{r=R} f_r(x)$, for both $(x - \lfloor x \rfloor) \geq 0$ and $(\lceil x \rceil - x) \geq 0$. Hence, only one direction of rounding can reduce the value of MAE.

Theorem 2: Given a predicted rating x, only rounding x to an integer can get the highest decrement of MAE compared with other numbers in $[\lfloor x \rfloor, \lceil x \rceil]$.

Proof: According to Equation (1) (2), as the signs of $Dec_{MAE}(x,\lfloor x \rfloor)$, $Dec_{MAE}(x,\lceil x \rceil)$ are all determined by the sign of $\sum_{r=1}^{r=\lfloor x \rfloor} f_r(x) - \sum_{r=\lceil x \rceil}^{r=R} f_r(x)$, the MAE value could be reduced by modifying the rating in only one direction, i.e. either increasing or decreasing x. In other words, if $\sum_{r=1}^{r=\lfloor x \rfloor} f_r(x) > \sum_{r=\lceil x \rceil}^{r=R} f_r(x)$, to decrease the MAE value, we should modify the original predicted rating to a candidate number x_0, where $x_0 \in (x, \lceil x \rceil)$. Therefore we can get the decrement on MAE value

$$Dec_{MAE}(x,x_0) = \frac{\sum_{r=1}^{r=R} f_r(x) * |x-r| - \sum_{r=1}^{r=R} f_r(x) * |x_0 - r|}{\sum_{r=1}^{r=R} f_r(x)} = \frac{\left[\sum_{r=1}^{r=\lfloor x \rfloor} f_i(x) - \sum_{r=\lceil x \rceil}^{r=R} f_r(x)\right] * (x - x_0)}{\sum_{r=1}^{r=R} f_r(x)} \tag{3}$$

Obviously, $Dec_{MAE}(x,x_0) < Dec_{MAE}(x,\lfloor x \rfloor)$. As can be seen in Equation (3), when x_0 is an integer, we get the largest decrement of MAE.

- **Rounding to the Optimal Integers**

In Section 4.1.1, we only consider rounding a predicted rating to its neighbor integer. However, sometimes this is not the optimal choice. For instance, in Fig. 3(b), for those samples with actual ratings of 1, the predicted ratings range from 0 to 5 with a mode of about 2.2. In this case, by only rounding the predicted rating to the neighbor integer (which is 2 for most samples) cannot achieve the best performance.

For this purpose, we modify Equation (1) (2) to calculate the decrement of MAE when rounding a given predicted rating x to each integer r_0, after derivation we get Equation (4) as follows:

$$Dec_{MAE}(x,r_0) = \begin{cases} \dfrac{\left[\left(\sum_{r=1}^{r=r_0} f_r(x) - \sum_{r=\lceil x \rceil}^{r=R} f_r(x)\right) * (x - r_0) + \sum_{i=r_0+1}^{i=\lfloor x \rfloor}\left(f_i(x) * (x - 2i + r_0)\right)\right]}{\sum_{r=1}^{r=R} f_r(x)}, & if\ r_0 \le \lfloor x \rfloor \\ \dfrac{\left[\left(\sum_{r=r_0}^{r=R} f_r(x) - \sum_{r=1}^{r=\lfloor x \rfloor} f_r(x)\right) * (r_0 - x) + \sum_{i=\lceil x \rceil}^{i=r_0-1}\left(f_i(x) * (2i - x - r_0)\right)\right]}{\sum_{r=1}^{r=R} f_r(x)}, & if\ r_0 \ge \lceil x \rceil \end{cases} \tag{4}$$

Theorem 3: Rounding a predicted rating to the neighbor integer is not always the optimal rounding approach.

Proof: Here, we take following as an example: given predicted rating: 3.2, and its probability distribution of each rating levels are:
$f_1(3.2) = 5/140$, $f_2(3.2) = 4/140$, $f_3(3.2) = 3/140$, $f_4(3.2) = 1/140$, $f_5(3.2) = 1/140$. According to Equation (4), $Dec_{MAE}(3.2,3) = 2/14$, $Dec_{MAE}(3.2,2) = 6/14$, $Dec_{MAE}(3.2,1) = 2/14$ we also get: $Dec_{MAE}(3.2,4) = -8/14$, $Dec_{MAE}(3.2,5) = -20/14$ and it's no doubt that we will make the biggest decrement of MAE when we round 3.2 to 2, rather than round to 3. Thus, it is necessary to consider rounding the predicted ratings to other integers besides its neighbors.

Assume the optimal rounding integer of predicted rating x is r_0, and $\sum_{r=1}^{r=\lfloor x \rfloor} f_r(x) > \sum_{r=\lceil x \rceil}^{r=R} f_r(x)$. We can calculate the decrement of MAE as follows:

$$Dec_{MAE}(x,r_0) = \left[\left(\sum_{r=1}^{r=r_0} f_r(x) - \sum_{r=\lceil x \rceil}^{r=R} f_r(x)\right) * (x - r_0) + \sum_{i=r_0+1}^{i=\lfloor x \rfloor}\left(f_i(x) * (x - 2i + r_0)\right)\right] \Big/ \sum_{r=1}^{r=R} f_r(x)$$

$$Dec_{MAE}(x,\lfloor x \rfloor) = \left(\sum_{r=1}^{r=\lfloor x \rfloor} f_r(x) - \sum_{r=\lceil x \rceil}^{r=R} f_r(x)\right) * (x - \lfloor x \rfloor) \Big/ \sum_{r=1}^{r=R} f_r(x)$$

Finally, we get the function to judge which integer is the optimal to the predicted rating x, the biggest value of Φ, the optimal integer of the predicted rating x is.

$$\Phi_{floor}(r_0,x) = \left[\sum_{r=1}^{r=r_0} f_r(x) - \sum_{r=\lceil x \rceil}^{r=R} f_r(x)\right] * (\lfloor x \rfloor - r_0) + \sum_{i=r_0+1}^{i=\lfloor x \rfloor}\left[f_i(x) * (\lfloor x \rfloor + r_0 - 2i)\right],\ if\ r_0 \in \{1, \mathrm{L}, \lfloor x \rfloor\} \tag{5}$$

4.2. RMSE Performance in Rounding Model

Compared to MAE, RMSE disproportionately penalizes large errors, so that, given a test set with four hidden items RMSE would prefer a algorithm that makes an error of 1 on three ratings and 0 on the fourth to one that makes an error of 2 on one rating and 0 on all three others, while MAE would prefer the second algorithm, Obviously, it is difficult to strictly define which metric is better to evaluate the accuracy of the predicted ratings.

As shown by Fig. 1, after rounding the predicted ratings, the values of RMSE increase for all rating level especially when training set ratio is larger than 0.5. In this section, we discuss the effect on RMSE values by rounding the predicted ratings. Similar to Section 4.1, two different rounding methods are considered.

- **Rounding to the Neighbor Integers**

First we consider rounding a predicted rating to the neighbor integers. According to our discussion above, for any predicted rating x , $x \in [\lfloor x \rfloor, \lceil x \rceil]$, if we round all predicted ratings with values of x to $\lfloor x \rfloor$, the decrement of RMSE will be:

$$Dec_{RMSE}(x, \lfloor x \rfloor) = \left(\sqrt{\sum_{r=1}^{r=R} f_r(x) * (x-r)^2} - \sqrt{\sum_{r=1}^{r=R} f_r(x) * (\lfloor x \rfloor - r)^2} \right) \Big/ \sqrt{\sum_{r=1}^{r=R} f_r(x)} \quad (6)$$

Similarly, if we round these predicted ratings to $\lceil x \rceil$, the decrement of RMSE will be:

$$Dec_{RMSE}(x, \lceil x \rceil) = \left(\sqrt{\sum_{r=1}^{r=R} f_r(x) * (x-r)^2} - \sqrt{\sum_{r=1}^{r=R} f_r(x) * (\lceil x \rceil - r)^2} \right) \Big/ \sqrt{\sum_{r=1}^{r=R} f_r(x)} \quad (7)$$

As the signs of Equations (6) (7) are determined by the sign of their numerators, the value of RMSE may not be reduced by either rounding x to $\lceil x \rceil$ or $\lfloor x \rfloor$.

- **Rounding to the Optimal Integers**

Second, we consider rounding the predicted rating to the optimal integers in order to reduce the value of RMSE. According to the formulas above, we calculate the decrement of RMSE when rounding the predicted rating to each candidate integer as follows:

$$Dec_{RMSE}(x, r_0) = \left(\sqrt{\sum_{r=1}^{r=R} f_r(x) * (x-r)^2} - \sqrt{\sum_{r=1}^{r=R} f_r(x) * (r_0 - r)^2} \right) \Big/ \sqrt{\sum_{r=1}^{r=R} f_r(x)}$$

The sign of $Dec_{RMSE}(x, r_0)$ is determined by the sign of its square root of the numerators. Thus, any modification may not reduce RMSE by rounding x to any integers.

4.3. Our Proposed Approach: PDPR_Neighbor

According to Theorems 1 and 2, to decrease the MAE value, we should round a predicted rating to either $\lfloor x \rfloor$ or $\lceil x \rceil$(Note that here we only consider the possible values in $[\lfloor x \rfloor, \lceil x \rceil]$). More specifically, we should round the predicted rating x to the integer $\lfloor x \rfloor$ when $\sum_{r=1}^{r=\lfloor x \rfloor} f_r(x) > \sum_{r=\lceil x \rceil}^{r=R} f_r(x)$, and to the integer $\lceil x \rceil$ when $\sum_{r=1}^{r=\lfloor x \rfloor} f_r(x) \le \sum_{r=\lceil x \rceil}^{r=R} f_r(x)$. Based on this finding, we propose our first rounding algorithm which only rounds a predicted rating to its floor or ceiling integer, called *PDPR_Neighbor*, as follows:

$$PDPR_{Neighbor}(x) = \begin{cases} BasicRound(x)\ , \\ \quad if\ x \notin [1, R] \vee |Diff(x)| \le Threshold \\ \left[\lfloor x \rfloor * (1 + sgn(Diff(x))) + \lceil x \rceil * (1 - sgn(Diff(x))) \right] / 2\ , \\ \quad if\ x \in [1, R] \wedge |Diff(x)| > Threshold \end{cases}$$

$$Diff(x) = \sum_{r=1}^{r=\lfloor x \rfloor} f_r(x) - \sum_{r=\lceil x \rceil}^{r=R} f_r(x)$$

$$Threshold = \min\left(\sum_{r=1}^{r=\lfloor x \rfloor} f_r(x), \sum_{r=\lceil x \rceil}^{r=R} f_r(x)\right) \Big/ 10$$

where $sgn(x)$ is sign function, as our rounding approach is under the assumption that the distribution of predicted rating in testing set keeps identical with the training set. So here, a threshold is set in our approach in order to avoid overfitting.

4.4. Our Proposed Approach: PDPR_Optimal

According to the example of Theorem 3's proof, the optimal integer of a predicted rating (3.2) can be determined by Equation (5). Thus, we propose our second algorithm which rounds a predicted rating to the optimal integer that makes largest decrement of MAE, called *PDPR_Optimal*, as follows:

$$PDPR_{Optimal}(x)=\begin{cases} BasicRound(x)\ ,\ if\ x\notin[1,R]\vee|Diff(x)|\le Threshold \\ \underset{r_0\in PossibleIntsSet(x)}{\arg\max}\left(\Phi_{floor}(r_0,x)*(1+sgn(Diff(x))+\Phi_{ceiling}(r_0,x)*(1+sgn(-Diff(x))\right), \\ \qquad if\ x\in[1,R]\wedge|Diff(x)|>Threshold \end{cases}$$

$$PossibleIntsSet(x)=\begin{cases}\{r\in Z|1\le r\le\lfloor x\rfloor\}\ ,\ if\ Diff(x)>0 \\ \{r\in Z|\lceil x\rceil\le r\le R\},\ if\ Diff(x)<0\end{cases}$$

A threshold is also set to avoid overfitting as our rounding approach is under the assumption that the distribution of predicted ratings in testing set must keep identical with the one in training set, while the distribution of predicted ratings can't completely consistent indeed.

The core part of above two round algorithms (*PDPR_Neighbor* and *PDPR_Optimal*) are all the calculation of each rating level's probabilistic distribution. We assume that the distribution of predicted ratings for each rating level follows Gaussian distribution. Compared with the algorithm *PDPR_Neighbor*, *PDPR_Optimal* rounds the predicted rating to the optimal integer in theory.

5. EXPERIMENT

Our approaches are tested on two different movie domains. The first one is the 100k-MovieLens rating data set with 100,000 ratings by 943 users on 1,682 movies. The ratings are integers ranging from 1 to 5. The second one is the EachMovie dataset with 2,811,983 numeric ratings by 72,916 users on 1,628 movies. The ratings range from 0.0 to 1.0 with a step size of 0.2. For EachMovie, we choose 100,118 ratings by the first 2936 users on 1,628 movies. We multiply all ratings by 5, and then plus one under the amplified ratings. Thus, the resulting ratings become integers ranging from 1 to 6.

5.1. The Necessity of Rounding Predicted Ratings

We compare the following algorithms with Basic Rounding approach: user-based kNN (using Pearson similarity and the neighborhood size of 30 which is suggested as the optimal value in literature [2]), Slope One [8], and Classical MF [9]. Besides, we also take the MAE of Rating Frequencies algorithm [5] as a reference, which predicted ratings are integers, to evaluate the rounding performance.

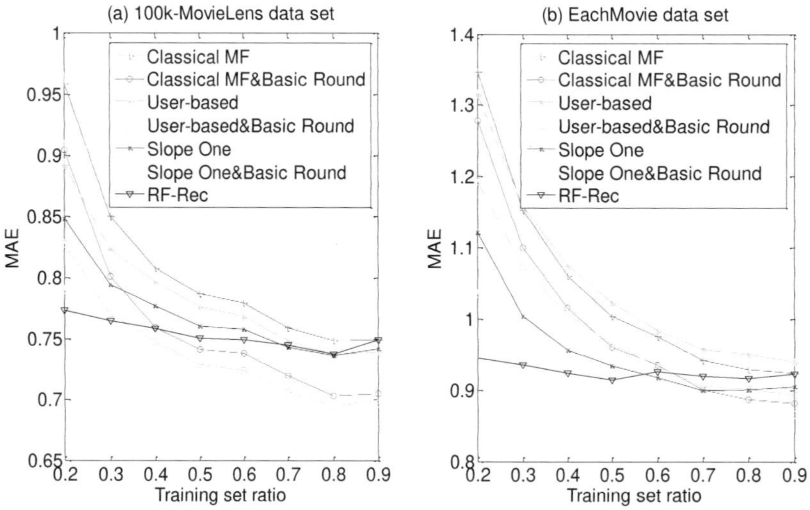

Fig. 6. The MAE values on MovieLens / EachMovie

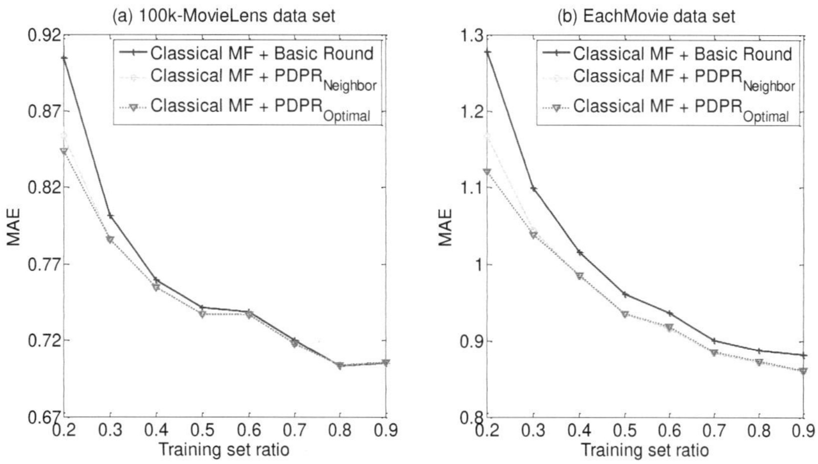

Fig. 7. The MAE values on two different data sets

Fig. 6(a), 7(b) shows the MAE values for different training set ratio from 0.1 to 0.9 on the MovieLens and EachMovie datasets respectively. As can be seen in Fig. 6(a), 7(b), for all different settings, the MAE values are significantly decreased after rounding the predicted ratings by Basic Round approach. This proves that rounding the predicted rating can decrease the MAE values for all tested recommendation algorithms. In particular, the decrement of MAE values is more inspiring on EachMovie data set.

5.2. The Effectiveness of Our Approaches

In this section, we conduct a series of experiments to compare the performance of our approach by changing the training ratio from 0.2 to 0.9 on MovieLens data set and EachMovie data set. From Fig. 7, it can be observed our proposed approaches achieve better performance on different training set ratios. Specifically, the MAE values those using our approaches are competitive with Basic Rounding. As large majority of absolute difference between original predicted ratings and actual ratings are less than 0.5, there even have not a pinpoint of difference between PDPR_Neighbor and PDPR_Optimal, while PDPR_Optimal have a higher robustness under some specified context.

6. Conclusion

In this work, we demonstrate the reason why rounding can bring different influences for MAE and RMSE metrics, and prove that rounding is necessary in post-processing of the predicted ratings and improving the performance and quality of the recommendation. Additionally, we also propose two rounding approaches based on the predicted ratings' probability distribution of each rating level, which can be used to round the predicted rating to the right integer rating and offer better predict accuracy. Extensive experiments on different data sets confirm the correctness of our analysis and effectiveness of our proposed approaches.

In the future, we will reform our rounding approach in order to improve the recall and precision of Top-K recommendation. And it is hoped that the rounding of predicted rating presented here will prove useful to the CF community as a reference scheme.

7. References

[1] Johns. Breese, David Heckerman and Carl Kadie. Empirical analysis of predictive algorithms for collaborative filtering. In Proceedings of the Fourteenth Conference on Uncertainty in Artificial Intelligence, 43-52, San Francisco, 1998.

[2] Jon Herlocker, Joseph A. Konstan, and John Riedl. An empirical analysis of design choices in neighborhood-based collaborative filtering algorithms. Information Retrieval, 5(4):287-310, 2002.

[3] Li Yong, Xu De-zhi, Zhang Yong and Xing Chun-xiao. MVC-based Incremental Reengineering Approach. Journal of Chinese Computer Systems, 29(3):469-472, 2008.

[4] Gediminas Adomavicius and Alexander Tuzhilin. Toward the next generation of recommender systems: A survey of the state-of-the-art and possible extensions. IEEE Transactions on Knowledge and Data Engineering, 17(6):734-749, 2005.
[5] Fatih Gedikli, Dietmar Jannach. Recommending Based on Rating Frequencies. ACM Conference on Recommender Systems, 233-236, Barcelona, 2010.
[6] Thomas Hofmann. Collaborative filtering via Gaussian probabilistic latent semantic analysis. ACM Special Interest Group on Information Retrieval, 259-266, Toronto, 2003.
[7] Rong Jin, Luo Si, ChengXiang Zhai and Jamie Callan. Collaborative Filtering with Decoupled Models for Preferences and Ratings. ACM International Conference on Information and Knowledge Management, 309-316, New Orleans, 2003.
[8] Daniel Lemire and Anna Maclachlan. Slope One Predictors for Online Rating-Based Collaborative Filtering. In Proceedings of the 5th SIAM International Conference on Data Mining, 471-475, Newport Beach, 2005.
[9] Yehuda Koren, Robert Bell, and Chris Volinsky. Matrix factorization techniques for recommender systems. IEEE Computer, 42(8):30–37, 2009.

Acknowledgments
Acknowledging assistance and encouragement from Data Platform Project Group (Door) , special work by technical staff or financial support from CFETS Information Technology (Shanghai) Co., Ltd.

IOP Conf. Series: Materials Science and Engineering **324** (2018) 012050 doi:10.1088/1757-899X/324/1/012050

Research on an innovative design model

Y Fu[1] and H Fang[1]
1.Art design College Guangdong University of Technology
Corresponding author: Yang Fu
E-mail: tcts@163.com

Abstract: The design methods of furniture are different from east to west; it has been the hotspot of the scholars. However, in terms of the theory of modern design innovation, neither the early creation theory, the modern design theory, nor the widely applied TRIZ theory can fully fit the modern furniture design innovation, so it is urgent to study the modern furniture design theory. This paper is based on the idea of TRIZ theory, using lots of literatures as data, and uses the method of statistical stratification to analyze and sort out the research of modern sitting equipment, and finally put forward the modern furniture design model, which provides new ideas and perspectives for the modern design of Chinese furniture.

1.Introduction

Furniture is one of the representatives of Chinese traditional products, is also the embodiment of ancient Chinese ideas of creation. It can be said that the development of furniture is accompanied by the development of the Chinese nation. Modern design furniture is the continuation of ancient Chinese furniture design. Chinese modern design began in the 1980s, which is subordinate to the traditional design of the creature. The design theory of Chinese modern creation originates from two aspects: one is inheriting the tradition of Chinese ancient creation, and continuing the basic design idea of "Unity of Methods and implements" and "Unity of Man and Nature". The second is from the Western "borrowlism" based on mathematics, physics, natural science evolved from modern design methods. These two aspects finally formed two kinds of thoughts of "the art theory of creation" and "the industrial theory of creation ", and influenced the development of Chinese modern design.

Although the two concepts of "art" and "industry" have different influences on the design of modern products, they all have the same characteristics in nature. First, both of them pay attention to the inheritance of culture. Creationism takes the traditional culture as the main body, embodies the humanistic spirit and artisan spirit. The design of science is standing on the "design culture". The difference is that the culture of "Creationism" refers to the essence and extension of traditional culture, while the culture of "Culturalism" refers more to the complex and multi-meaning modern broad sense culture. Second, both of them focus on the study of needs. The former emphasizes the social needs and emphasizing the practical; while the latter is emphasis on the relationship between people and things, by studying this relationship to derive a new demand, emphasizing that people's demands in different states. Third, both of them attach importance to the designer's knowledge and skills, but the former focus on the "folk art maternal theory", emphasizing the folk arts and crafts skills, and taking traditional court technology, literati and doctoral crafts and religious skills and other folk arts and crafts, as the main body and root of the traditional skills; The latter is based on the reorganization the knowledge of designers, with modern technology as a method, emphasizing the reorganization of resources[1]. Skills are the relationship between people and objects, providing services for different states of people.

In terms of the theory of "art" and "industry", the former is based on the traditional crafts, emphasizing the inheritance, which is an important method to study the ancient artifacts, but is lack of cohesion of modern design, and cannot guide the design and innovation of modern industrial production. While the latter is based on the modern, using modern methods specify the way of development for the modern design. However, it separated with the traditional design, ignoring the local experience, even ignoring the traditional and regional of culture.[2]

In the overall historical and cultural perspective, traditions should not be cut off and cannot be cut off. There are only a few intermittent records in the history about the development of technology and the continuation of the style. Meanwhile, under the "westernized" modern education, design has become a simple imitation of the western countries "advanced" mode of production, marketing and other forms.

2.METHODS

2.1Analysis of TRIZ method

TRIZ theory is a design theory which is used to solve the problem of design inheritance and innovation. Its founder Altshuller has always believed that innovation activities have their inherent laws, and not just based on cognitive science and creative thinking psychology theory that innovation activities only scientists and engineers' inspiration flashed.[3] Led by Altshuller, tons of research staff comes from research institutions in the Soviet Union, and its affiliated universities and enterprises, based on the 2.5 million natural sciences and technology patent documents, doing systematic and detailed analysis, and extracted 40 thousand patents as the key research, summarized and refined. And integrated the principles and rules of the multidisciplinary field, eventually found that the improvement of the product in any field, the change of technology and the innovation are all the same as the biological system. There exist the cycle of production, growth, maturity, aging and death. On the basis of this, the TRIZ theory is established, and the solving algorithm of invention is described. And after three stages of development, it has developed into a set of theoretical and practical solutions to the problem of the invention, which is based on the principle of technology system evolution.[4]

The TRIZ theory, however, has its scope of application as well as any other theory. The era of these 2.5 million patent documents selected by TRIZ theory is just before the high-tech revolution. At that time, regardless of new energy, new materials, or biotechnology are all in a weak period of development. However, the largest categories of innovation design were from the field of machinery and computer fields at that time. It can be inferred that the 2.5 million pieces of natural science and technology patent literature, mechanical and computer related literature should be in the majority, which is also matching the condition that the TRIZ theory is more applicable to the field of mechanical design and computer field. The vast majority of examples that are cited by Altshuller in introducing TRIZ theory are closely related with the mechanical technology case. However, for the chemical, electronics, medical science and other more abstract technical issues rarely involved.

In spite of this, the origin way of TRIZ theory that through a large number of investigation and selection, is worthy of our thinking and learning, which identify research points by some statistical methods in deepening research, summarizing and refining the principle and rule for the comprehensive multidisciplinary field, and use the algorithm to solve and ultimately determine the rules and methods of innovation design, and then continue to revise and improve, gradually be applied to the actual design.

2.2Furniture design literature statistics

The research method of modern furniture design is based on the research method of TRIZ theory. The author search traditional furniture, traditional furniture design, modern furniture, modern furniture design, four groups of key words by Guangdong University of Technology CNKI mirror database. The data from 2005 to 2015 is shown in Table , and the search type is subject search. The proportion of research on traditional furniture and modern furniture by modern scholars is basically the same, among them, the number of modern furniture design papers is higher than the number of traditional furniture design papers, but the study of furniture design is only a small part of the study of furniture.

Modern furniture has received more attention from the academic circle by comparing the circulation of traditional furniture and modern furniture articles from 2005 to 2015. In recent years, the circulation number of articles on modern furniture has been maintained at about 100 per year.

2.3Data selection method

The choice of data is the key problem in the research. In order to effectively pick up and deal with the data, the author uses the Analytic Hierarchy Process (AHP) which was proposed by Professor **T L Saaty** of University of Pittsburgh in the United States to analyze the literature in quantitative and qualitative way.[5] The literature is divided into different levels, to form a multi-level structure. At the same time, the judgment matrix is established by comparing each layer according to a certain criterion. The weight of the layer element for the criterion is obtained by computing the maximum eigenvalue of the judgment matrix and the corresponding eigenvector. On the basis, we can get the values of different ideas by calculating the combined weight of each level of elements for the overall objective, and taking them as the reference for the selection.[6]

According to this method, 832 articles of traditional furniture and 1215 articles of modern furniture are selected as the main data for this study. In these 2047 selected papers, journals are the largest, with 1653 articles. Conference papers are the least, only 18. The other includes 210 master's theses, 149 newspaper papers and 17 doctoral dissertations.

By calculating the weight value, we elected 17 doctoral thesis, 195 master's thesis, and 1532 journals as the final sample.

Table 1 Retrieval literature quantity

Subject Retrieval	Quantity
Traditional furniture	6149
Traditional furniture design	1162
Modern furniture	6854
Modern furniture design	2815

Table 2 Literature details

	traditional furniture	modern furniture	total
doctoral thesis	11	6	17
master's thesis	78	132	210
Journal article	633	1020	1653
conference papers	7	11	18
Newspaper article	103	46	149
total	832	1215	2047

2.4Research on Data Classification

Factor analysis can break down the original variables in the selected literature and derive the underlying "category" from which is more relevant to indicators. This paper will use this method to study the concerns of scholars in the furniture. The factor analysis principle assumes that there are p variables, $X_1, X_2, \cdots, X_p$ and the observations of p variables in n samples constitute an N × P-order raw data matrix[7]

$$X = \begin{bmatrix} X_{11} & X_{12} & \cdots & X_{1p} \\ X_{21} & X_{22} & \cdots & X_{2p} \\ \vdots & \vdots & \vdots & \vdots \\ X_{n1} & X_{n1} & \cdots & X_{np} \end{bmatrix}$$

In order to simplify the comparison between the variables, the standardized variables are set before performing factor analysis. We can assume that the standardized variable is Z1, Z2, ⋯, Zp. The p standardized variables can be represented by p new standardized variable factors F1, F2, ⋯, Fp. We use the equation (1) as follows:

$$z_j = a_{j1}F_1 + a_{j2}F_2 + \cdots + a_{jp}F_p,\ (j = 1, 2, \cdots, p) \quad (1)$$

In the formula a_{ij} $(i, j = 1, 2, \cdots, p)$ is the factor load matrix of matrix A. And choosing m factors with larger variance contribution (the cumulative contribution rate of m factors is above 85%).

At this time, equation (1) can be written as follows:

zj =aj1F1 +aj2F2 +⋯ +ajmF m +ej（j =1 , 2 , ⋯ , m (2)

In the formula, E1, E2 , ⋯ ,EP is the error term.

Using the above method, the 17 doctoral thesis have been divided into four categories (Furniture and life relationship, Culture and elements, Design theory, Style and decoration). From table 3, we can see their respective numbers.

In the same way, 195 master's theses are divided into 16 categories. The classifications and number of each category can be seen in table 4. It depicts their respective proportions in the total. Meanwhile, 1532 journal articles are divided into 29 categories. The classifications and number can be seen in Table .

Table 3 PhD thesis classification

Classification	Quantity
Furniture and life relationship	2
Culture and elements	4
Design theory	4
Style and decoration	7

Table 4 Master's thesis classification

Number	Classification	Quantity
1	Style and decoration	45
2	Furniture materials	22
3	Furniture structure style	21
4	Design theory	20
5	Literature Review	16
6	Culture and elements	14
7	Interior furnishings	12
8	Functional furniture	12
9	Special furniture	11
10	Ecological environment	5
11	Parts & Accessories	4
12	Furniture color	3
13	Furniture Marketing	3
14	Modern Technology Application	3
15	Human Engineering	2
16	Personalization	2

Table 5 Journal article classification

Number	Classification	Quantity
1	Literature Review	264
2	Culture and elements	159
3	Style and decoration	129
4	region and nationality	93
5	Furniture structure style	78
6	Furniture materials	63
7	Design theory	57
8	Chinese and Foreign Contrast	55
9	Functional furniture	55
10	Inheritance and innovation	54
11	Ecological environment	45
12	Special furniture	39
13	Furniture collection	38
14	Furniture Marketing	34
15	Personalization	33
16	manufacturing technology	32
17	Interior furnishings	32
18	Computer-aided	32
19	modern technology	28
20	Furniture designer	27
21	Furniture color	27
22	Parts & Accessories	24
23	Life style	22
24	Standardization	18
25	Human Engineering	17
26	Furniture evaluation	17
27	mechanical property	15
28	Parameterization	8
29	Designers cross-border	6

2.5Data research conclusion

Compared with these abundant papers, we can find that the scholars contribute a lot to the studies in furniture culture, the furniture structure style and the furniture style decoration, which are the hotspots in modern research. Furniture materials, design principles, functional furniture and other research directions have also been the favor of scholars. In addition, the research directions like inheritance and innovation of furniture, computer-aided design, and designer cross-border have also become new hot spots. In the sample, there are a large number of literature reviews on furniture development, the same classification also includes furniture collection, regional ethnic, Chinese and foreign contrast, furniture marketing, interior furnishings and other aspects. These cannot help us to study the traditional furniture design concerns, and should be excluded.

The remaining classifications are reintegrated into four directions according to their contents. These four aspects cover 23 research hotspots in the furniture literature ,they are design constraints, design source, design skills, design presentation, respectively.[8] At the same time, according to the nature of the design, design constraints are the impact from the outside; design sources are the user's needs; design skills are the designer's skills; design presentations are the presentation of the product.

3.DISCUSSION

3.1Design constraints: Potential

In the above data, the design constraints are made up of many aspects, such as culture and elements, style decoration, ecological environment, heritage, innovation, etc. These aspects are not visible, and we cannot see them with eyes, but they actually affected the design. These invisible constraints and influence can be called “potential” in Chinese culture. We often say that “Size up the situation”, “the wave of the future” and “momentum” in Chinese words. "Potential" is a kind of invisible influence for the surroundings, and it is also called as "field" in physics. Here, it is "the culture", it is "the environment”, it is the general name of what affects the design; it is a dimension of space and time.

3.2Design presentation: Pattern

In the above data, the design presentation of furniture contains structure style, materials, colors, special furniture, parts and accessories etc. Design presentation is a kind of external performance, and is the final pattern of furniture products. In the traditional culture, pattern represents the form of the objects existence. The Pattern of creation refers to specific conformation and state of existence, which is shown by the created things. It is a combination of design elements which mankind makes meaningful creation of subjective initiative.

3.3Design sources and skills: Principle

Design source and skills serves for the pattern, which are the embodiment of human factors. Design sources are the internal representation of the relationship of user and the furniture pattern, and design skills are the external representation of the relationship of designer and the furniture pattern. Traditional objects continue to inherit and innovation according to users’ requirement in different periods. Users’ requirements are divided into material needs and immaterial needs, which are corresponding to material factors and immaterial factors, at the same time, designers need to have knowledge and skills to meet material factors and immaterial factors.[9] This relationship between people and products is called “principle “in science of design.

The relationship between designers and users is changing all the time. In the primitive society, design appeared to meet human’s needs. Designers and users were the same person in the beginning when products related to human survival. These products’ designs are successful as they can play a role in the critical time, although their shapes are very simple.

3.4Relationship among the three

The French philosopher Sia had put the traditional artifacts into a multidimensional system to define. He believes that material culture and spiritual culture are all changing in a dynamic process and each of the traditional artifacts is bearing the change of the brand. [10]Creative culture always takes the human’s basic needs as a basis research spot. It not only concerns about the external structure of the material, but also includes various cultural phenomena in creation activity. It has a positive significance for the development of new life styles.

Today we refine architecture culture not only from the "pattern" of product, but also putting the "pattern" in the " principle" of the associated system, ultimately, we have to put the "pattern" and " principle" into the space of "potential" to interpret. "Pattern" shows “principle” and "principle" is used to explain product; and "principle" guides "potential" and "potential" is used to discuss people. "Interpretation of the Great Learning" refers to: "Things have their root and branches." "Principle" and "product" have been always developed simultaneously in Chinese culture. From the view of design point, the "principle" we talk about is the purposeful activity of human beings and the resulting of social phenomenon, a hidden structures which is behind the creative design and the fundament and causes "pattern" change. In a certain conditions, human purposeful activity must match up to the specific spatial "potential". It happens in behavioral interaction or information exchange, among people, or between human and everything[11]. In this process, people's consciousness has produced a certain meaning, and things have changed their "pattern" and it is consistent with the goal of science of design to create a richer human life style. Through the research on the external factors, we explore a new way of life which explores the cultural values of the dimensions; by adjusting the internal factors, create a suitable internal structure of external environment to manifest itself in the form of an object.

"Pattern" is the external display of "principle", which exists before things and determines existence of "pattern"; "principle" is an internal extension of "potential", which controls "principle" and gives the rule of "principle". Thus, it is not only to determine an object which is not another, but also as a link with the relationship between people and things, using a whole and dynamic vision to study the evolution and logic of design, as well as the relationship between the various elements in the system.

4.CONCLUSIONS

The furniture design model conforms to the traditional design features. Traditional design is based on three basic logic points: the first is the ruling systems logic, the second is the social relationship logic; and the third is the using logic of things. These three logical concepts are related to the three factors in the modern design: environmental factors, human factors and material factors. Design constraints are the modern extension of the traditional logic system, including culture, environment, style, genre, tradition, inheritance, and many other factors, which is also a modern continuation of the traditional system logic.

As Professor **G Z Liu** emphasizes that design is a kind of relationship[12], and emphasizes that "potential" determines the design expression and strategy. Modern design is based on the inheritance of the traditional design, the relationship among "potential", people and objects is just like the relationship among human, machine and environment in the early stage of industrial design.

The research of creation culture is proposed by summarizing the "traditional creation" and "modern industry" on the creation culture, it is a kind of theoretical research, which aims to excavate the material and spiritual relationship behind the ancient artifacts and summarizes the law of the development of the traditional artifacts, bringing inspiration for the modern design[13]. Science of design is a methodology, which can be considered as a research method that is like a kind of knowledge system for guiding practice. It also points out the research model of the lifestyle and the research methods are more diverse than traditional creation.

They belong to different research levels; one is "ancient research ", and another is "modern research", but the purpose of them is all for the future design. "Traditional and modern" just like "the East and the West", they are not antagonistic, but should be unified. In essence, they are consistent with the starting point, research methods and directions in design. They integrate the content on human lifestyle and creative intelligences in lives, through some research methods to develop the rules. As shown in figure 1, the furniture design model better reflects the relationships in modern furniture design. "Potential" is used to cover the constraints factors for design, "pattern" is used to cover the presentation factors for design, and "principle" is used to express the design sources and design skills which are reflected in human factor.

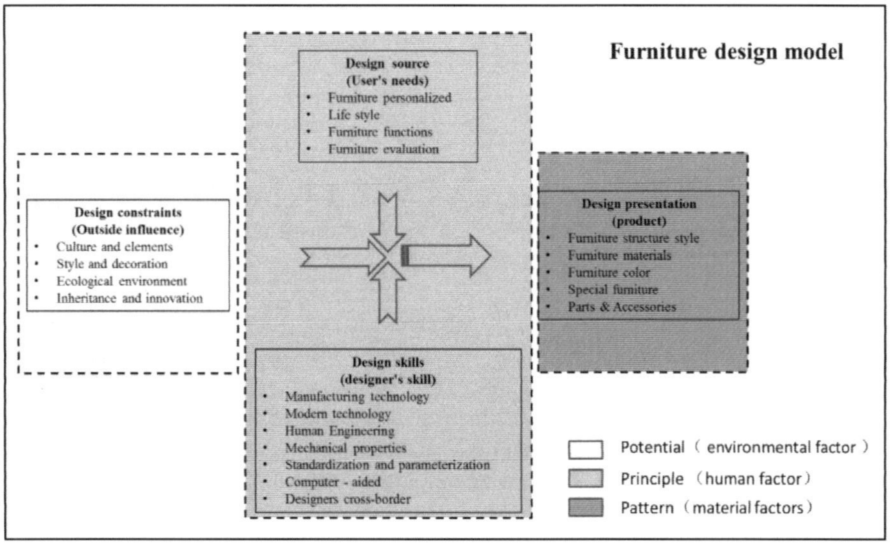

Figure 1 Furniture design model

---

[1] Nathanson, S., (1996) The culture of design, International Journal of the Legal Profession, 3(3), pp.301-313.
[2]Sanderson, H. W. (1936). Art and industry, Journal of the Royal Society of Arts, 84(4343),pp.380-381.
[3] Jiang, F., (2010) Application idea for TRIZ theory in innovation education, International Conference on Computer Science and Education pp.1535-1540.
[4] Zhang, J. Y., Guo, Y. L., Yang, S. C., & Wang, Y. W. et al., (2009). Innovation design of products based on triz theory, Journal of Machine Design, 26(2),pp.35-38..
[5]Braunschweig, T. Gijsbers, G., Janssen, W., Odame, H. H., & Meijerink, G. et al., (2013). Analytic hierarchy process, Linear Models in the Mathematics of Uncertainty, Springer Berlin Heidelberg, pp.93-117
[6]Vaidya, O. S and Kumar, S., (2006). Analytic hierarchy process: an overview of applications. European Journal of Operational Research, 169(1),pp.1-29..
[7] Mislevy, R. J., (1986) recent developments in the factor analysis of categorical variables, Journal of Educational & Behavioral Statistics, 11(1), pp.61–67.
[8]Yu, S ,(2001) Furniture design trends in the 21st century, Journal of Northeast Forestry University, 2(1), pp.82-84.
[9] Garner, S. W. (1992) the need for human factors in design and technology, Design & Technology Teaching, 24(2).pp.14-17
[10] Scott, K. (2000) Art and industry, Journal of Design History,Vol. 13 Issue 1, pp.1
[11] Veryzer, R. W and Mozota, B. B. D., (2005) the impact of user-oriented design on new product development: an examination of fundamental relationships, Journal of Product Innovation Management, 22(2), pp.128–143.
[12] Guanzhong, L.,(2010)Design——The Third Wisdom Preventing Our Future from Being Destroyed, Packaging Journal, 02(3), pp.9-13.
[13] Lu, M andYang, J., (2010) Notice of Retraction Inspiration of Chinese traditional culture to modern industrial design, IEEE, International Conference on Computer-Aided Industrial Design & Conceptual Design, Vol.2, pp.968 – 971.

IOP Conf. Series: Materials Science and Engineering **324** (2018) 012051 doi:10.1088/1757-899X/324/1/012051

Mobility and Position Error Analysis of a Complex Planar Mechanism with Redundant Constraints

Qipeng Sun[12] and Gangyan Li[1]

1 School of Mechanical & Electronic Engineering, Wuhan University of Technology,Wuhan 430070, China
2 Luoyang Institute of Science and Technology, Luoyang 471003,China
E-mail:sqpqq728@163.com

Abstract. Nowadays mechanisms with redundant constraints have been created and attracted much attention for their merits. The mechanism of the redundant constraints in a mechanical system is analyzed in this paper. A analysis method of Planar Linkage with a repetitive structure is proposed to get the number and type of constraints. According to the difference of applications and constraint characteristics, the redundant constraints are divided into the theoretical planar redundant constraints and the space-planar redundant constraints. And the calculation formula for the number of redundant constraints and type of judging method are carried out. And a complex mechanism with redundant constraints is analyzed of the influence about redundant constraints on mechanical performance. With the combination of theoretical derivation and simulation research, a mechanism analysis method is put forward about the position error of complex mechanism with redundant constraints. It points out the direction on how to eliminate or reduce the influence of redundant constraints.

1. Introduction

Currently, many closed mechanisms with redundant constraints have been created and have attracted much attention due to their merits, such as high stiffness, large workspace and simple structure [1,2]. As the uncertainty of redundant constraints in mechanical system are widespread, they have seriously effected on the mechanical performance of mechanism. And it often leads to the contradiction between a high precision of a mechanical system and low cost of product realization process [3-6]. And redundant constraint structure brings some problems such as poor flexibility, low control performance. However, redundant constraint structure is inevitable for the objective need of mechanical system to work normally.The accuracy of the position and posture of mechanism is one of the main performance indicators. It is directly related to the quality of work, especially in some cases when it is requiring higher precision, such as the existence of redundant constraint mechanism. The main representative methods to determine redundant constraints are screw theory method [2,7], linear transformation method [8] and screw algebraic method [9]. Screw theory method obtains the mobility by solving the movements spiral and analyzing spiral-order correlation. Linear transformation method only proposes calculation expression of the number of redundant constraints, while the constraint nature of redundancy judgment is not discussed.These methods can simultaneously determine the type and location of the redundant constraints, but they are too complex to learn and use for general mechanical designers.

2. Classification and Calculation of the Redundant Constraints

The existence of redundant constraints in planar linkage mechanism leads to a high sensitivity error, and causes a series of harmful effects on the mechanical work performance. In order to eliminate or reduce the redundant constraint and its harmful effects, it is important to judge correctly the existence of redundant constraints, including the number and type. This paper adopted the modified Grübler-Kutzbach[2, 7] formula:

$$M = 6(n-1) - \sum_{i=1}^{g} u_i + v \tag{1}$$

Where M is the prescribed DOF of mobility in the closed mechanism; n is the number of the links including base; g is the number of kinematic pairs; u_i is the ith constraint number of kinematic pair; v is the number of all the redundant constraints. It is known that, when v =0, Eq. (1) becomes Grübler-Kutzbach formula which can be used to solve DOF of the closed mechanisms without any redundant constraints. A formula for calculating the number of the redundant constraints of a closed mechanism is derived from Eq. (1) as below:

$$v = M - 6(n-1) + \sum_{i=1}^{g} u_i \tag{2}$$

Figure 1 presents a planar mechanism with DOF=1 , which shows that one of the constraints, in fact, does not work. It is referred to as "virtual constraint" or "repeat constraints". However ,When Eq. (2) is adopted to calculate the redundant constraints this mechanism, the result is: v=7(1-6 • 4+6 • 5=7). This indicates that 7 redundant constraints are present in the mechanism.Actually, it happens when the configuration of kinematic pair for some special requirements, or mechanical constraints that structure parameters must satisfy some special geometric conditions. In a situation of theological accuracy (mechanism without error), redundant constraints hardly effect the freedom and movement of the mechanism. According to the different constraint characteristics, redundant constraints are divided into 2 different kinds: the theoretical planar redundant constraints (v_1) and the space-planar redundant constraints (v_2).

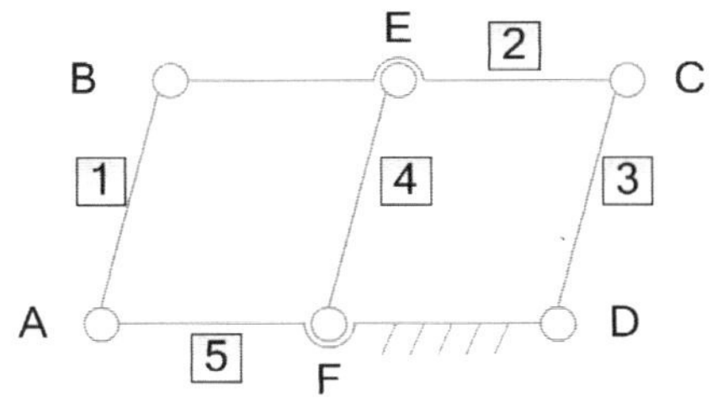

Figure 1. A planar mechanism with redundant constraints.

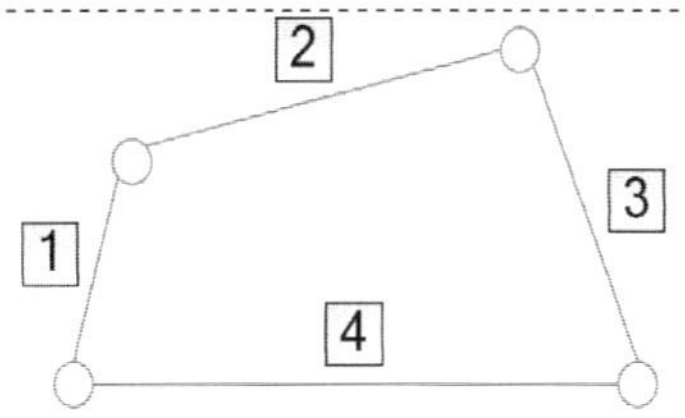

Figure 2. A planar 4-bar linkage.

In order to obtain more uniform movement of the mechanism, the EF rod provides a constraint to the parallelogram ABCD with DOF=1. With a condition of EF / / = AB, BE / / =AF, it does not affect the mobility of the original parallelogram mechanism. Therefore it is a redundant constraint and can be calculated by theological planar mechanism structure formula, namely the redundant constraint is v_1 = 1 (1 + 2·6-3·4 = 1). Therefore the redundant constraints in this mechanism are called the theoretical planar redundant constraints. The essence of being theoretical planar redundant constraints is that the trajectory of point E must be accurately equal to the circle whose center is F and radius is EF. Otherwise, it really can not move and becomes statically indeterminate structure. And the identity of the two tracks relies on AB = CD = EF, BC = AD, BE = AF. If anyone of them can not accurately meet the above conditions, it will cause movement disorders or make some component deformation. So this kind of redundant constraints requires a high structure dimensional accuracy. As shown in figure 2: A planar four-bar linkage, if the axis of the four revolute joins are parallel to each other, and perpendicular to the plane motion. We call it ideal planar mechanism. The theoretical planar redundant constraints can be calculated v_1 = 0 (1 + 2·4-3·3 = 0). It indicated that there is no theoretical planar redundant constraint in the mechanism. However, with the inevitable error from manufacturing, assembling and other factors, the mechanism can not satisfy the ideal movement conditions of a planar mechanism. Then it is supposed as a spatial mechanism, and its v_2=3(1-6·3+5·4=3) consisting of 1 translational redundant constraint and 2 rotational redundant constraint.

3. Redundant Constraints Analysis of Complex Mechanism

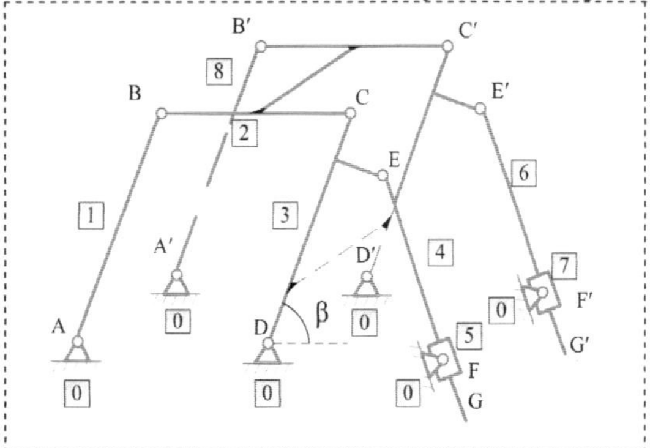

Figure 3. A complex mechanism

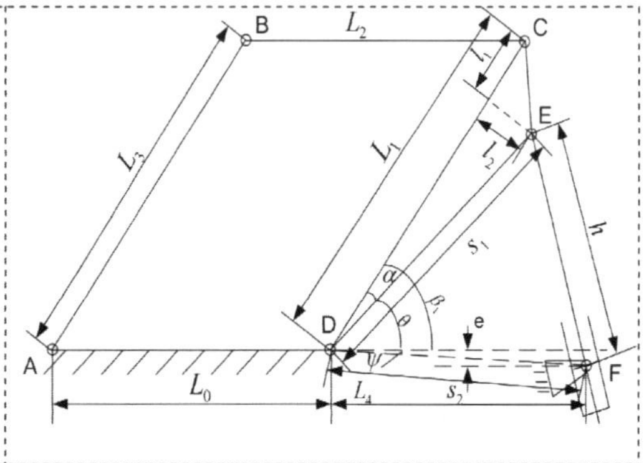

Figure 4. Planar six-bar mechanism.

The mechanism with repetitive structure can be regarded as two planar six -bar mechanism connected by bodies (2, 3). According to the principle of the redundant constraints analysis, the complex mechanism with redundant constraints is analyzed as below:

- Space - planar redundant constraints of two six-bar planar mechanism with DOF=1. According to Eq. (2), the number of space - planar redundant constraints of a planar six -bar mechanism (Fig.4)is 6 (1-6·5 + 5·7 = 6). So the total of space - planar redundant constraints v_2 = 6 · 2 = 12.
- Theoretical planar redundant constraints make the motion law consistent. The trajectory of points (B, B') is exactly the same and is expressed by function y = f (x). The independent variables is 1, so the theoretical planar redundant constraints v_1 = 1.

4. Position Error Analysis of the Complex Mechanism

According to the analysis of the redundant constraints, mechanism with redundant constraints (especially theoretical planar redundant constraints) is sensitive to the dimension error of the component. Therefore a higher requirement of position error on the mechanism is also put forward. The mechanism (figure 3) is so symmetrical, the movement form is consistent, that here we take a six-bar planar mechanism as research objective. The analysis model is shown in figure 4.

According to the geometric relations in figure 5, the expression of angle of L_1 is:

$$\beta_1 = \arccos\left(\frac{s_1^2+s_2^2-h^2}{2\cdot s_1\cdot s_2}\right)+\arccos\left(\frac{s_1^2+(L_1-l_1)^2-l_2^2}{2\cdot s_1\cdot (L_1-l_1)}\right)-\arccos\left(\frac{L_4^2+s_2^2-e^2}{2\cdot L_4\cdot s_2}\right) \tag{3}$$

Where Li is the length of ith bar; s_i and e are structural dimensions as shown in figure 5; h is elongation of the input bar. Eq. (3) reveals the relationship between the angle of CD bar and the elongation of the input components and the structural dimensions. It is inevitable that error exists in the process of manufacturing, assembling and working. At the same time there is an interaction between components and constraints. As a result, the output bar often fails to follow a theological motion trajectory and a position error is produced. It is supposed that the angle between BC and the horizontal direction is β_2, the position error calculation chart is shown in figure 5.

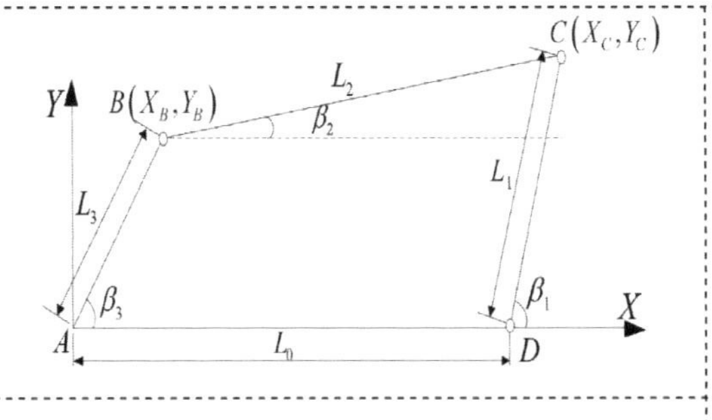

Figure 5. Position error calculation chart

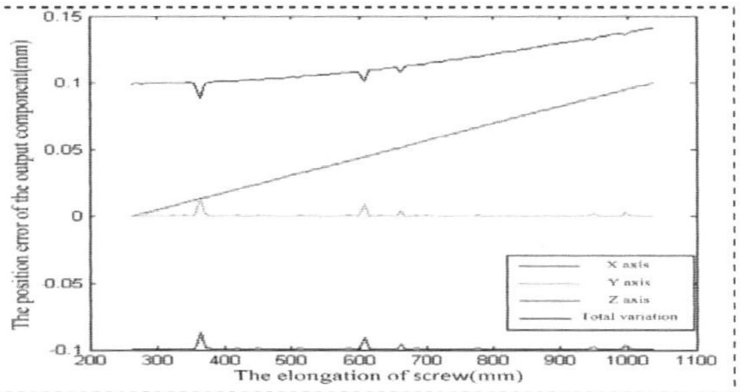

Figure 6. Curves of the position error.

Fig.5 presents a four-bar planar mechanism consisting of L_0、L_1、L_2、L_3. The bodies are connected by four revolute joints (A,B,C,D). The global, inertial reference frame XAY is established on the basis L_0. Where β_1 is the input Angle, and β_2 , β_3 are the output Angles. The vector equation is set up as below:

$$\begin{cases} X_B + L_2 \cdot \cos\beta_2 - L_1 \cdot \cos\beta_1 - L_0 = 0 \\ Y_B - L_3 \cdot \sin\beta_3 = 0 \\ X_C - L_1 \cdot \cos\beta_1 - L_0 = 0 \\ Y_C - L_1 \cdot \sin\beta_1 = 0 \\ X_B - L_3 \cdot \cos\beta_3 = 0 \\ Y_C - L_2 \cdot \sin\beta_2 - Y_B = 0 \end{cases} \tag{4}$$

Where β_1, β_2 , β_3 is the angle of CD, BC and AB rad; (X_B,Y_B) is the position coordinate of the point B, mm; (Xc,Yc) is the position coordinate of point C, mm; L_0, L_1, L_2, L_3 are structure parameters, mm. The partial derivatives of the parameters (β_1 ;L_0, L_1, L_2, L_3; β_2 , β_3)are derived from Eq.(4) as A,B and C :

$$\mathbf{A} = (L_1 \cdot \sin\beta_1, 0, L_1 \cdot \sin\beta_1, -L_1 \cdot \cos\beta_1, 0, 0)^T \; ;$$

$$\mathbf{B} = \begin{bmatrix} -\cos\beta_1 & \cos\beta_2 & 0 & -1 \\ 0 & 0 & -\sin\beta_3 & 0 \\ -\cos\beta_1 & 0 & 0 & -1 \\ -\sin\beta_1 & 0 & 0 & 0 \\ 0 & 0 & -\cos\beta_3 & 0 \\ 0 & -\sin\beta_2 & 0 & 0 \end{bmatrix} ; \mathbf{C} = \begin{bmatrix} 0 & 0 & 1 & 0 & -L_2 \cdot \sin\beta_2 & 0 \\ 0 & 0 & 0 & 1 & 0 & -L_3 \cdot \cos\beta_3 \\ 1 & 0 & 0 & 0 & 0 & 0 \\ 0 & 1 & 0 & 0 & 0 & 0 \\ 0 & 0 & 1 & 0 & 0 & L_3 \cdot \sin\beta_3 \\ 0 & 1 & 0 & -1 & -L_2 \cdot \cos\beta_3 & 0 \end{bmatrix} \tag{5}$$

Error column vector of input and structural parameters are as follows:

$$\mathbf{\Delta D} = (\Delta\beta_1)^T \; ; \; \mathbf{\Delta E} = (\Delta l_1, \Delta l_2, \Delta l_3, \Delta l_0)^T \tag{6}$$

According to the matrix equation, we get the expression of error of the output parameters as below:

$$\{\mathbf{\Delta F}\} = [\mathbf{C}]^{1}([\mathbf{A}]\{\mathbf{\Delta D}\} + [\mathbf{B}]\{\mathbf{\Delta E}\}) \tag{7}$$

Using MATLAB editor to solve vector equations to get the corresponding simulation curve. The simulation results are shown in figure 6.

As shown in figure 6, with the gradual increase of elongation, the position error values along the X-axis fluctuate around -0.1mm. The curve overall shows a fluctuant and the value changes little. When the screw elongation changes between 300-400 mm, the curve fluctuates no more than 0.014mm. The amount of change is so small that it is negligible. The change of the position error curve along the Y-axis is similar to that along with X-axis. It fluctuates around 0mm. With increasing elongation of the screw, the position error along the Z-axis direction shows a linear upward trend. The total amount of change in Z-axis direction is 0.1mm. This is the biggest factors which lead to the position error of the entire output mechanism.

5. Conclusion

The mechanism of the redundant constraints in a mechanical system is analyzed in this paper. According to the difference of applications and constraint characteristics, the theoretical planar redundant constraints and the space-planar redundant constraints are defined. And the calculation formula for the number of redundant constraints and method of type- judging are carried out. We find that space - planar redundant constraints are mainly sensitive to the error of axis of rotational pair and the theoretical planar redundant constraints are sensitive to the error of dimension. Respectively, a typical mechanism and a complicated mechanism are investigated as examples. With the above method, the number and type of redundant constraints are analyzed in order to further validate the rationality and practicality of the proposed method.

At the same time, a complex mechanism with redundant constraints is analyzed of the influence about redundant constraints on mechanical performance. A mechanism analysis method is put forward to the position error of complex mechanism with redundant constraints. After analyzing the driving mechanism and actuator, a mathematical model of position error is set up. The relationship of elongation of screw and the output linkage is obtained by simulation calculation. The simulation results will provide guidance for further research on how to eliminate and reduce impact of redundant constraints on the mechanism.

References

[1] Gogu, G., 2008, Structural Synthesis of Parallel Robots: Part 1: Methodology, Springer, New York

[2] Huang, Z., Liu, J., and Zeng, D., 2009, "A General Methodology for Mobility Analysis of Mechanisms Based on Constraint Screw Theory," *Sci. China*, Ser. E: Technol. Sci., **52(5)**, pp. 1337–1347

[3] Q. Zeng, K. F. Ehmann, Error Modeling of a Parallel Wedge Precision Positioning Stage. *Journal of Manufacturing Science and Engineering* **134**, 061005 (2012)

[4] S. Yun Cho, H. Keun Lee, C. Gook Park, Adaptive Redundant Inertial Measurement Unit/Global Positioning System Integration Filter Structure for Fault Tolerant Navigation. Journal of Dynamic Systems, *Measurement, and Control* **136**, 021009 (2013)

[5] Gan, D., Dai, J. S., and Liao, Q., 2009, "Mobility Change in Two Types of Metamorphic Parallel Mechanisms," Trans. ASME, *J. Mech. Rob.*, **1(4)**, p. 041007

[6] P. Haghighi, P. Mohan, J. J. Shah, J. K. Davidson, A Framework for Explicating Formal Geometrical and Dimensional Tolerances Schema From Manufacturing Process Plans for Three-Dimensional Conformance Analysis. Jo*urnal of Computing and Information Science in Engineering* **15**, 021010 (2015)

[7] LI Yanwen, WANG Lumin, LIU Jingfang, et al. Applicability and generality of the modified Grubler-Kutzbach criterion[J]. *Chinese Journal of Mechanical Engineering*, 2013, **26(2)**: 257-263

[8] ZHANG Yitong, LU Wenjuan, MU Dejun, et al. A novel mobility formula for parallel mechanisms expressed with mobility of general link-group[J]. *Chinese Journal of Mechanical Engineering*, 2013, **26(6)**: 1082-1090

[9] D. P. Bonny, M. L. Hull, S. M. Howell, Optimized design of an instrumented spatial linkage that minimizes errors in locating the rotational axes of the tibiofemoral joint: a computational analysis. *Journal of biomechanical engineering* **135**, 31003 (Mar 1, 2013)

IOP Conf. Series: Materials Science and Engineering **324** (2018) 012052 doi:10.1088/1757-899X/324/1/012052

The fluidized bed separator

R R Sharapov [1], E V Kharlamov [1] and V V Yadykina [2]

[1] Moscow State University of Civil Engineering, Yaroslavskoe shosse, 26, Moscow, Russia
[2] Belgorod state technological university named after V. G. Shukhov, Kostyukova str., 46, Belgorod, Russia

ABSTRACT
In the production of building materials produces a large amount of harmful substances, poisoning atmosphere and adjacent to the building materials site. One of the biggest polluting the ecology of cities adjacent to industry, is metallurgical industry. One of the most important technological processes is the enrichment, in which the rocks are extracted empty components. In the enrichment of iron ore in the production an increasing proportion consist the anthropogenic waste industries, which pollute the air basin and the huge territory around the factories that dumped these same wastes. This reduces as a space for human life and for the functioning and development of cities. These materials can be applied as construction materials, for example, in the preparation of fine-grained concrete as a mineral powder, and these materials can be applied in the construction of roads, housing, for receiving paint etc.

1 INTRODUCTION
Currently, mining enterprises are mainly used wet scheme of separeting raw materials and drum separators [1, 2]. For wet magnetic separation of strongly magnetic ores are used in drum-type separators PBM with a multipolar system of permanent magnets and is available in three versions: with direct-flow, counterflow and semi-counterflow baths [3, 4]. Direct-flow separators are used for materials with the grain size of 6 mm or less, counterflow ones – material fineness 2...3 mm and less, semi-counterflow ones – for material with particle size of 0.3 mm or less [5-7].

The above magnetic separators are separators for wet separation. They have the following negative drawbacks: the presence of water environment with a higher resistance compared to air; water environment has a strong opposition to the movement of magnetic and non-magnetic particles. As result the magnetic particles mix with nonmagnetic ones. When separating weakly magnetic materials is not observed magnetic flocculation (the formation of fortified units by the mutual attraction of the magnetized particles). It causes the need for feeding environment – water (waste water).

The source of waste water and recycled water are the drains dehydrating, washing machines and tailings. Polluting impurities are solids, hardness salts, heavy metal ions and organic substances. Untreated waste water that contains suspended impurities and aggregates, are the cause of the violation of the ecological system with all of its negative effects: shoal and dryen rivers, the vegetation dries, fades surrounding life [8, 9]. Turnover out a huge area, where merge slimes from the enrichment process of iron ore [10]. For the treatment of waste water and impurities has applied mechanical, chemical, physico-chemical and biochemical methods. All this leads to additional costs and cost of finished products.

An alternative of wet method is a dry method of separation. The lack of feeding environment provides significant water savings, there is no need of sewage treatment and eliminates the need for sludges, which occupy a considerable area and negatively affect to the atmosphere and lithosphere [11, 12]. Dry separation is the least energy intensive, as it is based on the natural property of magnetic attraction of ores.

2 METHODS

To improve the efficiency of dry separation tailings is possible due to the overlapping processes of magnetic extraction of ferrous particles and air stiring separeting the material [9, 12].

Based on these requirements we have developed and manufactured an experimental installation for studying the process of separation of a two component mixture. Photo of the current installation is shown in figure 1.

The experimental installation consists of the following main components: the belt conveyor 1, the conveyor unloading the magnetic particles 2, the scraper conveyor 3, the transporting tray 4, loading device 5, the stabilization system 6, the control panel 7, the discharge zone of the magnetic particles 8 and the zone of unloading non-magnetic particles 9. The belt conveyor is designed to extract from the mixture of the magnetic particles. Traction circuit inside the belt conveyor along the lower branches of the conveying tape hinged block of magnets. Belt conveyor set at an angle relative to the transporting tray. This situation creates a belt conveyor in relation to moving in the transported tray of a mixture of variable magnetic field from the minimum at the beginning of the extraction to the maximum at an end of the drum. Accordingly, the extraction process of the particles occurs during the period of movement of the mixture under the active section of the belt conveyor in the beginning to extract strongly magnetic particles and decreasing the magnetization of the particles.

Figure 1. Experimental installation

The transporting tray bottom made in the form of a box. The upper surface of the transporting tray made of permeable to air of the porous material. Passing through the porous material air forms a layer of compressed air between the surface of the porous material and the mixture, with the result that the air passes through the mixture creates the effect of fluidization. Separation of a mixture into fractions in the regime of fluidization increase the mobility of the particles. This increases the probability of their capture by the magnetic field, improves the recovery ratio of magnetic particles that reduces the possibility of occurrence of particles on the partition and coefficient their resistance movement of the particles.

3 RESULTS

Extracting the ability of the separator of the fluidized bed depends on the characteristics of the magnetic system, physical and mechanical properties of the particles of the separeted mixture and the geometric parameters of the working area. Recovery ratio of magnetic particles η is determined by the ratio of their volumes at the entrance to the active zone G (x_A) and on the way out of it (x_B) (figure 2):

$$\eta = 1 - \frac{G(x_B)}{G(x_A)}. \quad (1)$$

Extracting the particles in the separation zone causes aerodisperse stream of magnetic particles. They is moved along the working area of the separator fluidized bed at a speed v of the scraper conveyor. The concentration of magnetic particles in this stream are distributed across the height of the working area is very uneven: it sharply decreases from a maximum value in a boiling layer to a minimum value in the zone of deposition to the transport conveyer belt.

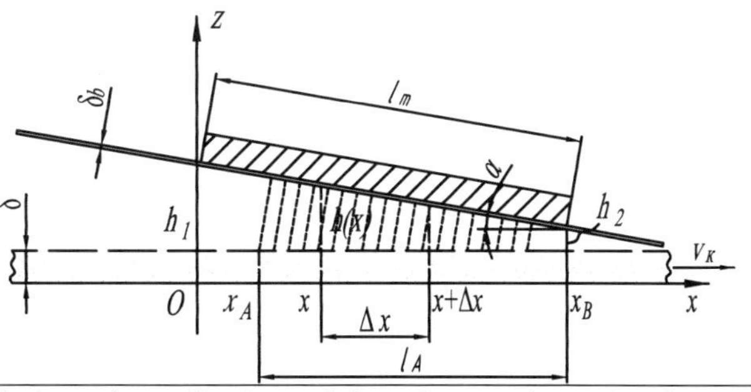

Figure 2. Calculation scheme for determining of extraction ratio of magnetite particles

Consider a part of the working area of the separator fluidized bed between the cross sections passing through the points x and $x+\Delta x$ (see figure 2). As a result of precipitation of magnetic particles, their concentration decreases from $C(x)$ to $C(x+\Delta x)$. Material balance equation of magnetic particles for the selected part of the working area:

$$G(x)-G(x+\Delta x)=\chi C(x)v_1 b\frac{\Delta x}{\cos\alpha}=\chi\frac{G(x)v_1\Delta x}{h(x)v\cos\alpha}. \quad (2)$$

Expressing the change volume magnetic particles through its differential, we obtain:

$$G(x)-G(x+\Delta x)=-(G(x+\Delta x)-G(x))\approx -dG(x), \quad (3)$$

and dividing the variables in equation (1), we get:

$$\frac{dG}{G}=-\frac{\chi v_1 dx}{v\,(h_1-x\mathrm{tg}\alpha)\cos\alpha}. \quad (4)$$

Integrated equation (4), we obtain:

$$\ln G\Big|_{G(x_A)}^{G(x_B)}=\frac{\chi v_1}{v\,\sin\alpha}\ln(h_1-x\mathrm{tg}\alpha\,)\Big|_{x_A}^{x_B}. \quad (5)$$

From relation (5) follows:

$$\frac{G(x_B)}{G(x_A)}=\left(\frac{h_1-x_B\mathrm{tg}\alpha}{h_1-x_A\mathrm{tg}\alpha}\right)^{\frac{\chi v_1}{v\sin\alpha}}. \quad (6)$$

Substituting expression (6) in the formula (1), we obtain a relation for the fractional extraction ratio of magnetic particles:

$$\eta(d)=1-\left(\frac{h_2}{h_2+l_A\mathrm{tg}\alpha}\right)^{\frac{\chi v_1}{v\,\sin\alpha}}. \quad (7)$$

For incoming in the formula (4) length of the active portion of the working area l_A we obtain:

$$l_A=-\frac{h_2-1{,}115\delta}{\mathrm{tg}\alpha}-\frac{1}{2c\sin\alpha}l_n\left(\frac{\rho g\cos\alpha}{\mu_0\chi\,cH_0^2}\right). \quad (8)$$

Parameter l_A and the precipitation velocity of particles on the transport conveyer belt v_1 (formula 10) takes the form:

$$l_A=-\frac{0{,}015-1{,}115\delta}{\mathrm{tg}\alpha}-\frac{\ln(0{,}183\cos\alpha\,)}{52{,}34\sin\alpha}, \quad (9)$$

$$v_1=4{,}472\cdot 10^{-4}d^2, \quad (10)$$

where the particle size d should be set in microns.

Taking into account formulas (9) and (10) the expression (7) can be rewritten in the form

$$\eta(d) = 1 - \left(\frac{0{,}015}{1{,}115\delta - \frac{\ln(0{,}183\cos\alpha\ )}{52{,}34\cos\alpha}} \right)^{\frac{4{,}472\cdot 10^{-4} d^2 \chi}{v\ \sin\alpha}} . \quad (11)$$

Comparison of theoretical and experimental values of the full coefficient of extraction of magnetic particles shows that the magnitude of the coefficient of heterogeneity of the distribution of their concentration in the working area of the separator fluidized bed χ depends mainly on the thickness of the partial mixture in the initial bound state δ. As a result of processing of experimental data obtained the following dependence:

$$\chi(\sigma) = 50\delta^2 - 0{,}15\delta + 0{,}0224 . \quad (12)$$

Substituting the formula (12) in (11) we obtain the final expression for the fractional extraction ratio of magnetic particles in an experimental fluidized bed separator:

$$\eta(d) = 1 - \left(\frac{0{,}015}{1{,}115\delta - \frac{\ln(0{,}183\cos\alpha)}{52{,}34\cos\alpha}} \right)^{\frac{4{,}472\cdot 10^{-4} d^2 (50\delta^2 - 0{,}15\delta + 0{,}0224)}{v\sin\alpha}} . \quad (13)$$

From formulas (4) and (5) we can see that the recovery ratio of magnetic particles depends on their physic-mechanical properties (density, size, shape, magnetic permeability), the characteristics of the magnetic system (its size, magnetic field strength, spacing of poles), the constructive-technological parameters of fluidized bed separator (dimensions, tilt angle of the magnetic system, the thickness of the partial layer of the mixture, the speed of movement of the scraper conveyor), but also properties of the air environment (temperature, viscosity).

To confirm the above presented experimental installation we conducted an experimental study. The used material is tails Lebedinsky mining plant. In the course of the experiment changed the following parameters: layer thickness δ of the separating material; moving velocity v of the separating material; the angle of inclination of a belt conveyor α and the magnetic field strength H.

4 CONCLUSIONS

Figure 3 shows the dependence of the effect of velocity of movement v of the separating material on the separation efficiency η under different values of the angle α of inclination of a belt conveyor, a constant layer thickness of the separating material δ = 8 mm and magnetic field H = 40 kApm Here we can see that all the dependences are of an extreme nature in the studied range of variation of the factors.

Analyzing graphs we can conclude that the maximum value of the separation efficiency η is achieved when the velocity of the separating material v = 0.02 mps, angle of installation of a belt conveyor α = 15° and η is 85.77 %.

With the aim of improving environmentally friendly production of materials metallurgical production under separation tailings with a high degree of effectiveness proposed scheme dry processing. It is proved that for these purposes can be applied by the magnetic principle of extraction of particles and airs tirring of separating material.

For conducting experiment we designed and constructed experimental installation of fluidized bed separator.

Experimental studies confirm that the proposed in article separator fluidized bed dry separation is more effective in comparison with traditional separators. Efficiency exceeds 90 %.

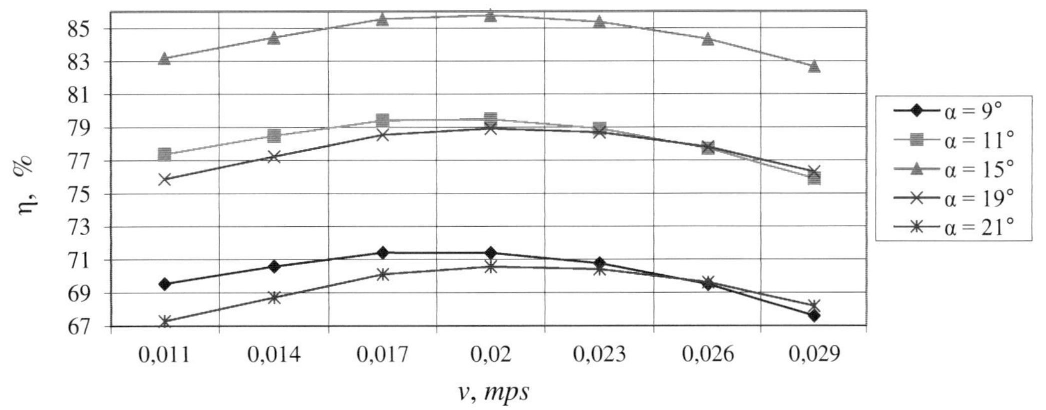

Figure 3. Dependence of the separation efficiency η from of the moving velocity of the separating material *v* under different values of the angle of inclination of a belt conveyor α, a constant layer thickness δ = 8 mm and magnetic field *H* = 40 kApm

REFERENCES

1. Bhagat R P and Pathak P N 1996 The effect of polymeric dispersant on magnetic separation of tungsten ore slimes *International Journal of Mineral Processing* V 47 № 3-4 pp 213-217

2. Kaljuvee T, Kuusik R and Veiderma M 1995 Enrichment of carbonate-phosphate ores by calcination and air separation *International Journal of Mineral Processing* V 43 № 1-2 pp 113-121

3. Lovás M, Murová I, Mockovciaková A, Rowson N and Jakabský S 2003 Intensification of magnetic separation and leaching of cu-ores by microwave radiation *Separation and Purification Technology* V 31 № 3 pp 291-299

4. Zhang H, Chen L, Zeng J, Ding L and Liu J 2015 Processing of lean iron ores by dry high intensity magnetic separation *Separation Science and Technology* V 50 № 11 pp 1689-1694

5. Simakov V A and Isaev V E 2015 X-ray fluorescence determination of magnetite iron in iron ores after preliminary magnetic separation *Inorganic Materials* V 51 № 14 pp 1438-1441

6. Turkenich A M, Lapshin E S and Dudnik V I 1997 Continuous barrier separation of weakly magnetic ores in the field of repulsive magnetic force *Magnetic and Electrical Separation* V 8 № 2 pp 81-87

7. Potapenko V E, Suvorova D I and Tyuryukhanova V V 1981 Beneficiation of magnesite ores by froth separation *Refractories and Industrial Ceramics* V 22 № 3-4 pp 146-148

8. Skorokhodov V F, Mesyats S P and Ostapenko S P 2011 Purification problem solution of waste water of industrial enterprises from multi-component contamination *Eurasian Mining* № 1 pp 58-60

9. Kharlamov E V, Sharapov R R and Yadykina V V 2015 On environmental friendliness increase for metallurgical production *International Journal of Applied Engineering Research* V 10 № 24 C 45108-45114

10. Melik-Gaykazov I V, Druz Yu G, Shchetkin V A, Komyagina Yu A and Skorokhodov V F 2012 The problem aspects of the protection of regional hydrosphere from the pollutted discharges of industrial waters of kovdor ore dressing and processing enterprise *Eurasian Mining* № 2 pp 58-60

11. Oshitani J, Franks G and Griffin M 2010 Dry dense medium separation of iron ore using a gas-solid fluidized bed *Advanced Powder Technology* V 21 № 5 pp 573-577

12. Sharapov R, Kapyrin P, Lozovaya S, Yadykina V and Agarkov A 2016 Research dedusting efficiency of the inertial hub with adjustable parameters *MATEC Web of Conferences 5. "5th International Scientific Conference "Integration, Partnership and Innovation in Construction Science and Education", IPICSE 2016"* 03007

Development of Semi-Automatic Lathe by using Intelligent Soft Computing Technique

S Sakthi [1], J Niresh [2], K Vignesh [3], G Anand Raj [4]

[1,2,3,4] Department of Automobile Engineering, PSG College of Technology, India.

E-mail: sakthisenthilkumar.95@gmail.com

Abstract: This paper discusses the enhancement of conventional lathe machine to semi-automated lathe machine by implementing a soft computing method. In the present scenario, lathe machine plays a vital role in the engineering division of manufacturing industry. While the manual lathe machines are economical, the accuracy and efficiency are not up to the mark. On the other hand, CNC machine provide the desired accuracy and efficiency, but requires a huge capital. In order to over come this situation, a semi-automated approach towards the conventional lathe machine is developed by employing stepper motors to the horizontal and vertical drive, that can be controlled by Arduino UNO -microcontroller. Based on the input parameters of the lathe operation the arduino coding is been generated and transferred to the UNO board. Thus upgrading from manual to semi-automatic lathe machines can significantly increase the accuracy and efficiency while, at the same time, keeping a check on investment cost and consequently provide a much needed escalation to the manufacturing industry.

1. Introduction

Lathe is a machine tool used principally for shaping articles of metal by causing the work piece to be held and rotated by the chuck while a tool bit is advanced into the work causing the cutting action[1]. The basic lathe that was designed to cut cylindrical metal stock has been developed further to produce screw threads, tapered work, drilled holes, knurled surfaces, and crankshafts as shown in Figure 1. The typical lathe provides a variety of rotating speeds and a means to manually and automatically move the cutting tool into the work piece. Some of the operations done by a lathe machines are tabulated below in Table 1.

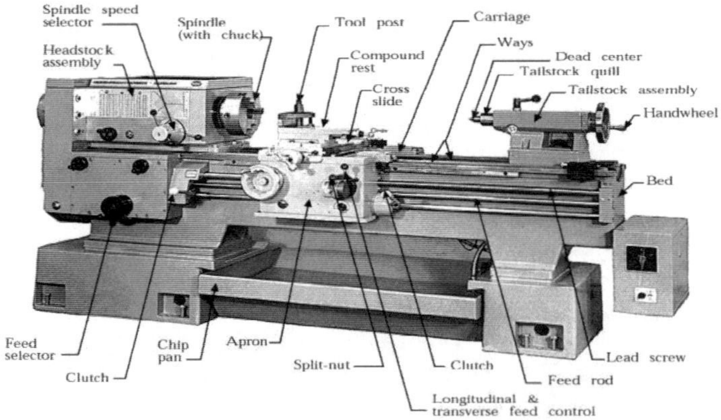

Figure 1. Conventional Lathe Machine [1]

IOP Conf. Series: Materials Science and Engineering **324** (2018) 012053 doi:10.1088/1757-899X/324/1/012053

Table 1. Operations done by Lathe Machine[6]

Operation	**Description**
Turning	To produce straight, conical, curved, or grooved work pieces.
Facing	To produce a flat surface at the end of the part or for making face grooves.
Boring	To enlarge a hole or cylindrical cavity made by a previous process or to produce circular internal grooves.
Drilling	To produce a hole by fixing a drill in the tailstock
Threading	To produce external or internal threads
Knurling	To produce a regularly shaped roughness on cylindrical surfaces

2. Soft-Computing Technique used

Arduino is an open-source prototyping platform based on easy-to-use hardware and software. Arduino boards are able to read inputs - light on a sensor or a finger on a button, and turn it into an output - activating a motor, turning on an LED. The user can tell the board what to do by sending a set of instructions to the microcontroller [4]. To do so the user should use the Arduino programming language (based on Wiring), and the Arduino Software (IDE), based on Processing. The Uno is a microcontroller board based on the ATmega328P [3].

2.1. Arduino UNO

The Uno differs from all preceding boards in that it does not use the FTDI USB-to-serial driver chip. Instead, it features the Atmega16U2 (Atmega8U2 up to version R2) programmed as a USB-to-serial converter. The Uno has a resettable poly-fuse that protects your computer's USB ports from shorts and overcurrent [2]. Arduino board is shown in the Figure 2.

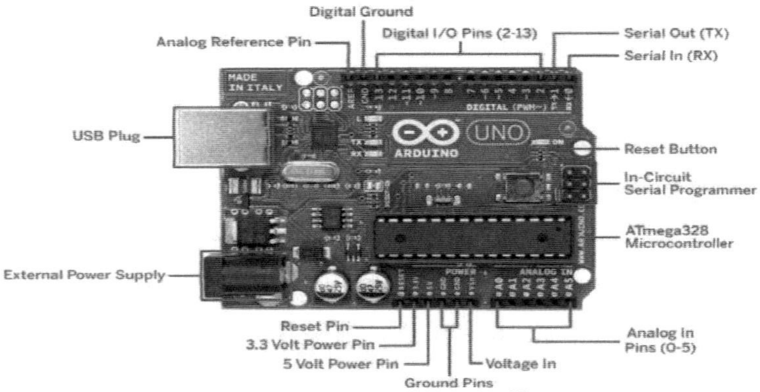

Figure 2. Arduino UNO[4]

3. Input Parameters

The specifications of micro-controller(Arduino UNO) & the lathe machine is tabulated in Table 2 & Table 3.

Table 2. Arduino UNO- specifications[4]

Microcontroller-Specification	
Microcontroller	ATmega328P
Operating Voltage	5V
Input Voltage	7-12V
PWM Digital I/O Pins	6
Analog Input Pins	6
DC Current per I/O Pin	20 mA
DC Current for 3.3V Pin	50 mA
Flash Memory	32 KB(ATmega328P)
SRAM	2 KB(ATmega328P)
EEPROM	1KB(ATmega328P)

Table 3. Lathe Machine-Parameters[5]

Lathe Machine	
Swing over machine bed	Min. 350mm
Swing over cross slide	Min. 190mm
Distance between centers	750-800mm
Turning Length	700-750mm
Width of bed	Mini.260mm
Normal Chuck diameter	160mm
Guide Length of carriage	350-365
Cross-slide travel	175-190
Width of cross-slide	140mm
Cross Travel	+= 10mm

4. Experimental Details

4.1 Principle Of Arduino Lathe :

Embedded System - is one that has hardware with software embedded in it as one of its important components.

4.2 Construction Of Arduino Lathe

The essential electronics used in the construction of arduino lathe is tabulate in Table 4 .

Table 4. Essential electronic parts used in the construction of Arduino Lathe

Components	Description
Motor	Motors are employed in both vertical & horizontal drive (i.e) Depth of cut & Lead Screw Movement/Rotation.
Rectifiers	Rectifiers are used to convert the AC into DC voltage.
Transformer	Transformers are used to rise or lower the voltage for motors & arduino board.
Relay Board	Relay acts as a switch, that is controlled by a small electromagnet which can be activated by a small signal. In other words, the relay is used to power on -high powered devices such as large motors with a signal from the Arduino.
Diode	Diodes acts like a low value resistor and allows the current flow.
Microcontroller	Microcontrollers are used to control the motor employed in feed & depth drive.

4.3 Working- Algorithm of Arduino lathe

The design phase of arduino lathe is developed using Solid Works software as shown in Figure 3. The coding is generated and compiled using Arduino IDE software as shown in Figure 4.The step by step working procedure of arduino lathe is tabulated in the Table 5.

IOP Conf. Series: Materials Science and Engineering **324** (2018) 012053 doi:10.1088/1757-899X/324/1/012053

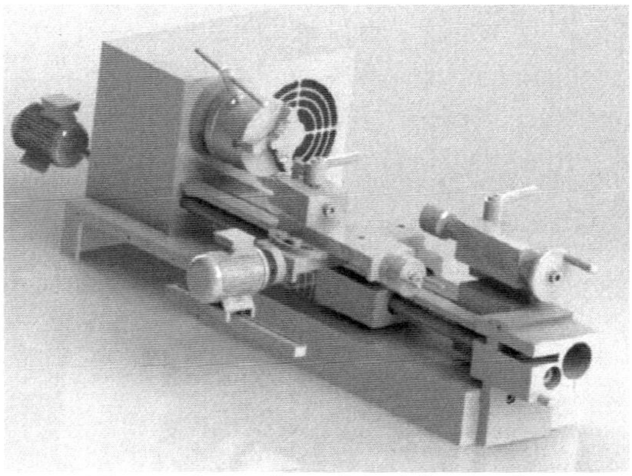

Figure 3. Design of Arduino Lathe

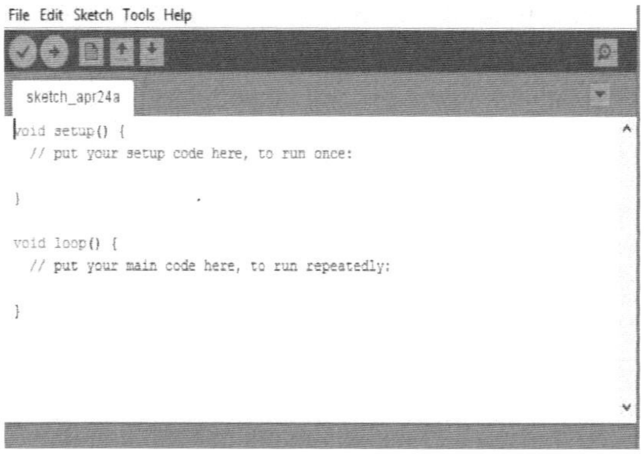

Figure 4. Arduino IDE software –Coding

Table 5. Working Algorithm for Arduino Lathe

Steps	Function
Step 1	The "Arduino Software IDE" is Initialized as shown in Figure 4.
Step 2	Based on the input parameters, the calculations are done for the required machining Process.
Step 3	With these results, the Arduino coding are generated for required Profile/Operation with specified Definition, Void Setup and Void Loop.
Step 4	Now the coding is compiled. So that, it ensures the coding generated is error free.
Step 5	These Coding are transmitted to the Arduino board by means of USB cable or accessed by means of its IP address.
Step 6	Mean while , the Work piece is mounted on the chuck of arduino lathe as shown in Figure 3.
Step 7	Now, the arduino reads the program and based on the coding generated it initiate the manufacturing process/operation.
Step 8	After the process gets over, if necessary the arduino can be reset for the new process/operation.

5. Formula used for Calculation :

The optimum cutting conditions (i.e. speeds, feed and depths of cut) depended upon the tool & work material for the required surface finish and the dimensional accuracy. The optimum cutting conditions used to determine the required spindle speeds for the proposed dimensional information about the work pieces to be machined .From the equation (1), (2),…(7), the formula required for calculation is achieved[6]. So that these results can be incorporated while generating Coding for desired lathe operation.

5.1 Determination of Power Requirements

The drive power required for the cutting motion depends upon the cutting speed and the cutting force. The power needed for the feed drive is also dependent upon the feed force and the feed rate. The cutting forces mainly depended upon the work material, the size of cut, tool geometry, and cutting speed and to extend the tool material. From the analysis of the forces in turning, it was found that the cutting forces in turning could be resolved into three components namely:

IOP Conf. Series: Materials Science and Engineering **324** (2018) 012053 doi:10.1088/1757-899X/324/1/012053

The tangential component : F_t-it is in the direction of the cutting velocity vector.

$$F_t = R_1.(F^x)_1.(d^y)_1 \; kgf \quad (1)$$

The radial component : F_r-is in the direction of the radius of the job.

$$F_r = R_2.(F^x)_2.(d^y)_2 \; kgf \quad (2)$$

The axial component : F_a-is in the direction of the longitudinal feed

$$F_a = R_3.(F^x)_3.(d^y)_3 \; kgf \quad (3)$$

Therefore the rate of work needed in turning approximately equals to the product of tangential force component and the cutting velocity.

The horsepower, needed for turning : $P = (F_t \text{ x } V_{max})/4500 \; hp$ (4)

The power rating, of the cutting drive : $P_{RC} = P / \eta_C \; hp$ (5)

The power rating, of the feed drive : $P_{RF} = (Q \text{ x } F) / \eta_f$ (6)

The feed force, for lathe operations : $Q = k.F_a + \beta(F_c + F_r + W_d)$ (7)

Where, η_C is the mechanical efficiency, β is the coefficient of friction between the carriage and the guide ways, k is a recommended factor, W_d is the weight of the parts being traversed, F_a,F_c and F_v are the components of the cutting force in the feed, cutting and radial directions respectively.

6. Conclusion

Therefore, the design of Semi-automatic lathe is successfully developed by the Solid Works Software. Which has the communion of both hardware(Arduino UNO) & software(Arduino IDE) resulting in the development of semi-automated approach towards the conventional lathe machine. As automation newly developed Arduino lathe is done by implementing some new features to the standard lathe, therefore the setup cost is increased, but when compared to the fully automated/CNC machine the setup cost is much low. The accuracy of the job manufactured in semi-automated lathe is high relatively, so the repeatability and dimensional stability of the manufactured part is achieved. Since the production rate is high, the semi-automated lathe machines will be very useful in mass production.

7. References

[1] Prakash N. Parmar , Prof. N. C. Mehta , Prof. Manish V. Trivedi, “*Investigation on Automation of Lathe Machine”,* International Journal of Emerging Technology and Advanced Engineering-IJETAE, ISSN 2250-2459, Vol.**4**, Issue 5, May 2014, page no. 524-529.

[2] Akshay R Sonawane, Arun Bhiva Rane and D. S. S. Sudhakar, “*Development Of A3-Axis Cnc Milling Machine With An Open Source Controller*”, IJRET, eISSN: 2319-1163 | pISSN: 2321, Aug-2017.

[3] S.Vigneshrao, “*Design Of 3-Axis Computer Numerical Control (Cnc) Router Using Stepper Motor*”, International Journal of Global Engineering (IJGE), E-ISSN: 2456-3099, Vol.**1**, Issue 3, March-2017.

[4] Kulkarni Bharat P, “*Arduino Based 3 axis PCB Drilling Machine*”, International Journal of Emerging Technologies in Engineering Research (IJETER). Vol. **4**, Issue 6, June (2016).

[5] Jahnavi Madireddy, "*Importance of Lathe Machine in Engineering Field and its usage*", Global Journal of Researches in Engineering: A Mechanical and Mechanics Engineering, Vol.**14**, Issue 6, Year 2014, ISSN: 2249-4596 , Print ISSN: 0975-5861

[6] Sundar Pandian and S. Raj Pandian, “*A Low-Cost Build-Your-Own Three Axis Cnc Mill Prototype*”, International Journal on Mechanical Engineering and Robotics (IJMER), ISSN (Print): 2321-5747, Vol.**2**, Issue-1, 2014.

Optimisation Of Cutting Parameters Of Composite Material Laser Cutting Process By Taguchi Method

S Lokesh[1], J Niresh[2], S Neelakrishnan[3], S P Deepak Rahul[4]

[1,2,3,4]Department of Automobile Engineering, PSG College of Technology, Coimbatore, India

lokeshsridhar96@gmail.com

Abstract: The aim of this work is to develop a laser cutting process model that can predict the relationship between the process input parameters and resultant surface roughness, kerf width characteristics. The research conduct is based on the Design of Experiment (DOE) analysis. Response Surface Methodology (RSM) is used in this work. It is one of the most practical and most effective techniques to develop a process model. Even though RSM has been used for the optimization of the laser process, this research investigates laser cutting of materials like Composite wood (veneer)to be best circumstances of laser cutting using RSM process. The input parameters evaluated are focal length, power supply and cutting speed, the output responses being kerf width, surface roughness, temperature. To efficiently optimize and customize the kerf width and surface roughness characteristics, a machine laser cutting process model using Taguchi L9 orthogonal methodology was proposed.

Keywords : Laser Cutting Machine, Composite Materials, Taguchi's Method, Minitab.

1. Introduction :

LASER cutting is a precise method of cutting a design from a given material using a CAD file to guide it. The parameters of laser cutting process play a main role on the cutting edge quality features. But, there are main obstacles that limit the laser extensive use in applications of routine material processing such as a beam size limitation, high cost of installation and replacement, additional and costly accessories, and need for skilled workers. Researchers studied the impact of different laser parameters on cut edge quality, kerf dimensions, roughness and surface finish. There are studies of the process parameters impacts, such as laser power, spot size and dimensions, and cutting speed on the resulting kerf size. Yilbas[1] studied the effect of laser power and oxygen gas pressure on the kerf width variation. Ghany and Newishy[2] indicated a positive relationship between laser cutting quality and the laser power, pulse frequency, cutting speed. Dubey &Yadava[3], Samant & Dahotre[4].There are machines capable of interpolating up to five axes, simultaneously. These type of machines are designed for cutting complex forms and their programming is much more complex than the 2D case. The cut can be executed in two dimensions (2D),for instance on metal sheets, or in three dimensions (3D) in the case of components that need to be cut in width, length and height such as tubes ICS-UNIDO2008[5]. The most widely used industrial lasers for the cutting of sheet metals are gaseous CO_2, solid state Nd:YAG and recently fibre. The gas laser CO_2 and solid state Nd:YAG laser have low beam power, better efficiency and good beam quality Choudhury & Shirley[6]. The fibre gains too much interest and development recently. There has been growing interest in recent years in the use of pulsed Nd:YAG lasers for precision cutting of thin sheet metals because of its high intensity, low mean beam power, good focusing characteristics, and narrow Heat Affected Zone (HAZ)[4]. The CO_2laser is usually more expensive than the Nd:YAG laser Sivarao et al.,[7]. CO_2 laser systems are available with reasonably a good beam quality at high output powers sufficient for thick-section metal cutting Wandera[8].

2. LASER CUTTING MACHINE:

A LASER cutting machineas shown in fig. 1. is a machine which is used to cut the materials with high accuracy and with very good surface finish. It is used to engrave many kinds of non-metal materials, such as acrylic, bamboo products, cloth, plastic, marble, glasses, woods, PVC. etc. This engraver can process the plexi glass products, acrylic display panel lenses, wood carving, wood percutaneous flowers, advertising products, crystal characters, packing boxes, models, toys and furniture. The table 1. Shows the specifications of LASER cutting machines.

Fig. 1, LASER CUTTING MACHINE

Table 1. machine parameters of LASER cutting machine

Parameter	Specifications
Max. Engraving speed	0-1000 mm/s
Max. Cutting speed	0-600 mm/s
Max. Cutting depth	max. 10mm / acrylic
Drive type	leadshine 3-ph stepper driver and motor
Laser type	Reci Glass Co_2 Laser Tube
Laser power	60w / 80w / 100w
Gross power	<1200 w

3. Material Used For Experiment:

Acrylic is a plastic manufactured using one or more derivatives of acrylic acid. Polymethyl Methacrylate acrylic, or PMMA, is one of the more widely used forms of acrylic due to its exceptional weatherability, strength, clarity and versatility. PMMA acrylic sheet exhibits glass-like qualities – clarity, brilliance, transparency, translucence – at half the weight with up to 10 times the impact resistance. Because it's thermoplastic and softens under extremely high temperatures, acrylic can be formed to virtually any shape. Incredibly durable, acrylic is a suitable solution for over a broad temperature range, and has superior weathering properties compared to other plastics.

4. EXPERIMENTAL DETAILS:

The Taguchi method[9,10] is a technique for optimizing a process or design using multiple parameters. A researcher should always fully understand the various experimental methods in order to properly apply them to individual studies to maximize both the efficiency and the result of a study. The complete Taguchi methods are actually comprised of three main phases, which are all intended to be conducted offline. These three phases include system design, parameter design, and tolerance design. The Taguchi parameter design stage, which is the phase used in study, is commonly referred to here. This phase requires that the factors are known ant that production should be in progress. The major goal of this phase is to increase the performance of the production process by adjusting the controlled factors. The three process parameters, cutting speed(A), feed(B), depth of cut(C) are considered in study. Based on Taguchi method, an L9 orthogonal array(OA) which has 9 different experiments was developed.

Table 2. level of parameters

Section	Factors/parameters name	Level 1	Level 2	Level 3
A	Cutting speed (mm/sec)	225	330	350
B	Feed (mm/rev)	0.05	0.075	1
C	Depth of cut (mm)	1	2	3

5. RESULTS AND DISCUSSIONS:

As referred earlier, in this work L9 orthogonal array was used to carry out the experiments and the experimental results were analysed using Taguchi method. In order to analyse the difference in the results, the signal to noise ratio(S/N ratio) analysis was done with cutting time, kerf widths and temperature as the outputs. The S/N ratio of each levels was calculated and the results were plotted by using Minitab software[11].The following table 3. shows that interference of various input parameter of focallength(8mm,10mm,12mm),cuttingpower(35watts,45watts,55watts),cuttingspeed(15mm\sec,25mm\sec,35mm\sec) in the output parameters such as cutting time, kerf width, temperature.

Table 3. variation of output parameters with respect to input parameters

Focal length (mm)	Cutting Power (watts)	Cutting Speed (mm/sec)	Cutting Time (sec)	Kerf width (mm)	Temperature (◦F)
8	35	15	10	0.28	400
10	35	25	6	0.2	400
12	35	35	4	0.25	400
8	45	25	6	0.28	460
10	45	35	4	0.2	460
12	45	15	10	0.25	460
8	55	35	5	0.28	520
10	55	15	10	0.2	520
12	55	25	6	0.25	520

5.1. PLOTS FOR SN RATIOS:

5.1.1. Cutting Time Graphs

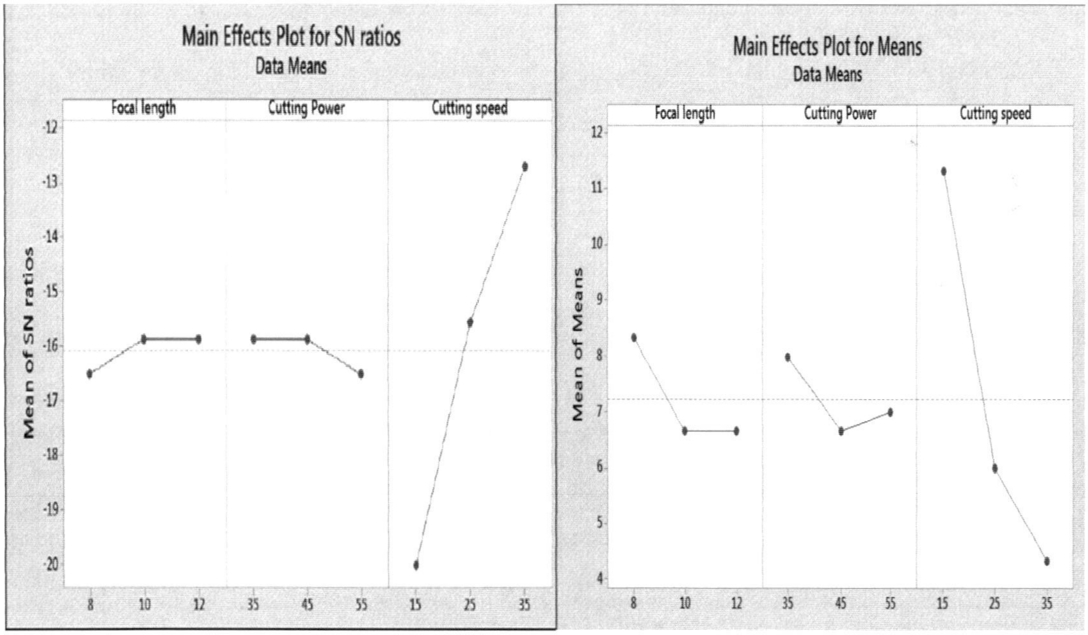

Fig.2. cutting time graphs

From the above graphs it is inferred that the cutting time decreases with increase in cutting speed whereas cutting power should be less for minimum cutting time, and the focal length should be moderate for optimum cutting time.

IOP Conf. Series: Materials Science and Engineering **324** (2018) 012054 doi:10.1088/1757-899X/324/1/012054

5.1.2. Kerf Width Graphs

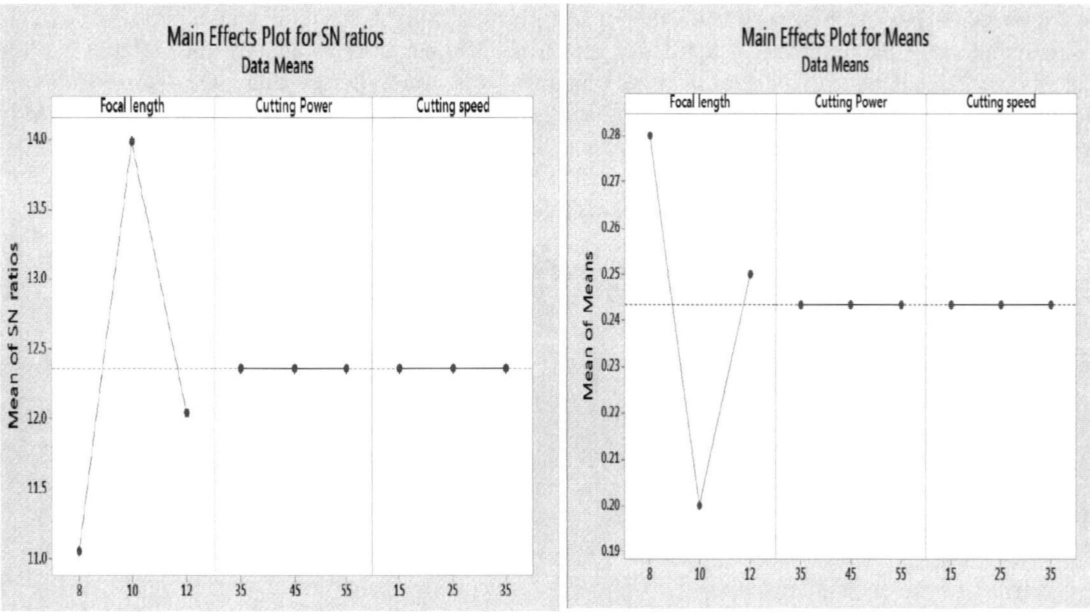

Fig.3. kerf width graphs

From the above graphs it is inferred that for obtaining good Kerf width the cutting power and the cutting speed should be as low as possible and the focal length have to be moderate.

5.1.3. Temperature Graphs

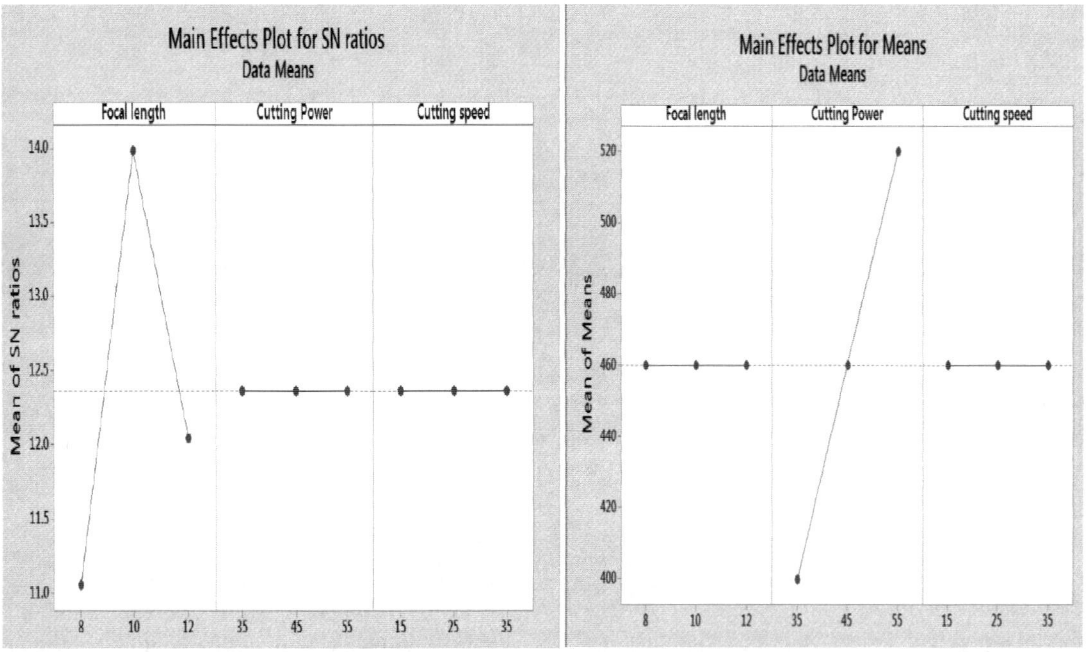

Fig.4, temperature graphs

Form the graph we came to know that for getting the optimum temperature level the focal length and cutting speed should be high and the power should be less.

6. Conclusion:

The experiment presented here is an overview of research work carried out in laser cutting process from the above discussion it can be concluded that Kerf width is best when power and speed are low and focal length is medium Temperature is best when power is medium, speed and focal length are maximum. Cutting time is best when power is minimum, speed is maximum and focal length is low.

7. References:

[1] R Kahraman, M Sunar, B Yilbas - *Journal of materials processing technology*, 20081
[2] Abdel Ghany, K. and Newishy, M., *Journal of Materials Processing Technology*, 168, 438-447
[3] AK Dubey, Vinod Yadava Int. *Journal of Machine Tools and Manufacture* (SCI Journal)
[4] Samant, A. N. & Dahotre, N. B. (2008a) *Ab initio physical analysis of single dimensional laser machining of silicon nitride. Adv. Eng. Mater.*, **10**, 978–981.
[5] ICS UNIDO (2008) Forum for Technology Transfer, Training Course on Technology Management.
[6] I.A.Choudhury, S.Shirley *Optics & Laser Technology* Volume **42**, Issue 3, April 2010.
[7] Sivarao et al., & Sharma &Yadhav *optimisation of parameters of LASER non linear cutting on stainless steel material.*
[8] C Wandera *Optimisation of parameters of Fiber LASER cutting*.
[9] Unitek Miyachi Group, *Welding Material Control,technical application* brief,**2**(1999), pp 1-5.
[10] G. Taguchi, in: *Introduction to quality engineering, Asian productivity organisation*, Tokyo 1990.
[11] Minitab user manual release 16 MINITAB Inc., State college, PA USA, 2003.

IOP Conf. Series: Materials Science and Engineering **324** (2018) 012055 doi:10.1088/1757-899X/324/1/012055

The selection of flying roller as an effort to increase the power of scooter-matic as the main power of centrifugal pump for fire fighter motor cycle

Himawan Hadi Sutrisno

Laboratory of Fire Material and Safety, Fire Safety Engineering Dept. Universitas Negeri Jakarta

himawansutrisno@yahoo.com
Himawan-hadi@unj.ac.id

Abstract. In densely populated settlements, fires often occur and cause losses. In some instances, the process of the occurrence of fires takes place so quickly that to reduce and avoid the occurrence of a fire disaster effort is required in accordance with the existing environmental condition. Fire fighter motorcycle by using motorcycle scooter-matic is considered suitable as one alternative to combating fire hazard in densely populated residential settlements. The use of motorcycle engines as the driving force of the pump often leads to unstable and not optimum power. Thus, the water spray on the centrifugal pump also becomes not maximum. To increase the engine power at scooter-matic engine idle rotation (700-2000 rpm), then the flying roller replacement with certain mass weight becomes an option. By selecting a 10 to 14 gram flying roller mass, the power analysis using a dynotest engine produces several variations. Of the calculation, the mass of a 14 gram flying roller provides a significant increase in motor power on the upper rotation. Meanwhile, on the lower power rotation using a flying roller with a mass of 10 grams provides an increase in power compared to a standard flying roller on a scooter matic motor engine. As a reference to the use of scooter-matic motor power as the pump power, the result of use of the flying roller with a mass of 10 grams becomes the best option.
Keyword: fire fighter motor cycle, dynotest, CVT

1. Introduction

The occurrence of fire disasters often comes and causes huge losses[1], especially in the big cities like Jakarta. Fire incidents recorded until the end of 2017 has reached 281 events with the period at the beginning of 2017 until mid of 2017 (source http://www.jakartafire.net/statistic). With a narrow road condition that does not allow fire trucks to enter the point of fire, the fire fighter motor cycle is one of the solutions that continues to be developed[2]. Under certain conditions, handling and countermeasure of fire hazard are still an interesting topic[3, 4].

This scooter-matic (scootic) is a motorcycle that uses automatic transmission, thus, it is very easy to drive. Automatic transmission or known as Continuously Variable Transmission (CVT) is a transmission that can make us feel comfortable driving because we only need to pull the gas without moving the transmission[5, 6]. by the shifting of paradigm aboutthis type, urban society preferable driving scooter matic than manual vehicle[7].In the design of fire fighter motor cycle manufacture[2], automatic scooter is a main option. In addition to providing ease in operation, this type of motor can utilize engine rotation as the driving force of the centrifugal pump. In a variety of needs, this type of vehicle can adjust the engine power generated to the need of the rider[8]. By replacing the flying roller with a certain mass weight. For example, when a motor requires a large torque at high speed, the installation of a flying roller with a mass weight of above a standard flying roller will trigger engine motor power. For the utilization of a scooter-matic engine as the main power of centrifugal pump for a fire fighter motor cycle, research of the flying roller which produces a big torque for lower rotation between 700rpm until 2,000rpm will be presented in this paper. In the following sub-chapter, we will describe the method of data retrieval and the result achieved.

2. Research Method

In this study, the selected flying roller is set with the weight of 10 to 14 grams. Each flying roller is selected, mounted on the CVT replacing the innate standard flying roller of the motor specification. Every installed flying roller, the torque of power generated is measured from 600 rpm to the highest rotation of 8,000 rpm. The flying roller used is a flying roller product of variations existing in the market. Specification of engine performance test describe in table 1 below.

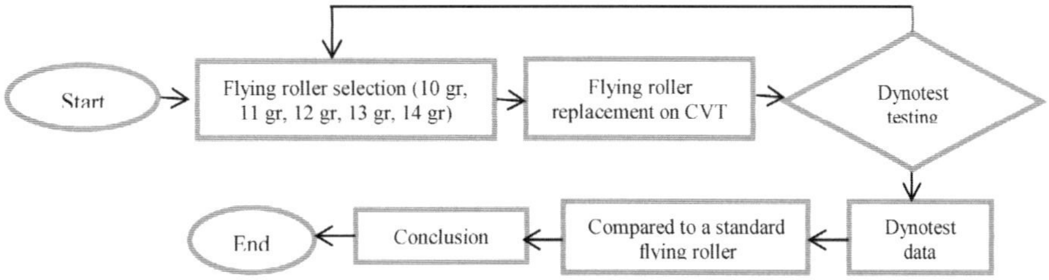

Figure 1 Flowchart

Meanwhile, the measuring instrument in the form of dynotest directly measures the performance of rotation produced by a scooter-matic motor engine. The measuring instrument data used is as follows:

Table 1. engine performance test specification

Dynotest Specification	
Type	Model 250i
L x W x H	246 x 213 x 167 cm
Diameter of tube	18 in / 45.72 cm
Max Horse Power	750 Hp
Maximum speed	200 Mph
Maximum torque	750 Ft/Lbs
Temperature	$0^0 - 70^0$ C
Hardware	*Dynoware* EX, CPU module, *Dyno* module,Atmospheric

At the time of torque measurement on the dynotest machine used, the position of the motor is placed on the measuring instrument and the engine is switched on at various engine rotation positions from low rotation to the highest rotation. Position the motor as shown below.

Figure 2. The position of motor power testing on the dynotest engine

Each type of the flying roller mounted on the CVT machine is tested 3 times with the engine heating time of 15 minutes.

3. Test Result

Data from the dynotest machine used, the engine power generated by replacing flying roller of various sizes of weight can be seen in the table 1 below.

Table 2. The Flying Roller Machine Power Data

IOP Conf. Series: Materials Science and Engineering **324** (2018) 012055 doi:10.1088/1757-899X/324/1/012055

RPM	POWER (HP)				
	I FR 10GR	II FR 11GR	III FR 12GR	IV FR 13GR	V FR 14GR
250	0.8	0.7	0.7	0.6	0.7
593	1.7	1.6	1.7	1.6	1.7
769	2.2	2.1	2.2	2.1	2.1
1000	2.9	2.8	2.8	2.7	2.7
1250	3.6	3.5	3.5	3.4	3.5
1500	4.3	4.0	4.3	4.0	4.1
1750	4.9	4.6	4.9	4.7	4.7
2000	4.7	4.9	5.0	4.9	4.9
2250	4.9	4.9	5.2	5.0	4.9
2500	5.4	5.2	5.5	5.4	5.2
2750	5.8	5.5	5.7	5.7	5.2
3000	6.1	5.8	6.0	5.7	5.2

The above table shows that scooter-matic motor engine power has a maximum value under different conditions. In the rotation between 700 rpm to 2,000 rpm, the power generated by a 10 gram flying roller has a relatively large power stability compared to the power generated by other flying rollers. But at this size, the maximum power is only at 6,750 rpm engine rotation. As for other flying rollers, the highest power is at 7,750 rpm rotation for the weight of 11 grams, 8,000 rpm for 12 grams, 8,750 rpm for 13 grams and 9,500 rpm for 14 flying roller. For more details, the difference in power generated at the lower rotation and the maximum power generated can be seen in Figure 3 below.

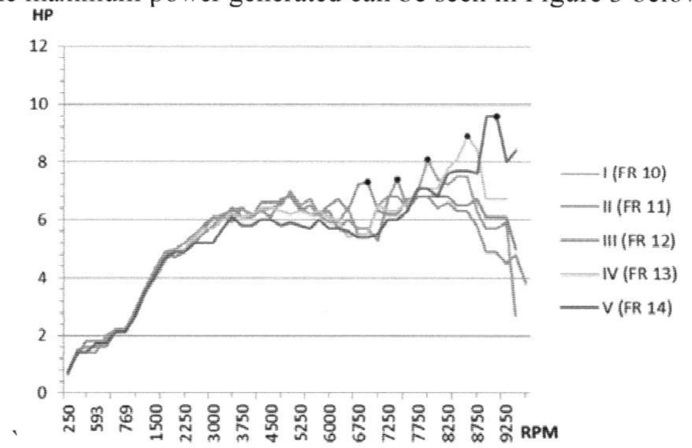

Figure 3 The Flying Roller Power Chart

As for the data about torque generated by a scooter-matic motor engine on each replacement of the flying roller is described in table 2 below. In each replacement of the flying roller, the data retrieved starts from the minimum engine rotation (250 rpm) to the maximum rotation (9,750 rpm).

Table 3 The Flying Roller Engine Torque Data

RPM	TORQUE (N.m)				
	I FR 10GR	II FR 11GR	III FR 12GR	IV FR 13GR	V FR 14GR
250	13.95	15.36	15.76	14.87	14.86
500	18.21	20.13	20.20	19.77	19.47
593	19.35	20.15	20.05	19.90	19.89
750	20.61	19.69	20.40	19.67	19.64
769	20.61	19.39	20.10	19.37	19.23
1000	20.38	19.11	19.98	19.23	19.11
1250	20.26	19.32	19.65	19.26	19.43
1500	20.41	18.86	20.08	19.06	18.94
1750	19.69	18.70	19.57	18.95	19.05
2000	16.63	17.26	17.59	17.36	17.25
2250	15.54	15.63	16.34	15.69	15.46
2500	15.24	14.83	15.65	15.33	14.75
2750	15.01	14.20	14.65	14.61	13.39
3000	14.34	13.77	14.18	13.53	12.35

Of the data above, at minimum rotation to 750 rpm rotation, the largest generated torque is recorded at the flying roller size of 11 grams, whereas in the 1,000 rpm to 2,000 rpm, the largest torque is generated by a flying roller size of 10 grams. At high rotation (5,000 rpm) to a maximum rotation of

9,750 rpm, the resulting torque decreases. This is due to the increase in the scooter-matic motor engine rotation. To illustrate the difference between each flying roller selected, the merging of data is presented on the following figure 4 chart.

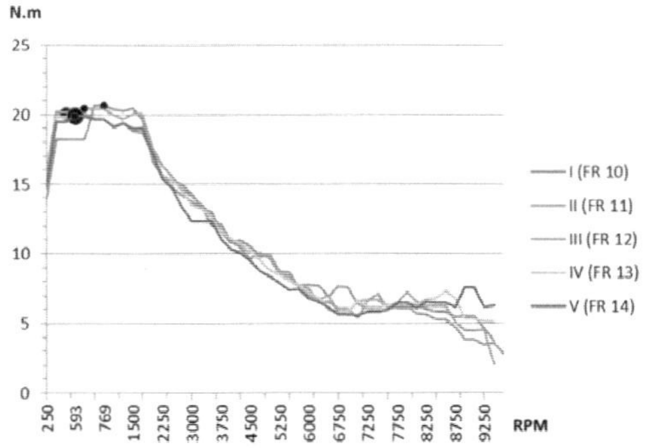

Figure 4 Chart of Resulting Increased Flying Roller Torque

Torque increase is generated only at 250 rpm rotation to 750 rpm rotation, while at 750 rpm to 2,250 rpm the resulting torque is relatively stable. On lap rotation of 2,250 rpm to maximum rotation of 9,750 rpm, the torque of all variations of the flying roller replacement has decreased.

4. Discussion

Engine power is useful when the motor is moving at a certain speed to achieve a higher and corrected speed against time. Meanwhile, torque is the ability of the engine to move or transfer the car or motor from idle to start.

Regarding to scootermatic engine utilization as the main driving force of the centrifugal pump for the fire fighter motor cycle, the greatest power and torque to drive the pump at 700 rpm to 2000 rpm are selected. From the results above, 10 gram flying roller has a great advantage of power and torque generated compared with the size of another flying roller.

In figure 5 below, we will illustrate the usage of scooter matic engine power as the driving force of the centrifugal pump. Then, the picture 6 is the initial test of centrifugal pump performance which is coupled with scooter matic machine.

Figure 5. usage of scooter matic engine as sentrifugal pump power[2]

Table 4. Specification of Sentrifugal pump

Centrifugal pump Specification	
Power	2,2 KW / 2,9 HP
Rotation	1450 rpm
pressure	3,2 bar (max)
Q max	500 l/menit
H max	18,5 m
Shead	9 m

Figure 6. initial test of centrifugal pump[2]

The centrifugal pump spesification which coupled with the scooter matic engine is as follows (see table 4 above) the selected pump has the minimum required power of .Hp. and the rotation at 1500 rpm. It is normal rotation for scooter matic, where the lower rotation is 500 rpm and 9000 rpm is maximum rotation. The performance of centrifugal pump as for fire fighter motor cycle can be seen in the table 5 below.

Table 5. preliminary test of centrifugal pump performance[2]

Motorcycle Rotation (rpm)	Pressure (bar)	Thrust everage (m)
750	1,05	1
1000	1,18	1
1250	1,31	3
1500	1,39	4
2000	1,69	6

5. Conclusion

By utilizing the scooter matic engine as the main driving force of the centrifugal pump in the fire fighter motor cycle, the replacement of the flying roller as an effort to increase the power and engine torque becomes very useful. But, the consequences of 10 gram flying roller to the engine effect need continously research.

References

[1] J. Wang, "Problems and solutions in the protection of historical urban areas," *Frontiers of Architectural Research,* vol. 1, pp. 40-43, 2012.

[2] H. H. Sutrisno and Triyono, "Designing a Firefighter Motorcycle as an Effort to Provide an Early Response to Fire Disaster: A Study Case in Jakarta Special Capital Region," *Journal Of Mechanical Engineering,* vol. SI, p. 12, 2017.

[3] W. J. H, J. H. Sun, S. M. Lo, L. J. Gao, and R. K. K. Yuen, "Statistical Analysis on the Temporal-spatial Characteristics of Urban Fires Under Typical Urbanization Features," *Procedia Engineering,* vol. 11, pp. 437-444, 2011.

[4] W. Helan, Z. Jiang, Y. Zheng, and T. Jun, "Study on Administrative and Educational Measures of China Urban Public Fire Safety Education," *Procedia Engineering,* vol. 84, pp. 151-165, 2014.

[5] W. Shabbir and S. A. Evangelou, "Efficiency analysis of a continously variable transmission with liniear control for a series hybrid electric vehicle," *Proceeding of the 19th Word Congress The International Federation of Automatic Control,* p. 6, 2014.

[6] D. Rockwood, N. Parks, and D. Garmire, "A continuously variable transmission for efficient urban transportation," *Sustainable Materials and Technologies,* vol. 1-2, pp. 36-41, 2014.

[7] J. Cai, G. Li, J. Li, and Y. Hu, "Study on effects of the secondary fuel manifold unfold pressure on engine starting performance," *Propulsion and Power Research,* 2017.

[8] A. Kompanakis, E. Siampis, E. Venelis, and S. Longo, "A Torque Vectoring Optimal Control Strategy for Combined Vehicle Dynamics Performance Enhancement and Electric Motor Ageing Minimisation," *IFAC-Paper Online,* vol. 11, p. 6, 2016.

IOP Conf. Series: Materials Science and Engineering **324** (2018) 012056 doi:10.1088/1757-899X/324/1/012056

Nonlinear system identification of the reduction nickel oxide smelting process in electric arc furnace

V Gubin[1] and A Firsov[2]

1 Candidate of technical sciences, Saint-Petersburg Mining University, Russia, gubin.vl@spmi.ru
2 Candidate of technical sciences, Saint-Petersburg Mining University, Russia, firs@spmi.ru

Abstract. As the title implies the article describes the nonlinear system identification of the reduction smelting process of nickel oxide in electric arc furnaces. It is suggested that for operational control ratio of components of the charge must be solved the problem of determining the qualitative composition of the melt in real time. The use of 0th harmonic of phase voltage AC furnace as an indirect measure of the melt composition is proposed. Brief description of the mechanism of occurrence and nature of the non-zero 0th harmonic of the AC voltage of the arc is given. It is shown that value of 0th harmonic of the arc voltage is not function of electrical parameters but depends of the material composition of the melt. Processed industrial data are given. Hammerstein-Wiener model is used for description of the dependence of 0th harmonic of the furnace voltage from the technical parameters of melting furnace: the melt composition and current. Recommendations are given about the practical use of the model.

1. Introduction

Most sulphide ores have traditionally been processed using rotary kiln electric furnace (RKEF) techniques to produce a nickel matte for electrorefining. RKEF process can include the next steps: pretreatment, drying, calcination in a fluidized-bed roaster, prereduction in a rotary kiln, smelting in electric arc furnace and, finally, electrorefining. The non-reduced calcine and the prereduced calcine from the RK charged into the electric arc furnace. Coke have been used as the reducing agent. [1, 2]

It is considered that optimal mass concentration of carbon in the reaction mass approximately equal 2.2%. It is minimize the consumption of energy for smelting. At the end of melting the carbon concentration must be equal to 0.04-0.12%. [3]

The quality control process is greatly influenced by the accuracy of determining the composition of the melt. Available methods of chemical analysis have a great time duration. As a rule, founder carries out a visual analysis of the sample to fast determinate the state of the metal in the furnace.

The process of smelting of a Nickel anode in an electric arc furnace as a control object is shown in figure 1. The model's inputs are mass flow of the components of the charge, having a pulsed characteristic. Perturbation are the change of the interelectrode distance, which depends on the position of the electrode and the height of the melt and slag; volatility of raw materials by composition. An automatic regulator of a known principle controls the position of the electrodes but, in fact, this parameter is external to the control system of technological mode of the furnace. Concentrations of melt components are considered as state variables. Outputs are electric characteristic – voltage and electrode's current, – and data obtained by visual analysis of the metal sample and the spectral analysis of the audio signal.

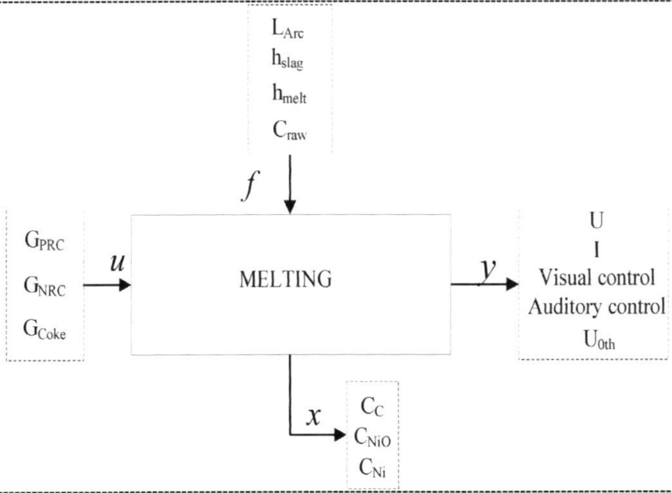

Figure 1 – The process of smelting of a Nickel anode in an electric arc furnace as a control object.

One of ways of the EAF process control is obtaining information about processes inside the furnace from the changing of electrothermal conditions in furnace. In the arc voltage AC may be present non-zero 0_{th} harmonic (U_{0th}), which depends on various melting conditions, such as temperature, the shape of the arc, resistance of the reaction mass. [4]

U_{0th} occurs due to electric arc mode, in which the density of the emission current of the first electrode can be more then density of the emission current of the second electrode. RMS-value of furnace's voltage during melting changes slightly because this is provided by automatic control of the position of the electrodes. At the same time, U_{0th} can be used as one of the parameters of the process for process control.

2. Experiment

Experimental (factory) research of process of reduction smelting was conducted in furnaces of the type RKZ-10.5. The experience was carried out measurement of phase voltages of the furnace, the measurement of current and voltage waveform at the output of the regulator the position of the electrodes. In the journal of melting was recorded charging time of all components of the charge.

ADC USB-6008 was used for data collection. Sampling frequency was 1000 Hz. Spectral analysis of a signal performed in real time in the program developed in the LabView development environment.

3. Result.

3.1. Data processing

This research shows that value of U_{0th} equals to about 4-6 Volts in the absence of disturbances during the period of carbonization, when the prereduced calcine and the coke was charged. The value of U_{0th} decreases significantly during charging. U_{0th} increases to 7-8 Volts by the end of the period of finishing, when the calcine was charged. [5, 6]

Figure 2 shows the measured value of U_{0th} for the one melt. The total number of recorded melts is 21.

IOP Conf. Series: Materials Science and Engineering **324** (2018) 012056 doi:10.1088/1757-899X/324/1/012056

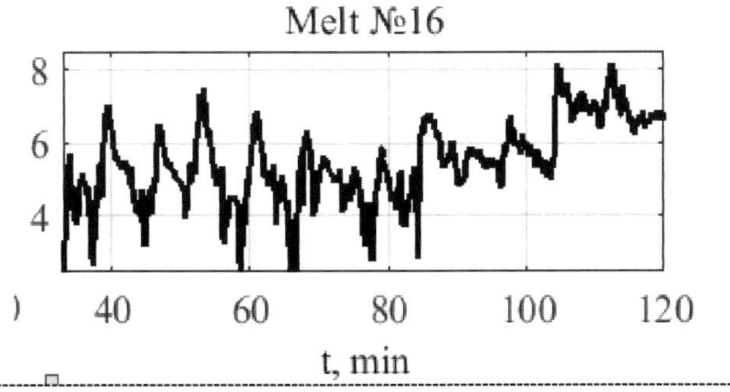

Figure 2 – values of U_{0th} during the melting.

3.2. Melting process modeling

The goal of the modeling was to determine the relationship between the ratio of the components of the charge and the value of U_{0th}. In the present paper, a block of nonlinear model was used; its structural diagram is shown in figure 3. The first block of the model is a dynamic model of material balance. Output of this block is theoretical material composition of the reaction mass.

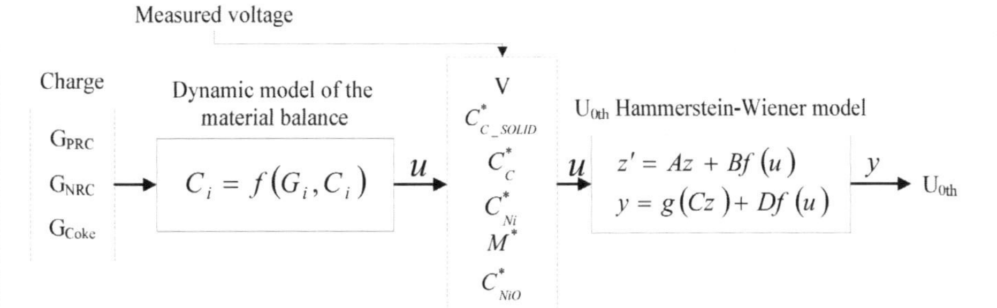

Figure 3 – Structure of the evaluation model of the U_{0th}.

The second part of the nonlinear block model is the simple cascade Hammerstein-Wiener model.

This model based on the calculated concentrations in the reaction mass and the measured values of voltage and U_{0th}. The model establishes a functional dependence between the value of U_{0th} and the composition of the reaction mass, and a current electrode. The resulting model has a high degree of convergence.

Hammerstein-Wiener Model belongs to the class of block-oriented models and describes a linear dynamic and a nonlinear static characteristic of the object of modeling.

The model can be described by the following equations:

$$x(t) = f(\alpha, u(t)) \quad (1)$$

$$r(t) = G(q^{-1}, \beta) \cdot x(t) \quad (2)$$

$$y_0(t) = g(\gamma, r(t)) \quad (3)$$

where $y_0(t)$ – model's prediction of the output $y(t)$, $G(q^{-1}, \beta)$ – a linear dynamical system parametrized by delay operator q^{-1}, α and γ – parameters of nonlinear functions f and g. All parameters can be presented as the vector of parameters $\theta = [\alpha, \beta, \gamma]$. [7]

List of model's inputs:

- u_1 the rms-value of the phase voltage, [0; 200] V;
- u_2 the concentration of solid carbon (coke), [0;10] % mass.;
- u_3 the concentration of carbon [0;2,5] % mass.;
- u_4 the total mass of furnace charge, [0; 28] ton;
- u_5 the concentration of a nickel [0;90] % mass.;

- u_6 the concentration of a nickel monoxide, [0;10] % mass.;

Model's output y is U_{0th}, [0;10] V.

The calculation of the coefficients of the model was carried out by the method of least squares. The target function was the residual sum of squares $\sum_{i=1}^{N}(y_i - \hat{y}_i)^2$. As a method of optimization was chosen as the Levenberg–Marquardt algorithm.

The model was reduced after training. The final model contains 3 inputs (u_1, u_2, u_5) and 1 output.

The model can be described by the state space equation:

$$\begin{aligned} z_{(k+1)\Delta t} &= Az_{k\Delta t} + Bf_{k\Delta t}(u_{k\Delta t}) \\ y_{k\Delta t} &= g_{k\Delta t}(Cz_{k\Delta t}) + Df_{k\Delta t}(u_{k\Delta t}) \end{aligned} \quad (4)$$

where $f_{k\Delta t}(u_{k\Delta t})$ – vector of functions of inputs, which are as follows:

$$x_i(t) = \begin{cases} x_i^L, \text{ при } u_i \le u_i^L \\ K(u_i - u_i^L) + x_1, \text{ при } u_i^L \le u_i \le u_i^H \\ x_i^H, \text{ при } u_i \ge u_i^H \end{cases} \quad (5)$$

The parameters of the function identified in the learning model are summarized in table 1.

Table 1. The parameters of the model

	x_i^L	x_i^H	u_i^L	u_i^H	K
u_1	- 3340	-17200	54,03	92,38	-361.41
u_2	13100	34200	15500	23800	2.54
u_5	8.15	3.14	0	0.05	-104.65

Vector $z_{k\Delta t}$ in the formula (5) is the vector of state variables, $y_{k\Delta t}$ – model's output. Function $g_{k\Delta t}$ has the following form:

$$g_{k\Delta t}(Cz_{k\Delta t}) = 9.9319 \cdot (10^{-5}) \cdot (Cz_{k\Delta t}) + 3.73 \quad (6)$$

State matrix, input matrix, output matrix and feedthrough matrix equal to:

$$A = \begin{bmatrix} 0{,}88 & 0 & 0 \\ 0 & 0{,}96 & 0 \\ 0 & 0 & 0{,}99 \end{bmatrix}, \quad B = \begin{bmatrix} 0{,}94 & 0 & 0 \\ 0 & 0{,}98 & 0 \\ 0 & 0 & 0{,}99 \end{bmatrix},$$

$$C = [-0{,}32 \quad 0.05 \quad -75.99], \; D = [0 \quad 0 \quad 0].$$

The resulting model has the following characteristics:

- mean squared error (MSE) equal to 0,54 V;
- mathematical expectation of the error equal to 0.004 V;
- the coefficient of determination equal to 0,52 V.

Figure 4 graphs the measured values of the U_{0th} and the values obtained during the simulation.

IOP Conf. Series: Materials Science and Engineering **324** (2018) 012056 doi:10.1088/1757-899X/324/1/012056

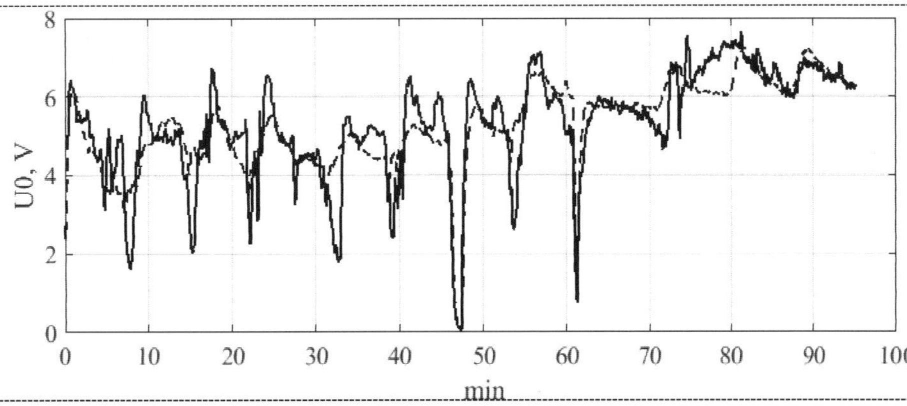

Figure 4 – A graph of the measured values of the U0th during the melting (solid line) in comparison with simulation results (dashed line).

4. Conclusions.

Zeroth harmonic of the AC voltage depends of the parameters of the arc and the properties of the melt, because furnace circuit include the arc and the melt, considered as a homogeneous reaction mass with variable conductivity. This statement can be used as the basis for determining the concentration of certain components of the melt on the value of the 0th harmonic.

The variability of the raw's composition makes inefficient use of open-loop automatic control system. Direct analysis of the composition of the melt can not be used for rapid adjustment of the charge composition.

In this regard, an interesting possibility to estimate the concentration of Nickel oxide in the melt by the proposed model. The feature of the model is that the input parameters are time-dependent functions of the mass flows of the components of the charge.

[1] Crundwell F K 2011 *Extractive Metallurgy of Nickel, Cobalt and Platinum Group Metals* (Oxford: Elsevier) p 610

[2] Voermann N, Gerritsen T, Candy I, Stober F and Matyas A 2004 Developments in furnace technology for ferronickel production *10th International Ferroalloys Congress* (Cape Town, South Africa)

[3] Paulson D. L and Anable W E 1977 Smelting prereduced nickel concentrate in an electric-arc furnace *US Dep. of the interior. Bureau of mines* vol 4 p 20

[4] Pedro A A 1993 About a nature of direct component of an electric arc phase voltage in the normal corundum furnace *Promyshlennaya energetika* vol 5 (Moscow: Energoprogress JSC) pp 28-31

[5] Gubin V V and Firsov A Yu 2015 The using of barrier effect of the arc for the control of reductive smelting of nickel monoxide *Natural and technical science* vol.6 (Moscow) pp. 463-468

[6] Pedro A A, Bazhin V Yu, Suslov A P, Firsov A Yu and Martynova E S 2017 Power control of the open electric arc for the electrode furnace *Steel* vol.7 (Moscow) pp. 21-23

[7] Giri F and Bai E 2010 *Block-oriented Nonlinear System Identification* (Springer) p. 426

IOP Conf. Series: Materials Science and Engineering **324** (2018) 012057 doi:10.1088/1757-899X/324/1/012057

Vehicle anti-rollover control strategy based on load transferring rate

W T Dai[1], H Q Du[2] and L Zhang[3]

1 Tianjin Sino-German University of Applied Sciences, Tianjin 300350
2 Tianjin Sino-German University of Applied Sciences, Tianjin 300350
3 Tianjin University of Technology and Education, Tianjin 300222
E-mail: duhuiqi@163.com

Abstract. When vehicles is drived on a low adhesion road or going on a high speed and sharp turn, it is prone to product some lateral stability problems, such as lateral sideslip or rollover. In order to improve the vehicle anti-rollover stability under these limited conditions, a SUV vehicle model with high mass center was built based on the software of CarSim and the rollover stability controller was designed using the static threshold value method for the lateral load transferring rate（LTR）. The simulations are shown that the vehicle anti-rollover stability under limit conditions is improved using the SUV model.

Key words: Vehicle; Anti-rollover; LTR

1. Introduction

At a low friction coefficient road or at a turning with high velocity, vehicle is easier to be out of control for the lateral stability. In order to improve the vehicle lateral anti-rollover stability, Chen proposed the Time-To-Rollover (TTR) concept to study the prediction system of vehicle rollover state[1]. To enhance the rollover control function of vehicles in extreme conditions, a vehicle dynamics model with 8-degrees rollover model was established, and an integrated control strategy was proposed[2]. To avoid automobile rollover accidents, a model based on an algorithm for rollover warning and a control strategy for rollover prevention were proposed. In the algorithm, the absolute value of the lateral Load-Transfer-Ratio was taken as the rollover index[3]. At present, some fuzzy controls, neural networks and the artificial intelligence control technology were introduced into the vehicle lateral stability control system[4]. However, in these lateral stability control, the yawing moment is often used as the controlled objective, it isn't fit to rollover condition. For instance, at a high adhesion coefficient road, a SUV with high mass center is easier to product vehicle rollover with a high-speed turning or emergency obstacle avoidance, because of the shift dynamic lateral load. In order to improve the rollover stability, a vehicle model with transmission, brake system and suspension system, etc. is built and the vehicle anti-rollover strategy under the limit conditions is studied.

2. Vehicle model

A SUV vehicle model with high mass center is built based on the software of CarSim as shown in Fig.1. The vehicle model is using a type of engine with the power of 200kw, a 6-speed manual transmission and an ABS model. Front suspension is the independent suspension, and rear suspension is the independent suspension.

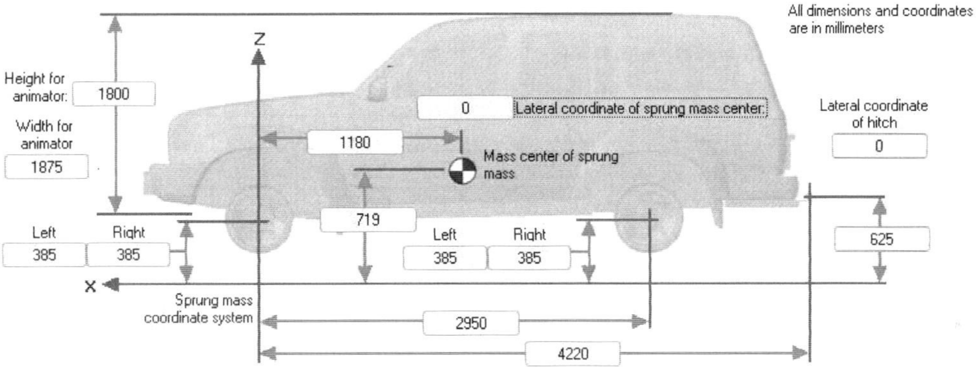

Figure 1. SUV vehicle model

The main parameters of the vehicle are as the followings: total mass = 1592 kg; Length/ width/ height = 3800/1875/1800 mm; mass center height = 1000 mm; wheelbase = 2950 mm; Front/ rear wheel tread = 1575/1575 mm; Moment of inertia around X = 614 kg•m^2 ; Moment of inertia around Y = 2488 kg•m^2 ; Moment of inertia around Z = 2488 kg.m^2.

3. Vehicle anti-rollover design based LTR

3.1. Vehicle anti-rollover principle

Roll force analysis at a vehicle turning is shown in Fig.2. In Fig.2, F_{zi} , F_{zo} - a vertical force for inside and outside wheel, t - wheel tread, h - vehicle mass center height, e - a distance from the roll center to mass center, h_g - a roll center height, a - lateral acceleration at mass center. In order to simplify the analysis, the automotive total mass and the sprung mass are considered being equal.

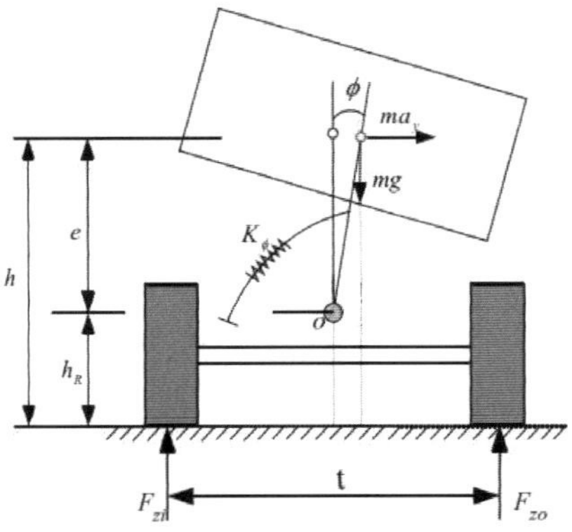

Figure 2. Roll force analysis at a turning

The body roll angle from the roll balance force around the center is shown as Eq.(1)

$$\phi = \frac{m \cdot a_y \cdot e}{K_\phi - m \cdot g \cdot e} \tag{1}$$

LTR (lateral load transferring rate) is used to express the vehicle lateral load transferring rate, namely the vertical load transfer between inside and outside wheel. The LTR from the force equilibrium equation around the center of the wheel track is shown as Eq.(2)

$$LTR = \frac{F_{zo} - F_{zi}}{F_{zo} + F_{zi}} = \frac{2}{t}\left(h \cdot \frac{a_y}{g} + e \cdot \phi\right) \tag{2}$$

When the LTR is transferring between inside and outside of wheel, the vehicle begin to roll. When $LTR \geq 1$, all of vertical load are sustained by the outside wheel, and the inside wheel is rolled over.

It is difficult to directly measure the vertical load of tire, and the LTR can be approximate expression by static roll coefficient R as Eq.(3):

$$R = \frac{2h}{t \cdot g} a_y \tag{3}$$

In the Eq.(3), the roll coefficient has a direct relationship with the lateral acceleration, and the R can be calculated by the lateral acceleration, then according to the static threshold to prevent the vehicle rollover.

When $|R| = 1$, the inside wheel is off the ground and vehicle happens a quasi static rollover. Considering the elastic deformation of the tire, the quasi static rollover threshold is generally decided as $|R| = 0.9$. The quasi static rollover threshold R is far more than the transient rollover threshold under the extreme conditions, therefore the $\hat{R}$ is chosen as the transient rollover threshold. When it is

$|R| \geq \hat{R}$, vehicle has the potential rollover tendency, the stability control is required.

3.2. Vehicle anti-rollover controller based LTR

The basic idea of anti–rollover stability control is to measure some information such as wheel velocity, lateral acceleration, vertical acceleration, etc. Using the lateral acceleration information to estimate the vehicle roll state and calculate the rollover coefficient. If the rollover coefficient exceeds the threshold, rollover controller will be activated. According to the longitudinal acceleration, the wheel cylinder pressure is calculated. By controlling the hydraulic regulator for brake system, the vehicle rollover stability is improved.

3.3. Brake wheel selection and cylinder pressure calculation

Anti-rollover control is mainly to limit the vehicle yawing motion to carry on under steer characteristics by the differential brake method. Because the vertical load on the inside wheel is much smaller than the load on the outside wheel, the braking force distribution inside the wheels should be less than the outside wheel. In order to make the vehicle have a high efficiency under steer tendency, usually only the front outside wheel is braked.

The brake wheel is determined according to the front wheel turning angle δ . The anti-rollover strategy is designed based on the brake force for front and outside wheel as shown in table 1.

Table 1. Control wheel for anti-rollover stability

Decision conditions		Control wheel
$\delta > 0$ (turn left)	$\lvert R\rvert \geq \hat{R}$	right and front wheel
$\delta < 0$ (turn right)	$\lvert R\rvert \geq \hat{R}$	left and front wheel
δ = any value	$\lvert R\rvert < \hat{R}$	No

The brake force for wheel is shown as in Eq.(4).

$$f_x = \begin{cases} 0 & |R| < \hat{R} \\ ma_x & |R| \geq \hat{R} \end{cases} \tag{4}$$

According to the Eq.(4), the braking force is transformed into the brake wheel cylinder pressure. When the controller is working, the wheel cylinder pressure for recovering vehicle stability is calculated. Through the hydraulic brake adjuster, the brake force is applied to the brake wheel to prevent the vehicle rollover.

4. Simulation analyses

For testing and verifying the controller, the simulations are measured under the typical hook mode. The hook condition is a kind of rapid reverse steering operation with sharp steering angle. Because in the hook conditions it is most likely to happen to rollover. It is often used to assess for the vehicle rollover performance.

The measure data under the controller are compared with the measure data without controller. Fig.3 is the input of steering wheel angle in hook test and Fig.4 is the vehicle driving state. It can be seen that the vehicle without controller happened rollover tendency, and the vehicle didn't happen rollover with the the lateral load transferring rate controller.

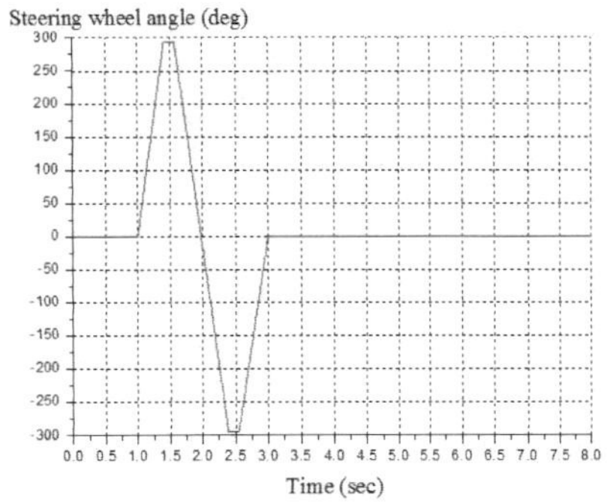

Figure 3. Steering wheel angle in hook test

Figure 4. Vehicle driving state

Fiures 4 are the measure data for the SUV model under the typical hook mode. As seen in Fig.4, the vehicle without controller happen rollover after 3.5s. The vehicle with the lateral load transferring rate controller is initially driving stability and the maximum vertical acceleration is 0.5g's, however after 4.3s it shows a rollover trend with the right and rear tire which is far from the ground. The vehicle with anti-rollover controller is driving stability and its maximum vertical acceleration is only 0.2g's, which is reduced by 60%.

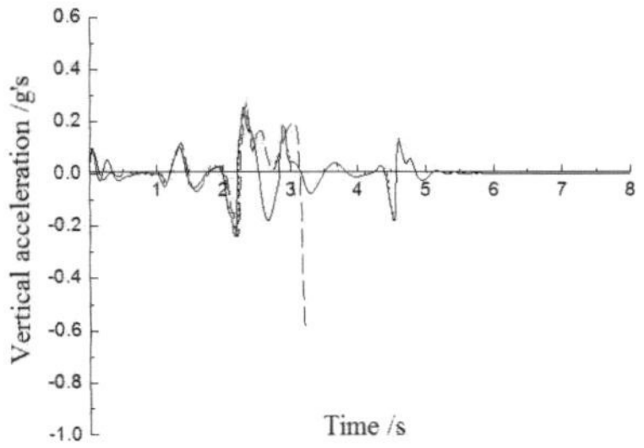

Figure 5. Vertical acceleration

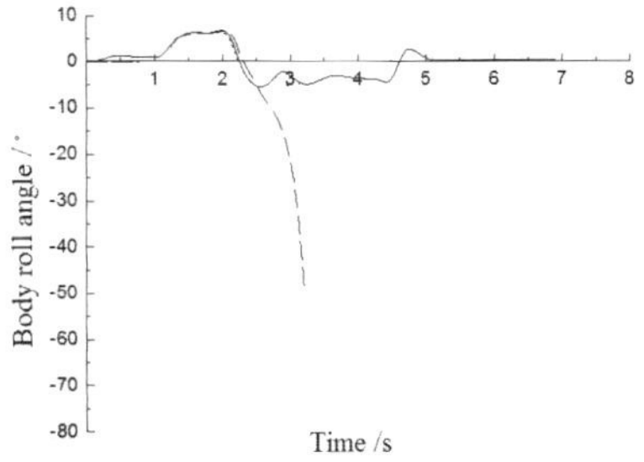

Figure 6. Body roll angle

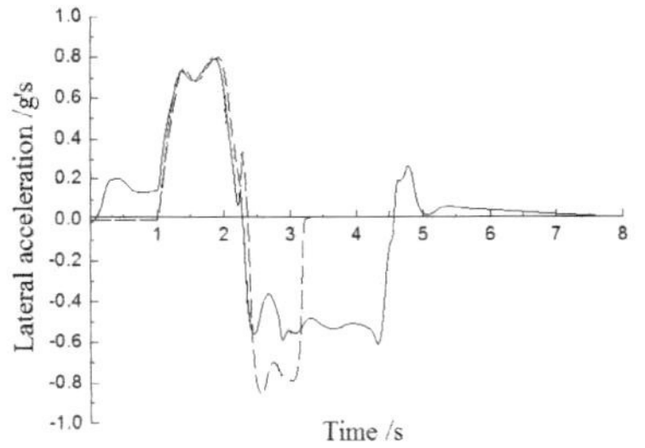

As seen in **Figure 7.** Lateral acceleration

Fig.6, the largest body roll angle of vehicle with the anti-rollover controller is reduced by 50% than the angle of vehicle without the controller, which greatly reduces the vehicle roll amplitude. As seem in Fig.7, the vehicle lateral acceleration with the anti-rollover controller was reduced by 30% than that without the controller. The vehicle lateral stability with the rollover controller is improved.

5. Conclusion

The SUV model with the suspension system, transmission system, braking system, etc. was established in the CarSim software. The validity of controller based on the lateral load transferring rate is verified using the SUV model. The simulations are shown that the vehicle anti-rollover stability under limit conditions is improved. In future, some new parameters will be considered to add the control strategy to improve the controller.

References

[1] Ou Jian, Chen xiangchuan, Zhou Xinhua, etc. Rollover control strategy based on vehicle stability control system. Journal of Southwest Jiaotong University, 2014, 49(2): 283-285.

[2] Jin Zhilin, Weng Jiansheng, Hu Haiyan. Rollover warning and anti-rollover control for automobiles. Journal of dynamicsand control., 2007(5), 17: 365-369.

[3] Huang Jieyan, Liu kun, Xiong Yi. Vehicle rollover warning and control system research. Light vehicle technology, 2011(2): 257-258.

[4]Ahmed Elmarakbi , Chandrasekaran Rengaraj , Alan Wheat-ley. The influence of electronic stability control, active sus-pension, driveline and front steering integrated system on the vehicle ride and handling

IOP Conf. Series: Materials Science and Engineering **324** (2018) 012057 doi:10.1088/1757-899X/324/1/012057

［J］． Global Journal of Researchesin Engineering Automotive Engineering，2013，13 （6）：446 —453.
[5]Zhang Lei． Robust analysis and optimization design of dou-ble front axle steering system［J］． SAE Paper，2013—01—9124.
[6]L. Mejía, V. Mata, F. Valero, et al. Dynamic Parameter Identification in the Front Suspension of a Vehicle:On the Influence of Different Base Parameter Sets. Multibody Mechatronic Systems, 2015, 25: 165-175
[7]Y. C. Chen, B. J. Zhang*, N. Zhang, M. Y. Zheng. A condensation method of the dynamic analysis of vertical vehicle-track interaction considering vehicle flexibility. Journal of Vibration and Acoustics, 137, 041010, 2015.8. (SCI, IDS: CL7OV)
[8]H.H Hong, L.F. Wang*, M.Y. Zheng, N. Zhang. Handling Analysis of a Vehicle Fitted with Roll-Plane Hydraulically Interconnected Suspension Using Motion-Mode Energy Method. SAE International Journal of Passenger Cars - Mechanical Systems. 2014, 7(1): 48-57. (EI, Accession number: 20142717903868)
[9]ZHENG Minyi, ZHANG Bangji*, ZHANG Jie, ZHANG Nong. Physical Parameter Identification Method Based on Modal Analysis for Two-axis On-road Vehicles: Theory and Simulation. Chinese Journal of Mechanical Engineering, DOI: 10.3901/CJME.2016.0108.004 (SCI, accepted January 8, 2016)
[10]KANG D O，HEO S J，KIM M S. Robust design optimization of the Mc Pherson suspension system with consideration of a bush compliance uncertainty ［J］. Proceedings of the Institu-tion of Mechanical Engineers，2010，224（6）：705-716.
[11]FENG C，DING N ，HE Y L . An integrated control algorithm of ABS and DYC for emergency braking on a-Split Road ［C］//Control Engineering and Communication Technology（IC—CECT）. New York：IEEE，2012 ：516-522.

IOP Conf. Series: Materials Science and Engineering **324** (2018) 012058 doi:10.1088/1757-899X/324/1/012058

Collapse Causes Analysis and Numerical Simulation for a Rigid Frame Multiple Arch Bridge

XinDai ZUO[1]

Research Institution of Highway, the Ministry of Transportation, CHINA
E-mail:330217252@qq.com

Abstract. Following the collapse accident of Baihe Bridge, the author built a plane model of the whole bridge firstly and analyzed the carrying capacity of the structure for a 170-tons lorry load. Then the author built a spatial finite element model which can accurately simulate the bridge collapse course. The collapse course was simulated and the accident scene was reproduced. Spatial analysis showed rotational stiffness of the pier bottom had a large influence on the collapse from of the superstructures. The conclusion was that the170 tons lorry load and multiple arch bridge design were the important factors leading to collapse.

1. Introduction

Baihe Bridge in Huairou District, Beijing was a reinforced concrete rigid-frame arch bridge, which had four spans and the single spans were 50 meters each. The rise to span ratio was 1/10 and the width of the bridge deck was11.5 meters. The substructures considered of solid piers and foundations of socketed piles. Fig.1 is the diagram of the span arrangements of Baihe Bridge.

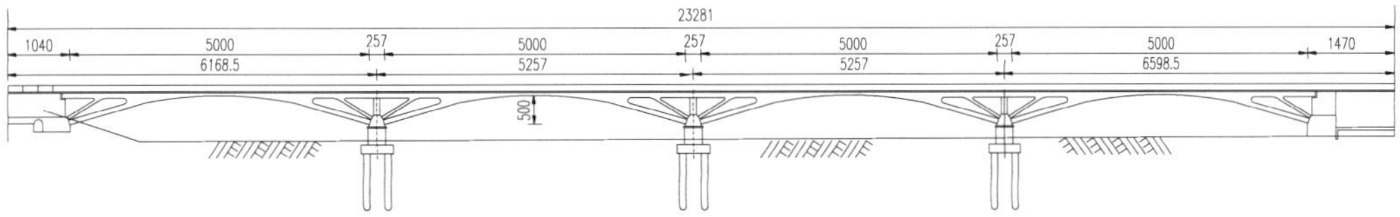

Fig 1. Span Arrangements of Baihe Bridge (unit.cm)

Baihe Bridge was built in 1988 and it has been operating for twenty years. In July 2011, when a 170 tons lorry which was severely overloaded moved to the first span of Baihe Bridge, the whole four spans collapsed successively. Fig.2 is the photo of the collapsed Baihe Bridge.

Fig 2. Baihe Bridge Collapsed

2. Analysis software and calculation contents

Bridge structure calculation software GQJS was used to build a plane finite element model of Baihe Bridge, allowing to analyze structural forces and load effects in a 170 tons lorry load.A general space finite element software was used to build a spatial finite element model in which geometric nonlinearity, material nonlinearity and contact nonlinear were considered[1]. In this part, the collapse process of Baihe Bridge was simulated and different rotational stiffness as a parameter influencing the collapse form was analyzed.

In the plane finite element model some factors such as the dead load, the heavy lorry load, construction phases, concrete shrinkage and creep were considered. The plane model included 161 solid elements and 150 series of reinforcement models. Wheel loads were arranged as the real situation in the first span. Fig.3 is the plane model of Baihe Bridge.

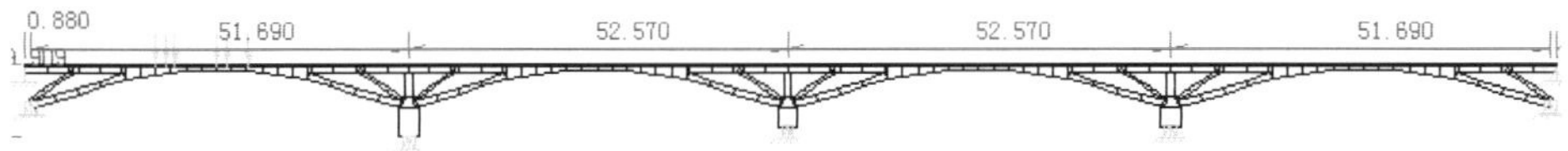

Fig 3. Plane Model of Baihe Bridge

The compression bearing capacity indicated that the major arch legs of the first span could not meet the design criteria of the 170 tons load. For example, the axial compressive strength in the right arch spring of the first span was 12.648MN, less than the axial load effect of 14.076MN. Other sections of the first span met design criteria but had a low degree of safety in compression and flexural capacity. From this we could already conclude that the main stress components had insufficient strength of the 170 tons load.

3. Numerical simulation of collapse process

3.1 Spatial finite element model

Using an adequate computer program a spatial simulation model was built in which geometric nonlinearity, material nonlinearity and contact nonlinearity were considered. The bridge deck, piers and block set walls were modelled as stratified shell elements, and the arch ribs were modelled as beam elements. Fig.4 is the spatial model of Baihe Bridge.

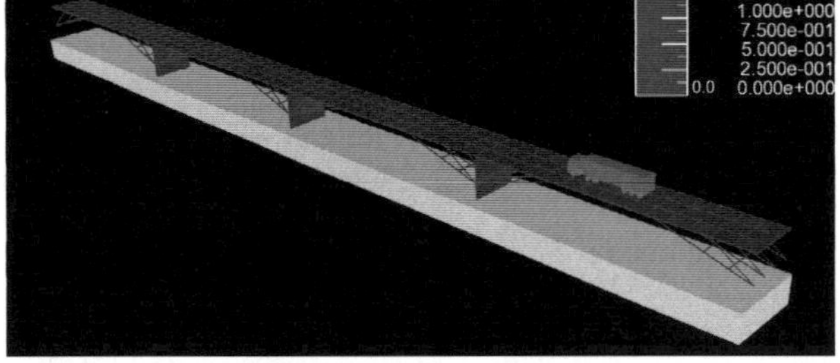

Fig 4. Analytical Model

3.2 Analysis cases

Two typical values of the rotational stiffness of the pier bottom were considered to simulate different constraint conditions according to some documentations[2][3], as showed in table 1.

Table 1. Analysis Case

Analysis case	Rotational stiffness of pier's bottom (×1011 Nmm/rad)
Case A	3.24
Case B	4.86

3.3 Analysis results

3.3.1 Analysis results in the case A. The whole four spans collapsed in the case A. Fig.5 lists dynamic printscreens of collapse process. Firstly, when the lorry transited to the center of the first span, the horizontal beams and arch ribs of the first span showed large plastic deformation and collapsed immediately, as indicated in a and b. Arch spring of the second span nearby the first span happened large deformation and collapsed immediately, as indicated in c and d. And then the third span and the forth span happened chain reaction and finally the four spans all collapsed, as indicated in e and f.

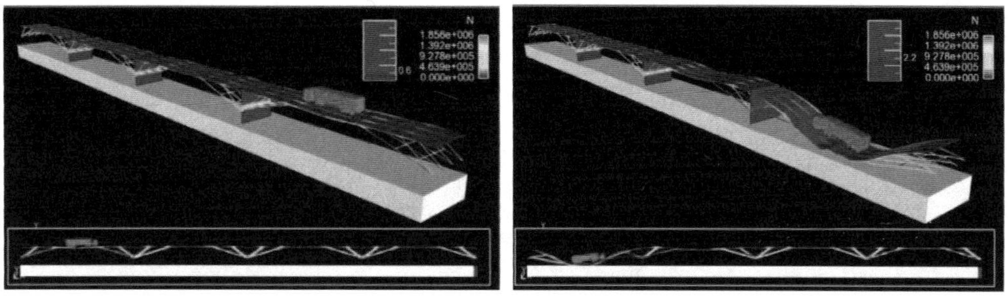

a: Large Plastic Deformation in the First Span **b:the First Span Collapsed**

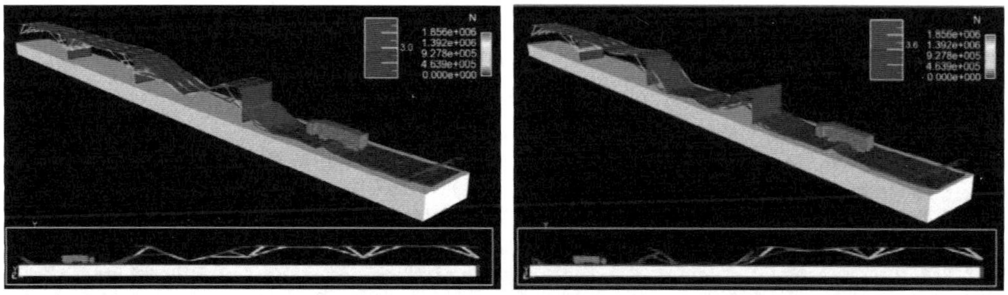

c:the Second Span Started to Collapse **d: the Second Span Collapsed Completely**

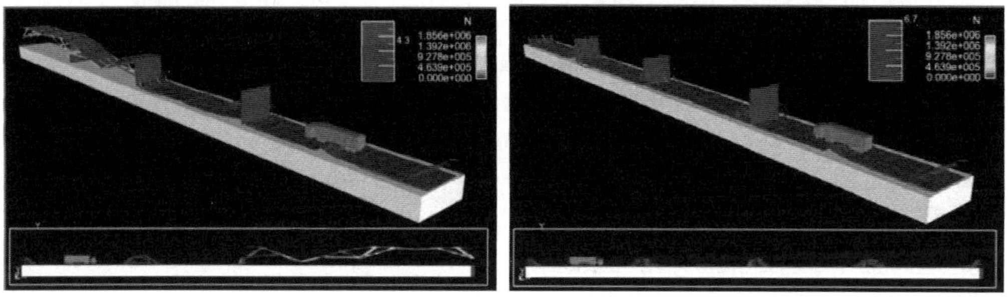

e: the Third Span Started to Collapse **f: the Forth Span Collapsed Completely**

Fig 5. Collapse Process in the Case A

IOP Conf. Series: Materials Science and Engineering **324** (2018) 012058 doi:10.1088/1757-899X/324/1/012058

3.3.2 Analysis results in the case B. If the rotational stiffness of the pier bottom equals 4.86×1011 Nmm/rad, progressive collapse ended in the second span. If the rotational stiffness was larger, the second span would not collapse. As indicated in Fig.6, first the arch spring of the first span showed plastic deformation, and with gradually increasing deformation, the first span collapsed, as indicated in a,b,c. Immediately the arch spring of the second span showed large plastic deformation, and started to collapse, as indicated in d,e,f. Further the second span collapsed and the arch ribs of the third span damaged to a certain extent. However due to the large enough rotational stiffness, the piers could bear the unbalanced horizontal force and further collapse ended.

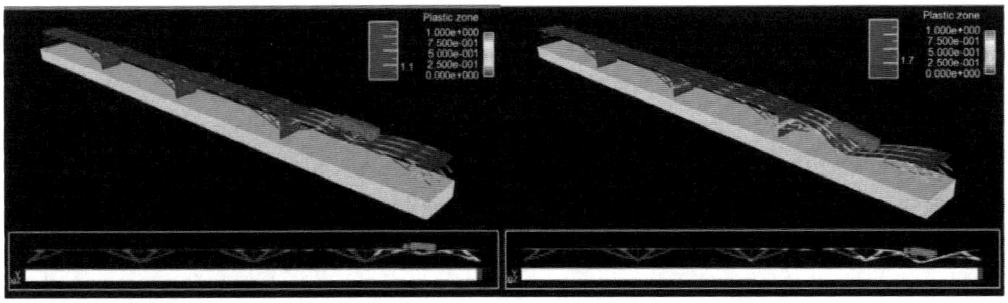

a: Large Plastic Deformation in the First Span **b:the First Span Stated to Collapse**

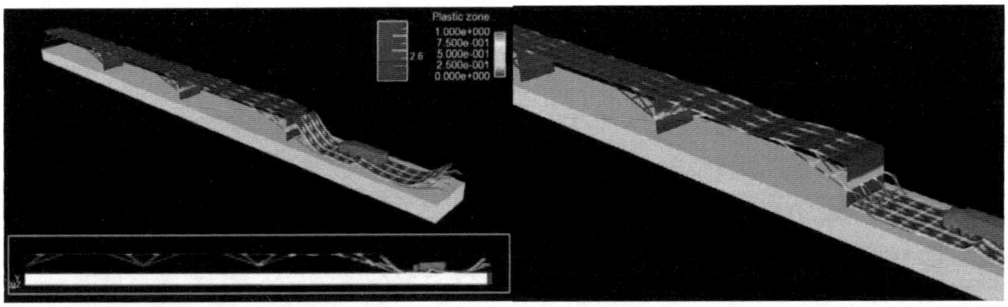

c:the First Span Collapsed Totally **d: Large Plastic Deformation in the Second Span**

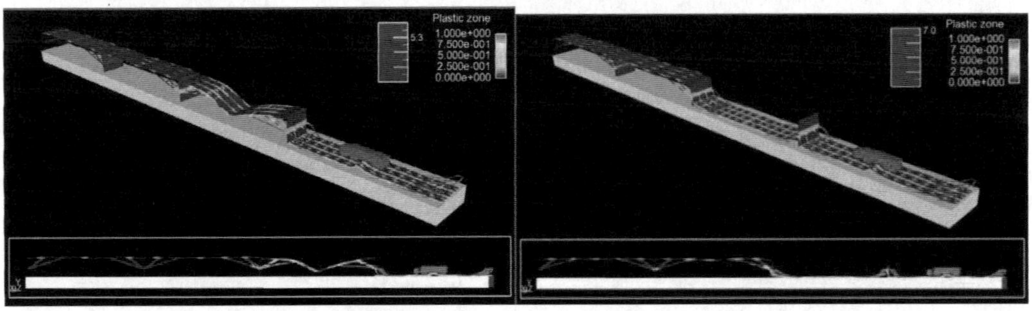

e: the Second Span Started to Collapse **f: the Second Span Collapsed Completely**

Fig 6. Collapse Process in the Case B

3.4 Influence of rotational stiffness on collapse forms

Fig.7,8 showed the displacement time history curve of the pier top in the case A and B. Displacement in the direction of the lower span takes positive values while negative value is taken to the opposite direction. The maximum horizontal displacement of the pier top is -230mm in the case A, while it equal -160mm in the case B.

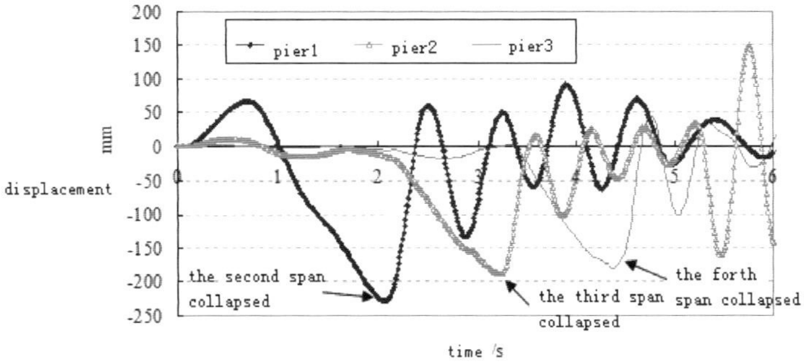

Fig 7. Displacement Time History Curve of the Top of Piers in the Case A

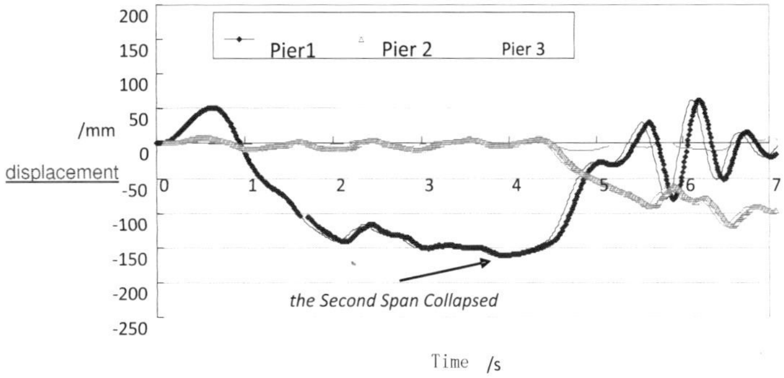

Fig 8. Displacement Time History Curve of the Top of Piers in the Case B

Fig.7 shows the horizontal displacement value of the top of pier1 which is positive at the beginning, subsequently there is a sharp shift towards negative value. As negative displacement increases to a certain value, the second span collapsed. When the horizontal displacement value of pier2 increased to a certain value, the third span collapsed, and so on, the chain reaction finally leading overall collapse. Fig.8 shows negative displacement of pier1 appearing a platform in the case B and later increasing lowly with time, finally the pier can't resist horizontal thrust from the second span and the second span collapsed. Because the rotational stiffness of pier2 is large enough to resist the horizontal thrust of the third span, progressive collapse ended.

3.5 Collapse cause analysis

The immediate cause of the bridge collapse was the overloaded lorry, because the 170 tons load was far beyond the design load. Different rotational stiffness of the pier bottom had a large influence on the collapse form. Baihe Bridge was designed as a multiple arch bridge and the piers were not designed for single direction thrusted piers. Because the rotational stiffness was insufficient, progressive collapse occurred in all four spans.

4. Conclusions

Many rigid-frame multiple arch bridges were built in the 1980s in China because of aesthetic value and for saving material. After thirty years operation this kind of bridges was insufficient to meet the existing design loads. Moreover, multiple arch design also implies a lot of questions. Baihe Bridge was a typical case. Through this paper engineers can understand this kind of bridge more clearly.

Aimed to this kind of bridge, we suggest to limit traffic load or strengthen and even dismantle this bridge type.

References to preprints.

[1] CHAO FENG, DANDAN SUN, "New Tutorial Examples & FAQ Resolution", China WaterPower Press, 2012, p.104-116.
[2] GUODING WANG, "Multiple Arch Calculation", China Communications Press, 2000, p.46-50.
[3] JIAN LEI, JIACHANG LIU, "Optimal Design Analysis of Rigid-Frame Multiple Arch Bridge", East China Highway, 2007, No. 4

Acknowledgement:
This wok was financially supported by the National Natural Science Foundation of China (51008145).

IOP Conf. Series: Materials Science and Engineering **324** (2018) 012059 doi:10.1088/1757-899X/324/1/012059

Development of integrated control system for smart factory in the injection molding process

M J Chung[1] and C Y Kim[2]

[1] Mechatronics Engineering Department, Korea Polytechnic University, Siheung, 15073, Rep. of Korea
[2] R&D Center, Rovix Co. Ltd, Seongnam, 13510, Rep. of Korea
E-mail: mjchung@kpu.ac.kr

Abstract. In this study, we proposed integrated control system for automation of injection molding process required for construction of smart factory. The injection molding process consists of heating, tool close, injection, cooling, tool open, and take-out. Take-out robot controller, image processing module, and process data acquisition interface module are developed and assembled to integrated control system. By adoption of integrated control system, the injection molding process can be simplified and the cost for construction of smart factory can be inexpensive.

1. Introduction

Goods made from plastic are used in the various area of living and industry. Injection molding process is used for production of plastic goods [1]. The injection molding machine used for injection molding process consists of hopper for supplying the plastic chip, heater for heating, screw for transferring melted material, mold for shape as showed in Figure 1 and Figure 2. This injection molding process has been conducted by manual.

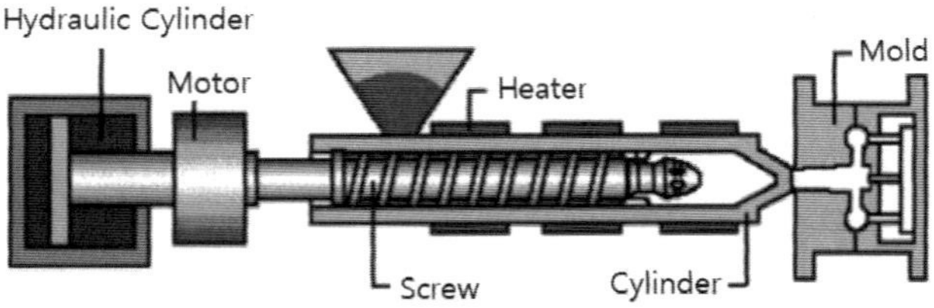

Figure 1. Configuration of the injection molding machine.

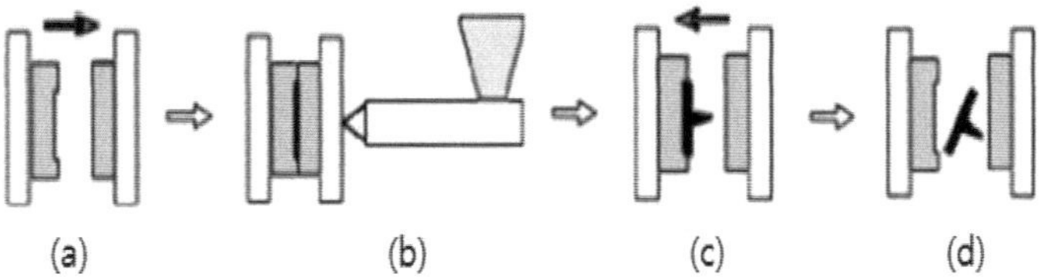

Figure 2. Injection molding has the process such as (a) tool close, (b) injection, (c) tool open, and (d) take-out.

Recently, the effort for saving of production cost by using the process improvement in the injection molding product, which can be achieved by adoption of smart factory, is increasing. Automation of production process, gathering of process data, and transfer to manufacturing execution system (MES) are required for the construction of smart factory [2]. Take-out robot, motion controller, mold monitoring device, and HMI HUB are used in the injection molding process to construct smart factory as showed in Figure 3. Process data generated in the injection molding process is transferred to MES via HMI HUB. By using the various peripheral devices such as dehumidifier, grinder, and refrigerator, the injection molding process is complex and cost for construction of smart factory is expensive.

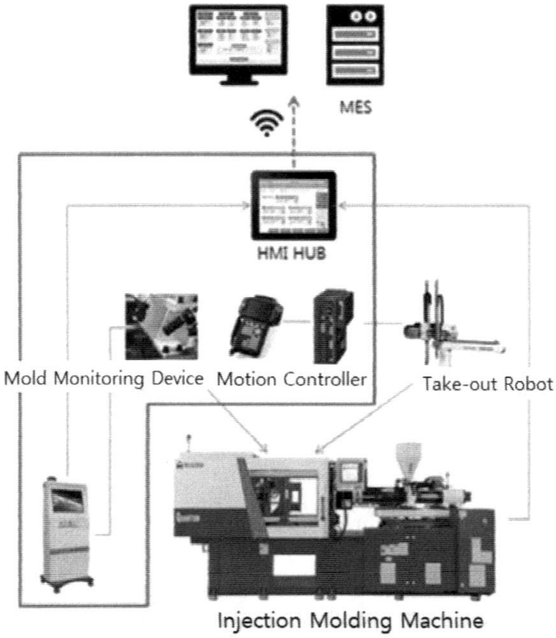

Figure 3. Configuration for construction of smart factory in the injection molding process.

In this study, integrated control system, which consists of take-out robot controller, image processing module, mold monitoring device controller, and process data acquisition interface module, is proposed to simplify the injection molding process and to reduce the cost for construction of smart factory.

2. Configuration of integrated control system

Integrated control system having the function such as take-out robot controlling, camera image processing, mold monitoring device controlling, process data acquisition, and data transferring, consists of take-out robot controller, image processing module, mold monitoring device controller, and process data acquisition interface module as showed in Figure 4.

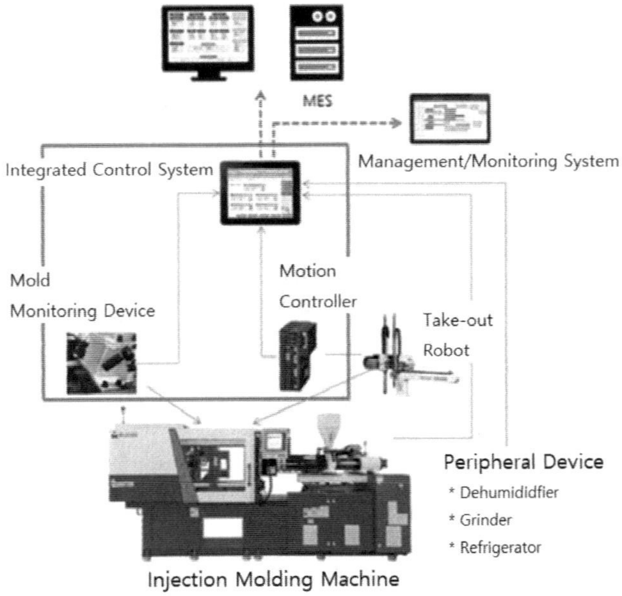

Figure 4. Configuration of integrated control system.

The process data generated by mold monitoring device, take-out robot, injection molding machine, and peripheral devices such as dehumidifier, grinder, and refrigerator are saved in the integrated control system and transferred to MES and process management and monitoring system.

3. Development of integrated control system

Integrated control system, which consists of take-out robot controller, image processing module, and process data acquisition interface module, is developed to simplify the injection molding process and to inexpensive the cost for construction of smart factory

3.1. Take-out robot controller

Take-out robot controller has functions such as 8-axis motor control, interpolation motion, point to point (PTP) motion, round motion. Figure 5 shows the developed take-out robot controller.

Figure 5. Developed take-out robot controller board.

3.2. Image processing module

Image processing module consists of camera, illuminator, and image processing algorithm. The mold status image captured by camera is transferred to the microprocessor, which has the image processing algorithm, of integrated control system and processed to catch the defect of the mold status. Figure 6 shows the captured image and processed image by developed image processing module.

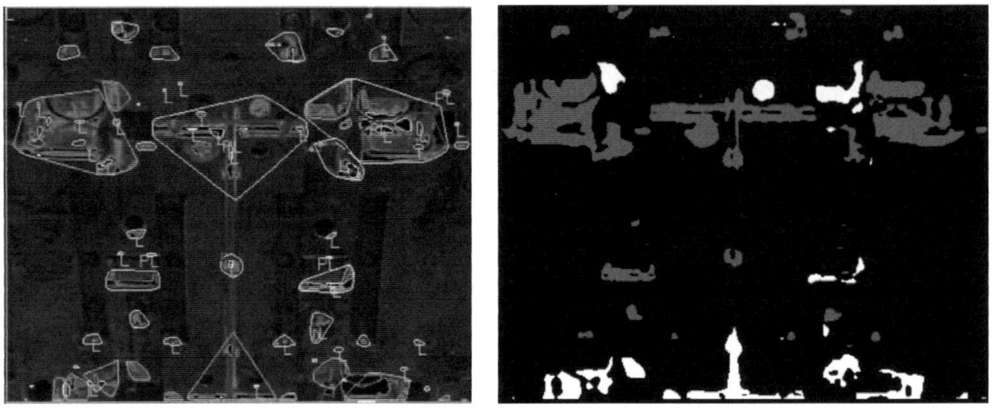

(a) (b)

Figure 6. (a) Captured image and (b) processed image by developed image processing module.

3.3. Process data acquisition interface module

Process data acquisition interface module has 8 analog interfaces, 32 each digital input and output, and serial port. Figure 7 shows developed process data acquisition interface module.

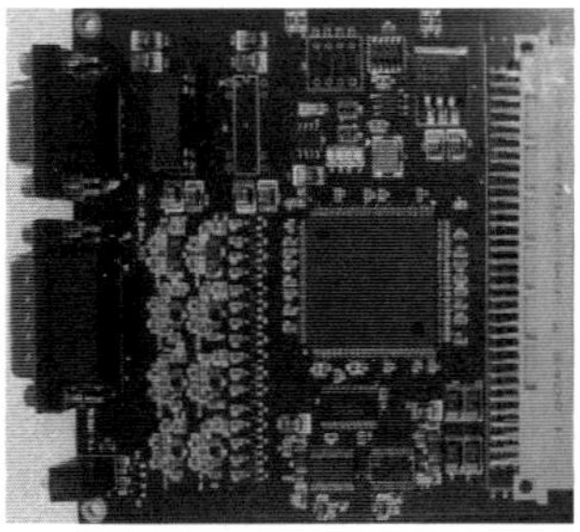

Figure 7. Developed process data acquisition interface module.

4. Conclusions

In this study, integrated control system having the function such as take-out robot controlling, camera image processing, mold monitoring device controlling, process data acquisition, and data transferring is proposed. By adoption of integrated control system, the injection molding process can be simplified and the cost for construction of smart factory can be inexpensive.

5. References

[1] Jung J T and Lee B K 2014 Numerical analysis of mold deformation including plastic melt flow during injection molding *Trans. Korean Soc. Mech. Eng.* A **38** 719-25

[2] Lee K T 2016 Smart factory technology trend and R&D road map *The Journal of Korea Institute of Electronics Engineers* **43** 16-24

Acknowledgments

Following are results of a study on the "Leaders in INdustry-university Cooperation +" Project, supported by the Ministry of Education and National Research Foundation of Korea.

IOP Conf. Series: Materials Science and Engineering **324** (2018) 012060 doi:10.1088/1757-899X/324/1/012060

Structure Optimization of Porous Dental Implant Based on 3D Printing

Fangqiu Ji[1,2],Chunyu Zhang[2*] and Xianshuai Chen[2]

1 School of Mechanical Engineering, Shanghai University of Engineering Science, Shanghai, 201620, China
2 GuangZhou Janus Biotechnology Co., ltd. Floor 1, Buliding A, No.1121 Haibin Rd., Nansha, GuangZhou, 511458, China
E-mail: fq.ji@gzjanus.com, cy.zhang@gzjanus.com

Abstract. In this paper, selective laser melting (SLM) technology is used to process complex structures. In combination with the theory of biomedicine, a porous implant with a porous structure is designed to induce bone cell growth. The mechanical strength advantage of SLM was discussed by observing the metallographic structure of SLM specimen with mechanical microscope and mechanical tensile test. The osseointegration of porous implants was observed and analyzed by biological experiments. By establishing a mechanical model, the mechanical properties of the bone implant combined with the jaw bone were studied by the simple mechanical analysis under static multi loading and the finite element mechanical analysis. According to the experimental observation and mechanical research, the optimization suggestions for the structure design of the implant made by SLM technology were put forward.

1. Introduction

With the aging of the society, the number of people who need to repair dental teeth is becoming larger and larger. In addition to the traditional method of denture installation, more people have chosen a more stable and beautiful dental implant. At present, the production process of dental implant maturity is precision investment casting. There are many defects in the implants produced by this method, and the common problem is that the mechanical properties of the implants can not meet the standard because of the incomplete casting. Second, this method does not take into account individual differences. Human oral and maxillofacial bones are divided into four types of bone, each type of bone structure and density are different. The implant produced by this method can not accurately control the elastic modulus and other data. In the interaction with human oral and jaw, it will cause the problem of stress shielding, thus affecting the normal growth of bone [1].

2. Production principle and characteristics of SLM Technology

Selective laser melting technology has made up for the shortcomings of the original 3D printing technology, and made it develop rapidly in many industrial fields. The production principle of SLM technology is to draw 3D modeling of implants by computer aided design software such as CAD, and then slice the 3D graphics into slices by software. At the beginning of processing, the SLM equipment controls the powder laying system to lay a layer of metal powder in the molding cavity, and the scanning system controls the laser beam to be burned on the powder according to the image information of the first slice. After the first layer of powder is processed, the system controls the piston to move the worktable downward, which is equal to the distance of a layer of powder. Then a powder is laid on the first layer of the processing, and then the second layers are processed and formed. Repeat the above steps until the entire workpiece is formed.

SLM technology not only has fast processing speed, but also has high machining precision. SLM technology is a kind of material processing method, so it can save a lot of precious raw materials in the process of production, thereby reducing the cost greatly. The most outstanding feature of SLM is that it can make complex shape and internal structure according to the work needs, so as to achieve better performance requirements. Therefore, the porous structure of the implant can be printed by its 3D, thereby changing the elastic modulus of the implant, and ultimately avoiding the problem of stress shielding. The porous structure implant can further optimize the structure design according to the finite element analysis structure.

3. Performance analysis of SLM specimen

Using German equipment SLM125 YLR-100-W, the commercial two grade pure titanium (CP-Ti Grade 2) powder was used as the raw material to produce 3D test piece. Then, the metallographic structure is observed by metallographic microscope, as shown in Figure 1. The metallographic structure of the part is shown in Figure 2. The metallographic structure of the parts produced by the traditional casting process is shown. It is not difficult to find that the Figure 1 is very fine acicular martensite, indicating that it contains higher strength and hardness. Figure 2 shows that it contains more martensite, because the martensite microstructure is decomposed into fine acicular $\alpha+\beta$ phase when heated in the $\alpha+\beta$ region, and the phase is easy to fracture due to shear mechanism [2]. The mechanical properties of the parts produced by SLM technology are better than those produced by the traditional casting process. Through tensile test, the yield strength of SLM specimen is 654 MPa, while the yield strength of commercial pure titanium forgings is 590 MPa[3]. It is shown that the strength of SLM specimen is better than that of forging.

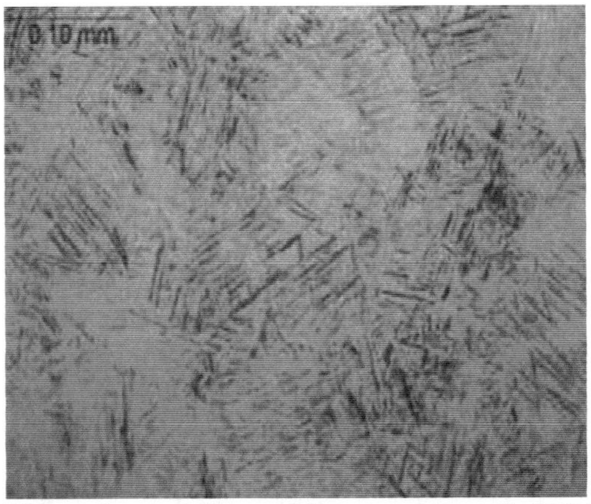

Figure1. metallographic structure of SLM workpiece Figure.

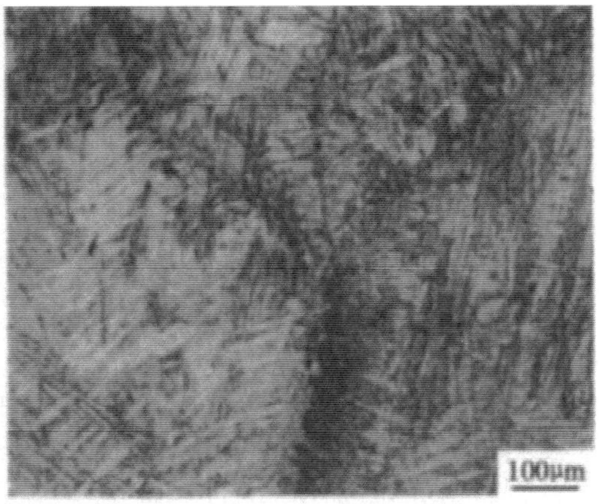

Figure2. metallographic structure of casting parts.

4. Design and experimental analysis of porous structure

In view of the biocompatibility of implant and human jaw, as well as to avoid stress shielding effect between the implant and the implant, and to improve the physical properties of implant. Based on the characteristics of SLM technology which can be used to manufacture complex structures, some porous structures are designed in this paper. Porous structure materials are widely used in biomedical materials because of their unique role in improving biocompatibility.

As shown in Figure 3, the implant is embedded in the outer part of the oral cavity and jaw using a porous structure design. Such a design is the use of porous structure can adhere to the growth of bone cells and promote bone cell ingrowth into the gap, so that the implant is more stable and jaw fixed. It is beneficial for the stress of the implant to be dispersed into the jaw and reduce the stress of the implant and the denture on it. In addition, the porous structure can adjust the density, strength and elastic modulus of the material by adjusting the porosity, so that the mechanical properties of the implant and the jaw match, and finally achieve the purpose of weakening or eliminating the stress shielding effect. Because of the improper treatment of stress shielding effect, the necrosis of the bone near the implant and the severe consequence of reducing its bearing capacity [4].

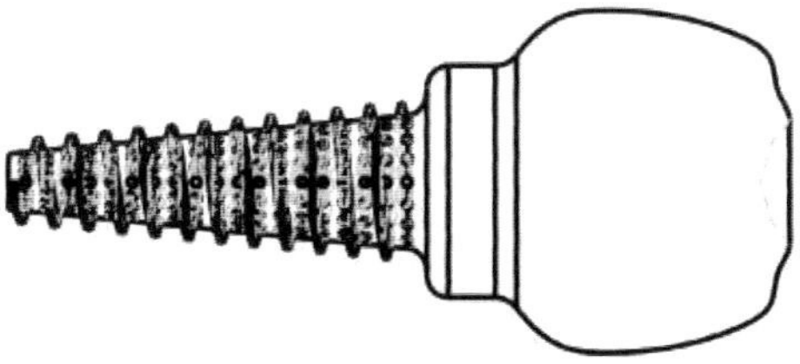

Figure 3. schematic diagram of implant structure.

At present, studies in the field of medicine show that the pore size of implants suitable for bone cell ingrowth is 100~600 μ m, considering the self recovery of capillary and jaw tissue, so the pore size is larger than 300 μ m[5]. Fukuda [6] found that the porous 500 μ m implant had the best effect on the ingrowth of bone cells through the implantation of different pores on the animal skeleton. Therefore, the aperture design of 500 μ m is reasonable, and the specific numerical value needs to be further determined. The laser spot of the equipment SLM125 is 87 μ m. Theoretically, the manufacturing accuracy is enough to produce the implant according to the design requirements.

According to the existing medical research, the porosity of human cortical bone is 5.36% ~ 14.1%, the pore size is 5~150 μ m, the compressive strength is 160 ~ 200MPa, and the elastic modulus is 3 ~ 30 GPa. The porosity of cancellous bone is 30% ~ 95%, the pore size is 20~1000 μ m[7], the compressive strength is 4 ~ 7 MPa, and the elastic modulus is 3 ~ 30 GPa. According to the jaw structure of different individuals, the implant can increase its biocompatibility with the jaw by adjusting the porosity. Therefore, the optimal scheme is to design the porosity distribution of the implant bone segment into two segments according to the different porosity distribution of the implant cortical bone and cancellous bone. At the same time, the implant elastic modulus approximate to the human skeleton can also effectively avoid stress shielding [8]. The elastic modulus of porous structure can be obtained according to the [9] formula of Nielsen model:

$$E = E_0 \times \frac{(1-\varphi)^2}{1+\left(\frac{1}{\rho}-1\right)\times\varphi} \tag{1}$$

In the formula, E is the elastic modulus of porous structure, E_0 is the modulus of elasticity of compact structure, φ is porosity, and ρ is shape factor. Assuming that the designed implant is the standard optimal shape, that is, the shape does not affect the structural strength of the implant, then ρ =1 is simplified (1) into a finite element:

$$E = E_0 \times (1-\varphi)^2 \tag{2}$$

It is known that the elastic modulus of SLM compact specimen is 110GPa, and the elastic modulus of porous structure should be designed from 3 to 30GPa according to the principle of similar elastic modulus. Therefore, the design range of porosity is 48% - 84%, which is similar to the porosity of human cancellous bone. Although it can not meet the approximate requirement of human skin bone porosity, similar elastic modulus can still avoid stress shielding. Therefore, the porosity of the cortical bone segment in the implant was further optimized when the porosity and elastic modulus could not be optimized simultaneously.

According to the design range of the porosity, the exact porosity value was selected randomly, and the pore size parameters of 500 micron m were used, and the standard rotational structure of the implant was made by SLM. Biological experiments were carried out and planted in the thigh of New Zealand rabbits. After eight weeks, the implant was scanned by histological section and CT scan, and the 3D reconstruction of CT scan was performed by software. As shown in Figure 4, the microstructure of the implant was reconstructed by microscopic CT, and the yellow was the new bone tissue. It can be seen that the bone tissue has entered into the internal pores, and the osseointegration is good.

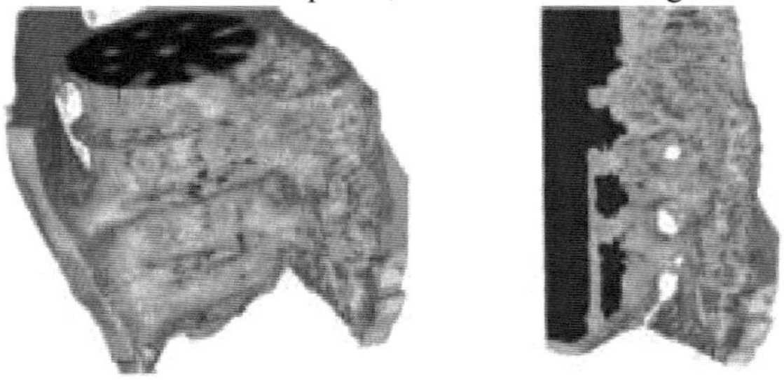

Figure 4. 3D CT reconstruction of implant tissue

5. Physical analysis of implant

5.1 establish the mechanical model

As shown in Figure 5, the anatomical structure of the dental implant after the implantation of the conventional dental implant is shown. Dental implant is composed of denture, implant foundation and abutment. Implant foundation is divided into cortical bone segment and spongy bone segment. Abutment is also called gingival segment. In the study, the design was used to simulate the force of the implant. Firstly, the mechanical analysis of the implant under the static multi loading condition was carried out. As shown in Figure 6, this is the schematic diagram of molar force simulation. In order to facilitate the calculation, it is simplified as shown in Figure 7. The length of the implant in the figure is a, and all the implants are implanted into the jaws. The length of the B part is the abutment of the implant, which connects the implant and the artificial denture. Kong and other [10] for long-term clinical trials, and from a biomechanical point of view, the diameter of the implant should be greater than 4.0 mm, length should be greater than 11.0 mm. According to the characteristics of Asian teeth, the preliminary design of abutment height B is equal to 3.0 mm, the artificial denture height h is 7.0 mm, and the pile foundation length a is 14 mm. The implant foundation is equivalent to a cylinder, the diameter of R_2 is 6.0 mm, and the implant abutment and the artificial denture are equivalent to a cylinder, and the diameter of the R_1 is 8.0 mm. The a segment is an implant foundation embedded entirely in the alveolar bone, which is considered to be surrounded by the homogeneous load of the alveolar bone. The B segment is embedded in the part of the periosteal flap, and its stress on the implant can be neglected due to its soft tissue characteristics.

When chewing the food, the molar plays a major role, which mainly bears the combined action of compressive stress, shear stress and torque. In the establishment of the mechanical model, it is assumed that the implant is in an ideal state, that is, the bone cell ingrowth into the implant and make it into an integral body with the jaw. Because the mechanical properties of the implant conform to the assumption of continuity, uniformity and isotropy, it is equivalent to the bar and then its mechanical analysis is carried out. By drawing the bending moment diagram of the implant, the dangerous section is determined as the surface L1 in figure 7. Studies have shown that normal people's molar will chew about 98 N to 225.4 N bite force [11], so F_1 take 225.4 N. Taking into account the masseter muscle plays a leading role in masseter muscle, and the masseter muscle is the control of the upper and lower jaw occlusion, so chewing teeth under the condition of the horizontal direction of the force F_2 is small, according to the oral physiology of Ohammed research, take F_2 120 N [12].

In this mechanical model, according to figure 7, the bending moment M1 is produced by the interaction of F_1 and F_2. According to the working characteristics of molars, the vertical load is regarded as uniform load, and its horizontal force is regarded as a point. Thus, the maximum torque and bending moment of the implant under the limit condition are obtained by using this model, and the formulas for calculating the torque M0 and the bending moment M1 are as follows:

$$M_0 = F_2 \times \frac{R_1}{2} \tag{3}$$

$$M_1 = \int_m^n \mathrm{qxdx} + F_2(b+h) \tag{4}$$

In the formula (m, n) for the F_1 interval, $q=F_1/R_1$, F_1 as a uniform load on the surface of the denture. By calculation, M_0 is equal to 0.48 N • m, and M_1 equals 1.43 N • m.

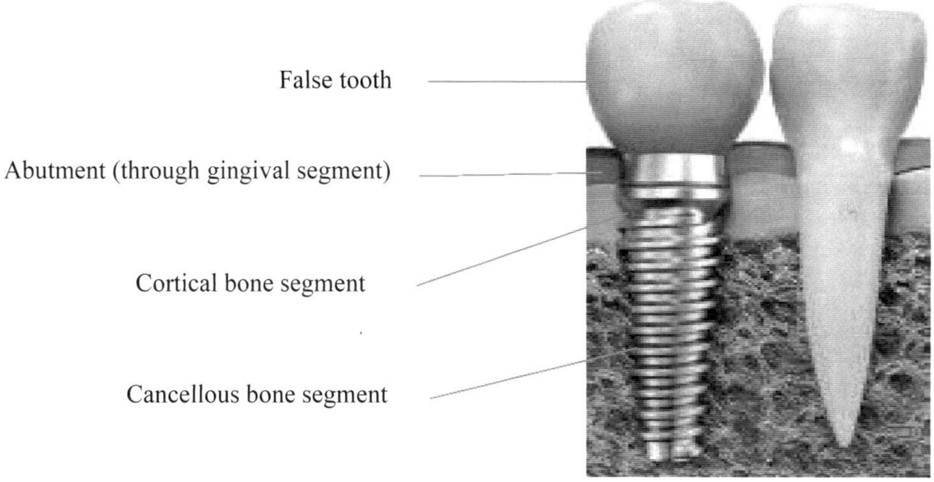

Figure 5. implant implant simulation

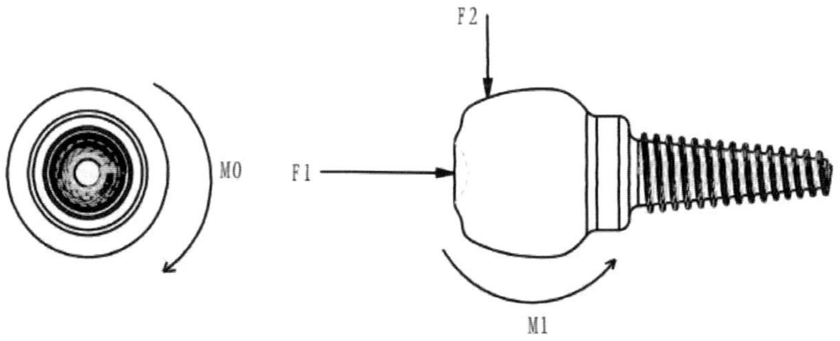

Figure 6. sketch of simulated force of molar

IOP Conf. Series: Materials Science and Engineering **324** (2018) 012060 doi:10.1088/1757-899X/324/1/012060

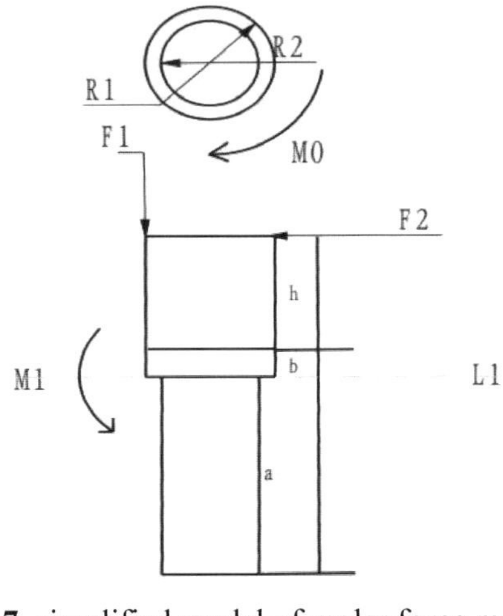

Figure 7. simplified model of molar force model

5.2 finite element analysis

The hypothesis of the mechanical model is that the implant and the jaw form a whole, ignoring the influence of the porous structure of the implant. However, the bending moment and torque are also present in the porous implant, and the data obtained from the model is similar to the true data in the porous structure when the implant and the jaw are in good osseointegration. After removing the simplified model of osseointegration, a finite element model for simulating the stress of implant with porous structure was established. The static multi loading mechanical model was calculated and the relevant data were imported into Ansys software to simulate the stress finite element analysis. Figure 8 is the deformation analysis diagram. According to the simulation results, the maximum deformation occurs at the abutment position of the implant. Figure 9 is the stress analysis diagram. According to the simulation results, the stress at the joint between the neck and the abutment of the pile is greatly influenced by the stress. The maximum stress of the implant at the dangerous interface is 35.1MPa, which is less than the yield strength of the compact structure SLM specimen. The relationship between the yield strength of porous structure and porosity and pore size remains to be experimentally investigated. The dangerous interface between the finite element and the mechanical model is similar to that of the dangerous interface, which indicates that the mechanical model is of higher accuracy. Through the stress distribution diagram, it is not difficult to find the bone graft part of the implant foundation, that is, the stress of the porous structure site is smaller, and less than 3.9 MPa. The experimental data will provide suggestions for further optimization of implant structure.

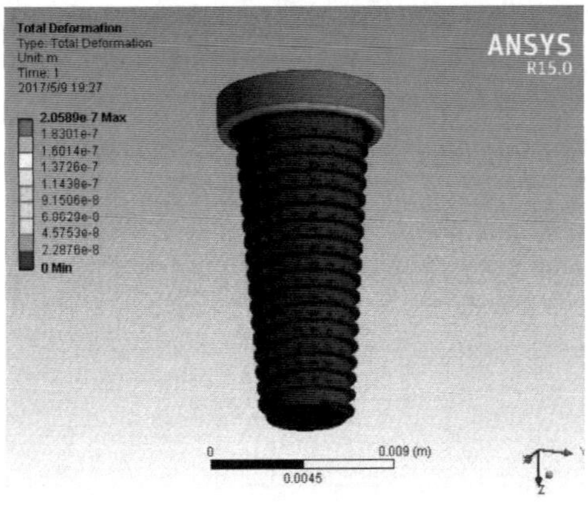

Figure 8. deformation analysis diagram

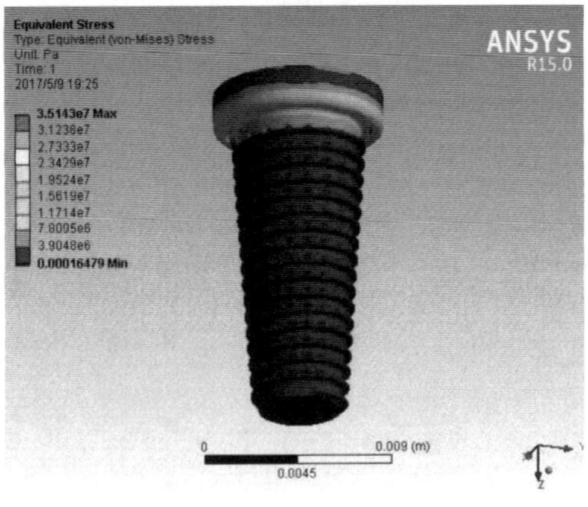

Figure 9. stress analysis diagram

6. Discussion

In this paper, the strength advantage of SLM technology for manufacturing implants was analyzed by observing the microstructure and tensile test of SLM pure titanium specimens. According to the theory of osseointegration, it is known that the porous structure contributes to the ingrowth of bone cells, and the porous structure can effectively avoid the problem of stress shielding. At the same time, controllable manufacturing of porous structure is another advantage of SLM technology compared with traditional manufacturing process.

According to the existing biomedical theories, it is demonstrated that the pores of 500 μ m are the optimal size for inducing bone ingrowth. According to the machining accuracy of the existing German equipment SLM 125, the machining dimension of 500 μ m porous structure pore is determined. In order to avoid stress shielding, it is necessary to adjust the elastic modulus of the implant close to the human bone tissue by adjusting the porosity of the implant. According to the Nielsen model and the elastic modulus distribution range of the human bone tissue, the design reference porosity of the implant was 48% - 84%. At the same time, the porosity of implant and jaw is beneficial to improve the biocompatibility of the implant, so the exact porosity value should be determined according to the specific porosity of the patient's jaw. The porosity distribution range of the implant is close to that of the cancellous bone, but the distribution of the porosity is different from that of the human skin. Therefore, the porosity meets the requirement of biocompatibility when the implant is implanted into the porous structure of cancellous bone segment. When designing the porous structure of cortical bone implants, only the design requirements of elastic modulus can be taken as the optimum design standard. Considering that the porosity of cortical bone is small, the elastic modulus of cortical bone is larger than that of Nielsen model. According to the reasons of stress shielding, it is known that the elastic modulus of implant is not greater than that of human bone tissue. Therefore, it is reasonable to design the porosity of implants larger than the porosity of cortical bone. Through biological experiments, the porous implant with pore size of 500 μ m and porosity of 48% - 84% was verified to have good biocompatibility, and a better implant jaw bone bonding plane was formed.

The formation of a good bone bonding plane is the prerequisite to ensure the initial stability of the implant, and the implant itself has enough strength and stiffness is the key to ensure the service life of the implant. Through theoretical analysis, the size and length of dental implants were determined. According to the analysis of oral anatomy, the load of implant in the limit condition was determined. By establishing a static multi loading mechanical model after implant implantation, it is assumed that the implant and jaw form a whole, ignoring the influence of the porous structure of implant. The mechanical properties of the implant are consistent with the assumptions of continuity, uniformity, isotropy, and then equivalent to the bar. Then the complex three-dimensional force is converted to two dimensions, and the key data of the moment and torque are obtained. The location of the dangerous interface of dental implant under static multi loading was determined.

The static load moment and torque obtained by the mechanics model of the implant in the conventional finite element simulation will be ignored, but it is real, and had better osseointegration in implant and jaw, model data and real data obtained in the porous structure is near. At the same time, the bending moment and torque are helpful to simplify the bone bonding model and remove unnecessary complex parts such as denture. The osseointegration model of porous dental implant and alveolar bone was established, and then the finite element analysis was carried out. The results of finite element analysis is the design of the porous structure of the strength and stiffness of the implant with normal work safety requirements, and to determine the similar position and static risk interface multi load mechanics model of the position, further validation of the static load mechanical model of high precision. According to the finite element analysis, the deformation of the implant occurs at the neck and base of the abutment, which is due to the fact that the implant can not transmit the force directly to the jaw in the gingival segment, which is consistent with the hypothesis of the static multi loading mechanical model. Therefore, when designing the neck and abutment of the implant, the stiffness should be increased as much as possible in the design requirements, that is to reduce the porosity and improve the elastic modulus. At the same time, this design also meets the requirements mentioned above to eliminate stress shielding and improve biocompatibility to reduce the porosity and increase the elastic modulus of implant cortical bone segment. According to the stress analysis diagram, most of the stress is concentrated in the connecting section between the implant neck and abutment, and the dangerous interface is between the bone graft section and the gingival segment. Therefore, in the design of porous implant, the stress concentration should be avoided and the density should be increased at the connection between the implant neck and abutment. It also shows that the two - segment implant has better structural strength than the one - stage implant. The stress of the porous structure of the whole cancellous bone segment is less than that of the human cancellous bone. Therefore, according to the different bone conditions, the intensity of individual design of porous structure is in line with the design requirements.

7. Summary

Through the existing biomedical theory, it is proved that porous structure with pore size of 500 μ m and porosity of 48% ~ 84% is proved to be beneficial to induce bone regeneration, and the rationality of the derivation is proved by biological experiments. The design of porous structure for porous implant according to different bone conditions is in line with the design requirements, but the connection between the implant neck and abutment should be designed with high density to eliminate stress shielding and improve the strength and stiffness of implant.

Reference

[1] Traini T, Mangano C, Sammons RL, et al. Direct laser metal sintering as a new approach to fabrication of an isoelastic functionally graded material for manufacture of porous titaniumdental implants [J]. Dent Mater,2008, 24(11):1525-1533.

[2] Xin S W,Zhao Y Q . Discussion on heat treatment and precipitation phase of titanium alloy[J].Metal heat treatment,2006,31(6):39-42.

[3] X.P. Li,J. Van Humbeeck,J.P. Kruth. Selective laser melting of weak-textured commercially pure titanium with high strength and ductility: A study from laser power perspective[J]. Materials & Design,2017,116:.

[4] Van der Stok J, Van der Jagt OP, Amin Yavari S, et al. Selective laser melting-produced porous titanium scaffolds regenerate bone in critical size cortical bone defects [J]. J Orthop Res, 2013,31(5):792-799.

[5] Polo-Corrales L, Latorre-Eateves M, Ramires-Vick JE. Scaffold design for bone regeneration [J]. J Nanosci Nanotechnal,2014,8(1):15-56.

[6] Fukuda A, Takenoto M, et al. Osteoinducation of porous Ti implants with a channel structure fabricated by selective laser melting [J]. Acta Biomaterialia,2011,7(5):2327-2336.

[7] Dabrowski B, Swieszkowski W, Godlinski D, et al. Highly porous titanium scaffolds for orthopedic applications [J]. Biomed Mater Res-applbiomater,2010,95B(1):53-61.

[8] Tan B S. Effect of Biomechanics on long-term retention of implants[J]. Chinese Journal of Prosthodontics,2014,15(02):122.

[9] Kang J F,Wang L. Microstructure design of 3D printing variable modulus metal prosthesis[J]. Journal of Mechanical Engineering,2017,53(05):175-180.

[10] Kong L, Li T, Wang Y,et al. Selection of optimal dental implant diameter and length in type IV bone: a three-dimensional finite element analysis[J]. Oral Maxillofac Surg,2009,38(10):1077-1083.

[11] Guo N S,Liu J P. The stress analysis of periodontal membrane of fixed abutment tooth with different thickness [J] .Chinese Clinical rehabilitation 2004,29(8):6352-6353.

[12] Ohammed Ibrahim M, Thulasingam C, Nasser KS, et al. Evaluation of design parameters of dental implant shape , diameter and length on stress distribution:a finite element analysis[J]. Indian Prosthodont Soc,2011,11(3):165-171.

Acknowledgments

This work is supported by the Science and Technology Program of Nansha, China (No. 2016GJ009).

Study of heat generation and cutting force according to minimization of grain size (500 nm to 180 nm) of WC ball endmill using FEM

J H Byeon[1], F Ahmed[1], T J Ko[1], D K lee[2], J S Kim[2]

1 School of Mechanical Engineering, Yeungnam University, 214-1, Dae-dong, Gyeongsan-si, Gyeongsangbuk-do, South Korea, 712-749
2. OSG Korea, Deagu Metropolitan City, South Korea

E-mail: tjko@yu.ac.kr

Abstract. As the industry develops, miniaturization and refinement of products are important issues. Precise machining is required for cutting, which is a typical method of machining a product. The factor determining the workability of the cutting process is the material of the tool. Tool materials include carbon tool steel, alloy tool steel, high-speed steel, cemented carbide, and ceramics. In the case of a carbide material, the smaller the particle size, the better the mechanical properties with higher hardness, strength and toughness. The specific heat, density, and thermal diffusivity are also changed through finer particle size of the material. In this study, finite element analysis was performed to investigate the change of heat generation and cutting power depending on the physical properties (specific heat, density, thermal diffusivity) of tool material. The thermal conductivity coefficient was obtained by measuring the thermal diffusivity, specific heat, and density of the material (180 nm) in which the particle size was finer and the particle material (0.05 μm) in the conventional size. The coefficient of thermal conductivity was calculated as 61.33 for 180nm class material and 46.13 for 0.05μm class material. As a result of finite element analysis using this value, the average temperature of exothermic heat of micronized particle material (180nm) was 532.75 ℃ and the temperature of existing material (0.05 ㎛) was 572.75 ℃. Cutting power was also compared but not significant. Therefore, if the thermal conductivity is increased through particle refinement, the surface power can be improved and the tool life can be prolonged by lowering the temperature generated in the tool during machining without giving a great influence to the cutting power.

1. Introduction

Density, specific heat, and thermal conductivity vary depending on the particle size of the material. As a result of analysis of the influence of the heating part of each material property (density, specific heat, thermal conductivity) through the preliminary experiment, it was found that the thermal conductivity has the greatest influence. In addition, the influence of the properties (density, specific heat, thermal conductivity) on the cutting force was not influenced. In this study, the finite element analysis is performed to investigate the change of heat generation and cutting power depending on the thermal conductivity which is changed by miniaturization of tool material.

Table 1. Result of measuring properties.

	Thermal conductivity	**Specific heat**	**Density**	**Thermal diffusivity**
Unit	$W/(m \times °C)$	$J/[kg \times °C]$	kg/m^3	mm^2/s
0.5 ㎛	46.913	190	14.439	18.1
0.18 ㎛	71.063	244	14.784	19.7

IOP Conf. Series: Materials Science and Engineering **324** (2018) 012061 doi:10.1088/1757-899X/324/1/012061

The material sizes of 500 nm and 180 nm were measured using the flash method, and the finite element analysis was performed using the measured values. Material property of table is table 1. The tool performance is evaluated by cutting force analysis, thermal analysis can be used to measure the maximum temperature of the tool and expect tool life to be prolonged at low tool temperatures.

2. FEM Set up

In this study, The tool is a ball-end mill which has a 2mm diameter and the workpiece is STD61 (HRC 52) which is used as a mold steel. The shape of the tool was fabricated using CAD(Fig.1). In the case of the analysis Fem is using by the commercial software "Advant Edge". the tool is rotated 720 ° (Fig.2) it means the cutting is happen 4 times. As the analysis condition, the mesh size of the workpiece set to 0.01mm. the mesh size of tool and contacting part(interface with workpiece and tool) define to 0.01mm. Cutting conditions is specified as down-milling, RPM 20000, FeedRate 800 (mm / min) cutting width 0.4 (mm) and cutting depth 0.4 (mm).

Cutter Diameter [Do]	2	mm
Core Diameter [Di]	1.5	mm
Number of Flutes	2	mm
Radial Rake Angle [a]	3	deg
Helix Angle [c]	29.5	deg
Radial Relief Angle [b]	30	deg
Edge Radius [r]	0.04	mm
Flute Radius [Rf]	0.08	mm
Width of Land [wol]	0.331	mm
Tool Length [ta]	5	mm

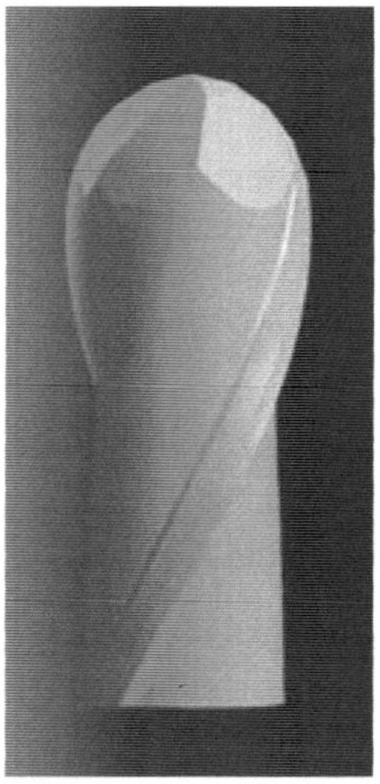

Figure 1. Modelling tool by 3D CAD

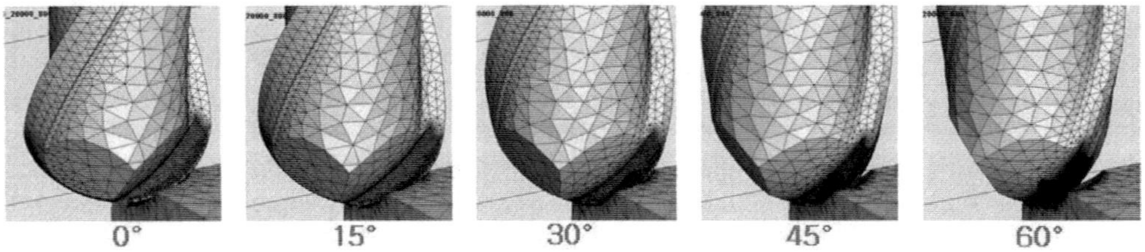

Figure 2. Example of FEM Boundary condition

3. FEM result

The temperature change according to the thermal conductivity is shown in the graph in Fig. 3, and the graph of cutting power and cutting force appears almost the same. Changes in thermal conductivity affect the tool temperature and have no significant effect on cutting power. Thermal analysis is performed by changing the temperature distribution at the cutting edge. Figure 4 shows the heat generated by the rotation angle. Indicates the temperature distribution change that occurs when cutting of the first blade proceeds. In a tool with a low thermal conductivity, thermal diffusion is not easy and the maximum temperature of the tool is increased due to heat concentration at the tip of the blade. Also, it is expected that the material with higher thermal conductivity will have more heat coming through the tool. However, as the tool receives the same amount of heat, the range of diffused heat spreads and the maximum temperature at the tool tip.

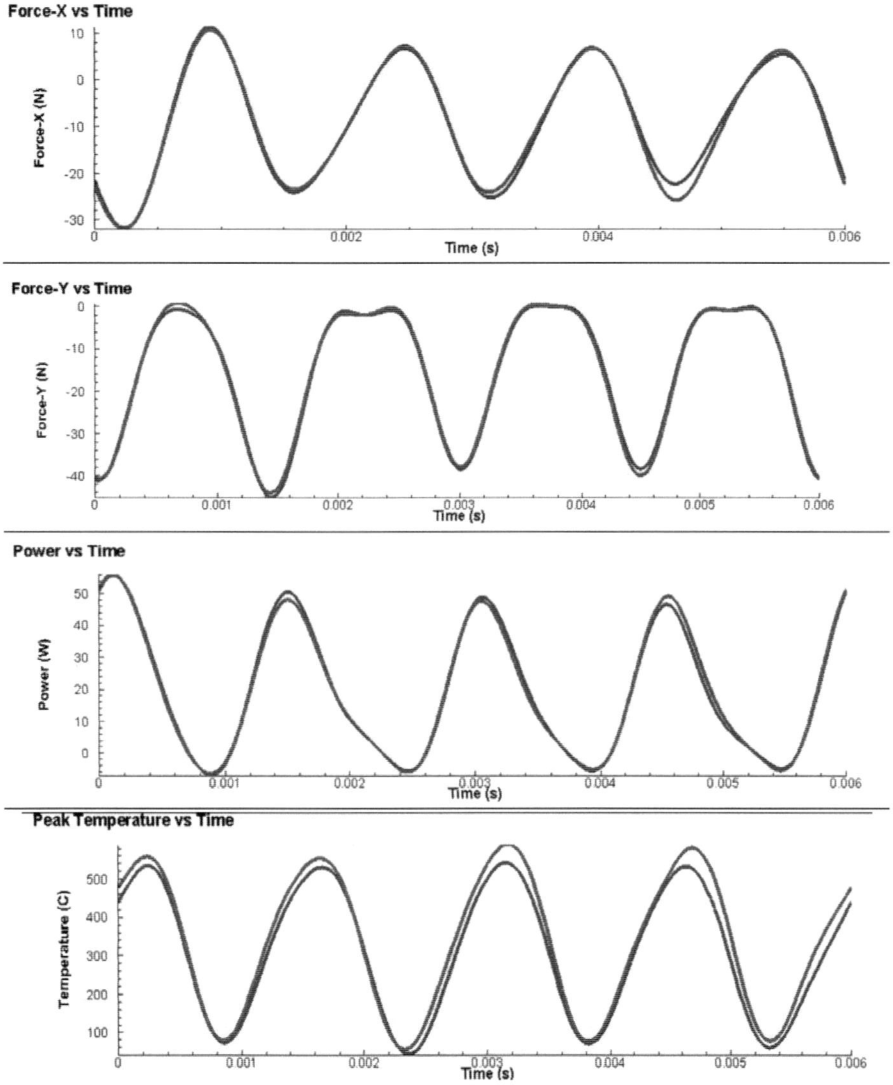

Figure 3. Cutting force & Cutting power & Temperature of tool [Red : 500nm ,Blue 180nm]

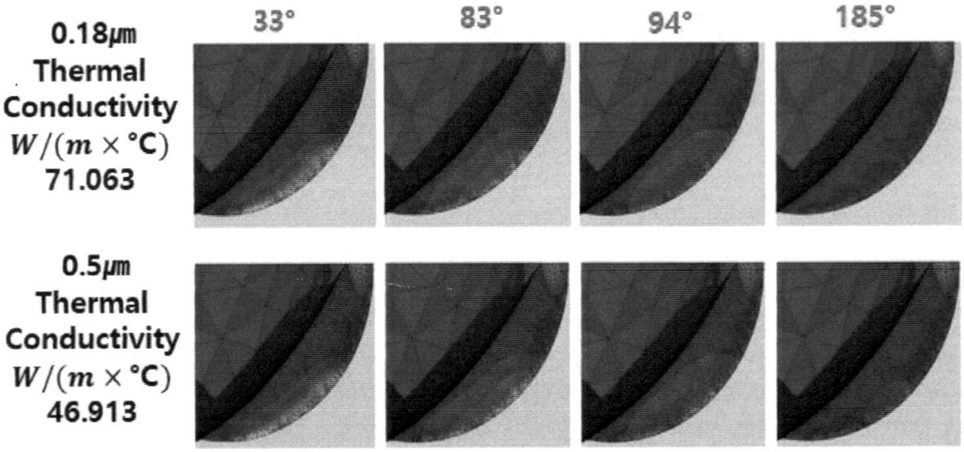

Figure 4. Heat transfer on blade

4. Conclusion

The change in thermal conductivity does not have a significant effect on the cutting force and cutting power. As a result of comparing the maximum temperature of the blade part, the temperature of 180nm tool with high thermal conductivity is low temperature. It means that at high thermal conductivity, the temperature distribution range is wide due to heat diffusion, and the maximum temperature of the blade is low. It also benefits on fast heat removal after cutting. Same as the result of the preliminary experiment.

References

[1] David K. Aspinwall et. al. "Cutting temperatures when ball nose end milling Y-TiAl intermetallic alloys" CIRP Annals- Manufacturing Technology 62 (2013) 75-78.

[2] Zhengwen Pu, Anshul Singh "High speed ball nose end milling of hardened AISI A2 tool steel with PCBN and coated carbide tools" Journal of Manufacturing Processes 15 (2013) 467-473

[3] Choi, S. Y., Kwon, D. G., Park, I. S., Wang, D. H., "A Study on the Cutting Forces and Tool Deformation when Flat-ended Pocket Machining", Journal of the Korean Society of Manufacturing Process Engineers, Vol.16, No.2, pp.28~33, 2017.

Acknowledgments

This work was supported by the Technology Innovation Program (or Industrial Strategic Technology Development Program(10067064, Development of tungsten carbide and PCD tools for high-quality finishing of high-precision parts and molds) funded By the Ministry of Trade, Industry & Energy (MOTIE, Korea).

IOP Conf. Series: Materials Science and Engineering **324** (2018) 012062 doi:10.1088/1757-899X/324/1/012062

Preparation of polyacrylonitrile nanofibrous membrane for fabrication of separator of lithium ion batteries

W U Arifeen[1], T Dong[1], R Kurniawan[1] and T J Ko[1]

1 School of Mechanical engineering, Yeungnam University, South Korea
E-mail: tjko@yu.ac.kr

Abstract. In this paper, the manufacturing process and morphology of nano fibrous membranes are discussed. These membranes are explored as separators in rechargeable lithium ion batteries. The function of separator is to allow the flow of ions while protecting the physical contact between positive and negative electrode. Therefore, the porosity, mechanical strength and thermal stability of separators possess significant importance. The separators are manufactured by electrospinning process and later the morphology is studied with the help of scanning electron microscope (SEM) images. The separator is prepared by polyacrylonitrile (PAN) and then exposed to the hot plate. The uniform, continuous and dense nano fibrous membrane is prepared with the help of electrospinning process providing the prevention of physical contact between electrode and stable enough to work in high temperatures leading to high performance lithium ion batteries separators.

1. Introduction

The lithium ion batteries have been utilized as the power source for most of the electronic devices and electric vehicles due to its higher energy density, longer service life, lower self-discharge rate and higher operational voltage than conventional batteries [1]. The separator is placed between negative and positive electrode to prevent the physical contact while allowing the flow of ions. Separator is either polymeric membrane or non-woven fabric mat depending on the fabrication process. The separator is microporous layer, chemically and electrochemically stable towards electrode and electrolyte, mechanically strong to withstand the high pressure during the assembly of the battery [2].

In this study, we demonstrated the experimental procedure, fabrication process and morphology from SEM images of nano fibrous membrane.

2. Experimental setup

2.1. Materials

N,N-dimethylformamide anhydrous (DMF), polyacrylonitrile (PAN, Mw = 150,000) were purchased from Sigma-Aldrich. Both the chemicals were used without further purification.

2.2. Preparation of electrospun nanofibrous membrane

10 wt. % concentration of PAN dissolved in DMF under the temperature of 35 °C and stirred it for 3.5 hours to get a complete phase change of the solution as shown in figure 1. The concentration lower than 7 wt. % results the formation of beads while more than 12 wt.% increases the thickness of nano fibrous membrane [3]. The solution was loaded to syringe having 9-gauge blunt end steel needle. The syringe was mounted to an 11 plus dual syringe pump (Harvard Apparatus). The feed rate was set to 0.04 ml/min. The high voltage of 17kV was applied to the solution through needle using high voltage power supply (NanoNC). The produced nano fibrous membranes were collected on an electrically grounded rotating drum having diameter of 9 cm and length of 20 cm (NanoNC DC90) covered with thin aluminum foil. The rotating speed of the drum was set to 150 rpm. The total time for producing a membrane was 15 min. The membranes were exposed to a heated plate at 50 °C for 4 minutes and pressed it with 70N weighed block.

Figure 1. Preparation of PAN solution

3. Results and discussion

3.1. Morphological structures

The SEM images of nano fibrous membranes show the continuity and uniformity. However, if we compare the compressed and heated membrane with non-compressed and non-heated membrane then it is clearly shown in the figures 2 and figures 3 that the former membrane is denser and closely packed than the later membrane respectively.

Figure 2 compressed and heated nano fibrous membranes

Figure 3 non-compressed non-heated nano fibrous membranes

The fiber diameter is ranging from 800 to 1000 nm and the diameter is uniform through out the membrane which is required to get uniformly distribution of porosity, strength and stability. The fiber diameter is shown in figure 4.

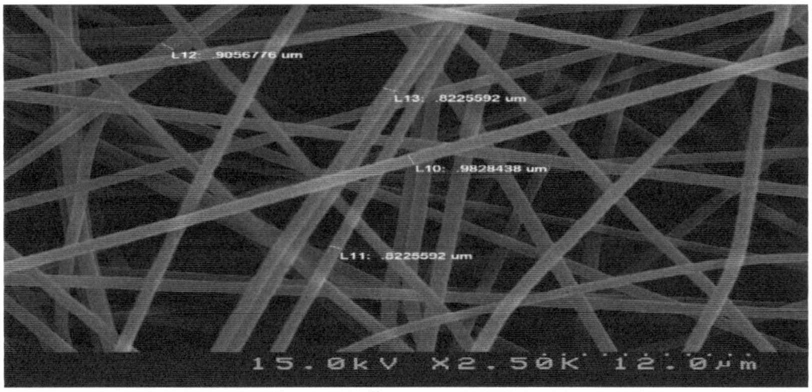

Figure 4 diameter of fibers

4. Conclusions

The thick and continuous membrane is prepared which can be used for fabrication of separators. The uniformity of the membrane is dependent on the PAN solution prepared which must be homogeneous. The compressed and heated membrane can be used in lithium ion batteries making it double layered or triple layered membrane. To making it double layered or triple layered the suitable binder can be used which must be chemically and electrochemically stable towards the electrode and electrolyte material. The low thickness is required for high power and high energy densities of about 25.4 µm [2].

5. References

[1] Wang H and Gao H 2016 A sandwich-like composite nonwoven separator for Li-ion batteries *Electrochim. Acta* **215** 525–34
[2] Zhang S S 2007 A review on the separators of liquid electrolyte Li-ion batteries *J. Power Sources* **164** 351–64
[3] Ma X, Kolla P, Yang R, Wang Z, Zhao Y, Smirnova A L and Fong H 2017 Electrospun polyacrylonitrile nanofibrous membranes with varied fiber diameters and different membrane porosities as lithium-ion battery separators *Electrochim. Acta* **236** 417–23

Acknowledgement
This research was supported by the National Research Foundation of Korea (NRF) grant funded by the Kore government (MSIT No. 217C000360).

IOP Conf. Series: Materials Science and Engineering **324** (2018) 012063 doi:10.1088/1757-899X/324/1/012063

Modal simulation analysis of novel 3D elliptical ultrasonic transducer

R Kurniawan[1], S Ali[1] and T J Ko[1]

1 School of Mechanical Engineering, Yeungnam University, 214-1, Dae-dong, Gyeongsan-si, Gyeongsangbuk-do, South Korea, 712-749
E-mail: tjko@yu.ac.kr

Abstract. This paper aims to present the modal simulation analysis results of a novel 3D elliptical ultrasonic transducer. This research aims to develop a novel elliptical transducer that works in ultrasonic and is able to generate a three dimensional motion in Cartesian space. The concept of the transducer design is basically to find a coupling frequency of the longitudinal-bending-bending mode. To achieve that purpose, the modal simulation analysis was performed to find a proper dimension of the transducer, thus the natural frequency of the 1st longitudinal mode is much closed with the two of natural frequency of the 3rd bending mode. The finite element modelling (FEM) was used to perform this work.

1. Introduction

The elliptical transducer development has increased gradually since the first transducer was proposed by Moriwaki et al. [1]. The Moriwaki's transducer consists of four piezoelectric plates which are located on surrounded beam surface [1]. The bending-bending vibration coupling mode is a fundamental principle of the Moriwaki's transducer. Moriwaki's transducer produced a vibration amplitude about 4 µm at 20 kHz.

A single excitation of the elliptical transducer was developed by Li et al., in which the transducer has a longitudinal-bending vibration mode because of an asymmetric transducer design [2]. An amplitude in tangential direction of 16 µm and 2 µm in normal direction can be generated in 22.5 kHz by the Li's transducer.

The elliptical transducer that has a perpendicular arrangement of the two of piezoelectric actuators was proposed by Kim et al. [3]. An amplitude of 2 µm and 1 µm at 18 kHz can be produced by Kim's transducer. A commercial elliptical transducer product is also available which made by Taga company [4].

A high performance elliptical transducer was proposed by Guo et al, in which the Guo's transducer is capable to generate amplitude approximately 8.79 µm and 7.28 µm at ultrasonic frequency of 28 kHz [5]. The Langevin's transducers with 60° inclination is proposed by Guo et al. which inspired from Kurosawa's ultrasonic motor [6]. The elliptical transducer that uses a bending-bending vibration coupling mode was proposed by Huang et al. [7]. Huang's transducer is capable to generate amplitude in both directions about 3.5 µm in vertical and horizontal at 19.3 kHz.

According to literatures above, there is no elliptical transducer that is capable to produce three-dimensional elliptical motion at ultrasonic frequency. The aim of this paper is to demonstrate the novel design of the elliptical transducer that is capable to produce three-dimensional elliptical motion at ultrasonic frequency. The FEM modal analysis has been carried out to analyse the dimensional parameters of the novel transducer associated to the longitudinal-bending-bending coupling frequency.

2. Design of 3D elliptical ultrasonic transducer

Fig. 1 shows the computer aided design (CAD) of the proposed 3D elliptical ultrasonic transducer in isometric view. In addition, the illustration of the piezoelectric actuator (PZT) polarity is also shown in Fig. 1. The proposed transducer consists of the conical horn, the conical back mass, the preload bolt, the PZT ceramics (full and half), and the steel thin plate. The conical horn is made by titanium alloy (Ti-6Al-4V) which has mechanical properties such as Young's modulus is 1.05 x 10^{11} Pa, mass density is 4430 kg/m^3, and Poisson's ratio is 0.31. The conical back mass is made by stainless steel (AISI 304) which has mechanical properties such as Young's modulus is 1.9 x 10^{11} Pa, mass density is 8000 kg/m^3, and Poisson's ratio is 0.29. The PZT ceramics made by SunnyTec Company was used in which has mechanical properties such as Young's modulus is 8 x 10^{10} Pa, mass density is 7700 kg/m^3, and Poisson's ratio is 0.32. The polarity of the PZT must be reversed for y-bending and z-bending vibration, meanwhile the polarity of the PZT is not reversed for longitudinal vibration as shown in Fig. 1. The steel thin plate is located between y-bending and z-bending half-PZT, which was determined by modal simulation analysis where the zero displacement has been known.

The dimensions of each length structure of the 3D elliptical transducer was determined using Langevin's law that is for longitudinal vibration based on a half-wavelength principle. After that, the dimensions of each length was adjusted by modal simulation to find a coupling frequency of longitudinal-bending-bending mode. Fig. 2 shows the dimensional parameters of the 3D elliptical ultrasonic transducer in which the thin steel plate is neglected. For instance, the working frequency was set about 25 kHz. L0 = 31.75 mm, L1 = 5.295 mm, L2 = 48.69 mm, L2-1 = 24.35 mm, L2-2 = 24.35 mm, L3 = 24.37 mm, L3-1 = 14.67 mm, D1 = 35 mm, D2 = 10 mm, and θ_B = 0° had been determined after calculation as preliminary dimension.

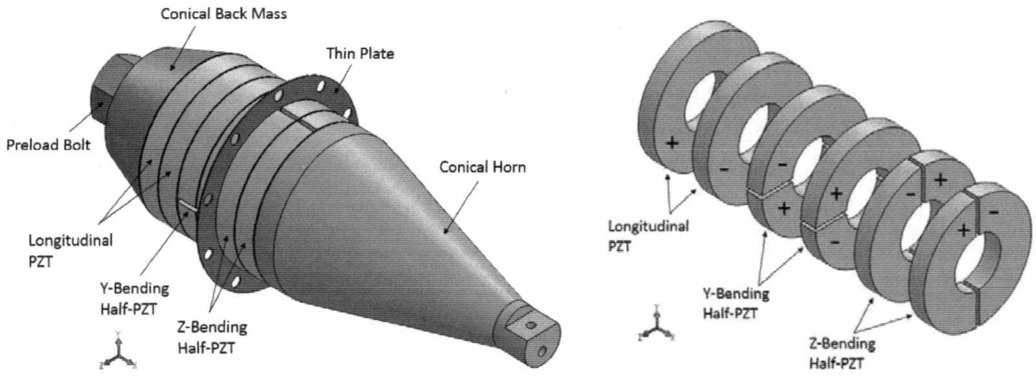

Figure 1. CAD model of the 3D elliptical ultrasonic transducer and the piezoelectric actuator polarity arrangement

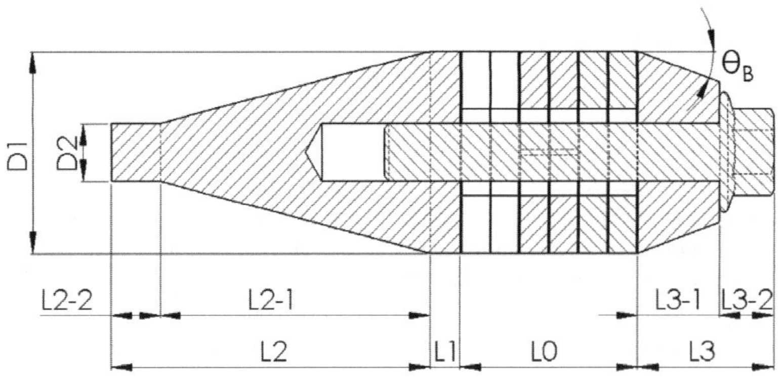

Figure 2. Dimensional parameters of the 3D elliptical ultrasonic transducer

3. Modal simulation results

Fig. 3 shows the effect of dimensional length associated with natural frequency of 1st longitudinal, 3rd y-bending, and 3rd z-bending vibration mode. As we can see, the simulation results showed that the coupling frequency was not found by changing parameter length of L3 and L1, although the natural frequency of both the 3rd y- and z-bending vibration mode are relatively closed. By increasing the parameter length of L2, it leads the similar frequency for both the 3rd y- and z- bending vibration mode. The parameter length L2-1 increases that leads the frequency of 1st longitudinal decreases, however the difference frequency between two 3rd y- and z- bending vibration mode remains exist. Finally, the coupling frequency can be found by increasing the value of angle θ_B, in the other words, the frequency of 1st longitudinal, 3rd y-bending, 3rd z-bending are similar or very closed each other.

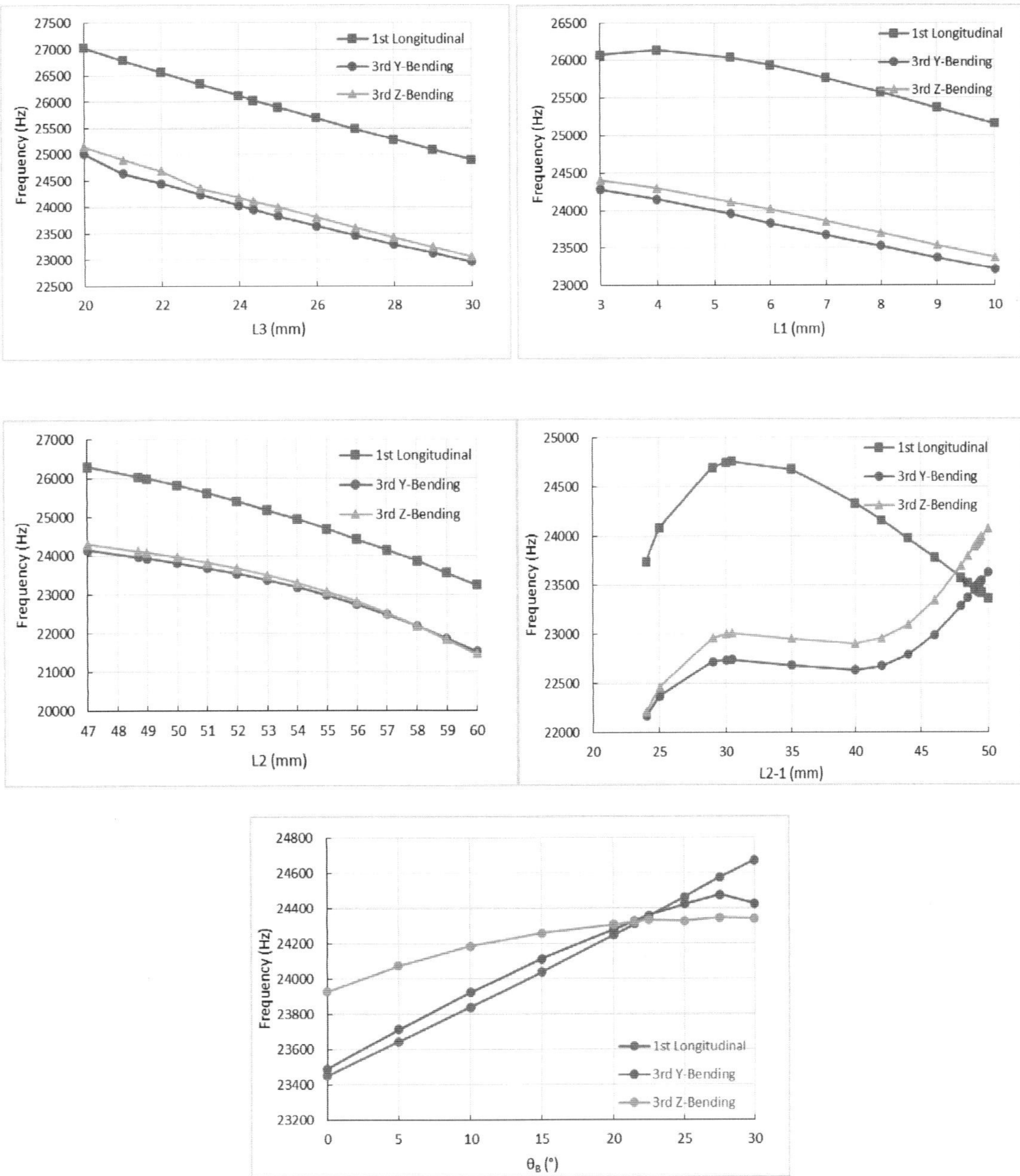

Figure 3. Effect of dimensional length associated with natural frequency of 1st longitudinal, 3rd y-bending, and 3rd z-bending vibration mode

IOP Conf. Series: Materials Science and Engineering **324** (2018) 012063 doi:10.1088/1757-899X/324/1/012063

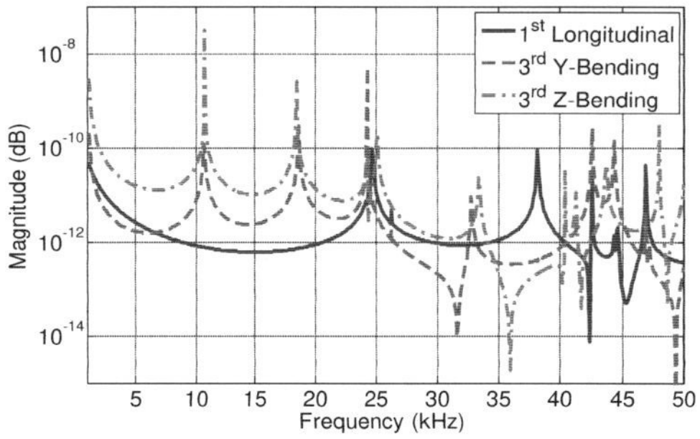

Figure 4. Frequency response function (FRF) graph of harmonic simulation result

(a) (b) (c)

Figure 5. Mode shape of the proposed transducer, **(a)** 1st longitudinal, **(b)** 3rd Y-bending, **(c)** 3rd Z-bending mode

Fig. 4 shows the FRF graph which indicates a closed frequency between 1st longitudinal and two of 3rd bending modes. Fig. 5 shows each mode shapes of the proposed transducer, 1st longitudinal, 3rd Y-bending, and 3rd Z-bending, respectively. As we can see, the blue colour indicates the lowest value of displacement. We can conclude that the thin steel plate can be placed between 3rd Y-bending and 3rd Z-bending half-piezo.

4. Conclusion

The modal simulation analysis have been carried out to find the coupling frequency of the proposed transducer. We can conclude that the coupling frequency can be found by adjusting the dimension of the proposed transducer.

References

[1] T. Moriwaki, E. Shamoto, Ultrasonic Elliptical Vibration Cutting, CIRP Ann. –Manuf. Technol. 44 (1995) 31-34.
[2] X. Li, D. Zhang, Ultrasonic elliptical vibration transducer driven by single actuator and its application in precision cutting, J. Mater. Process. Technol. 180(2006) 91-95.
[3] G. D. Kim, B.G Loh, Machining of micro-channels and pyramid patterns using elliptical vibration cutting, Int. J. Adv. Manuf. Technol. 49(2010) 961-968.
[4] C. Nath, M. Rahman, K.S. Neo, A study on ultrasonic elliptical vibration cutting of tungsten carbide, J. Mater. Process. Technol. 209(2009) 4459-4464.
[5] P. Guo, K.F. Ehmann, Development of a tertiary motion generator for elliptical vibration texturing, Precis. Eng. 37(2013) 364-371.
[6] M.K. Kurosawa, O. Kodaira, Y. Tsuchitoi, T. Higuchi, Transducer for high speed and large thrust ultrasonic linear motor using two sandwich-type vibrators, IEEE Trans. Ultrason. Ferroelectr. Freq. Control. 45(1998) 1188-1195.

Acknowledgments

This research was supported by Basic Science Research Program through the National Research Foundation of Korea (NRF) funded by the Ministry of Science, ICT and Future Planning (grant number NRF-2017R1A2B2003932).

IOP Conf. Series: Materials Science and Engineering **324** (2018) 012064 doi:10.1088/1757-899X/324/1/012064

Evaluation of rolling contact fatigue of induction heated 13Cr-2Ni-2Mo Stainless steel bar with Si_3N_4-ball

Ariyasu YADOIWA[1, 2], **Koshiro MIZOBE**[1,3], **and Katsuyuki KIDA** [1,4*]

[1] University of Toyama, Gofuku 3190, Toyama, 930-8555, Japan
[2] m1771227@ems.u-toyama.ac.jp
[3] kmizobe@eng.u-toyama.ac.jp
[4] kida@eng.u-toyama.ac.jp
[*]The corresponding author

Abstract. 13Cr % martensitic stainless steels were used in various industry, because they have excellent corrosion resistance and high hardness among other stainless steels. They are also expected as a bearing material, however, the research on rolling contact fatigue (RCF) is not enough. In this study, 13Cr-2Ni-2Mo stainless steels were quenched by induction heating and their RCF lives were evaluated. A Si_3N_4-ball was used in order to apply higher stress (P_{max} =5.6 GPa) than our previous tests (P_{max}=5.3 GPa), in a single-ball RCF testing machine. It was found that the basic life (L_{10}) was 2.20×10^6 cycles and Median life (L_{50}) was 6.04×10^6 cycles. In addition, Weibull modulus became higher than the previous tests.

Nomenclature
L_{10} : The basic life
L_{50} : The Median life
P_{max}: Hertzian maximum stress

1. Introduction

Stainless steels are used for various machine components, especially under the corrosive environment because they have excellent corrosion resistance. For these reasons, they are expected as one of bearing materials. However, it is necessary to process the heat treatment in order to improve durability for machine components. As the heat treatment method of 'Induction Heating (IH)' can harden only the surface for short time, it takes a low energy and cost. In the present study, the specimens of 13Cr-2Ni-2Mo stainless steel were quenched by the induction heating. This material obtains higher strength among other stainless steels by applying heat treatment [1-2]. Previous research reported induction heating improved the wear resistance [3]. According to Yuki and Sato [4], it was clear that induction heating has a positive effect on rolling contact fatigue (RCF). In our previous study [5], RCF life of the 13Cr-2Ni-2Mo stainless steel which were quenched by induction heating using a single-ball RCF system was investigated. However, most of the samples didn't generate flaking failure because RCF tests were performed by using a SUJ2-ball.

The purpose of this study is to evaluate the RCF life of 13Cr-2Ni-2Mo stainless steel bars which were quenched by induction heating based on only flaking fatigue data. In order to apply higher stress than the previous tests, RCF tests were performed by using the Si_3N_4-ball, which had higher hardness and Young's modulus than the SUJ2-ball.

2. Test method

2.1. Bar specimens and heat treatment

The bar specimens of 13Cr-2Ni-2Mo stainless steel bars were quenched by induction heating. Table 1 shows the chemical composition in weight percent of 13Cr-2Ni-2Mo stainless steel. Induction heating was carried out under the following conditions: Voltage of 10 kV, current intensity of 9 A and feed speed of 300 m/s. After the induction heating, bar specimens were tempered at 180 °C for 120 minutes.

Length of the bar specimen is 290 mm and a diameter is 20 mm. Mechanical properties of the specimen are Young` modulus of 199 GPa and Poisson's ratio of 0.30. Before measuring Vickers hardness, cross section of specimen bar was polished. The Vickers hardness was measured by a standard Vickers hardness tester (VMT-X7s, Matsuzawa Co., Ltd, Japan) under a load of 10 kgf and a loading time of 15 sec.

Table 1 Chemical composition of the specimens, 13Cr-2Ni-2Mo stainless steel (wt.%).

C	Si	Mn	P	S	Ni	Cr	Mo	N	Fe
0.18	0.29	0.18	0.022	0.001	1.03	13.12	1.85	0.07	R.

2.2. Rolling contact fatigue test

Figure 1 shows a single-ball RCF testing machine which was developed by Kida`s group in their previous work [6]. It can reproduce RCF of bearings by rotating a shaft while pushing up the ball with four bearings to apply a load. The diameter of a Si_3N_4-ball was 9.525 mm (3/8 inch). Mechanical properties of Si_3N_4 are Young's Modulus of 294 GPa and Poisson's ratio of 0.27.

Table 2 shows experimental conditions of RCF testing which were calculated using the mechanical properties. Based on the Hertzian contact theory, Hertzian maximum stress (P_{max}) were calculated [7]. Single-ball RCF tests were performed under the following conditions until flaking failures: rotating speed of 3000 rpm and Hertzian maximum stress of 5.6 GPa. Flaking failures and their maximum depths from the surface were observed by a Laser Confocal Microscope (LCM). It is noted that the present load was 833 [N], which was lither that in our previous work (877 [N]) [5].

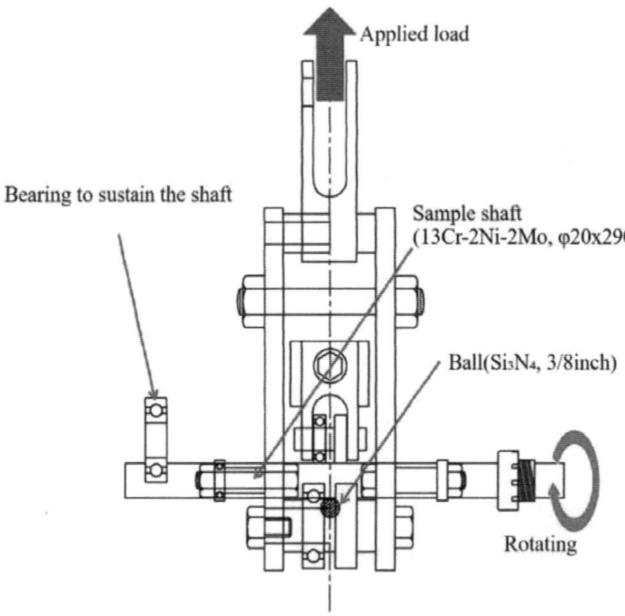

Figure 1 Single-ball RCF testing machine.

Table 2 Experimental conditions of rolling contact fatigue test.

Load [N]	Hertzian maximum stress P_{max} [GPa]	Semi-major radius of contact area [mm]	Semi-minor radius of contact area [mm]
833	5.6	0.303	0.233

3. Experimental result and discussion

3.1. Vickers hardness distribution

Figure 2 shows Vickers hardness distribution on the section of the 13Cr-2Ni-2Mo stainless steel bar. The Vickers hardness of surface was 576 HV, and the depth of heat affect zone is 0.8 mm from the surface.

Figure 3 shows an example of flaking failures. This was observed at 6.34×10^6 cycles. Based on the LCM observation results, it was found that the size of flaking failures in all samples is similar and the maximum depths from the surface is about 200 μm. Flaking failure occurred from heat affect zone in all samples.

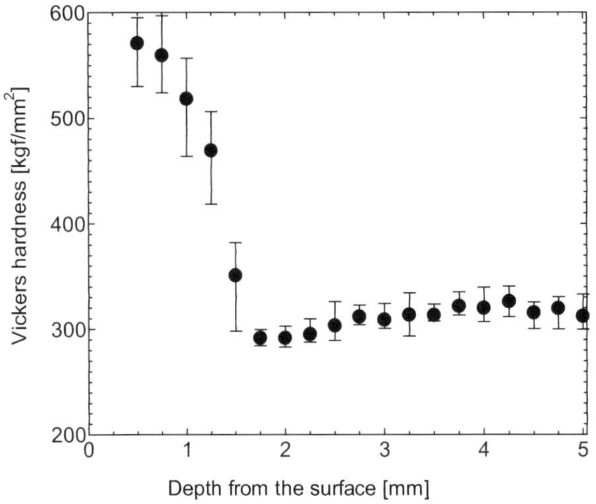

Figure 2 Vickers Hardness distribution.

Figure 3 Flaking failure at 6.34×10^6 cycles.

3.2. Weibull distributions

Reliability and fatigue life of bearings has been investigated based on Palmgren and Lundberg's investigation [8-9]. Based on this investigation, Ioannides & Harris [10] took into consideration the idea of 'fatigue' for life evaluation. Here, the life is evaluated based on the statistical data of RCF tests. In this study, the number of samples that generated flaking failures was 10. These datum were used to obtain Weibull distributions of RCF life. Based on Median rank method, the empirical datum are ranked in the Weibull analysis.

Figure 4 shows two Weibull distributions. One was calculated from the present results of RCF testing and the other from the previous data (Kida et al. [5]) for the same material. Significant difference of these data was P_{max}. In the previous study, RCF tests were performed under the following conditions: the ball was SUJ2 whose Young's modulus was 207 GPa and Poisson's ratio was 0.30; P_{max} was 5.3 GPa. Other conditions were the same as in the present tests. In addition, because the previous data includes the suspended data of the specimens which didn`t generate flaking failures, these were modified using L.G. Johnson's statistical method [11].

Flaking failures lives in all samples range from 1.0×10^6 to 2.0×10^7 cycles. The basic life (L_{10}) is 2.20×10^6 cycles and Median life (L_{50}) is 6.04×10^6 cycles. The present RCF life is lower than the previous data. This is because the present P_{max} is higher than the previous P_{max}. This result however indicates thatr the range of RCF life becomes smaller when P_{max} is higher. Furthermore, it is found that the Weibull modulus of RCF life of the present tests is 1.88, which is higher than the previous data. It is noted that the RCF life obtained under this experimental conditions is reliable because this Weibull modulus is sufficiently higher than the standard value (10/9) proposed by Palmgren and Lundberg [12].

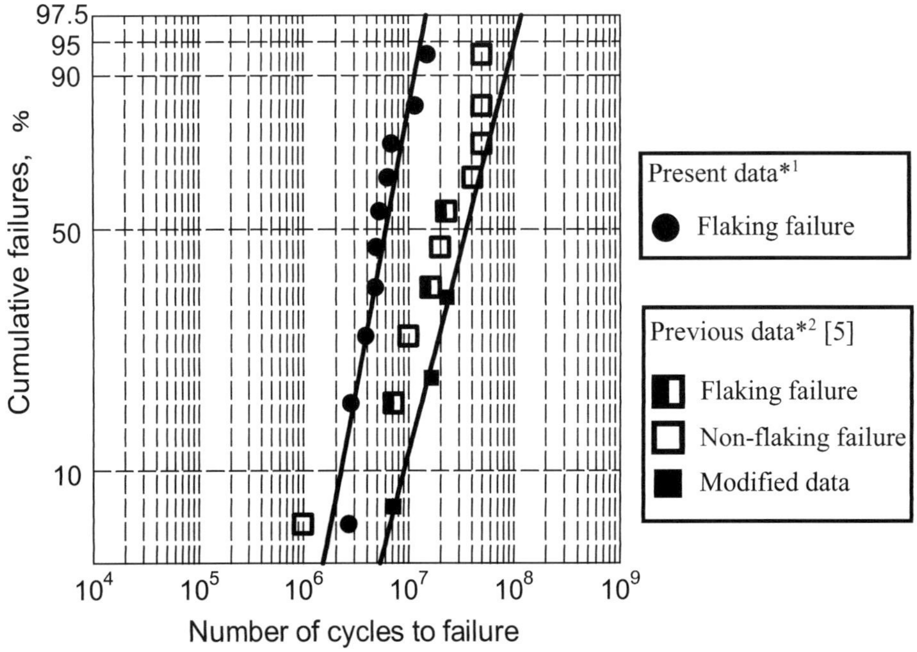

Figure 4 Weibull distribution of RCF life.

*1 RCF tests were conducted under the conditions shown in Table 2. Weibull modulus was 1.88 and the basic life was 2.20×10^6 cycles.

*2 RCF tests were conducted under the following conditions (P_{max} of 5.3 GPa, rotating speed of 3000 rpm and ♯46-lubricant conditions). Weibull modulus was 1.38 and the basic life was 9.01×10^6 cycles.

4. Conclusions

13Cr-2Ni-2Mo stainless steel bars were quenched by induction heating. Rolling contact fatigue tests of the bars were conducted using a Si_3N_4 single-ball RCF testing machine.

(1) Flaking failures generated in heat affected zone.

(2) The basic life (L_{10}) was 2.20×10^6 cycles and Median life (L_{50}) was 6.04×10^6 cycles.

(3) The present Weibull modulus of RCF life was 1.88, which was higher than the standard value (10/9).

IOP Conf. Series: Materials Science and Engineering **324** (2018) 012064 doi:10.1088/1757-899X/324/1/012064

References

[1] Mizobe K, Okamoto K, Shibukawa T, Kanemasu K, Kida K 2014 *Applied Mechanics and Materials* **563** pp.71-75

[2] Honda T, Santos E.C, Kida K, Shibukawa T 2012 *International Journal of Materials and Product Techology* **45** Issue 1-4

[3] Cao Y.J, Sun J.Q, Ma F., Chen Y.Y, Cheng X.Z, Gao X., Xie K. 2017 *Tribology International* **115** pp.108-115

[4] Yuki H, Sato M 2015 *NTN TECHNICAL REVIEW* **83** pp.75-80

[5] Kida K, Honda T, Seto Y, Santos E.C, Shibukawa T 2013 *Applied Mechanics and Materials* **300-301** pp.1372-1376

[6] Rozwadowska J, Kida K, Santos E.C, Honda T, Koike H, Kanemasu K 2011 *Advancced Materials Research* **418-420** pp.1613-1617

[7] Hertz H 1881 *Journal für die reine und angewandte Mathematik* **92** pp.156-171

[8] Lundberg G, Palmgren A 1947 *Mechanical Engineering Series* **1**

[9] Lundberg G, Palmgren A 1952 *Mechanical Engineering Series* **2**

[10] Harris T, McCool J 1996 *ASME, J.Tribology* **118** pp.297-310

[11] Johnson L 1964 Amsterdam, London, Newyork *The Statistical Treatment of Fatigue Experiments* pp.37-48

[12] Palmgren A, Lundberg G 1952 *Ingeniörs vetenskaps akademiens handlingar* **210**

IOP Conf. Series: Materials Science and Engineering **324** (2018) 012065 doi:10.1088/1757-899X/324/1/012065

Influence of the depth of heat affected zone on the fatigue strength and fracture surface in induction heated JIS SUJ2 bearing steel

K Egawa [1, 2], I Yoshida [1, 3], K Mizobe[1, 4], and K Kida[1, 5] *
[1]University of Toyama, Gofuku 3190, Toyama, 930-8555, Japan
[2]m1771203@ ems.u-toyama.ac.jp, [3]m1671239@ems.u-toyama.ac.jp,
[4]kmizobe@eng.u-toyama.ac.jp, [5]kida@eng.u-toyama.ac.jp
* The corresponding author

Abstract. Induction heating is used as one of surface hardening methods. Induction heating can control the depth of the heat affected zone. In this work, we prepared two kinds of induction heated bar specimens: one had a shallower heat affected zone; and the other had a thick heat affected zone. After the rotating bending fatigue tests, we observed the fracture surfaces and measured the value of retained austenite. We found that the depth of heat affect zone and the amount of retained austenite relate the fatigue strength.

1. Introduction

JIS SUJ2 bearing steel was used for mechanical parts which required high fatigue strength. For this reason, the effective strengthening method “quenching” is popular. In particular, induction heating is used as one of surface hardening methods. Induction heating can control the depth of the heat affected zone[1,2]. Kawasaki et al. investigated the mechanical properties of induction heated spring steel. They reported that induction heating improved the fatigue strength compared to that of furnace heated spring steel[3]. Komotori et al. prepared four kinds of hardened specimens and performed the rotating bending fatigue tests (RBT) in order to investigate the relation between the depth of heat affected zone and the fatigue strength. They reported that the fatigue limit related the depth of heat affected zone[4].

Yuki et al. investigated the influence of induction heating condition on the amount of retained austenite (RA). They reported that quenching temperature increased the amount of RA during induction heating process[5]. In generally, it is known that RA decreases the material hardness, but RA enhances the material toughness and resists the crack propagation[6,8,9]. Dommarco et al. reported the relation between the amount of RA and the contact fatigue life[10]. RA is effective in improving the fatigue strength.

In our previous work, we investigated the fracture surface of induction-heated JIS SUJ2 steel specimens. After the RBT, we observed the large inner cracks. We defined this crack as ”TRO crack”, and reported that TRO crack size increases with increasing the number of cycles to failure[7].

In this work, we prepared two kinds of induction heated bar specimens: one had a shallower heat affected zone; and the other had a thick heat affected zone. After the RBT, we observed the fracture surface and measured the RA. We investigate the influence of heat affected zone on the fatigue strength based on the stress-induced martensitic transformation.

IOP Conf. Series: Materials Science and Engineering **324** (2018) 012065 doi:10.1088/1757-899X/324/1/012065

2. Test method

2.1. Specimen (JIS SUJ2 steel) and heat treatment
We prepared JIS SUJ2 steel bar specimens. Table 1 shows the chemical compositions of the specimens. The specimens were spheroidized annealed [8] and were quenched by induction heating. The samples were heated up by the induction coil (For ϕ 10 mm) during the 300 rpm rotation. Immediately, the samples were cooled by water at room temperature. After that, Specimens were tempered at 180 °C for 60 minutes. We produced the 40 samples and divided these samples according to their depth of the hardened layer. The deeper samples are referred to as IH-deep and the shallower samples are referred to as IH-shallow.

Figure 1 shows the dimension of the RBT specimen. Stress concentration factor of breaking point was about 1.0[8].

Table 1 Chemical composition of the JIS SUJ2 steel bar specimens (weight %)

C	Si	Mn	P	S	Ni	Cr	Mo	Cu
1.00	0.17	0.39	0.019	0.006	0.06	1.38	0.02	0.12

2.2. Rotating bending fatigue test
Figure 2 shows the RBT machine which was developed by Kida's group[11]. The rotation speed was about 1600 rpm and the frequency of the fatigue tests was about 25 Hz. The stress amplitude was ranging from 386 to 746 MPa and stress ratio was -1. The target number of cycles was 1.0×10^8

Modified stress was calculated in order to evaluate the effect of the crack origin depth on low stress amplitude with following equation:

$$\sigma_m = \frac{r-d}{r}\sigma_n \qquad (1)$$

where σ_m is modified stress amplitude [MPa], r is radius of specimen [mm], d is depth of crack origin from the surface [mm] and σ_n is nominal stress amplitude [MPa].

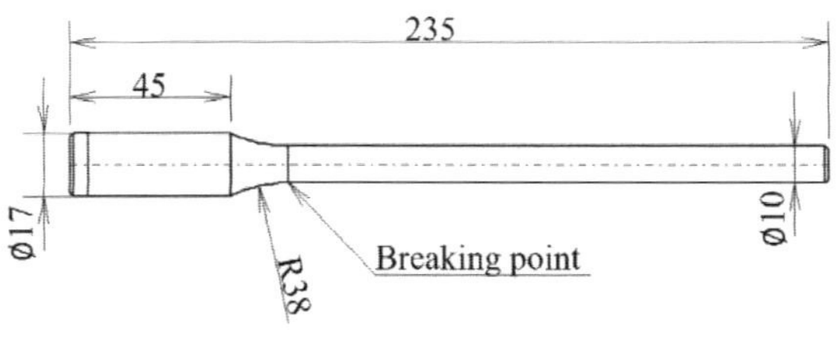

Figure 1 RBT specimen

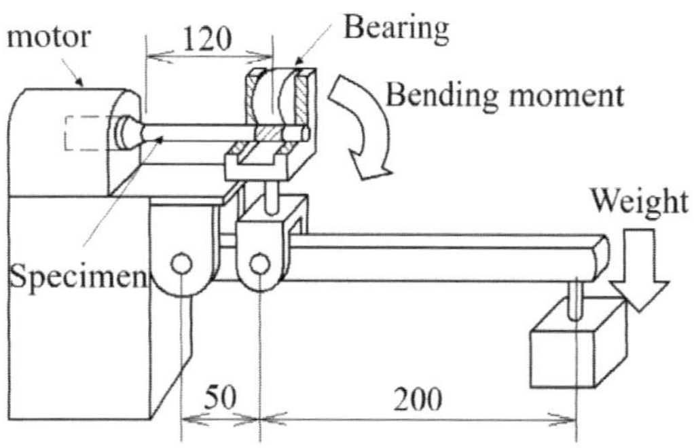

Figure 2 RBT machine

2.3. Fracture surface observation

After the RBT, fracture surfaces were observed by an optical microscope SHODENSHA TG140CCD. Vickers hardness distributions were measured by a MATSUZAWA VMT-X7s. The load was 10 kgf and the loading time was 15 seconds.

2.4. Retained austenite measurement method

We used a Bulker D8 DISCOVER with GADDS (2D detector) in order to measure the amount of RA. The measurement increments were 0.2 mm. Table 2 shows the test conditions of the RA measurement. We measured RA distributions on three sections: first and second are before and after the RBT and the last is fracture surface. Figure 3 shows the schematic illustration of their observation points. Figure 3(a) shows the observation points before the RBT. Specimens were cut and then polished by 1μm diamond paste. Figure 3(b) shows the observation points after the RBT. Specimen was cut and measured near the fracture surface. Figure 3(c) shows the fracture surface observation points after the RBT. We measured the amount of the RA at TRO crack and outside TRO crack in order to investigate the influence of crack propagation rate on the amount of RA.

Table 2 Observation conditions for RA measurement.

X-ray source	Cr Kα
Power	35 kV 40 mA
Diameter of collimator	0.1 mm
Detector distance	150 mm
Measurement angle	60 to 90 (2θ)
Scan time	20000 sec / point

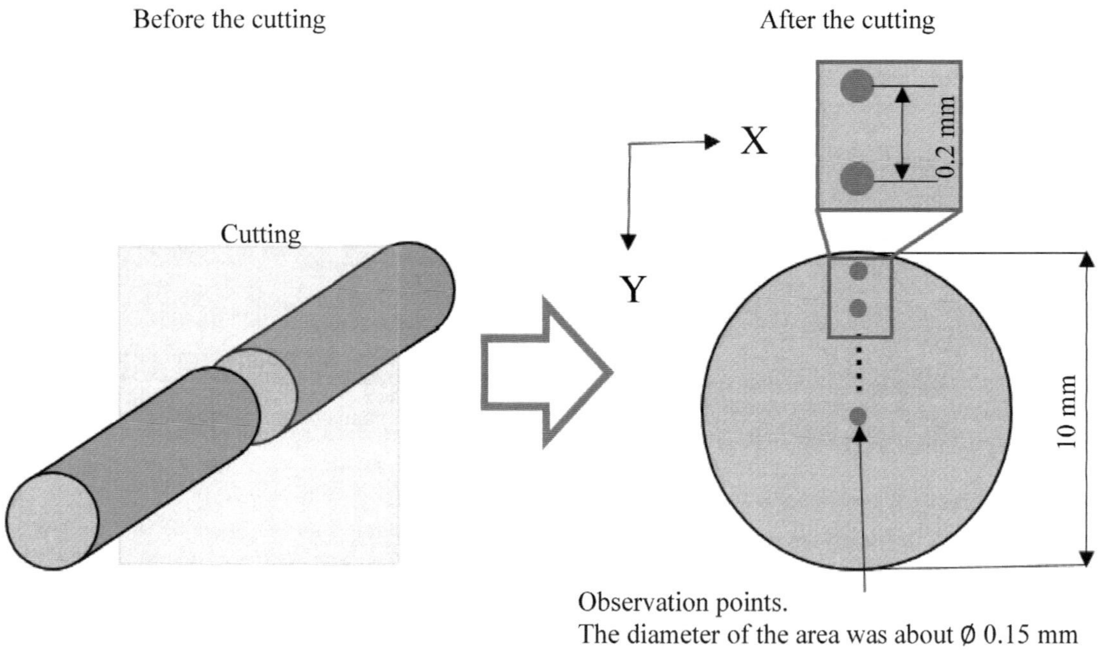

(a) Observation points before the RBT.

IOP Conf. Series: Materials Science and Engineering **324** (2018) 012065 doi:10.1088/1757-899X/324/1/012065

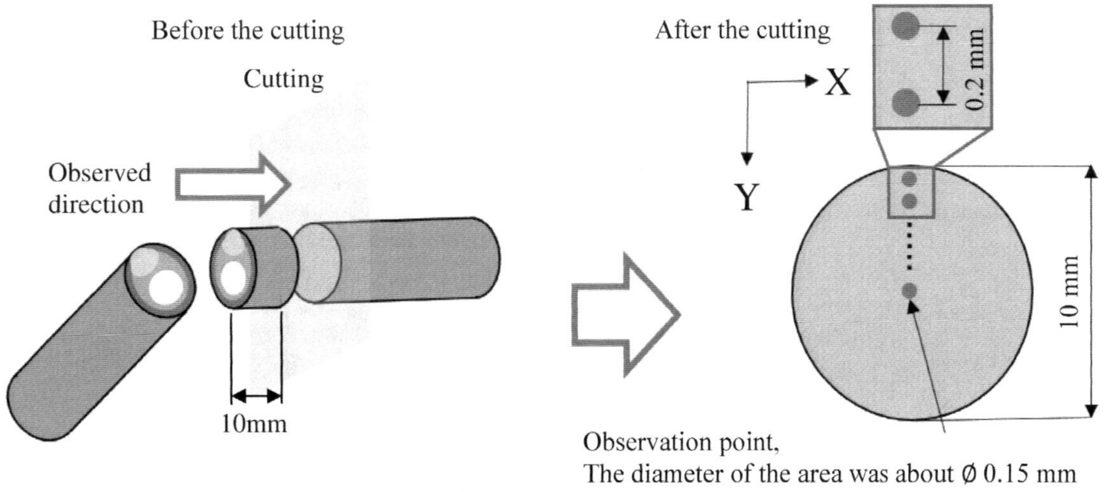

(b) Observation points after the RBT.

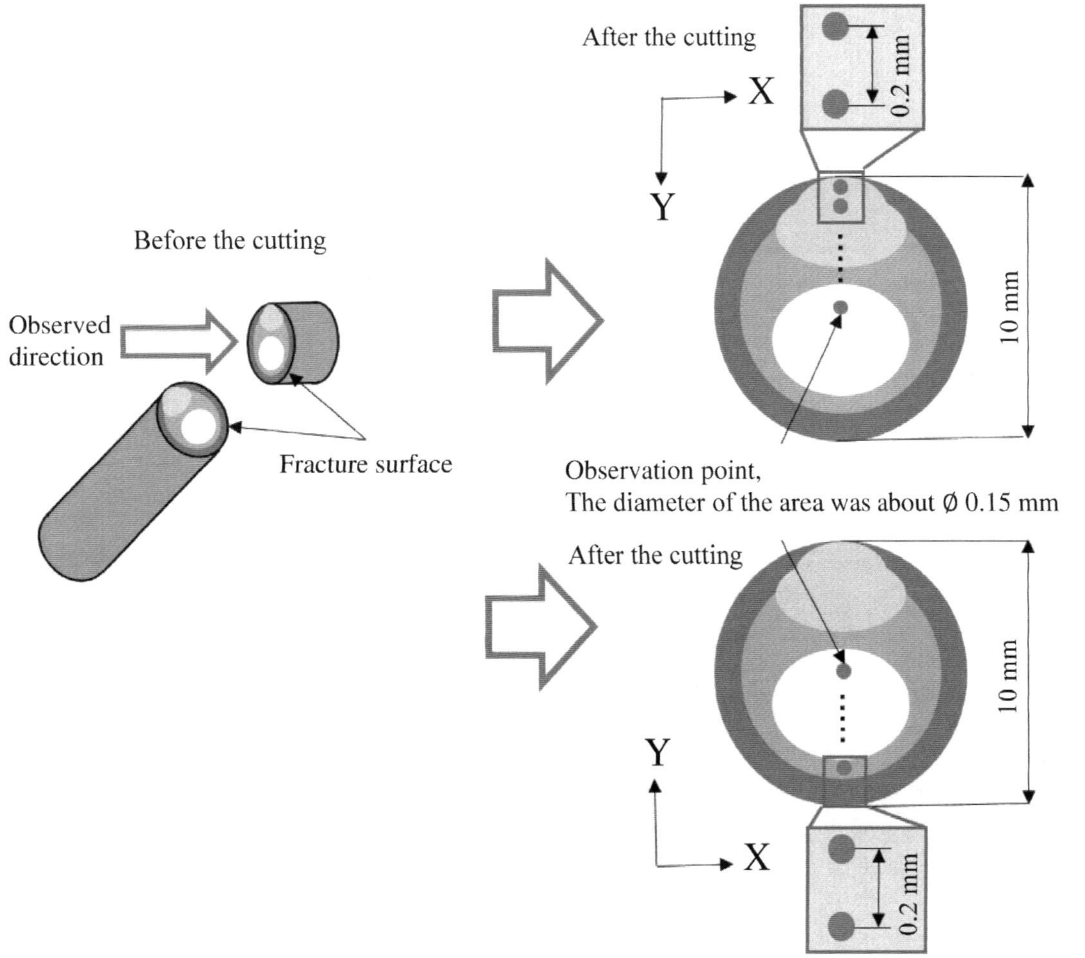

(c) Fracture surface observation points after the RBT.

Figure 3 Schematic illustration of RA measurement method.

IOP Conf. Series: Materials Science and Engineering **324** (2018) 012065 doi:10.1088/1757-899X/324/1/012065

3. Results and discussion

3.1. Rotating bending fatigue test and Vickers hardness distribution

Figure 4 shows Vickers hardness distributions of IH-deep and IH-shallow samples. Vickers hardness distribution of IH-shallow samples (○) is referred from our previous data[7]. Figure 5 shows the microstructures of the samples which were etched by nital. At heat affected zone, ferrite transformed into martensite. We defined the area where Vickers hardness is higher than 600 HV as "martensitic transformed area". Similarly, we also defined the area from 200 HV to 600 HV as "transition area" and the area below 200 HV as "ferrite area"

Vickers hardness of IH-deep and IH-shallow samples at 0.2 mm from the surface were 772 HV and 750 HV. The depth of martensitic transformed area of IH-deep were 1.1 mm and that of IH-shallow was 0.8 mm. Martensitic transformed area of IH-deep samples were larger than that of IH-shallow samples.

Figure 6 shows S-N diagram of nominal stress amplitude. S-N diagram of nominal stress amplitude of IH-Shallow samples (○) is referred from our previous data[7]. TRO crack can be seen in all specimens. The fatigue limit of IH-deep and IH-shallow samples were 427 MPa and 386 MPa. The fatigue limit of IH-deep sample was higher than that of IH-shallow samples under nominal stress amplitude. Figure 7 shows S-N diagram of modified stress amplitude. At the 10^8 cycles, we stop the test because it is fatigue limit. Therefore, we cannot calculate the modified stress. The fatigue strength of IH-deep samples was equal to that of IH-shallow samples under the modified stress amplitude.

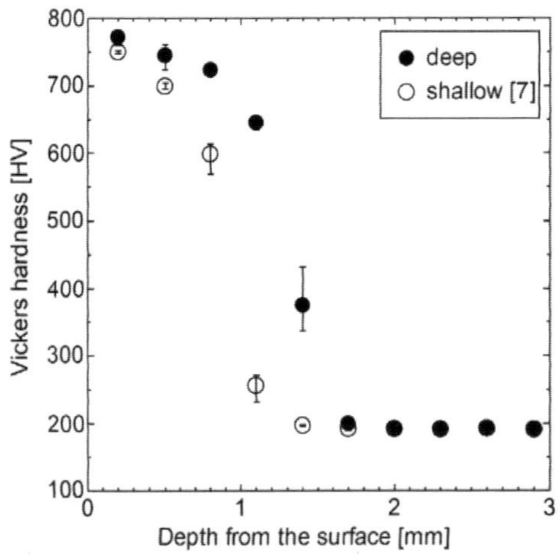

Figure 4 Vickers hardness distributions.

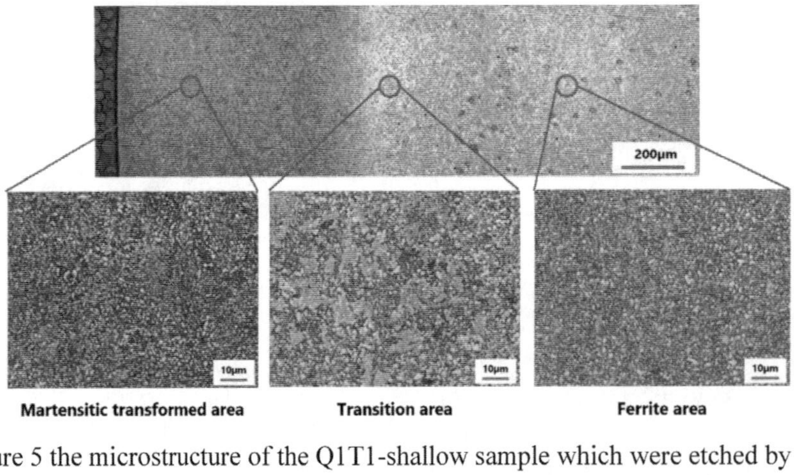

Figure 5 the microstructure of the Q1T1-shallow sample which were etched by nital.

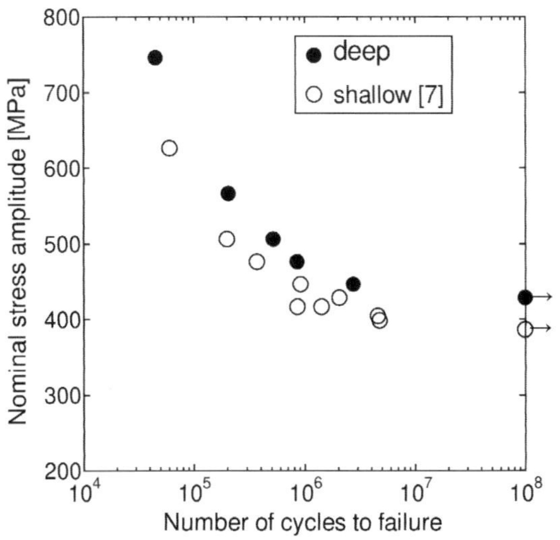

Figure 6 S-N diagram (nominal stress amplitude).

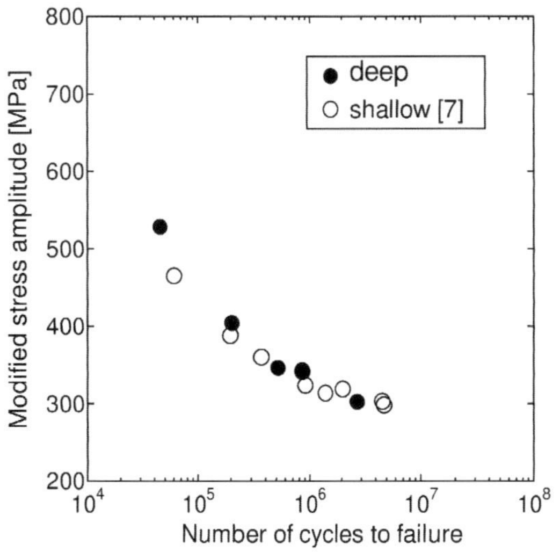

Figure 7 S-N diagram (modified stress amplitude).

3.2. Fracture surface observation

Figure 8 shows TRO crack area of IH-deep sample (σ_n=445 MPa, N_f=2.76×10^6). This TRO crack originated at the depth of 1.61 mm from the surface. Figure 9 shows TRO crack area of IH-shallow sample (σ_n=427 MPa, N_f=2.0×10^6). This TRO crack originated at the depth of 1.28 mm from the surface. Figure 10 shows one of the SEM images at the crack origin. The non-metallic inclusions can not be seen and the crack initiated from the boundary of the hardness layer. In this tests, all of the specimens were broken from the TRO cracks.

Figure 11 shows the relation between the TRO crack size and number of cycles to failure. TRO crack size of IH-shallow (○) is referred from our previous data[7]. TRO crack size of IH-deep samples was equal to that of IH-shallow at the same number of cycles to failure. Figure 12 shows the relation between aspect ratio and number of cycles to failure. Aspect ratio of IH-deep samples was equal to that of IH-shallow at the same number of cycles to failure. TRO crack size and aspect ratio increased with number of cycles to failure. This indicates that TRO crack size and aspect ratio did not relate the depth of martensitic transformed area.

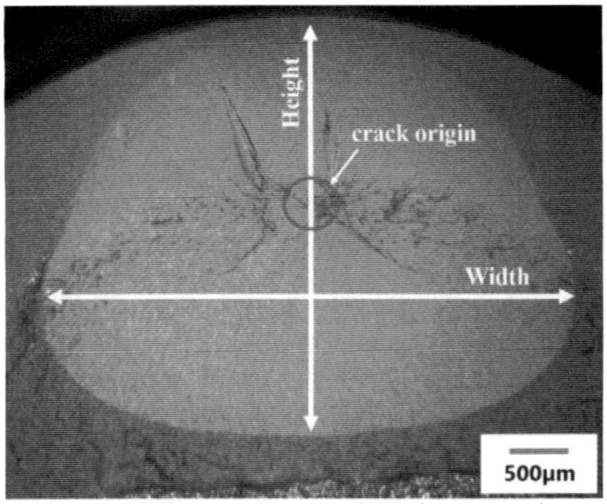

Figure 8 TRO crack area of IH-deep sample.
(σ_n=445 MPa, N_f=2.76×10^6)

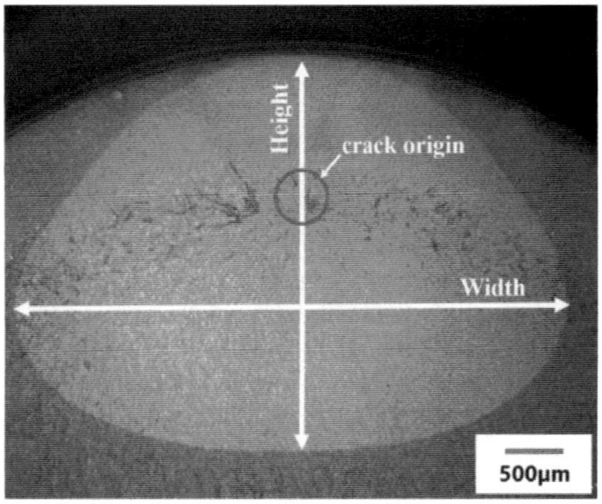

Figure 9 TRO crack area of IH-shallow sample.
(σ_n=427 MPa, N_f=2.0×10^6)

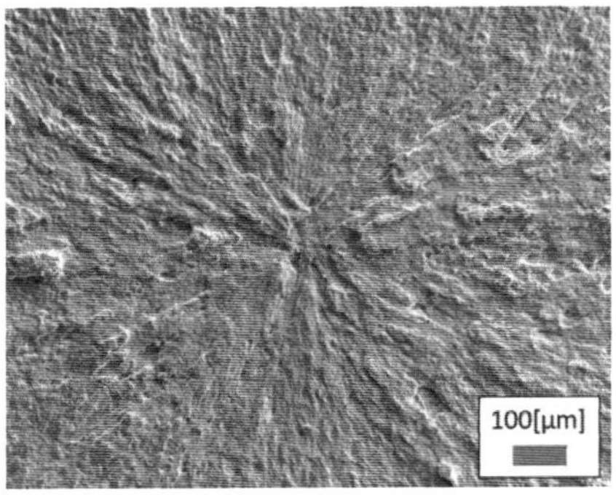

Figure 10 SEM image in crack origin

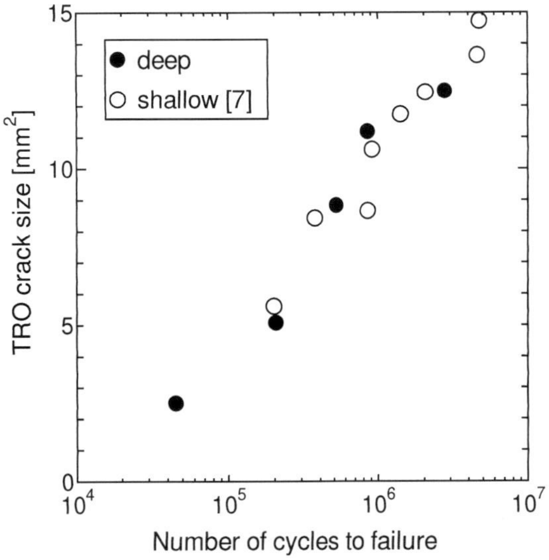

Figure 11 Relationship between TRO crack size and number of cycles to failure.

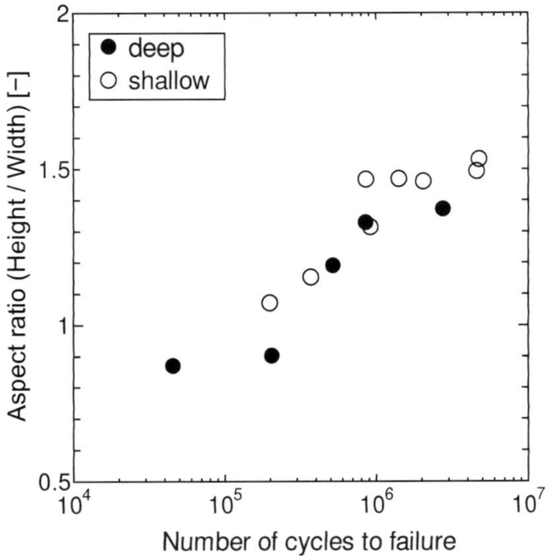

Figure 12 Relationship between aspect ratio and number of cycles to failure.

3.3. Results of retained austenite measurement

Figure 13 shows the RA distributions before the RBT. The value of RA of IH-deep and IH-shallow samples at 0.1 mm from the surface were 16.5 % and 7.2 %. At 0.7 mm from the surface, the value of RA of IH-shallow sample decreased to 0 %. At 1.3 mm from the surface, the value of RA of IH-deep sample decreased to 0%. The value of RA at martensitic transformed area of IH-deep sample was higher than that of IH-shallow sample. We suggest that the value of RA at martensitic transformed area related the depth of martensitic transformed area.

Figure 14 shows the RA distributions of IH-deep sample after the RBT. The value of RA after the RBT at 0.1 mm from the surface was 16.5 %, which is similar to the RA before the RBT. This means that the retained austenite did not decrease with only the bending moment. In both the TRO crack and the outside TRO crack, the decreasing values of RA were 1.4 % and 4.6 %. The values of both area were lower than that before the tests. This supports that the retained austenite did not decrease with only the bending moment without crack propagation. Figure 15 shows the RA distributions of IH-shallow sample after the RBT. The brief tendency of Fig. 14 is similar to that of Fig. 15. The value of RA after the RBT is the same as before the RBT and the value of RA at the fracture surface is lower than that before the test. These results concluded that the stress induced transformation occurred in both inside/outside TRO crack areas.

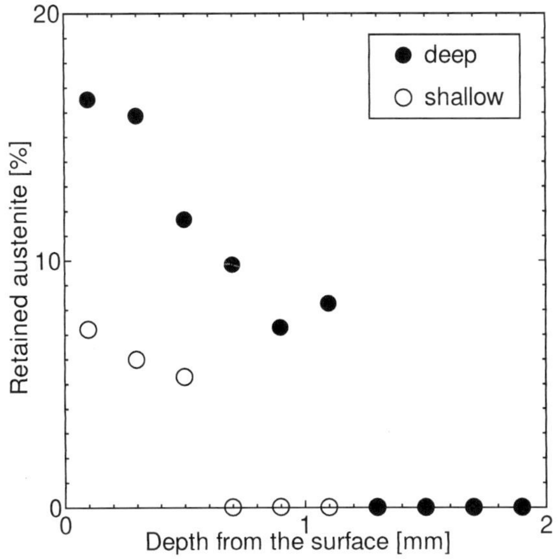

Figure 13 RA distributions before the RBT.

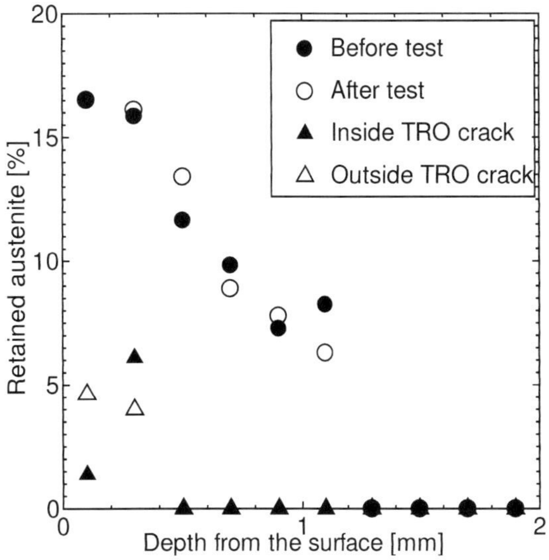

Figure 14 RA distributions of IH-deep sample after the RBT.

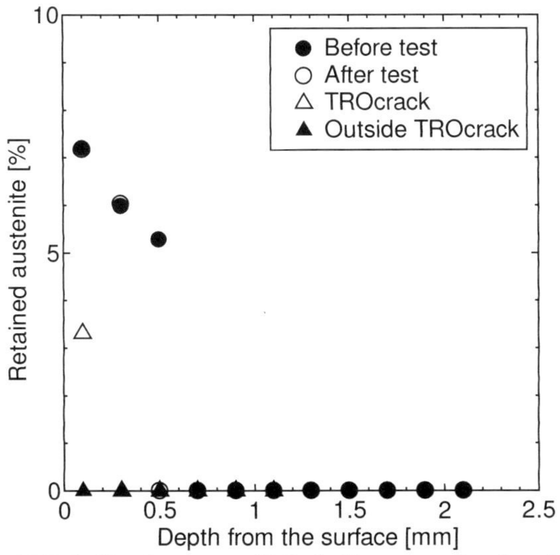

Figure 15 RA distributions of IH-shallow sample after the RBT.

4. Conclusions

We prepared two kinds of induction heated bar specimens, one was shallower heat affected zone (IH-shallow sample) and the other was thick heat affected zone (IH-deep sample). After the RBT, we observed fracture surfaces and measured the amount of retained austenite. The conclusions are as follows.

1. The fatigue limit of IH-deep sample was higher than that of IH-shallow samples under nominal stress amplitude.
2. The fatigue strength of IH-deep samples was similar to that of IH-shallow samples under the modified stress amplitude which included the effect of crack size and aspect ratio.
3. The value of RA at heat affected zone of IH-deep sample was higher than that of IH-shallow sample.
4. In both inside/outside TRO crack areas, stress induced transformation occurred.

References

[1] Savaria V, Bridier F, Bocher P 2016 *International Journal of Fatigue* **85** pp. 70-84
[2] Horikawa T, Nakamura H, Kawamura M, Kawasaki K, Misaka Y 1996 *Zairyou* **45**, No.4, pp. 437-442
[3] Kawasaki K, Chiba T, Takaoka N and Yazaki T 1987 *Tetsu-to-Hagane* **73** No. 16 pp 2290-2297
[4] Komotori J, Shimizu M, Misaka Y, Kawasaki K 2001 *International Journal of Fatigue* **23** pp 225–230
[5] Yuki H, Sato M and Ooki C 2017 *Tetsu-to-Hagane* **103** No.4 pp 194-200
[6] Diego-Calderón de I, Rodriguez-Calvillo P, Lara A, Molina-Aldareguia M J, Petrov H R, Knijf De D, Sabirov I 2015 *Materials Science and Engineering: A* **641** pp 215-224
[7] Yoshida I, Mizobe K and Kida K 2017 Materials Science Forum ISSN: 1662-9752 904 pp 24-28
[8] Christodoulou I P, Kermanidis T A, Krizan D 2016 *International Journal of Fatigue* **91** Part 1 pp 220-231
[9] GuanghuaYan LizhanHan ChuanweiLi XiaomengLuo JianfengGu 2017 Journal of Nuclear Materials 483 pp. 167-175
[10] Dommarco C R, Kozaczek J K, Bastias C P, Hahn T G, Rubin A C 2004 *Wear* **257** Issue 11 pp 1081-1088
[11] Mizobe K, Santos C E, Honda T, Koike H and Kida K 2012 *Int. J. Materials and Product Technology* **44** Nos. 3/4, pp 227-239

IOP Conf. Series: Materials Science and Engineering **324** (2018) 012066 doi:10.1088/1757-899X/324/1/012066

Natural Characteristics of The Herringbone Gear Transmission System

Jianxing Zhou, Wenlei Sun, Li Cao
School of Mechanical Engineering, Xinjiang University, Urumqi 830047, China
jianzhou82923@163.com.
wensunxj@163.com.
649823606@qq.com.

Abstract. According to the structure characteristics of herringbone gear transmission, a more realistic dynamic model of the transmission system is built in consideration of the inner excitation, herringbone gears axial positioning and sliding bearing etc. The natural frequencies of the system are calculated, and the vibration mode is divided into symmetric vibration modes and asymmetric vibration modes. The time history of system dynamic force is obtained by solving the dynamic model. The effects of the connection stiffness of left and right sides of herringbone gears and axial support stiffness on natural characteristics are discussed.

1. Introduction

Gear transmission is one of the most extensively adopted transmission modes; herringbone gear is favorably characterized by high bearing capacity, steady operation and small axial force[[1]], having been extensively applied to heavy load device, noise is associated with comfort and reliability of equipment directly. Therefore, proper establishment of transmission system analysis model to accurately analyze its dynamic characteristic shall be the basis to improve transmission system performance.

Scholars have made remarkable achievement in terms of dynamic characteristic of gear transmission system. Feng set up herringbone gear pair analysis model by three-dimensional finite element method to carry out analysis of contact behavior and modality of gear pair [[2]]. Wang established dynamical model for bending-torsion-axis coupling of herringbone gear based on lumped parameter method, then the influence of various excitations and gear tooth corrections on dynamic characteristic of herringbone gear are analyzed [[3]]. Wang built a 12-DOF nonlinear herringbone gear transmission model, the effect of the meshing stiffness and corner mesh impact on vibration characteristics under multi-load were studied [[4]]. Zhao proposed a parabola modification method for herringbone gears, and examples show that path of contact and transmission errors of optimized gears are preferable distinctly [[5]]. Considering that axial component of herringbone gear is identical in size and opposite in direction in theory, therefore, some researches on building of analysis model have ignored axial vibration of herringbone gear, which is reduced to spur gear, however, for herringbone gear with larger width-diameter ratio, such reduction is likely to bring about bigger error in calculation.

The paper has built the dynamic model of herringbone gear transmission, in which the axial positioning of herringbone gear and inner excitation are considered. The natural characteristic of herringbone gear transmission system is analyzed.

2. Dynamical model building

2.1. Dynamical model for herringbone gear transmission system

In the process of dynamical model building, left hand and right hand gears in gear reducer are taken as helical gear, two gears adopt axis section connection with stiffness, with bending-torsion-axis coupling vibration model for herringbone gear built based on the relationship shown in assembly drawing of gear reducer, as shown in fig.1, in which X and Y are transverse vibration respectively, Z is axial, p denotes input end, g indicates output end, k_{mi} in the fig represents meshing stiffness (in which i=1,2 indicates helical gear pair on left and right sides respectively). Every gear adopt sliding bearing support, in order to embody asymmetry of oil film stiffness[[6]], the paper adopts four stiffness coefficients to describe oil film stiffness of sliding bearing, in which k_{pix} and k_{piy} represent bearing support stiffness at input end; k_{gix} and k_{giy} represent bearing support stiffness at output end.

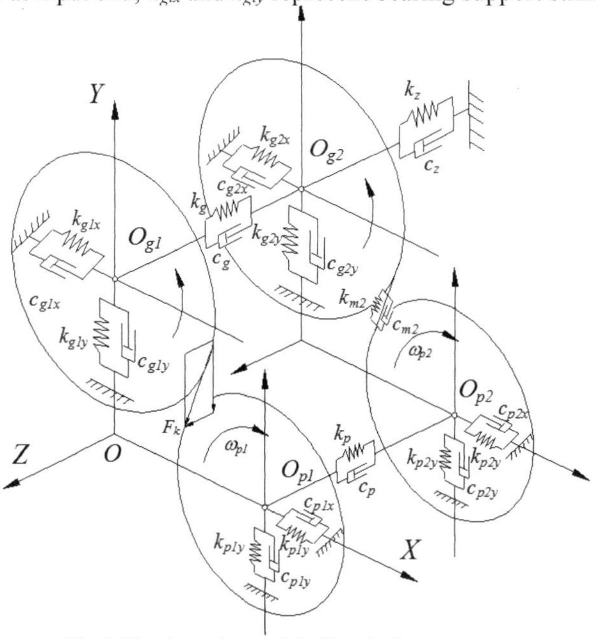

Fig 1. The dynamics model of herringbone gear system

The paper defines k_{p} and k_{g} as stiffness of intermediate connection axis section (namely intermediate tool withdrawal groove) of helical gear on left and right sides, which is abbreviated to intermediate connection stiffness. Since output end adopts axial positioning, therefore, elastic connection is added to g_2 along axial direction, k_{z} indicates axial positioning stiffness, with respect to physical model, k_{z} is axial stiffness of thrust face on bearing position.

Every gear has four degrees of freedom, namely translational degree of freedom along X, Y and Z directions and torsional degree of freedom around its center Generalized displacement vector of the system is expressed as

$$\{X\}=\{x_{p1}\ y_{p1}\ z_{p1}\ \theta_{p1}\ x_{g1}\ y_{g1}\ z_{g1}\ \theta_{g1}$$
$$x_{p2}\ y_{p2}\ z_{p2}\ \theta_{p2}\ x_{g2}\ y_{g2}\ z_{g2}\ \theta_{g2}\}^T$$

Mesh force and damping force of every gear pair are expressed as

IOP Conf. Series: Materials Science and Engineering **324** (2018) 012066 doi:10.1088/1757-899X/324/1/012066

$$
\begin{cases}
F_{K1} = k_{m1}\Big[\left(R_{p1}\theta_{p1} + x_{p1}\sin\alpha - y_{p1}\cos\alpha\right)\cos\beta_b + z_{p1}\sin\beta_b \\
\quad + \left(y_{g1}\cos\alpha - x_{g1}\sin\alpha - R_{g1}\theta_{g1}\right)\cos\beta_b - z_{g1}\sin\beta_b - e_1\Big] \\
F_{K2} = k_{m2}\Big[\left(R_{p2}\theta_{p2} + x_{p2}\sin\alpha - y_{p2}\cos\alpha\right)\cos\beta_b - z_{p2}\sin\beta_b \\
\quad + \left(y_{g2}\cos\alpha - x_{g2}\sin\alpha - R_{g2}\theta_{g2}\right)\cos\beta_b + z_{g2}\sin\beta_b - e_2\Big]
\end{cases}
\tag{1}
$$

Where: k_m is meshing stiffness of gear and cm is meshing damping of gear
Dynamic mesh force of gear pair is

$$F_{pg1} = F_{k1} + F_{c1} \qquad F_{pg2} = F_{k2} + F_{c2} \tag{2}$$

Model is converted into matrix form

$$[M]\{\ddot{X}\} + [C]\{\dot{X}\} + [K(t)]\{x\} = \{P(t)\} \tag{3}$$

Where: M , C , K —mass matrix, damping matrix and stiffness matrix
X —displacement vector
P —generalized force vector

2.2. *Herringbone gear meshing stiffness*

Herringbone gear pair is split into two symmetric helical gear pairs as shown in fig.2(a). One meshing period is divided into several parts, with finite element method adopted to calculate helical gear pair stiffness, in which meshing stiffness curve in one period is as shown in fig.2(b).

Table1. The system parameters

Number of teeth		Modulus (mm)	Pressure angle (°)	Helical angle (°)	Face width B (mm)
37	106	5	20	24	92

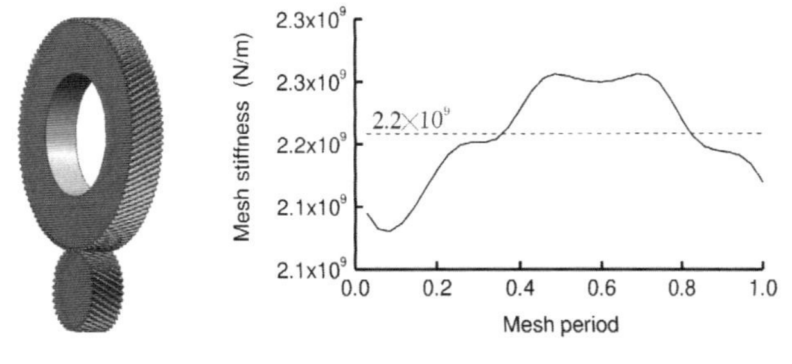

(a) Gear pair model (b) Gear mesh stiffness

Fig 2. The Gear pair model and mesh stiffness

Finite element method is adopted to calculate bending, torsional and axial stiffness of intermediate connecting axis to obtain that lateral bending stiffness of intermediate axis section of driving gear is $k_{px}=k_{py}=1.92\times10^9$N/m, axial stiffness is $k_{pz}=3.1\times10^9$ N/m, and torsional stiffness is $k_{p\theta}=1.89\times10^{10}$N·m/rad, while lateral bending stiffness of intermediate axis section of driven gear is $k_{gx}=k_{gy}=1.376\times10^9$ N/m, axial stiffness is $k_{gz}=1.1\times10^{10}$ N/m, torsional stiffness is $k_{g\theta}=2.7\times10^{10}$ N·m/rad.

3. Dynamic characteristic of herringbone gear transmission system

3.1. *Inherent characteristic of herringbone gear transmission system*

Inherent frequency of system is obtained by solution, as shown in table 2.

Table 2. The natural frequency of the gear system (Hz)

Order	Inherent frequency	Order	Inherent frequency	Order	Inherent frequency
1	269.59	6	653.46	11	1843.73
2	306.64	7	804.46	12	2108.46
3	359.60	8	967.76	13	2551.08
4	449.18	9	1507.08	14	9855.25
5	602.08	10	1714.18		

Mode of vibration of transmission system is as shown in fig.3, the abscissa in the figures corresponds to every degree of freedom of system in turn, namely degree of freedom in dynamical model $\{X_i\}$, ordinate indicates vibration amplitude relevant to every degree of freedom. It is observed from fig that mode of vibration of transmission system may be divided into two categories, one is symmetrical mode and the other one is asymmetrical mode, in which the first, third, fourth and sixth order mode of vibration is symmetrical mode of vibration for pinion gear p_1, p_2 and bull gear g_1 and g_2, other modes of vibration are not fully symmetrical, in which the second order is axial vibration mode of pinion gear p_1 and p_2, the fifth order is axial vibration mode of bull gear, the seventh order is lateral vibration mode of bull gear g_1 and g_2, the ninth and eleventh orders are lateral vibration mode of pinion gear p_1 and p_2, and the thirteenth and fourteenth orders are torsional vibration mode of gear.

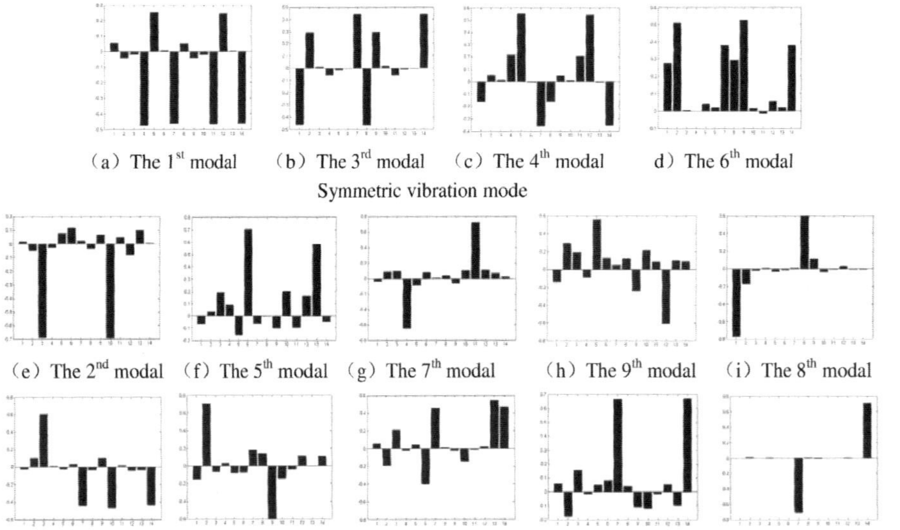

(a) The 1st modal (b) The 3rd modal (c) The 4th modal d) The 6th modal

Symmetric vibration mode

(e) The 2nd modal (f) The 5th modal (g) The 7th modal (h) The 9th modal (i) The 8th modal

(g) The 10th modal (k) The 11th modal (l) The 12th modal (m) The 13th modal (n) The 14th modal

Asymmetric vibration mode

Fig 3. The vibration mode of the transmission system

3.2. *Dynamic load of system*

Fig.4 shows time domain course and frequency spectrum of dynamic load of gear box bearing along direction Y, in which the mean value of dynamic load of bearing at input end is 8,192.3N, being on the contrary to positive direction defined in model and negative value; in addition, gravity action on gear at input end is consistent with dynamic load component on bearing along direction Y, being mutual superposition. Gravity of gear at output end is opposite to dynamic load component on bearing along direction Y, therefore they offset each other, with mean value smaller than that at input end, being 6385.4N.

For frequency spectrum, dynamic load on both ends generated larger peak value at mesh frequency and 2-multiple-frequency, since mesh frequency (1,233Hz) of gear pair and inherent frequency (1,507Hz) of lateral vibration of transmission system are similar, accordingly, fundamental harmonic is to directly transferred to bearing and become main part. The peak value at mesh frequency of bearing at input end is maximum, being 14.32N.

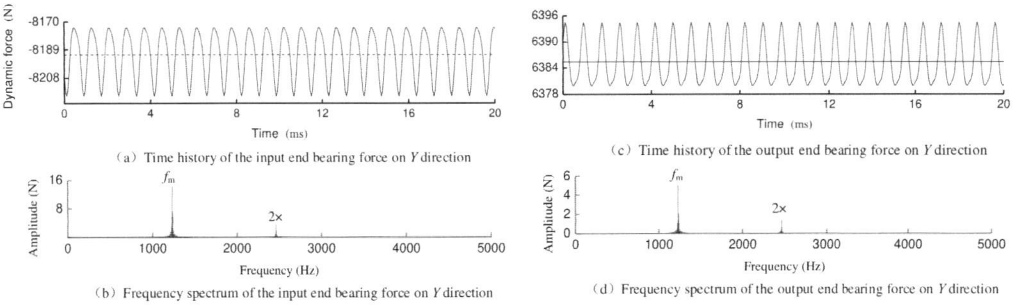

(a) Time history of the input end bearing force on Y direction
(b) Frequency spectrum of the input end bearing force on Y direction
(c) Time history of the output end bearing force on Y direction
(d) Frequency spectrum of the output end bearing force on Y direction

Fig 4. The bearing dynamic load of the gear system

4. Analysis of influencing factor of dynamic characteristic of system

4.1. Influence of stiffness of intermediate connection of herringbone gear on dynamic characteristic

Influence of intermediate connection stiffness on inherent frequency of transmission system is as shown in fig.5.

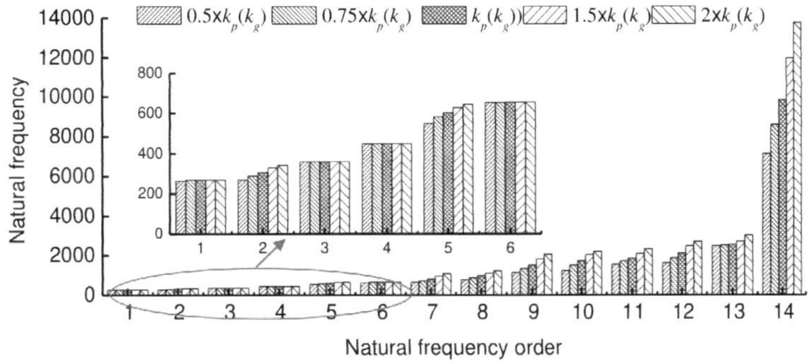

Fig 5. The natural frequency of the transmission system under different connection stiffness

Variation in intermediate connection stiffness has no influence on inherent characteristics of first order, third order, fourth order and sixth order, the reason for which is that the mode of vibration corresponding to inherent frequency of the four orders is symmetric vibration centering around intermediate connection axis, degree of freedom on both sides fail to subject to mutual coupling, accordingly, variation in intermediate connection stiffness has no influence on vibration characteristic. Other inherent frequencies are to show monotonous increase trend with the increase of intermediate connection stiffness.

4.2. Influence of axial positioning stiffness on dynamic characteristic

In herringbone gear transmission system, pinion gear is free of positioning in axial direction as a general rule; while, the entire system is subject to axial constraint by axial positioning on bearing thrust face of bull gear. The paper made a calculation of inherent frequency of transmission system in case that axial positioning stiffness is $0.5\times k_z$, $0.75\times k_z$, $1.5\times k_z$ and $2\times k_z$, from which it is observed that axial positioning stiffness only exerts an influence on inherent frequency (axial vibration mode of pinion gear) at second order and inherent frequency (axial vibration mode of bull gear) at fifth order of the system, having no influence on other inherent frequencies. The inherent frequency at second order and fifth order of system varies with axial positioning stiffness as shown in fig.6, from which it is observed that inherent frequencies at two orders are on the increase with the increase of positioning stiffness and that the difference is that positioning stiffness is to constrain pinion gear along axial direction in a indirect way by meshing spring after constraining bull gear, therefore, variation in stiffness is to exert insignificant influence on pinion gear, inherent frequency of axial vibration of pinion gear is less dependent on variation of positioning stiffness. Relatively speaking, inherent frequency at fifth order is the most sensitive to variation in axial positioning stiffness, with larger rate of variation.

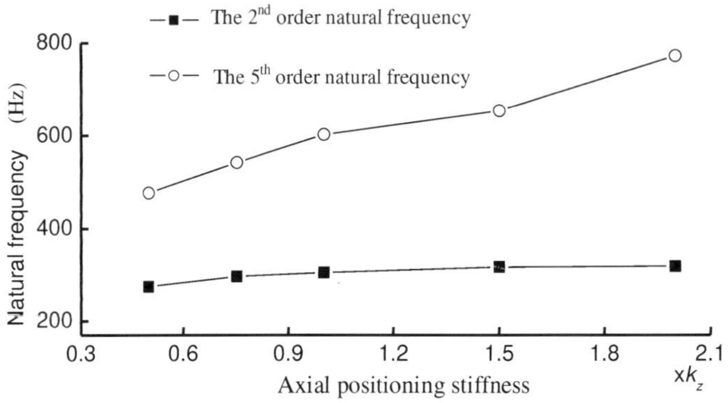

Fig 6. Relationship of the axial positioning stiffness in natural frequency

5. Conclusion

The paper set up dynamic model of the herringbone gear system, and natural characteristics of the gear transmission system are analyzed, some conclusions are found:

(1) the natural characteristic of transmission system may be divided into symmetrical modes and asymmetrical modes. The dynamic load of the bearing contains significant mesh frequency component.

(2) Variation in intermediate connection stiffness of herringbone gear is only to influence asymmetrical modes. With the increase of stiffness value, there is no influence on symmetrical modes, but other inherent frequencies are on the increase.

(3) Axial positioning stiffness of gear is only to influence axial mode of vibration in system. With the increase of axial positioning stiffness, fluctuation of dynamic load of axial support for gearbox is to gradually intensify.

ACKNOWLEDGMENTS

Thanks must be given to Xinjiang University for giving me a good surrounding, and Chinese National Natural Science Foundation (51665054) and Scientific and Technological Training Project (QN2016BS0082) for giving the author a good financial backing to finish the most important part of this research.

Reference

[1] Amendola J. B.. Single vs Double Helical Gears[J]. Turbomachinery International, 2006, 47(5): 34~38.

[2] Feng Zhigang, Wu Xinyue. Analysis on static and dynamic characteristics of herringbone gear in meshing[J]. Ship and Ocean Engineering, 2007, 36(5): 72~76.

[3] Wang Cheng, Fang Zongde, Zhang Molin.. Analysis of dynamics behavior for double helical gears transmissions[J]. Journal of Harbin Institute of Technology, 2011, 43(7): 122~126.

[4] Feng W., Zongde F., Shenjin L.. Dynamic characteristics of a double helical gear under multi-load[J]. Journal of Vibration and Shock. 2013,32(1):49~52.

[5] Ning Z., Pengyuan Q., Guili L.. Modification oprimization of double helical gears with high contact ratio[J]. Journal of National University of Defense Technology, 2015, 37(1):165~170.

[6] Bu Zhonghong, Liu Geng, Wu Liyan. Natural Characteristics Analysis on Herringbone Planetary Gear Trains with Slide Bearing Support[J]. Journal of Mechanical Engeering, 47(1): 80~88.

IOP Conf. Series: Materials Science and Engineering **324** (2018) 012067 doi:10.1088/1757-899X/324/1/012067

Research on the design of fixture for motor vibration test

W X Shen[1] and W S Ma[2] and L W Zhang[2]

1 Senior Production Editor, Hebei Jiaotong Vocational and Technical College, China
2 Production Assistant, Hebei Jiaotong Vocational and Technical College, China
E-mail: 595325836@qq.com

Abstract. The vibration reliability of the new energy automobile motor plays a very important role in driving safety, so it is very important to test the vibration durability of the motor. In the vibration test process, the fixture is very important, simulated road spectrum signal vibration can be transmitted without distortion to the motor through the fixture, fixture design directly affect the result of vibration endurance test. On the basis of new energy electric vehicle motor concrete structure, Two fixture design and fixture installation schemes for lateral cantilever type and base bearing type are put forward in this article, the selection of material, weighting process, middle alignment process and manufacturing process are summarized.The modal analysis and frequency response calculation of the fixture are carried out in this design, combine with influence caused by fixture height and structure profile on response frequency, the response frequency of each order of the fixture is calculated, then ultimately achieve the purpose of guiding the design.

The main performance of fixture for vibration durability test is passing the vibration energy from shaking table to the test piece, with little or no distortion in this passing process. Firstly the fixture need to clamp the test piece, and need to avoid resonant frequency in test frequency range in the passing process. The main influence factors of test result are system frequency、 centre of gravity、 quality and so on. The method of road spectrum excitation is used for Simulating vibration durability test of new energy automobile's motor on the test-bed, so researchers can investigate vibrate durability performance of motor system in a short time. The nature of the test is the simulation of vibrate reliability of new energy motor system in laboratory. Road spectrum excited vibration table is used to test motor vibrate durability in kinds of working-cases in real-time. The purpose of that is to simulate real working condition of motor , and to evaluate system vibrate reliability performance, to get its

Reliability quality.

According to the arrangement of the new energy vehicle's motor in the whole carload, currently the mainly two installations like fixtures are horizontal cantilever and base bearing type, are shown as below:

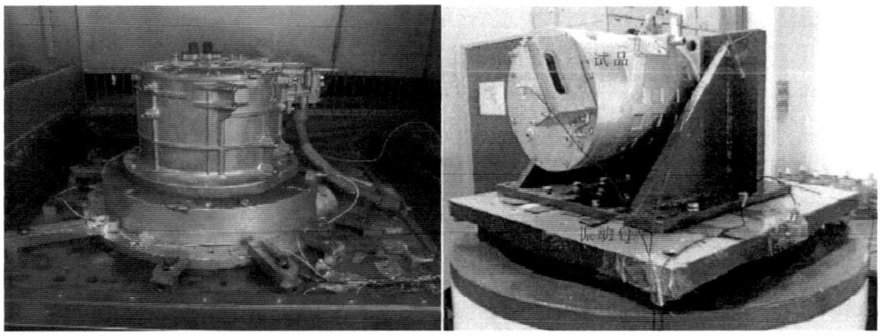

Figure 1. Horizontal cantilever and base bearing type fixtures.

Base bearing type fixture design scheme, it can be installed through adjusting the bolts location, (the bolts are on the vertical plate of the assembly), to meet requirements for different fixing bolt holes positions on the base for matching different test pieces; By adjusting location of waist-shape holes on the base of the assembly, the fixture can assemble with different screw holes which have different spacing on the vibration table , and finally meet genetic requirement of many types of motors. The fixture has the structure that its board and vertical plate are made as one body, also has many reinforce posts, it effective enhancement of the overall stiffness. For the whole structure, it has high vibrating basic frequency and big resistance, so it can avoid resonance in test. This kind of fixture can be installed easily, lots of preparing time before test can be saved . Horizontal cantilever type fixture design scheme, by adjusting the type-trough bolts' location which are on lower part of beam and upper part of the whole frame , finally meet genetic installation requirement of many types of test piece.Section steel and titled strutting pile are used in horizontal cantilever type fixture which can effectively increase the stiffness of the fixture.

The designing process includes the selection of material, weighting process, middle alignment process and manufacturing process . Material with big rigidity and big resistance should be chosen to manufacture fixtures for vibrate durability test. Fixtures made of big rigidity material has little influence to vibration and has wide frequency response, so it has more reliable performance for passing road spectrum vibration signal. Fixture's weight should be as light as possible,cause according to the formula $f_n = \left(\frac{k}{m}\right)^{\frac{1}{2}}$, only increase rigidity k, reduce mass m， the natural frequency of the whole fixture will increase. Otherwise , strength of fixture need to be considered, small strength can cause damage. When testing , the best condition is ,the center of gravity of test piece and fixture coincide in the central line of the vibration test-bed, which can avoid vibration of test-bed , the vibration will lead to vibrating waveform distortion. So fixture need to has symmetric and underslung structure, symmetric and closed shape such as cube、rectangular box、semisphere、cone and so on. Change of dynamic gravity center need to be considered also, especially elastic load even liquid load. Avoid jumpy change on cross section, that will reduce fixture's rigidity and fixture's natural frequency.

Main manufacture methods of fixture are global processing、screw joins、casting、welding、bonding and so on. Among all methods, global processing should be preferred firstly. Screw joins should be designed meticulously, decrease roughness of interactive surfaces , bolt pretension force should be bigger at least 10% than the calculated vibration pull-off force. For big fixtures which are used for high frequency , bolt space should be less than 80mm。Especially pay attention to shear stress of bolt or screw in manufacture. When use bolts made by aluminum alloy or Al-Mg alloy for connection ,coarse thread should be chosen .For the fixture which are frequently disassembled ,stainless steel screwed bushing or wire thread should be used. Epoxy resin can be used on bolts connecting part,it can improve connecting strength. When use horizontal smoothing board, When using horizontal slide, should pay special attention to the shearing force. The parts between the in fixture should be tightly connected to each other, and the holes should be filled with full expansion pin or filled with epoxy resin. The phenomenon of "hair bundles" also lead to waveform distortion. For varying complex surfaces、 thickness and cross sections, casting method is usually applied. The damping of the cast alloy is relatively large, so it is beneficial to reduce the resonance condition. After casting and grinding, the damping of the fixture will be reduced and it is disadvantageous to shock absorption. Welding jig is more convenient and low cost, but the quality of welding must be guaranteed, otherwise the welding position can be easily broken in vibration. Compared with the bolt connection fixture, the welding fixture has no "burr", which saves time than the casting method. For a small fixture ,bonding method is faster than welding method and save cost, is more commonly used with epoxy resin as the bonding material, when using screw connection, welding, epoxy resin adhesive connecting to make the same fixture, welding can get slightly higher resonance frequency , and the epoxy resin method get slightly lower resonance , welding and epoxy resin method get higher resonant frequency than the screw connection. In combination with the specific conditions, the fixture is test is connected by welding and bolt.

The modal analysis and spectrum calculation should be carried out to optimize the design of the fixture for the completion of the initial design for the more complex working conditions. The design optimization process of the base bearing fixture for the motor of the new energy vehicle is introduced below. As shown in the following figure, the resonant image of the six order response frequency after the optimized design.

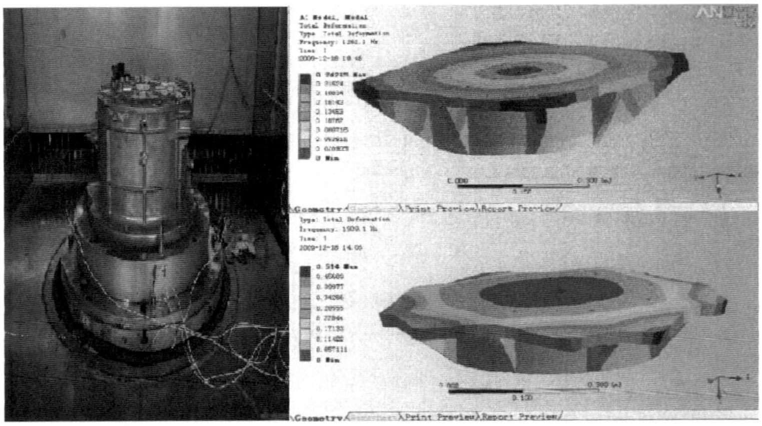

Figure 2. The resonant image of the six order response frequency after the optimized design.

Generally, when designing fixture, the strength and fatigue characteristics of materials are seldom considered. The high stiffness required by the fixture's high-frequency characteristics makes the fixture basically can not damaged by insufficient strength. The weight of fixture is often the most important parameter in fixture design. Therefore, aluminum magnesium alloy is chosen as fixture material and response spectrum need not be used in calculation. In the process of fixture design, combined with the structure and using requirements of the vibration test table, fixture mainly has symmetry design . Because the contact surface between the shaking table and the fixture is square, when designing the fixture surface for fixing test sample, the size should be designed as big as possible, so that the fixture can meet the requirements of the general test. So the designed fixture's outer diameter is 900mm, connected with the vibration test by bolts. At the same time, in order to reduce the weight, the weight reduction groove is used in the bottom area. The expanded clamp top surface is 1100 x 1100mm square, and 8 equidistance reinforcement are used to increase the stiffness of structure and increase the response frequency. In the process of design, the distribution direction of the connecting bolt of the fixed fixture is in the same angle as the direction of the reinforcing bar, so that the response frequency of the fixture can be improved. In Ansys modal analysis, for the convenient division of the grid, the connection part between model and test sample can be simplified, so that the calculation process can be simplified, but the effect to calculating result is not big.

In the design process, the first completed structure is a square with a diameter of 900mm at the bottom and a 1100×1100mm top square. The chamfering part of the fixture is about 120mm and the height is 300mm. After modal analysis, the first order response frequency is low, and it is necessary to be optimized. Therefore, the height of the fixture is changed to 200mm, and the other sizes remain unchanged. After analyzing the frequency response, we can see that, on the basis of meeting the using requirements, reducing the height of fixture has a very significant effect on improving its response frequency. However, the computer simulation analysis has limitations. Therefore, the measured frequency response data of the actual fixture in vibration endurance test is always slightly lower than the simulated value. At the same time, when using Ansys software to divide meshes, because of the limitation of computer hardware, In any case, the mesh can not be subdivided into the extent to which the test specimen can be completely reproduced, so the computation result always tend

IOP Conf. Series: Materials Science and Engineering **324** (2018) 012067 doi:10.1088/1757-899X/324/1/012067

to be bigger. So the real frequency of the real test specimen can not always meet the requirements of the high frequency test, so it is necessary to optimize the design again.

After lowering the overall height of the fixture, we can change the top surface to round shape with diameter of 1100mm, and the other data remain unchanged. The fixture at this time has double circular structure with a height of 200mm , a bottom diameter of 900mm, and a top surface diameter of 1100mm. The modal analysis is carried out again to the improved structure and the response frequency is calculated. The results show the response frequency of the optimized structure is greatly improved. Furthermore, after the grid division of the improved structure is further refined to 0.015, the response frequency of each order is reduced a little. More real data can be obtained. Through simulation analysis, we can think that the response frequency of the designed fixture tend to be stable. The first order frequency is 980Hz and six order frequency is 1474Hz ,this simulation results can meet the needs of the vibration endurance test of the new energy electric vehicle motor.

After the manufacture of the fixture for test is completed, scan the natural frequency of the fixture , then get the result that first natural frequency is 996Hz, which is slightly different from the modal analysis. When the fixture is in the simulation, cause the limited simulating capacity of Ansys software for the actual test pieces , the result always has deviation; And errors in machining process, such as surface roughness and parallelism, bolt hole position accuracy, also affect the results of the test; Otherwise human factors such as bolt fastening degree may also lead to deviation in results.

IOP Conf. Series: Materials Science and Engineering **324** (2018) 012067 doi:10.1088/1757-899X/324/1/012067

References

[1] Zheng S The design of fixture for vibration test [J] *Reliability Design and Process Control* 2006 ，24（5）p14-17

[2] Fu Z F Theory and application of modal analysis [M] *Shanghai Jiao Tong University Press* 2010

[3] Jin M Raymond L and Li D F Application of fixture for vibration test in the reliability test [J] *Environment and Reliability* 2004 p39-43

[4] Liu Z B and Liang Y Study on the vibration characteristics of the disc jig[J] *Mechanical Engineer* 2011 (7) p30-31

[5] Liang S G and Yin G F Modern mechanical optimization design method[J] *Chemical Industry Press* 2005 p5-6 p18-20

IOP Conf. Series: Materials Science and Engineering **324** (2018) 012068 doi:10.1088/1757-899X/324/1/012068

Heat rejection efficiency research of new energy automobile radiators

W S Ma[1] and W X Shen[2] and L W Zhang[2]

1 Senior Production Editor, Hebei Jiaotong Vocational and Technical College, China
2 Production Assistant, Hebei Jiaotong Vocational and Technical College, China
E-mail: 595325836@qq.com

Abstract. The driving system of new energy vehicle has larger heat load than conventional engine. How to ensure the heat dissipation performance of the cooling system is the focus of the design of new energy vehicle thermal management system. In this paper, the heat dissipation efficiency of the radiator of the hybrid electric vehicle is taken as the research object, the heat dissipation efficiency of the radiator of the new energy vehicle is studied through the multi-working-condition enthalpy difference test. In this paper, the test method in the current standard QC/T 468-2010 "automobile radiator" is taken, but not limited to the test conditions specified in the standard, 5 types of automobile radiator are chosen, each of them is tested 20 times in simulated condition of different wind speed and engine inlet temperature. Finally, regression analysis is carried out for the test results, and regression equation describing the relationship of radiator heat dissipation 、heat dissipation efficiency 、air side flow rate、cooling medium velocity and inlet air temperature is obtained, and the influence rule is systematically discussed.

New energy vehicles refers to the fuel cell electric vehicle (FCEV), compressed natural gas (CNG) vehicles, hybrid vehicles (HEV) which use non conventional vehicle fuel as power resource, the advanced technology of power control and drive power source are integrated for vehicle manufacturing. The driving system of new energy vehicle has larger heat load than conventional engine. Compared with traditional fuel vehicles, new energy vehicles have small noise, less exhaust emissions and can effectively protect the environment. At the UN climate change conference in 2015, countries and regions including Germany, Britain, Holland, Norway and the 18 states of the United States formed "zero emission vehicle alliance", promising that by 2050, the state will no longer sell fuel cars. In 2016, the number of new energy vehicles in China was nearly 1 million 90 thousand (pure electric account for more than 60%), which accounted for half of the global electric vehicle. On one hand, the new energy vehicles contain more power devices, has higher heat dissipation requirements; On the other hand, the new energy vehicle battery system for environment temperature requirements are more stringent, high or low temperature environment will significantly affect the vehicle mileage and battery life. Therefore, a higher requirement for the heat dissipation and heat dissipation efficiency of the new energy vehicle radiator is put forward.

In this paper, the heat dissipation efficiency of the radiator of the hybrid electric vehicle is taken as the research object, the heat dissipation efficiency of the radiator of the new energy vehicle is studied through the multi-working-condition enthalpy difference test. In this paper, the test method in the current standard QC/T 468-2010 "automobile radiator" is taken, but not limited to the test conditions specified in the standard, 5 types of automobile radiator are chosen, each of them is tested 20 times in simulated condition of different wind speed and engine inlet temperature. Finally, regression analysis is carried out for the test results, and regression equation describing the relationship of radiator heat dissipation、heat dissipation efficiency、air side flow rate、cooling medium velocity and inlet air temperature is obtained, and the influence rule is systematically discussed.

The heat dissipation efficiency of the radiator is the ratio of heat dissipation and heat absorption, the calculation method is shown as follows:

$$Qw = Gw \times Cpw(Tw1 - Tw2) \qquad (1.1)$$

Qw—— heat dissipation (kcal/hr)

Gw—— flow rate (kg/hr).

Cpw—— the specific heat of water (kcal/kg°C)= 1.0 kcal/kg°C

Tw1—— Enter water temp.(°C)

Tw2—— leaving water temp.(°C)

$$Qa = Ga \times Spa(Ta1 - Ta2) \qquad (1.2)$$

Qa—— heat absorption (kcal/hr)

Ga—— air flow rate (kg/hr)

Spa—— the specific heat of air (kcal/kg °C)= 0.24 kcal/kg °C

Ta1—— Inlet air temperature (°C)

Ta2—— Exit air temperature (°C).

Figure 1. Enthalpy difference test of radiator.

Figure 1 shows the radiator enthalpy difference test of a hybrid electric vehicle produced in Nanchang. The parameters of radiator enthalpy difference test mainly include 8 parameters, such as cooling medium velocity, air side flow, air side resistance and heat dissipation of the medium side. Table 1 shows the test result of the radiator with air side flow rate is 6000m^3/h.

Table 1. Test result of radiator sample1# when air side flow rate is 6000m^3/h.

	Indoor temp.	25.00			state		Test state		
Working condition	Indoor humidity				Start time		2016/6/26 15：05		
	Nozzles information	φ189×2；φ150×1			Test time length		4m 15s		
	Sampling interval	5s							
item	Channel name	1st group	2nd group	3rd group	4th group	5th group	6th group	7th group	average value
Measuring data	Indoor dry ball temperature(°C)	25.03	25.06	25.08	25.09	25.08	25.07	25.06	25.07
	Indoor relative humidity（%）	54.40	54.40	54.50	54.60	54.60	54.60	54.70	54.50
	Indoor atmospheric pressure（kPa）	100.4	100.4	100.4	100.4	100.4	100.4	100.4	100.4
	Face wind speed of radiator(m/s)	2.61	2.52	2.72	2.53	2.62	2.62	2.54	2.60
	Imported air density(kg/m^3)	1.16	1.16	1.16	1.16	1.16	1.16	1.16	1.16
	Inlet air temp. (°C)	25.03	25.06	25.08	25.09	25.08	25.07	25.06	25.07
	Outlet air temp. (°C)	48.36	48.29	48.20	48.18	48.10	48.06	48.00	48.17
	Nozzle pressure difference(Pa)	278.3	278.9	279.2	279.1	279.5	279.5	279.7	279.2
	Air resistance of radiator(Pa)	249.0	249.2	249.4	249.7	249.8	249.9	250.0	249.6
	Inlet water temp. (°C)	69.96	69.85	69.79	69.74	69.70	69.68	69.67	69.77
	Backwater temperature (°C)	60.32	60.26	60.20	60.20	60.12	60.11	60.09	60.19
	Water side resistance of	33.4	33.5	33.4	33.5	33.6	33.6	33.5	33.5

	radiator(Pa)								
	Volume flow(m^3/hr)	2.3880	2.3940	2.3900	2.3920	2.3950	2.3920	2.3900	2.3920
	mass flow(kg/hr)	1381.0	1385.0	1383.0	1384.0	1387.0	1384.0	1389.0	1385.0
	Medium density(kg/m^3)	978.89	979.72	979.95	980.12	979.84	979.69	980.26	979.78
	Radiator fan current(A)	0.00	0.00	0.00	0.00	0.00	0.00	0.00	0.00
	Radiator fan voltage(V)	70.0	69.9	69.8	69.7	69.7	69.7	69.7	69.8
Data processing	Inlet and outlet temperature difference(°C)	23.33	23.23	23.12	23.10	23.02	22.99	22.94	23.10
	Dielectric temperature difference°C	9.63	9.60	9.59	9.54	9.59	9.57	9.58	9.58
	Air flow(m^3/h)	5982.8	5988.4	5991.0	5990.1	5993.1	5992.4	5994.2	5990.3
	Medium flow(m^3/h)	3.7990	3.8070	3.8000	3.8040	3.8110	3.8040	3.8070	3.8050
	Air side heat flow	42473.2	42337.3	42170.6	42123.7	42015.2	41960.0	41891.2	42138.7
	Medium side heat flow	41631.7	41605.3	41529.4	41363.7	41602.5	41466.2	41538.8	41533.9
	Heat balance error(%)	1.98	1.73	1.52	1.81	0.98	1.18	0.89	1.44
	Heat transfer coefficient(W/(M^2 °C))	50.14	50.07	49.87	49.83	49.72	49.61	49.47	49.82

After many experiments, we found that the influence parameters of the radiator dissipation efficiency of the hybrid electric vehicle are mainly three parameters: the air side flow rate, the cooling medium velocity and the inlet air temperature. In order to discuss the three parameters' influence to the new energy vehicle radiator efficiency and heat dissipation, we will keep other parameters' value constant, Separately, only change three key parameters' value(change one parameter one time)：the air side flow、cooling medium flow rate and inlet temperature in radiator enthalpy difference test, in order to get the influence law of three key parameters and the radiator heat dissipation efficiency.

1. Influence of air side flow

For the radiator, the main thermal resistance is the thermal resistance of the air side. Therefore, the air side flow has great influence on the radiator of automobile, while in the actual vehicle driving process, the air side flow of the radiator is the sum of vehicle speed, natural wind speed and fan speed. We first set the test condition as the inlet air temperature is 25℃, cooling medium flow rate is 2.9m^3/h, the air side flow is 4000m^3/h, 4250m^3/h, 4500m^3/h, 4750m^3/h, 5000m^3/h, 5250m^3/h, 5500m^3/h, 5750m^3/h, 6000m^3/h in nine different conditions, the test results are shown in Table 2 (The data in the table for multiple measurements take the average value).

Table 2. Test data for different air side flow of radiators.

Air side flow (m^3/h)	4000	4250	4500	4750	5000	5250	5500	5750	6000
Heat transfer coefficient (W/m^2°C)	36.41	38.29	39.93	41.86	43.32	44.57	45.70	47.35	49.05
Dissipate heat efficiency(%)	37.1	36.0	34.7	33.5	32.5	31.7	30.9	30.3	29.7

The heat transfer coefficient (W/m^2 ℃) refers to the transferred heat of 1m thick material through the area of 1 square meters in 1 sec.,under stable heat transfer condition ,the temperature difference between the two sides of the surface is 1 degrees, the unit is W/m^2K（K can be replaced by ℃ here）.

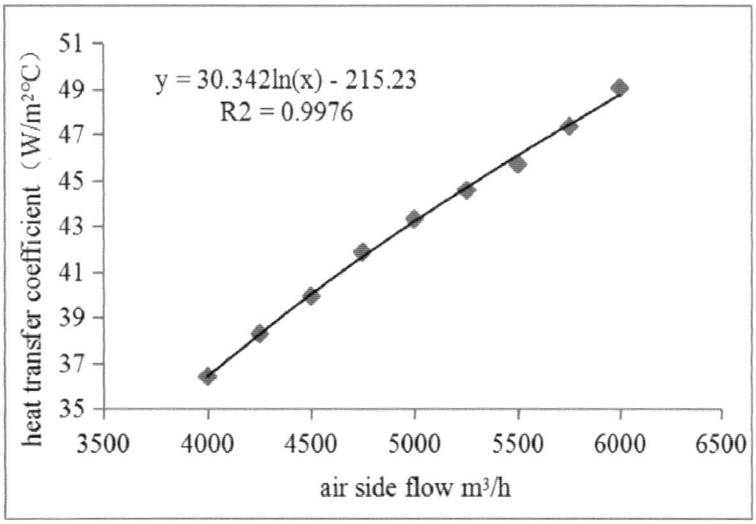

Figure 2. Regression equation diagram of air flow and heat transfer coefficient.

From Figure 2, it can be seen that with the increase of air flow, the heat transfer coefficient of the radiator increases significantly. This rule conforms to the common sense of the people. The increase of air speed will lead to the increase air volume fixed to the windward area per unit time, which will surely bring better cooling effect and increase the amount of heat dissipation. The regression equation of heat transfer coefficient and air side flow is $y=30.342\ln(x)-215.23$, in which y indicates the heat transfer coefficient (w/m^2), and X is the air flow (m^3/h). In the statistical analysis process of variable

IOP Conf. Series: Materials Science and Engineering **324** (2018) 012068 doi:10.1088/1757-899X/324/1/012068

linear regression， use the least squares method to estimate parameters, R square regression is the ratio of regression square and sum deviate square , means the proportion of the sum deviate square which can be explained by sum regression square, the bigger the better, the more accurate model we can get. The value of R square is between 0~1, the closer to 1, the better the regression fitting effect we get, generally think that more than 0.8 of the model fit is better. The ratio of difference squared sums means that the ratio of sum of deviate squares which can be explained by the sum of squares of regression. The more the ratio is, the more accurate the model is, the more significant the regression effect is. While the selected index equation fitting regression equation of R2=0.9976,means that the fitting degree is high, can meet the requirement.

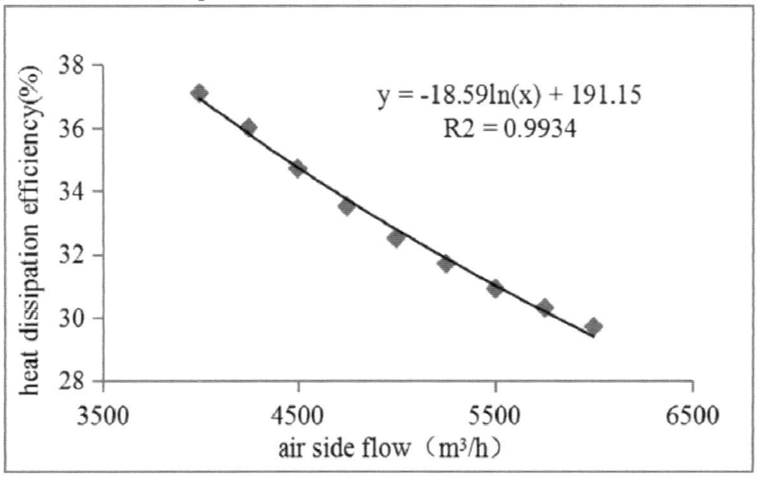

Figure 3. Regression equation diagram of air side flow and heat dissipation efficiency.

As can be seen from Figure 3, with the increase of air side flow or air side flow rate, the heat dissipation efficiency of radiator decreases gradually, and with the increase of air flow rate, the trend of decreasing heat efficiency is more and more gentle. This is different from our regular impression. Mainly because with increasing air velocity, wind resistance power lead to the increase in energy consumption, so the ratio of heat increased by the air which flowing through heat radiator to the heat emitted by the medium in the radiator get bigger and bigger, cooling efficiency increases gradually, the regression equation of the phenomenon described as y=-18.59ln (x) +191.15, where y represents the cooling efficiency,x represents said the air side flow. The R2 is 99.3%, and the curve fitting is very high.

2. Influence of flow velocity in cooling medium

The flow velocity of the cooling medium of the automobile radiator has a great influence on the heat transfer coefficient and the heat dissipation efficiency of the automobile radiator system. When the other conditions are constant, increasing the flow rate of the cooling medium can increase the heat dissipation of the system. However, the increase in the flow rate of the cooling medium increases the power consumption of the pump. With the increase of the flow velocity of the cooling medium, the heat transfer efficiency of the radiator increases gradually, but the when the velocity of the cooling medium increase, the increase degree of the heat transfer efficiency gradually decreases.

Table 3. Velocity test data of different cooling medium of radiator.

Velocity of the cooling medium (m^3/h)	3.0	3.5	4.0	4.5	5.0	5.5	6.0	6.5	7.0
Heat transfer coefficient (W/m^2°C)	37.44	40.86	44.38	47.69	50.62	54.91	58.70	60.57	62.38
Heat dissipation efficiency(%)	33.8	36.7	39.2	40.9	41.2	42.9	44.1	44.9	45.5

As shown in Table 3, keep the inlet temperature of 25 ℃ ,keep the air side flow 5000m^3/h, set coolant velocity as 3.0m^3/h, 3.5m^3/h, 4.0m^3/h, 4.5m^3/h, 5.0m^3/h, 5.5m^3/h, 6.0m^3/h, 6.5m^3/h, 7.0m^3/h nine kinds of conditions, the test results are shown in the following table (table data for repeated measurements of average value).

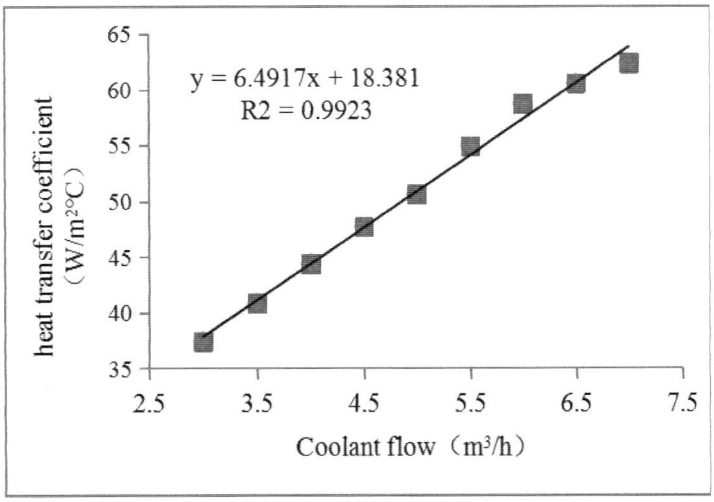

Figure 4. Regression equation diagram of flow velocity and heat transfer coefficient of cooling medium.

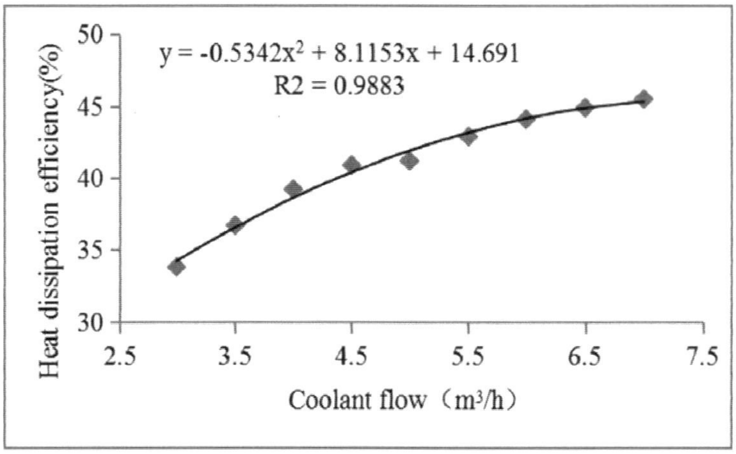

Figure 5. Regression equation diagram of flow velocity and heat dissipation efficiency of cooling medium.

As can be seen from figures 4 and 5, the heat transfer coefficient and heat dissipation efficiency of the radiator increases with the increase of the flow velocity of the cooling medium. The heat transfer coefficient, that is, the increasing trend of the heat dissipation is in the linear law, and is approximately proportional to the flow velocity of the cooling medium. The regression equation has two times polynomial form.$y=6.491x+18.38$， The Y shows the heat transfer coefficient, and the x is the cooling medium velocity, and the R2 is 99.2%, the curve fitting is very high。 The heat dissipation efficiency increases with the increase of the flow velocity of the cooling medium, but the increasing trend decreases gradually,and the regression equation has two times polynomial form.$y=-0.534x^2+8.115x+14.69$, the y indicates the efficiency of heat dissipation, and the x is the flow rate of the cooling medium. the R2 is 98.8%, and the curve fitting is very high.

3. Influence of air inlet temperature

The inlet temperature of the air side (cold fluid side) has a great influence on the heat dissipation of the radiator, and the bigger temperature difference between the cold and hot fluids means bigger heat dissipation.The inlet temperature of the radiator is very different during the different working conditions of the engine. So we set the radiator coolant flow as 2.9m^3/h, air flow 5000m^3/h, inlet air temperature was 25 ℃, 26.5℃, 28℃, 29.5℃, 31℃, 32.5℃, 34℃, 35.5℃ and 37℃, the test results are shown in the following table (the data in the table are repeatedly measured from the average value)

Table 4. Test data of different inlet air temperature of radiator.

Inlet air temperature（℃）	25.0	26.5	28.0	29.5	31.0	32.5	34.0	35.5	37.0
heat transfer coefficient（W/m^2℃）	43.32	42.72	42.07	41.50	41.42	40.17	38.24	38.15	37.25
Heat dissipation efficiency(%)	32.5	33.7	34.5	35.6	36.4	36.9	38.0	38.5	39.1

Through the test data of Table 4, it is easy to observe that with the increase of inlet air temperature,

IOP Conf. Series: Materials Science and Engineering **324** (2018) 012068 doi:10.1088/1757-899X/324/1/012068

the heat transfer coefficient and heat dissipation of automotive radiator are significantly reduced, but the heat dissipation efficiency is significantly increased. The air side (cold fluid side) entrance temperature has dual effects on the performance of the radiator: on the one hand, the air side entrance temperature decreases which increases the temperature difference between the cold and hot fluid ,and improve the heat dissipating capacity of the radiator, but on the other hand, the air side entrance temperature reduces the heat exchange efficiency decreases.

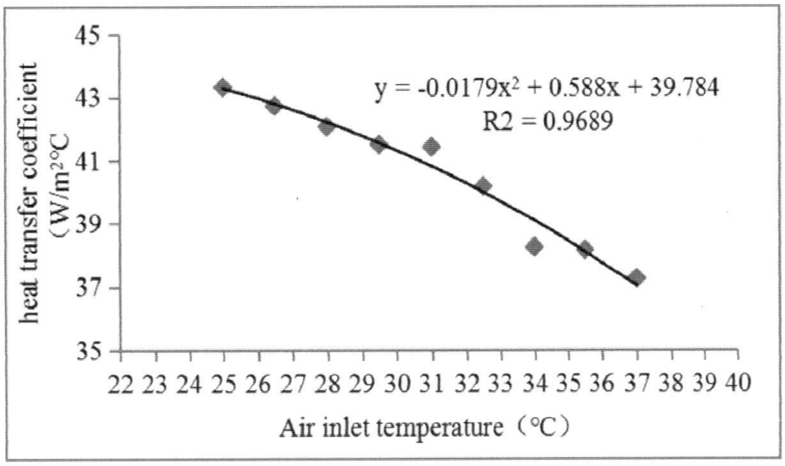

Figure 6. The regression equation diagram of the air inlet temperature and the heat transfer coefficient.

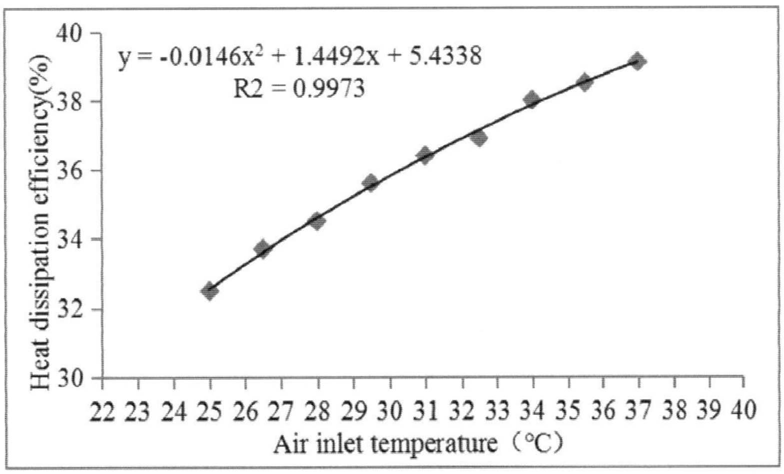

Figure 7. Regression equation diagram of air inlet temperature and heat dissipation efficiency.

As can be seen from Figures 6 and 7, the heat transfer efficiency increases with the increase of the heat transfer coefficient of the radiator as the inlet temperature increases. The regression equation of the heat transfer coefficient(the dissipated heat)which changes with the inlet air temperature is y=-0.017x2+0.588x+39.78, where y represents the heat transfer coefficient, and X indicates the inlet air temperature. the R2 is 96.8%, the curve fitting degree is very high. The heat dissipation efficiency increases with the increase of inlet air temperature, but the trend of increase decreases. The regression equation has two polynomial form, y=-0.014x2+1.449x+5.433, where y represents heat dissipation efficiency, and x represents coolant flow. The R2 is 99.7%, and the curve fitting is very high.

A summary of the full text and suggestions for follow-up work：

This paper systematically discusses the influence of air side flow, cooling medium velocity and inlet air temperature on radiator heat dissipation and heat dissipation efficiency, so as to provide reference for the design of new energy vehicle radiator. At the same time, the regression equation is established on the basis of a relatively small amount of data collection. So the applicability of the new energy automotive radiator except air mixed cars also need further verification.

IOP Conf. Series: Materials Science and Engineering **324** (2018) 012068 doi:10.1088/1757-899X/324/1/012068

References

[1] Won J P and Lee K Thermal flow study on the integrated all aluminum radiator and condenser [J]. *SAE* 01-01-1756

[2] Ma H G Prediction method of performance about heat transfer and resistance performance for vehicle radiator [J]. *Internal-combustion Engine Engineering* 2002 23(l) p33-36

[3] Chen J A Research about heat dissipation performance of vehicle tube type radiator [J] *Agricultural Machinery Journal* 2000 31(3) p84-87

[4] Sahnoun A and Webb R L Prediction of heat transfer and friction for the louver fin geometry[J] *Journal of Heat Transfer* 1992 114 p893-900

[5] Davenport C J Correlation for heart transfer and flow friction for heart transfer and flow friction characteristics of louver fin [J] *AICHE Symposium Series* 1983 225 p19-27

[6] Chou J P The effect of the flow no uniformity on the sizing of the engine radiator [J] *SAE* 02-25-1980

[7] Wu L P Performance research of vehicle tube type radiator [J] *Vehicle Engine* 2005 156(2) P62-66

[8] Zhang X Z Simulation study of louver fin vehicle radiator simulation study *Automobile Technology* 2005(1) p7-10.

[9] Alvon E Double pass versus single-pass radiators for automotive application [J] *SAE*11-06-1989

[10] Amir J Mohanlnlad and Hosni H etc. Thermal fluid characteristics of an automotive radiator used as the exothermal heat exchanger in an auto air conditioning system[J] *ASME* 2005 Heat Transfer 4 p315-323

[11] Saidi M H and Mozafari A A Experimental study of thermal performance and pressure drop in compact heat exchanger installed in automotive[J] *ASME* 2006 *Internal Combustion Engine Division Spring Technical Conf.* p551-557

[12] Patrick H E Study of thermal performance in engine compartment[J] *General Motor Aero/Thermal/Sealing Integration Center Report* Mareh1 2001

[13] Tamio K A Study for improving thermal effectiveness in automotive radiators [J].*SAE* 10-04-1994

[14] Zhu X and Liao Q. Analysis about flow resistance and heat transfer performance of tracked vehicle radiator [J] *Journal of Chongqing University* 2002(8) p40 -43

[18] Yu H Q and Wei Q Characteristic simulation of tracked vehicle radiator [J]. *Tractor and Farm Transporter* 2007(6) p68-70

IOP Conf. Series: Materials Science and Engineering **324** (2018) 012069 doi:10.1088/1757-899X/324/1/012069

Comparison of Models of Stress Relaxation in Failure Analysis for Connectors under Long-term Storage

Yilin Zhou[1], **Mengru Wan**[2]

1 Professor, Automation School, Beijing University of posts and Telecommunications, Beijing, China
2 Master student, Automation School, Beijing University of posts and Telecommunications, Beijing, China
E-mail: ylzhou@bupt.edu.cn

Abstract. Reliability requirements of the system equipment under long-term storage are put forward especially for the military products, so that the connectors in the equipment also need long-term storage life correspondingly. In this paper, the effects of stress relaxation of the elastic components on electrical contact of the connectors in long-term storage process were studied from the failure mechanism and degradation models. A wire spring connector was taken as an example to discuss the life prediction method for electrical contacts of the connectors based on stress relaxation degradation under long -term storage.

1. Introduction

The electrical connector is an interface between the conductive components in an equipment and those in other equipment. Its function is to transfer electrical energy or signal from one system to another. Electrical connectors are scattered in every parts of the system. Each connector contains several pairs or even hundreds of pairs of electrical contacts. According to statistics, 28% of the current system failures are caused by the failure of electrical connectors [1]. Most connectors under storage conditions are exposed to the air and are subject to various environmental factors such as temperature, humidity, atmospheric corrosion, etc. It involves failure mechanism such as stress relaxation of elastic materials, diffusion of the base material, and pore corrosion. The distribution of micro-pores on gold plating is discrete resulting in a relatively small probability of pore corrosion at small contact zones, especially for multiple pairs of contacts in parallel, so the effect of pore corrosion on electrical contact properties is ignored here. Most connector materials are plated with an intermediate layer, such as nickel, between the copper substrate and the gold plating on surface, which hinders the diffusion of the copper to the surface coating. Therefore, the process of electrical contact failure caused by the diffusion of the substrate material is very slow, which can also be neglected. Then, the failure mechanism of stress relaxation for elastic materials needs to be analysed and discussed. In this paper, three degradation models of stress relaxation were compared, the storage life of the wire spring connector was primarily estimated based on the accelerated experiments.

2. Failure analysis of stress relaxation of connectors

2.1. Mechanism and main influencing factors of stress relaxation

Stress relaxation is the contact pressure decreases with time under constant deformation. The process of stress relaxation is usually divided into two stages [2]. From a microscopic perspective, at first stage, the movable dislocation is remarkable while the driving force is high [3], so the speed of stress relaxation is fast. At second stage, the stress relaxation rate is small, because the dislocation and impurity atoms interact with the second phase particles, causing dislocations to proliferate, thus impeding dislocation movement.

The main factors affecting material stress relaxation are temperature, initial applied stress and material compositions. Temperature has the greatest effect on stress relaxation. Increasing temperature will change the microstructure of the material then will accelerate the loss of contact pressure [4], having an accelerating effect on two stages of stress relaxation. Stress relaxation is also affected by the size of the initial applied load. The more rapidly stress relaxation changes, the higher contact pressure is applied to the components. The relaxation rate is almost constant when the initial stress is between 30% and 90% of the yield strength at 0.2% residual strain [5]. After comparing the stress relaxation data of copper and copper alloys for electrical connectors, it was found that the combination of different alloying elements with pure copper increases its resistance to relaxation and its magnitude of anti-relaxation can vary depending on the type and amount of alloying elements added [5].

2.2. Model of stress relaxation of elastic materials

The common stress relaxation models are divided into empirical model and failure physical model, with time as the independent variable and residual stress as the dependent variable. Among the three models described below, the model a is the empirical model of stress relaxation and Larson-Miller parameter after extensive relaxation tests; the model b is a failure physics model based on the derivation of Arrhenius behaviour; based on the material parameters(E、η), a Maxwell linear elastic-viscous body physical model is proposed in the model c.

a. Empirical model of stress relaxation and Larson-Miller parameters

Stress relaxation can be characterized by permanent deformation or stress variation, when stress relaxation is characterized by deformation, the expression is [6]:

$$\lg\left(\frac{X}{X_0}\right) = aT(\lg t + C) + b \quad (1)$$

Where: X is the permanent deformation at time, X_0 is the initial deformation, T is the absolute temperature, t is the time, a and b are the constants to be determined, and the range of constant C is 16-20. At constant temperature, it can be further rewritten as:

$$\lg\left(\frac{X}{X_0}\right) = A\lg t + B \quad (2)$$

In the formula, A and B are parameters related to temperature, which can be obtained by the least squares method. When the stress relaxation quantity is characterized by residual stress, the expression is:

$$\lg\left(\frac{\sigma_0-\sigma_t}{\sigma_0}\right) = \mathrm{A}\lg \mathrm{t} + \mathrm{B} \quad (3)$$

Further rewritten as:

$$\lg\left(\frac{\sigma_t}{\sigma_0}\right)=A'\lg t+B' \tag{4}$$

Therefore, (2) (3) (4) can be considered as completely equivalent. The logarithm of relaxation amount is linearly correlated with the logarithm of time through the test data, regardless of being characterized by deformation or residual stress representation. The contact reed material of the test is divided into beryllium bronze and tin bronze strip [6], the $\lg(\frac{X}{X_0})$-lg t, which is fitted at four temperatures, is linearly related and is coincide with test data points.

b. The model of stress relaxation based on Arrhenius behaviour [7]:

$$t_1 = Ae^{\frac{Q}{kT}}\left[e^{\frac{1}{\beta}\left(1-\frac{\sigma_1}{\sigma_0}\right)}\right] \tag{5}$$

t_1 : time (in seconds) ; Q : activation energy ; k : Boltzman constant (1.38×10^{-23}J/K) ; T : absolute temperature ; A、β : temperature-dependent coefficient.

$$\frac{\sigma_1}{\sigma_0} = -\beta\ln t_1 + \beta\ln A + \beta\left(\frac{Q}{kT}\right) + 1$$

At a constant temperature, can be further converted to:

$$\sigma_t = A_0\lg t+A_1 \tag{6}$$

The model shows that at a certain temperature, the residual stress is linear with the logarithm of time. Whitley used the Bell Telephone Laboratories (BTL) to develop a method of determining the post-heat age relaxation of the hoop stress in copper wire wrapped around a nickel-silver [7].

c. MAXWELL linear elastic-viscous body model

Imagine the material's hysteresis elasticity and viscous behaviour as a combination of linear elasticity (following Hooke's law of elasticity) and linear viscosity (Newton's liquid damping cylinder) behaviour [8]:

$$\sigma = E\cdot\varepsilon_e \text{ and } \sigma = \eta\cdot\dot{\varepsilon}_\eta$$

Where σ is the normal stress, ε_e is elastic strain, η is the viscosity, $\dot{\varepsilon}_\eta$ is ideal line viscous strain rate.

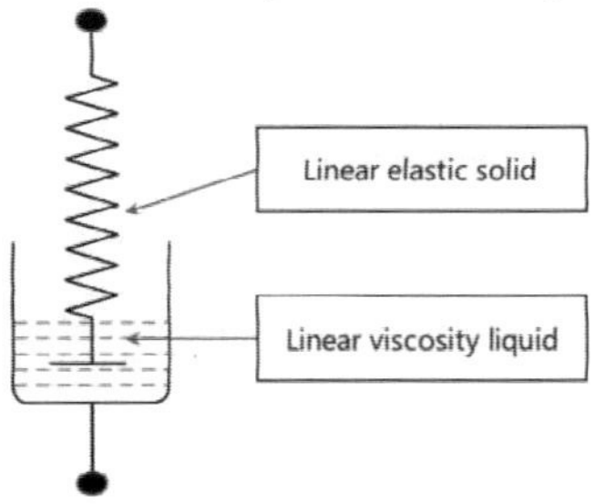

Figure 1. MAXWELL model for stress relaxation of beryllium bronze material

Based on the definition of stress relaxation, there is:

$$\dot{\varepsilon}=\dot{\varepsilon}_e+\dot{\varepsilon}_\eta=0$$

Substitute the stress-strain relationship:

$$E*\dot{\varepsilon}=\dot{\sigma}+\frac{E}{\eta}*\sigma=0$$

where $\dot{\varepsilon}$ is total strain rate, $\dot{\varepsilon}_e$ is ideal line elastic strain rate, $\dot{\sigma}$ is the stress rate, E is the ideal elastic modulus volume.

Let $t_R = \frac{\eta}{E}$, which is called the time factor (time-independent, temperature dependent), then

$$\frac{d\sigma}{\sigma} = -\frac{dt}{t_R}$$

After the integration : $\ln\sigma = -\frac{t}{t_R} + C_0$, C_0 is constant.

Consider when t=0, $\sigma = \sigma_0$, $\sigma_0 = E\varepsilon_0$, then : $C_0=\ln\sigma_0$,

$$\sigma = \sigma_0 \cdot \exp(-t/t_R) \quad (7)$$

$$\ln\frac{\sigma_t}{\sigma_0} = -\frac{1}{t_R} \cdot t$$

where: σ_0 is the initial stress, t is the stress relaxation time. Establish the relationship between residual stress and time.

Suppose: $\eta=\eta_0 \cdot \exp(-Q/KT)$,then $\ln\frac{\sigma_t}{\sigma_0} = -\frac{E}{\eta_0} \cdot \exp(Q/KT) \cdot t$

At a certain temperature, supposing $\frac{E}{\eta_0} \cdot \exp(Q/KT)$=C, then C is constant, So the above formula becomes:

$$\ln\sigma_t=\ln\sigma_0\text{-}Ct \quad (8)$$

The model expresses the stress relaxation behaviour of materials related to time and temperature through mechanical model. It is verified that the curve of bending stress relaxation of beryllium bronze strip accords well with the experimental data points, and the error is no more than 5%.

2.3. Comparison of test data fitting

A high temperature acceleration experiment was done on a wire spring connector of a beryllium bronze material, and the loss of contact pressure was characterized by separating force. From σ=F/S, where F is contact pressure, S is the contact area. It is assumed that in the process of stress relaxation, the area S is unchanged, and the residual stress σ_t in three models can be replaced with F. The test data were fitted under 144°C in three models, as shown in table 1.

Table 1. Test data fitting results

No. of Models	Type of models	Model(At 144°C)	R-squre
a	Empirical formula:	$\lg(\frac{F_t}{F_0})$=-0.2332lg$t$ -0.063	0.973
b	Based on Arrhenius behaviour:	$\frac{F_t}{F_0} = -0.2418\lg t + 0.759$	0.953
c	Based on linear elasticity - viscous body:	$\ln F_t = -0.006216t + 3.868$	0.705

3. DISCUSSION

3.1. Analysis of the model of stress relaxation

It can be seen from the experimental data that under the condition of high temperature, the fitting of model a is the highest. The Model b is the least and the model c has the worst fitting result. Obviously, model c is not suitable for this type of wire spring connector. Because the change of contact force under high temperature is remarkable, the decrease of contact force is not obvious when the temperature is low. The closer to the room temperature is, the more difficult to test. Therefore, the empirical formula fitted from a large amount of experimental data does not guarantee a high degree of reliability when extended to room temperature. Model b follows the Arrhenius behaviour and takes the parameter temperature and the initial stress mainly affecting stress relaxation of the connector into consideration. Theoretical deduction based on the failure physics makes it more rational for the prediction of stress relaxation. In summary, it is considered that model b has the best generality for predicting the stress relaxation behaviour of connectors.

3.2. Effect of stress relaxation on electrical contact properties

Connectors require low and stable contact resistance in applications. Thus, the contact resistance is used here as a degradation parameter for long-term storage of the connector to evaluate the storage life. For the convenience of discussion, if the two contact elements are the same material, the contact resistance expression on a contact pair is [9]:

$$R = R_c + R_f = \frac{\rho}{2}\sqrt{\frac{H}{F}} + \frac{\sigma_f H}{F} \tag{9}$$

Where R_c、R_f respectively represent the constriction resistance and the film resistance; ρ is resistivity of the conductor material；H is metal material's hard；F is contact pressure；σ_f is tunnel resistivity of the film.

When only the effect of stress relaxation on the increment of contact resistance is considered, the (9) is converted into:

$$\Delta R = \frac{\rho\sqrt{H}}{2}\left(\frac{1}{\sqrt{F_t}} - \frac{1}{\sqrt{F_0}}\right) + \sigma_f H\left(\frac{1}{\sqrt{F_t}} - \frac{1}{\sqrt{F_0}}\right)$$

If the area S is invariant during stress relaxation, contact force F replaces equation (6) where residual stress σ_t, to logarithmic after both sides:

$$\Delta R = \frac{\left(\frac{\rho\sqrt{H}}{2} + \sigma_f H\right)}{\sqrt{F_0}}\left(\frac{1}{\sqrt{A_0 lgt + A_1}} - 1\right)$$

Let K= $\frac{\left(\frac{\rho\sqrt{H}}{2} + \sigma_f H\right)}{\sqrt{F_0}}$, the above formula can be rewritten as:

$$\left(\frac{1}{\frac{\Delta R}{K} + 1}\right)^2 = A_0 \text{lgt} + A_1 \tag{10}$$

At 144 ℃, the parameter A_0、A_1in equation (6) can be fitted by the test data, which is 0.2418 and 0.759 respectively. Substituting equation (10) into the experimental data of the contact resistance increment and time (ΔR-t), the constant K can be fitted to 0.7385. After finishing the following formula, then:

$$\left(\frac{1}{\frac{\Delta R}{0.7385} + 1}\right)^2 = -0.2418 \text{logt} + 0.759 \tag{11}$$

The experimental data at 144 ℃ can be seen that the stress relaxation is divided into two stages, achieving a stable state after 16h, that is. Calculated by (11) formula, the contact resistance increases by 34.1% when the stress relaxation reaches stable state. Similarly, the increase in contact resistance at which the connector reaches the maximum amount of stress relaxation at different temperatures can thus be estimated.

3.3. Prediction and analysis of stress relaxation life

The following deduction is based on Model b for life prediction analysis. By Arrhenius theory, when the stress relaxation rate reaches a certain value, each test temperature and time follows the following relationship [10]:

$$\frac{1}{t_f} = Ae^{-\frac{Q}{RT}}$$

Take the logarithm on both sides and arrange it : $\ln t_f = \frac{Q}{R} \cdot \frac{1}{T} + (\ln 1 - \ln A)$

that is:

$$\ln t_f = \frac{Q}{R} \cdot \frac{1}{T} + B \tag{12}$$

By the equation (12), the logarithm of time has a linear relationship with the reciprocal of temperature. Through the acceleration experiments of different temperatures, the model is fitted first, and then according to the method in section 3.2, the corresponding time is obtained when the contact resistance increment reaches the threshold value. The linear expression of $\ln t_f$-$\frac{1}{T}$ can be obtained by fitting the relationship between temperature gradient and corresponding time according to (12). Thus, the life prediction of the elastic element on contact point of the connector due to stress relaxation is realized.

4. Conclusion

In this paper, the stress relaxation is analysed from mechanism and influencing factors, and the existing stress relaxation degradation models are summarized and compared. A wire spring connector with beryllium bronze material was tested using a series of high-temperature accelerated experimental data fitting in accordance with three types of degradation models. It was found that the empirical formula had the best fitting degree and the fitting effect of linear elastic-viscous body model was the worst. However, from the perspective of failure mechanism, it is considered that the model following Arrhenius behaviour is more suitable for describing the stress relaxation behaviour of elastic elements and has better generality under different temperature gradients. At 144°C, the contact resistance of the connector is increased by 34.1% when the stress relaxation reaches the stable value. Using high-temperature accelerated test can predict the contact resistance of the connector rises to a certain threshold value. Then the relationship between the different temperature and the corresponding failure time is obtained by fitting, and the extrapolation to room temperature can predict the storage life of the contact resistance due to stress relaxation under long-term storage condition of the connector.

Acknowledgement

The authors are grateful for the support of the National Natural Science Foundation of China (61674017) on this research.

IOP Conf. Series: Materials Science and Engineering **324** (2018) 012069 doi:10.1088/1757-899X/324/1/012069

References:

[1] Lin S D, Pan J and Chen W H 2010 *J. Electromechanical Components*. **29** pp 53-56

[2] Xiao X P and Liu R Q 2015 *J. Materials Review*. **29** pp 148-151

[3] Wang D J, Zhang W F and Fang X T 2013 *J. Advanced Materials Research*. **750-752** pp 675-678

[4] Milenko B and Valery V K 2007 *Electrical Contacts Fundamentals, Applications and Technology* (Boca Raton: CRC Press) pp 2-7

[5] Kenneth W H 1982 *Proc. Fifteenth Annual Connectors and Interconnection Technology Symp. (Philadelphia)* **11** pp 325-336

[6] Li M H 1985 *J. Electromechanical Components*.**1** pp 52-58

[7] Beach K L and Pascucci V C 2000 *Proc. of the Forty-Sixth IEEE Holm conf. (Chicago)* pp 1-17

[8] Zhang X M and Mao X P 2001 *J. The Chinese Journal of Nonferrous Metals*. **6** pp 988-992

[9] Wang S J 2017 *J. Journal of Mechanical Engineering*. **10** pp 180-186

[10] Fox A 1980 *J. Journal of Testing and Evaluation*. **3** pp 119-126

IOP Conf. Series: Materials Science and Engineering **324** (2018) 012070 doi:10.1088/1757-899X/324/1/012070

Inhibition effect of Arabic gum and cellulose acetate coatings on aluminium in acid/base media

S Alva, R Sundari, A Rahmatullah, and H Wahyudi

Department of Mechancal Engineering, Universitas Mercu Buana, Jakarta, Indonesia
E-mail: sagir.alva@mercubuana.ac.id

Abstract. Nowadays aluminium is broadly used for battery purpose due to its conductivity, non toxic and economic reasons. Arabic gum and cellulose acetate are used as potential inhibitors to hinder corrosion effect on aluminium plate immersed in a solution of hydrochloric acid or sodium hydroxide. This investigation has studied the corrosion rate in terms of different concentrations of acid or base media. The average inhibition efficiency in the interested concentration range of both HCl and NaOH (0.1 M – 3.0 M) for 1 x 1 cm^2 aluminium (Al) plate coated by 20% Arabic gum (AG) and 5% cellulose acetate (CA) with each thickness of 0.5 mm is found to be higher than 90%. The electrochemical behavior of corrosion effect is examined by cyclic voltammetric performance with respect to HCl or NaOH media. This investigation is useful especially for the study of Arabic gum and cellulose acetate utilized as polymer inhibitor in strong corrosive media.

1.Introduction

Increased aluminium utilization in many fields due to its conductivity, non toxic and low cost reasons has attracted strong interest to study the corrosion effect of aluminium in strong corrosive media such as acid or base media. Large utilities of aluminium batteries with its limitation in relative short life-time have generated ideas in seeking for materials that might overcome aluminium batteries obstacle. Carbon polymers conjugated with heteroatomic organic constituents (N, O, and S) assumed to possess characters related to strong inhibition effect [1]. In addition, the hydroxyl group of polysaccharides may increase polarity character of a polymer causing the polymer more dissolved in aqueous media.

Arabic gum classified as conjugated polysaccharides with glucoprotein originally came from East Africa assumed to possess good inhibition character related to its heteroatomic (N, O, and S) properties [1]. Besides, Arabic gum has other benefit properties in relation to its safety, high solubility in aqueous media and large molecular size. Several investigations were made to examine the role of Arabic gum in association with aluminium and mild steel corrosion in strong acid media, and even Arabic gum showed better inhibition character to aluminium rather than to mild steel [1].

Cellulose acetate as a synthetic material is an ester of acetic acid and cellulose, which has been largely used in photographic film, glues, and synthetic fibres. Cellulose acetate as its polysaccharide type with its hydroxyl group may enhance its polarity behaviour on reason of increasing aqueous solubility. Cellulose acetate membrane has been largely used in many industries on account of its availability, high flux, and hydrophilic character [2]. Recently, a derivate of cellulose acetate was produced to increase its resistance toward chlorine and biocompability [3].

On account of this matter, this investigation is interested in the study of inhibition effect of Arabic gum (AG) and cellulose acetate (CA) on aluminium corrosion in hydrochloric acid and sodium hydroxide solutions. The commercial aluminium 1100 plate (1 x 1 cm^2) is coated by AG as the first layer followed by CA as the second layer immersed in different concentrations of respective both HCl and NaOH solutions in the range of 0.1 M – 3.0 M. The electrochemical behaviour of corrosion – inhibition event has been investigated through cyclic voltammetric performance. Up to date, the Al corrosion using AG inhibitor has already been investigated, however, the utilization of double layer inhibitors with respect to AG and CA is assumed to be a new finding. Moreover, this study is intended to give benefit related to the knowledge of corrosion – inhibition process particularly involved with double layer inhibitor.

2. Material and Experimental

2.1 Materials

Materials used in this paper include commercial aluminium 1100 plate (1 x 1cm^2). Arabic gum and cellulose acetate purchased from Elnasr Ltd., and both electrolytes NaOH and NaCl purchased from Merck. Alcohol 70% was used for cleaning Al plate with a composition of 96.4% Al, 1.1% Mg, 0.7% Mn, 0.5% Fe, 0.2% Si, 0.2% Cu, and other undetected impurities.

2.2 Coating process of Al plate with polymer

The coating process is as follows:
1. To prepare an aqueous solution of AG (20%).
2. To paste the AG solution on Al plate (1x1 cm^2) and allowed to dry on air for 24h.
3. To prepare an aqueous solution of CA (5%).
4. To paste the CA solution as the second layer and dried on air for 24h.
5. To connect the bottom part of Al plate to a copper wire (15cm) and used as working electrode.

Each membrane polymer has thickness of 0.5 mm. A digital calliper was used to measure the thickness.

2.3 Determination of corrosion rate and inhibition efficiency

The determination of corrosion rate (CR) is derived from the Faraday's law expressed in equation (1),

$$CR = \frac{m}{tA} = \frac{iM}{zF} \qquad (1)$$

For aluminium, "m" is the mass of Al in mg, "t" is the time in sec, "A" is the surface area of Al in cm^2, "i" is the current density in $\mu A/cm^2$ if "CR" expressed in $mg/dm^2/day$. 1 F = 96 500 coulomb/mole and 1 coulomb = 1 ampere. sec., "z" = 3 for Al since it involved with 3 electrons transfer, "M" is the molecular weight of Al, i.e. 27 g/mole. Hence, "CR" in mm/y is defined by the current density "i" in

$\mu A/cm^2$ obtained from instrumental measurement multiply by a constant of 0.00327 and a factor of "M/zD" as described in equation (2) [4].

$$CR = 0.00327\,\frac{iM}{zD} \qquad (2)$$

"D" is the density of material, for Al = 2.7 g/cm^3.

The inhibition efficiency in this case is described as follows:

$$\text{Inhibition efficiency} = \frac{CRo - CR}{CRo} \times 100\%$$

CRo = corrosion rate without inhibitor

CR = corrosion rate with inhibitor

2.4 Cyclic voltammetry

The cyclic voltammetric performances are executed by a potentiostat of Corr Test Electrochemical Workstation using Corr Show application. A standard calomel electrode (SCE), platinum electrode, and aluminum connected to copper wire were used as reference electrode, auxiliary electrode, and working electrode, respectively. The current density, "i", was obtained from the polarization curve of current density vs. time and then determined the average current density during 60 sec. running. Then this average current density was used to determine the CR as stated in equation (2) above. Each measurement of CR was done for three replications. The measurements of current density were done for both Al plate immersed in HCl and NaOH. The cyclic voltammetry is an electrochemical curve of current vs. potential represented the oxidation reduction process in the electrochemical reaction of Al plate in the presence or absence of AG and CA for both aqueous solutions of HCl and NaOH, respectively.

3. Results and Discussion

3.1 Corrosion rate

All data of CRs related to a state using polymer inhibitor or free immersed both in HCl and NaOH aqueous solution in triplicate measurements, respectively, are not presented here due to space reason, however, the data results are shown in graphical presentations that are clearly to be understood. Generally, all the CR increased when concentrations of both HCl and NaOH increased (figure 1 and figure 2).

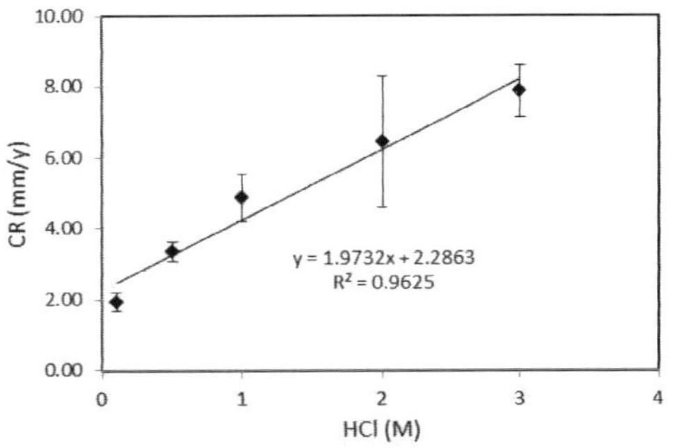

Figure 1. Corrosion rate of Al plate (1x1 cm^2) coated by 20% AG and 5% CA immersed in HCl.

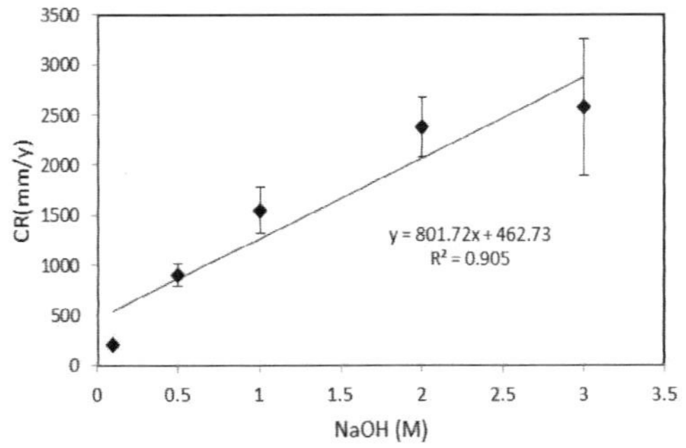

Figure 2. Corrosion rate of Al plate (1x1 cm^2) free of AG and CA immersed in HCl.

Regarding the values of CR in figure 1, the values of CR for Al plate in HCl are much lower in the presence of AG and CA (figure 1) compared to that in the absence of AG and CA (figure 2). Similar patterns are also shown in figure 3 and figure 4 when the Al plate immersed in NaOH. This is reasonable since AG and CA are polymers protected the Al plate from corrosive media (HCl and NaOH). The double layers of protecting polymers look strengthened the inhibition effect on Al plate corrosion. This phenomenon is justified by the high values of inhibition efficiency (Table 1) as the average values of inhibition efficiency for HCl and NaOH are found to be 97.13% and 95.79%, respectively.

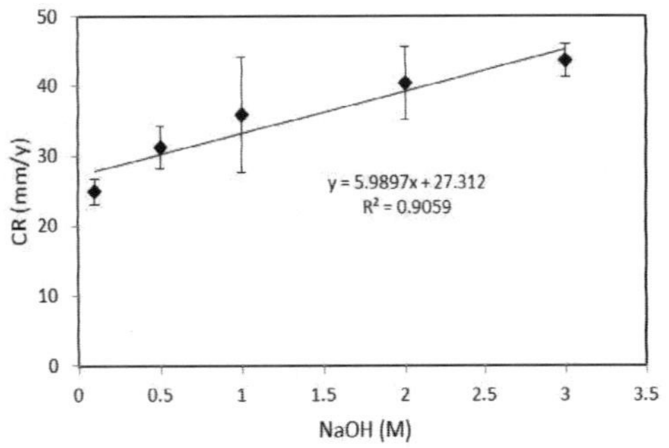

Figure 3. Corrosion rate of Al plate (1x1cm^2) coated by 20% AG and 5% CA immersed in NaOH.

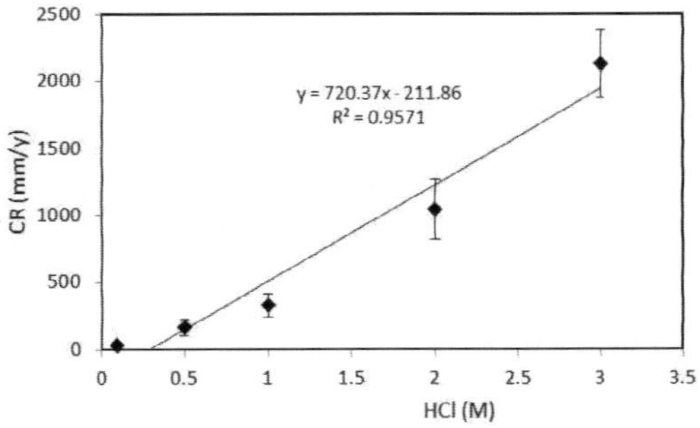

Figure 4. Corrosion rate of Al plate (1x1 cm^2) free of AG and CA immersed in NaOH.

3.2 Inhibition efficiency

Table 1. Inhibition efficiency of Al plate (1x1 cm^2) in the presence or absence of AG and CA immersed in HCl and NaOH, respectively. Triplicate measurement.

HCl aq.	Absence of Ag-CA	Presence of Ag-CA	Inhibition efficiency	NaOH aq.	Absence of Ag-CA	Presence of Ag-CA	Inhibition efficiency
0.1M	19.624	1.928	90.17 %	0.1M	210.083	24.914	88.14 %
0.5M	164.134	3.346	97.96 %	0.5M	898.720	31.259	96.52 %
1M	328.455	4.869	98.51 %	1M	1549.549	35.883	97.68 %
2M	1042.308	6.443	99.38 %	2M	2373.334	40.432	98.30 %
3M	2125.198	7.869	99.63 %	3M	2573.318	43.600	98.31 %
Average inhibition efficiency in HCl			97.13 %	Average inhibition efficiency in NaOH			95.79 %

3.3 Cyclic voltammetry

The electrochemical behaviour of Al plate corrosion in both HCl and NaOH are shown by their cyclic voltammetric performances (figures 5 – 8). Al corrosion by HCl in the absence of protective layers AG and CA occurred very rapidly particularly at higher acid concentrations that it means the redox reaction occurred so fast and as a result, the oxidation and reduction peaks changed very fast [5] causing a fluctuated response (figure 5) except at the lowest concentration of HCl (0.1M). According to Xhanari and Finsgar [6], at higher acid concentration (pH 3 – 4) chloride ion may dissolve Al through the formation of $AlCl_3$, which is soluble in water.

On the other hand, in the presence of AG and CA layers redox reaction between Al and water are almost totally hindered by those polymers coating Al plate and as a result, redox reaction almost apparently not occurred, it means there is no any oxidation or reduction peak as reflected by smooth lines of their CV performance (figure 6). In addition, in the presence of AG and CA the current yield is extremely low about in the order of 10^{-7} ampere/cm^2 (figure 6).

In the case of Al plate immersed in NaOH, the OH^- ion from NaOH enable to form hydrogen bond in water that may retard the migration rate of ions especially at lower concentration of NaOH [7]. This condition reflects the CV performance in NaOH (figure 7) that smooth lines observed for lower concentrations of NaOH (0.1 – 2M). Previous investigation reported a gray colour formed at the surface of Al when Al immersed in higher concentration of KOH as an indicator of the formation of Al_2O_3 layer [5].

As in the case of figure 7, the oxidation peak of the CV of NaOH 3M is apparently shown, it means that at higher concentration of NaOH (3M) the migration of ions in water is getting slower. Moreover, this investigation found the oxidation peak of the CV of NaOH 3M observed at about 1.0V (figure 7), and this case is in agreement with the facts finding obtained by the former report [8] using Al silicon coated by methyl cellulose. In the context of Al coated by AG and CA layers in NaOH media, the OH^- from NaOH may attack the acetyl group of CA resulted the acetyl group broke down from CA lowering the adherence of CA to AG [5]. Furthermore, the deacetylation process may reduce the protective function of CA that should be underlined in the case of using CA as protective layer for any metal in base media such as NaOH [5].

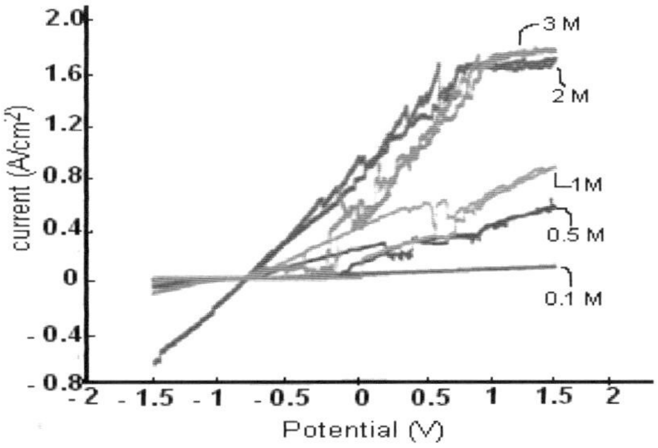

Figure 5. CV performance of Al plate free of AG and CA immersed in HCl (0.1M – 3.0 M).

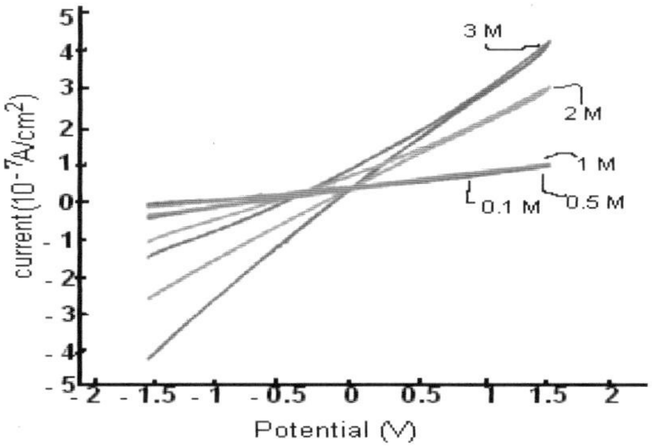

Figure 6. CV performance of Al plate coated by AG and CA immersed in HCl (0.1M – 3.0 M)

As in the case of Al corrosion in HCl media using AG and CA as protective layers, similar phenomenon with regard to the CV performance (figure 8) is also found with Al corrosion in NaOH using the same double layer polymer (AG and CA) as protective agent.

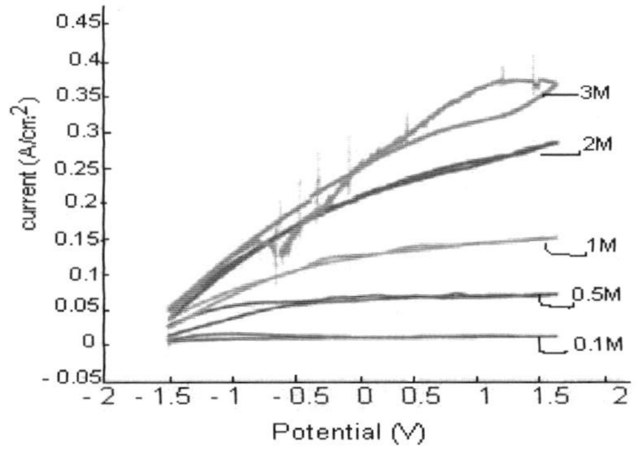

Figure 7. CV performance of Al plate free of AG and CA immersed in NaOH (0.1M – 3.0 M).

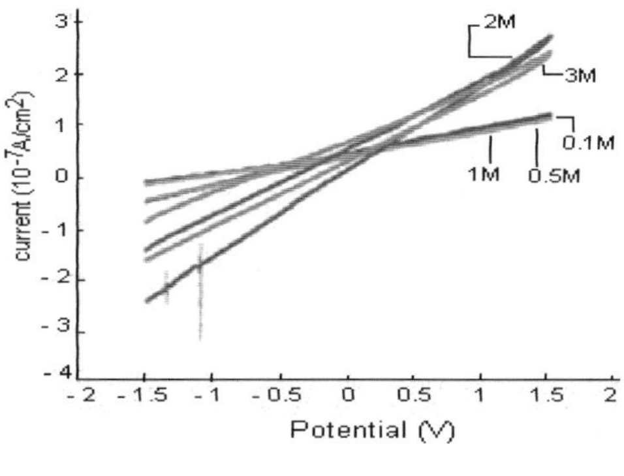

Figure 8. CV performance of Al plate coated by AG and CA immersed in NaOH (0.1M – 3.0 M).

The CV of figure 8 shows smooth lines performance for all concentrations of NaOH used (0.1-3M), which attributed no any redox reaction occurred during scanning process that it means the protective membrane layers (AG and CA) almost retarded the ions migration in aqueous solution. It should be noted that the current values in the CV of figure 8 is in the order of 10^{-7} ampere/cm^2 as the same order found with that in HCl (figure 6). This phenomenon is strengthened by the justification of its average inhibition efficiency (higher than 90%) for both HCl and NaOH as reported in Table 1.

4. Conclusions

1. Arabic gum and cellulose acetate have function as thin layer organic coating to prevent chemical reaction between Al and other substituent in corrosive media such as high concentration of strong acid.
2. The hetero-aromatic part (N, O and S) in polymer plays important role in reducing by-product reaction on Al surface. In addition, the polysaccharides in both polymers (AG and CA) increase polarity properties yielding better water solubility.
3. Arabic gum and cellulose acetate are members of organic coating group.

Acknowledgement

The authors send acknowledgement to Mercu Buana University Jakarta Indonesia for financial support through Mercu Buana Internal Funding.

References

[1] Abu-Dalo M A and Othman AA 2012 *International J Electrochem. Sci.* **7** 9303
[2] Baker and Richard W 2004 *Technology and Applications, California: Membrane Technology Research, Inc.*
[3] Tamborim S M, Dias S L P, Silva S N, Dick L F P and Azambuja D S 2011 *Corrosion Sci.* **53** 1571
[4] Fontana M G 1986 *Corrosion Engineering,* Mc Graw-Hill Book Company, Singapore
[5] Sari A K 2017 *Characterization study and corrosion rate of aluminium coated by cellulose acetate membrane,* Universitas Mercu Buana, Jakarta (unpublished article)
[6] Xhanari K and Finsgar M 2017 *Arabian J Chemistry* (article in press)
[7] Wijaya H F 2016 The *development of aluminium-air battery using Arabic gum polymer,* Universitas Mercu Buana, Jakarta (unpublished article)
[8] Eid S, Abdallah M, Kamar E M, El-Etre A Y 2015 *J Mater. Environ. Sci.* **6** 892

IOP Conf. Series: Materials Science and Engineering **324** (2018) 012071 doi:10.1088/1757-899X/324/1/012071

Analysis of Flexible Car Body of Straddle Monorail Vehicle

Zhong Yuanmu

CRRC QINGDAO SIFANG CO.,LTD., 88 jinhong east road, Qingdao, Shandong, 201804, China

zhongyuanmu@cqsf.com

Abstract. Based on the finite element model of straddle monorail vehicle, a rigid-flexible coupling dynamic model considering vehicle body's flexibility is established. The influence of vertical stiffness and vertical damping of the running wheel on the modal parameters of the car body is analyzed. The effect of flexible car body on modal parameters and vehicle ride quality is also studied. The results show that when the vertical stiffness of running wheel is less than 1 MN / m, the car body bounce and pitch frequency increase with the increasing of the vertical stiffness of the running wheel, when the running wheel vertical stiffness is 1MN / m or more, car body bounce and pitch frequency remained unchanged; When the vertical stiffness of the running wheel is below 1.8 MN / m, the vehicle body bounce and pitch damping ratio increase with the increasing of the vertical stiffness of the running wheel; When the running wheel vertical stiffness is 1.8MN / m or more, the car body bounce and pitch damping ratio remained unchanged; The running wheel vertical damping on the car body bounce and pitch frequency has no effect; Car body bounce and pitch damping ratio increase with the increasing of the vertical damping of the running wheel. The flexibility of the car body has no effect on the modal parameters of the car, which will improve the vehicle ride quality index.

1. Introduction

The straddle-type monorail vehicle is welcomed by all countries in the world because of its large capacity, safety, quickness, low noise, low pollution, less land occupation and adaptation to the large ramp. Ren Lihui[1] established a straddle-type monorail vehicle dynamics model, compiled a graphical simulation program, and the response characteristics of the straddle-type monorail vehicle passing through the curve and the joint of the track beam are simulated. Du zixue[2] analyzed the impact on vehicle passing performance by changing the vertical stiffness of running wheel. Urban rail vehicle body lightweight design led to lower car body stiffness, the problems caused by the flexible vibration of the car body can not be ignored. Therefore, thinking of the car body as an elastomer is closer to reality. Zhou Jinsong[3] adopted the modal superposition method to regard the flexible car body as the Euler beam to study the impact of vehicle body flexibility on vehicle ride quality, and installed in the car body dynamic vibration absorber to inhibit the car body flexible vibration. Based on the finite element analysis and multi-body dynamics simulation, a rigid-flexible coupled dynamic model of straddle-type vehicle with body flexibility is established in this paper. The influence of the flexible car body on the modal parameters of the car body and the power spectrum of the acceleration response is studied and analyzed, and its influence on the ride quality of the vehicle is analyzed.

2. Modelling

In this paper, combined with finite element analysis and multi-body dynamics simulation, a rigid-flexible coupled dynamic model of straddle monorail vehicle is established as shown in Fig.1. Flexible car body file contains the structure of the model and modal information.

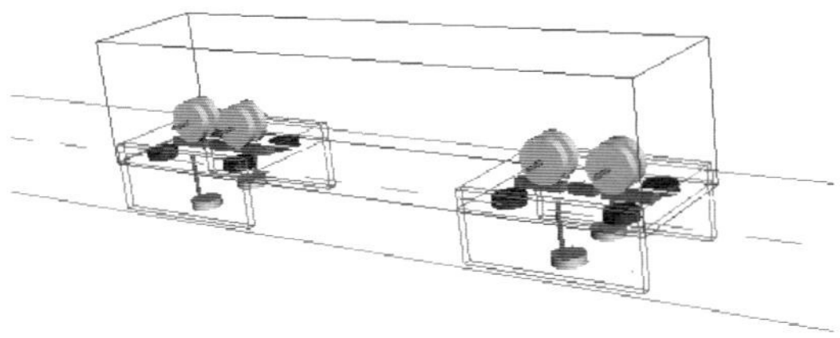

Figure 1. Rigid-flexible Coupling Dynamics Simulation Model

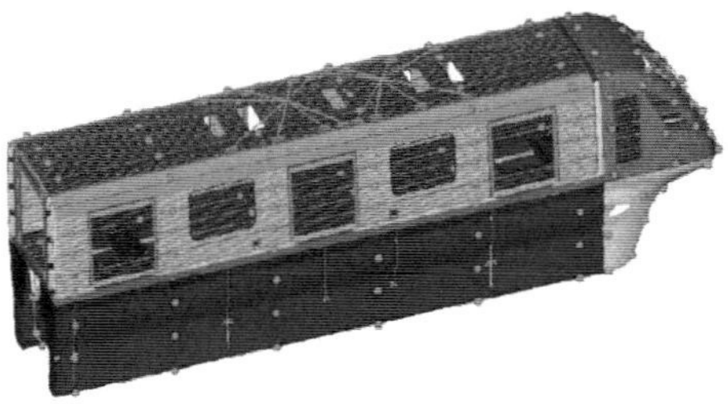

Figure 2. Car body Finite Element Model with Node Set

In the finite element model, a set of points that can describe the contour of the car body is established as shown in Fig.2. Then, the flexible car body file is imported into the established rigid car body model to replace the initial car body, and the relevant parameters are modified to obtain the complete rigid-flexible coupling model. The analysis of the coupling between rigid car body and flexible car body is helpful for further study on the influence of structure flexibility on vehicle dynamic performance.

3. Modal parameter analysis

Table 1 shows the comparison of the modal parameters of rigid car body and rigid-flexible car body of straddle monorail vehicle. As can be seen from the table, under the original parameters, the rigid car body and rigid-flexible coupling car body modal parameters has a little difference. But basically the same. The bounce frequency of the vehicle body is about 1Hz, damping ratio is about 18%; Pitch frequency is about 1.5Hz, damping ratio is about 23%. Modal parameters in a reasonable range.

Table 1. Modal parameters of the vehicle

Mode	Rigid car body		Rigid and flexible coupling car body	
	Natural frequency Hz	Damping ratio %	Natural frequency Hz	Damping ratio %
Car body bounce	1.149	18.1	1.150	17.8
Car body pitch	1.602	23.3	1.607	23.2
Frame bounce	3.328	16.5	3.368	16.6
Frame pitch	3.196	11.3	3.198	11.3
Car body yaw	1.234	8.3	1.207	7.9
Car body lower sway	1.283	16.3	1.276	15.8

3.1. The influence of the running wheel vertical stiffness on the modal parameters

By changing the vertical stiffness of the running wheel, the car body modal parameters can be calculated. Fig.3 shows the results of the frequency of the car body bounce and pitch. With all other parameters remaining the same, when the vertical stiffness of running wheel is less than 1 MN / m, the car body bounce and pitch frequency increase with the increasing of the vertical stiffness of the running wheel, when the running wheel vertical stiffness is 1MN / m or more, car body bounce and pitch frequency remained unchanged. Fig.4 shows the results of the damping ratio of the car body bounce and pitch. When the vertical stiffness of the running wheel is below 1.8 MN / m, the vehicle body bounce and pitch damping ratio increase with the increasing of the vertical rigidity of the running wheel; When the running wheel vertical stiffness is 1.8MN / m or more, the car body bounce and pitch damping remained unchanged.

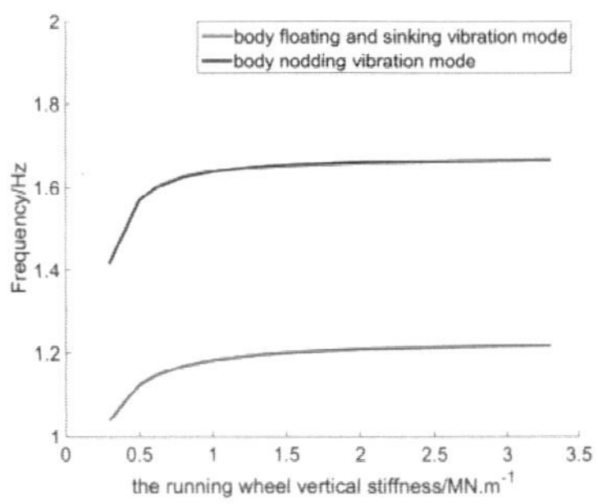

Figure 3. Modal Frequency of Car Body

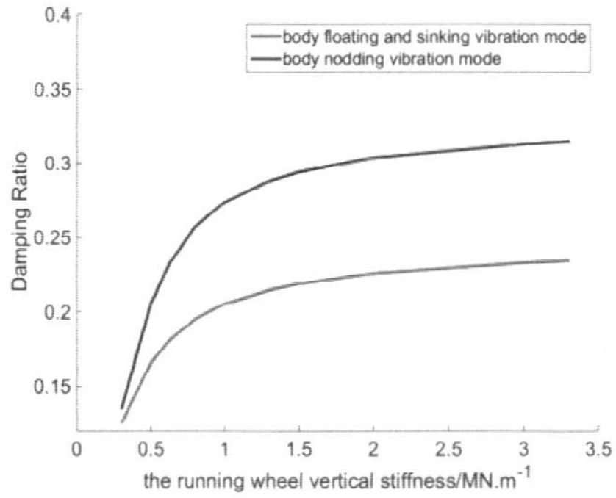

Figure 4. Modal damping Ratio of Car Body

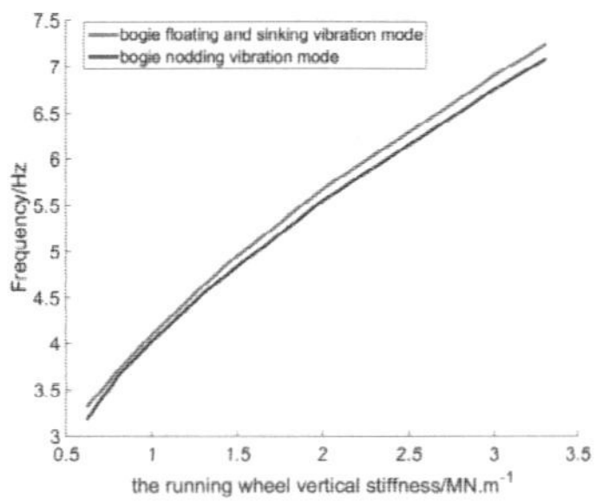

Figure 5. Modal Frequency of bogie

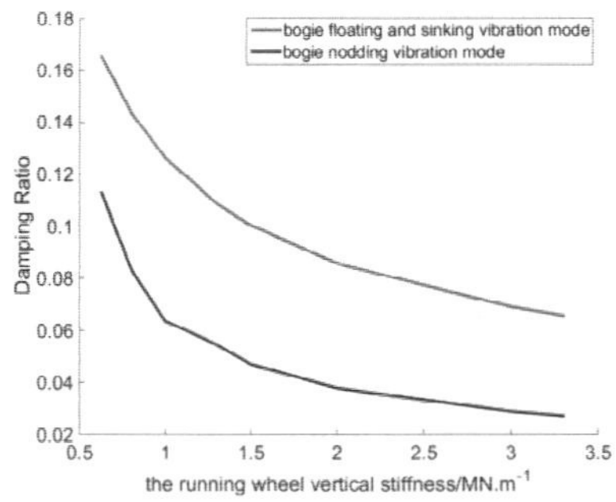

Figure 6. Modal damping Ratio of bogie

Fig.5 shows the results of the frequency of the bogie bounce and pitch. Bogie bounce and pitch frequency increase with the increasing of the vertical damping of the running wheel. The growth trend is almost linear. Fig.6 shows the results of the damping ratio of the bogie bounce and pitch. The bogie pitch and bounce frequency decrease with the increasing of the damping coefficient of the running wheel.

3.2. The influence of the running wheel vertical damping on the modal parameters

The vertical damping of the running wheel also has a certain impact on the modal parameters of the vehicle. Fig.7 shows the results of the frequency of the body bounce and pitch. The running wheel vertical damping on the car body bounce and pitch frequency has no effect. Fig.8 shows the results of the damping ratio of the car body bounce and pitch. Car body bounce and pitch damping ratio increase with the increasing of the vertical damping of the running wheel. The growth trend is almost linear.

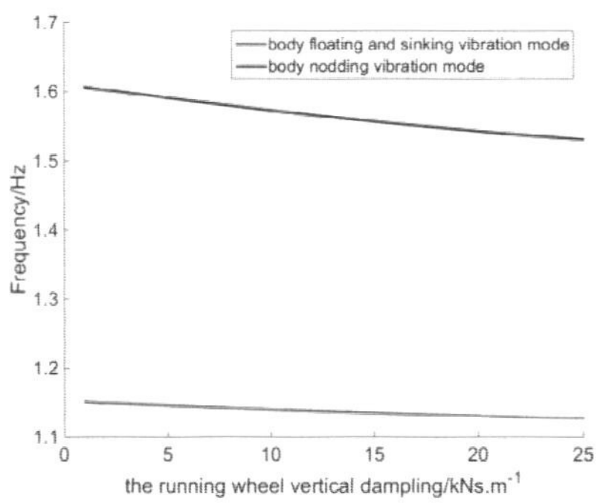

Figure 7. Modal Frequency of Car Body

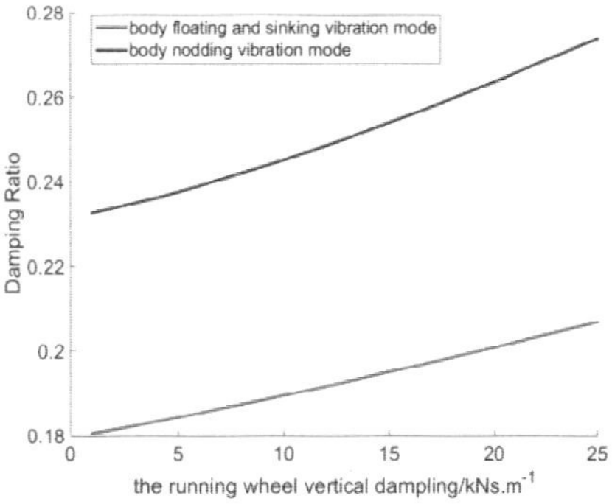

Figure 8. Modal damping Ratio of Car Body

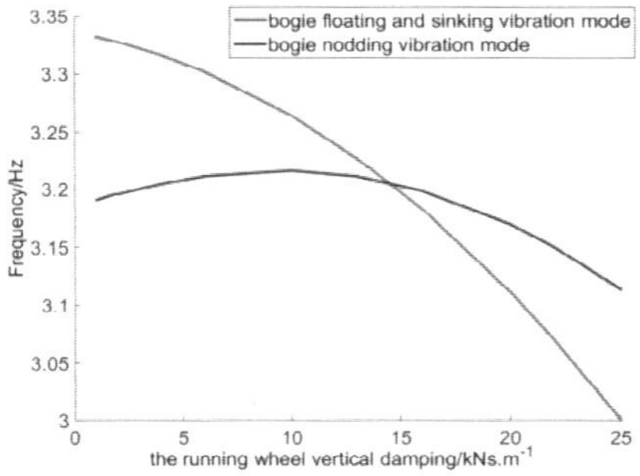

Figure 9. Modal Frequency of bogie

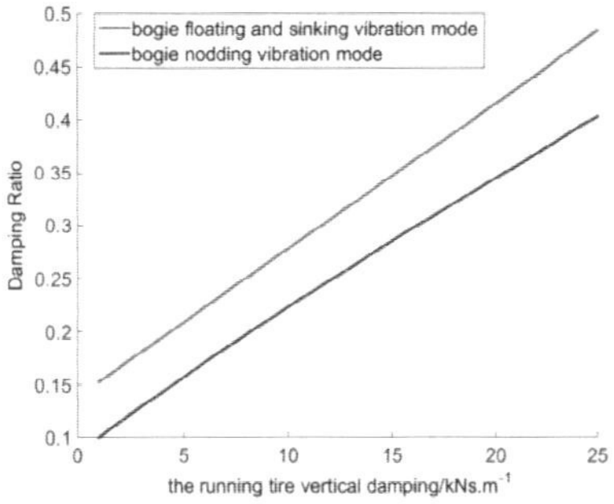

Figure 10. Modal damping Ratio of bogie

Fig.9 shows the results of the frequency of the bogie bounce and pitch. The bogie pitch and bounce frequency decrease with the increasing of the damping coefficient of the running wheel. When the vertical damping coefficient of the running wheel reaches 14 kNs/m, the frequency of the bogie bounce and pitch is equal to 3.2Hz. Fig.10 shows the results of the damping ratio of the bogie bounce and pitch. Bogie bounce and pitch damping ratio increase with the increasing of the vertical damping of the running wheel. The growth trend is almost linear.

4. Vehicle running quality analysis

Based on the rigid car body dynamics model and the rigid-flexible coupling dynamics model of the straddle monorail vehicle established above. The effects of different operating speeds on the ride quality of rigid car body and flexible car body are calculated respectively. Fig.11-13 shows the ride quality of rigid car body and flexible car body as a function of speed. We can see from the figure, rigid car body and flexible car body ride quality trend is about the same. And flexible car body quality index is slightly higher than rigid car body. In the 65-80km/h speed range, rigid car body and flexible car body quality index no more than 2.2. Passenger ride comfort will not be affected at this time.

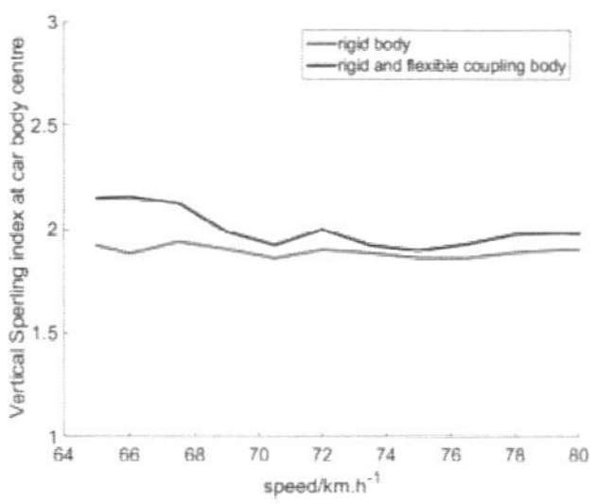

Figure 11. Ride quality at front of the car body

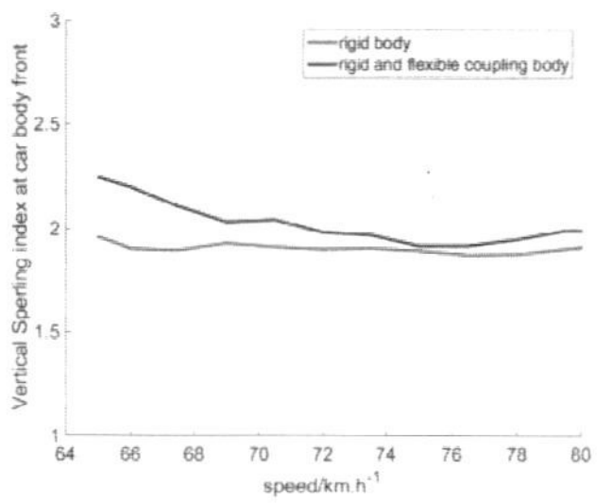

Figure 12. Ride quality at center of the car body

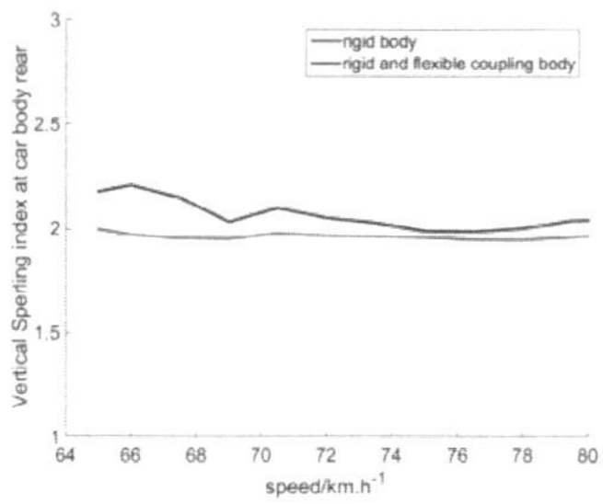

Figure 13. Ride quality at rear of the car body

5. Conclusion

Based on the finite element model of straddle monorail vehicle, a rigid-flexible coupling dynamic model considering vehicle car body's flexibility is established. When the vertical stiffness of running wheel is less than 1 MN/m, the vertical stiffness of the running wheel has a great influence on the modal parameters of the car body, and the car body bounce and pitch frequency increase with the increasing of the vertical stiffness of the running wheel. when the running wheel vertical stiffness is 1MN/m or more, car body bounce and pitch frequency remained unchanged; When the vertical stiffness of the running wheel is below 1.8 MN / m, the vertical stiffness of the running wheel has a great influence on the damping ratio of the car body, and the vehicle body bounce and pitch damping ratio increase with the increasing of the vertical rigidity of the running wheel. When the running wheel vertical stiffness is 1.8MN / m or more, the car body bounce and pitch damping ratio remained unchanged. The running wheel vertical damping on the car body bounce and pitch frequency has no effect; Car body bounce and pitch damping ratio increase with the increasing of the vertical damping of the running wheel. The flexibility of the car body has no effect on the modal parameters of the car, which will improve the vehicle ride quality index. In the 65-80km / h speed range, rigid car body and flexible car body quality index no more than 2.2. Passenger ride comfort will not be affected at this time.

References

[1] Ren L, Zhou J, Shen G. Dynamics Model and Simulation Study of a Straddle Type Monorail Car 2004 *China Railway Science*. 25(5), 26-32.

[2] Ren L, Zhou J, Shen G, Liu S. Hunting Stability of Straddle-type Monorail Car Based on Equation Eigenvalue 2003 *Journal of Tongji University*. 31(4): 469-472.

[3] Du Z, Zuo C. Impact of Running Wheel's Vertical Stiffness on Curve Negotiation Performance of Straddle-type Monorail Vehicle 2013 *Electric Drive For Locomotives*. (03), 40-42.

[4] Gong D, Zhou J, Sun W. Influence of under-chassis-suspended equipment on high-speed EMU trains and the design of suspension parameters 2016 *Proceedings of the Institution of Mechanical Engineers Part F Journal of Rail & Rapid Transit*. 230(8).

[5] Wang F, Zhou J, Ren L. Analysis on track spectrum density for dynamic simulations of high-speed vehicles 2002 *Journal of the China Railway Society*. 24(5), 21-27.

IOP Conf. Series: Materials Science and Engineering **324** (2018) 012072 doi:10.1088/1757-899X/324/1/012072

LCD denoise and the vector mutual information method in the application of the gear fault diagnosis under different working conditions

Zhang Xiangfeng[1], Jiang Hong[1]

[1]School of mechanical engineering, xinjiang university urumqi 830047, China

Corresponding Author: Jiang Hong, E-mail: onlyxjjh@xju.edu.cn.

Abstract. In this paper, the full vector LCD method is proposed to solve the misjudgment problem caused by the change of the working condition. First, the signal from different working condition is decomposed by LCD, to obtain the Intrinsic Scale Component (ISC)whose instantaneous frequency with physical significance. Then, calculate of the cross correlation coefficient between ISC and the original signal, signal denoising based on the principle of mutual information minimum. At last, calculate the sum of absolute Vector mutual information of the sample under different working condition and the denoised ISC as the characteristics to classify by use of Support vector machine (SVM) .The wind turbines vibration platform gear box experiment proves that this method can identify fault characteristics under different working conditions. The advantage of this method is that it reduce dependence of man's subjective experience, identify fault directly from the original data of vibration signal. It will has high engineering value.

1. INTRODUCTION

Machinery vibration signal contains a lot of physical meaning information while high-profile fault information is often submerged in it. The key for fault diagnosis is how to extract the effective fault information and it is also an important topic of the fault diagnosis at present. Empirical mode decomposition (EMD) and resonance demodulation method is often used for fault diagnosis .But the further development of the EMD was seriously affect because of the endpoint effect and modal aliasing , and its is restricted due to the frequency of the IMF has no physical meaning which restrict the application in the project. Resonance demodulation method has a good theoretical basis, but how to select its pass filter parameters is often difficult. Although many scholars use various methods such as spectral kurtosis method to achieve a rational selection of optimal band pass, the actual use is greatly limited.

Local characteristic-scale decomposition(LCD) is a new adaptive method for analyzing non-stationary signal and it developed in recent years. The basic idea of LCD is that: there is a sine curve which has a single modal and the partial matches between any maxima and adjacent minima, so the instantaneous frequency from LCD has physical significance. We can realize the mixed signal decomposition according to this approach to ensure that every ISC component contains a different vibration source information. Currently, LCD has been widely used in fault diagnosis of gears and bearings, Cheng Jun sheng[1] do the gear fault diagnosis in combination with LCD and envelope analysis ,the results showed that LCD method can be effectively applied to gear fault diagnosis, but he did not elaborate the reliability of the algorithm under different conditions . ZhengJingde[2] show that LCD has certain advantages over EMD, LMD and other methods in terms of the inhibition of the endpoint effect, the speed of calculation and the effect of decomposition by the experiment results.

Mutual information reflects the amount of information or statistical dependence of each other. The more correlation between two random variables, the greater the mutual information; otherwise, the smaller the mutual information.

Peng Suping [4] Implements the microseismic signal filtering and denoising by combined EMD and mutual information entropy. Usually, the fault signal and its ISC have high mutual information value, and therefore the value of the mutual information can be used as gear fault diagnosis features. The full vector theory is a method that uses the information fusion to carry out phase constant, amplitude of fusion of the signals in two directions, mainly to address the problem of shortage of single-channel information.

This paper puts forward a method to diagnosis gear fault under different working conditions by LCD combination with full vector mutual information. Comparing with the traditional method ,it have two advantages, on one hand, ensure the improvement of information, on the other hand, the fault feature can be obtained directly from the original vibration signal, without the subjective experience. Through analysis the full vector mutual information between samples and vibration signal of under different condition. The amplitude frequency modulation caused by different working conditions is reduced. Wind turbine gearbox vibration test proved this method can identify the fault characteristics under different conditions, it has a certain universality.

2. LCD and full mutual information vector feature extraction

Through the LCD decomposition and filtering Correlation coefficient ,we can obtained ISC signal containing vibration source information more. Calculation mutual information between sample and ISC signal as features for the classification. The steps of fault diagnosis method based on LCD and mutual information as follows:

(1)Provided n kinds of training sample under different conditions, referred to as $Class_1$, $Class_2$, ..., $Class_n$, using LCD algorithm decompose i-th training sample, and calculating the cross-correlation coefficient of ISC signal and the original sample, In cross-correlation maximum principle, select m ISC recorded as IS_i. A collection of total n ISC signal can be obtained by the above steps.

(2)Take the training sample of certain working conditions from the training sample set and denoted train1. Calculation mutual information of train 1 and Each ISC from IS1, sum absolute mutual information as Feature1 of train1; Calculation the followed mutual

IOP Conf. Series: Materials Science and Engineering **324** (2018) 012072 doi:10.1088/1757-899X/324/1/012072

information and get N features, then consist of the eigenvectors of the sample train1 with n feature, again and again, compute feature vectors of all the training samples.

(3) using the training sample feature vectors train SVM classifier

(4) select test sample randomly, calculate the feature vectors in accordance with step (1), (2). Pattern recognition by using the trained SVM classifier and get the state of test.

3. Experimental verification

3.1. Data source

Based on the above method, the gear box fault of the wind turbine vibration platform produced by American SQi company is identified. The experiment platform mainly includes DT9837 signal acquisition system, 608A11 acceleration sensor,PHC-100 CLUTCH magnetic particle brake and Gear Fault Bench.

Gear Fault Bench consists of speed regulating motor, two stage gear transmission. The main parameters of the experiment are: the high-speed shaft driving wheel gear is 36, 90, low speed shaft driving wheel teeth is 29,100, the sampling frequency is 10000Hz, the sampling time is 1S.Wind turbine vibration test platform shown in Figure 1.

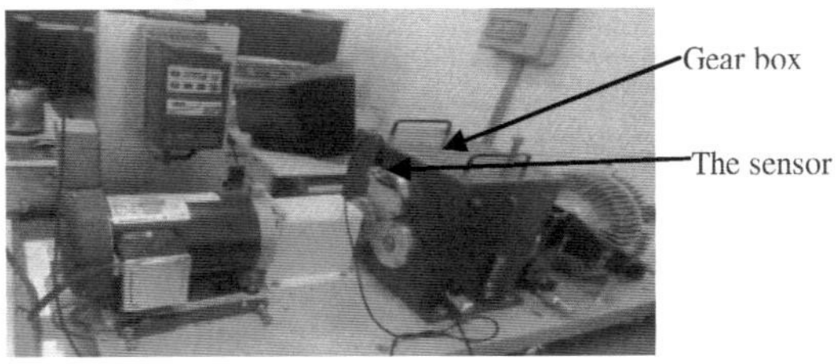

Fig.1 Wind turbines vibration experiment platform

3.2. RESULTS

The measured time domain signal of normal gear and tooth root crack is shown in Figure 2.

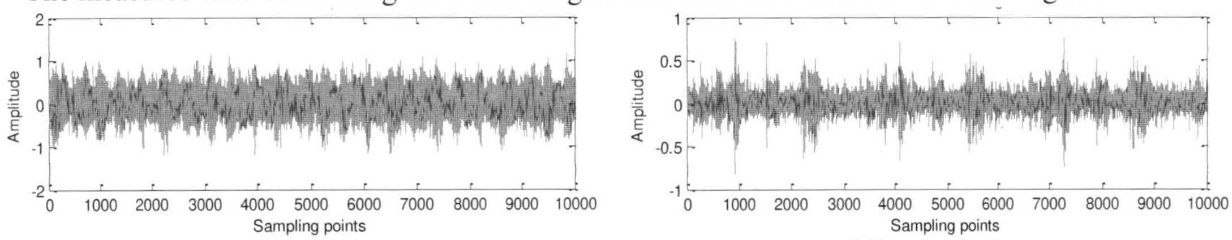

(a) Normal (b)Root crack

Fig.2 The time-domain signal of gear

In literature [7], the change of the load can cause the amplitude modulation phenomenon of the vibration signal, but no influence on frequency. While the change in the rotational speed caused FM phenomenon but no effect on the amplitude. LCD algorithm is mainly aimed at the frequency decomposition. Therefore, the condition of narrow load range(Load range 32 nm ~ 40 nm) and wide speed change(Speed range 180 r/min to 2000 r/min) are selected for experiment working. Randomly selected 50 groups of training sample and 100 groups of test sample.

The vibration signal of normal gear and gear of tooth crack under different working conditions is decomposed by LCD, The ISC components obtained are shown in Figure 3.

IOP Conf. Series: Materials Science and Engineering **324** (2018) 012072 doi:10.1088/1757-899X/324/1/012072

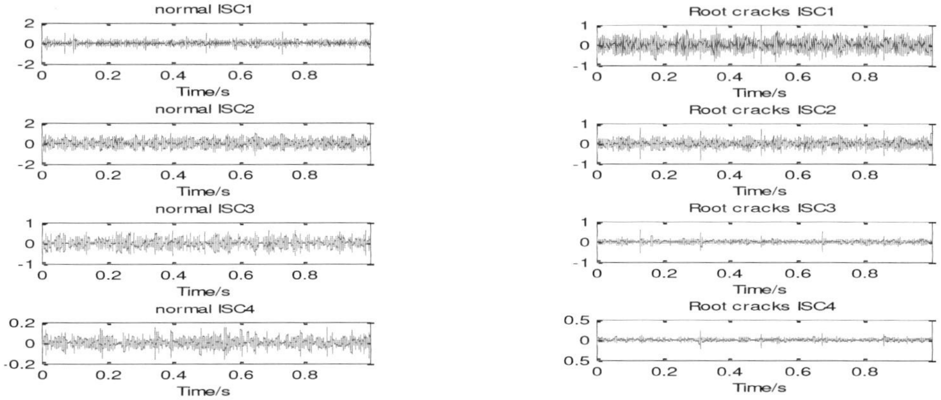

(a) Normal signal (b) Root cracks

Fig.3 ISC after signal decomposition

Compute the correlation coefficient of ISC and original signal ,the greater the coefficient, the more vibration signal component it contains. The correlation coefficient of ISC from normal/ tooth root crack and original signal are shown in table 1.

Tab.1 The correlation coefficient of the original signal and Normal/root crack

ISC	1	2	3	4	5	6
Normal	0.8937	0.2919	0.1112	0.0297	0.0605	0.0079
Root cracks	0.8893	0.3539	0.1410	0.0415	0.0110	0.0019

As shown in Table1, ,the correlation coefficient of the first 3 ISC components is greater than the mean value in all gear normal / root crack . That is to say, the first 3 ISC can fully express the signal characteristics. Through this method, on the one hand, noise reduction can be achieved, on the other hand the calculation of the total vector mutual information feature can reduced.

Record the filtered ISC of gear normal / root crack under n kinds of different conditions as Normal IS_n and Fault IS_n. The higher the N, the higher the coverage rate of the condition, and the greater accuracy of determination failure.

The gear normal / root crack information of vibration is used as the training sample, select s groups of signal from training samples denoted as $train_1 ... train_s$, and calculate mutual information with Normal IS_n and Fault IS_n. Part of full vector mutual information of are shown in Tables 2.

Tab.2 Sum of absolute full vector mutual information of the NIS_n/FIS_n and the R_s

	*NIS*1	*NIS*2	*NIS*3	*NIS*4	*NIS*5	*NIS*6	*FIS*1	*FIS*2	*FIS*3	*FIS*4	*FIS*5	*FIS*6
*NIS*1	0.4428	0.4545	0.2547	0.3574	0.3324	0.1368	0.0313	0.0492	0.0521	0.0521	0.0951	0.0657
*NIS*2	0.4204	0.5842	0.4478	0.3254	0.3124	0.3357	0.0863	0.0561	0.0656	0.0656	0.0811	0.0692
*NIS*3	0.4109	0.3562	0.2454	0.3241	0.4318	0.3349	0.0889	0.1821	0.1456	0.1456	0.0466	0.0721
*NIS*4	0.4014	0.3852	0.3612	0.3484	0.5231	0.2438	0.0693	0.0752	0.0851	0.0851	0.0812	0.0832
*NIS*5	0.4722	0.3484	0.3485	0.3241	0.3874	0.415	0.0385	0.0354	0.0645	0.0645	0.0684	0.0654
*FIS*1	0.0816	0.1024	0.0985	0.0657	0.0789	0.0763	0.2516	0.3452	0.3421	0.3241	0.3748	0.4325
*FIS*2	0.0808	0.0896	0.1245	0.0895	0.0854	0.1235	0.3688	0.3451	0.3631	0.2254	0.2894	0.4572
*FIS*3	0.0842	0.0778	0.0932	0.1024	0.0856	0.1034	0.3659	0.3841	0.3871	0.4621	0.3514	0.4327
*FIS*4	0.1153	0.1545	0.1679	0.1044	0.0867	0.1923	0.4829	0.3214	0.4324	0.5544	0.4512	0.4362
*FIS*5	0.1058	0.0864	0.0894	0.1865	0.0856	0.0942	0.5182	0.2842	0.3344	0.6854	0.5024	0.4216

From the tables, the absolute full vector mutual information of root crack gear and Normal IS_n substantially larger than the value corresponding to the normal gear ,description root cracks and normal IS_n is complementary distribution. While, The absolute value of full vector mutual information of normal gear and Fault IS_n substantially greater than the value corresponding to the tooth root crack, description normal gear and Fault IS_n is complementary distribution .Therefore, sum of the absolute full vector mutual information as feature of normal signal and root crack signals can distinguish different states.

Sum of absolute full vector mutual information of training samples with Normal IS_n as Feature1,with Fault IS_n as Feature2. Establish distribution of features With Feature1and Feature2,it is shown in figure 4.

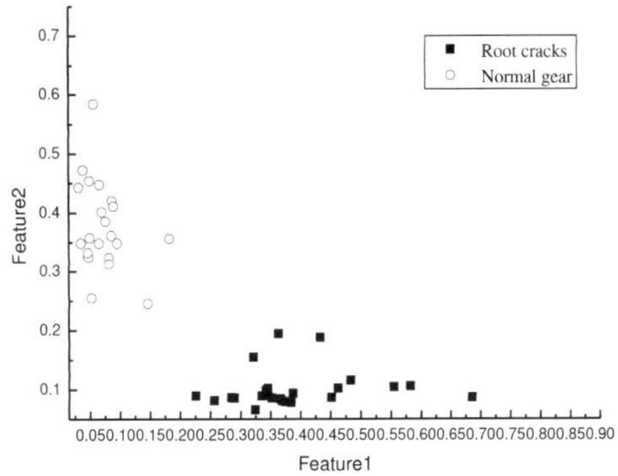

Fig.4 The characteristics distribution of the normal gears and tooth root crack

4. CONCLUSION

The following conclusions are obtained as follows:(1)The ISC contains some characteristic information of the weak fault of gear.(2) using the sum of absolute full vector mutual information as features, make up for the lack of single channel, at the same time , it avoid the impact of individual abnormal conditions. To compare with the direct mutual information ,its robustness and distinction extent are improved.(3)The method proposed in this paper can effectively extract fault feature of gear, and it does not depend on the subjective experience. It can solved gear fault diagnosis under different conditions.

References

[1] CHENG Junsheng ,YANG Yi,YANG Yu. Local Characteristic-scale Decomposition Method and Its Application to Gear Fault Diagnosis[J] Journal ofVibration Engineering, 2012, 48(9) :64—71.
[2] ZHENG Jinde. Research on Local Characteristic-scale Decomposition and Its Applications to Machinery Fault Diagnosis: [D], Changsha: Hunan University, 2014
[3] Dezhong Qi, Sanqiang Zhang, Mingyong Liu. An improved hierarchical genetic algorithm for collaborative optimization of manufacturing processes in metal structure manufacturing systems [J]. Advances in Mechanical Engineering(SAGE), 2017, 3(17): 40-44.
[4] HAO Wang-shen, WANG Hong-ming, et al. Gear fault feature extraction based on full vector permutation entropy[J],JOURNAL OF VIBRATION AND SHOCK
[5] LI Lingjun HAN Jie LI Pengyong HAO Wei CHEN Lei.Vector-bispectrum Analysis and Its Application in Machinery Fault Diagnosis[J]. Journal of Vibration Engineering, 2011, 47(17): 50-54.

The Optimisation of Processing Condition for Injected Mould Polypropylene-Nanoclay-Gigantochloa Scortechinii based on Melt Flow Index

M.H.Othman, M.S.Rosli, S.Hasan, A.M. Amin, M.Y.Hashim, O.M.F.Marwah and S.Y.M. Amin.

Department of Manufacturing and Industrial Engineering, Faculty of Mechanical and Manufacturing Engineering, Universiti Tun Hussein Onn Malaysia, 86400 Parit Raja, Batu Pahat, Johor Darul Takzim
hilmi@uthm.edu.my

Abstract. The fundamental knowledge of flow behaviour is essential in producing various plastic parts injection moulding process. Moreover, the adaptation of advanced polymer-nanocomposites such as polypropylene-nanoclay with natural fibres, for instance Gigantochloa Scortechinii may boost up the mechanical properties of the parts. Therefore, this project was proposed with the objective to optimise the processing condition of injected mould polypropylene-nanoclay-Gigantochloa Scortechini fibres based on the flow behaviour, which was melt flow index. At first, Gigantochloa Scortechinii fibres have to be preheated at temperature 120^{0}C and then mixed with polypropylene, maleic anhydride modified polypropylene oligomers (PPgMA) and nanoclay by using Brabender Plastograph machine. Next, forms of pellets were produced from the samples by using Granulator machine for use in the injection moulding process. The design of experiments that was used in the injection moulding process was Taguchi Method Orthogonal Array -$L_9 3^4$. Melt Flow Index (MF) was selected as the response. Based on the results, the value of MFI increased when the fiber content increase from 0% to 3%, which was 17.78 g/10min to 22.07 g/10min and decreased from 3% to 6%, which was 22.07 g/10min to 20.05 g/10min and 3%, which gives the highest value of MFI. Based on the signal to ratio analysis, the most influential parameter that affects the value of MFI was the melt temperature. The optimum parameter for 3% were 170^{0}C melt temperature, 35% packing pressure, 30% screw speed and 3 second filling time.

1. Introduction

It has been estimated that almost 170 million tons of plastics were produced worldwide during the year 2003 [1]. The use of thermoplastics has been continuously increasing in the last few years and penetrated the various fields, such as automotive and aerospace industry, household and sport appliances. Polypropylene (PP) composites have better material properties such as good strength, better stiffness, improve of ductility; good stability and thermal expansion; and low cost of production. Nevertheless, these properties depend on the matrix phase, the phase dispersion of fillers and strengthening mechanism, shapes and arrangements of filler particles and the bonding interface between filler and the matrix [2].

Polymer nanocomposites can be prepared by various processing techniques such as solution process, in-situ polymerization and melt blending. Each technique has an influence on the final characteristics and properties of the composites. The performance of this polymer system also can be upgraded to better properties with the presence of natural fibres as the filler material [3].

Direct intercalation by the molten polymer method of preparation of nanocomposites has greatest advantages over other methods as this method is environmentally benign due to the absence of organic solvents and is compatible with current industrial process, such as extrusion and injection moulding [4]. Most of these polymer nanocomposites were transformed into useful parts by using an injection moulding process. Unfortunately, it has been found difficult to control the properties of the material in order to manufacture consistently a part with no defect [5]. Hence, the synthesis of knowledge about the injection moulding processing might lead to solutions in producing good products. Therefore, the needs of optimising the processing condition shall be the priority in the injection moulding manufacturing process [6].

The increase of the resistance to flow and flow stability of the former system can be attributed to the improvement of the compatibility and interfacial adhesion between the filler and matrix as well as the dispersion of the filler in the matrix. Beside optimisation, studies about melt flow behaviour of polypropylene composites also need to be clarified [7].

2. Experimental

The experiment started from preparation of material. Then the injection moulding process was carried out according to the Taguchi Orthogonal Array. Flow behaviour evaluation was made after that, and optimisation through Signal to Noise Ratio was made at the end of the project.

2.1. Preparation of Materials

The polymer-nanocomposite are divided into two sections, the matrix and the fillers. For the matrix, the material that's been used in this research are polypropylene (PP) and for the fillers, the materials that been used are polypropylene-grafted-maleic anhydride (PPgMA), nanoclay (NC) and Gigantochloa Scortechinii (GS). Table 1 shows the compounding of these materials based on the formulation.

Table 1. Formulation of Polymer-Nanocomposites

No	PP	PPgMA	NC	GS	Label
1	84%	15%	1%	0%	0%GS
2	81%	15%	1%	3%	3%GS
3	78%	15%	1%	6%	6%GS

The material preparation starts from processing the GS which is a type of bamboo fibres. The fibres were chopped and refined to become short fibres. These fibres need to be pre-heated at 120°C. The process of mixing was made by using twin screw Brabender Lab-compounder KETSE 20/40 machine to make it into compounding. After that the mixture was formed into small pieces or it is called pallets with diameter less than 5mm by using Granulator SLM 50Fy machine before entering the injection moulding process.

2.2. Injection Moulding Process via Taguchi Orthogonal Array Method

In the injection moulding process, the sample that produce by using injection moulding machine usually influenced by the injection moulding parameters. Samples shall be produced by using an injection moulding process were prepared according to the Taguchi Optimisation Orthogonal Array Method, with selected formulation of polypropylene-nanoclay-Gigantochloa Scortechinii fibres. The injection moulding machine that used in this project was 0.7 tonne Nissei NP7F from Japan. The parameters that were considered were the melt temperature, packing pressure, screw speed and filling time. There were three levels and four parameters that had been analysed by using Taguchi Optimisation

Orthogonal Array Method based on level ($L_9 3^4$). The process parameter based on Orthogonal Array ($L_9 3^4$) was shown in Table 2.

Table 2. Orthogonal Array selected based on level ($L_9 3^4$)

No.	Melt Temperature (°C)	Packing Pressure (%)	Screw Speed (%)	Filling Time (s)
1	165	30	25	1
2	165	35	30	2
3	165	40	35	3
4	170	30	25	3
5	170	35	30	1
6	170	40	35	2
7	175	30	25	2
8	175	35	30	3
9	175	40	35	1

2.3. Flow Behaviour Evaluation

The analysis was carried out by comparison of the sample without using an injection moulding process and the sample after used the injection moulding process. The samples that have been produced in the injection moulding process was crushed using Granulator SLM 50Fy machine to become pellets. After that, the pellets were used to determine the flow behaviour of the melt polymer-nanocomposites, which is Melt Flow Index (MFI) by using Plastomer (Ceast type 6841) according the ASTM D1238 Standard.

2.4. The Optimisation of Processing Condition

In this research, signal to noise (S/N) ratio for melt flow behaviour properties were obtained and optimum level of the injection parameters was determined through S/N values to achieve maximum melt flow index result [6]. In this study, the larger the better was chosen to solve melt flow index based on the S/N ratio [8]. The calculation of this ratio was assisted by using Minitab 16 statistical software.

3. Results and Discussion

Table 3 shows the value of MFI based on the injection moulding process via Taguchi Method Orthogonal Array ($L_9 3^4$). The highest value of MFI was at formulation of 3% GS and the lowest value was at formulation 0% GS. When the value of GS increased from 0% to 3%, the value of MFI was also increased. However, when the value of GS increased from 3% to 6%, the MFI value was decreased.

Table 3. Value of MFI (g/10min)

Case No	0% GS	3% GS	6% GS
Before Moulding	15.35	18.77	18.44
1	16.90	19.25	18.08
2	16.15	20.39	18.34
3	16.33	20.08	18.63
4	17.32	22.07	20.00
5	17.78	21.28	19.33
6	17.29	20.46	19.31
7	16.75	19.35	17.97
8	16.57	20.70	18.77
9	15.70	21.35	20.05

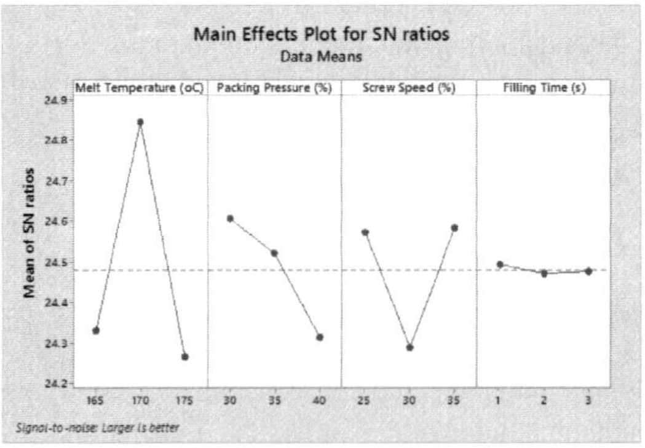

Fig. 1 Main Effect Plot for S/N ratio for 0% GS

For formulation 0% GS, the highest value for the MFI was in case number 5 which was 17.78 g/10min. The lowest value was in case 9 which is 15.7 g/10min compared with the MFI value without using an injection moulding process, which was 15.35 g/10min. When the injection moulding process was applied, the value of MFI increased in all case numbers from 1 to 9 compared to the value of melt flow index without injection moulding process.

As for 3% of GS, the highest MFI value was in case number 4, which was 22.07 g/10min. The lowest value was in case 1, which was 19.25 g/10min. This value was higher compared with the MFI value without using an injection moulding process (18.77 g/10min). When the injection moulding process via Taguchi orthogonal array method was applied, the value of MFI increased in all case numbers from 1 to 9 compared to the value of MFI before injection moulding process.

The analysis of MFI continued for 6% GS, whereby the highest value for the MFI was in case number 9 (20.05 g/10min) and the lowest value was in case 7 (17.97 g/10min). The value of melt flow index without injection moulding process was 18.44 g/10min. When the injection moulding process was applied, the value of MFI increased in case numbers 3, 4,5,6,8 and 9 but the value of MFI decreased in case number 1, 2, and 7 compared to the value of melt flow index without injection moulding process.

3.1. Optimisation of process condition based on Conceptual S/N Ratio Approach for Melt Flow Index (MFI) Results.

Figure 1, 2 and 3 shows the values of S/N responses at three levels for MFI based on formulation 0% GS, 3% GS and 6% GS. Based on the result for formulation 0% GS, the best combination parameter for optimising the value of MFI for 0%GS were 170°C for melt temperature, 30% of packing pressure, 35% of screw speed and 1 second for filling time. For the formulation 3% GS, the best combination parameter for optimising the value of MFI for 3%GS were 170°C for melt temperature, 35% of packing pressure, 30% of screw speed and 3 seconds for filling time. For the third formulation (6% GS) the optimum combination parameter for maximizing the value of MFI were 170°C for melt temperature, 40% of packing pressure, 30% for of screw speed and 1 second for filling time. Based on these findings, in can be rectified that by using the same orthogonal $L_9 3^4$, the data for melt temperature at level 2 usually provide the optimum level [9].

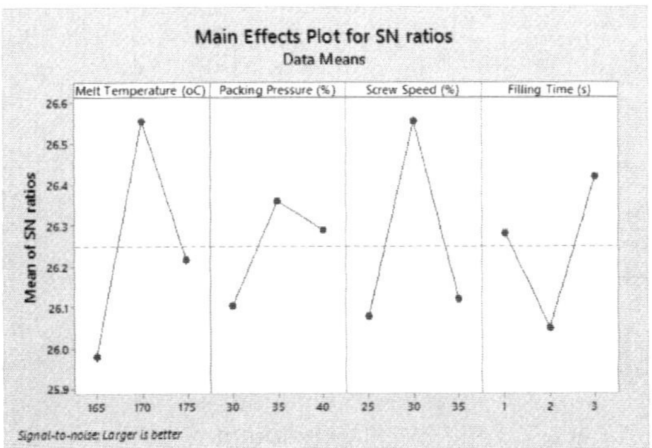

Fig. 2 Main Effect Plot for S/N ratio for 3% GS

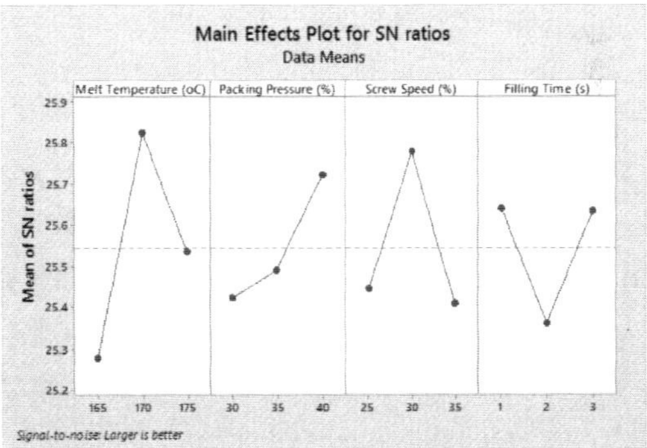

Fig. 3 Main Effect Plot for S/N ratio for 6% GS

4. Conclusion

In the conclusion, the highest value of MFI was found at formulation 3% GS which was 22.07 g/10min. The lowest value of MFI was at formulation 0% GS which is 17.78 g/10min. Based on the S/N ratio method, the optimum value for each formulation of the 0% GS, 3% GS and 6% GS were achieved. Based on these 3 formulations (0% GS, 3% GS and 6% GS), the melting temperature must at level 2 which was 170°C so that the optimum value of MFI can be achieved.

5. Acknowledgment
The highest appreciation dedicated to the Ministry of Higher Education Malaysia for sponsoring this research under the Fundamental Research Grant Scheme, FRGS (UTHM) Vot 1590 and Vot 1595. Exceptional recognitions to Universiti Tun Hussein Onn Malaysia for providing equipment and facilities, specifically at Advanced Material and Manufacturing Centre (AMMC) and Faculty of Mechanical and Manufacturing Engineering (FKMP).

References
[1] Achilias, D.S., Chemical recycling of polymer (methyl methacrylate) by pyrolysis. Potential use of the liquid fraction as a raw material for the reproduction of the polymer, Europe Polymer J, 43 (6), 2564-2575 (2007)
[2] H. Ismail, Komposit Polimer Diperkuat Pengisi dan Gentian Pendek Semulajadi. Pulau Pinang: Penerbit Universiti Sains Malaysia. (2004) 1-18.
[3] Albdiry, M., & Yousif, B. (2013). A critical review on the manufacturing processes in relation to the properties of nanoclay/polymer composites. Journal of Composite Materials, 47(9), 1093-1115
[4] Ray SS, Okamoto M. Polymer/layered silicate nanocomposites: a review from preparation to processing. Program Polymer Science 2003;28:1539–44.
[5] Dumitrescu, O. R., Baker, D. C., Foster, G. M., & Evans, K. E. (2005). Near infrared spectroscopy for in-line monitoring during injection moulding. *Polymer Testing*, *24*(3), 367–375.
[6] Ozcelik, B. (2011). Optimization of injection parameters for mechanical properties of specimens with weld line of polypropylene using Taguchi method. International Communications in Heat and Mass Transfer, 38(8), 1067-1072.
[7] Liang J.Z., Chen C.Y., Zou S.Y., Tsui C.P., Tang C.Y., Zhang S.D. (2015) Melt flow behaviour of polypropylene composites filled with multi-walled carbon nanotubes during extrusion, Polymer Testing, 45, pp. 41-46.
[8] Hasan, O., Tuncay E., Ibrahim, U. (2015). Application of Taguchi Optimization Technique in Determining Plastic Injection Molding Process Parameters for a Thin-shell Part.
[9] Ozcelik, B., Ozbay, A., Demirbas, E. (2010). Influence of Injection Parameters and Mold Materials on Mechanical Properties of ABS in Plastic Injection Molding, *International Communications in Heat and Mass Transfer*. ICHMT-02183; pp. 7.

IOP Conf. Series: Materials Science and Engineering **324** (2018) 012074 doi:10.1088/1757-899X/324/1/012074

Estimation of the duration life in thermomechanical fatigue using finite element post- processing

Z Achegaf, and A Khamlichi
National School of Applied Sciences at Tetouan, B.P. 2222 M'Hannech, Tetouan 93002, Morocco
achegaf.zineb@hotmail.fr, khamlichi7@yahoo.es

Abstract

In this study lifetime of structural parts subjected to thermomechanical fatigue, occurring under concurrent thermal and mechanical cycles is assessed. The special effect of phasing between thermal cycles and strain cycles was analysed. The methodology used was based on finite element post-processing analysis by specialized fatigue software package. A parametric study conducted on a standard specimen under various conditions of loadings had shown that thermomechanical fatigue is a complex phenomenon. In particular, it is not always more damaging than isothermal fatigue as usually believed.

1. Introduction

Thermomechanical fatigue (TMF) occurs in specific structures such as turbine blades and motor components as they are exposed to high combustion temperatures and high vibratory and friction loads. Material properties such as tensile strength, yield strength and Young's modulus decrease in general with increasing temperature. Fatigue lifetime properties are also temperature dependent. High temperature fatigue conditions imply that the fatigue load and temperature vary both as a function of time. As a result, fatigue damage accumulation is modified with increasing temperature. If thermal expansion is restricted, then thermal stresses are induced and fatigue will be affected also by structural aspect such as the actual boundary conditions. In addition to the cyclic strain, time and temperature are two more variables. The combined action of cyclic strains at high temperatures with time-dependent temperature variations constitutes the mean features of what it is called thermomechanical fatigue. The complexity of this problem scenario in practice is recognized to be considerable.Thomas et al.,[1] had shown through experiments that TMF testing is more severe that isothermal testing conducted at the maximum temperature. Fatigue damage is dependent on the phase relationship between strain and temperature. Bill et al.,[2]had found that lifetime of material specimens under in-phase (IP) thermomechanical cycling is well below isothermal fatigue lives obtained for a number of temperatures belonging to the temperature ranges that were investigated. Phase effects in the special case of metal matrix composites undergoing low cycle fatigue were discussed in [3].

There are two extreme thermomechanical situations; (IP) and out-of-phase (OP) cycles. Under IP cyclic loading, the maximum strain occurs at the same time as the peak temperature, and the minimum strain occurs at the minimum temperature. Under OP cycles, the maximum strain occurs at the minimum temperature, and minimum strain at the peak temperature. In many works reported in the literature, both in-phase and out-of-phase stress and temperatures cycles were found to be more damaging than isothermal stress cycles. Often, but not at all times, in-phase stress and temperature cycles were found to be more severe than out-of-phase cycles.

Models for the prediction of TMF lifetime are essential in order to anticipate when to change the fatigue sensitive components. Predicting lifetime can increase reliability and reduce serviceability costs. A large number of thermomechanical fatigue models were proposed in the literature, [4]. They can be classified into damage summation, strain-range partitioning, and strain energy partitioning. Each model was derived for particular use of a given material under specific ranges of temperature and cyclic strain. No versatile model could yet be used to predict in all circumstances lifetime under arbitrary thermomechanical loading.

In this work, TMF is analysed under various combinations of strain and temperature cycles by considering both IP and OP cycling. Finite element modelling of a standard test specimen is performed. Considering the Coffin-Manson fatigue model, the specialized fatigue post-processing software package, fe-safe,[5], was used to assess lifetime as function of the considered TMF loading.

2. Lifetime modelling in TMF

Different approaches have been introduced for the prediction of TMF lifetime, [6,7]. Rough phenomenological models relate measured total mechanical fields and lifetime without considering explicitly the different physical damaging mechanisms. Other approaches are cumulative damage based models where explicit consideration of the different damaging mechanisms is carried out. There are finally crack growth based models for which life is related to local inelastic strain at the crack tip. Gocmez et al. [7] have reviewed the classical fatigue criteria that were initially introduced for the uniaxial isothermal case. Since in reality engineering components are likely to experience multiaxial anisothermal loading, these early models had been extended subsequently. The concept of critical plane was introduced to incorporate isothermal multiaxial loading. Its extension to anisothermal loading needs yet to be achieved by taking into account temperature dependency of material parameters.

Energetic criterions based on energetic approaches showing a relation between dissipated energy and the number of cycles to failure were introduced by [8,9]. Gocmez et al.,[7] has extended this approach to take into account mean stress effect. However the application was focused on predicting life under uniaxial low cycle fatigue under OP-TMF. The generalization to multiaxial case is yet to be achieved.

From a practical engineering point of view, strain range partitioning based methods are more adequate. In these empiric approaches damage mechanisms are considered separately, [3,10]. The needed model parameters are to be identified from specific tests.In this work, the strain-life curves for isothermal fatigue analysis are defined under the following form

$$\frac{\Delta\varepsilon}{2} = \frac{\sigma'_f}{E}\left(2N_f\right)^b + \varepsilon'_f\left(2N_f\right)^c \qquad (1)$$

Where E is the Young's modulus, σ'_f is the fatigue strength coefficient, ε'_f the fatigue ductility coefficient, b the Basquin's fatigue strength exponent and c the Coffin-Manson fatigue ductility exponent.

To take into account TMF with varying phase relationship between temperature and strain cycles, parameters c and ε'_f are changed in the low cycle fatigue section of the strain-life curve according to the post-processing fatigue software, (Fe-safe, 2008). The modification is defined in terms of two thermal factors which are specific to a material: one for in-phase cycling and the second for out-of phase cycling. This is performed according to the modified life equation

$$\frac{\Delta\varepsilon}{2} = \frac{\sigma'_f}{E}\left(2N_f\right)^b + \frac{\varepsilon'_f}{\left(1-\psi_1\right)^c}\left(2N_f\right)^{\frac{c}{\left(1-\psi_2\right)}} \qquad (2)$$

With $\psi_1 = \left(1-tf_{1in}\right)\cos(\varphi)$ and $\psi_2 = \left(1-tf_{2out}\right)\sin(\varphi)$, where φ is the phase angel between the strain and temperature cycles. If $tf_{1in} = 1$and $tf_{2out} = 0$ 2out, then the isothermal damage curve would be retrieved. These two factors would normally be derived from plastic strain component of the strain-life curve. For each fatigue cycle, the actual phase relationship between strain and temperature is used to adjust the strain-life curve for any part of the analysed structure.

3. Numerical modelling of TMF

The specimen used in the numerical modelling is the standard CT16 specimen, figure 1. This specimen may be modelled in 2-D plane with a thickness B =16 mm, equal to the net thickness of the 3-D specimen. The other dimensions are the CT specimen width W = 32 mm and the initial crack length a = 0.25W. P designates the applied pressure on the outwards halves of the holes (Pa).

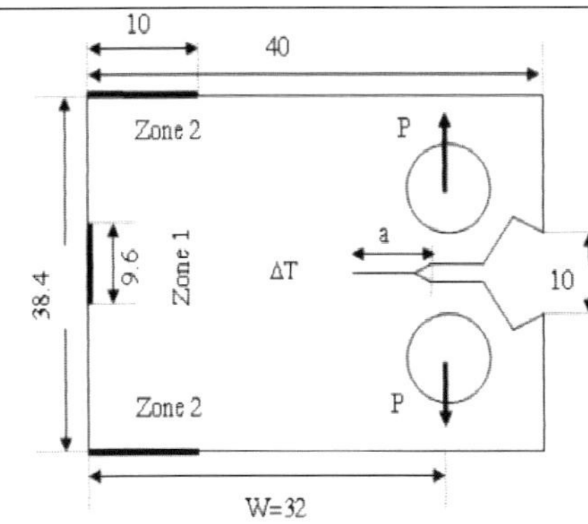

Figure 1. Geometric configuration of the specimen CT 16

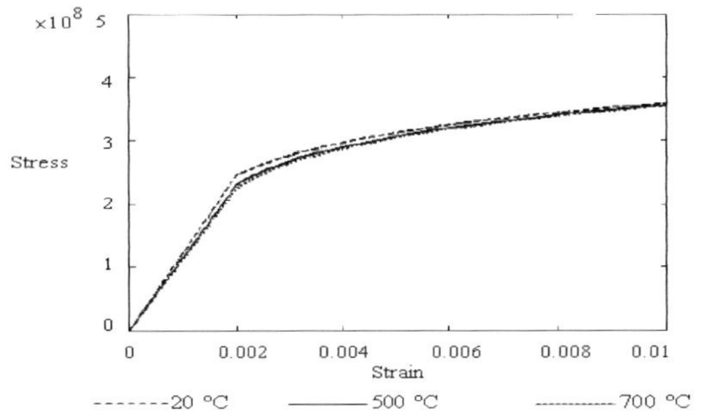

Figure 2. Elastic-plastic tension curves for 316 L steel as function of temperature

To obtain thermal stresses that do not vanish under uniform temperature field the specimen was fixed against horizontal displacement along the boundary zone 1 and against vertical displacement along the boundary zone 2. Thermal cycles are then considered to act simultaneously with elastic or elastic plastic strain cycles that are applied on the CT16 specimen. The material properties used are those of the stainless austenitic steel 316 L. Abaqus software is used to develop the finite element numerical model representing the test sample CT16 under the plane stress assumption. Mesh refinement is a crucial issue in the finite element simulation of fatigue. Abaqus option on adaptive remeshing integrating both energy and strain criteria were used.

The stress-strain curves for the thermomecanical fatigue of steel 316 L are defined in terms of the Young's modulus E, the strain hardening coefficient K and the strain hardening exponent n , as

$$\varepsilon = \frac{\sigma}{E} + \left(\frac{\sigma}{K}\right)^{\frac{1}{n}} \tag{3}$$

This steel was tested in the temperature range 20°C-700°C. Denoting u the Poisson's coefficient, σ_0 the yield stress and σ_m the ultimate strength, tension data of 316L steel are summarized in table 1.

Figure 2 presents the temperature dependent plastic curves of this material for the three temperatures: 20°C, 500°C and 700°C; plastic behaviour is shown to depend slightly on temperature.

The thermomecanical loading was selected according to either the in-phase mode between pressure P and temperature T, or to the out-of-phase mode between these two quantities.

Three different types of analyses were performed under Abaqus software: one cycle isothermal mechanical loading; one cycle IP-TMF loading; and one cycle OP-TMF loading.

Abaqus finite element results are transferred as inputs to fatigue post-processing software, [5]. Use is made here of the additional module TMF for thermomechanical fatigue solution. User can define load cycles to be analysed through many implemented mathematical functions and block operations. Fe-safe interpolates linearly the yield stress, the ultimate tensile stress and the endurance limit. Beyond the extremes of temperature interval the values at these extreme temperatures are used respectively.

For a TMF analysis, a minimum of 4 sets of data should be defined corresponding to at least 2 different temperatures and two different strain rates. Table 2 summarizes fatigue data for 316L steel. The TMF module allows converting isothermal fatigue life to an in-phase thermal cycle (0°) and an out-of-phase thermal cycle (180°C) in order to use the lifetime curve as defined by equation (2).

Table 1. Tension data for the 316 L steel in temperature range 20 °C - 700°C.

Temperature	E (MPa)	ν	σ_0 (MPA)	σ_m (MPa)
20 °C	193468	0.33	245.63	500
500 °C	152754	0.33	192.19	460
700 °C	142304	0.33	178.53	367

Table 2. Fatigue data for the 316 L steel in the temperature range 20 °C- 700 °C.

Temperature	K (MPa)	n	b	C	E_f'	σ_f' (MPa)	S value for $N_f = 10^4$	S value for $N_f = 10^6$
20°C	763.3	0.157	-0.0835	-0.5142	0.476	703.4	307.7	209.4
500 °C	848.41	0.206	-0.0903	-0.414	0.1272	582.6	238.2	157.2
700 °C	729.2	0.1954	-0.084	-0.4281	0.0821	848.42	195.2	138.6

4. Results and discussion

A parametric study is conducted on the CT16 specimen having the material data specified in section 3 under the following conditions: The applied pressure in Pa is varied in the set {9×10^6 ; 1.3×10^7; 1.5×10^7 ; 1.7×10^7 },the reference temperature in °C is varied in the set {200, 500 } ,the temperature gradient in °C takes one of the two values{5, 10 }.

Figure 3.a gives, at the reference temperature 200°C, the obtained results in terms of lifetime cycles for all the TMF loadings. Figure 3.b gives, in the same conditions as for figure 3.a, lifetime cycles but with the reference temperature equal to 500°C. Unlike what is largely known from literature dealing with TMF, figures 3.a and 3.b show that the isothermal loading is more damaging in some cases than the themomechanical loading.

A net distinction appears between elastic plastic regime{1.3×10^7; 1.5×10^7 ; 1.7×10^7 }and the elastic regime 9×10^6 Pa . One can notice that an elastic regime is recovered also with 1.3×10^7 Pa and $\Delta T = 10°C$ for the OP case. On the opposite, an elastic plastic regime was recovered with 6×10^6 Pa and $\Delta T = 10°C$ for the IP case. In the presence of plastic strains, only small differences appear between the isothermal, the in-phase and out-of-phase cycling. In some occasions the IP loading is more damaging, while in other cases the OP loading is more severe. This can also be verified when considering the themomechanical loading for which only elastic strains occur, 6×10^6 Pa.

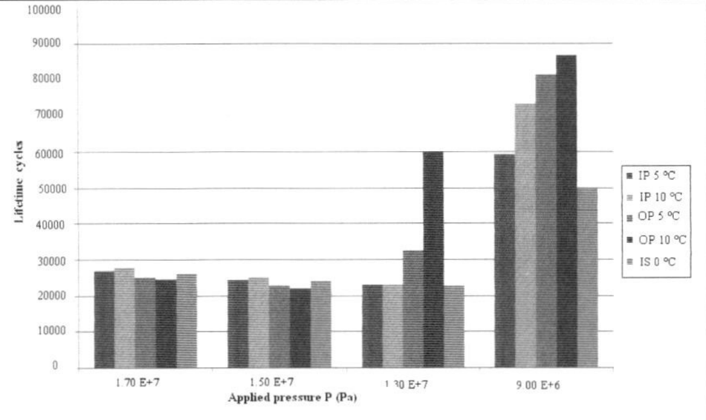

a.Reference temperature 200 °C

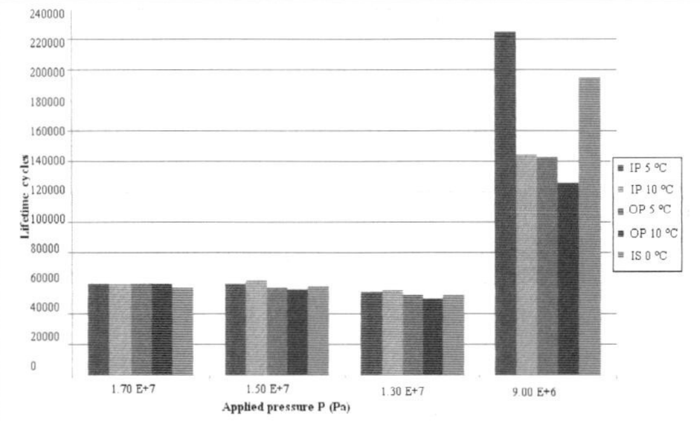

b.Reference temperature 500 °C

Figure 3. Life for TMF loadings as function of the applied mechanical load and thermal gradient;

5. Conclusions

Lifetime of the CT16 specimen with special boundary conditions and which was subjected to the action of various thermomechanical loadings was investigated by using finite element numerical modelling followed by fatigue pos-processing. Numerical simulations have shown that thermomechanical fatigue is not always more damaging than isothermal fatigue. Furthermore, no general conclusions could be drawn as to the relative influence of the in-phase cycles in comparison with the out-of-phase cycles.

Under plastic strains, only small differences appear between in-phase cycles and out-of-phase cycles. Considering both elastic and elastic-plastic regimes, comparison between these two scenarios has revealed that the out-phase mode can be more severe, but also less severe than the in-phase mode.

6. References

[1] Thomas, G., J. Bressers and D. Raynor (1982), Low-Cycle Fatigue and Life Prediction Methods. High Temperature Alloys for Gas Turbines, R. Brunetaud, Ed., D. Riedel Publishing Co., Netherlands, 291-317.

[2] Bill, R., M. Verrilli and G.R. Halford (1984), A Preliminary study of the Thermo-Mechanical Fatigue of Polycrystalline MAR M-200, NASA TP-2280, AVSCOM TR 83-C-6.

[3] Halford, G.R., S. Manson, B. Prillhofer, G. Winter and W. Eichlseder (1976), Life prediction of thermal–mechanical fatigue using strain range partitioning, thermal fatigue of materials and components. ASTM STP 612, 239-254.

[4] Zhuang,W.Z and N.S. Swansson (1998), Thermo-mechanical fatigue life prediction: a critical review, Airframes and engines division/aeronautical and maritime research laboratory, DSTO-TR-0609, Australia.

[5] Fe-safe (2008), Fe-safe User Manual, Safe Technology limited, Sheffield, England.

[6] Swanson, G.A., I. Linask, D.M. Nissley, P.P. Norris, T.G. Meyer and K.P. Walker (1986), Life prediction and constitutive models for engine hot section, Anisotropic materials program. NASA contractor report 174952.

[7] Gocmez, T., A. Awarke and S. Pishinger (2010), "A new low cycle fatigue criterion for the isothermal and out-of-phase thermomechanical loading", International Journal of Fatigue 32, 769-779.

[8] Skelton, R.P (1991), "Energy criteria for high temperature low cycle fatigue", Mater Sci Technol 7, 427–439.

[9] Skelton, R.P. (1993), "Cyclic hardening, softening and crack growth during high temperature fatigue", Mater Sci Technol 9, 1001-1008.

[10] Neu, R.W. and H. Sehitoglu (1989), "Thermomechanical fatigue, oxidation, and creep: part II. Life prediction", Metal Trans A, 20A, 495–508.

The Cross-Sectional Investigation of Oxide Scale FeCr Alloys and Commercial Ferritic Steel Implanted with Lanthanum and Titanium Dopants after Oxidation Test at 900°C

Darwin Sebayang[1], Hendi Saryanto[1, 2], Pudji Untoro[3], Tjipto Sujitno[4]

[1]Mercu Buana University, Jln. Meruya Selatan, Jakarta Barat, Indonesia
[2]Garuda Muda Institute, Jakarta, Indonesia
[3]Surya University, Jln. Boulevard Gading Serpong, Banten-Indonesia
[4]Centre For Accelerator and Material Process Technology. BATAN JL. Babarsari, D.I Yogayakarta, 55281, Indonesia

Email: hsaryanto@gmail.com & d_sebayang@hotmail.com

Abstract. The cross-sectional examinations of oxide scales formed by oxidation on the surface of FeCr alloys and Ferritic Steel that implanted with lanthanum and titanium dopants were observed and investigated. Scanning electron microscopy (SEM) coupled with energy dispersive X-ray spectroscopy (EDS) has been used to study the cross-sectional oxides produced by specimens after oxidation process. X-ray diffraction (XRD) analysis was used to strengthen the analysis of the oxide scale morphology, oxide phases and oxidation products. Cross-sectional observations show the effectiveness of La implantation for improving thinner and stronger scale/substrate interface during oxidation process. The result shows that the thickness of oxide scales formed on the surface of La implanted FeCr alloy and ferritic steel was found less than 3 μm and 300 μm, respectively. The oxide scale formed on the surface of La implanted specimens consisted roughly of Cr_2O_3 with a small amount of FeO mixture. The oxidation resistance of FeCr alloy that implanted with Lanthanum showed better quality than that of commercial alloy. The beneficial effect of La implanted sample on the oxidation resistance was clearly visible to improve the adherence of oxide scale. Cross-sectional observations revealed the effectiveness of La implantation for improving thinner and stronger scale/substrate interface during oxidation process which presents the homogenous structure with good adherence and a uniform inter-diffusion layer between substrate and oxide layer.

Keywords: Cr_2O_3, FeCr Alloys, Ferritic Steel, Crystallite Size, Lanthanum, Titanium

1. Introduction

The main problem with the use of conventional steel alloy is their inadequate high-temperature oxidation resistance [1]. Under SOFC operating conditions, ferritic steel form the oxide scale that is susceptible to spalling during normal fuel cell operation, in particular during thermal cycling encountered at start-up and shut down [2]. The extensive oxide losses from the surface or oxide spalling usually occur, which has become a serious challenge that leads to serious deterioration of the cell performances [3, 15]. Therefore, the alloys stabilization and oxidation performances at the high temperatures are essential to enhanced performance of the cell, chromium evaporation is known to be a major problem in SOFC stacks since volatile chromium is gets deposited on the cathode and poisons the oxygen electrode [4]. To enable practical application, two classes of technology can be approached to reduce the growth rate of the oxide scale, the first is to develop nanostructured alloys, and the other is to conduct surface treatment of the developed or available alloy with a reactive element [5,15]. FeCr alloys and ferritic stainless steels exhibit excellent properties as interconnect material in SOFCs [6]. The oxidation performance of FeCr used in high-temperature applications (such as Interconnector in SOFC) depends on their potential to form and maintain a Cr-rich protective oxide scale [7-8]. Technical requirements to be used as interconnect in SOFC should fulfill a number of specific requirements, i.e. a high oxidation resistance, a high electrical conductivity of the surface oxide scales, gas tightness, and a coefficient of thermal expansion matched to the electrolyte and the electrodes [9-10]. The role of the interconnect component in SOFC units is to electrically connect individual cells in series to form a multi-cell stack with higher power density. FeCr and Ferritic stainless steels have many desirable properties for use in solid oxide cell stacks and are the most common type of materials used as interconnect. When Cr based alloys and high-Cr ferritic steels are exposed to an oxidizing environment at elevated temperatures, an oxide layer is formed on the surface, acting as a barrier between the metal and the gas. Chromium increases the oxidation resistance of stainless steels by the formation of a chromia (Cr_2O_3) scale on the surface. Other oxides which are often encountered in the layers are iron oxides (wüstite FeO, magnetite Fe_3O_4, or hematite Fe_2O_3) [8,11-12].

Refer to our previous study [8,13-14] the cross-sectional investigation of FeCr alloy with smallest crystallite sizes can significantly reduces the growth of oxide scale and improve the high-temperature oxidation and corrosion resistance of FeCr alloy that oxidized at 900°C, However, the research did not investigate the characterization of oxide scale on a cross-sectional analysis and the growth mechanism of the oxide scale of implanted specimens. Therefore, the objective of this study is to investigate the cross-sectional characterization of oxide scale formed on the surfaces of implanted specimens compare with un-implanted specimens oxidized at 900°C. Oxidation kinetics, and as well as the evolution of oxide scale composition and microstructure are systematically studied. The mechanism of the formation of oxide scale on both of steels is discussed.

2. Specimens and Experimental Procedure

2.1 Specimens

The selected FeCr alloy with smallest crystallite size (38.51 nm) as the best oxidation results in our previous study [8] were selected and implanted with lanthanum and titanium dopants. Implantation of lanthanum and titanium dopants with nominal doses10^{17} ions/cm^2 was undertaken using 100 keV ion beam energy and 200 keV acceleration potential and the beam current density was maintained at 10 μA/cm^2. The depth concentration profiles of lanthanum and titanium dopants in to substrate of both type specimens were collective in Table. 1 [13-15].

Table 1. The depth concentration profiles of lanthanum and titanium dopants in to substrate of steels

Sample	**Depth profile (Å)**	
	Implanted with lanthanum	**Implanted with titanium**
FeCr alloys	297	706
Commercial FS	296	639

2.2. Oxidation Test

The samples of oxidation test were cut from the specimens to the size of 4x5x3 mm. The samples were subjected to oxidation tests in laboratory air at atmospheric pressure in a PROTHERM box furnace at 900°C for 100 h in laboratory air at the atmospheric pressure.

2.3. Preparation of cross-section specimens

Cross-section samples are mounted using "cold-mounting" resins. Samples were placed on a commercial sample holder (cylindrical), fixed through a drop of wax. Resin (plus hardener and accelerant) is then poured in the sample. After the sample and resin becoming a hard block, the samples were polished on the cross-sectional face up to 1500 grade SiC paper and final polishing using 0.05 μm diamond paste.

2.4. Microstructure Characterization

The oxidized alloys were investigated by a BRUKER D8 Advance X-ray diffraction (XRD) equipped with CuKα radiation and phase identification as referred to the ICDD PDF-2 database. The surface and cross-sectional morphology analyses of oxidized specimens were carried out with a JEOL JSM-6380 scanning electronic microscope (SEM) equipped with an energy-dispersive X-ray spectroscopy (EDS) analyzer. Furthermore, the best selected of oxidized specimens were analyses more detail using EDS line/mapping analysis.

3. Results and Discussion

3.1. Previous Results

The previous results on developed of FeCr Alloys, oxidation test, and parabolic rate were collected in Table 2 and Fig.1 [7, 8, 9, 10 & 11], where the parabolic rate constant (k_p) is determined by the mass gain (ΔW) per unit surface area of a specimen (A) and exposure time (t) [10, 13-15].

$$({}^{\Delta W}/_{A})^2 = k_p t \qquad (1)$$

Table 2. Mass gain change and parabolic rate constant (k_p) of FeCr alloys after oxidized at 900°C

Specimens	Crystallite size (nm)	Mass Gain (mg/cm²)	Parabolic Rate, k_p (mg²/cm⁴)
FeCr "A"	53.33	21.81	6.191
FeCr "B"	38.51	7.72	0.794
FeCr "C"	76.60	35.18	13.58

Table 1 show the mass gain curves of the studied alloys with different crystallite size oxidized at 900°C. Mass gain change (mg/cm^2) and parabolic rate (mg^2/cm^4) indicate that FeCr "B" as the smallest crystallite size was found as the lowest mass gain change and parabolic rate, which indicates as the best oxidation resistance.

3.2. Results of the current study

SEM morphologies of the cross-section of the oxide scale on the surface of implanted and the selected of un-implanted (FeCr "B") alloys after oxidized at 900°C for 100 h in air atmosphere are presented in Fig 1 below.

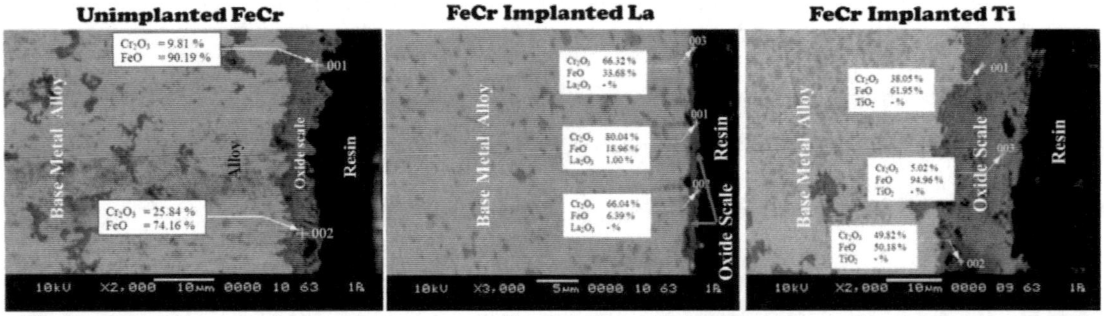

Figure 1. SEM morphologies of the cross-section of the oxide scale on the surface of implanted and un-implanted FeCr alloys oxidized at 900°C

Fig. 1 shows the cross-sectional morphology of oxide scale on the surface of implanted and un-implanted specimens oxidized at 900°C. A very thin oxide scale, less than 3 μm, had formed on the surface of FeCr alloy that implanted with lanthanum. While, the thickness of the oxide scale formed on the surface of

IOP Conf. Series: Materials Science and Engineering **324** (2018) 012075 doi:10.1088/1757-899X/324/1/012075

un-implanted FeCr alloy and FeCr alloy that implanted with titanium was formed about 8 and 15 μ m, respectively. According to EDX analysis, it can be seen that the oxide scale on the surface of FeCr alloy implanted with lanthanum was dominated by the presence of Cr_2O_3 about 60.05-80.04 %, which indicates that lanthanum implantation can protect the Cr_2O_3 layer, reduce the rate of Cr evaporation and reduce the growth of the oxide scale as evidenced by the thin layer of oxide scale. However, the oxide scale formed on the surface FeCr alloy that implanted with titanium dopant indicated covered and distributed evenly with the main composition of FeO approximately 50.18-94.96%. It can be noticed that titanium implantation ineffectively promotes Cr rich oxide. At the same time, the amount of Fe increased and diffused outwards, which caused the formation and rapid growth of FeO. Furthermore, Cr_2O_3 as the protective layer becomes non-protective, due to the formation of volatile Cr_2O_3 upon exposure to oxidizing atmospheres for extended periods.

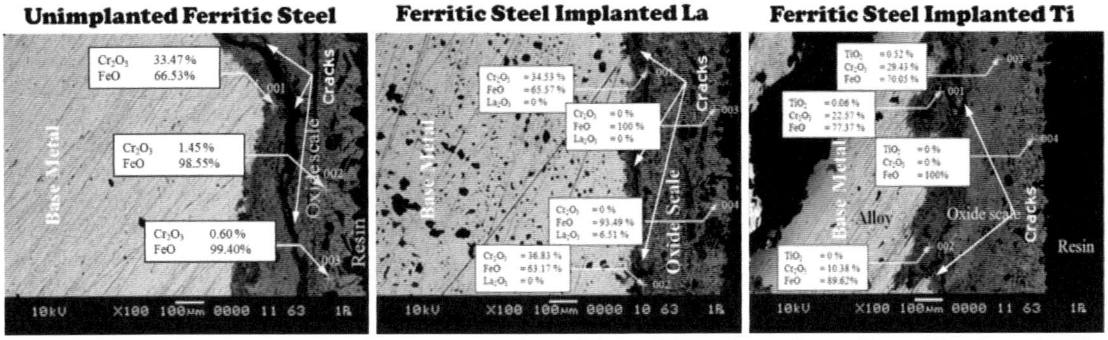

Figure 2. SEM morphologies of the cross-section of the oxide scale on the surface of implanted and un-implanted Ferritic Steel oxidized at 900°C

Fig. 2 shows the cross-sectional of oxide scale formed on the implanted and un-implanted ferritic steels oxidized at 900°C. The oxide scale on the surface of an-implanted ferritic steel is formed up to a scale thickness of 600 μm, where the oxide scales show mainly consisted of Cr_2O_3 about 33.47 and FeO 99.40 at%. However, the adherence between scale and a base metal reduced causes the occurrences of the largest crack along the scale. For the specimen that implanted with lanthanum show better oxide scale indicated by the thickness of the oxide scale was found about 300 μm thick, although the cracking also occurs but the presence of Cr_2O_3 as the protective layers is still significant in the range of 34.53-36.83 %. While, the presence of Cr_2O_3 on the oxide scale of FeCr that implanted with titanium was found in the range 10.38-29.43 at %. This phenomenon indicates that the implantation of lanthanum has influenced the formation stable layer of Cr_2O_3 and plasticity as well as protects it from evaporation and cracks on high temperature oxidation. The results of the cross-sectional analysis as described above have been confirmed by the results of the oxidation kinetics of both type specimens oxidized at 900°C as shown by following picture.

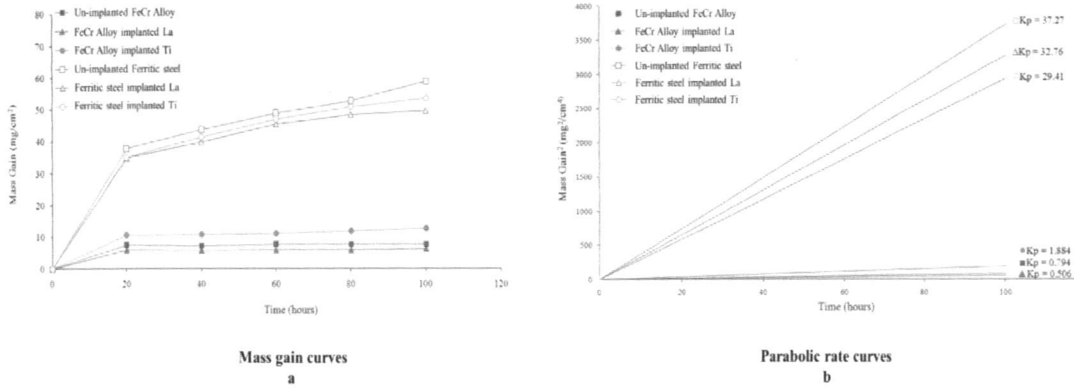

Figure 3. Mass gain curves and parabolic rate curves of un-implanted and implanted specimens oxidized at 900°C

Figs. 3 show the mass gain curves and parabolic rate curves of the implanted and un-implanted of FeCr alloys and Ferritic steel oxidized at 900°C. Mass gain curves and parabolic rate curves indicates that ferritic steels displayed higher mass gain trend and exhibits poor oxidation resistance. The beneficial effect of lanthanum implantation is clearly visible, the oxidation behaviour being improved. In fact, the mass gain and parabolic rate more lowest, if compared with the un-implanted and Ti implanted specimens. The effect of lanthanum implantation for specimen of FeCr alloy that implanted with lanthanum is significantly to reduce the mass gain change and parabolic rate from 7.73 mg/cm^2 and 0.794 mg^2/cm^4 to 6.29 mg/cm^2 and 0.506 mg^2/cm^4, respectively, in comparing with un-implanted FeCr alloy. Furthermore, The effect of lanthanum implantation for specimen of ferritic steel that implanted with lanthanum is significantly to reduce the mass gain change and parabolic rate from 59 mg/cm^2 and 37.27 mg^2/cm^4 to 49.7 mg/cm^2 and 29.41 mg^2/cm^4, respectively, in comparing with un-implanted ferritic steel. While, the effect of titanium implantation becomes ineffective to reduce the growth of oxide scale which indicates by the highest mass gain change and parabolic rate for both type of specimens. In order to strengthen the analysis result described above, XRD analysis were applied to identified the oxide phase of both type of specimens. The XRD patterns for specimens oxidized at 900°C for100 h are presented in Fig.4.

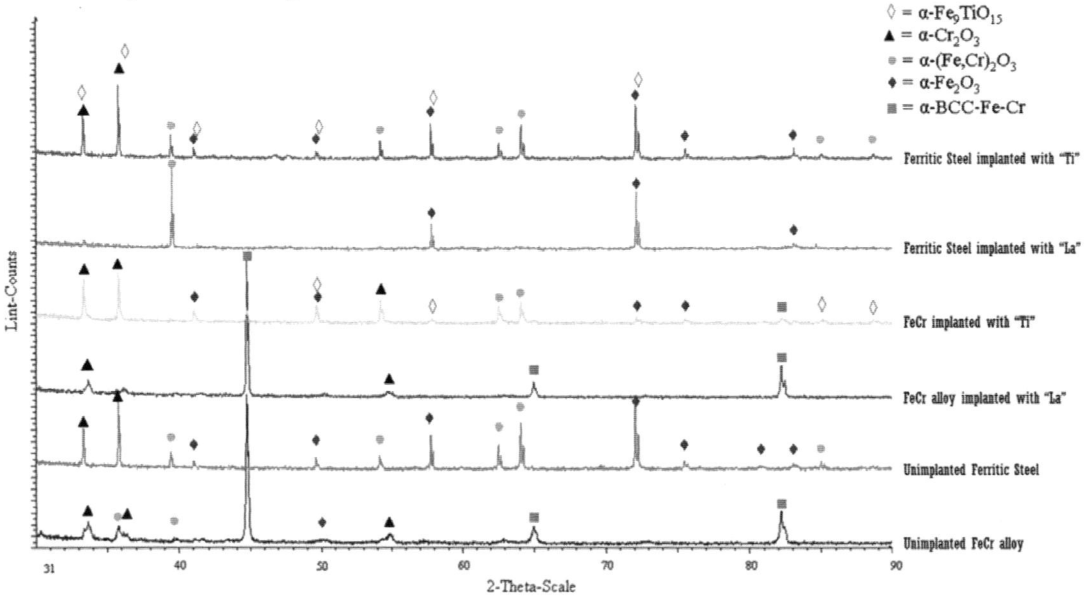

Figure 4. X-rays diffractograms of unimplanted and implanted specimens oxidized for 100 h at 900°C

Fig. 4 above shows the diffraction patterns of FeCr alloys and ferritic steels oxidized at 900°C. For unimplanted FeCr alloy, the signal peaks can be identified as α-(Fe,Cr)$_2$O$_3$, α-Fe$_2$O$_3$, and α-Cr$_2$O$_3$. The FeCr phase is also shown, but only on as the developed specimen which indicates the thin of oxide scale on the substrate surface. For the implanted specimens, the main oxide scale obtained were α-(Fe, Cr)$_2$O$_3$, α-Fe2O3, α-Cr$_2$O$_3$. The presence of the peak of the original alloy indicates that oxide scale was formed as a thin layer which dominated by the presence of Cr-rich oxide and small peaks of solid solution chromium iron oxide (Fe,Cr)$_2$O$_3$ [10, 13-15]. On the contrary, FeCr alloy that implanted with titanium dopant, all strong peaks were dominated by the presence of α-Fe$_9$TiO$_{15}$, hematite Fe$_2$O$_3$ and small peaks of solid solution chromium iron oxide (Fe,Cr)$_2$O$_3$, although the presence of the peaks of the original alloys still found with a small intensity which indicates that oxide scale was growing rapidly with the dominating presence of Fe-rich oxide due to ineffectively of titanium implantation. In order to understand more detail about the mechanism of oxide growth on the surface of specimens, EDX line/mapping analysis was used to verify the results in the formation of the oxide scale of FeCr and ferritic steel that implanted with lanthanum as the best oxidation results as shown in Fig.5 and 6.

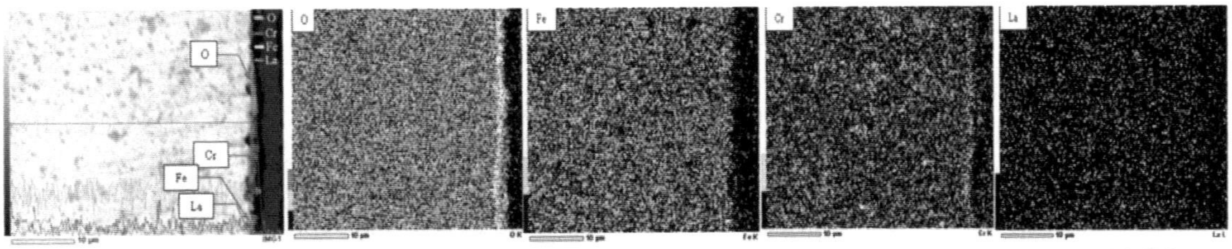

Figure 5. Cross-sectional EDS mapping analysis of FeCr implanted with La oxidized at 900°C

Fig.5 above shows that the scale formed on FeCr alloy that implanted with lanthanum consisted of corundum oxides layer of Cr_2O_3 wich confirmed by EDX mapping analyses which indicated by high presence of Cr rich oxide on the inner of oxide scale, while, the EDX mapping analyses indicates that the presence of Fe was found on the outer of oxide scale with small amount of Fe on the formation of FeO. However, EDX mapping analyses is not clear to detect the presence of Lanthanum on the oxide scale due to low doses of Lanthanum that implanted in to substrate of the alloy. This phenomenon indicates that the lanthanum implantation significantly to reduce the growth of oxide scale by reducing the Cr^{3+} ions diffuses to the surface of the specimens. where the enhancement of Cr diffusion guarantee the stable growth of the Cr-rich in the inner of the oxide scales, which significantly improve oxidation resistance, even a Cr_2O_3 scale may be formed at higher temperatures owing to the higher diffusion rate of Cr.

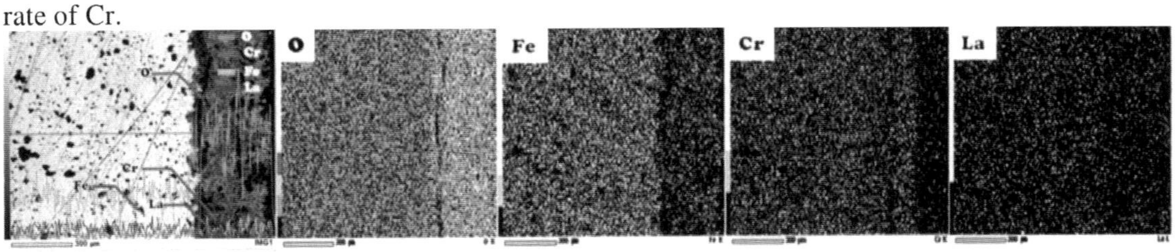

Figure 6. Cross-sectional EDS mapping analysis of ferritic steel implanted with La oxidized at 900°C

Fig.6 above shows that the Cr concentration in ferritic steel is still not sufficiently high to allow the formation of a more protective Cr_2O_3 layer, while the iron ions dissolve in and diffuse rapidly through the Cr_2O_3 scale and eventually an outer layer of fairly pure iron oxides will result. It may be caused by Fe and Cr ion transport is predominant then make the scale growth will proceed outwards, the oxygen transports through the FeO layer to the Cr_2O_3 layer and the Cr ions diffuse to the same layer; both of them react with each other, which make the oxide scale become thicker. Furthermore, due to the loose and incompact scale, thus causing the crack in the surface of the specimen which increase the rate of Cr evaporation. The oxygen may diffuse to the oxidation/alloy interface through the cracks and pores in the oxide scale. This phenomenon was confirmed by XRD results as shown in Fig. 3, where the result shows that the oxide phase that formed on the surface of ferritic steel that implanted with lanthanum is dominated by the presence of Fe_2O_3. It's also reinforcement by mass gain change and parabolic rate constant (k_p) as presented in Fig. 3. It can be noticed that a discontinuous Cr_2O_3 scale was formed, but could not protect the alloy from internal oxidation. At the same time, Fe became enriched and diffused outwards, which caused the formation and rapid growth of FeO, although it has been implanted with lanthanum due to the higher crystallite size effect.

4. Conclusion

Based on the results obtained in the oxidation of implanted and un-implnted FeCr alloys and ferritic steel at 900°C for 100 h the following conclusions can be formulated:

1. The oxidation resistance of FeCr alloy that implanted with Lanthanum showed better quality than that of commercial alloy. The beneficial effect of La implanted sample on the oxidation resistance was clearly visible to improve the adherence of oxide scale. Cross-sectional observations revealed the effectiveness of La implantation for improving thinner and stronger

scale/substrate interface during oxidation process which presents the homogenous structure with good adherence and a uniform inter-diffusion layer between substrate and oxide layer.

2. The specimens that implanted with lanthanum show better promoting the formation of Cr_2O_3 in large amount compositions which indicated that chromium preference diffused outward to form a Cr_2O_3 layer.
3. The oxide scale formed on the surface of FeCr alloys and ferritic steel that implanted with titanium dopant was thicker oxide scales indicating that significant increase in oxidation weight gain. It can be noticed that titanium implantation ineffectively promotes Cr rich oxide. At the same time, the amount of Fe increased and diffused outwards, which caused the formation and rapid growth of FeO. Furthermore, FeO nucleated and grew at the discontinuous areas such as the incompact Cr_2O_3 scale or its micro-defective areas; leading to some bulges and cracks.

5. References

[1] Kulekci, M. K. 2008. Magnesium and its alloys applications in automotive industry. *The International Journal of Advanced Manufacturing Technology*, *39*(9), 851-865.

[2] Cooper, L., Benhaddad, S., Wood, A., & Ivey, D. G. 2008. The effect of surface treatment on the oxidation of ferritic stainless steels used for solid oxide fuel cell interconnects. *Journal of Power Sources*, *184*(1), 220-228.

[3] Korsunsky, A. (Ed.). 2012. *Current Themes in Engineering Science 2010*. American Institute of Physics.

[4] Hui, R. 2013. Materials and Processing for Metal-Supported Solid Oxide Fuel Cells. *Materials for High-Temperature Fuel Cells*, 309-340.

[5] Saryanto, H., Deni, S. K., Untoro, P., Saleh, M. H., & Sebayang, D. 2010. Determination of Nanocrystalline Fe80Cr20 Powder based Alloys using Williamson-Hall Method. In *Advanced materials research* (Vol. 129, pp. 999-1003). Trans Tech Publications.

[6] Sachitanand, R., Sattari, M., Svensson, J. E., & Froitzheim, J. 2013. Evaluation of the oxidation and Cr evaporation properties of selected FeCr alloys used as SOFC interconnects. *international journal of hydrogen energy*, *38*(35), 15328-15334.

[7] Ali-Löytty, H. 2013. Microalloying Mediated Segregation and Interfacial Oxidation of FeCr Alloys for Solid-Oxide Fuel Cell Applications. *Tampereen teknillinen yliopisto. Julkaisu-Tampere University of Technology. Publication; 1114.*

[8] Saryanto, H., Sebayang, D., & Untoro, P.2017, May. Cross-Sectional Investigations of Oxide Scale Nanocrystalline FeCr Alloys after High-Temperature Oxidation Test at 900° C. In *IOP Conference Series: Materials Science and Engineering* (Vol. 196, No. 1, p. 012030). IOP Publishing.

[9] Hilpert, K., Quadakkers, W. J., & Singheiser, L. 2003. Interconnects. *Handbook of Fuel Cells.*

[10] Sebayang, D., Khaerudini, D. S., Saryanto, H., Othman, M. A., Sujitno, T., & Untoro, P. (2011). Effect of Nanocrystalline and Ti implantation on the oxidation behaviour of Fe80Cr20 alloy and commercial ferritic steel. In *Key Engineering Materials* (Vol. 474, pp. 2134-2139). Trans Tech Publications.

[11] Lessing, P. A. 2007. A review of sealing technologies applicable to solid oxide electrolysis cells. *Journal of Materials Science*, *42*(10), 3465-3476.

[12] Akanda, S. R. 2012. *Mechanical Characterization of Coating-Interconnect Interfaces and Anode-Electrolyte Interfaces for Solid Oxide Fuel Cells*. The Ohio State University.

[13] Sebayang D *et al* Oxidation Resistance of $Fe_{80}Cr_{20}$ Alloys Treated by Rare Earth Element Ion Implantation. Current Themes In Engineering Science 2010: *AIP Conference Proceedings.* 1394. No. 1 pp 90-102.

[14] Hendi, S., Darwin, S., Pudji, U, & Tjipto, S. 2009. Ion implantation process of Lanthanum and Titanium Dopants into a substrate of Fe80 Cr20.

[15] Sebayang, D., Khaerudini, D. S., Saryanto, H., Omar, B., Othman, M. A., Hamid, A &

Untoro, P. 2011. Oxidation Resistance of Unimplanted and Implanted of Nanocrystalline FeCr Alloy and Commercial Alloy with Lanthanum. *Journal of Advanced Microscopy Research*, *6*(4), 263-277.

IOP Conf. Series: Materials Science and Engineering **324** (2018) 012076 doi:10.1088/1757-899X/324/1/012076

Real-time parameter optimization based on neural network for smart injection molding

H Lee[1], Y Liau[1] and K Ryu[2]

1 PhD Student, Department of Industrial Engineering, Pusan National University, Busan, Korea.
2 Associate Professor, Department of Industrial Engineering, Pusan National University, Busan, Korea.
E-mail: hslee.yeeyeng85.kyryu@pusan.ac.kr

Abstract. The manufacturing industry has been facing several challenges, including sustainability, performance and quality of production. Manufacturers attempt to enhance the competitiveness of companies by implementing CPS (Cyber-Physical Systems) through the convergence of IoT(Internet of Things) and ICT(Information & Communication Technology) in the manufacturing process level. Injection molding process has a short cycle time and high productivity. This features have been making it suitable for mass production. In addition, this process is used to produce precise parts in various industry fields such as automobiles, optics and medical devices. Injection molding process has a mixture of discrete and continuous variables. In order to optimized the quality, variables that is generated in the injection molding process must be considered. Furthermore, Optimal parameter setting is time-consuming work to predict the optimum quality of the product. Since the process parameter cannot be easily corrected during the process execution. In this research, we propose a neural network based real-time process parameter optimization methodology that sets optimal process parameters by using mold data, molding machine data, and response data. This paper is expected to have academic contribution as a novel study of parameter optimization during production compare with pre - production parameter optimization in typical studies.

1. Introduction

Injection molding is a manufacturing process for producing plastics parts by injecting material into a mold. A wide variety of products are manufactured using injection molding, which vary greatly in their size, complexity and application. Injection molding process has two most common problems of all. The first problem is the derivation of optimal process parameters for initial injection molding machine setting considering various process variables in order to implement effective manufacturing after product design. In order to derive the optimal process variables, there is a process of continuously adjusting the process parameters in a trial-error manner which causes considerable loss. The second problem is the difficulty to change the value of initial process parameter during injection molding process, especially when defects occur during manufacturing. Currently, there is not yet method for quality improvement of this dilemma, and it is challenging to response to the occurrence of equipment failure and sudden production problems. In this research, we present a methodology of neural network based real-time parameter optimization in injection molding process and propose the detailed neural network structure. This study is expected to facilitate establishment of smart factories for injection mold, and to suggest a guideline for manufacturing innovation toward the fourth industrial revolution.

2. Literature research

Injection molding is the most widely used fundamental manufacturing process for plastic products. It is ideally suited for mass production of plastic parts with complex shapes requiring precise tolerance. In this process, hot polymer melt is forced into a cold empty cavity of a desired shape and is then allowed to solidify under a high holding pressure.

In order to improve the injection molding system, various researchers conducted injection process innovation and analysis. One of the representative researchers is one by Tuncay and Hasan[1]. They have studied the quality and process problems according to the characteristics of injection products by using neural network, Taguchi method, and Meta-heuristic method. In addition, the control parameters of injection molding machine were derived on the main characteristics such as defects (moldability, appearance defects, distortion, warping) of the injection mold, and proposed method for deriving the optimal parameters [2]. Lu and Khim[3] studied the effects of injection molding conditions on surface contours of a molded polycarbonate lens. In order to systematically analyze the effects of carious process parameters on the lens contours, statistical methods were used in conducting the experiments. Sadeghi[4] proposed a neural network model for predicting the quality or soundness of the injected plastic parts based on key process variables and material grade variations has been developed. The approach uses a backpropagation 4-2-3 net trained based on inputs/outputs data which were taken from simulation works carried out through a CAE software. In recent times, many techniques are being developed to solve multiple-input–multiple-output (MIMO) parameter- based problems, such as neural network, group method of data handling (GMDH) algorithm, fuzzy logic, etc. Nam [5] analyzed the temperature and pressure data inside the mold and controlled the process variables to optimize the quality using K-fold Cross Validation method. Gao [6] proposed a methodology for monitoring the quality of online produced products by using the Support Vector Regression method by utilizing the initial mold, melting temperature, and filling pressure values based on the design of experiments and simulation. Chen et al. [7] conducted a sequential analysis process using the Taguchi method, ANOVA, Response Surface Analysis method, and hybrid heuristic method (Genetic Algorithm + Particle Swam Algorithm) to derive manufacturing time, Many MIMO (Multiple Input and Multiple Output) studies have been conducted to optimize multiple quality levels. Taghizadeh et al. [8] derived initial process parameter settings for optimum quality by learning the mold temperature, melt temperature, settling value and degree of warping of the product using back propagation artificial neural network. The aforementioned researches focus on the reduction of defects by using historical data. However, we attempt to analyze the real-time data and to control the processing parameters in the real-time. The existing previous studies are mainly focused on the optimal process parameter setting right after product design. The studies are conducted on initial parameter optimization according to product and production characteristics by combining various statistical techniques and soft computing methods. It is essential to study the optimization in manufacturing field and shop floor by utilizing the technical factors and design principles of smart factory.

3. Real-time parameter optimization system

3.1. System framework

The proposed Smart injection molding system framework is shown in Figure. 1. The Smart injection molding system is a set of technologies that performs self-decision making, prediction, verification and control based on variable data generated during the process to optimize quality and production through data analysis, simulation, and control enhancement techniques. This system is able to conduct a self-learning of know-how and knowledge on characteristics of product quality and search the optimal initial parameter setting. Furthermore, this system can control the process parameters during production time using real-time data to prevent defects. The framework proposed in this study is inspired by six design principles of smart factory. The six design principles such as interoperability, virtualization, real-time capability, etc. are adapted in each module or functions. For example, interoperability is applied to communication between functions modules (Global decision maker and Local decision maker) and virtualization is applied to Digitalized factory model.

This framework is divided into three layers including planning layer, execution layer, and control layer. In this paper, we define and analyze the execution layer and the control layer of the injection mold system intensively.
In this paper, we focus on a data analysis framework for the control layer, which is a management area of the SCADA(Supervisory Control And Data Acquisition) system on the point of view of manufacturing support systems. The framework consists of four modules: Mold sensor, Machine controller, Vision sensor and Process parameter optimizer.
Mold sensor collects temperature data inside the mold. Machine controller acquires molding machine process parameter and internal cylinder temperatures. Vision sensor is constructed as a visual sensor system for collecting the shrinkage and warpage data of the final product produced. Process parameter optimizer derives optimal process variable values to minimize defects using the artificial neural network.
First, the machine controller acquires control log data (melting temperature, packing time,.etc) generated in the injection molding machine itself, and obtains the temperature (usually three or four zones) of the cylinder zone temperature inside the molding machine. This is replaced by the new setting value derived from the process parameter optimizer. The mold sensor acquires temperature data inside the actual mold to determine the difference between the control value of the actual equipment. A detailed description of process variable optimization is provided in the next section.

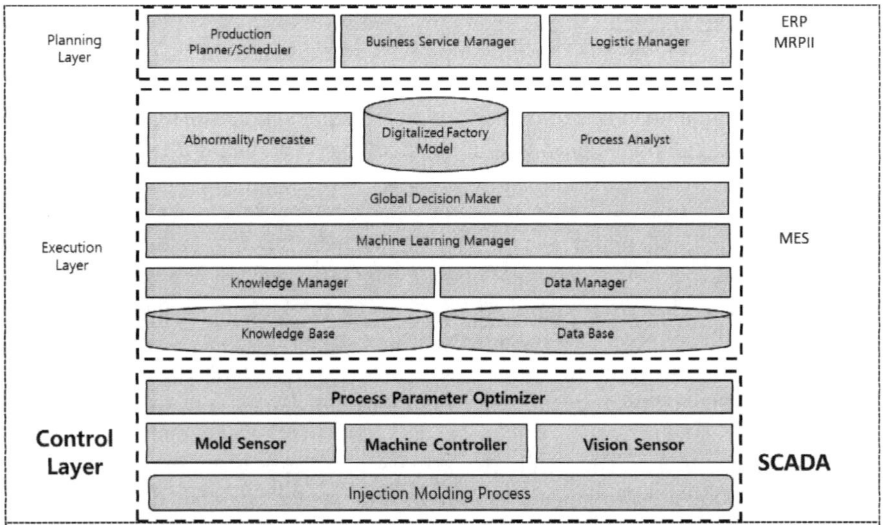

Figure 1. Framework of smart injection molding system.

3.2. Process parameter optimizer

The injection molding process has a mixed data in which continuous and discrete data. It is necessary to synchronize the type of data in order to learn and predict through the artificial neural network. In this study, continuous data is transformed into discrete form using Self-organizing maps (SOMs). SOMs were developed by Kohonen(1982) and are a peculiar class of artificial neural networks based on unsupervised competitive learning which consists of iteratively modifying the synaptic weights [9]. In self-organizing systems, the output neurons of the network compete among themselves to be activated, and an output neuron winning the competition is called a winner-takes-all neuron or a winning neuron [10]. The neurons, selectively tuned to discrete input patterns, are located on a one or two dimensional lattice in a topographic map, and they have intrinsic statistical features.
In this study, SOM is applied to extract the dynamic process parameter characteristics as the network input. The proposed methodology uses a sigmoid function for the backpropagation artificial neural network(BPNN) structure. Figure 2. is Architecture of the BPNN model. Table 1. is input and output variables of the proposed BPNN model.

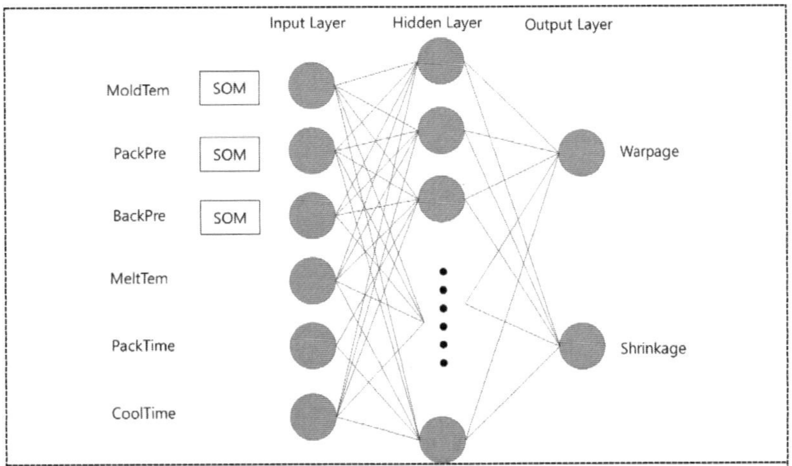

Figure 2. Architecture of the proposed BPNN model.

Table 1. Input & out variables of the proposed BPNN model

	Variables	Setting range	Data feature
Input	Mold temperature (℃)	20 - 80	Continuous
	Packing pressure (MPa)	30 - 175	Continuous
	Back pressure (MPa)	50 – 70	Continuous
	Melt temperature (℃)	180 - 280	Discrete
	Packing time (S)	1 - 18	Discrete
	Cooling time (S)	7 - 25	Discrete
Output	Warpage (mm)	0 – 1	Discrete
	Shrinkage (%)	0 - 3	Discrete

4. Conclusion

In this study, we proposed a system framework to optimize process parameters of injection molding system. The proposed system framework is divided into three categories. We focused on the function and data analysis methodology for the control layer. We proposed the artificial neural network structure of SOMs and BPNN based on the data obtained from the shop floor. The proposed artificial neural network is defined as a structure that supports learning of artificial neural network by discretizing continuous data. Furthermore, input and output variables of artificial neural network are defined based on existing literature. However, this study excluded the analytical process and the verification process of the experiment using real data. Therefore, it is expected that future work will be required to prove and verify the results of failure prediction and derivation of optimal variables using actual data.

Acknowledgments

This work was supported by the National Research Foundation of Korea(NRF) grant funded by the Korea government(MEST) (No. 2016R1A2B4014898).

References

[1] Erzurumlu T, Oktem H 2007 Comparison of Response Surface Model with Neural Network in Determining the Surface Quality of Moulded Parts *Materials & Design* **28** 459-465.

[2] Ozcelik B. Erzurumlu T 2006 Comparison of the Warpage Optimization in the Plastic Injection Molding using ANOVA, Neural Network Model and Genetic Algorithm *Journal of Materials Processing Technology* **171** 437-445.

[3] Lu X, Khim L 2001 a Statistical Experimental Study of the Injection Molding of Optical Lenses *Journal of Materials Processing Technology* **113** 189-195.

[4] Pattnaik S, Karunakar D, Jha P 2012 Application of Computer Simulation for Finding Optimum Gate Location in Plastic Injection Moulding Process *International Journal of Advanced Engineering Research and Studies* 159-161.

[5] Nam J 2016 Injection-moulded Lens form Error Prediction using Cavity Pressure and Temperature Signals based on K-fold Cross Validation *Proceedings of the Institution of Mechanical Engineers, Part B: Journal of Engineering Manufacture* 0954405416654421.

[6] Gao R X 2014 Online Product Quality Monitoring through In-process Measurement *CIRP Annals-Manufacturing Technology*. **63** 493-496.

[7] Chen W, Pen H, Shu C 2014 An Integrated Parameter Optimization System for MIMO Plastic Injection Molding using Soft Computing T*he International Journal of Advanced Manufacturing Technology* **73** 1465-1474.

[8] Taghizadeh S, Özdemir A, Uluer O 2013 Warpage Prediction in Plastic Injection Molded Part using Artificial Neural Network *Iranian Journal of Science and Technology, Transactions of Mechanical Engineering* **37** 149.

[9] Chen W C, Tai P H, Wang M W, Deng W J, Chen C T 2008 A neural network-based approach for dynamic quality prediction in a plastic injection molding process *Expert systems with Applications* **35** 843-849.

[10] Miller W T, Hewes R P, Glanz F H, Kraft L G 1990 Real-time dynamic control of an industrial manipulator using a neural network-based learning controller. *IEEE Transactions on Robotics and Automation* **6** 1-9.

IOP Conf. Series: Materials Science and Engineering **324** (2018) 012077 doi:10.1088/1757-899X/324/1/012077

Digital Twin concept for smart injection molding

Y Liau[1], H Lee[1] K Ryu[2]

1 PhD Student, Department of Industrial Engineering, Pusan National University, Busan, Korea.
2 Associate Professor, Department of Industrial Engineering, Pusan National University, Busan, Korea.
E-mail: yeeyeng85.hslee.kyryu@pusan.ac.kr

Abstract. Injection molding industry has evolved over decades and became the most common method to manufacture plastic parts. Monitoring and improvement in the injection molding industry are usually performed separately in each stage, i.e. mold design, mold making and injection molding process. However, in order to make a breakthrough and survive in the industrial revolution, all the stages in injection molding need to be linked and communicated with each other. Any changes in one stage will cause a certain effect in other stage because there is a correlation between each other. Hence, the simulation should not only based on the input of historical data, but it also needs to include the current condition of equipment and prediction of future events in other stages to make the responsive decision. This can be achieved by implementing the concept of Digital Twin that models the entire process as a virtual model and enables bidirectional control with the physical process. This paper presented types of data and technology required to build the Digital Twin for the injection molding industry. The concept includes Digital Twin of each stage and integration of these Digital Twin model as a thoroughgoing model of the injection molding industry.

1. Introduction

Injection molding is one of the most important and common manufacturing process to produce plastic parts in high volume and low cost. Products that manufactured using injection molding vary widely in the geometry complexity, size, materials and application in varies industry. Integration of smart manufacturing concept into injection molding known as smart injection molding which implement the Internet-of-Things (IoT) in injection molding that enables data collection, communication and analytics to achieve flexibility in process and improve production efficiency.

Injection molding industry covers three main stages, i.e. mold design, mold making and the injection molding process. The design of injection mold is an important stage because it is costly to do re-design and modification of mold in the future mass production. In current practice, mold design and making are perform either in-house or outsourcing. Mold designer gathers information on part, material and machine to determine the type of mold, layout of the mold in mold design. Once the design is completed, the mold designer will conduct the simulation using mold simulation software and the final design will send to mold making stage. Mold making stage consists of electrode design and manufacturing for mold insert, machining of mold components, assembly of mold and testing of the mold in machine. Tight tolerance and wide variety of processes involved increase the complexity of production planning. An efficient production planning is necessary in order to ensure the completion of mold on time. Finally, the production of plastic parts in injection molding process. It begins with production planning based on demands, machine and mold availability. During the injection molding process, personnel in the operation perform the control and monitoring as well as optimization of machine parameters setting and the inspection on molded parts. The production data such as parameter setting, machine and mold condition, and rejection will be input and updated from time to time to production planning for adjustment of schedule and maintenance schedule. Real-time control and monitoring is almost impossible because of the large volume of processing data required and difficulty to perform real-time inspection.

Every stage in injection molding industry affects each other on the performance of the production of molded parts. Various data in the entire injection molding industry, including mold design, mold manufacturing and the injection molding process need to be linked to achieve more efficient, smart and sustainable manufacturing environment. [1] Hence, the real-time bidirectional information flow between these stages is crucial to improve the performance from the aspect of quality, time and cost. In addition, a review report about the state of manufacturing conducted by General Electric (GE) in 2017 mentioned that there are not enough skilled people in manufacturing factories to fill manufacturing jobs and keep up with demand. This situation motivates the industry to shift towards smart manufacturing. The merging of Internet-of-Things (IoT) and the concept of Digital Twin enables the real-time data-driven manufacturing in various stages via convergence between physical and virtual production system. Therefore, this paper presented the application of Digital Twin concept in every stage of injection molding industry and the integration of each Digital Twin model that brings all the experts and data from each stage together to achieve complete and powerful insight in injection molding industry.

2. Concept of Digital Twin

Digital Twin plays important role in the 4th Industrial Revolution because it enables the combination of information technology and operation technology to create new value by link the preparatory production stage with real production. [3] The Digital Twin concept model introduced by Dr. Michael Grieves contains three main parts: physical products in real space, virtual products in virtual space and the connections of data and information that connects the physical and virtual space. [2] Digital Twin can be defined as a replication of real physical production system in digital model, which are used for system optimization, monitoring, diagnostics and prognostics using integration of artificial intelligence, machine learning and software analytics with large volume of data from physical systems.

Digital Twin model has five enabling components – sensors and physical assets from physical system, integration, data and analytics in the digital model. Data consists of data from enterprise and operational and environmental data from physical system obtained using sensors distributed in physical system. The sensors allow bidirectional data communication in real-time between physical system and virtual system using integration technology, which includes communication interface and security. Then the Digital Twin uses the analytics techniques to analyze the data for defined purposes and provides responsive action based on simulation result to physical asset for further action. [4] With the Digital Twin, the operation in a single machine as well in the interconnected system are visible to everyone who has authority to access the system, i.e. manufacturing, procurement, warehousing , transportation and logistics. Digital Twin allows the manufacturers to see how the physical system is performing by looking all the data in the manufacturing execution system. The analytics techniques in the Digital Twin model include what-if analysis and predictive analysis to simulate the real-time conditions in the physical system and predict the future state of the systems. The conceptualization, comparison and collaboration capability of Digital Twin enables us to conceptualize the manufacturing process visually, compare the option and outcome, and then finally collaborate with other manufacturing section.

General Electric (GE), Siemens and ANSYS introduced a series of platform to build the Digital Twin. Currently, this technology has been implemented in aerospace industry, wind-turbine power plant and automotive industry for performance improvement and operational inefficiency prediction. [5] For example, GE has built Digital Twins of jet engine that can achieve power optimization, monitoring and diagnostics of jet engine. DXC Technology built a Digital Twin for hybrid car manufacturing process to predict the performance of car before committing the changes in the manufacturing process.

3. Application of digital twin in injection molding mold design and manufacturing

This section presents the needs of Digital Twin at every stage in injection molding industry and how the Digital Twin connects with each other to build, operate and enhance the performance at each stage. Figure 1 shows the overview of information flow in the injection molding industry.

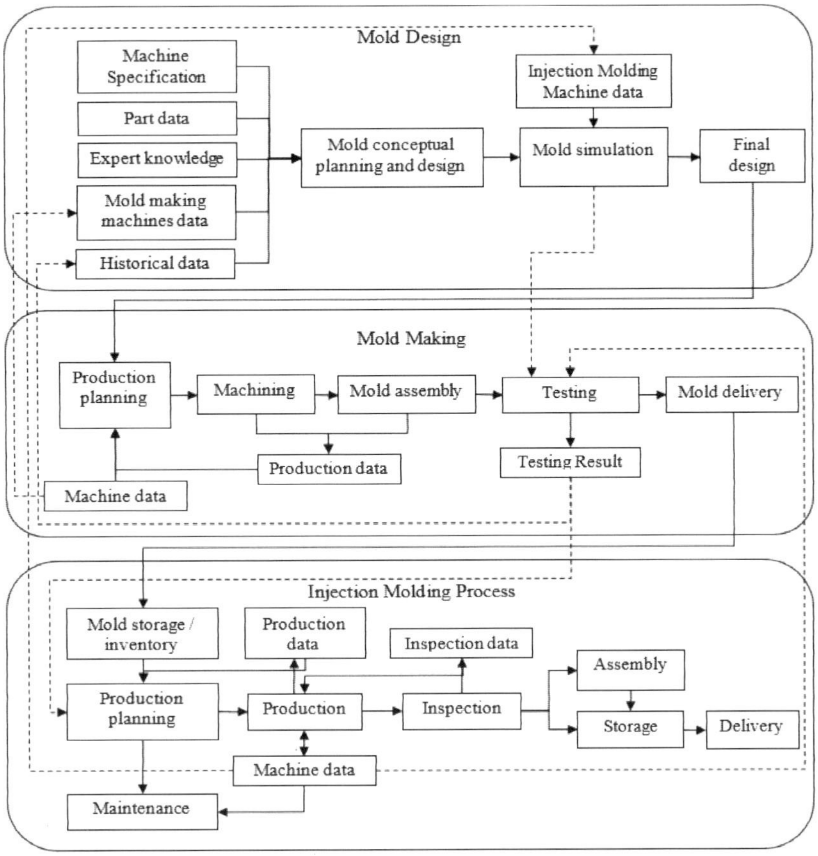

Figure 1. Overview of information flow in injection molding industry.

3.1 Injection mold design

Input information of part drawing and specification of material, molding machine specification and other tool specification such as type of mold, runner system, gate, use of robotics and estimated cycle time are necessary when design the mold. Mold designer has to be experienced enough to take all these considerations and the capability of mold making production to produce the designed mold. Mold design and simulation software such as Moldflow enables designer to design, simulate and analyze the mold designed and the part produced. The data input to simulation is based on the technical specification provided by the material supplier and machine manufacturer. However, in practice, the condition of a specific machine in production varies over time and this will cause the result from simulation differ from the real production. Therefore, in order to obtain accuracy and reduce resetting time during mold testing, the simulation has to be conducted using data from current condition of injection molding machine which to be used. These data can be obtained from the database of the Digital Twin model of the machine in the injection molding production. In addition, lead-time of mold design and making determine the lead-time of product to market. Continuous data input of machines for mold making process helps mold designers to gather the updated information on the machine condition and adjust design accordingly with the capability and availability of machining machines to avoid production delays due to the tool or machine breakdown during mold making. The final design of the mold will act as Digital Twin model of the mold and can be virtually installed in Digital Twin model of injection molding machine for future production planning and real-time process simulation.

3.2 Injection mold manufacturing

Once the mold design is completed, the detailed drawing of the mold will save in the database and mold-making production will retrieve the drawings for mold making process. Mold making mainly consists of machining of the parts, assembly and testing. Machining of mold components start with cutting of raw material, milling and lathe, wire-electrical discharge machining (WEDM), EDM and polishing. Each of the machine and tools has its own Digital Twin model. With the Digital Twin models of the machines, the overall physical production of mold making process can be visualized and simulate the production for production planning. During production, sensors distributed at each machine will send real time data to the Digital Twin models for production monitoring, quality prediction and prediction of machine condition for preventive maintenance scheduling. These data will update and store in each Digital Twin model and will be used for future mold design process.

In current practice, the production of mold making are done manually. However, because of the heavy weight of the mold, high variety number of components to be manufactured and assembled, application of human robot collaboration (HRC) systems is proposed in mold making process. Figure 2 shows the proposed framework of HRC systems in mold making. Simulation of the HRC systems is required to determine the collaborative task sequence, robot model selection and robot effector selection. Hence, the integration of the Digital Twin models of the human workers and the robots into the Digital Twin models of machines are necessary. Adaptive capability of robot to human working condition is one of the important issue in HRC systems. The Digital Twin models can analyze, simulate and predict the human worker's condition with the continuous input of human-robot unit data, and then input the result to robot for motion adjustment to adapt with the human worker's condition from time to time.

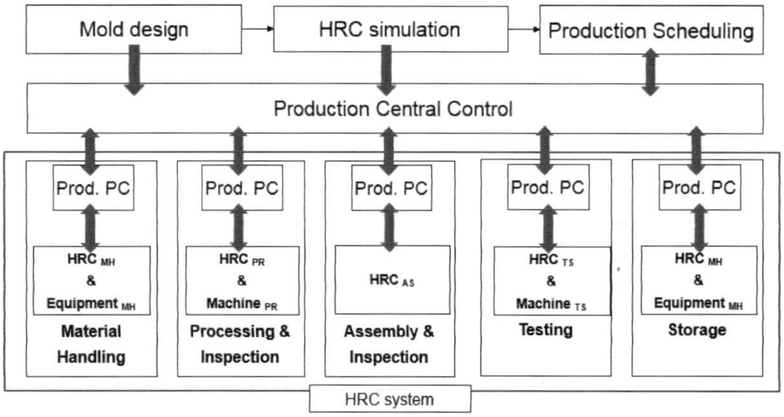

Figure 2. Mold manufacturing with human-robot collaboration.

4. Application of digital twin in injection molding process

Many research related to injection molding process focus on the optimization on process parameter and quality, but the trend of research is moving towards real-time process optimization, process monitoring, and prediction of quality defects. Application of Digital Twin concept in injection molding can help to achieve these objectives. Figure 3 illustrates the concept of Digital Twin in injection molding process.

Virtual production consists of the Digital Twin model of every single machine and equipment includes mold, material handling equipment and automated visual inspection system. Each of the machine is unique although they have the same machine specification and model because the machine condition will change when the production proceeds from time to time and this will cause instability in the performance of the molding machine. Besides, it is difficult to ensure the consistency of material supply. These problems lead to the inconsistency in quality of molded part. Before the production, the initial parameters are set according the result from mold testing in mold making stage. The Digital Twin models of automated inspection system also used to determine and select the corresponding image-processing algorithm and inspection equipment setup before the production starts. During the production, sensors distributed in each physical machine provides updated production data to manufacturing execution system and virtual models. The virtual model feedback the responsive action to physical production for production plan adjustment. The measurement of machine condition such as pressure and temperature corresponding to the machine parameter setting and the real-time result from automated inspection system are input to computational model of Digital Twin for the purpose of process optimization and quality prediction. These data can further analyze and generate maintenance schedule for mold and injection molding machine.

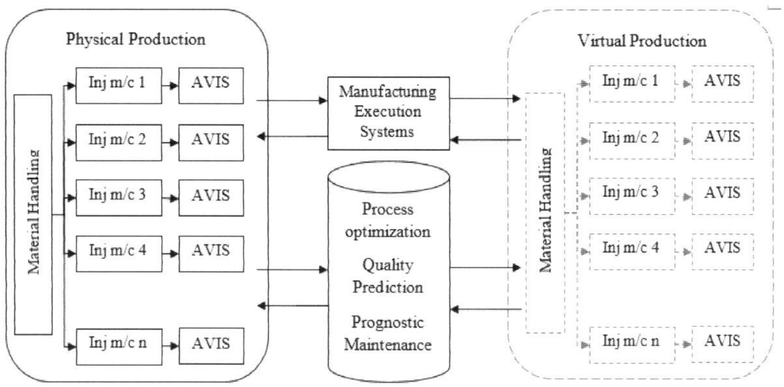

Figure 3. Digital Twin of injection molding process. (Inj m/c: injection molding machine; AVIS: automated visual inspection system)

5. Conclusion

This paper presented the application of Digital Twin concept in injection molding industry from mold design to mold making and finally injection molding process. Technology such as Internet-of-Things (IoT) and Cyber Physical System (CPS) play important role in Digital Twin concept. Some problems in realization of the Digital Twin in injection molding such as manual acquisition of data, implementation of sensors in injection molding machine, development of real-time automated visual inspection systems, optimization and prediction analytics model are yet to be solved to complete the Digital Twin application in injection molding industry.

Acknowledgments

This work was supported by the National Research Foundation of Korea (NRF) grant funded by the Korea government(MEST) (No. 2016R1A2B4014898).

References

[1] Tao F, Cheng J F, Qi Q L, Zhang M, Zhang H, Sui F Y 2017 Digital twin-driven product design, manufacturing and service with big data *The International Journal of Advanced Manufacturing Technology* 1-14.

[2] Grieves M 2014 Digital Twin: Manufacturing excellence through virtual factory replication http://www.apriso.com

[3] Uhlemann T H, Lehmann C, Steinhilper R 2017 The Digital Twin: Realizing the cyber-physical production system for Industry 4.0 *Procedia CIRP* **61** 335-340

[4] Parrott A, Warshaw L 2017 Industry 4.0 and the Digital Twin: Manufacturing meets its match *Deloitte University Press*

[5] Overton J, Brigham JC 2017 The Digital Twin: Data driven simulations innovate the manufacturing process. *White paper DXC Technology*

IOP Conf. Series: Materials Science and Engineering **324** (2018) 012078 doi:10.1088/1757-899X/324/1/012078

Mechanism design and optimization of a bionic kangaroo jumping robot

Y H Zhang, L Zheng, W J Ge* and Z H Zou

School of Mechanical Engineering, Northwestern Polytechnical University, Xi'an, China.

zhangyonghong@nwpu.edu.cn (Y H Zhang); zhenglei281@126.com (L Zheng); zouzhihua94@hotmail.com (Z H Zou); gwj@nwpu.edu.cn (W J Ge).

*Corresponding author to W J Ge, e-mail: gwj@nwpu.edu.cn

Abstract. Hopping robots have broad application prospects in the fields of military reconnaissance, field search or life rescue. However, current hopping robots still face the problems of weak jumping ability and load bearing. Inspired by the jumping of kangaroo, we design a Kangaroo hopping robot "Zbot", which has two degrees of freedom and three joints. The geared five-bar mechanism is used to decouple the knee and ankle joints of the robot. In order to get a bionic performance, the coupling mechanism parameters are optimized. The simulation and experiments show that the robot has an excellent jumping ability and load capacity.

1. Introduction

Jumping robot is a hot branch of robotics research field. Compared with wheeled, legged and tracked robots, they are more adaptive in mountainous, ravine and other complicated terrains with high jumping ability and discreteness of location. In addition, the jumping robot can effectively absorb the landing buffer energy and release it on the next hopping, to realize the reuse of partial energy. Therefore, jumping robot shows some advantages than other robots.

In 1991, the MIT developed the biokangaroo robot "Uniroo", which consists of a body, leg and tail. It has four joints: the tail and hip have collinear pivot axes at the hip, and four actuators for each joint. The robot and its boom together mass about 25kg, and it hops forward about 1m/s [1,2]. SH Hyon and his co-workers of Tohoku University, proposed a simple mechanical model of hind-limb named "KenKen" by mimicking dog hind limbs. Its articulated leg is composed of three links and two hydraulic actuators are adopted as muscles and linear springs as a tendon. It exploits an empirical controller based on characteristic dynamics of the model. The experiment showed that the robot was able hopping well in planar [3]. In 2013, the Festo Bionic Learning Network developed a hopping robot "BionicKangaroo", with a four-bar mechanism leg driven by pneumatic actuators. Elastic tendons connected with actuators in the leg kinematics are used to store the energy and to provide the power for hopping. In addition, the hip and tail joints are equipped with electrical drives to balance and stabilize the hopping cycle [4].

Miniaturization is another trend of hopping robot, because small robots have larger power mass ratio and more flexible characteristic jumping behaviour. Professor Paolo Dario developed a fast-long jumping robot "Grillo", at the taking-off phase, an escapement mechanism releases the drawn spring, and generates a power peak several times higher than that of using motor. In addition, The designer has developed a series of miniature jumping robots [5-7]. Based on the "Grillo", Micro Kovac of Federal Polytechnic Institute developed a 5cm 7g jumping robot. A four-bar leg linkage of robot is connected to the body on the ground link, which is extended via the input link using a torsion spring. The robot was able to jump up to 27 times than its own size. In their further research, the robot with additional wings can travel longer distances when jumping from an elevated starting position and reduce the kinetic energy on impact which needs to be absorbed by the robot structure [8, 9]. MSU jumper, developed by Zhao, can perform continuous jumping with four mechanisms. It charged the energy and performed the self-righting at the same time. The 23.5g robot could jump over 87cm high and 89cm away [10]. In 2016, Duncan W. Haldane, from the University of California at Berkeley, published on Science Robotics and presented a bionic galago robot "Salto". Using a series elastic actuator (SEA), Salto gains a surprising amount of power on the phase of taking-off. Salto was only 100g but can jump up to 1m every 0.58s averagely [11]. Xiaojuan Mo, the doctor of NWPU, China, proposed a jumping robot design scheme based on a four-bar mechanism to simulate the trajectory of the tibiae end of the locust and the knee joint rotation angle, which had been proved to be a good mimic of the movements of locusts jumping [12].

In this paper, we proposed a hopping robot model owning strong environmental adaptability, less freedom degree and strong load capacity. The hopping robot model imitates a kangaroo with thighs, lower legs and foot. In order to increase the kinematic flexibility, the ankle and knee joints are coupled through a geared five-bar mechanism. Two hydraulic cylinders are adopted to drive the robot's hip joint and ankle joint respectively. Secondly, the joint angles of the kangaroo are measured during the jumping, and the parameters of the coupled mechanism are optimized to imitate movement characteristics of kangaroo. Finally, the bionic performance and load-bearing capacity of robot are verified by both simulation and experiment.

2. Design of the mechanism

2.1. Design progress of bionic leg mechanism

Kangaroo hind-legs are very thick. It has a small body rotation angle while leaping in a large gait. It relies on the rapid swings of thighs, calves and feet to jump. Kangaroo can be seen as a chain of three degrees of freedom mode as figure 1 shows. This model requires three drivers, and the collaborative control is very difficult. Moreover, because of the existence of the hydraulic cylinder at the ankle joint, the lower body mass of the robot is too large. The body inertia at the time of taking off is relatively small, and the driving power is increased. Early our team's experimental studies have proved this mechanism is not effective.

Figure 1. Three-link jumping robot model.

The geared five-bar mechanism is richer in movement and force conversion features than the four-bar mechanism, and it can meet the requirements of a variety of motion trajectory. This paper develops a robot, "Zbot", of knee-ankle coupling based on a geared five-bar mechanism as shown in figure2. l_1,

l_2 and l_3 represent the lengths of Thigh, calve and foot of Zbot. Point P, N and C represent the hip, knee and ankle joints, and their corresponding angles are $\theta_k, \theta_i, \theta_j$. The parallelogram mechanism transmits the motion of the knee to the geared five-bar mechanism. The geared five-bar mechanism can be seen as a movement conversion device, which processed the knee movement with a certain law. Zbot retains the freedom of the hip, which can make the robot leg swing in the air so as to land smoothly. Zbot's knee and ankle are coupled by a mechanism, which not only reduces one degree of freedom of the mechanism but also improves the MA [13, 11]. MA is defined here as the ratio between the reaction force at the foot to the force applied by the actuator [11]. The foot made of silicon steel can increase the robot's power density when the robot takes off because it forms a series elastic actuator (SEA) with the driving cylinder.

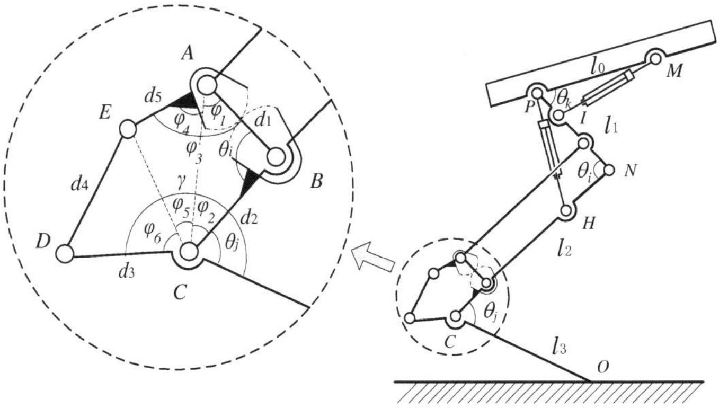

Figure 2. The geared five-bar mechanism.

2.2. Motion relationship between knee and ankle joints

The geared five-bar mechanism shown in Figure 2 is an essential part of the robot. It makes the ankle and knee angle has a certain relationship. The bar *AB, BC, CD, DE, EA* hinge end to end, the two gears at point *A* and *B* are fixed on the *AE* and *BC* respectively, the characteristics of the mechanism can be expressed by nine quantities: d_1, d_2, d_3, d_4, d_5, i, θ_{i0}, φ_{30}, γ. The d_1, d_2, d_3, d_4, d_5 represent the lengths of each linkage, i is the gear ratio. θ_{i0} and φ_{30} are the initial angle of θ_i and φ_3, Since *BC* and *EA* are connected through the gear pair, we can get the following equations:

$$\varphi_3 = \varphi_{30} + i \cdot (\theta_i - \theta_{i0}) \tag{1}$$

If d_{AC} is defined as the distance between point *A* and point *C* (The same below), the following equation can be drawn from the sine and cosine theorems of triangles.

$$d_{AC}^2 = d_1^2 + d_2^2 - 2d_1 d_2 \cos\theta_i \tag{2}$$

$$\frac{\sin\varphi_1}{d_2} = \frac{\sin\theta_i}{d_{AC}} \tag{3}$$

Equation (3) can be written in the following form.

$$\varphi_1 = \arcsin\left(\frac{d_2}{d_{AC}} \sin\theta_i\right) \tag{4}$$

And

$$\varphi_2 = 180° - \theta_i - \varphi_1 \tag{5}$$

In $\triangle ACE$, The following equation can be drawn.

$$\varphi_4 = \varphi_3 - \varphi_1 \tag{6}$$

$$d_{CE}^2 = d_5^2 + d_{AC}^2 - 2d_5 d_{AC} \cos\varphi_4 \tag{7}$$

IOP Conf. Series: Materials Science and Engineering **324** (2018) 012078 doi:10.1088/1757-899X/324/1/012078

$$\frac{\sin\varphi_4}{d_{CE}} = \frac{\sin\varphi_5}{d_5} \tag{8}$$

Equation (8) can be written in following form.

$$\varphi_5 = \arcsin(\frac{d_5}{d_{CE}}\sin\varphi_4) \tag{9}$$

And in $\triangle ECD$, The following equation can be drawn.

$$d_4^{\,2} = d_3^{\,2} + d_{CE}^{\,2} - 2d_3 d_{CE}\cos\varphi_6$$

$$\varphi_6 = \arccos(\frac{d_3^{\,2} + d_{CE}^{\,2} - d_4^{\,2}}{2d_3 d_{CE}}) \tag{10}$$

According to Figure 2 in the relationship between the angle can be drawn.

$$\theta_j = \gamma - (\varphi_2 + \varphi_5 + \varphi_6) \tag{11}$$

From Equation (3~11), the relationship between θ_i and θ_j can be derived.

$$\theta_j = f(\theta_i) \tag{12}$$

Equation (12) is the function between ankle and knee joints, which contains parameters: d_1, d_2, d_3, d_4, d_5, i, θ_{i0}, φ_{30}, γ . These parameters closely affect the knee-ankle movement. The value of each parameter corresponds to a movement relationship. A reasonable set of structural parameters would make the robot knee ankle joint motion characteristics achieve the best. The mechanism will be optimized in the next section to find out a set of structural parameters that make the jumping robot has bionic motion characteristics.

3. Biological observation and optimization of mechanism

3.1. Kangaroo morphology observation

This article directly regards the knee and ankle angle when kangaroo takes off as the optimization object. Firstly the key positions of the kangaroo are cut out during the whole cycle of kangaroo jumps as shown in figure 1. Then the knee angle and the corresponding ankle angles of each position are measured as shown in table 1. θ_{is} and θ_{js} represent the measured knee and ankle angle respectively.

table 1. Joint measurement of kangaroo's taking off

Kangaroo positions	Measured knee angles θ_{is} (°)	Measured ankle angles θ_{js} (°)
Position 1	95	82
Position 2	112	95
Position 3	120	125
Position 4	125	130
Position 5	132	152

3.2. Optimization of mechanism

The optimization of geared five-bar mechanism is to find out a set of parameter values that makes the movement of Zbot's knee and ankle imitate the corresponding movement of the kangaroo. If θ_{jk} represents the ankle angle value of the mechanism in position k, and θ_{jsk} represents the measured ankle angle value of the kangaroo in position k, the expression of the optimized object function can be written as follow.

$$\min f(\gamma, d_1, d_2, d_3, d_4, d_5, i, \theta_{i0}, \varphi_{30}) = \sum_{k=1}^{5}\left|\theta_{jk} - \theta_{jsk}\right| \tag{13}$$

Proportionally enlarging or reducing the size of the geared five-bar will not change the performance of mechanism as long as the initial posture of the mechanism unchanged. So according to

IOP Conf. Series: Materials Science and Engineering **324** (2018) 012078 doi:10.1088/1757-899X/324/1/012078

the overall size of the robot, we make d_1 equal to 54mm, θ_{i0} equal to 80°. The initial value and range of each parameter are determined by the size of the robot, which are shown in table 2.

Table 2. Optimized variables and their ranges.

Optimized variables	Range of optimized variables	Optimized results
d_1	54 mm	54 mm
d_2	50 mm ~ 80 mm	56 mm
d_3	20 mm ~ 80 mm	60 mm
d_4	20 mm ~ 80 mm	60 mm
d_5	20 mm ~ 50 mm	32 mm
i	1/3、1/2、1	1
θ_{i0}	80°	80°
φ_{30}	90°~ 120°	95°
γ	180°~ 250°	198°

The enumeration method is used for optimization. 1) The values of each independent variable are substituted into (12) to get the function between the ankle and knee. 2) The values of the knee angles of position 1~5 are respectively substituted into the function, and the mechanical ankle angles of each position are obtained. 3) Comparing these ankle angles with the corresponding measured ankle angles. the difference of two sets of ankle angles will be got. 4) Take the values of different independent variables and repeat the above process until the set of variables appear that make the difference minimum.

After the optimization, a best set of independent variables that make the bionic characteristic of the mechanism come to be obtained in table 2. The minimum value of the objective function is 18.9°. The deviation existing is caused by many reasons, such as the gap of each parameter and the discrete sampling points. The unavoidable error makes the optimization precision have a slight deviation, but the mechanism can still better imitate the position of kangaroo jumping.

By substituting the optimized results into (12), we can obtain ankle angle on the knee angle expression, and comparing optimized curve with the measured discrete point of knee-ankle joint in the coordinate, the figure 3 is obtained, which reveals the motion relationship of the knee and ankle during kangaroo jumping, The figure 3 shows, 1) The robot can mimic the kangaroo's jumping position to a large extent; 2) Entering a smaller rang of knee motion (80°~128°) can get a larger range of ankle motion(60°~152°), increases of ankle angle is about 1.9 times than that of knee, which shows that the robot has the same acceleration characteristics as the kangaroo while jumping.

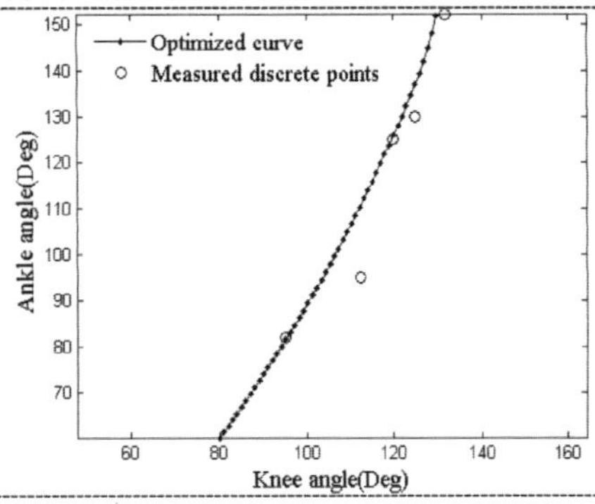

Figure 3. The optimized curve of the mechanism and the measured discrete points.

4. Kinematics simulation and experiments

4.1. Kinematics simulation

According to the results of optimization and the size of each robotic rod, a three-dimensional model of "Zbot" is established in SolidWorks and then imported into Adams. Materials properties, necessary markers, contacts and constraints are set up.

By planning trajectory and solving inverse kinematics of the robot, the displacement curve of the hip and knee cylinders can be obtained, and using them as the driving curve of the Zbot's hip and knee hydraulic cylinder. This article takes the robotic jumping at 3m/s in the direction of 45° as an example. The trajectory of the body center of mass and the robotic position in different moment are shown in Figure 4. As can be seen from the figure 4, Zbot jumps away from the ground at about 0.2 s, and the maximum height of the jumping is 0.5m. Zbot has a bionic kangaroo position and its body moves in a straight line when taking off, which in line with the biological characteristics.

Figure 5 shows the change in Zbot center of mass velocity. As can be seen from the figure 5, the body is evenly accelerated and the maximum velocity is about 3m/s that in line with expectations.

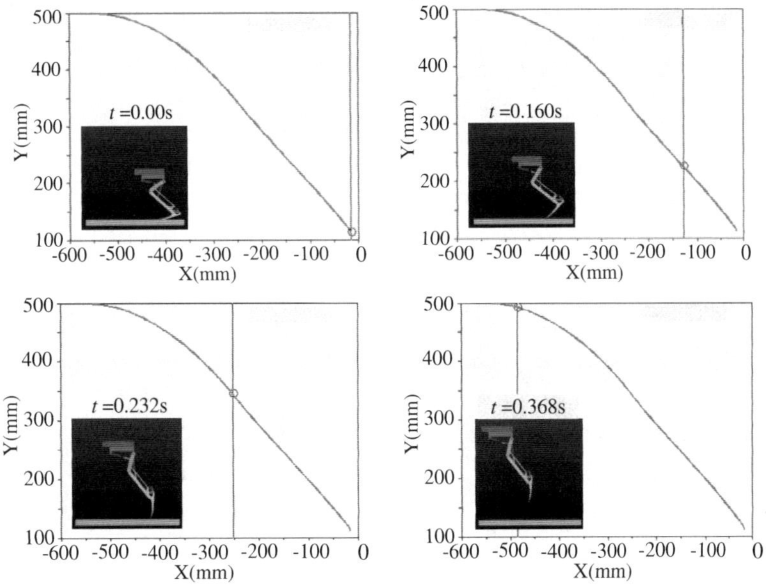

Figure 4. Zbot's taking off in different moment.

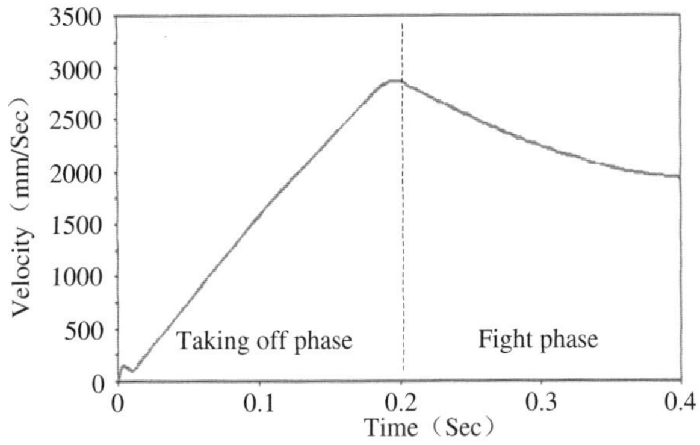

Figure 5. Centroid velocity of Zbot's taking off.

4.2. Experiments

Based on the results of the optimization of the mechanism, the prototype Zbot is fabricated as shown in figure 6. The overall size is 375mm × 380mm, and the material is mainly aluminum (6061). Zbot

IOP Conf. Series: Materials Science and Engineering **324** (2018) 012078 doi:10.1088/1757-899X/324/1/012078

weighs 4.5kg and is driven by two hydraulic cylinders at ankle and knee joints. In order to verify the robot's load-bearing capacity, ten weights (total 4kg) are fixed on the body of Zbot.

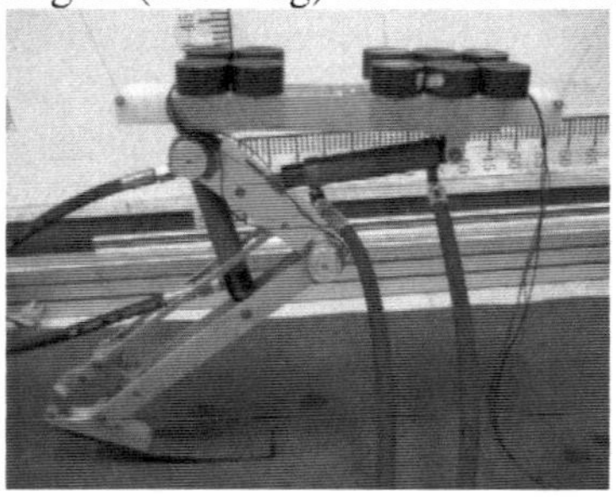

Figure 6. The prototype of Zbot.

Figure 7 is the video capture of robotic taking off, which shows that Zbot had a bionic position and had an excellent jumping ability under loaded conditions. The height it jumps is from 375mm to 880mm, which is 1.3 times than its own size. The horizontal distance it moves in half cycle is about 470mm. This performance is better than any other existing continuity jumping robot driven by hydraulic cylinders. This experiment proved that Zbot has good locomotion ability. Moreover, the weight it fixed on (4kg) approaches that of its own (4.5kg), which explains the robot has a good load capacity. It is worth mentioning that the load capacity of jumping robot is very important in future application, but few researches focus on this point. We will make continuous research on this. The jumping shown in Figure 7 is just one kind of hopping modes of Zbot. In addition, the robot can jump in different directions at different speeds by changing the motion of two hydraulic cylinders.

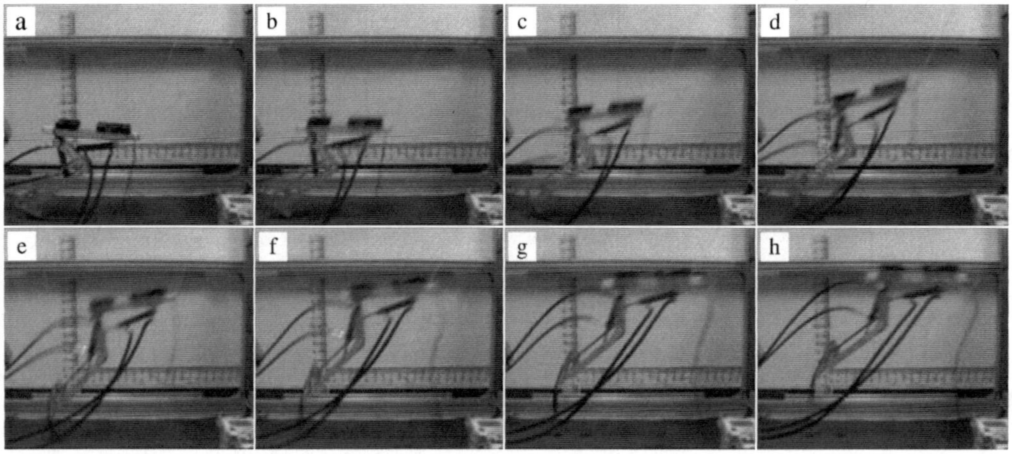

Figure 7. Video capture of Zbot's taking off.

5.Conlusion.

This paper presents a two-degree-of-freedom jumping robot "Zbot", which is inspired by the kangaroo jumping. The use of a geared five-bar mechanism to couple the robot's knee and ankle makes the robotic movement more flexible. The mechanism is optimized based on the position measurement of the kangaroo during take-off. The robot simulation and experiments were performed, and the results proved the feasibility of the mechanism.

Through the research stated in this paper, the following conclusions can be drawn: Zbot can not only imitate kangaroo jumping, but also has the ability of bear weight. It provides a new idea for the research of hopping robot.

In the future work, we will study the position adjustment of the robot during the fight phase and tail-based balance control, as well as the stability of the robot landing phase. We believe Zbot will have a very bright future in application.

Acknowledgments
The authors are very grateful to Dr. Zhao with building experimental platform. And authors would like to thank doctoral foundation at the University of China for financial support (fund number 20136102130001).

References
[1] Zeglin G U. Uniroo – A one legged dynamic hopping robot [D]. *Cambridge*, USA: MIT, 1991.
[2] Briggs R, Lee J, Haberland M, et al. Tails in biomimetic design: Analysis, simulation, and experiment[C]// *IEEE/rsj International Conference on Intelligent Robots and Systems*. IEEE, 2012:1473-1480.
[3] Hyon S H, Mita T. Development of a biologically inspired hopping robot-"Kenken"[C]// *IEEE International Conference on Robotics and Automation, 2002. Proceedings*. ICRA. 2002:3984-3991 vol.4.
[4] Graichen K, Hentzelt S, Hildebrandt A, et al. Control design for a bionic kangaroo[J]. *Control Engineering Practice*, 2015, 42:106-117.
[5] Scarfogliero, U, Stefanini, C, Dario, P. Design and Development of the Long-Jumping "Grillo" Mini Robot[C]// *IEEE International Conference on Robotics and Automation*. IEEE, 2007:467-472.
[6] Scarfogliero U, Stefanini C, Dario P. A bioinspired concept for high efficiency locomotion in micro robots: the jumping Robot Grillo[C]// *IEEE International Conference on Robotics and Automation. IEEE*, 2006:4037-4042.
[7] Scarfogliero U, Stefanini C, Dario P. The use of compliant joints and elastic energy storage in bio-inspired legged robots[J]. *Mechanism & Machine Theory*, 2009, 44(3): 580-590.
[8] Kovac M. Bioinspired Jumping Locomotion for Miniature Robotics[J]. *Epfl*, 2010.
[9] Kovac M, Fuchs M, Guignard A, et al. A miniature 7g jumping robot[C]// *IEEE International Conference on Robotics and Automation*. IEEE, 2008:373-378.
[10] Zhao J, Xu J, Gao B, et al. MSU Jumper: A Single-Motor-Actuated Miniature Steerable Jumping Robot[J]. *IEEE Transactions on Robotics*, 2013, 29(3):602-614.
[11] Haldane D W, Plecnik M M, Yim J K, et al. Robotic vertical jumping agility via series-elastic power modulation[J]. *Science Robotics*, 2016, 1(1): eaag2048.
[12] Mo X, Ge W, Wang S, et al. Mechanical Design and Dynamics Simulation of Locust-Inspired Straight Line Four-Bar Jumping Mechanism[C]//*Mechanism and Machine Science:* Proceedings of ASIAN MMS 2016 & CCMMS 2016. Springer Singapore, 2017: 429-442.
[13] T. J. Roberts, R. L. Marsh, Probing the limits to muscle-powered accelerations: Lessons from jumping bullfrogs. *J. Exp. Biol*. 206, 2567–2580 (2003).

Economic Optimization Analysis of Chengdu Electric Community Bus Operation

Wang Yidong, Cai Yun, Tan Zhengping, Wan Xiong
School of Automobile and Transportation, Xihua University, Chengdu 610039 China
E-mail: 453152438@qq.com

Abstract. In recent years, the government has strongly supported and promoted electric vehicles and has given priority to demonstration and popularization in the field of public transport. The economy of public transport operations has drawn increasing attention. In this paper, Chengdu wireless charging pure electric community bus is used as the research object, the battery, air conditioning, driver's driving behavior and other economic influence factors were analyzed, and optimizing the operation plan through case data analysis, through the reasonable battery matching and mode of operation to help businesses effectively save operating costs and enhance economic efficiency.

1. Introduction

At present, many scholars at home and abroad have done research on the economy of new energy vehicles. These studies can be divided into two categories in genera: the first category is economic evaluation from the point of energy conservation and emission reduction or energy utilization, such as Yang Feng and other scholars, using the "energy rate" as an indicator, selecting Toyota RAV4 electric cars and fuel cars as the evaluation object and analysising the economy of both, it was concluded that the energy consumption cost of pure electric vehicle is superior to that of traditional fuel vehicle under the same condition. Wang Zhenpo and other scholars put forward that the energy utilization ratio is a comprehensive evaluation parameter for the power performance and energy consumption economy of pure electric vehicles, which provides relevant theoretical basis for energy - saving optimization of electric vehicles. The second type is economic evaluation and analysis based on the flow of vehicle lifetime funds. For instance, Zeng Ming[1] and other scholars set up the evaluation model of the economic benefits and environment of HEV from the entire life cycle of the automobile, Which concluded that HEV will be more competitive in the future, and they made some suggestions for the future development of electric vehicles. Wan Jian[2] and other scholars made a dynamic analysis of the economics of new energy vehicles and established an economic analysis model based on the time value of funds. They also made an economic comparison and analysis of the three buses BEB, BEB and CDB. These scholars have made a good theoretical basis for the economic research of pure electric vehicles and which also have good reference value, but these lack the research on the operating economy of pure electric public vehicles. Therefore, this paper combines with the actual situation of Chengdu wireless charging pure electric community bus operation and analyzes its economy, and it is very important and valuable to economically optimize the operation plan.

2. Economic influence factors of pure electric community bus

The difference between the economic evaluation of a pure electric community bus and a traditional fuel bus mainly lies in the following aspects: First, during the driving process of the traditional fuel buses,the time of refueling is short and the number of refueling times is less, while the electric community buses need to be charged, the charging time will affect the electric buses normal departure frequency, and they need to be recharged many times every day, therefore , in the case of the same amount of transport , the total number of pure electric buses will be increased with respect to the total use of conventional fuel buses. Secondly, the pure electric bus needs to build more charging facilities, the costs of pure electric bus maintenance are relatively high, and the power battery is very expensive. Third, the energy consumption evaluation of pure electric buses is different from that of traditional vehicles under different operating conditions and different speeds, the driver's operating level, the weather environment, traffic control, load and so on will have some impacts on energy consumption.

2.1 The impacts of power battery on the economy of electric bus

The only source of energy for electric community buses is the battery. On the one hand, the discharge of the battery is largely affected by temperature. On the other hand, the air conditioner of the electric bus consumes a large amount of energy. The opening of the air conditioner seriously affects the driving range of the bus. Ambient temperature for air conditioning requirements are different, buses don’t have to open the air conditioner in some seasons, such as most of the time in Chengdu in spring and autumn buses don’t open the air-conditioned. Therefore, whether the air conditioning turned on and power battery performance are important factors affecting the economy of electric buses.

In recent years, the rapid development of pure electric vehicles, lithium-ion batteries are widely used, from the power considerations, the more mature lithium-ion batteries are mainly two kinds, namely the ternary materials and lithium iron phosphate batteries. Comparatively speaking, the energy density of ternary material is more high, the technology and safety of lithium iron phosphate battery is relatively good. Considering all aspects, the city buses mainly use the lithium iron phosphate batteries. At present, the price of lithium-ion battery is still high, so that the price of pure electric vehicles enhance a lot. From the market conditions in 2015, the domestic price of lithium iron phosphate batterry is about 2000 yuan / Kwh, only the cost of the battery accounts for 40% to 50% of the cost of the vehicle, there is still a lot of technical space in terms of cost.

2.2 Economic impact of air conditioning on electric buses

In electric vehicles, the total system energy consumption of air conditioning accounted for more than 60% of the total auxiliary system, which has a great impact on the economy of electric vehicles, therefore, electric vehicles on the air conditioning energy efficiency requirements are very high. In theory, the driving range of electric vehicles when turning on the air conditioners is related to the speed of the electric vehicles, and the driving range increases as the speed increases. The reason is relatively simple, assuming that driving 100km distance, If the speed of 50 km / h goes through the entire journey and the air conditioning takes 2 U, the whole journey of 100km/h is reduced by half as time.

However, in actual vehicle driving, the higher the driving speed is, the longer the driving range is. However, the higher the speed is, the lower the power consumption per unit air conditioner is, but the greater the driving resistance (wind resistance, rolling resistance, etc.) the higher power consumption of per mileage overcoming the driving resistance.If the electric car turn on the air conditioner, then the driving range will greatly reduce. Therefore, in the economic evaluation of electric vehicles, the opening or closing of the air conditioning has a great impact on the results.

2.3 The impact of driving behavior on energy consumption

Because the electric vehicles are motor-driven, electric community buses do not have idling conditions and their operating characteristics can be reduced to acceleration, deceleration and uniform speed. The operation change of the driver in the constant speed process is less ,transient characteristics of the vehicle is not obvious, thus the uniform process of driving behaviors have little effect on energy consumption, The main effect of energy consumption is the behaviors of acceleration and deceleration, and the classic acceleration or deceleration of the pure electric community bus is the process of entering and leaving the station. During the process of outbound, the driver has an expected value for the final speed of the station. The expected speed has a certain relationship with the route. The higher the expected value, the faster the driver accelerates and the faster the station exits. The more energy is consumed, Process. The distance between bus stops is fixed. The higher the proportion of high accelerations is, the more the number of quick brakes is, the greater the fluctuation of positive and negative accelerations are, and the worse the driver's driving habits are, then the poorer the energy consumption performance is.

3. Economic optimization analysis of operational electric bus

The model of the study is a pure electric bus produced by Chengdu Shudu Bus Company. The electric bus is loaded with 30 people and the battery type is lithium iron phosphate battery. The nominal capacity of the battery pack is 39.27 Kw•h, and the charging mode is wireless charging. The operation route is Chengdu 1058 community bus operation route, and 1058 is the first commercial bus line of wireless charging community bus in China. There are 2 charging stations in the whole line. One is set at the starting point and the other is set at the intermediate station.

3.1 Test data collection

Pure electric buses are much more expensive than conventional fuel buses. Therefore, the actual cost of a vehicle is an important factor in its analysis of operating costs. In the case of meeting the operating requirements,the minimum amount of vehicle input and the lowest battery cost should match. In order to draw the best economic plan, we need to test and record the trial operation of 1058 pure electric community bus firstly. Testing parameters include mileage, driving time, the remaining power of car recorder, the controller temperature, power meter readings, charging time, data records shown in Table 1. Full load during the test, but do not stop by the passengers.

Table 1 .Test data for pure electric Community bus

Times	Driven mileage	Travel time	Meter reading before charging	Meter reading after charging	Remaining battery charge
1	22km	1h56min	163.7Kw•h	178.1Kw•h	70%
2	22km	1h30min	178.1Kw•h	196.8Kw•h	61%
3	18km	1h14min	196.8Kw•h	212.5Kw•h	66%
Times	Charging time	Driving laps	road condition	Charging station to the starting station	Terminal back to the charging station
1	36min	4	Off-peak time	4min	10min
2	41min	4	Peak time/Off air conditioning	2min	7min
3	41min	3	Peak time/On air conditioning	3min	8min

3.2 Calculation of optimal vehicle configuration number

3.2.1 power consumption calculation of peak time and Off-peak time

- Peak time without air conditioning running power consumption per kilometer of electricity = (reading after the second charge - reading after the second charge) / second mileage = (196.8-178.1) /22≈0.85Kw•h/Km
- Off-peak time air conditioning run per kilometer consumption of power grid = (after the third charge reading - the third charge after reading) / third mileage = (212.5-196.8) /18≈ 0.87Kw•h/Km
- Off-peak power consumption per kilometer peak energy = (after the first reading of the charge - the first charge before reading) / first mileage = (178.1-163.7) /22 ≈ 0.65Kw•h/Km

3.2.2 Calculation of turnover time from peak time to off-peak time

(1) Calculation of stop time

Chinese Automotive Technology Center statistics, the cost of our typical city road bus idle time for the entire circulation time of 27.9%, because the electric vehicle is driven by a motor, so the electric community bus does not exist idle condition, operating characteristics can be simplified as acceleration, deceleration, speed.1058 bus route for clockwise loop running. Don't wait for the red lights, this paper considered idle time is equal to stop time. In Table 1, the average one-way travel time is 25min, the total idle time (stop time) “n” can be obtained by the following formula:

$$27.9\% = \frac{n}{n+25}$$

Calculated by the formula can be “n = 9.7min”, therefore, one-way travel time:

- one-way time=one-way travel time+Total stop time=9.7+25=34.7 min
- stop time=Total stop time/Number of intermediate stations= （9.7×60） /17=34 s

(2) Calculation of turnover time

The departure interval of Chengdu public transport morning and evening peak time is about 5min / trips, The departure interval of off-peak is about 10min / trips. Bus 1058 runs at 34.7min in one-way travel (lap time). Peak-time stop time of the start and termius is 5min. Therefore:

- Turnover time of peak time=one-way time+peak time stop time of the start and termius=34.7+5=39.7min

- Turnover time of off-peak time=one-way time+off-peak time stop time of the start and termius=34.7+10=44.7min

In order to facilitate the calculation, the turnaround time will be rounded by 18s, that is, the turnaround time of each bus peak will be 40 minutes and the peak turnaround time will be 45 minutes. In other words, it takes 40 minutes for each bus to start another bus after the rush hour. A turnaround time 40min departure situation shown in Table 2.

Table 2. Distribution of electric buses at peak time

Time（min）	0	5	10	15	20	25	30	35	40
Departure bus	1st	2nd	3rd	4th	5th	6th	7th	8th	1st

(3) At least the required number of buses

Each peak time lasts 2 hours. According to Table 2 above, only 8 community buses are required to operate for 3 laps per vehicle to meet the normal operation of the line. Due to less traffic at off-peak time, departure interval will be longer, and 8 community buses are sufficient to operate.

3.3 Minimum battery capacity ratio

3.3.1 Each car daily power consumption calculation

The total distance of 1058 community bus is 5km, starting from the charging station to the origin station and from the terminal back to the charging station total mileage is 3km. Driving laps at peak time = (60×4) /40≈6 laps, driven mileage at off-peak time = (60×10) / 45≈13 laps. Air conditioning power consumption per kilometer is about 0.2Kw•h

- Driven mileage at peak time =Driving laps at peak time×One-way distance=6×5=30Km
- Driven mileage at off-peak time =Driving laps at off-peak time × One-way distance=13×5≈65km
- Daily consumption of electricity per car per day = Driven mileage at peak time × Energy consumption per mileage during peak time of air conditioning+ Driven mileage at off-peak time ×Energy consumption per mileage during off-peak time + the mileage of returning to the station × energy consumption during off-peak time=30×0.87+65×0.67+3×0.67=71.66Kw•h

3.3.2 Charging time of Average per unit

The 1058 Shudu bus pure electric community bus nominal voltage is 377.6V, the nominal discharge capacity is 104mA • h, it can be calculated that the Battery nominal charge = nominal voltage × nominal discharge= 377.6V × 104mA ·h= 39.27kw ·h.

- The first test battery power consumption = (1-70%) × 39.27Kw • h = 11.78Kw • h = the first charging amount
- The second test battery power consumption = (1-61%) × 39.27Kw • h = 15.32Kw • h = the second charging amount
- The second test battery power consumption = (1-61%) × 39.27Kw • h = 15.32Kw • h = the second charging amount

The charge time of per battery unit can be obtained by the Table 1:

- The first test per unit charging time = first charging time / first chargeing amount = 36 × 60 / 11.78≈183.36s / Kw • h.
- The second test per unit charging time = second charging time / second charging amount

= 41 × 60 / 15.32≈160.57s / Kw • h.

- The third test per unit charging time = the third charging time / third charging amount = 41×60 / 13.35≈184.27s / Kw • h.
- Charging time of Average per unit = total charging time / total charging amount = (36 + 41 + 41) × 60 / (11.78 + 15.32 + 13.35) = 175.03 s / Kw•h.

3.3.3 Daily recharging time and recharging amount

According to the calculation in 3.3.1, the total daily electricity consumption of this community bus is 71.66Kw • h, the intermediate stations stop for 34s and the recharge time is 30s.Recharge time every day = Recharge time in middle station + Recharge time in starting station = Laps count per day × Recharge time in middle station per lap + Recharge time in start station at peak time + Recharge time in start station at off-peak time = 19 × 30 + 6 × 5 × 60 + 12 × 10 × 60 = 9570s.

Therefore, the battery charging amount per car = daily recharging time / average per unit charging time = 9570 / 175.03 = 43.38Kw•h.

3.3.4 The minimum battery capacity.

The minimum battery charge = Total daily battery charge - Daily battery replenishment = 60.42-43.38 = 17Kw • h.

Because the battery can not be completely discharged, lithium iron phosphate battery's depth of discharge is about 70%, the battery nominal charge = 17 / 0.7 = 24.3Kw•h.

Because the battery capacity will gradually decay during its used process . In the case of room temperature and standard discharge rate, when the battery standard discharge capacity decays to the nominal capacity of 70% or 80% of the time as the battery end of life time, this paper decay to the nominal capacity of 75% as the battery end of life time, then The actual nominal battery capacity of a pure electric bus is determined as: 24.3 / 0.75 = 32.4 Kw•h.

The test car had a nominal battery pack capacity of 39.27kw•h when it left the factory,but according to the calculation of this paper, actually only 32.4Kw•h to meet operational needs, reducing the 6.87Kw • h, lithium iron phosphate battery price of 2 ~ 5 yuan / wh, take the price of 3 yuan / wh, then each bus can reduce the investment of 6.87 × 1000 × 3 = 20.61 million yuan, the entire line 8 community bus can reduce input 2.0611 × 8 = 165,444 yuan.

The parameter of the single-cell lithium iron phosphate battery for the 1058 electric bus is voltage 3.2V, nominal discharge capacity 104AH, actual nominal capacity (Kw • h) = (battery voltage (V) × battery capacity (AH)) / 1000 , thus, battery voltage = 32.4 × 1000 ÷ 104 = 311.5V. The number of required single cells = battery voltage / cell voltage = 311.5 / 3.2 = 97, only 97 sections of 3.2V / 104AH small capacity battery in series can meet the normal operation of electric community bus every day.

4. Conclusion

Having based on the analysis of three economic factors such as power battery, air-conditioner and driver's behavior of pure electric community bus and the actual test collection data, the best economical operation plan is analyzed , the optimal vehicle configuration of the bus line and the battery capacity of daily operation, as well as the power charging during operation, are calculated, and the minimum battery capacity ratio is obtained. The result shows that the optimal number of vehicles in this bus line is 8, the minimum bus voltage in each community bus is 311.5V, and the single cell capacity is 104A•h, which is composed of 97 3.2V / 104AH small capacity cells in series to meet the normal operation of pure electric community bus every day, significantly reducing the battery capacity than the original car, reducing the cost of enterprise investment.

References

[1] Zeng Ming, Shi Hui.Calculation model and analysis of full life cycle cost of hybrid electric vehicle [J] .Modern Electric Power, 2014,31 (1).
[2] Wan Jian, Ji Jinzhang.Analysis of new energy bus economy considering the time value of funds [J] .Journal of Highway and Transportation Research and Development, 2015,32 (3).
[3] Ding Nan nan. Optimum Matching and Energy Efficiency Analysis of Power System in Urban Electric Bus [D]. Wuhan University of Technology, 2014

Acknowledgments

Supported by Project of Science and Technology Support Program of Sichuan Province （2016GZ0017）， Key Laboratory of Automotive Engineering in Sichuan Province（szjj2014-072）， Chengdu Science and Technology Project（2016-RK00-00151-ZF）.

Prediction on the charging demand for electric vehicles in Chengdu

Cai yun[1], Zhang wanquan[2] ,You wei[3], Mao pan[1]
1.School of Automobile and Transportation, Xihua University, Chengdu 610039 China
2. The General Assembly Department Of China Aerospace Science and Technology Corporation Seventh Academy, Chengdu 610100 China
3. The Seventh Design Department Of China Aerospace Science and Technology Corporation Seventh Academy, Chengdu 610100 China
E-mail: 29682131@qq.com

Abstract. The development of the electric vehicle charging station facilities speed directly affect the development of electric vehicle speed. And the charging demand of electric vehicles is one of the main factors influencing the electric vehicle charging facilities. The paper collected and collated car ownership in recent years, the use of elastic coefficient to predict Chengdu electric vehicle ownership, further modeling to give electric vehicle charging demand..

1.Introduction

Electric vehicle charging and switching facilities are an important facility for electric vehicles and are the main factors influencing the development and promotion of the electric vehicle, however, electric vehicle charging demand is one of the main factors which affect the effect of charging station layout, only by accurately predicting the charging demand, can rational layout of the electric vehicle charging station be achieved. There are many kinds of predicting methods for electric vehicle charging demand. In this paper, a systematic method is used to predict the charging demand of electric vehicles in Chengdu.

2.Prediction Idea

The elastic coefficient method is used to predict the number of cars in Chengdu city by 2020, and then establishes the model and combines the current situation to analyze and predict the annual car ownership and charging demand of the private cars, taxis and buses and electric vehicles.

3. Prediction of the car ownership

3.1 The Introduction of the Methods

The Elastic Coefficient Method is an indirect prediction method which is based on the prediction of the development and change of a factor and predicts the development of another

IOP Conf. Series: Materials Science and Engineering **324** (2018) 012080 doi:10.1088/1757-899X/324/1/012080

factor by the elastic coefficient. The Elasticity Coefficient Method is an indirect prediction method for the development of another factor on the basis of the change of one factor. From a mathematical point of view: $E = X \bullet f'(x) \div f(x)$, E means elastic coefficient, that is the limit of ratio $x \bullet f'(x) \div f(x)$ of the relative increment of the function $f(x)$ and the relative increment of the independent variable x is the elastic coefficient of the function $f(x)$. It is seen that the elastic coefficient of the function $f(x)$ is the ratio of the relative change of function and relative change of independent variable, which correctly reflects sensitivity of the changes between them, elastic coefficient $|E| = 1$ shows the relative change between the two and is called a single elasticity, the elastic coefficient $|E| = 1$ means that the relative change of Y is greater than the relative change of X, which is called elasticity, the coefficient of elasticity means that the relative change of Y is less than the relative change of X, which is called inelasticity.

Based on the definition of the elastic coefficient method here, we take the current GDP data change and predict the growth rate of car ownership, and the formula is as follows:

$$R = Rgdp \bullet E$$

$$N_n = (1 + R) \bullet N_{n-1}$$

In the formula, R- Annual growth rate of car ownership

R_{gdp} -GDP Growth rate

E- Elastic coefficient

N_n - car ownership in N year

N_{n-1} - car ownership in N-1 year

3.2Prediction of the car ownership

Although in the new energy vehicles plan of Sichuan province and put forward our electric car development goal, but because of the charging infrastructure, and the limitation of charging technology, for the development of it will be hard to achieve our goals. Therefore, this paper combines relevant theoretical models and analyses the actual status quo to predict the car ownership of Chengdu in the future.

The number of automobile grew slowly due to the slow development of the national economy a few years earlier ago, however, in recent years, the rapid economy develops rapidly, the growth of car ownership has become rapidly, showing explosive trend. But according to the slow growth of GDP, it will make domestic car ownership tend to saturation in the next few years, the growth rate will gradually decline in the future. According to the GDP growth prediction of the National Statistical Bureau respectively 9%, 8% and 7% during "11th Five-

Year" to "12th Five-Year" and "13th Five-Year" .The following table 1 is car ownership of our country from 2008 year to 2004 year, table 2 is GDP situation of our country from 2008 year to 2014 year.

Table1. Car ownership of our country from 2008 year to 2016 year

Year	Car ownership /units	Annual increase/ten thousand units	Annual growth rate/%
2008	5099.6	741.0	17.0
2009	6280.6	1181.0	23.1
2010	7801.8	1521.5	24.2
2011	9356.3	1554.5	19.9
2012	10933.1	1576.8	16.8
2013	13700.0	2766.9	25.3
2014	15400.0	1700.0	12.4
2015	17200.0	1800.0	11.6
2016	19400.0	2200.0	12.7

Table 2. The situation of GDP from 2008 year to 2016 year

Year	GDP/Hundred million yuan	Annual growth/hundred million yuan	Annual growth rate/%
2008	316751.7	27744.7	9.6
2009	345629.2	28877.5	9.2
2010	408990.3	63361.1	18.3
2011	484123.5	75133.2	18.3
2012	534123.1	49999.6	10.3
2013	588018.8	53895.7	10.1
2014	640000.0	51981.2	8.8
2015	689052.0	49100.0	7.6
2016	744127.0	54900.0	7.9

According to table 1 and table 2, the growth rate of the car ownership of China and the annual growth rate of GDP in China can be obtained by combining formula 1-1. The elastic coefficients of the over years are shown in Table 3 below.

Table 3. The elastic coefficient of over years

Year	2008	2009	2010	2011	2012	2013	2014	2015	2016
Elastic coefficient	1.7	2.5	1.3	1.1	1.6	2.5	1.4	1.5	1.6

Combining the formula and principle of the elasticity coefficient, taking into account elastic coefficient in 2009 and 2003 is too large to considered, the average value of the elastic coefficient were 1.4 from 2008 to 2014, So the elasticity coefficient of 2015 is regarded as 1.4. Because this paper considers to predict car ownership in the future, the growth rate of GDP slowly decreases, while car ownership has gradually become saturated value, which there is a transitional phase, elastic coefficient must gradually reduce, so the elasticity coefficient in 2016 is regarded as 1.2 , the elastic coefficient should be less than 1 and be regarded as 0.8 from

2017 to 2020, that is the growth rate of the car ownership is 0.8 times the annual growth rate of GDP . According to the resulting elasticity coefficient and GDP growth rate, based on 2014, car ownership of 2015 to 2020 could be predicted table 4.

Table 4. Car ownership in Chengdu from 2012 to 2020

Year	Growth rate of car ownership /%	GDP growth rate/%	Elastic coefficient	Vehicle population /hundred million units
2014	16.8	12.0	1.4	336.1
2015	18.0	12.0	1.5	396.5
2016	12.32	7.7	1.6	445.3
2017	9.8	7.0	1.4	488.9
2018	7.7	7.0	1.1	526.5
2019	5.6	7.0	0.8	555.9
2020	5.6	7.0	0.8	587.0

According to predicted national car ownership, combining with the proportion of electric car ownership which is predicted by the Ministry of Industry 0.28% and 2.2% in 2015 and 2020. Here, the paper predicts the number of electric vehicles in Chengdu according to the three high, middle and low ratio schemes . Among high, low and middle ratio schemes, 2012, 2015 and 2020, respectively, 0.6%, 2%, 9%, 0.3%, 1.2%, 6%, 0.1%, 0.3% and 3%. Taking into account the electric vehicle due to national policy and its various aspects will certainly present increasing trend of rapid growth during 2014 to 2020 and fully consider the study and take the proportion of electric vehicles from 2012 to 2020. The prediction is shown in Table 5 below.

Table 5. car ownership of the vehicle car over years

Year	situation	2012	2013	2014	2015	2016	2017	2018	2019	2020
ownership	/	240.0	308.2	336.1	396.5	445.3	488.9	526.5	555.9	587.0
different rate	high/%	0.6	0.9	1.3	2	2.5	3.5	5	7	9
	middle/%	0.3	0.6	0.9	1.2	1.8	2.6	3.4	4.6	6
	low/%	0.1	0.14	0.2	0.3	0.5	0.9	1.5	2.2	3
number	high	1.44	2.77	4.36	7.90	11.13	17.11	26.33	38.91	52.83
	middle	0.72	1.84	3.02	4.76	8.01	12.71	17.90	25.57	35.22
	low	0.24	0.43	0.67	1.19	2.23	4.40	7.90	12.22	17.61

Taking into account the actual situation, choosing the data of the low scheme to calculate. According to the introduction of the development stage, in the early stage ,there is the development of the bus, taxi and other fields, the slow development of private cars; but for the long-term, private cars lies in a stage of rapid development and will show explosive growth, while buses and taxis is steadily developing. Based on this situation, this paper hypotheses taxi, bus and private cars accounted for the proportion of electric car ownership as in Table 6 shown in 2014 to 2020, taking into account the actual situation, the paper chooses electric vehicle ownership to predict low scheme to calculate the number of private cars and taxis, table 7 can be obtained.

Table 6. Proportion of cars in the different market

Year	2014	2015	2016	2017	2018	2019	2020
Private car	20%	40%	60%	78%	87%	90%	92%

Taxi	30%	25%	18%	10%	7%	4.5%	4%
Bus	50%	35%	22%	12%	8%	5.5%	4%

Table 7. The number of the Private car, taxi and bus

Year	2014	2015	2016	2017	2018	2019	2020
Vehicle car	6700	11800	22000	42400	75500	118400	172700
Private car	1340	4720	13200	33072	58890	106560	158884
Taxi	2010	2950	3960	4240	5285	5328	6908
Bus	3350	4130	4840	5088	6040	6512	6908

4 Prediction of Charging Demand

The charging demand for electric vehicles is to prepare the layout of the electric charging station, so the prediction of its demand should firstly consider the layout and facilities of the charging station.

4.1 Predicting Premise

1) Now, the average mileage of batteries for electric vehicles is 180 kilometers , But as time goes on, the relevant experts predict that in the next 5 years, electric vehicle mileage will reach 300 km or more, but because the mature time of the technology is full of uncertainty, and Considering the deep discharge of the automobile, the service life of the battery will be affected .

2) the main charging mode of the private car is regular charging, which is supplemented by charging in the public fast charging station. So supposing that the car is charged at the public quick charging station, once from Monday to Friday and once at the weekend.

3) Taking into account the battery life problems and high depreciation costs, and now the technology and other issues, the construction of the public rapid charging station is based on the principle that the main battery replacement is considered, DC fast charge is supplemented , that is, when the vehicle comes to charge, the battery is replaced under normal conditions, and secondly, the fast charge is selected. This will not only reduce the battery load, but also improve the utilization rate of vehicles.

4.2 Normal Charging Demand

Normal charging is usually aimed at private cars. Now, in order to solve the energy and environmental problems, the government has vigorously promoted the development of electric vehicles as a means of green transportation. So as to introduce a variety of policies to promote the popularization of electric vehicles, so for the private users of electric vehicles, the construction of charging facilities should be funded by the government to build charging facilities. For private electric vehicles, charging facilities should be equipped with charging piles in accordance with the ratio 1:1.

The formula for charging demand is as follows:

$$Q=N*S\ mileage/S\ available\ mileage$$

In the formula, Q- Actual charging demand of every day

N- the number of the charging cars

S Mileage- average daily mileage

S Available mileage -Battery mileage available

Regular charging mostly aims at the private cars. The average daily mileage is 60 kilometers, and the battery mileage is 150 kilometers. Therefore, a private car charge 0.4 times a day / day.

4.3 Battery replacement charge demand

For buses, environmental sanitation vehicles and vehicles of other groups usually charge by the battery replacement. If the bus is full of electricity, available mileage is 260 kilometers, while the bus in Chengdu run 230 kilometers per day on average. Therefore, the bus is full of electricity, basically it can run one day, that is, the bus charge per day is 1 times / day. Battery replacement mode is more suitable for buses and other special vehicles to charge, the size of the battery replacement station to meet the jurisdiction of the vehicle battery supply of the second day within the station.

4.4 Electricity demand of public fast charging station

The main service object of public fast charging station is the taxi and the private car that needs temporary power supply. According to the survey, the daily mileage of taxis is about 400 kilometers, while the available mileage of the taxi is 150 kilometers, and the number of charging is 2.67 times per day. Since taxis are sometimes required to maintain and attendance rates are unlikely to be 100%, the attendance rate is 90%. The formula for charging demand for taxis can be derived by formula 3-1.

$$Q=N*S\ mileage/S\ available*C$$

For private cars, the normal charging mode is chose in most cases, and only when the power needs to be temporarily added, the public fast charge can be chose. According to the hypothesis, private cars need temporarily public recharging from Monday to Friday, at least once a week, or 2 times a week at public charging from Monday to Friday . Considering the outside factors such as private car temporary maintenance, the attendance rate of private cars is also 90%. Therefore, the formula for charging electricity demand of private car in public fast charging station is as follows:

$$Q=N*q*C$$

In the formula：Q-charging demand（car/day）；

q-average daily charging demand of every car(from Monday to Friday1/5times every day，weekends 1/2times every day）；

N-the number of the charging private car ；

C-attendance rate；

From the above, you can know that the charging demand on the weekend is significantly greater than Monday to Friday charging demand, In order to avoid too many traveling owners on the weekends and waiting in line too long when charging, when planning the scale of the charging facilities, the requirements of the weekend should be considered to meet the vehicle's electricity demand as a target construction. Due to the driving characteristics of the normal charging service objects, it is possible to build AC charging piles in the special-purpose parking areas and public parking places, and the construction of battery replacement stations in fixed parking areas is suitable for group cars. In view of the location of these two kinds of charging facilities, the existing parking spaces and parking lots can be assured, while the public fast charging stations can not be assured and need to be calculated by prediction. Therefore, this thesis regards the public fast charging station as the main object of the study.

For the taxi, according to the formula 3-2, the formula of the taxi charging demand can be gained: Q=N*2.67*90%; according to the formula 3-3, the formula of the private car charging demand can be gained : Q=N*0.5*90%, charging demand of the taxi and the private electric cars are given from 2014 to 2020 in table 8.

Table8. The charging demand of the private car and taxi from 2014 to 2020

IOP Conf. Series: Materials Science and Engineering **324** (2018) 012080 doi:10.1088/1757-899X/324/1/012080

Year	2014	2015	2016	2017	2018	2019	2020
The charging demand of the private car/car/day	603	2124	5940	14310	29558	47952	72275
The charging demand of the taxi/taxi/	4830	7088	9515	11207	12699	14225	16599

5.Conclusion

This paper combines the development of electric vehicles in Chengdu. In the last few years, the number of car ownership has been collected to predict the number of electric vehicles in Chengdu's 13th five-year plan, and its products are predicted. The model is further established to predict the charging demand of one of the main factors influencing the layout of electric vehicle charging stations. A good setup is provided for the distribution of charging stations for electric vehicles.

References

[1] TONG Fang,LAN Fengchong,CHEN Jiqing. Analysis of Factors Affecting the New Energy Vehicle Development and Ownership Forecast[J]. Science and Technology Management R esearch,2016(17):P112-116

[2]Fu danhua, Guo zhongye. Study on Optimization of site selection for new urban energy vehicle charging pile[J]. Industrial & Science Tribune,2017(11):P75-82

Acknowledgments

Supported by Project of Science and Technology Support Program of Sichuan Province （2016GZ0017）， Key Laboratory of Automotive Engineering in Sichuan Province（szjj2014-072）， Chengdu Science and Technology Project（2016-RK00-00151-ZF）.

IOP Conf. Series: Materials Science and Engineering **324** (2018) 012081 doi:10.1088/1757-899X/324/1/012081

Research on the Applicable Method of Valuation of Pure Electric Used vehicles

Cai yun, Tan zhengping, Wang yidong, Mao pan
School of Automobile and Transportation, Xihua University, Chengdu 610039 China
E-mail: 29682131@qq.com

Abstract. With the rapid growth in the ownership of pure electric vehicles, the research on the valuation of used electric vehicles has become the key to the development of the pure electric used vehicle market. The paper analyzed the application of the three value assessment methods, current market price method, capitalized earning method and replacement cost method , in pure electric used vehicles, and draws a conclusion that the replacement cost method is more suitable for pure electric used car. At the same time, the article also conducted a parametric correction exploration research, aiming at the characteristics of pure electric vehicles and replacement cost of the constituent factors. Through the analysis of the applicability parameters of physical devaluation, functional devaluation and economic devaluation, the revised replacement cost method can be used for the valuation of purely used electric vehicles for private use.

1.Background

In response to the global environment of energy security and environmental pollution, the development of pure electric vehicles has become an important measure for all countries to save energy and reduce emissions. With the improvement of people's awareness of environmental protection, the improvement of charging infrastructure, the implementation of subsidy policy of private individuals buying electric cars, pure electric vehicles in the market continue to be popular. In recent years, the number of electric vehicles also increased rapidly, pure electric used car market transactions more and more. Combined with the theory of asset valuation, commonly used asset valuation methods include current market price method, Capitalized earning method and replacement cost method, etc. These methods have been well applied in the evaluation of traditional fuel vehicles, and the details of the applicable parameters and models have been discussed. However, the application of evaluation methods of pure electric used cars is still blank. As a result, it is very important to study the method of EV evaluation, accurately assess the value of pure electric used vehicles, and Choose what method to evaluate. It plays an important role in the development of pure electric used car market and will indirectly affect the development of the entire automotive industry.

IOP Conf. Series: Materials Science and Engineering **324** (2018) 012081 doi:10.1088/1757-899X/324/1/012081

2.The Development of Pure Electric Used Car Market

2.1 Sales of pure electric used cars

In 2016, the total global sales of new energy vehicles (EV and PHEV) were 774,000, 40% increasing over the same period and accounting for 0.85% of global vehicle sales. In all countries, sales in China, the United States, and the European market have increased rapidly. The total number of sales in the three major markets is 702,100, accounting for 90.7% of the global market.

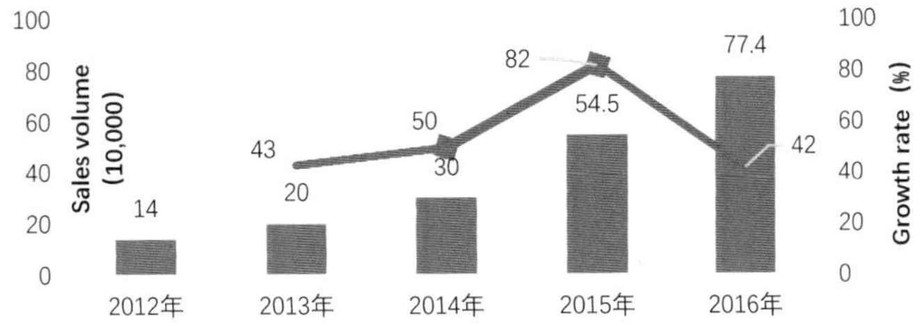

Figure 1 : Global new energy vehicle sales and growth rate in 2012-2016

In recent 3 years, China's new energy vehicles have maintained a rapid growth trend, with an average annual growth rate of 30%. In 2016, the production and sales of new energy vehicles in China were 517,000 and 507,000, creasing 51.7% and 53% respectively. The production and sales of pure electric vehicles were 417,000 and 409,000, accounting for 80.6% of the total output of new energy vehicles in China. In the first 11 months of 2017, China sold 604,300 new energy vehicles.

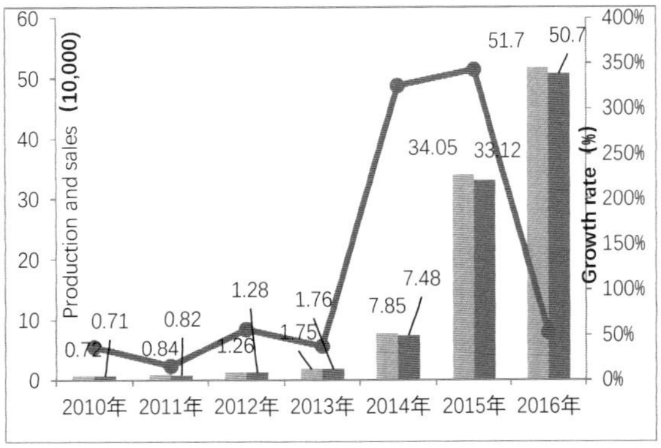

Figure 2 : The production , sales and growth of new energy vehicles in China in 2010-2016

2.2 Problems of pure electric used car industry

With the increase in sales and ownership of new energy vehicles, it has been more than four years since the beginning of 2012 in major cities. This time is close to the average change cycle of most car owners for 6-8 years. In the mainstream used car website, there are already some pure electric car sources, but the price of the same type of car is very different. And from the time of posting information, many pure electric cars were unmanned in the months after they were released, and the value of the pure electric used cars was low relative to the new ones. From the entire electric used car industry, there are three main problems. First, due to the lack of pure electric used car value assessment industry standards and the difference of the components between the pure electric car and the traditional fuel car, we cann't evaluate pure electric cars based on the industry standard of conventional fuel cars. Second, due to the lack of professional practitioners, most of the used car practitioners in our country do not understand the related knowledge of pure electric cars and it is hard to make scientific judgments on the state of the three electric core components. Third, due to the lack of viable pure electric used car value assessment method, there are valuation methods that apply to traditional fuel vehicles on the market. For purely electric vehicles, there is a lack of accepted methods in this field.

3.Applicability analysis of used vehicle evaluation method

3.1 Analysis of the applicability of current market price method

The current market price method is a method to determine the price of the vehicle to be assessed by comparison. It uses the market survey to select one or several vehicles that are the same or similar to the vehicles to be assessed as the reference to analyze the structure, function, performance, old and new, regional differences, trading conditions and transaction prices of the reference objects, Comparison, find out the difference between the two and reflected in the price difference, after adjustment, calculate the price of the old motor vehicle. The prerequisites for application include four aspects. The first is to have an active, fair market with sufficient reference to choose. The second is that at least one recent, comparable, already traded, or already priced reference vehicle has not been concluded. Third is the reference material compared with the evaluation vehicle, the technical parameters, such as the data can be collected, and the price factors are clear and can be quantified. The fourth is to have a relatively sound car trading information system.

For pure electric used cars, this is a new field that comes with the development of electric cars. Because the application conditions of the current market price law are unable to be realized, there is no sufficient reference, no index comparison, quantitative differences, and no fair, sound automobile trading information system. Therefore, if the current market price method is applied on pure electric used cars, it will bring great difficulty to the evaluation.

3.2 Analysis of the applicability of Capitalized earning method

Capitalized earning method, also known as income recovery method, refers to an asset valuation method that determines the value of the vehicle to be assessed by estimating the expected future return on the assets being evaluated. The essence of this method is to convert the future earnings of the automobile into the present value of the asset and use it as the revaluation value of the asset to be evaluated. The basic theoretical formula can be expressed as the revaluation of used cars equals the sum of the expected annual return of the car. The calculation formula as follows:

$$P=\sum_{t=1}^{n}\frac{A_t}{(1+i)^t}=\frac{A_1}{(1+i)^1}+\frac{A_2}{(1+i)^2}+\ldots+\frac{A_n}{(1+i)^n}$$

In the formula, P represents the evaluation value. At represents the expected future income of the future t income period and the residual value of the vehicle is negligible in general estimation. N represents the number of years of income. i represents the discount rate. t represents the period of return, usually by year.

Remaining economic life span refers to the period from the base date of assessment to the time when vehicles reach the end of life, according to "car scrapped standards" to determine the remaining life of the car economy life. For example, the average life span of the fuel taxis is

IOP Conf. Series: Materials Science and Engineering **324** (2018) 012081 doi:10.1088/1757-899X/324/1/012081

calculated in 8 years, and the electric taxis are calculated for 6 years due to the particularity of their battery pack.

Capitalized earning method is more suitable for the pure electric vehicle in the operation of the investment. This method is not applicable to most of the current pure electric cars are used by families for their own use. Therefore, capitalized earning method can only be used to evaluate the value of a few operating pure electric vehicle, and can not be applied to the value evaluation of a non-operating pure electric vehicle.

3.3 Analysis of the applicability of replacement cost method

The replacement cost method refers to the total cost required to re purchase a new state of the vehicle under the current conditions, minus the difference between the old depreciation of the valuation vehicle and the current price of the vehicle being assessed. The old devaluation of the vehicles being assessed includes physical devaluation, functional devaluation and economic devaluation.

$$P = B - D_1 - D_2 - D_3$$

In the expression: P represents the evaluation value of the vehicle being assessed. B represents the reset cost. D1 represents substantial devaluation. D2 represents functional devaluation. D3 represents economic devaluation.

The replacement cost method is practical because of the replacement cost of used cars and all kinds of old devaluation fully being considered. The replacement cost method is used to evaluate the value of used cars, mainly determining the full cost of replacement cost, all kinds of old depreciation, or estimating the depreciation situation by determining the new rate.

In the above discussion of evaluation methods, only the replacement cost method is applicable to the vast majority of pure electric used cars. Because the replacement cost method takes full consideration of the loss of the vehicle, the results of the evaluation tend to be more fair and reasonable. However, due to the different structure of pure electric vehicles and traditional fuel vehicles, the different market environment, so the value of its influencing factors are different. The total weight of individual assembly parts is larger than that of the total vehicle value. Therefore, when using the replacement cost method to evaluate the value of the pure electric second-hand vehicle, the original parameters should be improved based on the replacement cost method. This can be more accurate to determine the new rate of pure electric vehicles.

4.Analysis of influencing factors of replacement cost method

The pure electric vehicles have some characteristics that battery components are of great value and the loss of battery, motor and electric control is difficult to be measured directly. Therefore, there are many factors that affect their value.

The replacement cost method can analyze the factors influencing the value of pure electric vehicles from three aspects: entity depreciation, function devaluation and economic devaluation

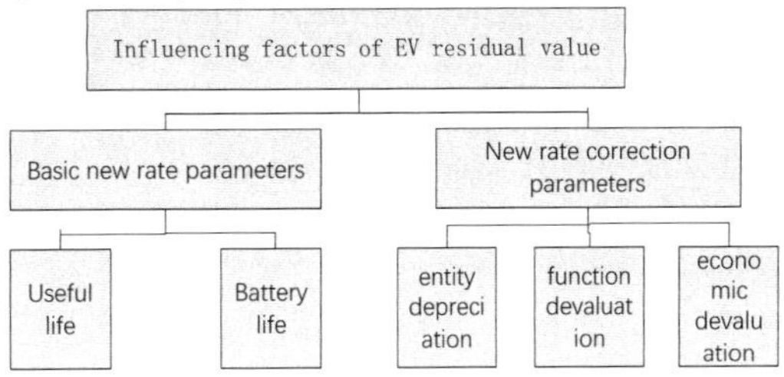

IOP Conf. Series: Materials Science and Engineering **324** (2018) 012081 doi:10.1088/1757-899X/324/1/012081

Figure 3: Factors affecting the residual value of pure electric vehicles

4.1 Correction analysis of entity devaluation parameters

Due to the limitation of endurance, pure electric cars are mainly used for short distance transportation in working life at present, and the working conditions are mostly urban roads. Therefore, the working conditions of pure electric cars can not be considered.

(1) vehicle condition

Physical loss is more obvious in pure electric vehicles, Solid loss, for example, the battery life of a pure electric car will decline with the increase of charge and discharge times. The detection of pure electric vehicle is divided into 6 main aspects: internal detection, lighting detection, body appearance and chassis detection, battery system detection, drive motor system detection, road test detection.

(2) Maintenance condition factors

The maintenance is different from the comprehensive physical parameters of the vehicle. The comprehensive physical parameters of the vehicle are aimed at the parts of the whole vehicle, while the maintenance status reflects the level of vehicle maintenance and maintenance. The timely maintenance of vehicles can increase the new rate of vehicles. Maintenance records can be determined by the improvement of the maintenance records of the 4S store.

(3) working nature

For pure electric vehicles, their working nature is the same as that of traditional cars, mainly for private use, business (business) and operation. The loss of pure electric cars is also different with different working properties. Based on this situation, different coefficients are required for different work properties.

(4) accident conditions

For vehicles, whether there is a serious accident has a great impact on the value of the vehicle. Because the vehicle repaired by overhaul is usually very poor even though it has been restored to its service performance, so it is difficult to restore the original technology and the possibility of expansion of the fault location.

4.2 Correction analysis of function devaluation parameters

The factors that affect the functional devaluation mainly include the stability of the new car price and the rate of the upgrading of the model. In China, the pure electric car is still in its development stage, and has not yet formed its scale. If we want to make technological breakthroughs and reduce costs, it will be very difficult for new models to replace, so the price of new cars will hardly change much. The influence factors of this part are often related to the market share, which can be considered in combination with the factors of economic devaluation.

4.3 Correction analysis of economic devaluation parameters

Economic devaluation is the impact of macro environment on the value of used cars, such as market factors, national policies, etc. For pure electric used cars, besides the brand value and supply and demand factors, the factors that affect their economic depreciation should include factory quality assurance policy and market share.

(1) Brand value

The impact of the brand on the residual value of a pure electric vehicle has nothing to do with the vehicle's own car condition. At the present stage, the electric vehicle brand in our country is less. We can divide it into three levels, one is the joint venture brand, the other is the high awareness of the independent brand, and the third is the general brand in the independent brand.

(2) Manufacturers warranty policy

In the same brand awareness of the models, its manufacturers warranty policy also affects the evaluation of pure electric cars. In the quality assurance policy of the manufacturer, BYD is a life-long warranty of the electric core, and the whole vehicle enjoys a super long warranty of 6 years /15 million kilometers. Bei qi new energy vehicles can be replaced by the old new, providing a vehicle for 3 years /12 million kilometers, the core components for 5 years /15 million kilometers of quality assurance services. According to these different quality assurance policies, different coefficients are chosen.

(3) market share

In the assessment of pure electric used cars, the market share has a certain impact on the value of the car. For cars with high market share, their used cars will be favored by consumers.

(4) supply and demand

In the used car market, the used cars with large supply and demand tend to have a relatively high evaluation price.

5. Conclusion

Through the application analysis of the existing value evaluation method in pure electric used car evaluation, replacement cost method is more feasible, due to more factors and more reasonable. At the same time, in the application of replacement cost method to evaluate the pure electric second-hand vehicle, we can modify the three main factors of entity depreciation, functional devaluation and economic depreciation. On the basis, the revised reposition cost method can be used to evaluate the value of pure electric used cars for private use.

Reference lists

[1]Zhou you. Study on the value evaluation method of used cars based on the replacement cost method [D]. Liaoning University of Technology, 2014

[2]GAO Wei, DENG Zhaowen, YU Yang. Improvement of Production Costs Method for Second-Hand Car and Its Evaluation System Design[J]. Tractor & Farm Transporter,2015 (6) :P32-35

Acknowledgments

Supported by Project of Science and Technology Support Program of Sichuan Province (2016GZ0017), Key Laboratory of Automotive Engineering in Sichuan Province (szjj2014-072), Chengdu Science and Technology Project (2016-RK00-00151-ZF).

IOP Conf. Series: Materials Science and Engineering **324** (2018) 012082 doi:10.1088/1757-899X/324/1/012082

Research on the Value Evaluation of Used Pure Electric Car Based on the Replacement Cost Method

Tan zhengping, Cai yun, Wang yidong, Mao pan
School of Automobile and Transportation, Xihua University, Chengdu 610039 China
E-mail: 29682131@qq.com

Abstract. In this paper, the value evaluation of the used pure electric car is carried out by the replacement cost method, which fills the blank of the value evaluation of the electric vehicle. The basic principle of using the replacement cost method, combined with the actual cost of pure electric cars, puts forward the calculation method of second-hand electric car into a new rate based on the use of AHP method to construct the weight matrix comprehensive adjustment coefficient of related factors, the improved method of value evaluation system for second-hand car

1. Background

The dual pressure of energy and environment has prompted China to vigorously develop new energy vehicles, making more and more new energy vehicles flowing into the market. According to the data of China Automotive Industry Association, by the end of October 2017, the number of new energy vehicles in China was 1 million 580 thousand, of which 1 million 143 thousand were pure electric vehicles, accounting for 72.3% of the total amount of new energy vehicles. The increasing number of new energy vehicles will activate the second-hand car market for new energy vehicles, especially the second-hand car market for pure electric vehicles.

At present, the research on the evaluation method of used car is mainly aimed at the traditional car. Chen Junyi and other based on the basic principle of replacement cost method, using the least square method of fuzzy evaluation method to calculate the comprehensive adjustment coefficient weight to evaluate the value of second-hand car. Yang Yayi and otherset up a second-hand car loss evaluation model to determine the loss degree of second-hand cars by using the grey relational degree theory to make a comprehensive analysis of the second-hand car loss evaluation model. Lin Yilin and other use the analytic hierarchy process (AHP) and Delphy scoring method to calculate the weight coefficient of each index factor in the value evaluation, so as to determine the remaining value of the used car. However, compared with traditional cars, the value components of new energy vehicles have changed significantly. The value of batteries, motors and electronically controlled vehicles has accounted for more than 50%, especially the depreciation of battery consumption is no longer synchronized with vehicle. New energy evaluation methods are needed for second-hand new energy vehicles, but there is no research on the valuation methods of second-hand new energy vehicles, so it is necessary to carry out this research. In this paper, a used pure electric car is taken as an example to carry out the study of its value evaluation method.

2.overview of the replacement cost method

Used cars are a class of products of fixed assets machine equipment. Like other assets assessment, the general theory of asset assessment should also be followed. The value evaluation methods of second-hand cars include the current market price method, the present value method of income, and the replacement cost method.

The replacement cost method refers to the total cost of re buying a new state of the vehicle under the current conditions, minus the value of the old depreciation of the estimated vehicle as an evaluation method for the current price of the vehicle being assessed. The old devaluation of the vehicles being assessed includes physical devaluation, functional devaluation and economic devaluation. In the actual evaluation, the substantive depreciation is characterized by the new rate of the automobile, and the functional devaluation and the economic devaluation are characterized by the comprehensive adjustment coefficient. The calculation formula is as follows:

$$P = B \times C \times I$$

In the formula, P- evaluation value; B- replacement cost; C- into a new rate; I- comprehensive adjustment factor.

3.Analysis of the value characteristics of a used pure electric car

3.1 characteristics of cost composition

The cost composition of pure electric car has changed greatly compared with the traditional car. Taking several pure electric cars listed in China's domestic market as an example, the cost of battery pack occupies a large proportion in vehicle cost, as shown in Table 1. As can be seen from the table, the general battery group accounts for about 40% of the manufacturer's guide price, and some even more than 50%.

Table 1. several major domestic models of the battery pack costs

model	manufactu	battery	the proportion of battery costs
Beiqi	22.69	30.4	50.2%
Roewe E50	23.49	22.4	35.7%
Beiqi	19.2	25.6	50.0%
JAC iEV5	17.98	23	47.9%

3.2 battery life has a great influence on the value of the whole vehicle

In China, there are detailed scrapped requirements for all kinds of commercial vehicles. Although there is no age requirement for non business vehicles, the industry has been established. In general, the life of vehicles is 15 years. But for pure electric cars, battery life is affected by existing technical barriers, charging times, charging speed and environment. The service life is not equal to the life of vehicle itself. And the cost of battery is high in the cost of pure electric car, so the impact of battery life shortage on vehicle value evaluation should be considered separately.

3.3 pure electric car purchase subsidies vary from time to time

In order to promote the development of electric vehicle technology, China has implemented the subsidy policy of new energy vehicles since 2010, and the amount of subsidies has been decreasing year by year. The subsidy standard has increased year by year, and the subsidy standard is shown in Table 2. It can be seen from the table that the 2017-2018 year subsidy standard dropped by 20% on the basis of 2016, and the 2019-2020 year subsidy standard dropped by 40% on the basis of 2016. The subsidy policy is adjusted year by year for the second hand pure electric car, its replacement cost will change every year. Moreover, subsidies for pure electric cars are different in different cities of China. Some cities are supplemented by 50% on the basis of state subsidies, some of them are 40%, some even do not subsidize them.

Table 2. 2013-2020 pure electric car national subsidy standards (million)

Mileage (km)	2013	2014	2015	2016	2017	2018	2019	2020
80≤R<150	3.5	3.325	3.15	-	-	-	-	-
100≤R<150	-	-	-	2.5	2	2	1.5	1.5
150≤R<250	5	4.75	4.5	4.5	3.6	3.6	2.7	2.7
R≥250	6	5.7	5.4	5.5	4.4	4.4	3.3	3.3

4. Study on value evaluation of second hand pure electric car based on resetting cost method

Because the second-hand electric car market is not mature, the cost compared with the traditional car changed greatly, and the influence on vehicle battery life value, Car Buying subsidies vary from time to time, so the evaluation method used in this paper to pure electric car based on the replacement cost method.

4.1 determination of the cost of resetting

Pure electric cars enjoy the national and local financial subsidies, and the intensity of subsidies varies every year. Therefore, the replacement accounting method to determine the replacement cost of vehicles must be improved, and the actual subsidies that need to be repurchased should be considered. The calculation expression is as follows:

$$B = B_1 + B_2 - S$$

In the formula, B- the cost of replacement; B_1 - the market price of the new vehicle purchased; B_2 - the sum of the taxes paid in one go; S - the sum of the state and local financial subsidies at the time of the renewal.

It is especially pointed out that, because the technology of electric vehicle is still in the development stage, and the product is improving continuously, if the product is updated, the market price of the new vehicle should be the price of updating the product, that is, using the replacement cost.

4.2 determination of the new rate

The new rate is a parameter that characterizes the physical devaluation of the vehicle. Due to the cost composition of a pure electric car, the determination of its new rate is divided into two parts, that is, the new rate of the battery and the new rate of the other components of the vehicle.

4.2.1 calculation method for new rate of battery formation

In the actual evaluation of the inspection, inspection personnel due to the lack of experimental instruments and data statistical analysis of professional skills, often can not be a car battery test and set up a model to predict the residual life, so relatively accurate estimates of the residual life of the battery in great difficulties. This paper presents a method to calculate the battery life through the charge and discharge times. That is to say, the number of batteries charged is estimated by mileage and combined with the number of battery charging and discharging provided by the manufacturer, and then the remaining life of battery (the number of remaining charges) is estimated. According to research, the users of pure electric cars usually use only 60-80% of battery power, and they will be charged. This paper takes 80% for calculation, so the calculation formula of the number of battery use is as follows:

According to research, pure electric car users generally only 60-80% of the battery power, it will charge. This article takes 80% when calculating, so the calculation formula of times of battery usage is:

$$M0 = \frac{Mileage\ traveled}{recharge\ mileage \times 80\%}$$

$$\boldsymbol{C_B = M_0/M}$$

In the formula, M_0- The number of battery has been charged; M- The number of battery charge and discharge provided by the manufacturer; C_B- The battery into a new rate.

4.2.2New rate calculation method for complete vehicle formation

It is estimated that at present, most of the pure electric car batteries account for about 40% of the total value of the vehicle. Therefore, the new rate of the new rate is 40% in the new rate of the new rate of the used pure electric car, and the weight of the new rate of the other components of the vehicle is 60%.

$$C=60\%C_O+40\%C_B$$

In the formula, C- Vehicle into a new rate; C_O - New rates of other components; C_B- Battery into a new rate.

The new rate of the other components of the vehicle refers to the calculation method of the new rate of the traditional automobile, which can be calculated by the equal speed depreciation method or the accelerated depreciation method according to the situation.

4.3 Determination of comprehensive adjustment coefficient

4.3.1 The establishment of a comprehensive adjustment coefficient model

With the use of theory of surplus value and actual vehicle, the second-hand electric car comprehensive adjustment coefficient is the residual value of parameters and usage parameters of the two part, the surplus value parameters of two levels, the use of parameter setting a level, establish a comprehensive adjustment coefficient model, as shown in figure 1. The formula of the comprehensive adjustment coefficient is as follows:

$$I = W \times K$$

In the formula, I- comprehensive adjustment factor; W- residual value parameter; K - usage parameter.

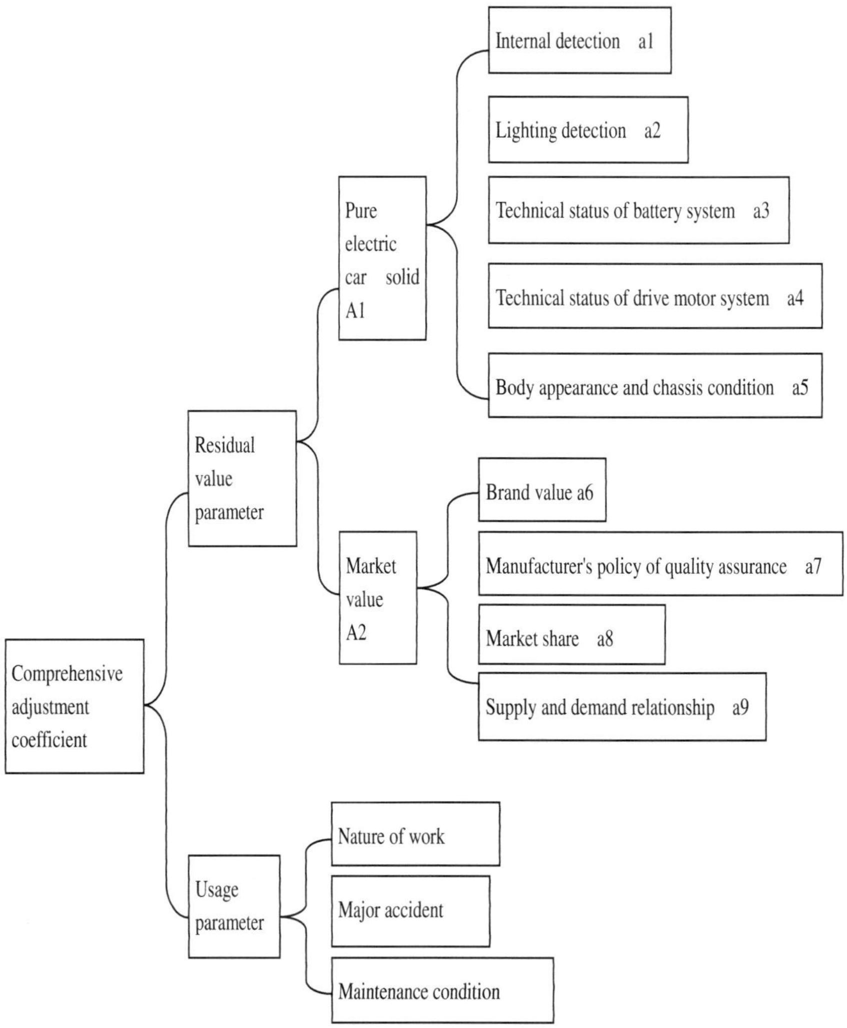

Figure 1. comprehensive adjustment coefficient

4.3.2 Residual value parameter model

The analytic hierarchy process (AHP) is used to construct the W - A judgment matrix according to the expert advice. The eigenvector is obtained after the specification:

$$W^{*} = [0.83 \quad 0.17]$$

$$W = \begin{bmatrix} 1 & 5 \\ 1/5 & 1 \end{bmatrix}$$

For the importance degree of every factor of the pure electric car entity, the weight is determined according to the proportion of the cost of each part of the car to the vehicle value, and the A1 a judgement matrix is established according to the weight. The characteristic vector is obtained as follows:

$$A_1^{*} = [0.0209 \quad 0.0371 \quad 0.3772 \quad 0.2970 \quad 0.2678]$$

$$A_1 = \begin{bmatrix} 1 & 1/3 & 1/9 & 1/5 & 1/7 \\ 3 & 1 & 1/7 & 1/3 & 1/5 \\ 9 & 7 & 1 & 5 & 3 \\ 5 & 3 & 5 & 1 & 1/5 \\ 7 & 5 & 1/3 & 5 & 1 \end{bmatrix}$$

According to the weight of the factors of market value, the expert scoring method is adopted, and the opinion sets of expert scoring results are collected. After transforming the weighted average, the A2 a judgment matrix is constructed, as shown in the formula. The characteristic vector is obtained as follows:

$$A_2^* = [0.4869 \quad 0.1395 \quad 0.0705 \quad 0.3031]$$

$$A_2 = \begin{bmatrix} 1 & 4 & 5 & 2 \\ 1/4 & 1 & 3 & 1/3 \\ 1/5 & 1/3 & 1 & 1/4 \\ 1/2 & 3 & 4 & 1 \end{bmatrix}$$

To sum up, the weight distribution of the remaining value parameters is obtained, as shown in Table 3. Therefore, the calculation formula of the residual value parameters is as follows:

$$W = A_1 + A_2$$

$$A_1 = a_1R_1 + a_2R_2 + a_3R_3 + a_4R_4 + a_5R_5$$

$$A_2 = a_6R_6 + a_7R_7 + a_8R_8 + a_9R_9$$

In the formula, W - Residual value parameters; A_1- Pure electric car physical factors; A_2- Market value factors; a_i - Residual value factors corresponding weights; R_i - Factors corresponding to the score value.

Table 3 .weight distribution of residual value parameters

Level a	Level A		Level a Total weight
	Pure electric car physical factors A1	Market value A2	
	A1 = 0.83	A2 = 0.17	
Internal (electrical instrumentation and interior)	0. 0209	–	0. 017
Lighting assembly	0. 0371	–	0. 031
Battery system	0. 3772	–	0. 313
Drive the motor system	0. 2970	–	0. 247

Body appearance and chassis assembly	0. 2678	–	0. 222
Brand Value	–	0. 4869	0. 083
Manufacturers warranty	–	0. 1395	0. 023
market share	–	0. 0705	0. 012
Market supply and demand	–	0. 3031	0. 052

4.3.3 Use case parameter model

For the nature of K1, K2 and K3, these three factors are still obtained through expert score, and K matrix is shown in formula 12. Specification to obtain their feature vectors: K^*=(0.614&0.268&0.118), the nature of the work K1, a major accident K2, maintain the weight of these three factors maintenance status K3 values were 0.614, 0.268, 0.118. The value of the use parameter can be obtained by combining the weight value combined with the score.

$$K = K_i R_K$$

In the formula, K_i - the weight corresponding to the usage parameter factor; R_k- the score value corresponding to the factor.

$$\mathrm{K} = \begin{bmatrix} 1 & 3 & 4 \\ 1/3 & 1 & 3 \\ 1/4 & 1/3 & 1 \end{bmatrix}$$

5. Conclusion

In view of the problems of EV subsidies, battery cost and technology status, this paper will improve the replacement cost method of second-hand car evaluation, and form a systematic replacement cost evaluation method for second-hand ev. However, as the market of mature used pure electric car has not yet formed, this method still needs to be revised and perfected through a large amount of actual transaction data.

References

[1] Chen Junyi, Wang Hongyan. An improved comprehensive adjustment method based on fuzzy evaluation method to determine the value of used car [J]. Shanghai car, 2009 (07): 18-23.

[2] Yang Yayi, Zhang Furen. Analysis of damaged degree of used car based on combined model [J]. Journal of Chongqing Jiaotong University (NATURAL SCIENCE EDITION), 2012,31 (04): 869-872.

[3] Lin Yilin, Chen Rongzhang. Research and Discussion on the residual value of used car based on analytic hierarchy process (AHP) [J]. Shanghai automobile, 2010 (12): 50-54.

Acknowledgments

Supported by Project of Science and Technology Support Program of Sichuan Province（2016GZ0017）, Key Laboratory of Automotive Engineering in Sichuan Province（szjj2014-072）, Chengdu Science and Technology Project（2016-RK00-00151-ZF）.

IOP Conf. Series: Materials Science and Engineering **324** (2018) 012083 doi:10.1088/1757-899X/324/1/012083

The study on the Layout of the Charging Station in Chengdu

Cai yun[1], Zhang wanquan[2], You wei[3], Mao pan[1]
1.School of Automobile and Transportation, Xihua University, Chengdu 610039 China
2. The General Assembly Department Of China Aerospace Science and Technology Corporation Seventh Academy, Chengdu 610100 China
3. The Seventh Design Department Of China Aerospace Science and Technology Corporation Seventh Academy, Chengdu 610100 China
E-mail: 29682131@qq.com

Abstract. In this paper , the comprehensive analysis of the factors affecting the layout of the electric car, considering the principle of layout of the charging station .Using queuing theory in operational research to establish mathematical model and basing on the principle of saving resource and convenient owner to optimize site number. Combining the theory of center to determine the service radius, Using the Gravity method to determine the initial location, Finally using the method of center of gravity to locate the charging station's location.

1.Introduction

Electric vehicle charging and changing equipment is an important supporting facility of electric vehicles. It is one of the main factors that affect the development and promotion of the electric vehicles. Only by improving the supporting facilities can the development of electric vehicles be promoted. Chengdu city is one of the 5 pilot city of third group of ten city 1000 project, but the development of the electric vehicles in Chengdu city is slow and stagnant due to the imperfect development of charging facilities . In order to promote the development of electric vehicles in Chengdu City, this paper combines the existing research results at home and abroad and puts forward its own opinion, theoretical model is established and applied to the practice, which can provide the reference for the location of charging facilities of the electric vehicles.

2. The Planning of Building the Charging Facilities Station

The research of this thesis is planned on the basis of the standard charging station and calculates it based on the data about the car ownership and charging demand of Chengdu City in the paper Prediction on the charging demand for electric vehicles in Chengdu. First of all, this paper should confirm the service capacity and scale construction of the standard charging stations and use the mathematical models to plan and confirm the number of construction of charging stations.

2.1 Scale of the standard charging station

Combining the scale of the gas station to confirm the scale of the standard charging station.[1]The formula of the service capability is shown as the following:

$$C = M \bullet \mu \bullet t$$

In the formula, C- the maximum service capability of the standard charging station, set/ day M- the sets of the facility(Battery replacement 12 sets, DC charging 4 sets).μ- the service rate (theoretically, 4units/h, actually, 3units/h).t- the working time (theoretically, 24h , actually, 16h)

Considering the actual situation of the gas station, the scale of the standard charging station can be obtained as shown in Table 1.

Table 1 .The scale of the standard charging station

Charging mode	The set of the DC charging	The set of the battery replacement	Maximum service capability	Actual service capability
Standard units	4	12	1440units/day	768 units/day

2.2 Queuing Model Construction and Solution

2.2.1 The Analysis and Description of the Problem

The electric car that is lack of electricity and needs to be charged on the road, this process is a random process. According to the experience, the process of charging demand for electric vehicles meets the three major characteristics of the Poisson distribution introduced above. So the paper can use Poisson flow to describe the law of the electric vehicle charging demand. When two customers arrive at the service station, the time interval T satisfies the time negative exponential distribution,which can be described. According to the principle of queuing theory[2] and the full analysis of the arrival of electric vehicles at the service station, this paper makes the following hypothesis:

(1) The charging time of each electric vehicle follows a negative exponential distribution;

(2) Each electric vehicle charging and system capacity limits service system;

(3) The number of electric vehicles waiting to be charged can not exceed a certain number, so that the number of electric cars waiting in line can not exceed 5 units ;

(4) Queuing time can not exceed a certain period of time, or the waiting time is too long, it will affect the owner's trip. The waiting time is not more than 15 minutes or 0.25 hours.

2.2.2 The Building of the Model

The queuing model has mainly loss model, waiting model, mixed model, according to the above analysis of the charging system, this paper is standard multi-service waiting model M/M/C, that is Poisson input, negative exponential distribution, waiting queue model of C service desk, service in accordance with the rules first-come -first -served and customer source is infinite. When the customer arrives at the service desk, the time interval follows the Poisson distribution of the parameter, and the service time obeys the negative exponential distribution of the parameter. The average service rate of every table cloth assisting each service desk is $c\mu(n \geq c)$ or $n\mu(n < c)$, $\rho = \frac{\lambda}{c\mu}$ is average utilization rate of the service institutions, that is service intensity. When $\rho = \frac{\lambda}{c\mu} < 1$, the system does not end up in an infinite queue. The system transition state diagram of model M/M/C is shown as the system state transition diagram of the following 1. M/M/C model:

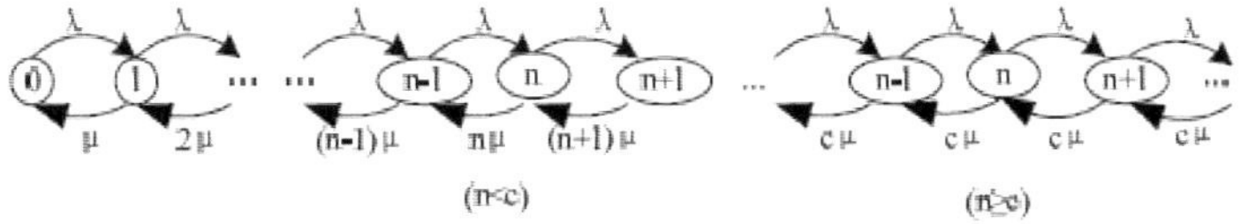

Fig1:The system state transition diagram 1M/M/C model

In the state transfer system state diagram, 0 indicates that there is no electric vehicle charging station and charger idle, transferring state 0 to state 1 represents a car arrives at the service institution, 1 charger work, others idle, the transferring rate is μP_1, similarly, The analysis of state n shifts to state n+1,there are 2 cases: first is when $n \le C$, the N service desk work, state transferring rate is $n\mu P_n$,second is when $n > C$, C electric cars are served, N-C electric cars waiting in line, the state transferring rate is $C\mu P_n$.

The steady-state equilibrium equation of the system is:

$$\mu P_1 = \lambda P_0$$
$$(n+1)\mu P_{n+1} + \lambda P_{n-1} = (\lambda + n\mu)Pn, (1 \le n \le c)$$
$$c\mu P_{n+1} + \lambda P_{n-1} = (\lambda + c\mu)P_n, (n > c)$$

In the formula, $\sum_{n\to 0}^{\infty} P_n = 1$ and $\rho = \frac{\lambda}{c\mu} \le 1$, therefore, the probability of the system state can be obtained by recursion:

$$P_0 = \left[\sum_{n=0}^{c-1} \frac{1}{n!}\left(\frac{\lambda}{\mu}\right)^n + \frac{1}{c!}\left(\frac{\lambda}{\mu}\right)^c \left(1 - \frac{\lambda}{c\mu}\right)^{-1}\right]^{-1}$$

The operating index of the system is:

$$L_q = ((c-1)!(c\mu - \lambda)^2)^{-1} \lambda\mu\left(\frac{\lambda}{\mu}\right)^c P_0,\ L_s = Lq + \frac{\lambda}{\mu},\ W_q = \frac{L_q}{\lambda},\ W_S = \frac{Ls}{\lambda}$$

In the formula, PO is the probability that the system state is 0 at the moment T, and λ is the average arrival rate for the electric vehicle to the charging station , Unit is the unit per hour, and the formula is as follows;

$$\lambda = \frac{B_0 \bullet K}{N \bullet T}$$

In the formula, B_0-total demand for vehicles , unit is car number/ day

K-Service vehicles account for 0.9 of the total amount of services in the whole day

N-number of the urban construction sites

T-the actual time for the service at the charging station is 16 hours

The total charge charging demand and the number of the urban charging station are the factors which influence city electric vehicle arrival rate, while the demand is gradually increasing with the development of electric vehicles, so the city can adjust number of charging μ is the average service rate of vehicles per hour, the average service rate of the fast charging station is 3 vehicles per hour, The average service rate of the battery charging station is 4 units / hour, taking into account the influence of external factors and convenient calculation , the paper will value the average service rate of 3 vehicles per hour.

Because the private car demand is bigger on weekends, in order to avoid queuing on weekends and charging time too long , which will influence the residents' travel. Here, the charging demand of the residents weekend needs to be ensured, so meeting the charging demand of 2015 year is regarded as an aim of building station of the paper, according to the standard of charging station service capacity is768 vehicles per day,

In order to save energy and capital, the minimum number of stations in Chengdu is planned, and the number of stations, average service capacity, service intensity and service level are as following scale of station construction of table 2.

Table 2. scale of station construction

The minimum number of stations	Average arrival rate	Service intensity	Average queue length	Average waiting time
12	43.2	0.91	5.97	0.14h

Because the average queue length is 5.97 and greater than 5 does not meet the prerequisites, the minimum number of stations should be increased, by gradually increasing the number of stations to gain the most appropriate number of building station to meet the service requirements as shown in table 3.

Table 3 .the scale of building station in the future

The minimum number of stations	Average arrival rate	Service intensity	Average queue length	Average waiting time
13	39.8	0.83	4.37	0.11h

Because the average queue length is 4.37, less than 5, and the average waiting time is 0.11 hours, less than 0.25 hours, the most suitable station scale is 13 station. With the same method, the paper can get the station scale of public fast charging station in the coming years, as shown in table 4.

Table4 .the number of the fast charging station in the future

Year	2015	2016	2017	2018	2019	2020
The number of the station	13	20	33	56	81	116

3 The Planning of the Location

The idea of planning the location of the charging station is: firstly, to use the central ground theory to divide the service area and reduce the overlap of facilities between the service areas and avoid unnecessary waste of resource allocation. And then make use of gravity model approach to firstly choose primary site in each primary service area, but the primary site in

IOP Conf. Series: Materials Science and Engineering **324** (2018) 012083 doi:10.1088/1757-899X/324/1/012083

practice is not necessarily can be achieved, so you can choose n possible locations in the coordinates near the possible sites as an alternative site, finally focus on the model to determine the final location.

3.1 The Central Ground Theory Determining the Service Radius

According to the central gravity theory, the paper regards 2015 as one -level central place and 2020 as 2- level center place. The same level is mutually excluded into a honeycomb distribution; the different levels are nested with each other, and the higher level contains the lower level.

Formula based on service radius:

$$R_{\max} = V \bullet t \bullet u$$

In the formula, $R_{\max}$ -Maximum radius of service km

V- Average speed of the car km/h(30km/h)

t- the maximum time of the finding the center place(10min)

u- influencing factors(0.8)

So $R_{\max}$ =4km, According to the formula of regular hexagon area is 41.52km, the center ground area is 41.52km . So the paper gets the lowest level central area and the one-level center place. The main research object of this paper is the public fast charging station, and the main service objects for public fast charging station are the traffic. Taxi distribution in the urban areas is more scattered, so the paper only needs to study the object, that is , service areas of universities, hospitals, shopping malls, traders Hotel. The user distribution of tourism scenic spot as the research object to represent the center area, the number of users is evenly distributed according to the size of the area , for private cars is evenly distributed according to the size of the residential area. The Tianfu Square is used as the coordinate origin, and the center of the one- level center is shown in figure 2.

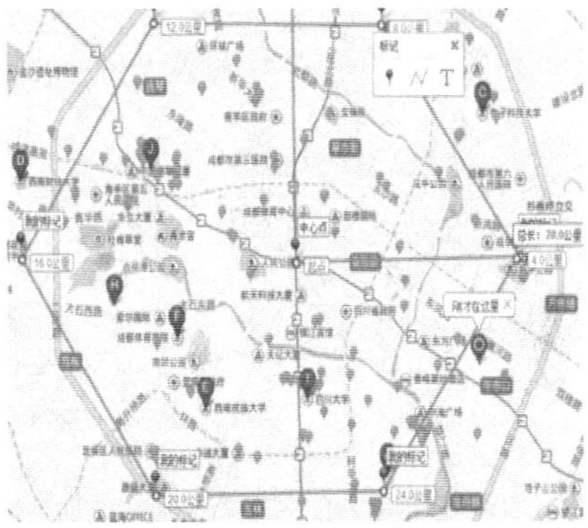

Figure 2. Central service scope centerss

3.2 To Determine the Primary Site of the Center with Gravity Model Method

Here, this paper respectively calculates one place to meet the one-level central level, that is, to accord with the charging demand in 2015, and to meet the two -level center place, that is, to meet of the coordinates of the charging demand in 2020 . First calculate the coordinates that meet the one- level center place, and then take the demand and coordinates of each point into the formula of gravity model method:

Figure 3, the alternative location the residential area

$$C_x = \frac{\sum D_{ix} V_i}{\sum V_i}$$

$$C_y = \frac{\sum D_{iy} V_i}{\sum V_i}$$

The primary site is(-0.83, -1.15), exactly falls in the residential area, operability is very low in Figure 3, the alternative location the residential area, so combining the gravity point near the gas station which has built to choose the location to realize resource utilization. In the area near the gas station, there are the oil gas station (Gaosheng Bridge), Sinopec gas station (Qingshui River gas station), Sinopec (hongpailou gas station), regarding three gas stations as alternative sites. As shown in: Figure 1 is Sinopec gas station (Qingshui River gas station); 2 is Sinopec (hongpailou gas station); 6 is Petro China gas station (Gaosheng Bridge Road).

3.3 Center of Gravity Method to Confirming the Final Location

The gravity center model is designed to solve the location problem of two or more facilities, Make the demand point to the weighted distance of its nearest facility point and the Tc minimum, and the formula of the model is as follows:

$$\min Tc = \sum_{i=1}^{m} E_{ij} d_{ij}, (j = 1,2,\bullet\bullet\bullet n)$$

In the formula, n- the number of the optional sites

m- the number of the demand point

E_{ij} - the demand intensity of the demand point i to the facility point j

d_{ij} - the shortest distance of the demand point i to the facility point j,

According to Euler formula is $d_{ij} = \sqrt{(X_i - x_j)^2 + (Y_i - y_j)^2}$.

Regarding the minimum distance between user and service points as the location index, according to the formula 3-2 of the gravity center method, making the required data applied into the formula to get the shortest distance of the alternative sites: S Qingshui River =1987.5,S Hongpailou =2013.2 , Tgaoshengqiao =1635.4. Therefore, finally the paper determines Petra China(Gaosheng Bridge Road gas station).

The location of the charging station in 2015 has reached the most central location, and the others are scattered from the center to the surrounding area and can be gained the location in the same way. With the increase of the number of electric cars, in order to meet its needs, the construction of charging stations will be increased every year. The two-level center places will be filled with Chengdu by 2020 as shown in Figures 4 and 5.

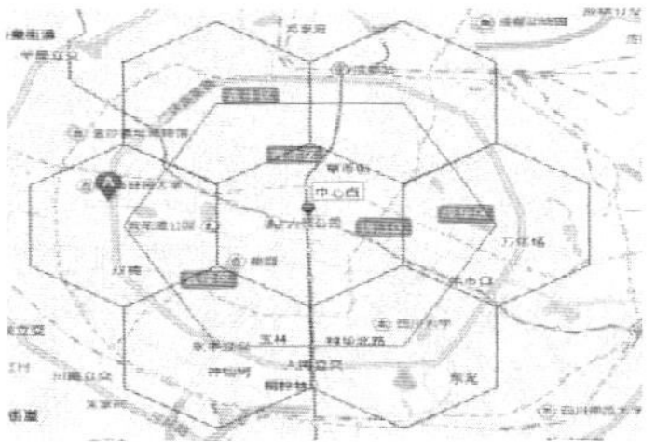

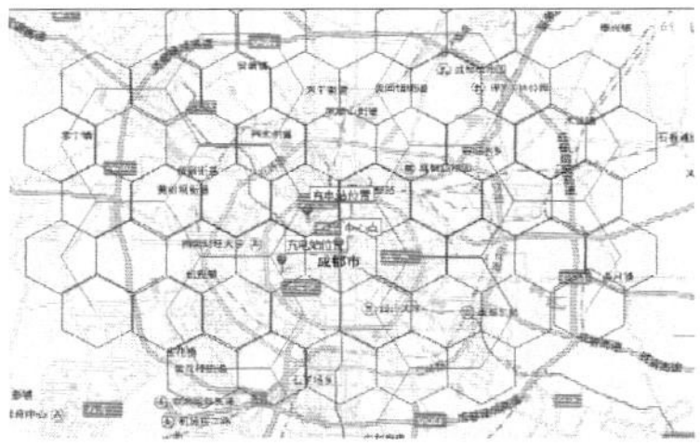

Figure 4 the situation of building station in 2016 Figure 5 the situation of the building station in 2020.

Electric vehicle charging system is an important basic support system of electric vehicles, and also is an important link in the process of commercialization and industrialization of electric vehicles. In this aspect, it is necessary for the government to introduce corresponding policies to compulsorily promote, which helps to effectively accelerate the development of electric vehicles. For example, the government can introduce a policy for Sinopec, PetroChina and the state grid and other large state-owned enterprises and electric vehicle batteries businesses joint "three policy" and unify battery model as possibly as; the government can be free to construct AC charging pile in the area or parking lot, let private car etc normally charge from Monday to Friday and reduce traffic congestion situation.

4. Conclusion

(1)The paper combines the related theory of choosing the location to study the plan of the layout of the charging station of vehicle cars, the areas that are worth affirming are as follows:

(2) After determining the standard charging station model, the queuing theory is applied to optimize the number of stations.

(3) Using the central gravity theory to determine the service radius, using gravity model approach to determine the first location, the central gravity method determines the final location. The combination of the three methods achieves the scientificalness and accuracy of site selection.

(4) But the location of the electric vehicle charging station involves many models, constraints and many uncertain variables. It is a very complicated project. This thesis is based on the present urban construction layout of Chengdu city. With the development of economy, if the layout of the city is changed into satellite layout, the prerequisite of the study will be changed.

References

[1] Wang Derong, Wang Zhiyuan, Li Bixin. Research on Location Decision Method of City Gas Station Based on Fuzzy Matter Element Analysis [J]. Journal of Chongqing University of Science and Technology (natural science edition). 2008 (03) :P55-60

[2] Tong Dai, Yin Guofu, He Taibi, etc. Preliminary Study of Optimization Methods for the Natural Gas Fueling Station [J]. Journal of Xihua University (natural science edition). 2002 (04) 17-20

Acknowledgments

Supported by Project of Science and Technology Support Program of Sichuan Province （2016GZ0017）， Key Laboratory of Automotive Engineering in Sichuan Province（szjj2014-072）， Chengdu Science and Technology Project（2016-RK00-00151-ZF）.

The Development on the Ultra-hard Honing Oilstones in Preparation and Application

Z. C. Zhou*, J. Du, S.Y. Gu

School of Mechanical and Electronic Engineering, Suzhou Vocational University, Suzhou, 215104, China

E-mail: zhouzhengcun@126.com

Keywords: ultra-hard honing oilstones, CBN oilstone, diamond oilstone, the preparation and application of ultra-hard honing oilstone

Abstract. Physical, chemical and mechanical properties of some abrasives such as diamond, CBN are described in this paper. The preparation and application of ultra-hard oilstones such as diamond and CBN are also discussed. They are widely applied in honing and promote great development of honing technology. However, the standardizing binder compositions of diamond and CBN can not easily obtained. In addition, the precision requisition of grinded parts is different from part to part and the grinded material kinds are diverse. Therefore, the oilstone performance is difficult to match reasonably the grinded parts. Those factors result in the complication of the oilstone preparation.

1. Introduction

Honing is an important processing method to obtain precise and ultra-precise machine parts, especially the hole-parts, e.g. cylinder liner, fluid cylinder. The property of oilstones is a key factor. It influences the processing precision and quality of the parts, but also processing efficiency and oilstone consumption. The property parameters of oilstones are included in types of abrasive, hardness, particle sizes [1]. There are many kinds of grinding materials. Diamond and Cubic Boron Nitride (CBN) belong to super-hard abrasives. SiC and Al_2O_3, etc, have low hardness compared with CBN and diamond [2]. The hardness of oilstone is regarded as the jointing firmness of abrasive with binder, which is also called as holding force of binder with abrasive. The surface of oilstone can not be easily removed when the holding force is large [3]. The hardness of oilstone should be adapted to that of the finished material and flexural strength should be meted basic requirements. The particle sizes of abrasives are also related to the roughness of the finishing surface of parts. Therefore, many factors should be considered when the oilstone is chosen, which include the material, roughness, dimensional precision, shape of finished parts. In this paper, the abrasive characteristics of the honing oilstone are described. The study development of the preparation and choice of oilstone is also discussed.

IOP Conf. Series: Materials Science and Engineering **324** (2018) 012084 doi:10.1088/1757-899X/324/1/012084

2. The common characteristics of the ultra-hard abrasives

The oilstone can be produced in common using the abrasives such as diamond, CBN, Al_2O_3, SiC and so on. They have not only different hardness, but also different physical and chemical properties. The diamond and CBN belong to the ultra-hard abrasives.

2.1 The characteristics and properties of the diamond

The diamond is the highest in hardness in all materials now, whose chemical composite is carbon containing little impurity. The natural diamond contains the impurities such as N, Al, Si, Ca, Mg, and so on. There are graphite, catalyst metal, pyrophyllite and N in the synthetic diamond [4].

The diamond is isometric system crystal with face-centered cubic lattice. The lattice constant (a) is 0.35670 nm. The Moh's hardness of diamond is 10 and micro-hardness is up to 10000~101000 MPa. The high hardness of diamond is related to large bond energy. The wear resistance of diamond is about 140 times as big as Al_2O_3.The natural diamond has an elastic modulus of 900 MPa and a bending strength of 210~490 MPa and a compressive strength of 2 GPa. For the synthetic diamond, elastic modulus is between 740 and 1050 GPa. The bending strength is 300MPa and a compressive strength 2 GPa. The diamond has large heat capacity and good thermal conductivity. The coefficient of heat conductivity is 0.35cal/(cm s °C) and specific heat is 0.12 cal/(g °C). The linear expansion coefficient is $0.9 \sim 1.18 \times 10^{-6}$/ °C (0~100 °C). The melting point is above 3550 °C. The diamond will burn if it is heated to the temperatures of 850 °C~1000 °C in air. The diamond can be dissolved in the salt molten mass such as monarkite, potassium nitrate, etc though it can not be dissolved in sour and alkali.

2.2 The characteristics and properties of CBN

CBN is cubic crystal consisting of N and B. It has not been found in natural world [5]. Its hardness is next only to the diamond. However, the thermal stability and chemical inertness of CBN are substantially superior to those of the diamond. CBN is transformed from hexagonal boron nitride (hBN) under high temperature and pressure [6-7]. Boron nitride possesses two like-diamond structures formed by the SP^3 hybridization in addition to the two structures of hBN and rBN. The former two structures are CBN with zinc blende structure and WBN with wurtzite-type.

CBN and diamond are similar in their structures ad their lattice parameters are very close. The lattice of diamond is 0.3567 nm and CBN 0.3615nm. Their bodings are the same that are covalent bond formed by the hybridization along the tetrahedron. The diamond contains only the covalent bonds between C and C while the CBN consists of the covalent bond between B and N.

It is known that the main mechanical properties are strength, micro-hardness and toughness for the CBN single crystal. The Moh's hardness of CBN is 9.8 and its micro-hardness is 7000~10000 MPa. The compressive strength of CBN is 0.8 GPa and the linear expansion coefficient is $2.1 \sim 2.3 \times 10^{-6}$/ °C. Its melting point is above 3300 °C. CBN has extremely excellent chemical stability for the acids. It is not corroded in all the acids such as hydrochloric acid, sulphuric acid, nitric acid, phosphoric acid, hydrofluoric acid, aqua regia and so on. CBN doesn't apparently interact with Fe and C, which provides the extreme superiority for the grinding iron and steel.

3. The preparation and application of ultra-hard honing oilstone

The development of diamond and CBN materials promotes the jumping of honing technology [8].The diamond honing oilstone is manufactured using metallic binder. A binder with high binding strength and good wear resistance is used to prepare the diamond honing oilstone in order to improve the cutting action and machining efficiency [9]. Li Yan-ping et al [10] discussed the new development of honing oilstone. The diamond and CBN honing oilstone can be used to hone all kinds of materials, which are steel, hard alloy, cast iron, superheat resisting alloy, electrodeposition, overlay, ceramics and glass. The diamond honing oilstone is shaped using cold or hot compression technology. The metal mold is used when the cold compression is adopted. A large pressure of 400~500MPa is needed and mold lifetime is long. The graphite mold is used when the hot compression. The shaping pressure is small that is 15~20 Mpa. The strength loss is small and mold consuming is large when the sintering temperature is low and the shaping pressure is small. Shu Zhi et al developed a honing oilstone with new binder and summarized a perfect composition, technology and manufacturing method. The main compositions of the binder are Fe, Q_{663}, Cu, Ni, Sn, Ti, Cr, TiC, WC and ceritics [11]. Zou Wenjun et al [9] introduced the preparation of the honing oilstone to machine the plateau honing cross hatch of the engine cylinder liner. The 170/200 mesh diamond is used to prepare the honing oilstone and the honing testing is completed. The honing oilstone hardness is up to HRB100.2 and the bending strength is 231.3MPa when the compositions consist of 50%Cu, 20%Fe, HP1 addition agent 10%, HP2 addition agent 15%, HP3 addition agent 5%. The optimal sintering technology is that the sintering is kept for 4 mins at 760 ℃ and 3MPa. Wang Yanhui et al [12] introduced the surface treating technology and application of the ultra-hard diamond and CBN honing stone. The new development of surface treating technology is also summarized in Ref. [12]. The plated-Ti diamond and CBN by the evaporating in vacuum is widely applied in all kinds of tools with metal and ceramic binders in order to avoid falling abrasive particles and to prevent the abrasive particles from the thermal etching of the binder. The diamond volume is decreased and the cost of the binders is also reduced when the plated-Ti diamond is used. The combine plated-Ti/Ni diamond can further improve the efficiency of the plated-Ti diamond. The lifetime is prolonged and the sharpness is improved ten percent. The plated-Al_2O_3 ultra-hard abrasive particles are suit for the tools whose binder is resin or ceramic. The plated-Si diamond and CBN possess large binding force and good thermal-resistance. Especially, they have good compatible binding force with Fe-based and ceramic binders and play an important role in the Fe-based and ceramic binders Li Zhihong et al [13] manufactured the CBN tool material with the ceramic binder. The results showed that the suitable sintering temperature of vitrified bond CBN grinding tools should be lower than 800 °C. The strength of vitrified bond CBN grinding tools was obviously affected by the difference of thermal expansion coefficient between vitrified bond and CBN abrasives. C1 bond was more suitable for the manufacturing of vitrified bond CBN grinding tools. Within the sintering temperature range, relatively higher sintering temperature was beneficial to the strength of the bond bridge and the holding strength between bond and CBN particles.

Ichida et al [14-16] deals with the grinding characteristics of newly developed ultrafine-polycrystalline CBN (CBN-U). The results show that the grinding ratio is around 10 times higher than that with the conventional CBN (CBN-B) abrasives. Grinding forces in grinding with CBN-U abrasives are reduced by 20-30% compared with those in grinding with CBN-B abrasives. The CBN-U is suitable for the applications with a high dimensional accuracy in creep feed profile grinding for nickel-based super-alloys, because it gives less profile wear, and hence better form retention, than conventional CBN abrasive. The fracture strength of CBN-U grain is about 1.6 times higher than that of the representative conventional polycrystalline CBN grain (CBN-W5). The grinding ratio in grinding with the CBN-U grains is around 8 times higher than that in grinding with the CBN-W5 grains. Ichida et al [17-18] also investigated the formation mechanism of finished surface in ultrahigh-speed grinding with cubic boron nitride wheels. It have been confirmed that the roughness of the ground surface decreases with an increase in grinding speed, and this decrease is mainly due to the reduction of the swelling ratio with increasing grinding speed. Recently, Tao et al [19] investigated the preparation and mechanical properties of diamond honing oilstone. The diamond honing oilstone was fabricated by hot-pressing at 550–650 °C and 25 MPa pressure for 4 min, using Cu–Sn based alloys as binder metal and uncoated or W-coated diamond grains as abrasive material. The experimental results show that the interfaces between the diamond grains and metal matrices of all the oilstone samples are smooth and no defects are observed in the metal matrices. The bending strength and rockwell hardness of the honing stones increase with the sintering temperature increasing from 550 °C to 650 °C, and the bending strength decrease with the increase of diamond grains faction. The minimum grinding ratio is obtained as the diamond was W-coated, which can be attributed to the improved interfacial bonding derived from the W coating.

4. Conclusions

Ultra-hard honing oilstones with diamond and CBN are widely applied in honing and promote great development of honing technology. However, the standardizing binder compositions of diamond and CBN can not easily obtained. In addition, the precision requisition of grinded parts is different from part to part and the grinded-material kinds are diversity. Therefore, the oilstone performance is difficult to match reasonably the different grinded parts. Those factors result in the complication of the oilstone preparation.

Acknowledgement

This work is sponsored by Qing Lan Project of Jiangsu in China.

References

[1]Liu M,Wang D H 2001 *Machine Manufacturing* (10)11 (in Chinese)
[2]Li Z H, Li L 2012 *China Venture Capital* (10) 31 (in Chinese)
[3] Li Y J 1989 *Grinding Machine and Grinding* (3) 59 (in Chinese)
[4] Cahn R W, Haasen P, Kramer E J 1994 *Materials Science and Technology. In: Structure and Properties of Ceramics* Vol. 11, ed. by M. V. Swain (Weinheim, New York, et al.)
[5]Wang G Z 2005 *Superhard Material Engineering* 17 (63) 41 (in Chinese)
[6]Wentorf R H 1957 *J. Chem. Phys.*26 (4) 956
[7]Hirano S, Yamaguchi T, Naka S 1981 *J. Amer. Ceram. Soc.* 64 734
[8]Malkin S. 1985 *Annals of the CIRP* 34 (2)557
[9] Zou W J,Han P, Peng J et al 2001 *Diamond & Abrasives Engineering* 30 (1) 29 (in Chinese)
[10]Li Y P, Wang S X, Sun J S, Zhou J M 2010 *Tool Technology* 2010 44 12 (in Chinese)
[11]Shu Z, Xu H S 2006 *Exploration Engineering* (6) 57(in Chinese)
[12]Wang Y H 2009 *Diamond & Abrasives Engineering* (1) 5 (in Chinese)
[13]Li Z H,Yuan Q M,Yang Z F 2003 *Acta Material Composite Sinica*, 20 (5)39
[14]Ichida Y, Sato R, Morimoto Y, Inoue Y 2006 *JSME International Journal* 49 (1) 94
[15]Ichida Y. 2008 *Diamond & Related Materials* 17 1791
[16]Ichida Y. 2011 *Int. J. Machining and Machinability of Materials* 9 (3/4)197
[17]Ichida Y, Sato R, Morimoto Y , Ohsawa Y and Fredj N B 2006 *JSME International Journal* 49 (1)100
[18]Ichida Y, Sato R, Morimoto Y 2006 *JSME International Journal* 49 (1) 106.

IOP Conf. Series: Materials Science and Engineering **324** (2018) 012084 doi:10.1088/1757-899X/324/1/012084

[19]Tao J N, Zhang X Z, Liu G W, Xu Z W, et al 2017 *Materials Science Forum* 893 354.

IOP Conf. Series: Materials Science and Engineering **324** (2018) 012085 doi:10.1088/1757-899X/324/1/012085

Analysis of the influence of the milling machine throw-out plate surfacing and structure design on throwing material

Yuanbin Fang[1,2], Hongwei Sha[3], Yunmin Yu[4] and Bing Chen[4]

[1] XCMG Jiangsu Xuzhou Engineering Machinery Research Institute, Xuzhou, J S 221004, China
[2] State key Laboratory of Intelligent Manufacturing of Advanced Construction Machinery, XCMG Construction Machinery Co., Ltd, Xuzhou, J S 221004, China
[3] XCMG Xuzhou Construction Machinery Co. Ltd, Xuzhou, J S 221004, China
[4] XCMG Road Machinery Corporation, Xuzhou, J S 221004, China

fybflying@163.com, 18362801@qq.com, yym88088@126.com, chenbing18560@163.com

Abstract. Material composition, hardness and wear properties of the throw-out plate improved are analysed on a road milling machine. At the same time, analyse the tissue and performance of Fe based alloy named Fe60 cladding layer using the plasma surfacing method. And the original and improved throw-out plates are analysed throwing material effect by the dynamic analysis. Then the throw-out plate samples are verified. The results show that Fe60 powder is selected as surface strengthening material. By the improved structure, the hardness of the throw-out plate increases from 14.6HRC to 57.5HRC, and the wear resistance increases from $0.452g^{-1}$ to $16.393g^{-1}$. At the same time, it increases from 3263 to 3433 to fall into the collecting material number of milling machine. It provides important guidance for structure design and process design of the milling machine throw-out plate.

1. Introduction

Road milling machine is a kind of common asphalt pavement maintenance equipment. Through the design of the rotor tool, the depth and efficiency of the milling asphalt pavement are ensured. The waste material is thrown to the conveyor through the throw-out plate, and the effect of the throwing material affects the collecting efficiency. The scraper flattens the leftover waste and the wavy pavement of the milling surface, and collects it. Abrasive wear by in the process of collecting waste, surface sharp shock and bump will lead to failure of the scraper wear. The distance between the milling road and the scraper becomes large to result in material collecting effect becoming worse. At the same time, power consumption is large, and vibration intensifies. Pavement roughness is poor and the trimming performance is bad. Seriously affect the efficiency and performance [1,2]. Throwing material effect is very important to the fatigue life of the scraper. Therefore, improving structure and enhancing the wear resistance of the throw-out plate to improve the material receiving effect has a direct impact.

With the rotor scraper design of the milling machine continues to mature, milling depth and efficiency have greatly improved, which put forward higher requirements of throwing material efficiency. Gao Lu evaluated different texture pavement surfaces, which were tested in different milling drums, forward speeds and cutting depths. The forward speeds were positively associated with both skid resistance and macro texture [3]. Bae A. developed building information modeling to optimize milling quantity and pavement quality, which enabled optimization of milling and paving options [4]. The structure of the throw-out plate that is attached to the spiral structure of the rotor scraper is mainly rectangle. The research on improving the wear resistance helps to improve the throwing material effect. Plasma surfacing uses compressed plasma arc as heat source to melt the filler material and some base metal that form the deposited layer. Compared with common surfacing methods such as manual arc and argon arc surfacing, the plasma surfacing has the advantages of high production efficiency, low dilution rate, good process stability, easy realization of automation, and stable quality of welding layer. Filled materials commonly used in powder, including iron based, nickel based, cobalt based, copper based, metal-ceramic composite materials, and so on. Each material has its own characteristics and scope of application. The mechanical properties of iron-based materials vary widely. The toughness and wear resistance can achieve a good match and meet the requirements of different work conditions, and the price is cheap. Shchitsyn Y. found that the plasma surfacing of steel under reverse polarity current was feasible, and the microstructure of the surfacing layer was studied [5]. Sidorov S. found the thickness of the surfacing layer was a factor affecting the service life of these components. Surfacing was the most effective way to react with the abrasive medium under working conditions. The composition had certain allowable value for the thickness of the surfacing layer [6]. More research is to study the surfacing, but there is little research on the application of technology and structure together, especially the analysis of tne throw-out plate surfacing and structure design of throwing material.

Based on the wear resistance of the pavement milling machine throw-out plate are analyzed, it obtains the surfacing material. At the same time, the performance and microstructure of the surfacing layer are analyzed. Then, the influence of the original and improved structure on the throwing material effect is simulated by the dynamic simulation. Finally, the samples of the throw-out plate are verified.

2. Performance analysis of the original and improved plate

2.1. Performance analysis of the original plate.

The throw-out plate material is ordinary carbon steel. To study the hardness and wear resistance of the original throw-out plate shown in Figure 1, chemical composition of the throw-out plate is analysed using the ARL-3460 spark spectrometer. The results of the test are shown in table 1.

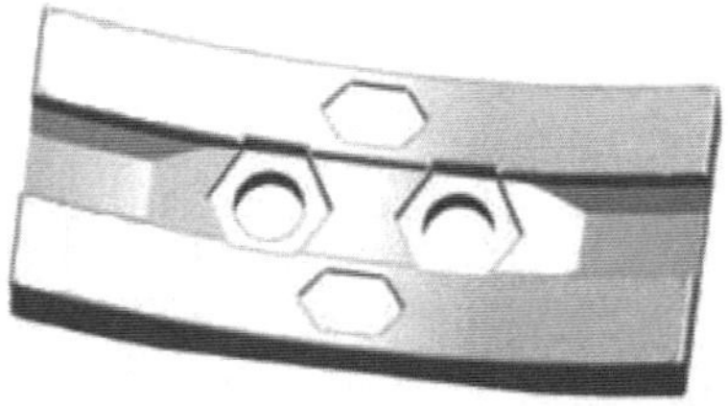

Figure 1. A diagram of the original throw-out plate

Table 1. The chemical composition of the throw-out plate.

Chemical composition	C	Si	Mn	P	S	Cr	Ni	Cu
Content	0.06	0.47	0.46	0.905	0.015	0.51	0.20	0.30

The hardness of the throw-out plate adopts THRP-150D digital display Rockwell hardness tester. The results of the test are shown in table 2.

Table 2. The hardness of the throw-out plate.

The test sample	1	2	3	Average value
Hardness value (HRC)	14.2	14.1	15.5	14.6

The wear resistance is used by pin-disc wear tester. The results of the test are shown in table 3. From table 2 and table 3, the original throw-out plate is low in hardness and poor in wear resistance. In the form of structure, the slotting in the center of the throw-out plate not only ensures the bolt assem-

IOP Conf. Series: Materials Science and Engineering **324** (2018) 012085 doi:10.1088/1757-899X/324/1/012085

bly, but also realizes the rule surface to ensure the effect of the throwing material. From the performance analysis, the hardness and wear amount are poor, and it needs other processes to be improved.

Table 3. The wear performance of the throw-out plate.

The sample	Weigh before test (g)	Weigh after test (g)	difference (g)	loss (g)	Wear resistance (g^{-1})
1	9.596	7.403	2.193	2.213	0.452
2	9.506	7.273	2.233		

2.2. Performance analysis of the improved throw-out plate.

To increase the wear-resistance of the throw-out plate, the structure is improved as shown in Figure 2.

Figure 2. A diagram of the improved throw-out plate r

According to the working condition of the throw-out plate and the characteristics of the iron base material, the plasma surfacing is adopted. Fe60 powder is selected as the surface strengthening material. The above-mentioned analysis method is used to analyze the composition, hardness and wear resistance of the surfacing layer. The chemical composition of the surfacing layer is shown in table 4.

Table 4. The chemical composition of the surfacing layer.

Chemical composition	Fe	C	Mn	Si	P	S	Ni	Cr	Cu	Mo	V	Co
Content	81.68	0.33	0.54	1.46	0.04	0.01	0.86	14.10	0.04	0.18	0.13	0.11

The hardness of the surfacing layer is shown in table 5.

Table 5. The hardness of the surfacing layer.

Fe60	Nominal hardness	Actual hardness
Hardness value (HRC)	60	57.5

The wear performance of the surfacing layer is shown in table 6.

Table 6. The wear performance of the surfacing layer.

The sample	Weigh before test (g)	Weigh after test (g)	difference (g)	loss (g)	Wear resistance (g^{-1})
1	11.213	11.154	0.059	0.061	16.393
2	11.125	11.062	0.063		

From table 5 and table 6, it is known that the wear resistance of the improved throw-out plate is raised from 14.6HRC to 57.5HRC, and the wear resistance has been greatly improved.

To study the microstructure of iron base alloy powder deposited layer, metallographic preparation is made. Metallographic grinding surface is made up of cross section of surfacing layer. The corrosion agent is selected by ferric chloride, hydrochloric acid, and alcohol solution, and the corrosion time is about 10-20s. The microstructure of the metallographic microscope is observed by DMI5000M inverted metallographic microscope after the preparation of the sample. The results are shown in Figure 3.

From Figure 3, the Fe60 powder surfacing layer contains more Cr and B elements, and the hard phase is mainly chromium carbide and boron carbide. The surfacing layer is mainly dendrite structure. The hardness is 57.5HRC, and wear resistance is good.

IOP Conf. Series: Materials Science and Engineering **324** (2018) 012085 doi:10.1088/1757-899X/324/1/012085

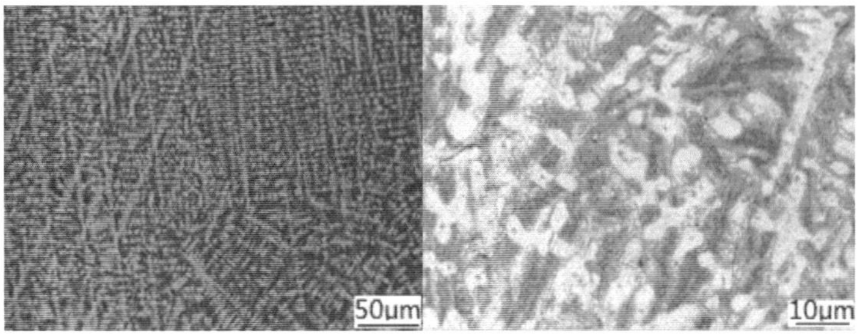

×200 ×1000

Figure 3. The microstructure of Fe60 powder surfacing layer

3. Result analysis and discussion

3.1. Comparison and analysis of the throwing material effect.

To investigate the influences of the throw-out plate structure design of throwing material effect on milling machine, the throwing material effect of the original and improved throw-out plate is simulated by the dynamic analysis. The same size discrete element is adopted to simulate the road surface. The angle of the throw-out plate is in conformity with the actual situation, and the angle with the milling drum is 90 degrees. The rotation speed of the milling drum matches the actual situation. The effect of the throwing material is evaluated by calculating the number of pavement materials falling into the simulated collecting orifice. Select the shed, detachment and completely out of three times into collecting quantities of materials to obtain the results as shown in figure 4, and n is defined as the quantity of material.

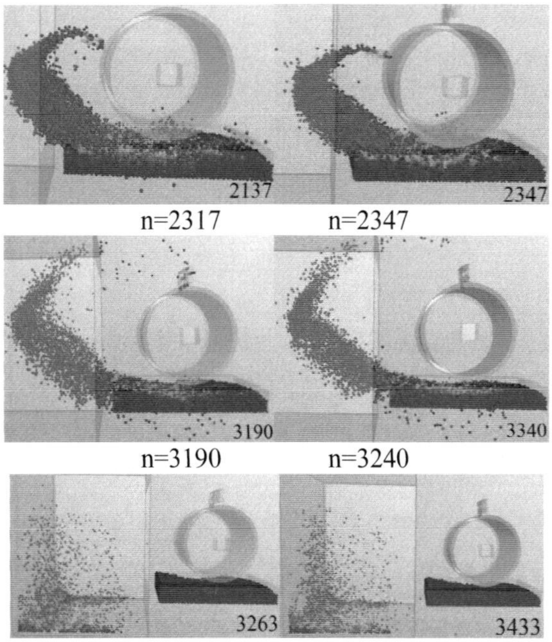

n=2317 n=2347

n=3190 n=3240

n=3263 n=3433

Figure 4. The throwing material effect simulation of the throw-out plate

From figure 4, the throwing material effect of the improved throw-out plate is better. The quantity of material is 3433, and the original is 3263. Without considering the material size, the adhesion conditions between the material and the throw-out plate, under the existing conditions, the aggregate number is similar, and the material spilled in the shape of the air distribution are basically the same, but the throw-out plate is still a small amount of material the adhesion in the groove in the run-up to the above. Therefore, the surface shape of the throw-out plate is not the key factor that affects the throwing material effect, but the improved structure can improve the throwing material effect.

3.2. *Test verification.*

On the above research results, the process parameters are used to test the sample of the throw-out plate shown in Table 7. The samples of the throw-out plate are shown in Figure 5. The plasma spray welding equipment is used to fix the throw-out plate on the turntable by using the tooling. After setting the turntable rotation speed and the surfacing process parameters, the automatic process is carried out.

Table 7. The process parameters of the plasma spray welding.

Lonized gas (L/H)	Non-arc current (A)	Arc voltage (V)	Oscillating speed (mm/min)	Left stay (s)
0.3	36	26	1400	0.3
Powder gas (L/H)	Non-arc current (A)	Arc voltage (V)	Rotational speed (mm/min)	Right stay (s)
0.2	21	22	6	0.3
Protection gas (L/H)	Speed (mm/min)	Supply (g/min)	Distance (mm)	Width (mm)
0.75	3.5	27	300	20

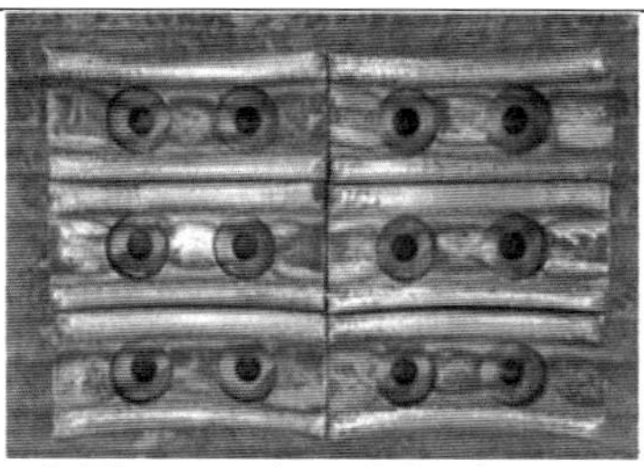

Figure 5. The samples of the throw-out plate

4. Conclusions

Based on a road milling machine original and improved throw-out plate, material composition, hardness and wear properties are analyzed. Using the method of plasma surfacing, material composition, surfacing layer hardness, wear properties and microstructure of the improved throw-out plate are analyzed. The throwing material effect of the original and improved plate is simulated and analyzed. The conclusions are as follows. The throwing material surface adopts the plasma surfacing, and Fe60 powder is as surface hardening material. Both hardness and wear resistance greatly improve. At the same time, the improved structure of the throw-out plate is conducive to enhance the throwing material effect of the milling machine. In this study, there is no comparative study on other iron based materials and materials which are cheap and can improve the wear resistance, when surface hardening materials are used. Only Fe60 is analyzed. Therefore, it is necessary to further optimize and analyze the surfacing materials of the throw-out plate.

References

[1] Haga T, Tsuge H, Ishihara T, Kumai S, and Watari H 2014 *Key Eng. Mater.* **611-612** 623-628
[2] Hao Y, Yu H F, Liu Z, Zhou Y L, and Mu X M 2014 *Coal Sci. Technol.* **42**(11) 67-72
[3] Gao L, de Fortier Smit A, Prozzi J, Buddhavarapu P, Murphy M, and Song L. 2015 *Constr. Build. Mater.* **102** 602-610
[4] Bae A, Lee D, and Park B. 2016 *Can. J. Civil Eng.* **43**(10) 886-896
[5] Shchitsyn Y, Belinin D, Neulybin S, Ol'shanskaya T, Shchitsyn V, and Simonov M. 2017 *Metallurgist* **61**(7-8) 607-612
[6] Sidorov S, Khoroshenkov V, Lobachevskii Ya, and Akhmedova T. 2017 *Metallurgist* **60**(11-12) 1290-1294

IOP Conf. Series: Materials Science and Engineering **324** (2018) 012086 doi:10.1088/1757-899X/324/1/012086

Research to Assembly Scheme for Satellite Deck Based on Robot Flexibility Control Principle

Tao Guo, Ruiqin Hu, Zhengyi Xiao, Jingjing Zhao, Zhikai Fang

Beijing Institute of Spacecraft Environment Engineering. No.104 Youyi Rd. Haidian District. Beijing. China

E-mail: guotao102@163.com

Abstract. Deck assembly is critical quality control point in final satellite assembly process, and cable extrusion and structure collision problems in assembly process will affect development quality and progress of satellite directly. Aimed at problems existing in deck assembly process, assembly project scheme for satellite deck based on robot flexibility control principle is proposed in this paper. Scheme is introduced firstly; secondly, key technologies on end force perception and flexible docking control in the scheme are studied; then, implementation process of assembly scheme for satellite deck is described in detail; finally, actual application case of assembly scheme is given. Result shows that compared with traditional assembly scheme, assembly scheme for satellite deck based on robot flexibility control principle has obvious advantages in work efficiency, reliability and universality aspects etc.

1. Introduction

Docking assembly of the deck is critical quality control point in final satellite assembly process, and important safety control point. Satellite deck generally has 2 kinds of assembly modes: the first one is process that high-mass deck assembly is finished with the help of 4-6 DOF pure mechanical deck overturn trailer vehicles; the second one is process that small-mass deck assembly is finished by relying on manual lifting and support of operators completely. It is difficult to control assembly accuracy of traditional deck assembly mode, and in the mode, assembly efficiency is relatively low, and especially in docking assembly under full- capsule state, cable and equipment bruise or locating pin bush clamping stagnation may appear easily, which affects development quality and progress of satellite to a certain extent.

Industrial robot is one of main equipment in flexible assembly. Foreign robot assembly is widely used in the fields of aviation, aerospace, weapon and missile etc., and is in the leading position in technical research[1]. The theoretical research on robot assembly in large spacecraft component has also been widely carried out in China[2-6]. However, there are few studies on project implementation scheme of large spacecraft component assembly.

Combining with investigation to domestic and overseas satellite assembly mode, this paper proposes assembly project scheme for satellite deck based on robot flexibility control principle to overcome several disadvantages in assembly docking process, such as operation instability, excessive human factor dependency and experience-based judgment etc., reduce collision, bruise and clamping stagnation risks in docking assembly process, and improve assembly quality and safety of docking assembly link.

2. Project scheme

Auxiliary assembly system based on robot flexibility control principle is shown in Figure1, and mainly consists of mobile platform, industrial robot, end effector and six-dimensional force sensor etc. Main functions of each part are as follows: mobile platform is used for rapid transport of industrial robot at different final assembly stations and support and fixation to robot; industrial robot and supporting control cabinet are used to finish adjustment to pose required and docking in deck assembly process; end effector provides mechanical interfaces with six-dimensional force sensor, acceleration sensor and deck, and realizes accurate positioning and reliable clamping simultaneously; six-dimensional force sensor is important sensor of intelligent robot, and it can detect all force information of three-dimensional space (Cartesian coordinate system) simultaneously, i.e. 3 force components and 3 moment components.

Project scheme is a kind of assembly scheme with participation of people. Operator shall finish state confirmation before deck assembly and in assembly process, robot will replace docking overturn function of original overturn trailer vehicle, and support function of original overturn trailer vehicle is realized through relatively light and simple support structures.

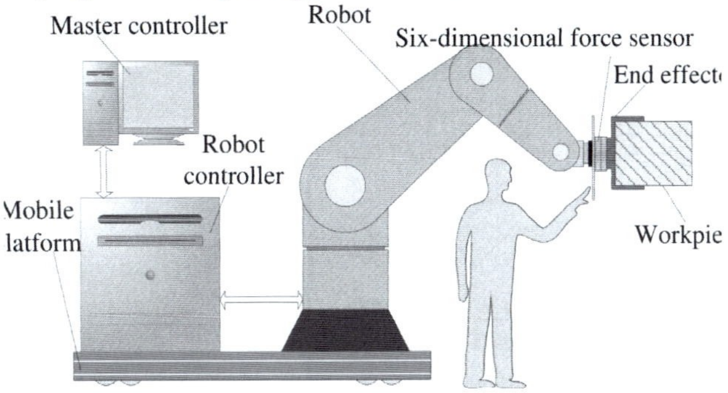

Figure 1. Diagram on System Composition

3. Flexibility control principle

3.1. End force perception

Six-dimensional force sensor is generally installed between robot and load, load gravity is imposed to six-dimensional force sensor, and to obtain acting force of the external to load through six-dimensional force sensor, effect on gravity of load and inertia force must be made. If there is no external force imposed to load, information on force and moment measured by six-dimensional force sensor is completely caused by load gravity. Under the condition, multigroup measured data of six-dimensional force sensor can be obtained by controlling robot to make load under different space postures, and least square method shall be utilized then to solve load weight and gravity center coordinate[3,7].

Schematic diagram on effect of load gravity in six-dimensional force sensor coordinate system is shown in Figure 2, six-dimensional force sensor coordinate system is space rectangular coordinate system with 3 coordinate axes, i.e., X, Y and Z, and load gravity is G, coordinate of load centroid in six-dimensional force sensor coordinate system is (x, y, z), and acting component force of load gravity G in X, Y and Z respectively is G_x, G_y, G_z, and applied moment of load gravity G to X, Y and Z respectively is M_{gx}, M_{gy}, M_{gz}, and according to relationship between force and moment, there is:

$$\begin{cases} M_{gx} = G_z \times y - G_y \times z \\ M_{gy} = G_x \times z - G_z \times x \\ M_{gz} = G_y \times x - G_x \times y \end{cases} \quad (1)$$

Measure and obtain multigroup six-dimensional force sensor data by selecting numerous different load poses, to solve and obtain coordinate of load centroid in six-dimensional force sensor coordinate system, （x， y， z） and size of load gravity G.

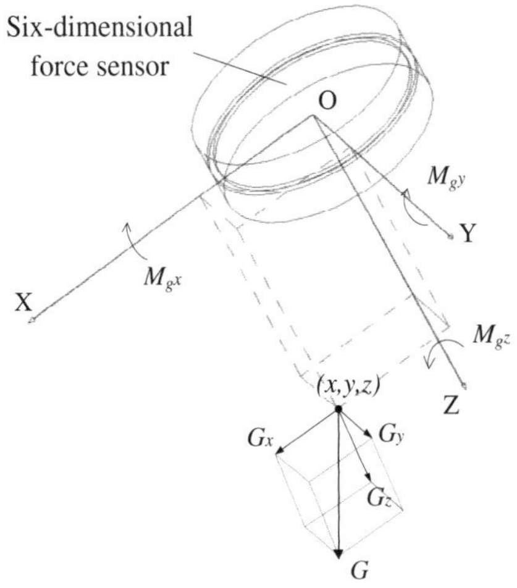

Figure 2. Schematic Diagram on Effect of Load Gravity in Six-dimensional Force Sensor Coordinate System

Assumed that 3 force components measured by six-dimensional force sensor under the condition that external force is imposed to load are F_x, F_y, F_z, and 3 moment components are M_x, M_y, M_z.

In flexibility control process, direction of load gravity in six-dimensional force sensor coordinate system changes with change of posture at end of mechanical arm. Through calibration to placement posture of mechanical arm, included angles, α, β, γ ,between gravity direction and axis X, Y and Z of six-dimensional force sensor coordinate system can be obtained in real time through mechanical arm control system, and projections of load gravity in 3 coordinate axes of six-dimensional force sensor coordinate system can be calculated as follows:

$$\begin{cases} G_x = G \times \cos\alpha \\ G_y = G \times \cos\beta \\ G_z = G \times \cos\gamma \end{cases} \tag{2}$$

G_x, G_y, G_z are parts generated because of load gravity in force component F_x, F_y, F_z measured by six-dimensional force sensor, and substitute them and load gravity coordinate obtained, （x， y， z） into formula (1), to obtain M_{gx}, M_{gy}, M_{gz}, which are parts generated because of load gravity in moment component data M_x, M_y, M_z measured by six-dimensional force sensor, thus obtaining:

Components of external force in 3 coordinate axes are:

$$\begin{cases} F_{ex} = F_x - G_x \\ F_{ey} = F_y - G_y \\ F_{ez} = F_z - G_z \end{cases} \tag{3}$$

Components of external moment in 3 coordinate axes are:

$$\begin{cases} M_{ex} = M_x - M_{gx} \\ M_{ey} = M_y - M_{gy} \\ M_{ez} = M_z - M_{gz} \end{cases} \tag{4}$$

3.2. Flexible docking control

Schematic diagram on flexible docking is shown in Figure 3. When 2 docking surfaces approach to each other under unparallel condition, point contact or line contact will occur firstly, keep that contact force is within safety range, and adjust pose of workpiece installed according to force and moment information generated through contact to make 2 installation surfaces approach to each other continuously and turn to parallel until it is impossible to make approach under the premise that safety contact force is guaranteed, which means that installation surface docking is finished.

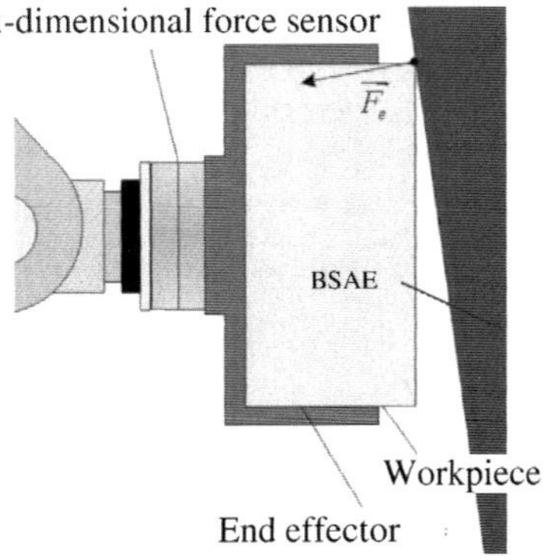

Figure 3. Schematic Diagram on Adjustment to Flexible Docking Pose of Mechanical Arm

Considering actual installation demand, in docking adjustment, in direction perpendicular to installation surface, it is required that contact force shall be kept within safety range, and in direction parallel to installation surface, workpiece shall still move along installation surface. Define robot base coordinate system in installation surface, which is recorded as BASE, and define axis Z perpendicular to installation surface with external direction being BASE, and after contact between workpiece and installation surface, external force imposed to workpiece, $\overrightarrow{F_e}$, is perpendicular to installation surface.

Flexible docking control flow chart of mechanical arm is shown in Figure 4. In control, information on external force/ moment imposed to load is obtained in real time, and judgment control is made according to different external effect information:

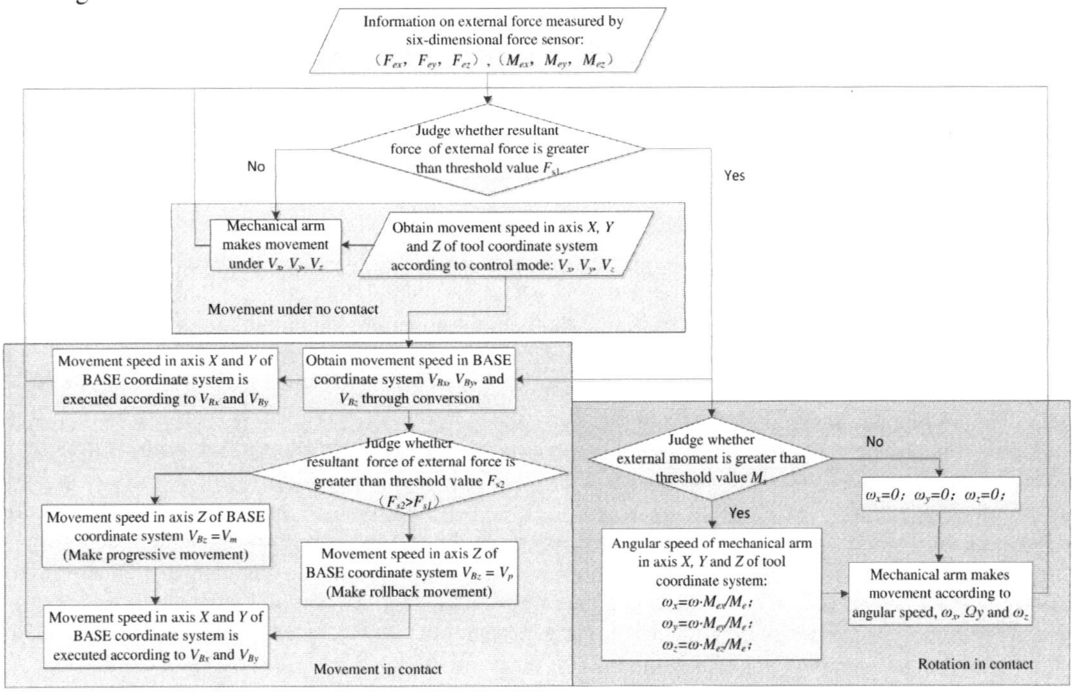

Figure 4. Flexible Docking Control Flow Chart of Mechanical Arm

3.2.1. Movement *under no contact:* judge whether resultant force $|\vec{F_e}|$ of external force is greater than threshold value F_{s1} preset, and if $|\vec{F_e}| < F_{s1}$, it is considered that there is no contact, and control movement of mechanical arm towards installation surface according to control strategy in free space; if $|\vec{F_e}| > F_{s1}$, it is considered that there is contact, control mechanical arm according to control strategy for force/position to realize flexible docking of workpiece.

3.2.2. Movement *in contact:* if $|\vec{F_e}| > F_{s1}$, it is considered that there is contact, and speed of BASE coordinate system in axis *X, Y* and *Z* shall be controlled independently. Convert coordinate system speed of robot tool for input control to component of BASE coordinate system in axis *X* and *Y* to obtain speed of BASE coordinate system in axis *X* and *Y*. Speed of BASE coordinate system in axis *Z* shall be fed back and controlled according to size of contact force. When $|\vec{F_e}| > F_{s2}$, workpiece will make rollback movement, and when $|\vec{F_e}| < F_{s2}$, workpiece will make progressive movement.

3.2.3. Rotation *in contact:* if $|\vec{F_e}| > F_{s1}$, it is considered that there is contact, and judge whether external moment $|\vec{M_e}|$is greater than threshold value preset, M_s; if $|\vec{M_e}| < M_s$, it is considered that no rotation is required, and if $|\vec{M_e}| > M_s$, then convert angular speed component of mechanical arm according to moment component, and adjust posture of workpiece.

4. Implementation flow

Logic control chart on robot auxiliary assembly flow system is shown in Figure 5.

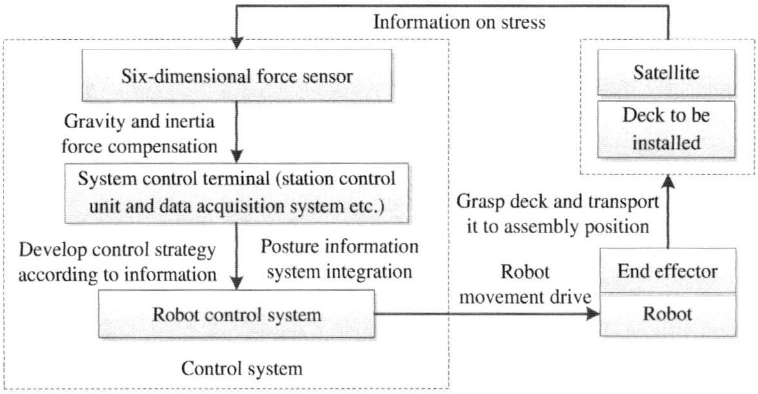

Figure 5. Logical Relationship Chart on Assembly Flow System

Utilize robot and six-dimensional force sensor to finish mass detection of deck at different poses, transmit mass detection data to upper computer through Ethernet and upper computer calculates weight of deck to be installed and centroid position, which means obtaining gravity compensation parameter of deck to be installed correctly.

Make manual traction to realize movement of end effector towards target location, large six-dimensional force sensor measures external force imposed to deck in real time, and develop mixed control strategy for force/ displacement to form control command.

Lower computer gains mixed control strategy for force/ displacement of upper computer and control command through RSI (Robot Sensor Interface), and robot drives motor to make movement according to direction of force/ moment imposed. Control robot to make continuous and compliant movement through real-time measurement feedback, to realize compliant force control in movement process.

Above process is coarse positioning process on large stroke (away from satellite) of deck to be installed. When deck to be installed approaches to target location, switch movement mode of robot to fine adjustment mode, so that robot can be operated under lower speed. Compliant force control mode will be adopted for compliant assembly of deck until docking requirements are met.

5. Application case and effect

5.1. Key point design

Compared with large-batch production line assembly, satellite deck assembly is characterized by small batch, and uncertain assembly station etc. Therefore, assembly scheme for satellite deck based on robot flexibility control principle must adapt to different assembly sites and working conditions of different assembly. In practical project application, following problems must be considered:

- Support structure used to match with robot must be light and convenient, and reduce demand to site space as much as possible;
- Robot must be provided with movable working platform, which is convenient for adjustment to relative pose relationship between robot and satellite;
- Height of movable working platform shall be proper, and ensure that movement envelope of robot can cover opening and closing route of deck;
- Choose model of sensor according to the maximum force and the maximum moment born by 6 DOF sensors in deck assembly process, and the maximum force and the maximum moment value of sensor must be greater than 1.2 times of practical application condition;
- In flexible docking process, line speed of deck in movement cannot exceed 6mm/s, and angular speed in rotation cannot exceed 1°/s, and ensure that operator has enough response time to stop robot under accident.

5.2. Application case

At present, assembly scheme for satellite deck based on robot flexibility control principle has been successfully applied to satellite. Full load weight m_1 of deck is 80Kg roughly, and the maximum size of deck cannot exceed 1500mm×1000mm. Design simple support structure as shown in Figure 6 by combining with practical application condition. Support structure consists of 2 parts, respectively being overturn support and holder. Overturn support is directly connected to deck, and robot finishes opening and closing of deck by grasping end effector on overturn support; holder can support overturn support and deck. Total weight m_2 of overturn support and holder does not exceed 20Kg, and load force arm L of centroid of composite solid constituted with deck to 6 DOF sensors does not exceed 500mm. Therefore, the maximum moment M imposed to 6 DOF sensors is as follows:

$$M=(m1+m2)\times L<(80+20)\times 0.5=500Nm \quad (5)$$

Model of 6 DOF sensors is determined as Omega190 of ATI Company (SI-3600-700, the maximum force is 3600Kg and the maximum moment is 700Nm). Model of robot is determined as KUKA-210 (the maximum bearing is 210Kg). Final assembly process of deck is shown in Figure 7.

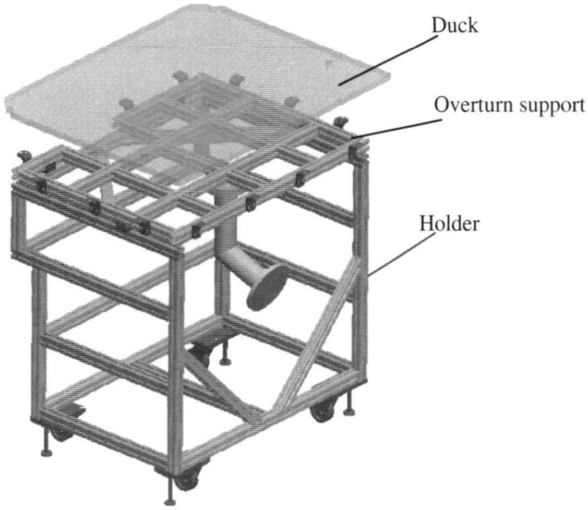

Figure 6. Schematic Diagram on Support Structure

Figure 7. Assembly Process of Deck

5.3. Application effect

Compared with traditional deck assembly method, robot auxiliary assembly scheme has obvious advantages in work efficiency, reliability and universality aspects etc., which is shown in Table 1.

Table 1. Advantage of Robot Assembly Scheme

Aspect	Traditional method	Robot Assembly
Work efficiency	4-5 operators are required in deck overturn docking process, and operation time is 0.5~1 hour with low work efficiency.	3-4 operators are required in deck overturn docking process, and operation time is 0.5 hour roughly and work efficiency is improved obviously.
Reliability	Assembly process is rigid assembly, which may cause deck bruise and dowel clamping stagnation easily, and has high requirement to skill level of operator, with relatively bad reliability.	Assembly process is flexible docking, which avoids dowel clamping stagnation basically, and has low dependency to operator with high reliability.
Universality	Overturn trailer vehicle is developed through customization and can only be used for certain satellite.	Robots are general goods, and can be used for assembly process of other satellite.

6. Conclusion

Aimed at problems existing in satellite deck assembly process, assembly scheme for satellite deck based on robot flexibility control principle is proposed in this paper, and in virtue of end force perception and flexibility control technology of robot, high-efficiency and high-reliability installation of satellite deck can be realized. Assembly scheme for satellite deck based on robot flexibility control principle has certain universality, which can provide reference for assembly process of other similar large spacecraft component.

7. Reference

[1] Zhang Naifeng, Zhang Zhixian and Tao Weiqian 2012 *J. Robot Technique and Application*, **6** 9-11

[2] Wei Yuee, Zhang Chengli, Meng Wei, Xiong Tao and Hu Ruiqin 2014 *J. Aeronautical Manufacturing Technology*, **21** 147-152

[3] Zhang Lijian, Hu Ruiqin and Yi Wangmin 2017 *J SensorActa Automatica Sinica*, **43** 439-447

[4] Buren, Sun Gang, Hu Ruiqin, Fu Hao, Zhang Lijian and Tang Laiying 2014 *J. Spacecraft Environment Engineering*, **31** 430-435

[5] Qiu Tiecheng, Zhang Man, Zhang Liwei, Zhao Lu and Xu Yiliu 2012 *J. Spacecraft Environment Engineering*, **29** 579-585

[6] Xiong Tao 2011 *J. Aeronautical Manufacturing Technology*, **22** 6-39

[7] Kubela T, Pochyly A, Singule V and Flekal L 2012 Mechatronics: Recent Technological and Scientific Advances (Berlin: Springer-Verlag) 429-437

IOP Conf. Series: Materials Science and Engineering **324** (2018) 012087 doi:10.1088/1757-899X/324/1/012087

Wear behavior of AA 5083/SiC nano-particle metal matrix composite: Statistical analysis

Amir Hussain Idrisi[1], Abdel-Hamid Ismail Mourad[2], Dinu Thomas Thekkuden[1], John Victor Christy[1]

[1] **Research Scholar, Department of Mechanical Engineering, UAE University, PO. Box 15551 Al-Ain, UAE**
[2] **Professor, Department of Mechanical Engineering, UAE University, PO. Box 15551 Al-Ain, UAE**
E-mail: ahmourad@uaeu.ac.ae

Abstract: This paper reports study on statistical analysis of the wear characteristics of AA5083/SiC nanocomposite. The aluminum matrix composites with different wt % (0%, 1% and 2%) of SiC nanoparticles were fabricated by using stir casting route. The developed composites were used in the manufacturing of spur gears on which the study was conducted. A specially designed test rig was used in testing the wear performance of the gears. The wear was investigated under different conditions of applied load (10N, 20N, and 30N) and operation time (30 mins, 60 mins, 90 mins, and 120mins). The analysis carried out at room temperature under constant speed of 1450 rpm. The wear parameters were optimized by using Taguchi's method. During this statistical approach, L27 Orthogonal array was selected for the analysis of output. Furthermore, analysis of variance (ANOVA) was used to investigate the influence of applied load, operation time and SiC wt. % on wear behaviour. The wear resistance was analyzed by selecting "smaller is better" characteristics as the objective of the model. From this research, it is observed that experiment time and SiC wt % have the most significant effect on the wear performance followed by the applied load.

Keywords: Metal matrix composite, 5083 Al Alloy, SiC nano reinforcement, wear test, ANOVA, Taguchi method.

1. Introduction

Aluminium alloys play a crucial role in various engineering fields due to their superiority in quality to weight proportion and other mechanical properties [1]. These alloys are used as matrix phase in aluminium matrix composites. These composites achieve enormous interest and are widely utilized in various applications of making automobile and aerospace components because of their different essential properties like light weight, bring down cost, high stiffness, easy fabrication and high dimensional stability which aided to mould the material to desired shape and size. The metal matrix composites (MMCs) are most commonly used in engineering applications due to their ease of fabrication, lower costs, recyclability and isotropic properties [2].

Aluminium alloy reinforced with particulates are a most promising way to enhance the wear properties of any metal matrix composites (MMCs). There are numerous reinforcements which have been utilized for improving the mechanical characteristics including wear properties of aluminum based alloys, but the fillers such as silicon carbide [3–9] and alumina [10–12] to aluminium alloys were also reported to enhance the tribological properties of AMCs. For determining the better reinforcement aluminum matrix composites reinforced with SiC and Al_2O_3 are compared for wear performance. The analysis of results showed that SiC particles were more effective than Al_2O_3 particles for the enhancement of wear resistance of aluminum matrix composites due to its property of high hardness [13, 14]. Given results expressed that SiC preferably can be used as reinforcement. It was also reported that temperature widely affects the tribological properties of the AMCs. Martin et al. [15] conducted wear test on fabricated composite and stated that at the temperature range of 20-200 Al alloy reinforced with 15 vol% SiC particles showed the transition from mild wear to sever wear with the increase in temperature. Similarly, Muratoglu et al. [16] also studied the effect of temperature on wear behavior of Al-Cu alloy reinforced with 25 vol% SiC. The study demonstrated that wear resistance decreases with increase in temperature for the base metal as well as reinforced composite material.

Various combinations of 5083 Al alloy and SiC have been utilized to improve the tribological properties of aluminum matrix composite. Gargatte et al. [17] fabricated AA 5083 reinforced with SiC particles using stir casting technique, where the different volume fraction of SiC particles (3, 5 and 7 wt%) were used for synthesis. The wear test was conducted on the pin-on-disc testing machine to examine the wear behavior of the aluminum alloy and its composites. It was observed that the wear rate decreased by increasing the reinforcement percentage of SiC particles. Al-5083 reinforced with 3, 5 & 7 wt % shows lesser wear rate compared with pure Al-5083 alloy. Similarly, a dry sliding wear test was performed on 5083 Al alloy composite with 10 wt.% SiC by using the pin-on-disk machine. For testing normal load (10N, 20N, 30N) was applied for constant time of 30 minutes at different sliding distance (754 m, 1131 m, 1508 m, 1885 m) and the sliding velocity (0.42 m/s, 0.63 m/s, 0.84 m/s, 1.04 m/s). It was found that applied load has the highest influence on the wear rate by Ravindra et al. [18]. Furthermore, a wide range of SiC weight percentage (5%, 10%, 20% and 25%) utilized as reinforcement by Manoj Singla et al. [19] for fabrication of composite and reported that 20% weight fraction of SiC particles has minimum wear under different applied load. The structure and properties of these metal matrix composites can be varied by the size as well as the type of the reinforcement and nature of bonding [20]. Aluminium matrix composite strengthened with micron or nano-sized particles enhanced mechanical and physical properties in composite materials. The distribution of nano-sized reinforcing particles also affects the morphology and interfacial characteristics of nano-composites as reported by Xiaoxin Y et al. [21].

For analysis of wear results statistical technique used by many researchers. S. Basavarajappa et al. [22] used Taguchi's method for investigating the influence of wear parameters like as normal load, sliding speed and sliding distance on dry sliding wear of the composites. Prasanta [23] used L27 Taguchi orthogonal design for optimization of four process parameters, viz., concentration of reducing agent, concentration of nickel source solution, bath temperature, and annealing temperature. The wear characteristics of electroless Ni–P coatings sliding is used as the response. It is observed that annealing temperature and bath temperature have the most significant influence on wear characteristics of electroless Ni–P coating. Wear performance was also analyzed by sachin [24] for the composite and the matrix alloy size using a linear factorial design approach. The linear regression equation for the wear rate of the alloy and composite when tested against the SiC and Al2O3 emery paper expressed in term of sliding distance, applied load and abrasive size. It was observed that the wear rate increased with increasing abrasive size, applied load and sliding distance for SiC emery paper, whereas the wear rate increased with abrasive particle and applied load, and it decreased with sliding distance for Al2O3 paper. In the light of discussed literature, an attempt is made to study the effect of applied load, experimental time and SiC wt % on wear behavior of the Al/SiC nano-composites using Taguchi design of experiments. The analysis of variance (ANOVA) was employed to determine the influence percentage of various factors and its interaction on the wear of the composites.

2. Material and method

2.1 Materials

Al 5000 arrangement is widely utilized as a part of marine and aviation applications considering their dominant corrosion resistance, formability, and great welding qualities. However, because of low strength and poor wear resistance, the utilization of this arrangement is limited. Al 5083 alloy used as a base metal for this work. It is a high Mg-Al composite with light-weight and broadly utilized for the marine and automobile applications. The composition is shown in Table 1. The aluminium alloy was strengthened by mixing nano-particles of SiC through stir casting process [25].

Table 1. Al 5083 alloy chemical composition

Element	Zn	Ti	Si	Cu	Fe	Cr	Mn	Mg	Al
% Present	0.25	0.15	0.4	0.1	0.4	0.05-0.25	0.4-1.0	4.0-4.9	Balance

IOP Conf. Series: Materials Science and Engineering **324** (2018) 012087 doi:10.1088/1757-899X/324/1/012087

2.2 Basic concept of Design of experiments (DOE)

The system of laying out the states of analyses including various elements was first proposed by the Englishman, Sir R.A. Fisher [26]. This technique is known as the factorial design of experiments. Taguchi developed technique for directing the design of experiments which depend on all around characterized rules. This method utilizes a set of arrays called orthogonal arrays. These standards exhibit stipulate the technique of performing the minimum number of tests which could give the full data of the considerable number of variables that influence the execution parameter [27]. The crux of the orthogonal arrays strategy lies in picking the level combinations of the input design factors for each analysis.
The design of an experiment involves the following steps:

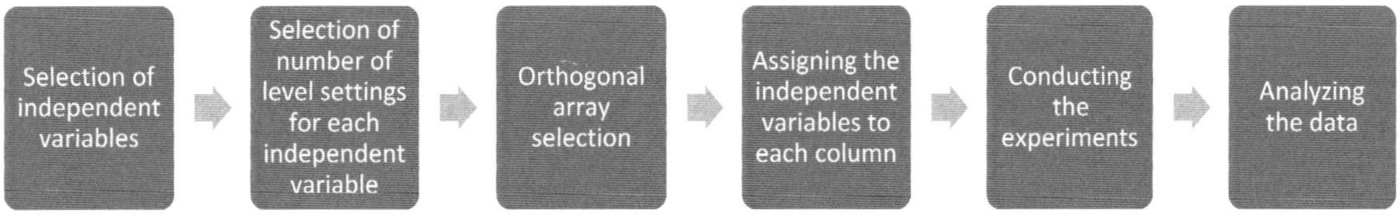

Figure 1. Steps for design of an experiment

2.3 Plan of experiments

There are numerous standard orthogonal arrays which can be implied for number of independent design factors and levels. The L27 orthogonal array is implied for understanding the impact of 3 autonomous elements each having 3 levels of value. This array has 27 rows and 13 columns. There are three parameters (applied load, time, and SiC weight percentage) which were used for performing the wear analysis furthermore three levels of these parameter were selected. The levels of these factors decided for experimentation are given in table 2.

Table 2. Variables and their levels

Level	Applied Load (N)	Experiment time (mins)	SiC (wt%)
1	10	30	0
2	20	60	1
3	30	90	2

3. Results and discussion

3.1 Results for Statistically analysis of wear rate

The experiments were directed according to the orthogonal array. For determining the effect of each variable on the output, the signal-to-noise ratio or the SN number, needs to be evaluated for each conducted experiment. The SN ratios were also obtained for various combinations of parameters by using MINITAB 17 as shown in table 3. The experimental values were transformed into S/N ratios for measuring the quality attributes.

Table 3. Results of L27 Orthogonal array for AA 5083 composite reinforced with nanoparticles

Sr. No.	Applied Load (N)	Experiment Time (mins)	SiC wt %	Wear (gms)	SNRA1
1	10	30	0	0.02235	33.01445
2	10	30	1	0.02293	32.79192
3	10	30	2	0.01392	37.12722
4	10	60	0	0.05312	25.49484
5	10	60	1	0.04853	26.27979
6	10	60	2	0.03486	29.15345
7	10	90	0	0.08356	21.56003

8	10	90	1	0.07247	22.79683
9	10	90	2	0.05455	25.2641
10	20	30	0	0.02924	30.68045
11	20	30	1	0.02686	31.41788
12	20	30	2	0.01584	36.0049
13	20	60	0	0.06621	23.58153
14	20	60	1	0.05243	25.6084
15	20	60	2	0.03719	28.59148
16	20	90	0	0.10146	19.8741
17	20	90	1	0.07666	22.30862
18	20	90	2	0.057	24.8825
19	30	30	0	0.03772	28.46857
20	30	30	1	0.03275	29.69577
21	30	30	2	0.01659	35.60307
22	30	60	0	0.07634	22.34496
23	30	60	1	0.06044	24.37351
24	30	60	2	0.03983	27.99579
25	30	90	0	0.10954	19.20855
26	30	90	1	0.08913	20.99952
27	30	90	2	0.06194	24.16058

3.2 Minitab analysis of the Taguchi design

The Influence of the control process parameters such as load, experiment time and SiC weight % on wear has been analyzed. The ranking of process parameters is obtained by Minitab using the signal to noise ratios and means for different parameter levels for wear.

Table 4. Response Table for Signal to Noise Ratios

Level	Applied Load (N)	Experiment time (mins)	SiC wt(%)
1	28.16	32.76	24.91
2	26.99	25.94	26.25
3	25.87	22.34	29.86
Delta	2.29	10.42	4.95
Rank	3	1	2

Table 5. Response Table for Means

Level	Applied Load (N)	Experiment time (mins)	SiC wt (%)
1	0.04514	0.02424	0.06439
2	0.05143	0.05211	0.05358
3	0.05825	0.07848	0.03686
Delta	0.01311	0.05423	0.02754
Rank	3	1	2

Table 4 and table 5 demonstrate the impact of elements on the different levels of trial and contribution to the general wear of Al Alloy composite. It can be comprehended that time is more critical as it has

IOP Conf. Series: Materials Science and Engineering **324** (2018) 012087 doi:10.1088/1757-899X/324/1/012087

the first rank for the fabricated composite component. The impact of SiC wt% on wear behaviour is more than the applied load however not as much as experiment time for SiC nanoparticles.

3.3 Residual Analysis

The main effect plot for mean wear and S/N ratios showed in figure 2(a) and figure 2(b) respectively. The significance of each parameter is determined from the inclination of the main effects plot. It is clear from the main effect plot for wear and main effect plot for S/N ratio that experiment time is a most significant parameter, while SiC wt% also has some considerable effect. These plots were generated based on the condition that "smaller is better" so for minimizing the wear, applied load and experiment time level should be low whereas SiC wt% level must be high. It is also observed from figure 2(a) and figure 2(b) that the parameter for which the line has the higher slope will have the most significant effect.

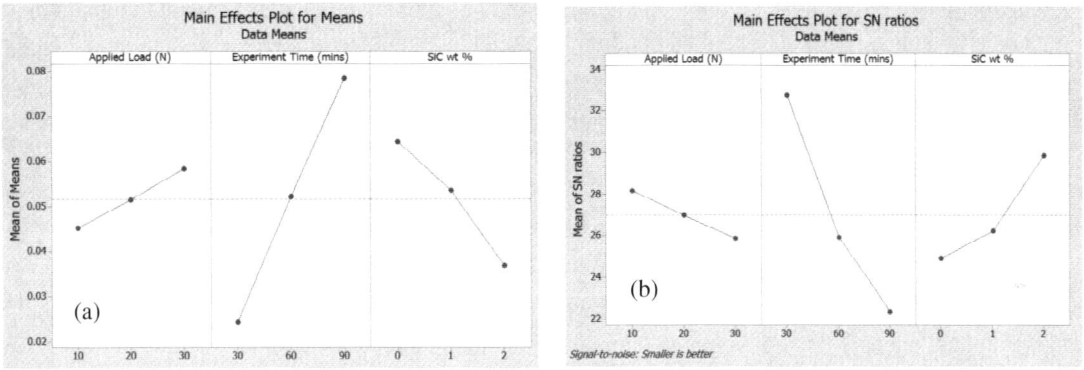

Figure 2. Main effects for plot for a) Means b) SN Ratio

The interaction plot between the process parameters such as applied load, experiment time and SiC wt % are also shown in figure 3. The non-parallel lines of parameter play a vital role in the study of an interaction plot means. If the lines of an interaction plot are not parallel, it suggests that there is nominal interaction occurred and if the lines interact each other, then strong interaction occurred between the parameters. Figure 3 shows that there is a substantial interaction between the experiment time and SiC wt.% while moderate interaction between the applied load and experiment time and between applied load and SiC wt.%.

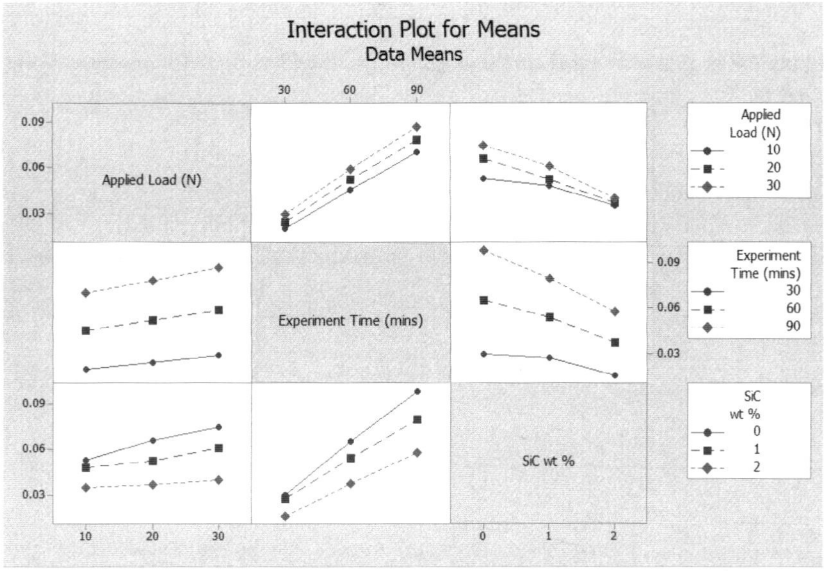

Figure 3. Interaction effect plot of process parameters on wear

3.4 Results of ANOVA in Minitab

The experimental results were analyzed by analysis of variance (ANOVA) which is used to investigate the influence the wear parameters like applied load, sliding distance and sliding speed. By performing the analysis of variance, it can be decided which factor dominate over the other and the percentage contribution of that particular independent variable. The analysis of variance results for the wear rate for three factors varied at three levels and interactions of those factors. The analysis showed in table 6 was carried out for the confidence level 95%.

Table 6. Analysis of variance of nanocomposite

Source	DF	Seq SS	Contribution	Adj SS	Adj MS	F-Value
Applied Load (N)	2	0.000774	4.23%	0.000774	0.000387	137.19
Experiment Time (mins)	2	0.01324	72.40%	0.01324	0.00662	2347.2
SiC wt %	2	0.003464	18.95%	0.003464	0.001732	614.16
Applied Load (N)*Experiment Time (mins)	4	0.000041	0.23%	0.000041	0.00001	3.65
Applied Load (N)*SiC wt %	4	0.000223	1.22%	0.000223	0.000056	19.75
Experiment Time (mins)*SiC wt %	4	0.000521	2.85%	0.000521	0.00013	46.23
Error	8	0.000023	0.12%	0.000023	0.000003	
Total	26	0.018286	100.00%			

It can be observed in table 6 that for aluminium metal matrix composites, that time and SiC wt % has the highest influence on wear

3.5 Multiple linear regression model

A multiple linear regression model is developed using statistical software "MINITAB 17". This model gives the relationship between an independent / predicted variable & a response variable by fitting a linear equation to observed data. The generated regression equation established a correlation between the significant terms obtained from ANOVA analysis namely applied load, experiment time and SiC wt%. The regression equation developed for wear calculation in weight (grams) as given in equation (1)

Wear (gms) = -0.01585 + 0.000699 Applied Load (N) + 0.000998 Experiment Time (mins)
+ 0.00751 SiC wt % + 0.000006 Applied Load (N)*Experiment Time (mins)
- 0.000413 Applied Load (N)*SiC wt % - 0.000217 Experiment Time (mins)*SiC wt % ....(1)

From above equation, it is observed that the experiment time has significant role in the wear weight (grams) followed by SiC wt %.

4. Conclusions

From the Taguchi experimental design method results, it is found that the experiment time has the highest influence on wear rate followed by SIC wt % then the applied load for Al-5083 composite reinforced with SiC nano-particles. It is also clear from the ANOVA results that the contribution of applied load, experimental time and SiC wt % is 4.23%, 72.40%, 18.95% respectively in the analysis. it can be concluded from this study that for minimizing the wear, load and time level should be low whereas SiC wt% level must be high.

IOP Conf. Series: Materials Science and Engineering **324** (2018) 012087 doi:10.1088/1757-899X/324/1/012087

5. References:

[1] A. Baradeswaran, A. Elayaperumal, "Effect of Graphite Content on Tribological behaviour of Aluminium alloy - Graphite Composite", European Journal of Scientific, 53(2): 163-170.
[2] M. S. Zedalis and P. S. Gilman, "Aluminum based metal matrix composites," ed: Google Patents, 1990.
[3] Idrisi, Amir Hussain, and Abdel-Hamid Ismail Mourad. "Fabrication and Wear Analysis of Aluminium Matrix Composite Reinforced by SiC Micro and Nano Particles." In ASME 2017 Pressure Vessels and Piping Conference, pp. V06AT06A033-V06AT06A033. American Society of Mechanical Engineers, 2017..
[4] E. Candan, H. Ahlatci, H. Cimenoglu, "Abrasive wear behavior of Al–SiC composites produced by pressure infiltration technique", Wear 247 (2001) 133–138.
[5] Idrisi Hussain, A., V. Singh Dev, and Vipul Saxena. "Development and testing of Al5083 alloy reinforced by SiC particles." International journal of scientific research engineering and technology 2, no. 11 (2014): 697-704.
[6] S. Won, U. Jong, S. Won, D. Keun, K. Ogi, "Heat treatment and wear characteristics of Al/SiCp composites fabricated by duplex process", Compo-sites Part B—Engineering 34 (2003)737–745.
[7] Idrisi Hussain, A., Deva, S. "Development and testing of metal matrix composite by reinforcement of SiC particles on Al 5XXX series alloy", International Journal of Scientific Research Engineering & Technology (2014), 697-704.
[8] Y. Sahin, "Abrasive wear behavior of SiC/2014 aluminum composite, Tribology International", 43(2010)939–943.
[9] S. K. Ghosh, P. Saha, "Crack and wear behavior of SiC particulate reinforced aluminium based metal matrix composite fabricated by direct metal laser sintering process", Materials Design 32 (2011) 139–145.
[10] L. J. Yang, "The transient and steady wear coefficients of A6061 aluminum alloy reinforced with alumina particles", Composites Science and Technology 63 (2003) 575–583.
[11] A. M. Al-Qutub, I. M. Allam, T. W. Qureshi, "Effect of sub-micron Al2O3 concentration on dry wear properties of 6061 aluminum based composite", Journal of Materials Processing Technology 172 (2006) 327–331.
[12] A. M. Al-Qutub, I. M. Allam, M. A. Abdul Samad, "Wear and friction of Al–Al2O3 composites at various sliding speeds", Journal of Materials Science 43 (2008) 5797–5803.
[13] Hosking FM, Folgar Portillo F, Wonderlin R, Mehrabian R., "Composite of aluminium alloys: fabrication and wear behaviour", Journal Material Science, 17 pp.477–98
[14] Lee CS, Kim YH, Han KS, Lim T. 1992, "Wear behaviour of aluminium matrix composite materials", Journal Material Science, 27, pp793–800.
[15] Martin A., Martinez M. A., LLorca J. (1996), "Wear of Sic - reinforced Al - matrix Composites in the Temperature Range 20-2000C", Wear 193, 169-179.
[16] Muratoglu M., Ak s oy M. (2000), "The Effects of Temperature on Wear Behaviors of Al–Cu alloy and Al–Cu/SiC composite", Materials Science and Engineering A282, 91–99.
[17] Gargatte, Sourabh, Rahul R. Upadhye, Venkatesh S. Dandagi, Srikanth R. Desai, and Bhimappa S. Waghamode. "Preparation & Characterization of Al-5083 Alloy Composites." Journal of Minerals and Materials Characterization and Engineering 1, no. 01 (2013): 8.
[18] Ravindra Singh Rana, Rajesh Purohit, Anil kumar Sharma, SaraswatiRana, "Optimization of Wear Performance of AA 5083/10 Wt. % SiC_pComposites Using Taguchi Method", Procedia Materials Science 6 (2014) 503 – 511.
[19] Manoj Singla, Lakhvir Singh, Vikas Chawla, "Study of Wear Properties of Al-SiC Composites", Journal of Minerals & Materials Characterization & Engineering, Vol. 8, No.10, pp.813-819, 2009.
[20] Sanjay Kumar Thakur, Brij Kumar Dhindaw, "The influence of interfacial characteristics between SiCp and Mg/Al metal matrix on wear, coefficient of friction and micro hardness", Wear 247 (2001) 191–201.
[21] Xiaoxin Y, Tao L, Yongda Y, Haibo W, Guoyi T, et al. (2015), "Enhanced grain refinement and micro hardness of Ti-Al-V alloy by electropulsing ultrasonic shock", Journal of Alloys and Compounds 621: 66-70.

[22] Basavarajappa, S., G. Chandramohan, and J. Paulo Davim. "Application of Taguchi techniques to study dry sliding wear behaviour of metal matrix composites." Materials & design 28, no. 4 (2007): 1393-1398.
[23] Sahoo, Prasanta. "Wear behaviour of electroless Ni–P coatings and optimization of process parameters using Taguchi method." Materials & Design 30, no. 4 (2009): 1341-1349.
[24] Sahin, Y. "Wear behaviour of aluminium alloy and its composites reinforced by SiC particles using statistical analysis." Materials & design 24, no. 2 (2003): 95-103
[25] V. N. Gaitonde, S. R. Karnik, M. S. Jayaprakash, "Some Studies on Wear and Corrosion Properties of Al5083/Al2O3/Graphite Hybrid Composites", Journal of Minerals and Materials Characterization and Engineering, 2012, 11, 695-703.
[26] Fisher R. A. 1951. Design of Experiments, Oliver & Boyd, Edinburgh, UK
[27] Taguchi, G., Konishi, S., 1987, Taguchi Methods Orthogonal Arrays and Linear Graphs: Tools for Quality Engineering, American Supplier Institute, Dearborn.

IOP Conf. Series: Materials Science and Engineering **324** (2018) 012088 doi:10.1088/1757-899X/324/1/012088

Large-Scale 3D Printing: The Way Forward

Hamad Al Jassmi[1], Fady Al Najjar[2], Abdel-Hamid Ismail Mourad[3]
[1]Department of Civil Engineering, College of Engineering, UAE University
[2]Department of Mechanical Engineering, College of Engineering, UAE University
[3]Department of Computer Science and Software Engineering, College of Information Technology, UAE University
Email: h.aljasmi@uaeu.ac.ae

Abstract. Research on small-scale 3D printing has rapidly evolved, where numerous industrial products have been tested and successfully applied. Nonetheless, research on large-scale 3D printing, directed to large-scale applications such as construction and automotive manufacturing, yet demands a great a great deal of efforts. Large-scale 3D printing is considered an interdisciplinary topic and requires establishing a blended knowledge base from numerous research fields including structural engineering, materials science, mechatronics, software engineering, artificial intelligence and architectural engineering. This review article summarizes key topics of relevance to new research trends on large-scale 3D printing, particularly pertaining (1) technological solutions of additive construction (i.e. the 3D printers themselves), (2) materials science challenges, and (3) new design opportunities.

Keywords: Additive Construction; Large-Scale 3D Printing

1. Introduction

The hype of large-scale 3D printing was largely triggered by a Chinese Company called 'Winsun', who claimed in the public domain that it was able to print "10 full-sized houses in a day", which only costed $5,000 each, using a 'top-secret' cement recipe made "entirely from recycled materials" [1]. WinSun used Contour Crafting (i.e. gantry-based large-scale 3D printing; see Figure 3), a technology pioneered by Professor Khoshnevis from the University of South California [2]. Despite that Contour Crafting was first announced in 2004, it has only received enormous wide-spread media coverage in 2014 through WinSun. Nonetheless, WinSun's claims in the media were confronted by severe criticisms and doubts on whether the technology is mature enough to go that far [3], especially that the Chinese company did not release any details about the technical characteristics of the used materials [4]. Regardless of WinSun's credibility, the company was surely successful at making large-scale 3D printing a 'hot-topic', and managed to convince governments about it.

Following that hype, the Egyptian Government signed a deal for printing 20,000 single-storey housing units with WinSun [5], the UAE government signed a contract to print the 'Office of the Future' with WinSun, and more recently, Saudi officials studied the use of 3D printing to build 1.5 million housing units with WinSun [6]. Above all, the government of UAE has made a serious stride towards expediting the spread of technology by announcing its target to have 25% of UAE's construction built using large-scale 3D printing, as well as signing a MoU with Autodesk with an investment of 100 million U.S. dollars to boost large-scale 3D printing start-ups. Simultaneous to governmental initiatives, large-scale 3D printing has been dramatically grasping the interest of scientists and technical specialists, as indicated by the evolution of publications during the past five years (Figure 1).

Figure 1 summarises a review of publications relevant to large-scale construction-oriented 3D printing as till September 2015, which the authors refer to as "additive construction". Based on our track of the topic's evolvement, year 2016, which was not included in [7] review, experienced an even more dramatic increase in the number of journal and conference proceeding articles, as more scientists began investigating the technology. Attention to large-scale 3D printing is expected to further grow at an exponential rate.

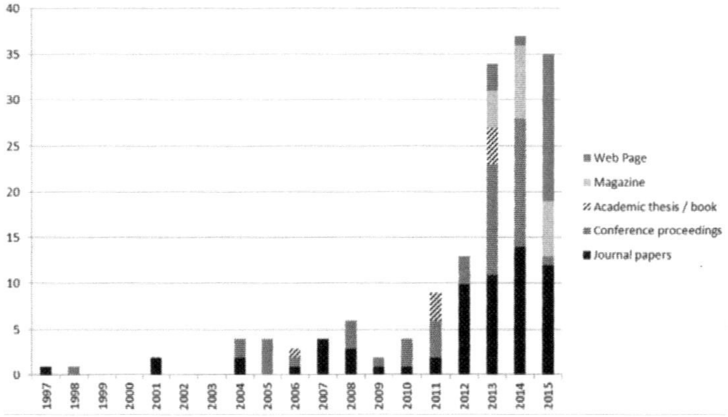

Figure 1. Evolution of number and type of publications relevant to 'large-scale 3D printing' during the last twenty years [7]

2. Overview

Large-scale 3D printing is a special type of *additive manufacturing* that specialises on erecting large-scale, heavy, and often permanent structures. Some authors prefer using the term *additive construction*, in reference to the principle of *additive manufacturing* scaled up for construction [7]. *Additive manufacturing* itself is defined by ASTM as "the process of joining materials to make objects from 3D model data, usually layer upon layer, as opposed to subtractive manufacturing methodologies" [8]. This is what has recently became known as *3D printing*.

3D printing has a relatively long history of development, traced back to the 1980's at which time it was used for rapid prototyping of products in manufacturing. The technology advanced to an extent that it may now replace conventional subtractive manufacturing, and has become accessible to the public. Today a 3D printer is priced as low as $250. Throughout the past 3 decades, numerous approaches to additive manufacturing were invented, including stereolithography, selective laser sintering, fused deposition modeling, and inkjet powder printing. Each is outlined in the Appendix.

While these technologies has been successful for small scale purposes, scaling-up 3D printing for the use of construction, and to replace conventional building methods, is yet a challenge.

According to Labonnote et al. [7], research efforts in large-scale 3D printing has been devoted to three main aspects:

1. technological solutions of additive fabrication (i.e. the 3D printers themselves),
2. material science challenges,
3. and new building design opportunities

The following sections will review current state of research in each of these aspects and will discuss areas of further investigation and potential improvements.

3. Large-scale 3D printing technologies

Research on large-scale 3D printing technologies could be classified as either gantry-based, robotic-arm based, or swarm-based.

3.1 Gantry Solutions

A gantry-based solution is simply a scale-up of additive manufacturing to additive construction. In short, it is "a giant 3D printer", with an actuator controlled in translation to any direction along the X, Y, Z-axes in cartesian coordinates [7]. The most noticeable and noteworthy gantry-based developments are D-Shape and Contour crafting:

- *D-Shape* is a process similar to the inkjet powder bed method (refer to Appendix) where a binder is selectively sprayed on the printing material [9]. When the binder contacts the powder printing material, the solidification process starts and a new layer is added. It takes 24 hours for the solidification process to be completed. D-Shape is a factory gantry-based powder-bed 3D printer that has the ability to print architectural structures with dimensions up to 6 × 6 × 6 m (Figure 2). The products printed using D-shape technology can have a compressive strength up to 235–242 MPa [9]. Surplus powder that is not a part of the structure acts as a support to the structure while the solidification process takes place. Once the printing of an object is completed, it is removed out of the loose powder bed [10]. The surplus powder can be reused for other structures as well. D-Shape technology was mainly designed for factory (off-site) production of structural elements having complex geometries. However, using the technology to print on-site is currently being studied, where construction material available on site, like sand, and binder materials can be used [11].

Figure 2. D-Shape 3D Printing Technology.

- *Contour crafting (CC)* is the most recent and the most promising gantry-based 3-D printing technology developed for the construction industry [9]. CC is an in-situ process where the material (usually cement-based paste) is extruded successively layer by layer [10]. The CC system consists of a gantry system and a nozzle. The gantry system used in CC is very alike to that used in precast concrete fabrication. However, on-site employees are–theoretically–not needed in contour crafting as the process is fully automated, unlike in precast concrete fabrication where on-site employees are required during production to make sure that the discharge system is working properly. A set of actuated and computer-controlled trowels are used to trowel the printed material extruded from the nozzle. This is done to ensure a smooth and accurate planar and free-form surfaces. One of the main issues related to contour crafting is maintaining a uniform level of viscosity. A uniform level of viscosity is required for an improved structural strength and a smoother surface finish [9].

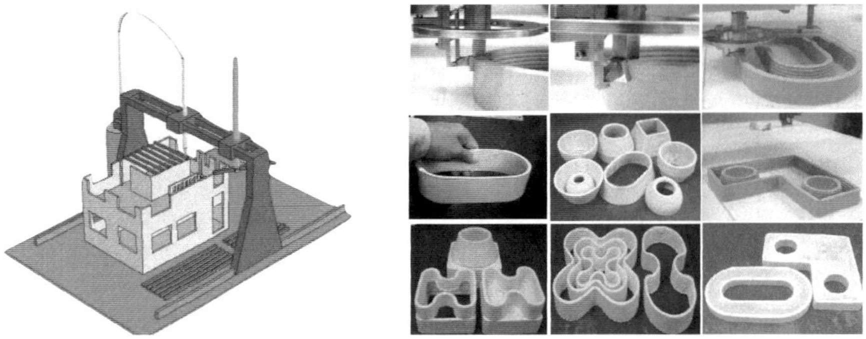

Figure 3. (left) initial Contour Crafting conceptualisation, and (right) first realisation attempts of 2.5D printed concrete shapes [2]

3.2 Robotic Arm Solutions

Although D-Shape and Contour Crafting were successful in scaling up additive manufacturing for the use of construction, they yet suffer from significant limitations. D-Shape is criticised for being unreliable due to the difficulty of hydrating cementitious powder in a satisfactory and regular manner; It is also unpractical as one has to eventually remove the powder and clean the product, adding significant time and effort to the process.

On the other hand, Contour Crafting is criticized for being limited to vertical extrusion. Hence CC does not literally yield 3D, but rather 2.5D typologies (vertical extensions of a planar shape) [11]. As in small-scale 3D printing techniques, the extrusion path in CC uses 3D-to-2D slicing software. The software slices a 3D shape into layers of constant thickness, which are then extruded one up onto the other. Each layer is then made of a contour line and a filling pattern that looks like a honeycomb; whereas the filling density is adjusted on the basis of given requirements. The drawback of such technique is apparent when attempting to print a cantilevered structure; i.e. a structure that sways horizontally from the base layer, as shown in Figure 4 (left). Clearly, weak interfacial zones (red) between the printed layers (grey) are created [11].

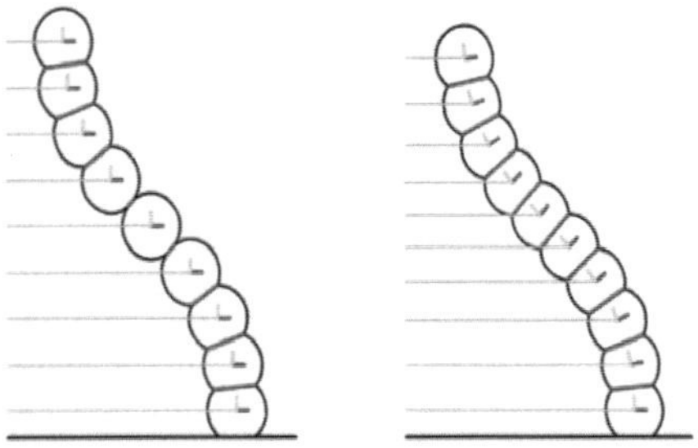

Figure 4. The Tangential Continuity Method (TCM) implemented used a 6-axis robotic arm by Gosselin et al. [11]

In response to the aforementioned limitations, [11]invented the Tangential Continuity Method (TCM) illustrated in Figure 4 (right). The method makes use of the increased degrees of freedom of a 6-axis robotic arm to generate a building path that are, accordingly to [11], truly three dimensional. The paths are made of non-planner layers with locally varying thicknesses (grey). The obvious advantage of such strategy is to increases areas of contact between two layers (red), hence avoiding the geometrical gaps between the layers; This eventually yields more efficient and mechanically sound constructions [11].

As aforementioned, the invention of the Tangential Continuity Method (TCM) would have not been possible without the use of the increased degrees of freedom offered by 6-axis robotic arms. For this purpose, and for their high software and hardware versatility and flexibility, robotic arms has now become the dominant tool for large-scale 3D printing research and development. Industrial robots like Kuka or ABB are accompanied with accessible software and scripting languages that ease the task of planning and controlling robotic trajectories. Designers and artists has further developed workflows to bridge the gap between digital and material worlds, with an aim to automate the materialisation of CAD/CAM (computer-aided design and computer-aided manufacturing) files. Automatic 'File-to-factory' is not a barrier anymore [12], but at the same time it is a subject of further research and development.

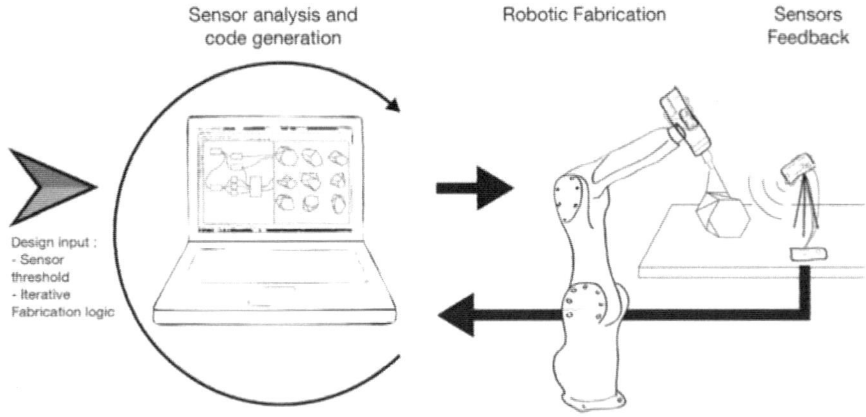

Figure 5. Ideal robotic-arm fabrication (or 3D printing) workflow [12]

An ideal workflow for robotic-arm fabrication is as following (Figure 5):

- Design and model the product using CAD parametric design software; for this purpose *Rhinoceros3D* is often used. *Grasshopper3D* plug-on is simultaneously used to visually program *Rhinoceros3D* models and build generative algorithms (or generative art). Through *Rhinoceros3D+Grasshopper3D* (Figure 7), the design is flexibly generated based on adjustable parameters and algorithmic rules.
- Convert the *Rhinoceros3D+Grasshopper3D* CAD model into robotic instructions through CAM plugins, e.g. KUKA|prc. The plugin generates accurate simulation of robotic trajectories, which allows the designed to virtually explore the potential and limitations of robotic fabrication processes.
- Extrude and receive real-time feedback from the end-effector, using micro-controllers (e.g. Arduino) and sensors to control material-specific parameters. This becomes especially crucial when extruding non-static materials such as clay, polymer or concrete. Precise sensor feedback is needed to enable tracking, fine-tuning and synchronisation between material, machine and design [12]. Control parameters include: (1) environmental conditions (e.g. air temperature, relative humidity), (2) sheer viscosity at the extrusion point, (3) feed rates, and (4) displacement speed. All these parameters will help predicting the final deposition location of the extruded material. Plenty of feedback sensors could be used for this purpose, such as thermometers, water flow meters, cameras (through image-pixel based analysis), fast laser scanners, Infrared (for light

sensitive material), and more. Such feedback may thence feed an artificially intelligence logic to adapt the extruders behaviour on real-time basis (refer to Figure 6).

To sum up, the current main challenge confronting large-scale 3D-printing research is the design of a smart end-effector that intelligently bridges digital and fabrication worlds, while finding optimum material properties under optimum environmental set-ups. Clearly, all this requires competences in material science, mechatronics, artificial intelligence and structural engineering. Section 3.2 reviews material science research in particular.

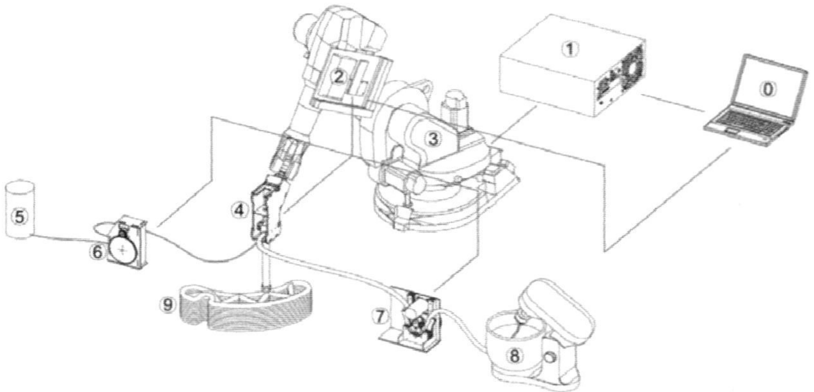

Figure 6. Concrete-type 3D-printing system components consisting of (0) System command, (1) Robot controller, (2) Printing controller, (3) Robotic arm, (4) Printhead, (5) Concrete accelerating agent, (6) Peristaltic pump for accelerating agent, (7) Peristaltic pump for premix, (8) Premix mixer, and (9) 3D printed object.

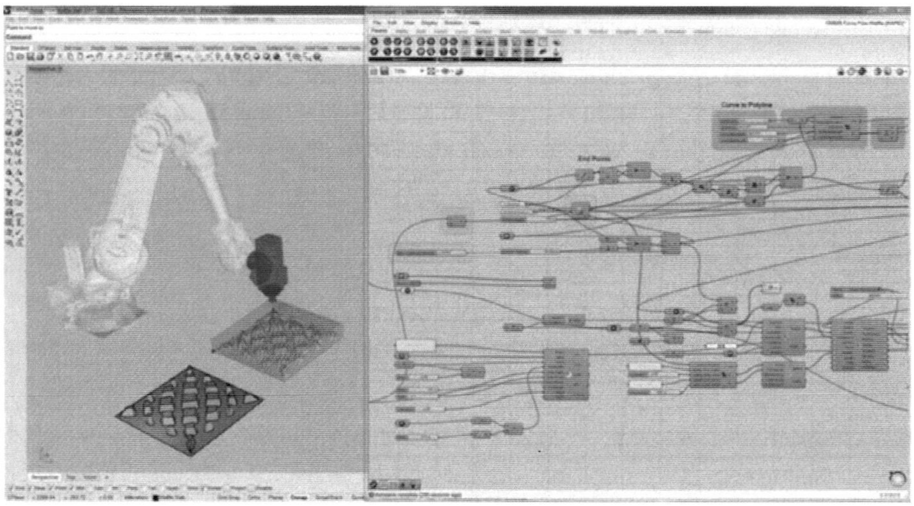

Figure 7. Grasshopper plugin for simulating robotic trajectories translated from a CAD model [13]

3.3 Swarm Solutions

Swarm approach is a radically different solution. It rejects the use of a single giant printer, in favour of the collective performance of several and smaller mobile robots: "swarm" [7]. In 1997, Pegna [14] has envisioned the approach as constructing large structures by an army of mechanical 'ants', spilling one grain of sand at a time. The motive is to extend the concept of additive construction to areas where human exposure, payload and transportation activities are dangerous [14]. Ceccanti et al. [15] proposed a swarm-based solution to build moon habitats; and argued that a smaller wheeled printer controlled by a rover responsible for collecting and depositing the regolith would be more efficient than D-shape in that case.

Swarm approach was implemented by a research team from MIT who introduced the concept of building architectural structures using a swarm of single, independent robots. The research team envisaged the robots' capability to build and then climb tabular structures, which successively form single threads within a larger network constituting the developing structure. Similarly, students in California College of the Arts in San Francisco designed autonomous robots to carry out construction work in hostile environments, which used sawdust to perform additive construction [16]. The application of swarm approach depends highly on the robots' ability to climb. Moreover, the ability of robots to communicate within a swarm is key to the success of this approach, which requires real-time sensor feedback to support algorithmic decision making processes [7].

Sasa Jokic and Petr Novikov from the Institute of Advanced Architecture of Catalunya created a group of three small robots called "mini-builders" for executing large scale structures [17]. Mini-builders constitute three types of robots:

- Foundation Robot, which uses tracks for movement, and a line follower sensor put in front of the robot for positioning. This robot places the base layer so that the grip robot could proceed with the additive construction process.
- Grip Robot, which is self-attached to the structure and fastened between four rollers. Each one of the rollers is connected to a steering actuator to allow the robot to position itself within and above the structure. This method allows the robots to carry out horizontal and curved printing of components. Material curing times are accelerated by the use of heaters.
- Vacuum Robots, which are used to strengthen the shell printed by the previously mentioned robot types. Those robots consist of a vacuum generator and a suction cup, can travel over surfaces at any angle and reinforce the shell by attaching themselves to it and printing extra layers over it.

Figure 8. Grip Robot [17].

Another interesting swarm-based large-scale 3D printing uses flying robots, which are intended to overcome height limitations. Hunt et al. [18] has tested the practicality of "aerial additive construction" through a flying quadcopter robot capable of depositing polyurethane expanding foam in mid-flight.

Limitations include flight stability, materials availability, battery life of quadcopter and regulation restrictions.

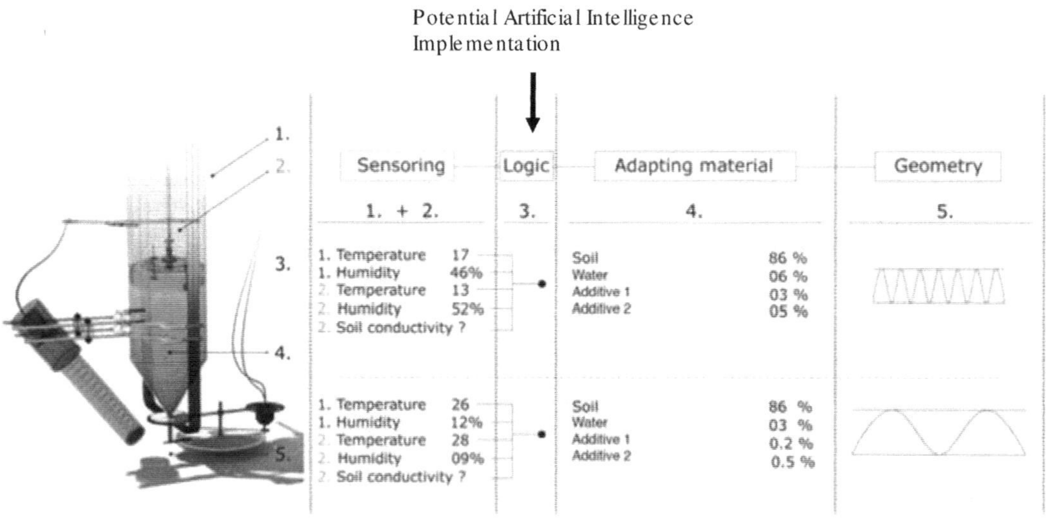

Figure 9. 3D printing process informed by sensors [12]

4. Material science challenges

Materials used in 3D printing (or additive manufacturing in general) can either be solid-like, viscous-like, powder-like or liquid-like. 3D processes based on solid-like materials involve layering the material in solid form. The different layers are bind by a number of techniques, like using a glue-like material to hold the successive layers together. 3D processes based on viscous-like materials involve extruding a viscous liquid from a printing nozzle. The materials are cured after extrusion in order to solidify. Contour Crafting, Fused Deposition Modelling and Inkjet processes are examples of 3D processes that are based on viscous-like materials. 3D processes based on powder-like materials involve the transformation of a material from a powder to a solid state using melting or sintering. D-Shape, Laser Sintering, Power Binding Printing, Selective Laser Melting and Selective Laser Sintering are all processes that are based on powder-like materials. 3D processes based on liquid-like materials involve the transformation of a material from a liquid to a solid state, such as Stereolithography where a light source is used for curing [7].

4.1 Concrete-type Materials

Most recent large-scale 3D printing research efforts has focused on concrete-type materials (e.g. Contour Crafting). For cementitious (or any viscous-like) 3D-printing to succeed, the extruded material should meet specific, sometimes contradictory requirements. Lim et al. [10] identified four extrusion-based process-related characteristics that need to be considered and optimised. These are namely: (1) pumpability, (2) printability, (3) buildability and (4) open time. *Pumpability* is the ease with which a material travels through the printing system. *Printability* is the ease with which a material is deposited from an extruder. *Buildability* is the resistance of the wet deposited material to deformation when the upper layers are printed on it. Finally, *open time* is the time in which the previous properties remain within the acceptable limits. Balancing the printability and buildability has always been known as the most critical property. This is because, the adhesion between layers which affects the bond strength, is a function of the time interval between layers. Hence, the process should be designed such that the individual layers are able to adhere with one another (an open time that is short enough), and yet which still have time to cure (an open time that is long enough) [10,19–21]. This can be done by calibrating input parameters, including: the rate of extrusion, the thickness of layers, linear speed, part diameter, number of layers and the output parameters, including: layer width, part height, vertical profile and surface roughness [22].

Recently Feng et al. [23] studied the mechanical behaviour of a concrete-type 3D printed element, and revealed through microscopic observations, mechanical test of cubes and small beams, and Finite Element modelling, the effect of the printing process on the structural behaviour. Their study indicated that the layer-on-layer printing process led to an apparent orthotropic behaviour that was relevant to compression strength and elastic modulus, but not to failure mode. Their results confirmed that due to the anisotropic nature of material distribution, all extrusion-based processes are likely to create components that are strongly anisotropic [21], and that this will have a significant effect on the load-bearing capacity of the construction in question. The same conclusions were reached by Le et al. [20] who measured the effects of voids that appeared between deposited filaments on the orthotropic compressive and flexural strengths of extruded concrete. Moreover, FE analysis conducted by Feng et al. [23] showed that the printing direction has a significant influence on the load bearing capacity of the structure.

For these reasons, one may not rely on the structural performance of 3D printed elements, as their mechanical properties may be highly unpredictable. Although some researchers attempted to add glass fiber to the printing concrete to increase its strength, the material was yet too brittle to be used as load bearing components, or components that span horizontally such as slabs and staircases [9]. Nonetheless, the material was proven reliable enough to print moulds in which conventional concrete are poured into (Figure 10). The 3D printed moulds in that case becomes part of the final product; as such a great advantage is obtained through the elimination of de-moulding [9]. Above that, the most attractive advantage of 3D printed moulds is the ability to produce complex 3D shapes in comparison to casting processes [11]. More about geometric advantages and design opportunities are discussed in Section 3.3.

Figure 10. 3D printed moulds [9]

IOP Conf. Series: Materials Science and Engineering **324** (2018) 012088 doi:10.1088/1757-899X/324/1/012088

4.2 Beyond Load-bearing Properties

As mentioned above, the state-of-the-art of large-scale 3D printing is largely dominated by concrete-type research efforts, with an immense focus on the bearing capacity of the printed material. However, a sound building material cannot merely depend on good load-bearing capacity. Other essential qualities of materials that should be investigated include durability, water vapour diffusion resistance, thermal properties, and/or fire-resistant properties. Additionally, it is required that the selected materials does not pose risk on the heath, safety and environment [24]. Material storage is also considered an issue as large-scale construction requires large volumes of material. Thus, one should also consider that the material used for 3D printing is inexpensive and have lightweight to facilitate storage [25]. This challenge can be addressed by using locally-derived materials that does not require transportation or storage. For instance, some authors proposed an "off-grid" 3D printer that can work using sand as the raw material for building.

Striding beyond sole bearing capacity investigations would probably entail moving out of the box to explore materials other than concrete for large-scale 3D printing applications. Labonnote et al. [7] argue that recent large-scale 3D printing efforts were largely influenced by the heritage of construction in which buildings are composed of combination of materials, each exhibit specific properties serving different requirements. For example, a wall may be composed of multiple materials including timber studs to provide adequate load-bearing properties, mineral wool to provide good insulation properties, and vapour barriers to provide good water vapour diffusion resistance properties.

Current large-scale 3D printing efforts seem to follow the same heritage of focusing on uni-functional material properties. This makes it unlikely to significantly favour large-scale 3D printing ahead of traditional construction methods. For this reason Labonnote et al. [7]except that today's standard use of extruded concrete in large-scale 3D printing will be disqualified in favour of other materials and/or processes. The authors argue that future opportunities for the application of material science in large-scale 3D printing are two-fold:

- Development of innovative materials that exhibit an optimal combination of all essential properties. Such inventions would increase the efficiency of construction through the reduction of material usage and the elimination of the extra work and time devoted for installing numerous materials. An example of such material science invention is developed by the Engineering Excellence Group Sydney [7]. The group invented an approach to print moulds made of wax, which they refer to as FreeFab Wax (Figure 11). The FreeFab is a replacement to the conventional timber-made, steel-made, or more recently concrete-printed moulds. FreeFab Wax exhibits both load-bearing properties, as well as sustainability properties. The wax is 3D printed to yield the required tolerance and surface finish; thereby, it is placed after use in a recycling station that melts the wax off the finished element and collects it for re-use again in other mould printing processes [24]. Figure 14 presents a conceptual overview of the FreeFAB Wax printing and recycling process.
- Investigation of large-scale 3D printing processes that support the extrusion of non-homogeneous Functionally Graded Materials (FGMs) whose properties smoothly vary along an axis (e.g. Figure 12). FGMs are inspired by the natural formation of bones and teeth in which the microstructure of a composite material varies from one material to another through a specific gradient. These would facilitate the production of versatile building materials. FGMs have already been investigated for the use of small-scale 3D printing [26] and should be also considered for large-scale 3D printing.

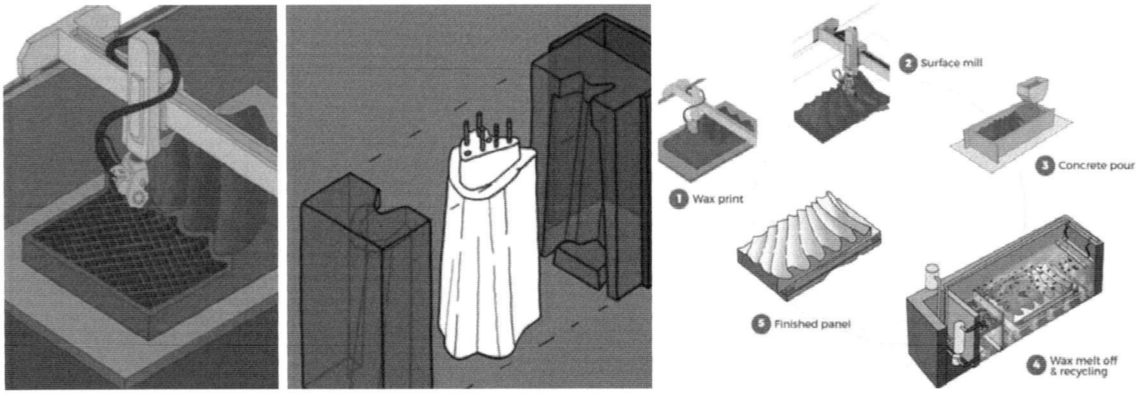

Figure 11. A conceptual illustration of the FreeFAB wax mould fabrication and recycling process [27]

Figure 12. FGM 3D-printed structures by MIT Medial Lab [28]

5. New building design opportunities

In the field of architectural design, a significant trade-off exists between the will to save cost, which entails producing simplistic and straight walls, and between the desire to produce novel and non-standard buildings, which entails producing customised moulds and formwork for one-time use, thus generating large amounts of waste materials and unforeseen construction delays. Large-scale 3D printing offers unlimited potential for producing complex geometries that can't be identified by basic geometrical notions and were only accomplished for limited projects [29]. Now that it is mathematically proved by Anatasiou et al. [30] that any 3D structure can be created by means of layer by layer. Structures can be made into almost any shape [15] regardless of their difficulty, expenses or possibility [14].

5.1 Smart Design and Building Performance Optimization

Furthermore, large-scale 3D printing provides the opportunity of optimising topologies at no extra costs. Optimised components results in material savings and reductions in weight [31] and aids in other types of optimisation such as those linked to thermal or acoustic properties. Gosselin et al. [11], for instance, experimented on the production of complex 3D printed concrete elements to attaint ultra-high performance buildings. An example of their work is illustrated in Figure 14. The research team compared the shown complex 3D printed wall (Figure 13a) against a typical conventionally-built wall (Figure 13b). They found that a gain in thermal insulation performance of 56% was obtained in comparison to a conventional wall, only by optimising the geometry of the internal reinforcement. The wavy shell results in reducing the amount of heat (i.e. heat flux $\Phi^{\rightarrow}$) flowing through the whole structural element. Additionally, the authors argue that the wavy shell is structurally efficient as it contains straight bars acting as a stable truss, suited to support loadings; the shell also acts as a doubly-corrugated reinforcement, enhancing the element's flexural rigidity [11].

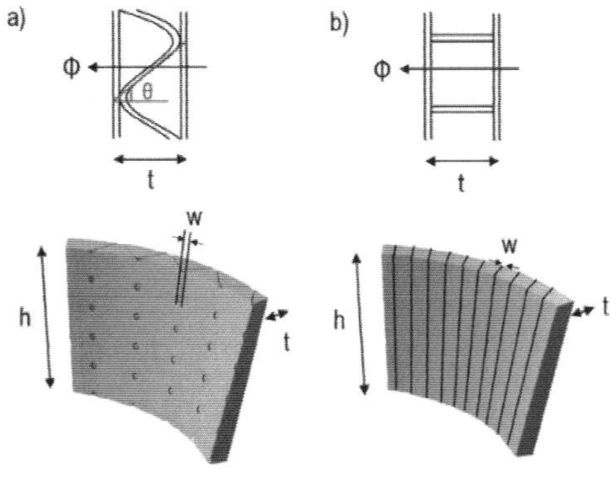

Figure 13. (a) complex 3D printed wall with wavy internal reinforcement, verses (b) wall with conventionally built internal reinforcement [11].

Moreover, large-scale 3D printing provides a better and more flexible methodology than mass customisation for meeting customer and/or architectural requirements for individualised products. Mass customisation depends on the use of diverse combinations of pre-assembled units and/or hindered differentiation methods. However, Strauss [21] offers a controlled individualisation level. Gardiner [24] considered flexibility regarding individualisation as a possible future niche market for large scale 3D printing. It actually represents a great interest to architects as 3D printing may be a true game-changer for architectural design and for future rich cooperation between engineers and architects. 3D printing's added value exceeds the possibility to embed "increased functionality" [32]. In fact, it can also contribute to the building of "smart structures" which embrace various properties such as solar responsively, technologies for monitoring and maintenance of buildings, self-healing materials and seismic forces dissipation. More potentials of large scale 3D printing were investigated like façade construction and predicted innovative smart applications [21].

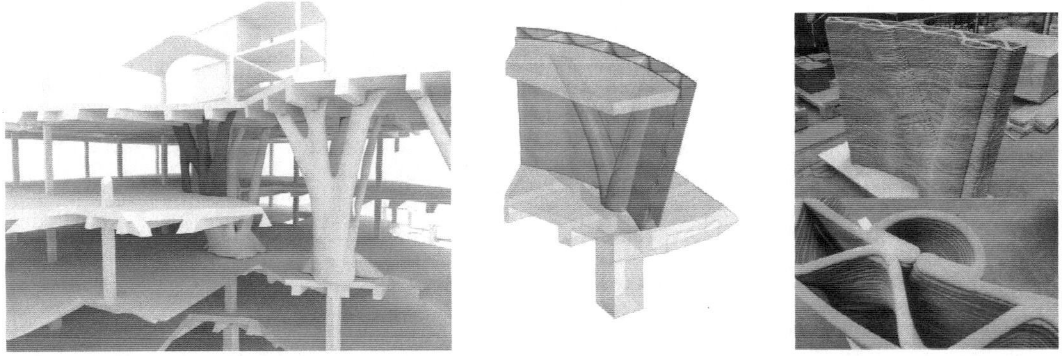

Figure 14. Architectural context for the multi-functional wall element (left), CAD model for the multi-functionnal wall (centre) and the 3D printed high-performance concrete multi-functionnal wall (right), designed and constructed by Gosselin et al. [11]

6. Conclusions

The articles reviewed three core inter-disciplinary research themes pertaining large-scale 3D printing: (1) technological solutions of additive construction (i.e. the 3D printers themselves), (2) material science challenges, and (3) new building design opportunities.

Improving the 3D printing technology: State-of-the-art large-scale 3D printing technologies could be classified as either gantry-based, robotic-arm based, or swarm-based. Gantry based solutions served well for materialising the concept of large-scale 3D printing, but yet suffer significant drawbacks. Swarm based solutions seem promising as for approaching unreachable construction zones, but are too conceptual so far. There is a great potential for robotic-arm solutions to push the future of large-scale 3D printing forward, given their well-established workflows, aside with their high versatility and flexibility. The main challenge is to design a smart robotic-arm end-effector that intelligently bridges digital and fabrication worlds, while finding optimum material properties under optimum environmental set-ups.

Tackling Material Science Challenges: Most recent large-scale 3D printing research efforts has focused on concrete-type materials. For cementitious (or any viscous-like) 3D-printing to succeed, the extruded material should meet specific, sometimes contradictory requirements. These are namely: (1) pumpability, (2) printability, (3) buildability and (4) open time. Recently employed large-scale 3D printing systems were successful in achieving a right equilibrium between these requirements, and hence attained satisfactory load-bearing capacity parameters for full-scale real-world structures. Nonetheless, one may not rely on the structural performance of 3D printed elements, as their mechanical properties are yet highly unpredictable. This is due to the effect of voids that appear between deposited filaments on the orthotropic compressive and the flexural strengths of the extruded concrete. Nonetheless, till day the printed material were proven reliable enough to develop moulds in which conventional concrete are poured into. 3D printed moulds have a significant advantage over conventional moulds in their ability to produce complex 3D shapes. Despite this advantage, 3D printed elements is unlikely to significantly become favoured ahead of traditional construction methods, unless researchers stride beyond the uni-functional material heritage of construction towards the innovation of multi-functional material. A multi-functional material is that which not only exhibits one property (e.g. load-bearing), but rather multiple properties.

Exploring New Design Opportunities: Designers are often confronted with a trade-off between cost savings, which entails producing simplistic and straight walls, and between novel and non-standard buildings, which entails producing customised moulds and formwork for one-time use, thus generating large amounts of waste materials and unforeseen construction delays. Current large-scale 3D printing state-of-the-art allow constructing complex geometries that were not possible in the past. Future research may investigate this opportunity aside with its implications on the future of building design, not only from an aesthetic perspective, but also from the perspective of (thermal, acoustic, etc.) performance of structures. Also the value of large-scale 3D printing may exceed “increased functionality”; It can also contribute to the building of "smart structures" which embrace various properties such as solar responsively, technologies for monitoring and maintenance of buildings, self-healing materials and seismic forces dissipation. All in all, large-scale 3D printing would create a paradigm change in the field of building design. Future architects are expected to become robotic-aware, in which they would consider robotic arm constraints for the design of a given building element.

7. Acknowledgement

This work was financially supported by the Research Affairs Office at UAE University Grant No. 31N196.

8. References

[1] S. Griffiths, Giant 3D printer creates 10 full-sized houses in a DAY: Bungalows built from layers of waste materials cost less than £3,000 each., DailyMail. (2014). http://www.dailymail.co.uk/sciencetech/article-2615076/Giant-3D-printer-creates-10-sized-houses-DAY-Bungalows-built-layers-waste-materials-cost-3-000-each.html (accessed December 28, 2017).

[2] B. Khoshnevis, Automated construction by contour crafting—related robotics and information technologies, Autom. Constr. 13 (2004) 5–19.

[3] J. Brandon, Don’t Believe the Hype: WinSun Unveils Bland Multi-Story 3D Printed Buildings, SolidSmack. (2015). https://www.solidsmack.com/fabrication/dont-believe-hype-winsun-unveils-bland-multi-story-3d-printed-buildings/ (accessed January 2, 2018).

[4] I. OBERTI, F. PLANTAMURA, Is 3D printed house sustainable?, in: Proc. Int. Conf. CISBAT 2015 Futur. Build. Dist. Sustain. from Nano to Urban Scale, LESO-PB, EPFL, 2015: pp. 173–178.

[5] N. Davison, 3D-printed cities: is this the future?, Guard. (2015). https://www.theguardian.com/cities/2015/feb/26/3d-printed-cities-future-housing-architecture (accessed January 2, 2018).

[6] Alec, Saudi Arabia plans to 3D print 1.5 million houses with WinSun's construction 3D printing tech, 3Ders. (2016). https://www.3ders.org/articles/20160802-saudi-arabia-plans-to-3d-print-15-million-houses-with-winsuns-construction-3d-printing-tech.html (accessed January 2, 2018).

[7] N. Labonnote, A. Rønnquist, B. Manum, P. Rüther, Additive construction: State-of-the-art, challenges and opportunities, Autom. Constr. 72 (2016) 347–366.

[8] The American Society for Material and Testing, Standard Terminology for Additive Manufacturing Technologies, in: ASTM Int., West Consholhocken, United States, 2009.

[9] P. Wu, J. Wang, X. Wang, A critical review of the use of 3-D printing in the construction industry, Autom. Constr. 68 (2016) 21–31.

[10] S. Lim, R.A. Buswell, T.T. Le, S.A. Austin, A.G.F. Gibb, T. Thorpe, Developments in construction-scale additive manufacturing processes, Autom. Constr. 21 (2012) 262–268.

[11] C. Gosselin, R. Duballet, P. Roux, N. Gaudillière, J. Dirrenberger, P. Morel, Large-scale 3D printing of ultra-high performance concrete–a new processing route for architects and builders, Mater. Des. 100 (2016) 102–109.

[12] A. Dubor, G. Camprodom, G.B. Diaz, D. Reinhardt, R. Saunders, K. Dunn, M. Niemelä, S. Horlyck, S. Alarcon-Licona, D. Wozniak-O'Connor, Sensors and workflow evolutions: developing a framework for instant robotic toolpath revision, in: Robot. Fabr. Archit. Art Des. 2016, Springer, 2016: pp. 410–425.

[13] J.B. Gardiner, S.R. Janssen, FreeFab, in: Robot. Fabr. Archit. Art Des. 2014, Springer, 2014: pp. 131–146.

[14] J. Pegna, Exploratory investigation of solid freeform construction, Autom. Constr. 5 (1997) 427–437.

[15] F. Ceccanti, E. Dini, X. De Kestelier, V. Colla, L. Pambaguian, 3D printing technology for a moon outpost exploiting lunar soil, in: 61st Int. Astronaut. Congr. Prague, CZ, IAC-10-D3, 2010.

[16] B.B. Millsaps, 3D Printing Robots Capable of Building in Outlying Areas, 3D Print. (2015). https://3dprint.com/32903/swarmscapers-robots/ (accessed January 2, 2018).

[17] Institute of Advanced Architecture of Catalunya - IAAC, Minibuilders: small robots printing big structures, (2017). http://robots.iaac.net/ (accessed January 2, 2018).

[18] G. Hunt, F. Mitzalis, T. Alhinai, P.A. Hooper, M. Kovac, 3D printing with flying robots, in: Robot. Autom. (ICRA), 2014 IEEE Int. Conf., IEEE, 2014: pp. 4493–4499.

[19] A. Perrot, D. Rangeard, A. Pierre, Structural built-up of cement-based materials used for 3D-printing extrusion techniques, Mater. Struct. 49 (2016) 1213–1220.

[20] T.T. Le, S.A. Austin, S. Lim, R.A. Buswell, A.G.F. Gibb, T. Thorpe, Mix design and fresh properties for high-performance printing concrete, Mater. Struct. 45 (2012) 1221–1232.

[21] H. Strauss, AM Envelope: The Potential of Additive Manufacturing for facade constructions, TU Delft, 2013.

[22] B. Khoshnevis, R. Russell, H. Kwon, S. Bukkapatnam, Crafting large prototypes, IEEE Robot. Autom. Mag. 8 (2001) 33–42.

[23] P. Feng, X. Meng, J.-F. Chen, L. Ye, Mechanical properties of structures 3D printed with cementitious powders, Constr. Build. Mater. 93 (2015) 486–497.

[24] J. Gardiner, Exploring the emerging design territory of construction 3D printing-project led architectural research, RMIT University, Melbourne, Australia, 2011.

[25] E. Barnett, C. Gosselin, Large-scale 3D printing with a cable-suspended robot, Addit. Manuf. 7 (2015) 27–44.

[26] S. Chianrabutra, B.G. Mellor, S. Yang, A dry powder material delivery device for multiple material additive manufacturing, Univ. Southampton, UK. (2014).

[27] J.B. Gardiner, S. Janssen, N. Kirchner, A Realisation of a Construction Scale Robotic System

for 3D Printing of Complex Formwork, in: ISARC. Proc. Int. Symp. Autom. Robot. Constr., Vilnius Gediminas Technical University, Department of Construction Economics & Property, 2016: p. 1.

[28] Functionally graded printing, MIT Media Lab. (2012). http://matter.media.mit.edu/tools/details/3d-printing-of-functionally-graded-materials#prettyPhoto[3d-printing-of-functionally-graded-materials]/1/ (accessed January 2, 2018).

[29] I. Vizotto, Computational generation of free-form shells in architectural design and civil engineering, Autom. Constr. 19 (2010) 1087–1105.

[30] A. Anastasiou, C. Tsirmpas, A. Rompas, K. Giokas, D. Koutsouris, 3D printing: Basic concepts mathematics and technologies, in: Bioinforma. Bioeng. (BIBE), 2013 IEEE 13th Int. Conf., IEEE, 2013: pp. 1–4.

[31] H. Valkenaers, D. Jansen, A. Voet, A. Van Gysel, E. Ferraris, Additive manufacturing for concrete: a 3D printing principle, in: Proc. 14th Euspen Int. Conf., euspen, 2014: pp. 139–142.

[32] R. Soar, Additive manufacturing technologies for the construction industry, Rapid Manuf. An Ind. Revolut. Digit. Age. (2006) 249–273.

IOP Conf. Series: Materials Science and Engineering **324** (2018) 012089 doi:10.1088/1757-899X/324/1/012089

Preparation and properties of UHMWPE microporous membrane for lithium ion battery diaphragm

ZHAO Changsong, HE Jianyun*[1], LI Jiawei, TONG Jinge, XIONG Jinping*[2]

Beijing University of Chemical Technology, Beijing 100029, China
E-mail: [1]jyhe2009@163.com,[2]xiongjp@mail.buct.edu.cn

Abstract. Based on the principle of heat - induced phase separation (TIPS).The ultrahigh molecular weight polyethylene (UHMWPE) microporous membrane was prepared. The effects of material formula and preparation process on the structure and properties of UHMWPE microporous membrane and the internal relationship between the process parameters and the performance of microporous membrane were studied. The forming process of microporous membrane was optimized and the UHMWPE microporous membranes with different properties were prepared and assembled into the half-battery and the full battery. The key electrochemical properties of lithium ion batteries, such as ion conductivity and cycle performance, were tested. The properties of lithium ion battery was explored. Results shown that the UHMWPE microporous membrane met the requirements of power lithium ion battery.

1. Introduction

UHMWPE microporous membrane has excellent mechanical properties and chemical stability, It has big potential application prospect[1] in the field of lithium ion batteries. However, The preparation process of microporous membrane based on the TIPS of UHMWPE is relatively complex, and the research on the process factors affecting microporous membrane structure is less. Therefore, in this paper, based on the principle of thermal induced phase separation (TIPS) the forming technique of preparation of UHMWPE microporous membrane system was studied, and the influence of average molecular weight of UHMWPE, material proportion and preparation technology on the properties of UHMWPE structure was studied. The UHMWPE change rule of crystalline morphology in the process of UHMWPE microporous membrane preparation was explored, the relation between process parameters and product performance was studied, the preparation technology of UHMWPE microporous membrane and the associated properties that UHMWPE microporous membrane and the performance of lithium ion battery using properties such as ionic conductivity and cycle performance were also researched.

Lithium ion battery has high working voltage, large energy density, no memory effect, long cycle life and green environmental protection, and many other advantages[2-3]. In recent years, it has become one of research hotspots in the field of new energy materials. Lithium ion batteries are used in aerospace, communications, medical, transportation, and many other areas of electronics， Lithium ion battery is gradually replace the traditional lead acid and nickel metal hydride batteries， It plays an important role[4-5] in car instead of petrochemical energy as a driving force.

2. Experimental

2.1 *Materials*

UHMWPE number average molecular weight are： 2.0×105、5.0×105、8.0×105、1.1×106、1.3×106、2.0×106， purchase from Mitsui chemicals company in Japan; Liquid paraffin oil） 70# ,kinematic viscosity was 70mm2/s（40℃）, lash point: 200 ℃ or higher, Bought in hangzhou petrochemical co., LTD; Antioxidant 1010 hindered phenols, bought in ciba company .

2.2 *Preparation* of UHMWPE Membrane

Put the UHMWPE (powder), antioxidants, liquid paraffin oil (LP) in mechanical premixed according to certain proportion within the agitator, Then after the premixed material was added to the synthetic twin screw extruder (L/D = 60:1, ϕm = 40, Nanjing KeYa company) , Material in the extruder melt mixing, plasticizing completely, and then after T die stretch film to the rapid cooling roller,By controlling the quenching of roller temperature, producing UHMWPE slabs under different cooling rate; Then with adjustable stretch ratio of homemade tensile testing machine (8.0 x8.0 maximum tensile rate),the UHMWPE thick slices respectively in different stretching temperature, into the ventilation conditions, for different ratio of tensile. Then with a homemade extraction machine stretch film to extraction respectively, formed with microporous stretch film.F inally ,with the heat setting machine will microporous stretch film for heat setting , get the UHMWPE microporous membrane.

3 .Results and Discussion

3.1 The *effect* of UHMWPE formula and diluent on the microporous membrane structure and performance

This study selects the liquid paraffin oil as diluent of UHMWPE. The formula of UHMWPE and diluent are the important factors which affect the structure morphology and properties of lithium ion battery using UHMWPE microporous membrane. Only at high temperatures UHMWPE and liquid paraffin can form the thermodynamics single-phase. This study was carried out at melt mixing experiment with different UHMWPE mass fraction (10%, 20%, 30%, 10%, 20%) , When the UHMWPE mass percent concentration is 10% and 20%, The melt strength of extruded UHMWPE thick sheets is lower. However, if the UHMWPE mass percent concentration is too high, the UHMWPE is difficult to disperse in the UHMWPE- diluent system, the mixing effect is poor, and the extruded sheet surface is rough. Figure 2 is the microstructure of UHMWPE microporous membrane prepared at UHMWPE mass fraction of 30%, 40%, 50%,. When UHMWPE mass percent concentration was 30%, pore diameter distribution is uniform and permeability is good. With the increment of UHMWPE mass fraction, the amount and the size of the pore gradually reduced. When UHMWPE mass fraction increased to 50%, microporous basically is closed. This is because, with the increasing of mass fraction UHMWPE the interaction between molecules enhanced and the mutual entanglement degree of UHMWPE macromolecular chain increased, therefore the tight interlaced network formed. Experiments show that when UHMWPE mass fraction is 30%,the aperture size of UHMWPE microporous membrane is uniform, porosity is moderate.

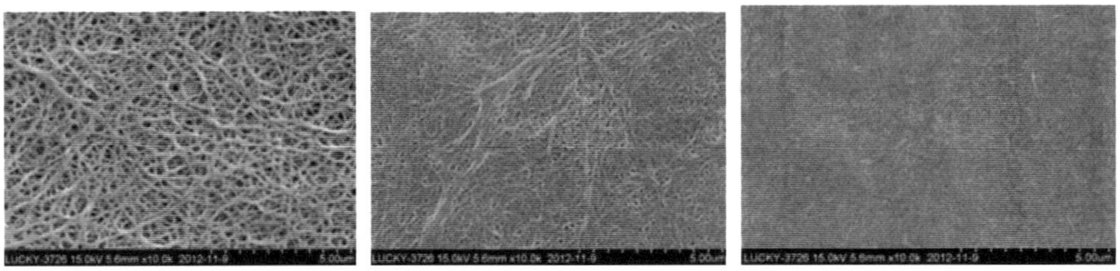

(a) (b) (c)

Fig.1 The surface morphology of the UHMWPE microporous membrance
at different ratios between UHMWPE and Oil: (a) 20% (b) 30% (c) 40%.

3.2 The *effect* of stretching temperature on the microporous membrane morphology and porosity

The porosity of Lithium ion battery with microporous membrane are the key indicators of UHMWPE microporous membrane, it directly influence the battery capacity of the lithium-ion batteries and its safety. High porosity, can accommodate more electrolyte, upgrade the capacity of the battery.

Under the condition of the UHMWPE mass fraction of 20%. with stretching ratio 6x6 and stretching speed 10 mm/s, at 70°C,90°C,110°C and 130°C respectively, UHMWPE thick slices were stretched. Figure 5 is the micro topography of UHMWPE microporous membrane under different tensile temperature. Experiments show that the tensile temperature is lower than 70 °C, UHMWPE slabs stretching is difficult, stretch film is easy to break, When the tensile temperature reaches the fastest crystallization temperature 110°C, It is good for stretching, under this temperature, the membrane pore size distribution is uniform, moderate porosity was 42%. However, when the tensile temperature is much higher than the fastest crystallization temperature, the porosity is low. Therefore, at 90-110°Cstretching temperature with 6x6 longitudinal transverse stretching ratio, good mesh structure of the microporous membrane was obtained. Figure 6 is the relationship of the microporous membrane pore ratio and stretching temperature.

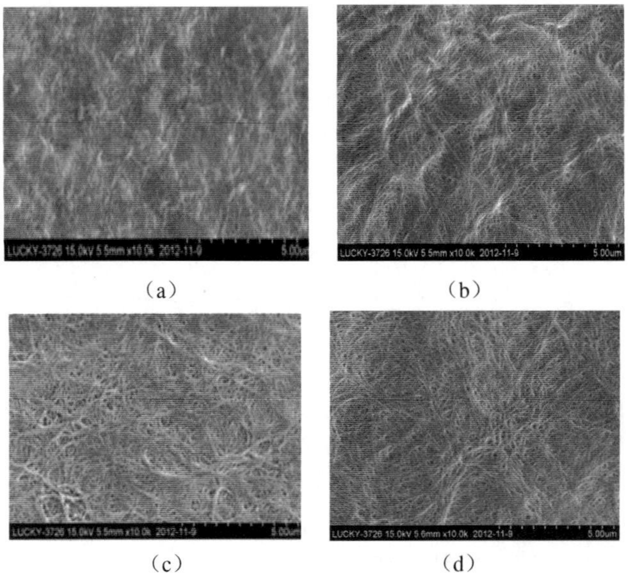

(a) (b)
(c) (d)

Fig.2 Morphology of UHMWPE films with different stretching temperature

（a） 70℃, （b） 90℃, （c） 120℃, （d） 130℃

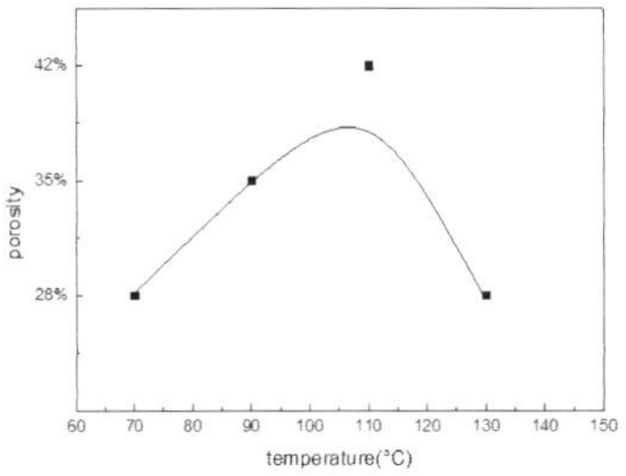

Fig.3 the microporous membrane of porosity with different stretching temperature

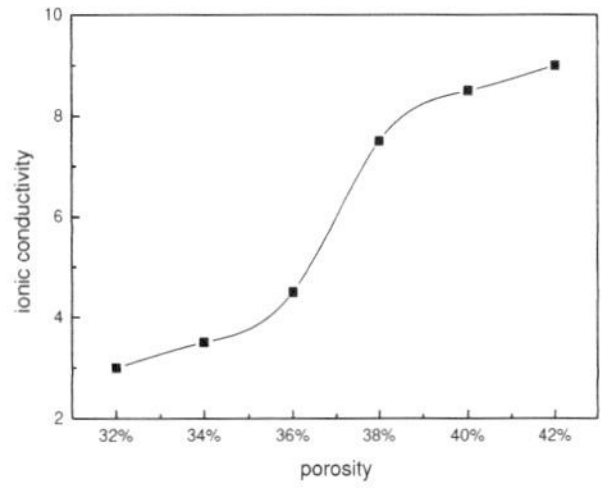

Fig.4 The relationship betweenIonic conductivity and permeability of the microporous membrane

3.3 The *influence* of UHMWPE microporous membrane on ionic conductivity of Li ion battery

The ionic conductivity of UHMWPE microporous membrane affects the discharge characteristics and capacity characteristics of lithium ion batteries. It is an important index to measure the electrical properties of UHMWPE microporous membranes. The porosity of UHMWPE microporous membrane has an important influence on the ionic conductivity of Li ion battery. Samples of UHMWPE microporous membranes with porosities of 33%, 36%, 38% and 42% were assembled into button half cells and tested for ionic conductivity. Fig. 7 is the relationship between the porosity of the UHMWPE microporous membrane and the ionic conductivity. As can be seen from the diagram, the ionic conductivity increases with the increase of porosity of UHMWPE microporous membrane，Because as the porosity of UHMWPE microporous membrane increasing, the capacity of microporous membrane to absorb electrolyte increases.

4. Conclusion

（1） using UHMWPE - liquid paraffin oil system produce UHMWPE microporous membrane, when UHMWPE mass percent concentration is 30%, the UHMWPE has good dispersion in the system of microporous membrane and the pore aperture size is moderate, pore size distribution is uniform. The microporous membrane can meet the performance requirements of the lithium ion battery.

（2） The drawing process have important influence on the microstructure and physical properties of UHMWPE microporous membrane. When the longitudinal transverse stretching ratio is 5x5, stretching temperature is 90-110 °C, the microporous membrane has good mesh structure, pore size distribution is uniform, the connectivity is good, porosity is moderate, microporous membrane thickness is uniform, which can basically meet the performance requirements of lithium ion battery.

（3） when the porosity is 38% - 38%, Lithium ion battery has good ionic conductivity.

5. Reference

[1] Jingqiang Zhang, Shuangqiang Chen, Xiuqiang Xie, Katja Kretschmer, Xiaodan Huang, Bing Sun, Guoxiu Wang. Porous Poly(vinylidene fluoride-co-hexafluoropropylene) polymer membrane with sandwich-like architecture for highly safe lithium ion batteries[J]. Science Direct:Journal of Membrane Science 472, (2014)133-140.

[2] Xiaosong Huang, Jonathon Hitt. Lithium ion battery separators: Development and performance characterization of a composite membrane[J]. SciVerse ScienceDirect: Journal of Membrane Science 425-426 (2016) 163-168.
[3] S.Y. Xiao, Y.Q. Yang, M.X. Li, F.X. Wang, Z Chang, Y.P. Wu, X. Liu. A composite membrane based on biocompatible cellulose as a host of gel polymer electrolyte for lithium ion batteries[J]. ScienceDirect: Journal of Power Power Sources 270, (2014)53-58.
[4] Shiying Xiao, Faxing Wang, Yaqiong Yang, Zheng Chang and Yuping Wu.An environmentally friendly and economic membrane based on cellulose as a gel polymer electrolyte for lithium ion batteries [J].ROYAL SOCIETY OF CHEMISTRY:RS. Adv, 2014, 4, 76.
[5] Nishi Y. The development of lithium ion secondary batteries [J].Materials Science and Engineering:R.Repeports,2011, 33(4:) 109-134.

Research on non-contact scanning measurement of pressure vessel head

W J Ding, C H Jiang, X L Liao, S P Sheng, Y L Liu
Hangzhou Institute of Special Equipment Inspection, Hangzhou, China

Email： 2420673693@qq.com

Abstract. This paper gives firstly a brief analysis of the harm of the shape deviation of pressure vessel head and disadvantages of traditional head mechanical contact measurement. On the basis of these, a non-contact laser-scanning measurement is presented. And then this paper introduces its principle, and analyzes in detail existing errors in measuring the head contour. Finally, Simulated defect measurement results are presented in the paper.

1.Introduction

Head is one of the main components of the pressure vessel, which is to ensure the sealing of pressure vessel. Strength failure, instability can lead to the overall safety problems of pressure vessels. The influence of different shapes of head on large pressure vessels is obviously, especially for the strength and stability of large pressure vessels.

The stress concentration of the head is closely related to its shape, and the larger head shape deviation may cause more serious stress concentration. Sometimes, the stress intensity where the inner wall of the small head appeared transition to straight edge by round (R) can be nearly 4 times the normal stress, it is obviously that the geometric deviation of the shape of the head can not be ignored for the safety of pressure vessels [1]. Check of the head parameters, especially the shape deviation, is an important work in the manufacture of pressure vessels [2].

At present, the measurement of the inner surface shape and the important parameters of the head geometry are mainly based on the mechanical contact measurement, including the whole model test and head inspection coordinate measuring instrument[3]. The former way was to detect the offset distance between the inner surface of the head to model; the latter way was to formula design of head curve theory coordinate detector, point contact head surface to obtain the measured data by using the detector, in order to determine the shape and quality of the inner surface of the head. These testing methods essentially use mechanical contact measurement, there are many problems in testing such as consumption, low measurement accuracy, low efficiency and poor operability, etc [4].

Therefore, this paper proposes a laser non-contact scanning method for head parameter detection, detection of important parameters of head by laser non-contact measurement and scanning motor, in order to overcome the shortcomings of traditional detection methods.

2. The pressure vessel head non-contact laser-scanning measurement

The method proposed in this paper complete the inspection of the head parameters, by using the laser scanning detection device located in the head end fixing to the support rod, and distance detection and rotating scanning detection of the motor. Here, this paper selects the detection of inner surface profile curve of head as an example to introduce the method.

3. The positioning and the axis position of head scanning measurement

One of the key parts in the head scan detection is the localization and determination of the axis position. According to the standard, the deviation of the head of the pressure vessel is the deviation of the section profile curve and the standard curve of the axis of the head. So capturing the position of the axis of the head is also the key element of the measurement. And the positioning is to find the shape of the head detection or zero starting point. Figure 1 is the schematic diagram for overlooking the scanning detection mechanism to detect the head. As shown in Figure 1, the scan detection mechanism includes a support bar (2 in Figure 1a), a locking device (3 in Figure 1a) and a distance detection module and a multi dimension scanning module (4 in Figure 1a). During the detection, the supporting rod used in arbitrary head (1 in Figure 1a) on the end, the lower end of the support rod on both sides respectively is fixed on the edge of the arc concave head with the locking device. The distance detection module and multi dimension scanning module is arranged in the support bar below, and is located in the arc concave surface of head, as illustrated in Figure 1b. Figure 2 is the assembly sketch map of distance detection module and the multi dimension scanning module. There, distance detection module (8 in Figure 2) is used to realize the detection of the distance among the points of the head, and the multi dimension scanning module (5, 6, 7 in Figure 2) provides the distance detection module and the rotation of the two dimensions.

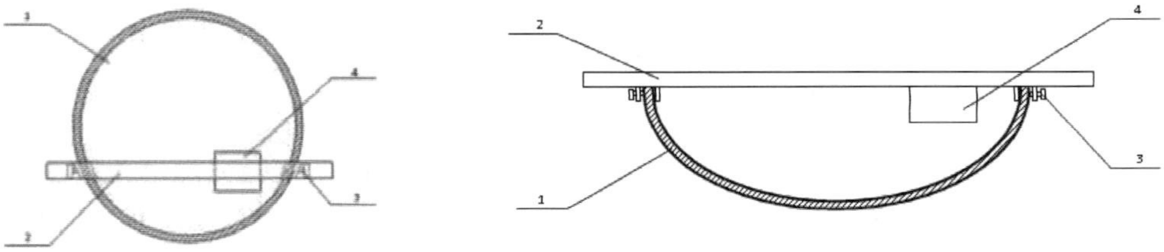

Figure 1. Schematic diagram of scanning detection mechanism

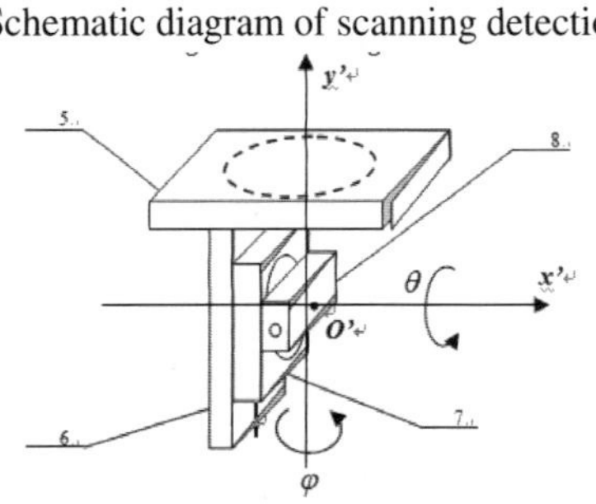

Figure 2. Schematic diagram of distance detection and multi-dimension scanning module

The method determination of the axis position is shown in Figure 3. *O'* is distance detection module (3 in Figure 7) for the measurement reference point of the device itself. When measuring, the first multi dimension scanning module direction is rotated to the zero position. Here is the definition of the zero distance detection module detects the direction (emergent light) parallel to the head m(contact surface supporting bar and head end). The direction angle of the multi dimension scanning module is arbitrary. In the distance detection module, The measure distance from the reference point to the end face of the head is *O'A*. Similarly, the scanning module drives the distance detection module to rotate 90 degrees for two times in the same direction, and the distance between the detection module and the end face of the head is detected, the distance *O'A'* and the distance *O'B'* were measured respectively. At the moment, the position of the center of the end face *O"* to the reference point of the distance detection module *O'* can be determined. The angle between the *O'O"* direction and the initial direction of the distance detection module, *O'A* direction, is φ_o, this angle is determined by the following formula:

$$\varphi_o = \arctan\frac{O'B - O'B'}{O'A - O'A'} = \arctan\frac{b_1 - b_2}{a_1 - a_2} \qquad (1)$$

In the formula, the distance values of *O'A*, *O'A'*, *O'B*, were a_1、a_2、b_1 and b_2. Here, the *O"* which is the center of the end face of the head, which is the real center of the surface profile of the head, but also the origin of the head inspection.

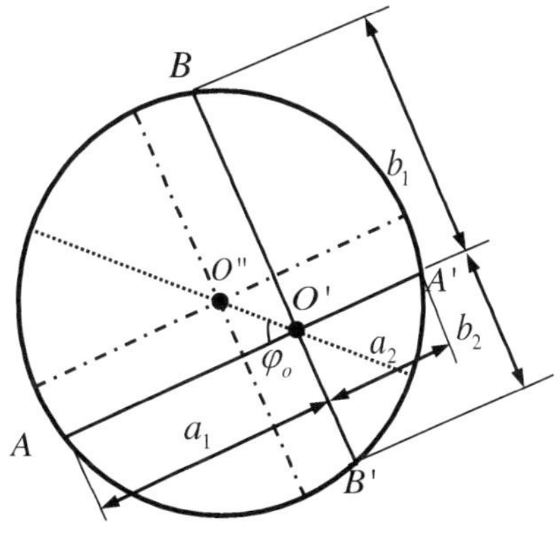

Figure 3. Determination of end face scanning direction of head

We can through the multi dimension scanning module rotate angle φ_o along the direction to complete positioning. When the direction of rotation is in place, the distance detection module is in position zero, that is, the distance detection module detects the direction is parallel to the end face of the head, At the same time, the direction is also through the *O"*, which is the center point of the inner surface profile curve of the head. At the moment, a point on the head which is detected by the distance detection module is the starting point of the head detection. Thus, the positioning of the head detection is completed.

IOP Conf. Series: Materials Science and Engineering **324** (2018) 012090 doi:10.1088/1757-899X/324/1/012090

4. Detection of the deviation of the inner surface of the head

The positioning and the determination of the central axis make the distance detection module be detected in the initial state. As shown in Figure 4, the shape error of the surface profile of the head is detected. Multi dimension scanning module direction at position zero. Direction scanning direction center of the end head face of *O"*. At the moment, the multi-dimensional scanning module rotates along the direction (the scanning surface), the distance detection module can detect the actual shape of the contour curve of the inner surface of the head.

At the time of detection, distance detection module detects along the starting point from the head start scanning head detection to another relative face. In this paper, the measurement of the shape error of the inner surface profile is based on the vertex of the head as the reference point of the error calculation. The point is that the vertical head end line intersection with the inner surface contour line of the head which go through the center point of the head end *O"*. It is also the point *O* in the graph. The vertical is the vertical axis. Let's make a coordinate system with the reference point *O* as the center, the detection of the coordinates of the surface of the head in the Figure 4 scanning process can be determined by the formula:

$$\begin{cases} x_c = \frac{1}{2}\sqrt{(a_1 - a_2)^2 + (b_1 - b_2)^2} - d\cos\theta_c \\ y_c = h_H - d\sin\theta_c \end{cases} \quad (2)$$

In the formula, d is the distance to the point C of the surface head Inner surface which is detected by the distance detection module. a_1、a_2、b_1 and b_2 respectively in the detection is the distance value of *O'A*, *O'A'*, *O'B*, *O'B'*. θ_c is the angle when the distance detection module is transferred to the point C in the direction of the multi dimension scanning module. Its positive or negative lies on the right hand rule in the coordinate system of Figure 2. Its angle size is equal to the angle $\angle O''O'C$. Its range meets the following formula:

$$\arctan\left[(h_H - p_b)/(p_a + \frac{1}{2}\sqrt{(a_1 - a_2)^2 + (b_1 - b_2)^2})\right] \le \theta_c \le \arctan\left[\pi - (h_H - p_b)/(p_a - \frac{1}{2}\sqrt{(a_1 - a_2)^2 + (b_1 - b_2)^2})\right] \quad (3)$$

The variables p_a and p_b are the long half axis and short half axis of the head. h_H is the actual distance between point *O'* of the test module reference to the point *O* of the head vertex along the y direction, or the actual distance between the center point *O"* of the head face and point *O* of the head vertex in the formula (2) and formula (3).

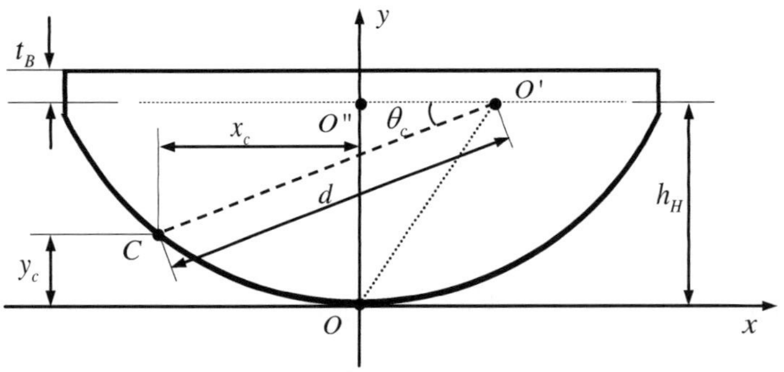

Figure 4. Scanning of inner surface profile of head

The actual distance between the center point O'' of the head face and point O of the head vertex is determined as shown in Figure 6. In the figure, the curve A is the theoretical outline of the head, and the curve B is the actual contour line of the head. The distance detection module is driven by the multi dimension scanning module to start from the detection start point, and after turning the angleθ_H along the direction, the outgoing light of the distance detection module just passes through the point O_A of the head vertex. Angleθ_H to meet the next type:

$$\theta_H = \arctan\left[(H - t_B) / \frac{1}{2}\sqrt{(a_1 - a_2)^2 + (b_1 - b_2)^2} \right] \quad (4)$$

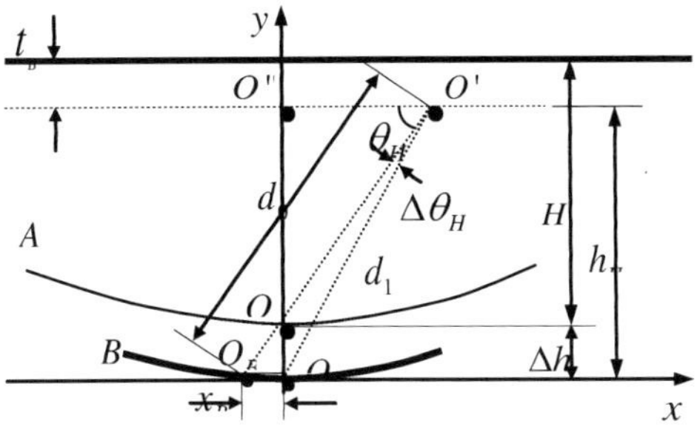

Figure 5. The determination of the actual distance between the head end face center and the head vertex

In the formula, the variable H is the distance from the head end to the head end of the head. The distance is also the total depth of the head. t_B is the distance from the detection module to the point O' of the end face. It is not consistent between the actual contour line of the head and the theoretical outline of the line A, assuming that both deviate Δh from the diagram in the curve B. At the moment, the light to reach the end face of the head is not the point O of the head vertex, but a point O_B near the top of the head, after the distance detection module rotation angle θ_H. Thus, it still need an angle $\Delta\theta_H$ to revise, after the distance detection module rotates angle θ_H. This angle $\Delta\theta_H$ satisfies the following formula:

$$\Delta\theta_H \approx (d_0 \cos\theta_H - \frac{1}{2}\sqrt{(a_1 - a_2)^2 + (b_1 - b_2)^2}) / d_0 \quad (5)$$

In the formula, d_0 is the actual detection after distance the distance detection module rotation angle θ_H from its reference point O' to the head vertex. Range detection module from the detection of the initial point began to rotate angle θ_H+$\Delta\theta_H$, its emergent light reach the point O of the head vertex. Set the detected distance as d_0, it can be obtained:

$$h_H = d_1 \cos(\theta_H + \Delta\theta_H) \quad (6)$$

By the formula (2), the distance detection module can detect the actual contour curve of the inner surface of the head. Finally, the shape error of the inner surface of the head can be determined by comparing with the theoretical curve of the inner surface of the head.

The definition of shape error is shown in Figure 7 (a). The curve B in the figure is the actual contour of the head. Making a theoretical contour curve of the head with the actual head contour vertex O as a reference point, which is curve A in the diagram, the center point of theoretical contour is point $O^{\#}$. Here, the shape error of the internal contour of the head is defined as the distance d-value from the point to the theoretical contour curve to the theoretical profile curve of the head in the normal direction. In the figure, the shape error of a point C_B（θ_c）on the contour curve is the difference between C_BO_f and C_AO_f, which is ε. Among them, C_BO_f is the normal direction of point C_A on the theoretical contour curve, the shape error value is obtained by the following formula:

$$\begin{cases} \varepsilon(\theta_c)=\left\{\left(\frac{1}{2}\sqrt{(a_1-a_2)^2+(b_1-b_2)^2}-d\cos\theta_c-x_{C_A}\right)^2+\left(h_H-d\sin\theta_c-y_{C_A}\right)^2\right\}^{1/2} \\ x_{C_A}=a\cdot b/\sqrt{b^2+a^2\tan^2\alpha} \\ y_{C_A}=a\cdot b\cdot\tan\alpha/\sqrt{b^2+a^2\tan^2\alpha} \end{cases} \quad (7)$$

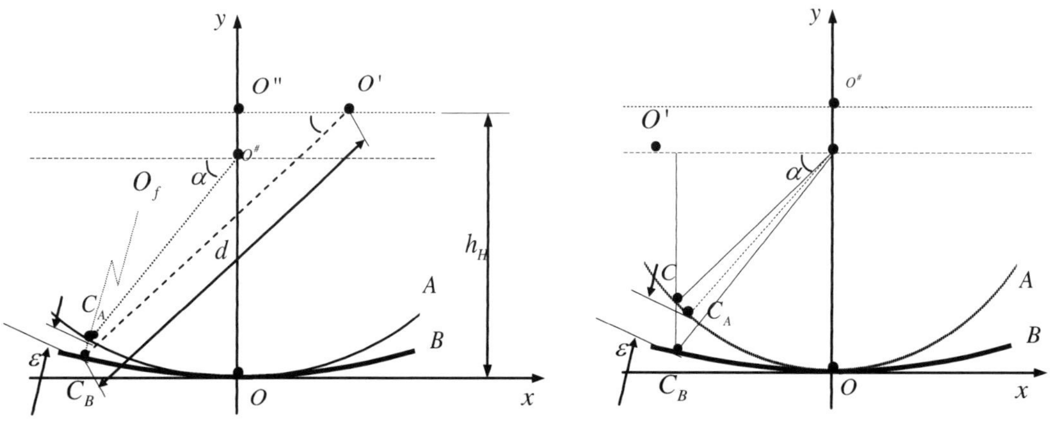

(a) Definition of shape error (b) The determination of the point on the theoretical curve in error calculation

Figure 6. The Definition and calculation of shape error

In the formula, d is the detection module rotation to the angle θ_c, the detection module reference point O' to the head of the actual contour line corresponding to point C_B detection distance. The parameters α or C_A on the approximate theoretical curve are determined as shown in Figure 7 (b) shows. Through the actual contour curve point C_B to do the y axis parallel line and with the theoretical contour curve intersection point C, the parameter α for the half of the sum of angle $O'O^{\#}C$ and $O'O^{\#}C_B$ ($\alpha=$（$\angle O'O^{\#}C+\angle O'O^{\#}C_B$）/2).

5. Actual measurement of head shape deviation based on non-contact laser scanning

In order to realize the non-contact laser scanning measurement and error analysis, this paper has carried out a field test on two different types of head equipment company in Hangzhou.

1）Elliptical head. Nominal parameter:

Shape: ellipse; diameter: 2800mm; total depth: 750mm; straight edge: 50mm.

First of all, we used the principle of this paper to test the parameters of the header 3 times.

Secondly, we carried out point by point actual measurement of the head with a ruler, and compared the instrument test data with the measured data. The test data are shown in table 1.

Table 1. Comparison of instrument test and ruler measurement in head shape errors

	measure	1st	2nd	3rd	σ
Maximum internal deviation (mm)	8	10.3	9.2	9.8	0.551
Maximum external deviation(mm)	-1	-2.3	-1.2	-2.5	0.7

From Table 1, three measurement data by Instrument and measurement data contrast. The shape of the uncertainty of the maximum degree of 0.7mm in repetitive aspect; in accuracy aspect, the maximum deviation of the maximum deviation of 2.3mm; maximum deviation of the maximum 1.5mm.

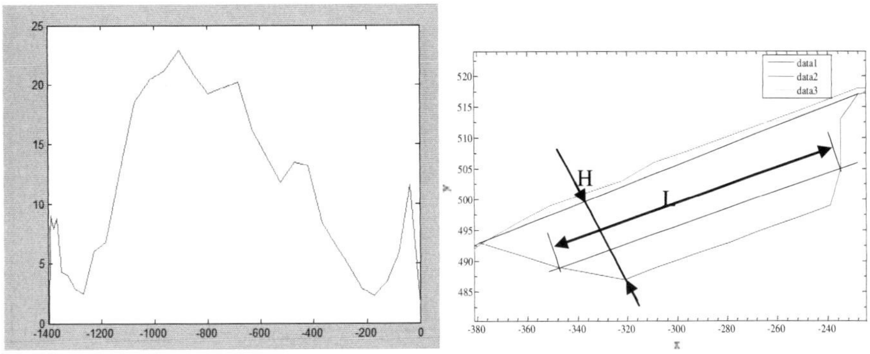

(a) Local amplification test curve (b) Selection of measurement direction

Figure 7. The test of block shaped objects placed on the inner surface of the head

We placed an iron round cake shaped object at the surface of the head at random, block shaped objects are tested, and tested the block shaped objects, local test curves are shown in Figure 7 (a). The block shaped objects in actual size are as follows: Length: 111.5mm; thickness: 15.5mm.

In Figure 7 (b), the direction of the thickness of H is taken as the direction of the radial direction (the direction of the measurement is directed toward the end of the end face center). The length of the L measuring point from block shaped object points on both sides and the actual laser scanning test are shown in Figure 7 (b). The measured actual dimensions are:

Length: 114.1mm; deviation: -2.6mm;

Thickness: 16.44mm; deviation: -0.94mm.

According to the actual test, this method of non-contact laser scanning measurement can be used to measure the shape of the head and the correct the parameters of measurement, and it is in accordance with the theoretical deviation analysis. In addition, from the theoretical analysis and the actual test, in order to increase the accuracy of the non-contact laser scanning device scanning, reduce the deviation, it have the following measures:

1) Using high precision laser distance measuring equipment. The accuracy of the equipment directly determines the follow-up scanning accuracy;

2) Before testing, it is needed to accurately determine the central axis, that is, the distance between the laser range finder and the center of the end face of the end face of the head. This deviation, which is mainly the deviation between the two centers of the end face of the vertical head, will affect the accuracy of the subsequent scanning. This deviation,

which is mainly the deviation between the two centers of the end face of the vertical head, will affect the accuracy of the subsequent scanning.

3) Mechanical assembly stability. Due to the influence of the precision of the mechanical device or part in the production, the actual size and design size will be biased. There will be some deviation in the coordination of the mechanical structure, and the cumulative effect of the mechanical assembly errors will affect the detection accuracy of the whole device[5].

6. Conclusion

The transitional zone between the straight side section and the semi ellipsoid of the head is a sensitive area, the slight decrease of the corner radius will lead to the rapid increase of the numerical value of the shape system, and the maximum stress and the circumferential compressive stress on the outer surface of the transition region also increase rapidly. For internal pressure forming head, it means that the possibility of damage to both strength and stability will rapidly increase. So, it is very necessary to detect the deviation of the header of the pressure vessel[6]. Comparing with the traditional contact measurement, the non-contact laser scanning measurement in this paper has the following advantages:

1) Eliminates the links of making and using the model. The test is simple and convenient, also the deviation caused by the test is reduced.
2) It can automatically find header center section to achieve data collection and processing of data, as well as the visual display of data;
3) In addition to the detection of the shape of the inner surface of the head, it is also easy to detect other parameters of the head, including roundness, depth, diameter, etc. At the same time, the head of various shapes can be detected, which can increase the convenience of detection, which is of great significance for the prevention of the failure of the container.

References

[1] Xinlin Teng. Discussion about the Safety Inspection and Quality Supervision of Pressure Vessels[J]. China New Technologies and Products, 2011, (14): 41-43.

[2] MA Li MIAO Cunjian ZHU Xiaobo. Research on Prediction Method of Plastic Deformation for Cold Stamping Formed Standard Elliptical Head Made of Austenitic Stainless Steel. JOURNAL OF MECHANICAL ENGINEERING, 2015, 51(6): 19-25.

[3] Mingzhi Liu. The Coordinate Measuring Machine for Head Test [J]. Petro-chemical Equipment, 1991, 20(1): 37-38.

[4] Zhang Jiangyu. Measure- ment of Shape Tolerance for Elliptical Head. CPVT. 1994, 1,11(1)：83-85.

[5] LIN Jinghuan，DAI Yong，SHENG Shuiping，CHEN Haiyun. New universal laser scanning head detector[J]. Laser Technology. 2014, 9, 38(5): 651-654.

[6] XIAN Chaozuo, Yuexin, HE Jianhua. Discu- ssion on the allowable deviation of the shape of the head [J]. CHINA PETROLEUM MACHINERY. 2004，32(4): 43-45.